Moon

Viorel Badescu (Ed.)

Moon

Prospective Energy and Material Resources

 Springer

Editor
Viorel Badescu
Polytechnic University of Bucharest
Candida Oancea Institute
Bucharest
Romania

ISBN 978-3-642-27968-3   e-ISBN 978-3-642-27969-0
DOI 10.1007/978-3-642-27969-0
Springer Heidelberg New York Dordrecht London

Library of Congress Control Number: 2012930487

Printed on acid-free paper

Springer is part of Springer Science+Business Media (www.springer.com)

To those named lunatics since
they have the same ideas as everybody
but ten years earlier.

Foreword

Lunar Exploration and Utilization: Why and How?

Jules Verne was the first one who wrote a fascinating book on his dream to conquer the Moon, but certainly before and after there have been many enthusiastic people dreaming about that. Only a bit more than a decade after the first man orbited earth, NASA started an extensive moon programme culminating in putting men on the moon, but other nations like Russia (UdSSR), Japan, China and India managed at least to put robotic probes on or around the moon.

There are a number of rationales to explore the Moon, emplace a robotic village and then an international lunar base (see International Lunar Exploration Working Group (ILEWG) – http://sci.esa.int/ilewg/) - a global civilisation imperative, and scientific, political, socio-economical, technological rationales. We have to learn from results and lessons from Apollo & Luna missions, and from the more recent robotic missions, as well as from the International Space Station collaborative experience. The Moon is a natural and large space station offering assets, knowledge, a bench for enhancing technical and science skills, and resources for industries, and citizens at large. It will rely on the development of advanced technologies, in a context of effective and affordable cost, sustainability and work share between nations and stakeholders.

Apollo Background

It began on July 1960 as a continuation of Mercury Program and its maximum goal was to send manned missions to the Moon and return back safety to the Earth, this efforts strongly supported by a powerful speech given to congress by US President Kennedy on 25th May 1961. Within the program 15 successful missions took place and sadly 2 setbacks, Apollo 1 suffered a fire during a pre-flight test killing 3 astronauts and Apollo 13 where an explosion in one oxygen tank required abortion of the mission but fortunately the crew returned safely back to Earth. Apollo 2, 3, 4, 5 and 6 had as main goal test flight systems in automatic launches. Apollo 7 and 9 were manned missions orbiting the Earth and Apollo 8 and 10 were manned missions that orbited the Moon. 5 missions landed on the Moon, from Apollo 11 to Apollo 17, except Apollo 13. Apollo missions 15 and 17 are especially interesting from a geological field work point of view. Apollo 15 was the first mission where the Astronauts field work was supported with a Lunar Roving Vehicle and Apollo 17 was the first mission to include a geologist. We recall their main features as they are relevant to inspire, dimension, and adapt for long duration stays and EVAs for the next phase of human return.

Technology Challenges

The trip to the Moon can nowadays be considered as a standard human spaced activity, provided that a spacecraft big enough to carry the team and payload can be built within the financial constraint all space agencies have to tackle. That is if the mission is restricted to be completed within a lunar day. To survive a lunar night, which is on areas most interesting for exploration up to 14 earth days long, significant energy resources are required, basically eliminating batteries as electrochemical energy storage right away, but clearly point towards fuel cells or nuclear sources, which maybe excluded for cost or political reasons. Another challenge is the amount of radiation bombarding humans and sensitive equipment. And the environmental temperatures are very low. How can sufficient food be produced to supply the explorers?

Renewed Lunar Exploration, SMART-1 Mission and Other ESA Activities

Europe entered the lunar adventure with SMART-1, the small mission for advanced research and technology. It was launched in 2003 from Kourou on an Ariane 5, and orbited the Moon from November 2004 for nearly two years until its impact in September 2006. SMART-1 tested a solar electric propulsion motor –slow but efficient, and new techniques, observing the Earth for the long journey, and discovering the views of the dark side, poles and unexplored regions of the Moon.

SMART-1 carried miniature instruments that measured the Moon in the invisible (infrared or X-rays) to map the mineral and chemical elements. SMART-1 mapped the polar regions and discovered peaks of eternal light, ideal sites for future manned bases. Finally SMART-1 ended its mission with a controlled impact observed from Earth. SMART-1, Europe's first lunar mission demonstrated technologies for future science and exploration missions.

SMART-1 launched a new international exploration, opening up future collaborations with the missions in lunar orbit: the Japanese Kaguya, Chinese Chang'E1 probe for which ESA provided a control station and receiving data, the Indian Chandrayaan probe -1 carrying three instruments inherited from SMART-1 probe, Lunar Reconnaissance Orbiter LRO and the LCROSS polar impactor launched in 2009.

ESMO, the European Student Moon Orbiter, is developed as education and inspiration activity with a network of universities for launch in 2013. ESA has conducted a set of generic lunar studies, and design concepts for lunar landers (LEDA, EuroMoon, Lunar Exploration Study LES3, Moon-NEXT). It is now performing industrial studies for a mid-class ESA Lunar Lander, with precision landing near peak of light.

ESA has conducted a comparative lunar architecture assignment in collaboration with NASA with a focus on the development of an European logistic cargo lander launched by Ariane.

International Lunar Missions and Roadmap

Japan launched its lunar probe (Selene/Kaguya) in September 2007 just before China's and India's lunar missions. Kaguya is JAXA's first large lunar explorer that carried a comprehensive suite of scientific instruments with successful operations until June 2009.

CNSA in China is building up a space program with high ambitions. Among the main targets are a robotic program for exploring the Moon and human spaceflight. In particular, the success of its three manned Shenzhou missions has encouraged China to envisage on an own permanent manned spaceflight system in the future and possibly a Chinese lunar landing. In October 2007, China launched its first lunar probe, Chang'E1, as the first mission of the China Lunar Exploration Program (CLEP), with participation of ESA for mission support and ground station control and data collection. The Chang'E2 orbiter was launched in October 2010. A taikonaut in 2008 performed China's first extravehicular activity (EVA). Chang'E phase II and Chang-e phase III are planned for the future. Chang-e phase III will be accompanied by a sample return mission.

In India ISRO developed new programs and launched Chandrayaan-1 in October 2008 to the Moon, as the first Indian planetary mission. International instruments on this mission include C1XS, SIR2 and SARA delivered by ESA, M3, mini-SAR by NASA, and RADOM from Bulgaria. A second robotic lunar mission in collaboration with Russia - Luna-Glob is in the planning stage for launch after 2013.

In summary, the main space powers - the United States, Russia, Europe, Japan, Canada, China and India - are developing new capabilities to fulfil their exploration ambitions. Space actors have the potential to complement each other in collaborative efforts to explore Moon and beyond. The development of new capabilities of rising space actors China and India will allow a global exploration program with a higher frequency and diversity of space missions.

Field tests on earth preparing for lunar exploration include the investigation of geological and geochemical techniques like drilling of cores and sampling, remote control field rovers, cameras, instruments, evaluating crew operations, simulations and EVAs and many other factors. Worldwide sites in extreme environments are used to perform these steps for robotic and human exploration. Numerous programs are currently undertaken worldwide including various stakeholders.

ILEWG has developed with task groups on technology and a lunar base, technical pilot projects that organised and coordinated field campaigns in Utah desert research station, at Eifel volcanic park, Rio Tinto, and other sites, in collaboration with ESA, NASA and academic/industrial partners. The goals of ILEWG field campaigns include: 1) testing instruments, rovers, landers, EVA technologies, habitat and field laboratory; 2) performing field research in geology and sample analysis, exobiology; 3) studying human factors and crew aspects; 4) outreach and students training.

De facto, the space agencies and stakeholders have developed precursor science and robotic missions, and advanced on the roadmap for lunar and global exploration as recommended by ILEWG (Fig. 1).

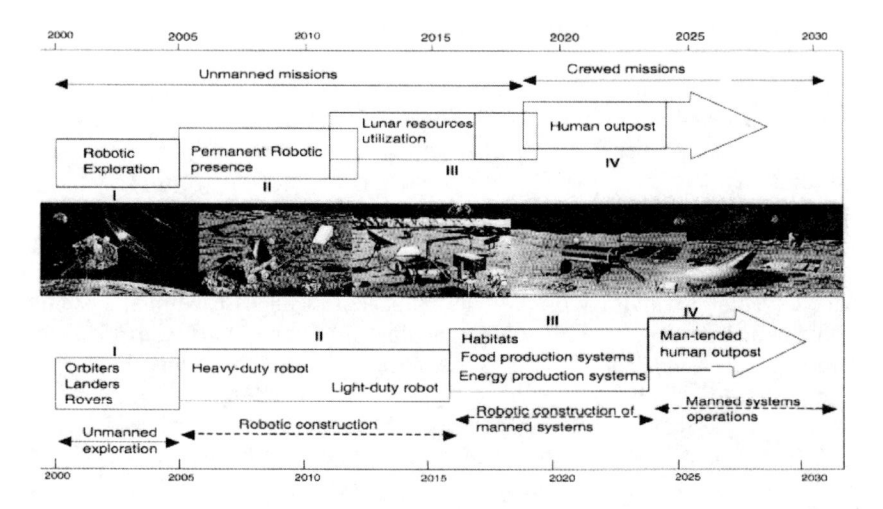

Fig. 1 Roadmap for lunar and global exploration as recommended by ILEWG

Science Enabling Exploration

Relevant science questions from the community include (see http://sci.esa.int/ ilewg):

- What does the Moon tell us on processes that are shaping Earth-like planets (tectonics, volcanism, impact craters, erosion, space weathering, volatiles)?
- What is the present structure, composition and past evolution of the lunar interior?
- Did the Moon form in a giant impact and how? How was the Earth evolution and habitability affected by this violent event, and by lunar tidal forcing?
- How can we return samples from large impact basins as windows to the lunar interior and as record of the early and late heavy bombardment?
- What can we learn on the delivery of water and organics by comets and asteroids from sampling cores of the lunar polar ices deposits? Are they prebiotic ingredients in lunar soils or ices?
- How to find and return samples ejected from the early Earth (and possibly the oldest fossils) now buried under a few meters of lunar regolith?
- How to use most effectively the Moon as platform for astrophysics, cosmology and fundamental physics, compared to Earth or space based laboratories?
- How to use an international robotic village (as recommended by ILEWG) to provide the measurements to fulfil these scientific objectives?

The ILEWG recommendations are clearly expressed in Beijing and Sorreto Declarations:

"Recognizing the importance of the geophysical studies of the interior of the Moon for understanding its formation and evolution, the necessity for a better monitoring of all natural hazards (radiation, meteorites impacts and shallow moonquakes) on the surface, and the series of landers planned by agencies in the period 2010-2015 as an unique opportunity for setting up a geophysical network on the Moon, we recommend the creation of an international scientific working group for definition of a common standard for future Moon network instruments, in a way comparable to Earth seismology and magnetism networks. We encourage interested agencies and research organizations to study inclusion of network instruments in the Moon lander payload and also piggyback deployment of a Moon Geophysical and Environmental Suitcase." (International Conference on Exploration and Utilization of the Moon (ICEUM) 8-11, Beijing 2006 - http://sci.esa.int/iceum8)" and "To address outstanding lunar science questions remaining to be resolved (relating to mineralogy, geochemistry, interior structure, gravity, topography, polar regions, volatiles, environment protection) as well as the scientific investigations that can be performed from the Moon as a platform (astrophysics, solar physics, Earth observations, life sciences) (ICEUM9, Sorrento, 2007 - http://sci.esa.int/iceum9)".

Towards a Lunar Robotic Village and an International Lunar Base

The next decade will see landers and vehicles that will allow for a global robotic village. From 2020 onwards manned missions could prepare an International Lunar Base. Space agencies and other stakeholders are collaborating on the development and operation of robotic villages and lunar bases, with six areas of activity:

- For science, to conduct research to better understand our Earth and solar system history, observe the cosmos from the moon, and adapt the terrestrial life on the moon;
- For technical innovation, to develop robots, techniques for using local resources and support life (bacteria, plants, animals, humans and biosphere) that will reduce risks and prepare the future;
- For our civilization, to learn how to extend human presence and sustaining an international manned lunar base, which can be a refuge for our biosphere or a stepping stone out of the cradle land;
- For a global partnership, based on international collaborations and challenging to develop peaceful village inhabited robotics and bases;
- For economic growth, with skilled jobs and activities that benefit the Earth;
- To engage the public and inspire the next generation, to innovate, to undertake and fulfil our destiny as explorers.

Infrastructure Development and the Lunar Role in Human Expansion into the Solar System

The ICEUM community has a recommended goal – to establish a permanent human presence on the Moon within 20 years. The benefits from this will include new energy and materials resources, new human support technologies, and new professional and societal opportunities for people from all over the World, and an opportunity to demonstrate cooperation between people of all nations. ILEWG should:

- Prepare a roadmap with schedule and milestone events for lunar exploration and development and encourage space agencies to undertake missions and technology development projects that support the roadmap.
- Examine existing national proposals for lunar exploration to find inconsistencies, synergies, and missed opportunities. Make recommendations for improving or enhancing the missions to the national space agencies.
- Continue to identify and quantify benefits to humans on Earth from exploration and development of the Moon.
- Support efforts to diminish the cost of transportation to space.
- Promote government - industry partnerships and commercial initiatives for lunar exploration, development and utilisation.
- Encourage programs to develop infrastructure technologies - power, habitation, closed life support, transportation of materials and people, communications, etc. - that are required for a self-sufficient settlement on the Moon.
- Promote the development and demonstration of lunar resource utilisation technologies.
- Identify one or more critical projects of economic/societal interest, which can focus world-wide attention on lunar development and can be implemented soon as a pathfinder activity for international participation, for example, the production of the first photovoltaic device from lunar materials).
- Promote the refinement of space laws that will define the rules, conditions, and constraints on lunar development.
- Encourage participation and free thinking about lunar possibilities by young people around the World and provide means for their new ideas to be incorporated into space agency programs. Promote an international competition for new ideas, with the prize being the naming of a feature on the Moon.
- Encourage expansion of ILEWG to a world-wide basis, not simply those Nations now undertaking or planning lunar exploration missions.
- Educate the world-wide public, particularly the youth, about the potential for humans to live permanently on the Moon.
- Include consideration of the environmental effects to the Moon and the Earth in the implementation of lunar exploration and development.
- Promote innovative astrobiology initiatives using the Moon, new concepts for scientific advancement, and new uses of lunar programs for infotainment.

- Support the concept of the "Lunar Explorers' Society" and assist in organising a founding convention.
- ILEWG to support or promote the development of international standards for hardware and software to be utilised on the Moon.

The Lunar Global Robotic Village

ILEWG supports the goals of a comprehensive series of surface elements including landers and rovers at the poles and other key sites. The model envisages the deployment of landers from different countries that are developed and operated in co-ordination. The rovers should perform complementary cooperative robotics and exploration tasks, and demonstrate enabling technologies. Such a program will initiate and enhance international collaboration, as well as science, commercial and public engagement opportunities. Various infrastructure assets such as telecom, power generation and energy storage can be shared by the international partners. The planning and development of a "global lunar robotic village" will encourage and stimulate the peaceful and progressive development to investigate the Moon, and foster international cooperation between nations, space agencies and private companies, (ICEUM7-9, Toronto 2005 - http://sci.esa.int/iceum7).

The rationale for and possible implementation of a global lunar robotic village have been discussed by ILEWG community, with a phased approach with orbital reconnaissance, small landers, a network of landers for science (of, on and from the Moon) and exploration. Then advanced robotic precursors to human missions with deployment of large infrastructures and resource utilisation would conduct operations imminent to human arrival, during and between human early missions. Possible elements of the Global Robotic Village are discussed in ILEWG volumes (ICEUM4 2000 pp. 219-263, pp. 385-391 - http://sci.esa.int/iceum4, ICEUM9 2007 pp. 82-190 - http://sci.esa.int/iceum9). The rationale for lunar sample return is described in (ICEUM9 2007 pp. 59-60 - http://sci.esa.int/iceum9). Priority areas include the south pole Aitken basin impacts melts (as a probe of lower crust and upper mantle material, and a constraint on the chronology of giant bombardment), samples from polar volatiles, and from the youngest lunar volcanic units in Procellarum. Lunar samples can be returned with automatic missions such as Moonrise, Chang'EIII, or from human missions to the surface.

The Moon's proximity to Earth allows lunar sample return to act as a testbed for robotic technologies enabling sample return from more distant planetary bodies. Between 1969 and 1972 the six Apollo missions that landed astronauts on the Moon returned a collection of over 2,000 soil samples (in total 382 kg). However, Apollo samples were collected from a relatively small, equatorial area of the Moon that consists of conditions that are atypically for the Moon. Therefore many follow-on lunar sample return options have been evaluated in the last decade to bring back samples from other locations on the Moon, and take them into terrestrial laboratories to perform a full suite of investigations such as mineralogical, lithological, geochemical and geo-chronological analyses that are not possible to conduct via in-situ exploration.

Technology and Human Robotic Partnerships for an International Lunar Base

Planetary science stands to be a major beneficiary of human space exploration. Human exploration of the Moon facilitates landing, operating and maintaining massive and complex scientific equipment as well as large-scale exploratory activities such as drilling. Human exploration can enable the intelligent and efficient collection of samples in large quantities, covering different location and wider geographical areas. Human exploration of the Moon allows for increased opportunities for serendipitous discoveries. Furthermore it takes advantage of the fact that human beings can work intelligently and quickly, make sense of complexity and are able to troubleshoot unforeseen problems with inherent flexibility. Whereas robots are expendable, environmentally robust, continuously present, and characterized by physical durability, they suffer from limited intellectual capability, a slow data rate and power constraints. Overall, however, robotic exploration is comparatively cheap—both in terms of cost and risk. Humans, in turn, are mechanically flexible, able to communicate and can handle difficult terrain. They can easily adapt to different situations, are intellectually flexible, but they require life support, and need to sleep and eat. Overall, human exploration is expensive. In order to make the best use of each system's advantages, a sensible long-term exploration roadmap will envision that robots and humans should explore in synergistic partnership on the Moon.

An international lunar base design requires the knowledge of many different disciplines, e.g. engineers, architects, industrial designers and medical personnel. ILEWG and Lunar Exploration Analysis Group (LEAG) have worked since a decade on concepts for a lunar base as an important milestone in their roadmap. The development of a lunar base as example for international cooperation has been identified on political level by the Beijing Declaration (2008 - http://sci.esa.int/ iceum8).

An important element for lunar outpost architecture is the habitation module. Its configuration is an important element of the outpost architecture definition, and it is function of the environmental requirements, of the radiation shielding approach, of the transportation/operation constraints and of the distribution of functions between habitat elements (separated or integrated in the habitation core). Looking at the different solutions analyzed in the past and at the several trade-offs performed on the radiation protection options, the cylindrical module option will probably be the initial adopted one. Additional deployable volumes will allow extending the internal volumes with limited impacts in mass and in transportation volumes. Radiation protection can be provided by bags filled with regolith.

Finally as stated by the participants in the ILEWG/LEAG/SRR 2008 conference:

"We, reaffirm our commitment to international lunar exploration, from the analysis and integration of current lunar orbiter data, to the development of lunar landers and rovers, the build up of a global robotic village, and the preparation for human settlements and international lunar bases" (ICEUM10, Cape Canaveral, 2008 http://sci.esa.int/iceum10).

* * *

A book on the exploration and utilization of Moon resources thus comes at the right moment. Such a book will contribute to the dissemination of existing knowledge and will support the efforts of numerous groups devoted to Lunar research. It thus is a welcome initiative of Springer Verlag and a highly appreciated effort by Prof. Viorel Badescu to collect thirty chapters, written by selected experts in the field.

Max Schautz & Bernard H. Foing
European Space Agency - ESTEC & International Lunar Exploration Working
Group

Preface

The Earth has limited material and energy resources. Further development of the humanity will require going beyond our planet for mining and use of extraterrestrial mineral resources and search of power sources. The exploitation of the natural resources of the moon is a first natural step on this direction. Lunar materials may contribute to the betterment of conditions of people on Earth but they also may be used to establish permanent settlements on the Moon. This will allow developing new technologies, systems and flight operation techniques to continue space exploration. Also, lunar materials are attractive as feedstocks for large scale space-based industrialization since the cost to get materials off the Earth is much larger than to get them off the Moon.

The lunar soil consists mostly of oxides of metals and silicon. The oxygen, which is bound in molecular silicates and metal oxides, makes up roughly 40% of lunar soil. On the Moon there are no big free metal ores. However, significant quantities of free iron granules exist in the soil thanks mostly to asteroid craters. The lunar South Pole's area contains significant quantities of volatile compounds such as hydrogen, sulfur dioxide, carbon dioxide, formaldehyde, ammonia, methane, methanol, mercury and sodium.

The Moon's soil is fine powder and this ensures cheap mining and mineral processing. Metals can be extracted by space-based processing and products to make from semi-processed lunar materials are "lunarcrete", fiberglass, various glass-ceramic composites, and oxygen. Some of the existing substances, such as the atomized metal powders, are substitutes for hydrogen. They can be used to produce fuel propellants for hydrogen-oxygen rocketry. One expects the first shelters on the Moon will be inflatable structures imported from Earth but building structures may be constructed from ceramics and metals created in-situ.

One of the most important resources on the Moon is water. It exists in higher concentrations than anticipated a decade ago. Water may be used for drinking, producing hydrogen fuel and oxygen for breathing, making Moon exploration and exploitation easier. The propellant produced in this way could be supplied in orbital refueling stations to spacecrafts departing Earth, extending considerably their range.

The Moon is rich in helium-3, which comes from the outer layer of the sun and is blown around the solar system by solar winds. The total amount of helium-3 on the Moon can reach one to five million tons. Helium-3 is considered a long-term, stable, safe and clean material to be used by the new range of fusion nuclear power stations. However, there are about 15 tons of helium-3 on Earth while in the world about 100 tons will be needed every year. Also, the classic Buck Rogers propulsion system based on helium-3 fusion energy, requiring less radioactive shielding and lightening the load may be the key to future space exploration and settlement.

The Moon does not contain large enough quantities of chemicals such as carbon, potassium and phosphorous needed for some future human activities like agriculture. They have to be imported from Earth but once the first crop is obtained, then the chemicals can be reused in a natural cycle.

Power may be generated on Moon in various ways. Solar cells may be manufactured from lunar materials while nuclear power may be generated by using uranium mined on the Moon. Fuel cells may be used, based on the hydrogen and oxygen obtained from local materials.

A number of public agencies and private companies already expressed their interest in the exploitation of Moon resources. In 2007 NASA reported work on a new space vehicle destined to fly the astronauts to the moon around 2020, to deploy a permanent base. Recent economic events seem to postpone this initiative. China has announced plans to map in detail Moon surface and interest in exploiting helium-3 reserves. A Chinese lunar based might be established by 2024. The Russian company RSC Energia analyze a plan to land on the Moon in three missions, i.e. a lunar fly-around mission, a circumlunar orbit injection with automatic landing of the lunar module and the final manned landing on the Moon. An industrial transportation system to support regular missions to the Moon and lunar mining operations of isotope helium-3 is envisaged to operate after 2025.

The UN Treaties state that the Moon and its minerals are the common heritage of humanity. Therefore, the exploitation of lunar resources requires an appropriate legal framework which would likely demand joint international co-operation.

This book presents an inventory of the Moon's energy and material resources for prospective human use. One investigates the advantages and limitations of various systems supplying manned lunar bases with energy and other vital resources. The present book collects together recent proposals and innovative options and solutions. It is a good starting point for researchers involved in current and impending Moon-related activities.

The book is structured along logical lines of progressive thought and may be conceptually divided into four parts. The first part contains six chapters and deals with a detailed presentation of material resources on Moon. After the introductory Chap. 1, presenting a survey of Moon's geological resources, a group of two chapters (2 and 3) refer to helium and water on Moon. Chapter 4 refers to existing information about lunar surface/subsurface obtained from microwave exploration. The last two chapters in the first part present lunar minerals and their resource utilization with reference to energy production and lunar agriculture (Chap. 5) and the lunar holes and lava tubes as material resources (Chap. 6), respectively.

The second part of the book contains fifteen chapters and deals with technological aspects related to material and energy resources existing on Moon. Chapters 7 and 8 present existing technology to be used for in situ oxygen and water production while Chap. 9 refers to the utilization of lunar regolith for habitation applications. The next two chapters focus on lunar drilling, excavation and mining (Chap. 10) and on the transportation, handling and processing of lunar regolith (Chap. 11). Chapter 12 presents the past, present and future of power systems used for Moon surface exploration. The usage of lunar resources for energy generation is presented in Chap. 13 while Chap. 14 refers to perpetual sunshine, moderate temperatures and perpetual cold as lunar polar resources. The available solar energy on Moon is treated in Chap. 15 while Chap. 16 refers to photovoltaic power generation on Moon. Chapters 17, 18 and 19 deal with fuel cell power system options, cooling systems and heating systems, respectively. The last two chapters in this part refer to deployable lunar habitation design (Chap. 20) and usage of composite materials for inflatable construction on Moon (Chap. 21).

The third part of the book consists of two chapters dealing with legal matter. Chapter 22 refers to the natural resources of the Moon and the legal regulation while Chap. 23 shows details about the property status of lunar resources.

The fourth part of the book consists of seven chapters referring to technological aspects in the perspective of a more distant future. Chapter 24 describes a telecommunication and navigation services in support of lunar exploration and exploitation while Chap. 25 shows the details of a laser power beaming architecture for supplying power to the lunar surface. Several aspects related to building the first lunar base are treated in Chap. 26. Advanced systems concept for autonomous construction of lunar surface structures are presented in Chap. 27. Chapter 28 introduces a new concept of managing the land surface on Moon – the bacillithic cratertecture. Chapter 29 presents several considerations concerning the modern lunar management, with focus on the private sector. The fourth part ends with a short visionary Chap. 30 showing the role that the Moon is expected to play in the Solar System exploration.

More details about the thirty chapters of the book are given below.

Chapter 1 by Jennifer E. Edmunson and Douglas L. Rickman shows that the Moon's geologic resources lie in the regolith, naturally concentrated minerals, and large-scale physical features. Likely formed by a giant impact experienced by the proto-Earth, the consolidated, molten Moon crystallized into a layered structure. Major impacts into the hardened layers exposed and remobilized materials. Smaller impacts have "gardened" the upper surface for billions of years. The resulting regolith is a mixture of glasses, minerals, and rock fragments. Space weathering creates agglutinates, implants solar particles, and changes the properties of the uppermost surface of the Moon over time; these processes collectively contribute to the regolith's local maturity. Chemically, mixtures of high calcium plagioclase, pyroxenes, and olivine describe the bulk of the material, with other phases such as sulfides and oxides as minor or trace constituents. Impact basins are the dominant morphological features on the surface, resulting in both smooth areas and extremely rugged terrain. The $1.5°$ axial tilt of the Moon keeps some deep craters permanently shadowed, with temperatures plunging to 29K. Volcanic features are also significant, such as rilles and lava tubes.

Chapter 2 by Kimberly R. Kuhlman and Gerald L. Kulcinski shows that the Moon is the most promising source of large amounts of ^{3}He needed for future fusion fuel, and ^{3}He appears to be correlated with titanium-rich lunar regoliths. The parabolic differential equation for diffusion was combined with solar wind data and a description of the lunar environment to create a computational model of helium implantation and diffusion in the lunar regolith. The diffusivity of helium isotopes in ilmenite, the lunar mineral thought to retain the most helium, was measured experimentally. Samples were implanted with ^{4}He and ^{3}He at solar-wind energies using plasma source ion implantation (PSII). Stepwise heating with mass spectroscopy of the evolved He isotopes demonstrated release behavior similar to that of the Apollo 11 regoliths. Moderate fluences of hydrogen were seen to retard ^{3}He release, while a high fluence of hydrogen had little effect. The helium diffusion in ilmenite was characterized by four distinct activation energies, $E_1=0.26$ eV, $E_2 \cong 0.5$ eV, $E_3 \cong 1.5$ eV and $E_4 > 2.2$ eV. These energies are characteristic of diffusion through two amorphous layers, detrapping from oxygen vacancies and constitutional vacancies, respectively. The diffusivity of ^{3}He was seen to be at least a factor of 10 higher than the diffusivity of ^{4}He, which was determined to be 9×10^{-24} cm^2/s at 400K with activation energy of 1.4 eV.

Chapter 3 by Larissa V. Starukhina focuses on the possible existence of water on Moon. The results of the remote sensing experiments aimed on search of water on the Moon are discussed: (1) measurements of hydrogen content by the neutron spectrometers on Lunar Prospector and LCROSS, (2) recent observations of the absorption bands near 3 μm in reflectance spectra of the lunar surface, and (3) radar observations of polar regions of the Moon and Mercury. All the results can be explained without presence of water, which is supported by model calculations. The hydrogen content measured on the Moon, including permanently shadowed areas, can be accounted for by trapping and retention of solar wind protons on radiation defects in the lunar soil particles, retention mechanisms at low temperatures being up to blistering (accumulation of the implanted gas in pores) and physical adsorption. Infrared absorption near 3 μm is most probably due to chemical trapping of solar wind protons in OH-groups. Strong, highly depolarized radar echoes from polar and lower latitude craters of Mercury can be explained by decrease of the dielectric loss of silicate material with temperature. All the alternative explanations overcome the main difficulty of water ice hypothesis, namely, its inconsistency with local regional variations of the physical parameters observed in all three types of experiments. Global and local variations of the parameters are due to exponential dependence $\sim \exp(-U/kT)$ of outgassing rates and dielectric loss on temperature T and activation energies U characteristic of surface material. In addition to these factors, regional variations of the depth of 3 μm absorption in reflectance spectra should be correlated with variations of surface brightness. Thus, at present, neither of the remote sensing results can be considered as reliable proof for water on the Moon. For more confident interpretation of the observations, many laboratory simulations are required.

Chapter 4 by Ya-Qiu Jin reminds that China had successfully launched its first lunar exploration satellite Chang'E-1 (CE-1) on 24 October 2007 at lunar circle orbit ~200 km. A duplicate CE-2 at lower orbit ~100 km was also launched in 1 October 2010. A multi-channel microwave radiometer, for the first time, was aboard the CE-1 (and CE-2) satellite with the purpose of measuring the microwave thermal emission from the lunar surface layer. There are four frequency channels for CE-1 microwave radiometer: 3.0, 7.8, 19.35 and 37.0 GHz. The measurements of the multi-channel brightness temperature, T_B, are applied to invert the global distribution of the regolith layer thickness, from which the total inventory of ^{3}He (Helium-3) stored in the lunar regolith layer can be estimated quantitatively. The chapter reports an overall work by the author's laboratory for this mission, in both passive and active microwave remote sensing of lunar media, including the microwave emission modeling and numerical simulation of T_B from lunar surface media, data validation and retrieval of lunar regolith layer thickness from multi-channel CE-1 T_B data, and evaluation of global inventory of ^{3}He in lunar regolith, and diurnal temperature change of the cratered lunar surfaces. Radar echoes from the cratered lunar surface are also numerically calculated. Employing a back projection (BP) algorithm, SAR image of lunar surface at L bands is numerically simulated. Then, high frequency (HF) radar echoes from the cratered lunar surface/subsurface are studied for simulation of radar echoes imaging. Their dependences on the physical properties of lunar layering structures are analyzed.

Chapter 5 by Yuuki Yazawa, Akira Yamaguchi and Hiroshi Takeda refers to some Lunar minerals and their utilization. After looking at the varieties, global distribution and usefulness of lunar minerals, the authors introduce utilization of plagioclase, the most abundant minerals of the highland. This topic is also related to the science themes addresses by the authors. The formation of plagioclase highland crust form the lunar magma ocean is the most important questions to be answered. Instead of establishing multi purpose permanent bases, the authors are assuming a small base for construction of a solar power station to solve energy crises on Earth, and small agricultural dome to glow green vegetables for short human stay. The chapter covers utilization of extreme calcic plagioclase (anorthite $CaAl_2Si_2O_8$) and focus on the utilization of Si, Al and Ca in this mineral. Also a plan is presented to use silicon extracted from the lunar calcic plagioclase to produce a Solar Power Satellite station (SPS). The authors consider Ca-plagioclase as the most important mineral for soil development at the highland base by utilization of humic substances, especially fulvic acid for lunar agriculture.

Chapter 6 by Junichi Haruyama, Tomokatsu Morota, Shingo Kobayashi, Shujiro Sawai, Paul G. Lucey, Motomaro Shirao, and Masaki N. Nishino reminds that the Japanese lunar orbiter SELENE recently discovered three large holes in the lunar maria at the Moon's Marius Hills region, at Mare Tranquillitatis, and at Mare Ingenii. Each of these holes has near vertical or overhanging walls, tens of meters in both diameter and depth. Detailed observations of these features by SELENE and LRO indicate "sub-lunarean" voids, probably lava tube caverns, which seem associated with the holes. The inner walls and floors of the holes could provide numerous clues to understanding lunar volcanism, the history of solar activity, lunar dynamos, and dust transportation mechanisms. Temperature extremes, cosmic ray radiation, and micrometeorite hazards at the floors of the holes are expected to be more moderate than at the lunar surface; therefore, the bottoms of lunar holes are also promising candidates for constructing future lunar bases. As similar holes and possible lava tubes have been found on Mars as well, exploring the Moon's holes and probable associated lava tubes will provide lessons for exploring those of Mars. The unique characteristics of these penetrating lunar holes present opportunities for good science and engineering endeavors, including possible sites for lunar resources and settlements.

Chapter 7 by Carsten Schwandt, James A. Hamilton, Derek J. Fray and Ian A. Crawford begins by showing that there is a renewed interest in space exploration with the targets of establishing a permanent human settlement on the moon and launching manned missions to Mars. It is a general consensus that extraterrestrial in situ resource utilisation (ISRU) will play a crucial role in these scenarios. Amongst the main objectives are, firstly, the generation of oxygen on the moon from locally occurring materials in order to sustain human life and produce an oxidant for burning rocket fuel during space travel and, secondly, the extraction of metals so as to be able to create a lunar infrastructure. This chapter discusses ISRU on the moon in view of economic constraints, lunar geology, and scientific and engineering aspects. A summary is given of the various oxygen generation techniques that have been selected by NASA for in-depth assessment, and a particular emphasis is placed on the Ilmenox process. This process has been derived from the FFC-Cambridge process, which is capable of removing oxygen from metal oxides via electrochemical processing in molten salts. The unique, and experimentally proven, benefit of the Ilmenox process is that, through the utilisation of a specially developed inert anode, it has become possible to produce simultaneously oxygen and metal from lunar regolith with an excellent yield and at a fast rate. Importantly, and in contrast to some other suggested ISRU processes, the Ilmenox process can be used to extract oxygen from the calcium and aluminium-rich *anorthositic* regoliths typical of the lunar highlands including the lunar poles, and is therefore not restricted to the iron-rich basaltic regoliths of the lunar maria. This greatly enhances its potential versatility as an ISRU process in support of future exploration activities.

Chapter 8 by Yang Li, Xiongyao Li, Shijie Wang, Hong Tang, Hong Gan, Shijie Li, Guangfei Wei,Yongchun Zheng, Kang T. Tsang and Ziyuan Ouyang shows that in-situ water and oxygen production is one of the key technologies for the construction and maintenance of lunar bases. By investigating the potential processes of water and oxygen production and analyzing the property of ilmenite, reducing ilmenite by hydrogen with microwave heating is considered to be an appropriate process of water and oxygen in-situ production on the Moon. To comprehend the in-situ produced process, reaction temperature, reaction time, qualities and grain size of ilmenite were discussed. With traditional heating experimental analysis, the efficiency of water production could be improved by increasing the reaction temperature and reaction time. But, the energy consuming would also increase. Grain size of ilmenite shows an important effect. Most of the ilmenite grains were surrounded or aperture packed by iron and titanium after reaction, which indicate the ilmenite has been partly reduced. As the simulated experiment shows, 20g ilmenite reduced at 1000°C is the best process. It provides the maximum water production per unit energy consuming with a water production ratio 1.25%.

Chapter 9 by Eric J. Faierson and Kathryn V. Logan shows that the use of in-situ planetary resources will provide a sustainable infrastructure for human colonization of celestial bodies such as the Moon. This chapter discusses a geothermite reaction between lunar regolith simulant and aluminum powder that has potential for in-situ resource utilization (ISRU) applications. The term "geothermite" is used to refer to chemical reactions that exhibit thermite-type behavior, and incorporate unrefined minerals and glass reacting with a reducing agent. Geothermite reactions have been initiated in a standard Earth atmosphere and a vacuum environment. The reactions were performed using various mass ratios of lunar regolith simulant and aluminum in order to establish a range of mass ratios that would allow repeatable reaction completion. Optimizing the mass ratio is also important to conserve the quantity of aluminum used in the reaction. Unreacted regolith simulant, as well as reaction products, were analyzed using X-ray diffraction (XRD) in order to identify chemical species formed during the reaction. Data from XRD analysis, combined with thermodynamic data, were used to propose thermodynamically favorable chemical reactions that could occur within the geothermite reaction. Scanning electron microscopy (SEM) and energy dispersive spectroscopy (EDS) were utilized to analyze the microstructure and chemical composition of unreacted regolith simulant and products of the geothermite reaction. Compressive strength measurements were performed on selected reaction products. In addition, proof of concept for the fabrication of Voissoir dome elements utilizing a reusable crucible was demonstrated.

Chapter 10 by Kris Zacny begins by reminding that the analysis of lunar samples returned by the US Apollo and the Soviet Luna missions revealed that the regolith contains a lot of elements, including oxygen that can be used to sustain human presence on the Moon. In addition, the most recent discovery of large deposits of hydrogen suggests there is a substantial amount of water-ice on the Moon. Water-ice can be mined and split into hydrogen and oxygen and used to generate electricity in fuel cells and sustain human presence by providing needed oxygen. Hydrogen and oxygen can also be used as rocket fuel and oxidizer for the journey back to Earth, other destinations in the Solar System, or for refueling the Earth observing satellites and the International Space Station. Past decades of lunar exploration have given us enough knowledge to convince us that the Moon is worth continuing exploration. Our technology has also been steadily improving and the success rate of landing on the Moon or Orbiting the Moon has gone from 0% in 1958 to 100% since 1990. This is especially encouraging as we prepare ourselves for the next decade of exploration. A topic of In-situ Resource Utilization (ISRU) is quite broad. It covers all the aspects of assuring sustainable human presence on the Moon, including mining and processing of local resources. The focus of this chapter is on mining and excavation technologies that will enable human exploration of the Moon and possibly lunar tourism. A number of examples of past, present and future lunar excavation technologies are presented.

Chapter 11 by Otis R. Walton shows that the wide size distribution and large fine-fraction of lunar regolith create challenges for handling and transport; unique features of the lunar environment contribute to additional concerns. Very fine lunar dust particles, some of them electrostatically charged, will affect any mechanical equipment on the moon. Fine dust will adhere to all surfaces. The hard vacuum leads to higher cohesion/adhesion forces. Under reduced gravity the behavior of fine granular solids changes to that of a more cohesive material - a factor of four reduction in gravity-level is enough to dramatically change the apparent character of a powder from free-flowing to cohesive, in the same size apparatus. The well-known Geldart classification of the fluidization characteristics of powders also changes at reduced gravity-levels, causing the boundaries between classifications to shift to larger particle sizes. Any equipment utilizing gravity to function (including hoppers, conveyors, and fluidized beds) will operate more slowly, and will need to be larger in size than comparable terrestrial designs. Proof-testing under appropriate conditions will be required for robust designs.

Chapter 12, by Simon Fraser reminds that the major milestones in lunar exploration have been made in the 1960s and 1970s, where first fly-by probes and orbiters were sent to the Moon. These first missions were soon followed by missions impacting on the lunar surface – either intentionally or unintentionally – before the first soft landing succeeded in 1966. The U.S. space program sent the first humans onto the Moon in 1969, whereas the Soviet program sent two surface rovers and a total of three sample return missions onto the Moon. Based on a discussion of the power systems of these past lunar surface exploration systems, future power system options are discussed. Power generation and energy storage requirements will become more and more challenging with the high-power lunar surface exploration applications considered in the future. This includes the next-generation crew landing craft, the Altair, as well as the power system options for all the other stationary and mobile mission elements required in the future lunar surface exploration and exploitation. Launch mass reduction is always an issue with space missions, and thus also in lunar exploration. In-Situ Resource Utilisation and In-Situ Propellant Production are innovative concepts aiming at a reduction of Earth launch mass by utilizing in-situ-produced materials and propellants. In case of the Moon, the in-situ production of oxygen from regolith is considered a technically feasible and very promising option not only for the Earth-return vehicle, but also for power generation options utilizing fuel cell technology, for instance. In addition, the recent discovery of hydrogen in shaded polar craters theoretically enables the in-situ production of hydrogen and oxygen. These could be used as propellants for ascent and Earth-return vehicles as well as being the enabler for advanced energy storage options in high-output power applications. Energy storage and power generation options are discussed alone as well as in the context of In-Situ Resource Utilisation in this chapter.

Chapter 13, by Alex Ignatiev and Alexandre Freundlich shows that the ubiquitous presence of solar energy in space begs for its significant utilization for the supply of energy for use by humankind in space. To that end, the striving of man to move into the cosmos will require extensive amounts of energy, principally in the form of electrical energy. The Moon presents a unique opportunity to generate the extensive amounts of energy that will be required by humankind in the future through the use of its indigenous resources. The Moon possesses in its regolith all of the chemical elements needed to make silicon solar cells. Furthermore, the Moon's surface is at an ultra-high vacuum condition, thereby allowing for the direct use of vacuum deposition technique without vacuum chambers to fabricate thin film devices such as solar cells on the lunar surface. As a result, the application of thin film growth technology in the vacuum environment of the Moon can yield the fabrication of thin film silicon solar cells directly on the surface of the

Moon for the development of a lunar power system as well as for the supply of electrical energy to space.

Chapter 14 by James D. Burke shows that lunar polar illumination conditions, with sunlight nearly uninterrupted and always horizontal, lead to a thermal situation that is much more benign than anywhere else on the Moon. Power supply and environmental control for a lunar habitat are as a result much more easily provided. Permanent shadow in crater bottoms causes temperatures there to be as low as 40 K, enabling efficient heat rejection systems. These thermal conditions constitute a natural resource independent of any frozen volatiles that may be found in those crater bottoms. Because the favorable topographic and illumination combination exists in only a limited region near each pole, it will be necessary to devise a practical policy and legal structure to assure sustainable and equitable use of the resource.

Chapter 15, by Xiongyao Li, Wen Yu, Shijie Wang, Shijie Li, Hong Tang, Yang Li, Yongchun Zheng, Kang T. Tsang and Ziyuan Ouyang shows that solar radiation is an exterior energy source of the Moon and represents a key resource with respect to returning to the Moon. It controls the variation of lunar-surface temperature during the lunation, and changes the thermal radiation properties of the lunar surface. In lunar Earth-based exploration, orbital exploration, and manned and unmanned lunar surface activities, solar radiation is an important factor which should be considered. This chapter introduces a method to estimate the solar irradiance and topography shadow on the Moon. Based on the theoretical model, the change of solar irradiance on the Moon was discussed. To comprehend the solar radiation on the flat lunar surface, a lunar-surface effective solar irradiance real-time model in terms of the relationship between lunar-surface effective solar irradiance and solar constant, solar radiation incidence angle and the Sun-Moon distance has been proposed. The theoretical erroneous percentage of this model is less than 0.28% in 100 years from 1950 to 2050. And the result of numerical simulation showed that the solar irradiance on the lunar surface would change form 1321.5 to 1416.6 $W \cdot m^{-2}$ in 2007, averaging 1368.0 $W \cdot m^{-2}$. The maximum effective solar irradiances on the lunar surface are estimated to be 37.3 $W \cdot m^{-2}$ and 38.0 $W \cdot m^{-2}$ at the South and North Poles, respectively. By considering the effect of lunar topography, a topography shadow mode parameterized with selenographic coordinates, elevations, and time is introduced. With high spatial resolution topography data, the solar illumination conditions at any selenographic coordination could be estimated by this model at any date and time. With the analysis of topography shadow, it was shown that the maximum topography shadow range was less than 180 km at incidence direction.

Chapter 16 by Trivandrum Eswaraiyer Girish and Sujatha Aranya deals with some aspects of the photovoltaic power generation on the moon considering the variations in the lunar physical environment as well as the current space solar cell technology. The authors have calculated the relative power expected from Si, GaAs, Gr III–V multijunction and CIGS solar cells when operated under extreme solar proton irradiation conditions and high lunar daytime temperature. 2–3% annual solar cell degradation is most probable in the lunar radiation environment. To improve the Si solar cell performance on the moon one can adopt techniques such as solar cell cooling, sunlight concentration and proton radiation shielding. The authors also discussed the prospects of photovoltaic power generation during the lunar night and on the polar regions of the moon using very large LILT solar cell arrays.

Chapter17, by Simon Fraser shows that fuel cells have played an important role in spacecraft power system design since the 1960s, when the first fuel cell power plant was operated aboard the Gemini 5 spacecraft. Fuel cells were also installed in the Apollo Command and Service Module. Plans to utilize fuel cells in the descent stage of the Lunar Module were cancelled due to various problems in the development phase. The relatively short surface stay and the low average power consumption made an all-battery-powered solution possible, and thus prevented the first lunar surface application of a fuel cell power plant with the Lunar Module. Fuel cell development was continued after the end of the Apollo Program, and improved versions of the Apollo-era alkaline fuel cell power plant are still used in the Shuttle Orbiters today. Terrestrial fuel cell technology has seen a strong increase in research & development since the early 1990s, mainly driven by advantages in conversion efficiency compared to internal combustion engines as well as the zero-emission operation if hydrogen is used as fuel. Space fuel cell technology strongly benefits from these terrestrial developments and is now considered for wide range of challenging applications in space, planetary and lunar exploration. One of these applications is the projected new crewed surface landing element of the Constellation Program, the Altair. Altair, previously known as the Lunar Surface Access Module, will be designed for a longer surface stay than its predecessor, the Apollo Lunar Module, and will also have a significantly higher electrical energy consumption. Fuel cells are therefore considered as primary candidate for the Altair power plant. Many more power generation and energy storage applications will become available as the mission planning of future robotic and crewed surface exploration elements proceeds. Possibilities and limitations of fuel cell technology in lunar surface exploration and exploitation applications are discussed in this chapter.

Chapter 18, by Simon Fraser shows that the diurnal temperature on the lunar surfaces ranges normally between a minimum temperature of the order of above 100 K and a maximum temperature of the order of up to 400 K. Recent investigations have revealed that polar-adjacent regions having a low elevation, such as the floors of impact craters, may never be directly illuminated, whereas regions of high elevation may be permanently illuminated. These regions are very interesting with respect to future in-situ resource utilization and surface base location, and are particularly challenging with respect to the environmental conditions faced. Lunar days and nights are very long due to the sidereal revolution period of 27.32 terrestrial days, which further contributes to the complexity of thermal management and control. The principal goal of spacecraft thermal control systems is to maintain equipment temperatures in specified ranges (usually room temperature) during all mission life. This guarantees optimum performance when the equipment is operating and avoids damage when the equipment is not operating. A number of different methods and technologies for thermal management and control have been applied in the past, or are currently being considered or developed. Heat pumps, whether mechanically- or thermally-driven, are considered in a number of different mission profiles and scenarios, as they provide the possibility to release heat at sink temperatures exceeding source temperature. Considering that radiator size and mass may become a limiting factor, particularly but not only with mobile mission elements, this is a very interesting and relevant factor. The objective of this chapter is to discuss the possibilities and limitations of heat pumps in space mission thermal control system applications in general, and with respect to future lunar surface exploration and exploitation applications in particular.

Chapter 19, by Viorel Badescu refers to the problem of how much heat at a given temperature can be converted into heat at another temperature on Moon. First, the thermal heat efficiency, represented by the heat gain factor, is calculated by using non-equilibrium thermodynamics of discrete systems. For performing this calculation, an irreversible Jaynes engine is introduced and compared with conventional heating by heat conduction. The profit of heat supply and the obtained higher stationary temperature of the heated room by using a Jaynes engine are calculated for various Moon thermal environments. An endoreversible approach is also used. It allows more detailed calculations. Comparison to the conventional heating demonstrates that energy saving is possible on the Moon by changing the traditional heating technology. Next, the chapter refers to the problem of how much heat is necessary for oven operation on Moon. An endoreversible thermodynamics approach is used. The optimum operation temperatures and the oven heat gain factor depend on three parameters incorporating both design and thermal factors. The quantities associated with reversible operations are recovered in the limiting case of infinitely large conductance. In the more realistic case of finite sizes one shows that the equipartition principle acts for the conductances of the heat engines and heat pumps driving the endoreversible oven, but does not act at the level of the system. The results show that reducing the present day energy consumption by an order of magnitude for oven operation on Moon is, in principle, possible.

Chapter 20 by Sandra Haeuplik-Meusburger and Kursad Ozdemir shows that, taking into account the limited knowledge of 'Moon Living Conditions', which have been acquired mainly through the Apollo missions and late orbital inspections, the advanced phase of lunar exploration still lies ahead as a major task. Advanced exploration means extended capabilities and mission durations on the lunar surface. To sum up the goals and scope, the humans boldly looks forward to be on the Moon and furthermore to expand their research and the living and working possibilities there. Deployable structures come into play, providing efficiency in multiple segments of the lunar exploration missions. Ranging from space suits to collapsible habitation systems, deployable elements of lunar human exploration present an array of assets that may play crucial roles in future expeditions, just as the deployable lunar rover did for the Apollo missions. Within the aforementioned perspective, this chapter provides insight to designed and applied deployable habitation systems for human lunar exploration.

Chapter 21 by Alexey Kondyurin reminds that despite the Moon is the nearest celestial body to our Earth, only short term stays have been possible till now. In order for long-term missions to be possible, large pressurized constructions are needed. The 15-20 m^3 Altair habitat planned in the Constellation Program and the 6.65 m^3 pressurized crew compartment volume realized in the Apollo program are insufficient. Hundreds of cubic meters per crew member are required for living area, working area, greenhouse with sufficient plants and animals for food, air and water recovery and storage. There is only one way to get sufficient volume of the pressurized crew compartment: an inflatable construction. The soft shell of an inflatable construction can be prepared on Earth, folded and transported in a small container to the Moon. Then the shell is inflated to a sufficiently bigger volume than the container. After that a chemical reaction rigidization process in shell material has to be started to get durable frame. The chapter discusses experimental and theoretical results to answer: What are problems associated with curing a composite material on the Moon? How can we cure a liquid composition in vacuum? How can cosmic rays help with curing? How curing can be done in lunar frost? A cylindrical construction with diameter of 10 m and length of 80 m with the internal volume of the habitat is about 6300 m^3 is calculated. This is about 1000 times bigger than the pressurized crew compartment in Apollo Moon module, and about 300 times bigger than the planned Altair habitat in Constellation Program.

Chapter 22 by Lotta Viikari reminds that the Moon does not fall under national sovereignty of any state. Lunar resource activities are thus governed by public international space law. Unfortunately, the international agreements pertaining to such activities are relatively old and cannot answer the demands of the modern space sector. The Moon Treaty of 1979 was an attempt to regulate resource activities on the Moon to the benefit of all humanity: it declares lunar resources to be the Common Heritage of Mankind and envisages the establishment of an international regime as soon as exploitation is about to become feasible. However, as the number of actors interested in outer space and having the resources to explore it multiplied over the decades, the process of producing space law became increasingly complicated. Consequently, there has not been much progress in developing the legal regime for the exploitation of lunar resources since 1979. If celestial resource activities are pursued, complex questions of law, politics, economics and technology cannot be avoided. The problematic issues in the use of the natural resources of the Moon include the role of private actors; environmental concerns; the possibilities of developing countries; the purpose of use of the resources; interests of future generations; etc. At the moment, pressure is rapidly building on the international community to establish a more acceptable and in every aspect more distinctive legal regime applicable to the exploitation of space resources and, in particular, one under which they can also be utilized for commercial purposes. In all likelihood, various models of regulation will be needed, depending on the particular circumstances of the resources. This chapter introduces first some questions of general nature related to various approaches to lunar resource activities. Next, it examines the historical development of the relevant legal regulation. The Moon Treaty is studied in detail, as well as reactions of the international community to the Treaty – including the eventual political failure of the Moon Treaty. The current status and future prospects of the Moon Treaty are considered after a short assessment of the so-called Space Benefits Declaration of 1996. It becomes obvious that the current law of outer space is unable to offer much for the practical needs of the space sector in lunar resource utilization. Equally obvious is that an innovative, flexible and open-minded approach is needed in looking for solutions to the shortcomings of the present regulatory framework.

Chapter 23 by Virgiliu Pop shows that the issue of ownership of planetary resources is crucial in an era where the Moon is seen as a watery treasure chest, capable of quenching the NewSpace entrepreneurs' thirst for rocket fuel. This chapter offers a legal and ideological overview of the issues at stake in the field of exploitation of planetary resources; in lay terms, it aims to offer an answer to the question "Who owns the Moon?" After critically analyzing and dismantling with legal arguments the trivial issue of sale of extraterrestrial real estate, the chapter proceeds to examine the commons regime – currently in place on the Moon according to the Outer Space Treaty – and its two challengers – the Moon Agreement's common heritage of mankind regime on the left of the ideological spectrum, and the frontier paradigm on the right. The chapter concludes that the most pertinent arguments incline the balance towards marketable property rights and towards legal norms supporting these.

Chapter 24 by Marco Cenzon and Dragoş Alexandru Păun tries to offer a structured approach to the design of a future lunar telecommunications and navigation service network with the aim of obtaining a sustainable global exploration effort. Based on available information regarding the strategies of space agencies worldwide, the lunar exploration effort has been divided into several phases. The number of users and their particular needs will drive the system architecture and a generational approach must be taken to ensure that the needs of a growing number of users and their complexity will be met. The NASA Space Communication Architecture Working Group (SCAWG) predicts that by the year 2030, a single spacecraft would be capable of transmitting in an 8 hour interval a quantity of data equivalent to the entire quantity present in the Planetary Data System today. A set of requirements has been drafted to meet mission necessities and was separated into three categories: Communication, Navigation and Common Requirements. The technical analysis presented has been largely derived from the project work activity performed during the fourth edition of the SEEDS activities. The service architecture is based on an incremental constellation of lunar satellites able to relay data and to transmit the navigation signal. These satellites are supported by a series of Moon and Earth surface elements such as orbitography beacons, Earth orbit satellites and ground antennas. Locally on the lunar surface, the vehicles and fixed elements will perform inter-element communication and short-range navigation by means of dedicated on-board equipment.

Chapter 25 by Henry W. Brandhorst describes a concept for the commercial infrastructure that must be in place before access to the lunar resources can become economically feasible. An efficiently-functioning infrastructure will encompass a range of transportation nodes and the means to supply energy supply to the lunar surface on a full time basis. The architecture described uses conventional chemical propulsion to LEO and electric propulsion (EP) for the transit to GEO and LLO. Fuel supply depots in LEO are utilized to supply fuel for the electric propulsion vehicles transiting to LLO. EP vehicles use lightweight, radiation-tolerant, high voltage solar arrays that enable multiple transits through the Van Allen radiation belts with minimal damage. Power is supplied to the lunar surface via infrared laser power beaming to III-V-based single junction solar cells. Detailed orbital mechanics studies were run to assess the number of satellites and their orbits that would enable full time energy at any site on the lunar surface. This finding enables energy-rich bases to be established over the entire surface. Thus the combination of an integrated transportation system and an energy supply approach enables an economically feasible approach to full lunar resource utilization.

Chapter 26 by Werner Grandl suggests how humanity can build an initial Lunar Base by using current technologies, minimizing costs and providing maximum safety for astronauts. The design study proposes a modular Lunar Base built at least of six cylindrical modules. These light-weight modules are made of aluminium sheets and trapezoidal aluminium sheeting to provide stiffness and structural redundancy. The proposed double-shell structure allows to fill the cavity between outer and inner hull with lunar material (regolith) as an in situ resource for shielding. The initial Stage 1 can be inhabited by eight astronauts. The station can be enlarged by stages, finally becoming an "urban structure" for some dozens of inhabitants. For launching -as an existing rocket- the Ariane 5 launcher is assumed. To land the modules on the lunar surface we propose the Teleoperated Rocket Crane, which is assembled in lunar orbit. This vehicle can land, move and assemble the modules on the lunar surface. The entire landing and assembling process is operated by remote control from a manned Lunar Orbiter. Human transport to lunar orbit, crew exchange, supply of resources for the astronauts and fuel for vehicles are beyond the scope of the chapter.

Chapter 27 by Haim Benaroya, Stephen Indyk and Sohrob Mottaghi shows that manned exploration calls for a return to the Moon, requiring the construction of structures for long term habitation, challenging the engineer and the astronauts that must erect the structure. Given the costs associated with bringing material to the Moon and the difficulties of lunar construction, what is the best way to erect habitable structures on the lunar surface? The authors briefly review some key studies of the role that ISRU can play in the manned exploration and settlement of the Moon and beyond. In addition they outline a "grand vision" for the design of a layered manufacturing machine that operates under primarily solar power and that can construct a generic structure in advance of astronaut arrival.

Chapter 28, by Magnus Larsson and Alex Kaiser argue that *bacillithic cratertecture* – an architecture resulting from the "deliberate displacement of a terrestrial microorganism in an extraterrestrial context" – could be a way of simplifying and accelerating the construction of lunar habitats while limiting cost. Based on a terrestrial architecture that pushes its analogous space architecture towards the realms of microbial geotechnology and beyond, consiliently bridging the biological, digital, and architectural worlds in order to create a foundation for future strategies that could be based on biological computing and programmable built environments, *Moon Dune* is a forward-thinking scheme that conjures up new possibilities for advanced *in-situ* resource utilisation (ISRU) strategies – the idea of putting local raw materials to practical use as humanity expands further into the solar system. By working within the lunar regolith itself, the authors seek to minimise the amount of material that has to be brought to the Moon, as well as the amount of material that needs to be brought back. Harnessesing the emergent behaviours of bacterial cellular agents, the project utilises a radically modular design strategy that opens up a much wider array of possible constructed forms and spatial arrangements, while potentially offering a resource-efficient way of sustainably transporting and utilising biologically manufactured, renewable building materials at the molecular scale. Engineering strains of the approximately five nonillion bacteria on Earth, only a fraction of which have been studied, the *Moon Dune* architecture uses microbial construction workers to create the theoretical material *bacillith*, a potentially reconfigurable, computational biomaterial that could eventually form the foundation for a spectacular design strategy falling somewhere between *in silico* and *in vivo*: sculpting crater habitats out of ready-made lunar buildings through biologically programmable processes based on biochemical precipitation within the granular medium of the regolith. The result could be networks of *habitable biological machines* stretching out across a 4.6-billion-year-old extraterrestrial construction site. The advantages of such a strategy are potentially vast: once a safe and non-contaminating technology is in place, rather than coping with the enormous costs of delivering building materials or entire compounds to the Moon, vials of microbes – a renewable resource that can be grown *in situ*, could be transported instead. Rather than carrying materials, we could carry microorganisms, tools with which to sculpt the existing material masses, bacterial *micronauts* programmed into lunar bases by a new generation of free-thinking space architects inspired by recent advances in synthetic biology.

Chapter 29 by Mike H. Ryan and Ida Kutschera deals with the fundamentals of modern Lunar management, focusing on the private sector. "Business as usual" is a phrase that is commonly used for a wide range of business activities familiar to most managers. While the phrase can have pejorative overtones, the sense is that managerial practice has a range of norms and common considerations. This perception is not particularly accurate when considering special management settings in very remote locations or in extreme circumstances. Traditional management discussion has centered on examples of these locations such as Antarctica or in extremely remote duty stations. The problems of isolation and environmental extremes coupled with complications of transportation make managing in these places much more difficult and substantially different. To some extent, they require adjustments of management practice and additional considerations with respect to personnel. Current contemplation of long-duration private-sector space activities that will eventually include living and working on the Moon necessitate extension of our understanding of what it means to work in remote and hostile environments. This chapter examines many of the issues and problems that prospective managers and employees must consider prior to initiating commercial activity on the Moon.

Chapter 30 by Haym Benaroya presents a short fictional history of the path humanity took in its manned exploration and settlement of our Solar System. This year – the year 2169 – is a celebration of the day the first men landed on the Moon, where the first steps were taken in the expansion of our civilization to the Moon and beyond. These past 200 years prove beyond a doubt that nothing is beyond humanity's reach – intellectually or financially. *Apollo* was our first foray into the Solar System. Space was and is the world's *West* – as in *Go West* for new opportunities for freedom and limitless economic growth. Much of what we have accomplished in the Solar System during the past 200 years had been predicted, but there were surprises. It was all fantastic; we are still in our infancy in space. We are successful today because of our parents who made their vision a reality. Our children's success tomorrow depends on the efforts of their parents to build upon that earlier vision.

The book allows the reader to acquire a clear understanding of the scientific fundamentals behind specific technologies to be used on Moon in the future. The principal audience consists of researchers (engineers, physicists, biologists) involved or interested in space exploration in general and in Moon exploration in special. Also, the book may be useful for industry developers interested in joining national or international space programs. Finally, it may be used for undergraduate, postgraduate and doctoral teaching in faculties of engineering and natural sciences.

Viorel Badescu

Acknowledgments

A critical part of writing any book is the review process, and the authors and editor are very much obliged to the following researchers who patiently helped them read through subsequent chapters and who made valuable suggestions: Dr. Angel Abbud-Madrid (*Colorado School of Mines, Golden, CO, USA*), Dr. Manuela Aguzzi (*Altec S.p.A., Italy*), Prof. Bulent Akinoglu (*Middle East Technical University, Ankara, Turkey*), Prof. Elisabeth Back Impallomeni (*University of Padua, Italy*), Prof. Nanan Balan (*University of Sheffield, UK*), Prof. Haim Baruh (*State University of New Jersey, NJ, USA*), Dr. Carsten Baur (*European Space Agency, Noordwijk, The Netherlands*), Prof. Leonhard Bernold (*School of Civil & Environmental Engineering, UNSW Sydney NSW, Australia*), Prof. William R. Bigler (*Louisiana State University, Shreveport, LA, USA*), Dr. Giorgio Borriello (*Aviospace S.r.l., Torino, Italy*), Prof. Iver H. Cairns (*University of Sydney, Australia*), Mr. Richard B. Cathcart (*Geographos, Burbank, CA, USA*), Dr. Feng Chen (*Institute of Geochemistry, Guiyang, China*), Dr. Jeremy Curtis (*UK Space Agency*), Dr. Peter A. Curreri (*NASA-Marshall Space Flight Center, Huntsville, AL, USA*), Dr. Jason DeJong (*University of California, Davis, CA, USA*), Prof. Alexandru Dobrovicescu (*Polytechnic University of Bucharest, Romania*), Dr. Christopher Dreyer (*University of Colorado, Boulder, CO, USA*), Dr. Jeniffer E. Edmunson (*Marshall Space Flight Center, Huntsville, AL, USA*), Dr. William C. Feldman (*Planetary Science Institute, New Mexico, USA*), Dr. Bernard Foing (*European Space Agency*), Mr. Joseph Friedlander (*Shave Shomron, Israel*), Dr. Rupert Gerzer (*Institute of Aerospace Medicine, Cologne, Germany*), Dr. Paul S. Greenberg (*NASA Lewis Research Center, Cleveland, OH, USA*), Prof. Viktor Hacker (*Technical University Graz, Austria*), Prof. Kari Hakapää (*University of Lapland, Helsinki, Finland*), Dr. Daniel Hernandez (*Devil-Hop Consultancy, France*), Dr. J. Marvin Herndon (*Transdyne Corporation, San Diego, CA, USA*), Dr. Hugh Hill (*International Space University, Strasbourg, France*), Dr. Joe T. Howell (*NASA Marshall Space Flight Center, USA*), Dr. Sharon A. Jefferies (*NASA Langley Research Center, USA*), Prof. Mikhail Kreslavsky (*University of California, Santa Cruz, CA, USA*), Dr. Xiong-Yao Li (*Lunar & Planetary Science Research Center, Guiyang, China*), Dr. Yang Liu (*University of Tennesses, Knoxville, TN, USA*), Dr. Thomas B. McCord (*The Bear Fight Institute, Winthrop, WA, USA*), Dr. Carole A. McLemore (*NASA, Marshall Space Flight Center, Huntsville, AL, USA*), Dr. Wendell Mendell (*NASA Johnson Space Center, Houston, TX, USA*), Dr. Erwin Mondre (*Wien, Austria*), Dr. Hiroko Miyahara (*The University of Tokyo, Chiba, Japan*), Prof. Mark Nagurka (*Marquette University, Milwaukee, WI, USA*), Dr. L. E. Nyquist (*NASA Johnson Space Center, Houston, TX, USA*), Dr. Galina S. Nechitailo (*Institute of Biochemical Physics, Moscow, Russia*), Dr. Sarah Noble

(*Planetary Science Division, NASA, Washington, DC, USA*), Dr. Makiko Ohtake (*Japanese Space Agency, Sagamihara, Kanagawa, Japan*), Dr. Mark J. O'Neill (*Entech Solar Inc, Fort Worth, Texas, USA*), Dr. Stephen Ransom (*SRConsultancy, Stuhr, Germany*), Prof. Radu Rugescu (*Polytechnic University of Bucharest, Romania*), Dr. Max Schautz (*European Space Agency, Noordwijk, The Netherlands*), Dr. Christian M. Schrader (*NASA, Huntsville, AL, USA*), Prof. Scott W. Sherman (*Texas A&M University, Corpus Christi, TX, USA*), Dr. Larissa Starukhina (*Astronomical Institute of Kharkov National University, Kharkov, Ukraine*), Dr. Christopher Stott (*International Institute of Space Commerce, Isle of Man, UK*), Dr. Matthew Stuttard (*Astrium Ltd., Stevenage, Hertfordshire, UK*), Prof. Paul van Susante (*Colorado School of Miner, Golden, CO, USA*), Dr. Berin Michael Szoka (*TechFreedom, USA*), Dr. Paul Todd (*Techshot, Inc., Greenville, IN, USA*), Prof. Rainer Wieler (*Institut f. Geochemie und Petrologie, Zürich, Swiss*), Prof. MasamichiYamashita (*Institute of Space and Astronautical Science, Kanagawa, Japan*), Dr Kris Zacny (*Honeybee Robotics Spacecraft Mechanisms Corporation, Pasadena, USA*), Dr. Limin Zhou (*East China Normal University, Shanghai, China*), Dr. Yongliao Zou (*National Astronomical Observatories, Chinese Academy of Sciences*).

The editor, furthermore, owes a debt of gratitude to all authors. Collaborating with these stimulating colleagues has been a privilege and a very satisfying experience.

Contents

1 A Survey of Geologic Resources

Jennifer Edmunson and Douglas L. Rickman

Marshall Space Flight Center,
Huntsville, AL, USA

1.1 Introduction

This chapter focuses on the resources available from the Moon itself: regolith, geologically concentrated materials, and lunar physical features that will enable habitation and generation of power on the surface. This chapter briefly covers the formation of the Moon and thus the formation of the crust of the Moon, as well as the evolution of the regolith. The characteristics of the regolith are provided in some detail, including its mineralogy and lithology. The location of high concentrations of specific minerals or rocks is noted. Other ideal locations for in situ resource utilization technology and lunar habitation are presented.

This chapter is intended to be a brief review of current knowledge, and to serve as a foundational source for further study. Each concept presented here has a wealth of literature associated with it; the reader is therefore directed to that literature with each discussion. With great interest in possible manned lunar landings and continued study of the Moon by multiple satellites, the available information changes regularly.

1.2 Formation and Evolution of the Moon

Given the similarities in composition, the nearly-identical oxygen isotopic signature of the Earth and Moon, and the angular momentum of the Earth-Moon system, the Moon likely formed when an oblique, "Giant Impact" occurred between the proto-Earth and a Mars-sized impactor (e.g. Hartmann and Davis 1975). Debris from this impact, made of material from both the impactor and the proto-Earth, was lofted into orbit around the Earth (e.g. Canup and Asphaug 2001) and may have undergone diffusive equilibration (e.g. Pahlevan and Stevenson 2007).

Accretion of this debris, by gravity, contained sufficient energy to melt the Moon and form it into a sphere. At this stage, the Moon had a magma ocean.

There are two widely recognized models for the formation of the lunar crust. The first involves crystallization of minerals during the slow cooling of the magma ocean. This model, termed the "lunar magma ocean" model, claims that dense minerals, such as olivine and pyroxene, sank to the bottom of the lunar magma ocean as they formed, while the mineral anorthite (Ca-rich plagioclase), which was less dense than the residual relatively Fe-rich surrounding magma, floated to the top and formed the lunar highlands crust (e.g. Smith et al. 1970; Wood et al. 1970). In the second model, the "postmagma ocean" model, the magma ocean cooled, but did not form an anorthositic (anorthite-rich) crust (Longhi 2003). Early convection and density adjustments caused heating and partial melting of the upper layer of the Moon; the liquid produced during partial melting separated into a less dense anorthositic parent liquid and a denser pyroxenite (pyroxene-rich) liquid. The less dense anorthositic liquid then reached the surface and formed the anorthosites. Although there remains some debate as to which model is more likely, both explain the presence of an anorthositic crust on the Moon.

Both formation models agree upon the need for a density overturn in the Moon after crust formation. The density of cumulates (a term used to describe accumulated minerals) increases with decreasing depth because iron-rich minerals form after less dense magnesium-rich minerals. Thus, the denser minerals form after the less dense cumulates have already crystallized and settled to the bottom. For cumulate pile gravitational stability, the densest cumulates must be at the bottom. As a consequence, cumulates overturned in the mantle. Recent articles pertaining to the lunar magma ocean and crustal formation on the Moon call upon this process to explain the array of geochemical sources observed for lunar samples (e.g. Parmentier et al. 2002; Elkins-Tanton et al. 2011).

1.3 Impacting the Lunar Surface

After crystallization of the anorthositic crust, the Moon was subjected to intense bombardment by meteorites; the lunar surface was shaped by over four billion years of impacts. The resulting broken rock that separates the Moon's bedrock from space, the lunar regolith, evolved from repeated impacts, including those from micrometeorites. The repeated impact "tilling" of the regolith gave rise to the term "gardening", which describes the impact-driven process by which fresh surface material is exposed and other surface material is buried on the Moon. This process also gave rise to the lunar "soil"; collections of regolith particles less than 1cm in size are referred to as soils (McKay et al. 1991). Figure 1.1 shows some attributes of the regolith.

Fig. 1.1 Regolith components and physical properties. (a) This large boulder is a part of the lunar regolith. Astronaut Harrison Schmitt for scale. NASA photograph AS17-140-21496. (b) In situ regolith. NASA photograph AS11-40-5878. (c) A portion of Apollo sample 76335 (a magnesian suite troctolite). (d) Scanning electron microscope image of an agglutinate. Photo credit D. S. McKay, NASA Johnson Space Center. (e) Orange and black volcanic glass beads collected during the Apollo 17 mission. NASA photograph S73-15171. Beads range in size from 20 to 45 microns. (f) Lunar core sample 10004. NASA photograph S69-40945. Approximately half of the core section is unfilled due to a core design flaw and the dry, uncompressible property of the regolith. A large rock aided in keeping the remaining regolith in place

One of the largest crater basins in the Solar System is on the Moon; it is called the South Pole-Aitken Basin. This crater basin is also one of the oldest in the Solar System. According to Wilhelms (1987), the average crater size decreased as the age of the Moon increased, indicating a decrease in the size of the impactors and the flux of impactors over time. That is not to say small impacts were not part of the early bombardment of the Moon, but that they were covered by larger (or later) crater ejecta. The impactor size was likely a complete continuum, from the largest to the smallest size possible; in addition to the large impact craters observable from Earth, the Moon has extremely small, even microscopic, craters - these are from micrometeorites (typically 1 mm or less in size). Figure 1.2 is a cross-section of a micrometeorite impact crater on the surface of Apollo 16 sample 62236.

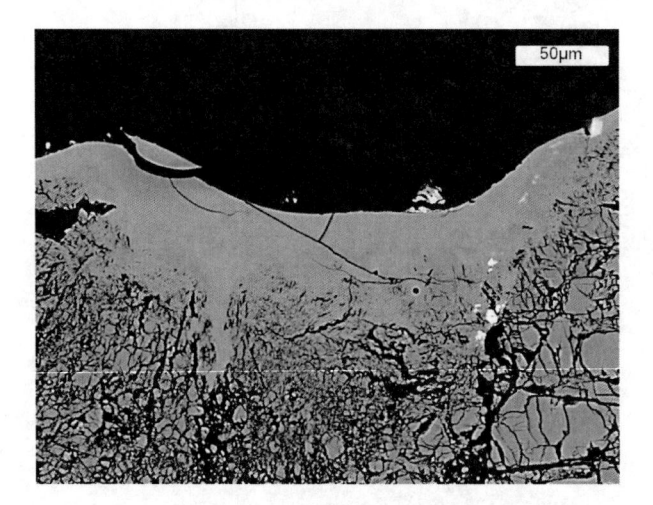

Fig. 1.2 Scanning electron microscope image of a micrometeorite impact crater cross-section in an Apollo 16 sample (62236, 58). Black indicates edge of the thin section. Gray is plagioclase. Note the broken appearance of plagioclase in the majority of the sample, and the smooth glass of the impact crater site. White phases are pyroxenes.

The Moon experiences hypervelocity impacts (e.g. Hörz et al. 1991; French 1998). That is, an impactor will hit the lunar surface traveling at speeds greater than 3 km/s (Hörz et al. 1991). At impact, the lunar surface is compressed at the impact site. A shock wave travels outward from the impact site, pushing all surrounding rocks outward and, depending on the energy, vaporizing the impacting body. A portion of the Moon will also be vaporized and more of it will be melted. If the melt is small enough it will chill rapidly and make impact melt glass. The shock wave throws out near-surface material (ejecta) which blankets the surrounding area. In most cases, the uppermost impacted rocks are melted. Rocks surrounding the impact site that are not melted will rebound as a result of the rarefaction wave, but will be brecciated (broken) in the process as they do not have sufficient elastic properties. Craters are filled with vertically ejected material and portions of the crater walls that exceeded the angle of repose. This includes any original ejecta on the original crater rim. The final crater, known as the apparent crater, appears shallower than the original crater. Impacts with more energy will cause uplift in the center of some craters; the central peak forms as a result of concentrated upward motion of a rarefaction wave (e.g. Dence 1968). These high-energy, central uplifted craters are called complex craters, whereas those without central peaks are called simple craters.

Micrometeorite impacts create tiny simple craters, and are also responsible for the formation of lunar agglutinates. Agglutinates are aggregates of other regolith particles on the lunar surface, including other agglutinates. These aggregates are held together by vesicular (hole-bearing) impact melt glass. Agglutinates are typically less than 1 mm in size and can constitute approximately 60% of studied mature lunar soils by volume (Papike et al. 1998). In addition, the abundance of agglutinates increases with decreasing grain size and increasing maturity of the regolith (e.g. McKay et al. 1974; Taylor et al. 2001). Maturity, in the lunar sense, describes how long the surface layer has been exposed to space; this refers to space weathering. Micrometeorites can also create vapor coatings on lunar materials. These coatings do not have the same chemical characteristics as the host sample, therefore they must have been deposited and not created in situ (e.g. Keller and McKay 1993, 1997; Hapke 2001; Taylor et al. 2010). This type of space weathering is also apparent on the surface of Mercury (e.g. Cintala 1992; Hapke 2001).

1.4 Space Weathering

Impacts are included in a group of processes referred to as "space weathering". In addition to impacts, space weathering describes the effects of cosmic ray, solar wind, and ultraviolet (UV) radiation exposure. These particular processes influence the exposed surface, but the effects can be observed in samples placed at depth due to gardening.

Galactic cosmic ray particles have energies of 100 MeV to 10 GeV, and can change the isotopic and elemental chemistry of the surface with which they interact. Neutron capture reactions can create higher concentrations of specific isotopes (e.g., ^{150}Sm from ^{149}Sm) on a planet's surface; this effect decreases with depth (e.g. Lingenfelter et al. 1972). Scientists have used this technique to determine the age of exposure and the depth at which the sample was exposed to cosmic rays. The cosmic ray particles can also produce lighter elements and isotopes, such as ^{15}N used by Mathew and Marti (2001) to calculate the rate of spallation (the expulsion of atoms or atomic nucleus components) at the lunar surface.

The Moon's lack of an atmosphere allows very concise study of solar wind particles and their implantation. Solar wind describes the abundant, low-energy (1 keV) particles (i.e., protons, electrons, and alpha particles to a lesser extent, as well as light volatile elements) being emitted by the Sun and implanted, typically to a depth of a few tens of nanometers, into the lunar surface (e.g. Lucey et al. 2006 and references therein). Solar energetic particles, those accelerated to 1-100 MeV, can be implanted up to approximately 1 cm in depth. Much of the solar-implanted material remains implanted, but some is able to diffuse out of the rock. Thus, solar-implanted material is not used to determine the age of exposure of the lunar surface.

Ultraviolet radiation exposure and its effect on minerals is a relatively new field of study. Wong et al. (2010) noted degradation of forsterite during exposure to UV radiation and observed a shift in the peak center and shape of Raman spectra. Wong et al. (2010) claim that UV radiation causes the crystal lattice order to become disordered, and may in fact liberate O_2 gas. Future work is planned to further quantify the effects of UV radiation on minerals.

Lunar geologists use maturity as an index used to describe how long a regolith surface resided at or near the surface of the Moon. Mature lunar regolith has evidence of space weathering; changes in isotopic and elemental composition, micrometeorite impacts, a darkening of the albedo, and the formation of nanophase iron. The maturity of the lunar regolith can indicate its grain size, from the counteracting properties of comminution (breaking apart, destruction) by impact and agglutination (welding together of particles with impact melt glass); this process was studied in detail by McKay et al. (1974) and Lindsay (1975). The magnetic properties of lunar materials are also influenced by space weathering. Superparamagnetic nanophase iron is formed by space weathering; it carries a characteristic electron spin resonance of $g = 2.1$ (Hapke et al. 1975). This material also lowers the albedo of lunar regolith, in particular the fine fraction (<25 µm) when compared to powdered lunar rocks of analogous composition (Hapke et al. 1975; Pieters et al. 1993). Space weathering induces a higher absorption coefficient (the change in albedo), an absence of absorption bands, and a decrease in the absorption with increasing wavelength (Hapke et al. 1975). A way to quantify the maturity of lunar soils by using their optical (spectral) characteristics, abbreviated OMAT, was presented by Lucey et al. (2000). This particular measure of maturity compares the reflectance ratio of 950 and 750 nm to a hypothetical dark red "hypermature" end-member composition. Another way to express maturity is the use of I_s/FeO (Morris 1978). In this measure, the intensity (I_s) of the ferromagnetic resonance of the sample at $g = 2.1$ is divided by the concentration of FeO in the sample. Immature samples have a ratio of 0 to 29, submature from 30-59, and mature is greater than or equal to 60. Note that maturity does not depend on the type of rock at the lunar surface.

1.5 Characteristics of the Lunar Regolith

1.5.1 Physical Description

The lunar regolith is composed of mineral fragments, impact melt glasses and glass beads, agglutinates, and localized occurrences of volcanic glass beads. The depth of the regolith is estimated to be between 4 and 5 m in mare areas, and 10 to 12 m in highland areas (McKay et al. 1991); and has been determined, using radar and optical data, to average between 5 and 12 m globally (e.g. Shkuratov and Bondarenko 2001). In contrast to all terrestrial geologic materials, the lunar regolith is not effectively sorted in either size or shape. The size of lunar regolith particles varies from submicron to stadium-sized boulders on the surface. However, most of the regolith (up to 90% in some measurements) is below 1 cm (Heiken et al. 1974). Carrier (2003) compiled measurements from the sub-centimeter fraction and determined a particle size distribution (Figure 1.3). In general, the shape of lunar regolith particles is very angular because there is no observed widespread weathering process on the Moon equivalent to that on the Earth (i.e., wind and water-involved chemical and mechanical breakdown processes). Some particles, including impact-generated and volcanic glass beads and some agglutinates, have rather spherical or smooth shapes. Quantitative lunar particle shape data is lacking, although a few have attempted to characterize them (e.g. Carrier et al. 1991; Liu et al. 2008; and references therein). Due to the majority of the particles being elongated, and the mechanisms such as moonquakes that pack the particles, preferred orientation of particles has been noted in Apollo core samples (e.g. Carrier et al. 1991). The packing of these particles gives rise to the density of the regolith. The average bulk density of the regolith was determined to be 1.5 g/cm^3 and includes void spaces; the specific gravity of the lunar regolith averages 3.1 g/cm^3 and does not include void spaces (Carrier et al. 1991, and references therein). Carrier et al. (1991) determined that, if in situ lunar regolith was excavated, the top 3 m would remain vertical, and a 60° slope would survive to a depth of approximately 10 m. If lunar regolith was excavated and compacted, a 45° slope would survive. If the regolith was simply dumped on the surface, the angle of repose would be near 40°.

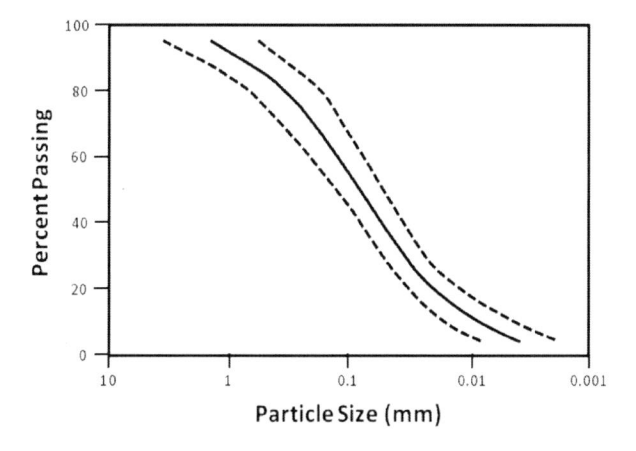

Fig. 1.3 Particle size distribution for lunar regolith samples < 10 mm in size (redrawn from Carrier, 2003). Solid line indicates calculated average size of lunar regolith particles. Dashed lines indicate one sigma variation of the average particle size. Percent passing indicates the relative amount of material passing through a sieve with a particular mesh size.

1.5.2 Particle Textures

There are numerous textures observed in lunar regolith particles. The following list explains some general geologic terms for textures in regolith particles. These terms are explained here to provide a way for simulant users and regolith examiners to understand some of the geologic literature on the Apollo and simulant samples.

- Holocrystalline: entirely crystalline (this does not necessarily mean the particle is a single crystal, although that may be the case). An example would be a brecciated mineral or rock clast (piece).

- Holohyaline: entirely glass. Examples include impact melt glass, agglutinate glass, and most volcanic glass beads.
- Hypocrystalline: contains both crystals and glass. Examples include agglutinates and some impact melts, as well as rocks created from the shock lithification (making into a rock) of regolith (these are called regolith breccias).

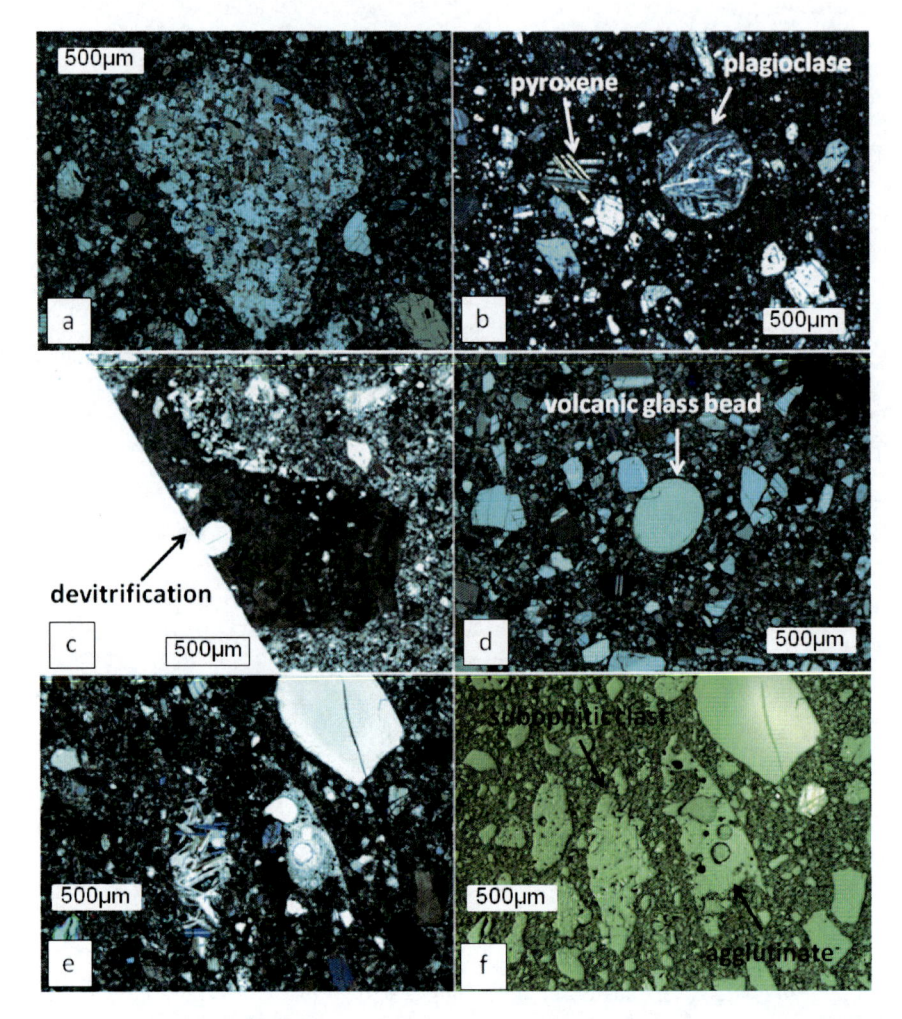

Fig. 1.4 Some particle textures for Apollo 16 regolith. (a) Poikiloblastic, coarse granulite (center clast). Photo taken in plane polarized light. Thin section 60009, 6020. (b) Intersertal plagioclase sphere with a pyroxene grain to left showing exsolution lamellae (two pyroxene compositions at different crystallographic orientations). Picture taken in crossed nicols. Thin section 60009, 6020. (c) Spherulitic texture as a result of devitrification. Photo taken in plane polarized light. Thin section 68001, 6031. (d) Holohyaline volcanic glass bead. Photo taken in plane polarized light. (e) Subophitic clast, to the left of a hypocrystalline agglutinate, picture taken in plane polarized light. (f) Reflected light image of (e), showing the size of the subophitic clast and agglutinate.

Some of the textures observed in lunar regolith particles have more specific textures. Figure 1.4 illustrates some common textures observed in Apollo 16 regolith core samples.

- Poikiloblastic: a metamorphic texture in which a single crystal of one mineral encloses many smaller grains of other minerals.
- Granulitic: a completely recrystallized rock that retains none of its original texture.
- Intersertal: unaligned plagioclase crystals that are relatively large in size making up a substantial portion of a rock.
- Spherulitic: fibrous crystals that radiate from a point and are clustered in spherical groups. Can be a sign of devitrification (the formation of crystals in hot glass).
- Subophitic: pyroxene and plagioclase grains of similar size, with pyroxene encompassing the ends of some plagioclase grains.

1.5.3 Chemical Composition

A review of the chemistry, mineralogy, and petrology of lunar soil samples < 1 mm in size was published by Papike et al. (1982). Table 1.1 shows data from this reference, collected initially by instrumental neutron activation analysis.

Table 1.1 Compositions of Apollo soils (< 1 mm) in oxide weight percent (Papike et al. 1982)

	10084, 1591	12003, 464	14163, 778	15271, 27	64501, 122	72501, 15
SiO_2	41.3	46.9	47.3	46.0	45.3	45.2
TiO_2	7.5	2.3	1.6	1.5	0.37	1.4
Al_2O_3	13.7	14.2	17.8	16.4	27.7	20.1
Cr_2O_3	0.290	0.387	0.200	0.350	0.090	0.230
FeO	15.8	15.4	10.5	12.8	4.2	9.50
MnO	0.213	0.195	0.135	0.162	0.056	0.120
MgO	8.0	9.2	9.6	10.8	4.9	10.0
CaO	12.5	11.1	11.4	11.7	17.2	12.5
Na_2O	0.41	0.67	0.70	0.49	0.44	0.44
K_2O	0.14	0.41	0.55	0.22	0.10	0.17
Total	99.8	100.8	99.8	100.4	100.3	99.7

Note: Apollo sample number (5 digits) is followed by split number.

Note there are several conventions in such a table that are standard practice in geology. The elements reported are referred to as the major, rock forming elements. Other elements are certainly present and could even be nearly as abundant as some of the elements reported here. The compositions are presented as oxide weight percent, not element weight percent; this does not imply that these specific oxides exist on the Moon and are not intended to represent oxide minerals or species. For the actual chemistry see the following section on mineralogy. The total will not sum to 100% for several reasons, such as unmeasured components, non-stoichiometry, and time sensitivity of the technique; the totals can be imprecise individually or collectively by several percent. Iron is shown as $Fe^{2+}O$, although it may actually be in more than one valence state (e.g., Fe^0 and $Fe^{3+}_2O_3$) on the Moon.

1.5.4 Mineralogy and Constituent Phases

The vast majority of the Moon's surface is composed of a few silicate minerals (plagioclase feldspar, pyroxene, and olivine). In these minerals, the silicon is in tetrahedral coordination with four oxygen atoms.

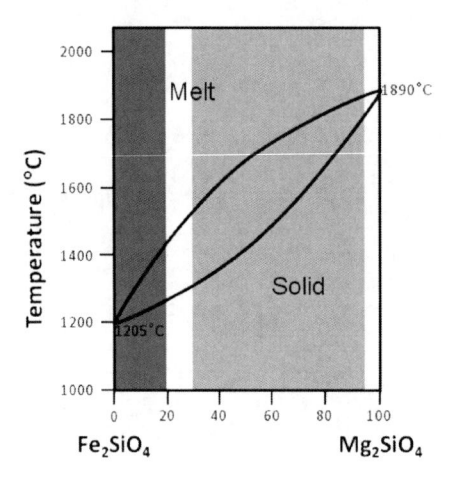

Fig. 1.5 The olivine composition-temperature phase diagram at 1 atmosphere, redrawn from Klein and Hurlbut (1993). The solidus (bottom, below this curve everything is solid) and liquidus (top, above this curve everything is melt) are shown. Between the solidus and liquidus curves is a mixture of crystals and melt. A discussion of phase diagrams, tying melt composition to the composition of early- and late-forming crystals, as well as the temperature of melting and temperature of crystallization, can be found in Hess (1989). Lunar olivine compositions have a great range, from approximately 30% Mg_2SiO_4 in Fe-rich mare basalts to approximately 95% Mg_2SiO_4 in magnesian suite samples (e.g. Papike et al. 1998; represented by the light gray area). However, some extremely Fe-rich olivines (0-10% Mg_2SiO_4, represented by the dark gray area) have been found in extremely late stage crystallization mineral assemblages (Papike et al. 1998).

The mineralogy of lunar regolith dictates most of its properties, such as melting temperature, abrasivity, strength, and trace element distribution. For example, the olivine mineral group melts between 1200°C (pure Fe olivine, fayalite) and 1890°C (pure Mg olivine, forsterite) at one atmosphere pressure, depending on its composition (Figure 1.5).

Because olivine's composition can range between 100% Fe_2SiO_4 and 100% Mg_2SiO_4 (both Fe and Mg are +2 cations with similar ionic radii), this range is called a "solid solution". Estimates for the hardness of fayalite on the Mohs hardness scale are approximately 6.5 (approximately 700 on the Knoop hardness scale), whereas the hardness of forsterite is estimated to be 7 (approximately 850 on the Knoop hardness scale). Olivine has an affinity for other +2 cations (such as platinum-group elements, Cr^{2+}, Mn, Ca, and Sr), as well as small +3 cations (such as V^{3+} and the heavy rare earth elements). Vanadium^{2+} also substitutes into the structure because the lunar environment is exceedingly reducing (e.g. Papike et al. 2005).

Plagioclase is another mineral group, and is the most abundant crystalline, and shocked diaplectic glass, component of the lunar crust. In fact, the lunar highlands are dominated by plagioclase, and a significant amount of plagioclase can also be found in lunar mare. Plagioclase is the name given to the solid solution between anorthite $(CaAl_2Si_2O_8)$ and albite $(NaAlSi_3O_8)$. This solid solution employs "coupled substitution". That is, $Ca^{2+} + Al^{3+}$ are substituted by $Na^{1+} + Si^{4+}$ in the plagioclase structure. The composition of the mineral is often denoted by molar percent of calcium; thus, An_{100} has 100% Ca (anorthite) and An_0 has 0% Ca (albite). Figure 1.6 presents the temperature regimes for the plagioclase series.

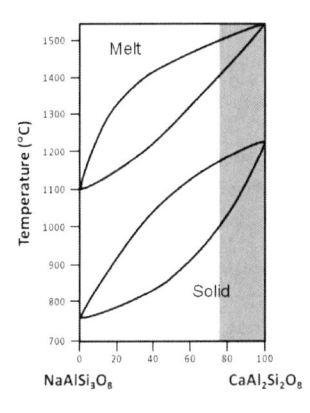

Fig. 1.6 The plagioclase temperature-composition diagram (Deer et al. 1992), indicating solidus and liquidus temperatures for anhydrous conditions at 1 bar pressure (top; Bowen 1913) and 5 kbar H_2O conditions (bottom; Yoder et al. 1975). Highlighted area indicates compositions of lunar plagioclase (Papike et al. 1976). There is currently a debate as to the quantity of water on the Moon or in its interior, from anhydrous to relatively H_2O-rich (c.f. Sharp et al. 2010; and McCubbin et al. 2010; respectively). Pressures studied for anorthosite genesis range from 1 bar to 47 kbar (Longhi 2003)

Note that plagioclase encompasses the minerals albite (An_{0-10}), oligoclase (An_{10-30}), andesine (An_{30-50}), labradorite (An_{50-70}), bytownite (An_{70-90}), and anorthite (An_{90-100}). Most of the lunar highland plagioclase is approximately An_{95} (Heiken et al. 1991) due to a lack of sodium on the Moon, whereas bytownite and anorthite are both found in mare basalts (e.g. Karner et al. 2004). Minor and trace elements found in plagioclase minerals include Sr, which substitutes readily for Ca, and light rare earth elements. The element Eu, which can have both 2+ and 3+ ionic charges, tends to substitute in significant abundance into the Ca^{2+} site of the plagioclase structure as Eu^{2+}. This particular substitution leads to a "positive Eu anomaly" in plagioclase rare earth element patterns. The hardness of well-crystallized plagioclase is 6 on the Mohs scale (approximately 550 on the Knoop scale). The hardness of lunar plagioclase may frequently be less (Cole et al. 2010) presumably due to shock-induced features such as fractures or diaplectic glass formation.

Pyroxene is the final mineral group that is a major component of the lunar crust. Pyroxene has four end-member compositions: enstatite ($MgSiO_3$), ferrosilite ($FeSiO_3$), diopside ($CaMgSi_2O_6$), and hedenbergite ($CaFeSi_2O_6$). Pyroxene can be almost any composition between the four end-members, although cooling conditions and temperature/pressure changes can result in exsolution (the formation of two minerals from a crystal of a single composition, Figure 1.4b). Figure 1.7 shows the pyroxene quadrilateral, with the four end-member compositions, and compositions of lunar pyroxenes. The variations in the composition (and crystallographic structure) of pyroxene by major and minor elements cause variations in the hardness, from 5-6.5 on the Mohs hardness scale (approximately 320 to 850 on the Knoop scale). Minor and trace elements normally found in pyroxenes are Mn, Sr, Cr^{3+}, and heavy rare earth elements.

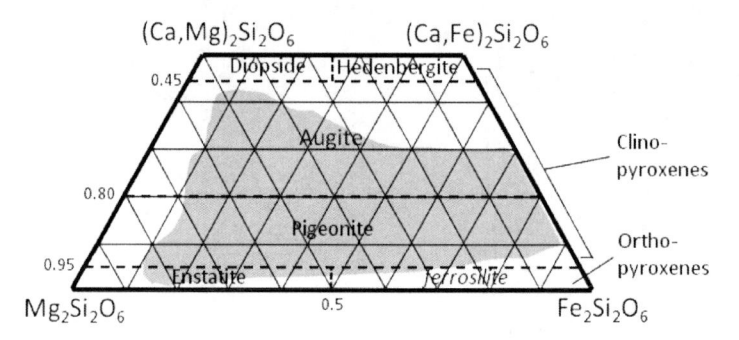

Fig. 1.7 Pyroxene Ca-Mg-Fe quadrilateral showing mineral names (Morimoto 1988) and approximate range of lunar pyroxenes according to Heiken et al. (1991)

Glass is a ubiquitous component of regolith and can be the dominant phase. There are two different sources of glass in the regolith. By far, the most dominant source of glass is impact. The melt produced during impact typically solidifies quickly, creating impact melt glass. Large amounts of impact melt, often produced in large-scale impacts, can cool slowly enough to crystallize. Not only is impact melt produced during large-scale impacts (such as the impact melt sea in the crater Copernicus), impact melt is produced in small-scale micrometeorite impacts as well. This glass produces lunar agglutinates. The vesicular nature of agglutinates makes them incredibly fragile. Impact-produced glass is found globally on the Moon. In contrast, volcanic fire-fountaining, the release of molten material balistically, is a highly localized source of glass in the regolith. The molten material ejected from lunar volcanic vents solidifies quickly in the cold, low-gravity environment. In special cases this resulted in spherical or near spherical glass beads, and the color of the glass beads reflects the composition of the beads. Specific deposits of colored glass beads were observed in significant quantities by the Apollo 15 (green glass) and Apollo 17 (orange and black glass) crews. The difference between these glasses is their composition; green glasses are rich in Mg and Fe and poor in Ti, while orange and black glasses are rich in Ti. Orange and black glasses can be distinguished by their crystalline component (the composition is the same, but the black glasses contain tiny crystals formed because of a decrease in the cooling rate).

One additional component of the lunar regolith is nanophase iron, located in agglutinates and within amorphous rims on lunar minerals exposed to space weathering. Nanophase iron is metallic iron globules. These globules can be between a few nm up to several hundred nm in size (e.g. James et al. 2002). Nanophase iron forms by the reduction of Fe in coordination with O in mineral structures and glasses. There are many proposed methods of nanophase iron formation. One involves the reduction of Fe simply due to the thermal energy of an impact (Sasaki et al. 2001). Another is the reduction of Fe during micrometeorite impact vaporization and vapor deposition (Hapke et al. 1975; Keller and McKay 1993). The role, if any, of nanophase iron formation due to saturation of the lunar surface by highly reducing H in the solar wind, remains to be determined.

There are several minor or trace mineral phases present in the regolith that might become significant under some conditions, including sulfide minerals such as the most abundant lunar sulfide, troilite (FeS). Sulfur is frequently considered a deleterious constituent in manufacturing conditions. Troilite may be especially worrisome, as it is unstable in the presence of H_2O; it spontaneously decomposes into hydrated iron oxides and sulfuric acid. In similar manner, there are mineral phases containing phosphorus and halogen elements as primary constituents. There are also trace quantities of a number of other elements normally avoided on Earth, specifically Sb, As, Se and similar elements. The minerals and phases containing these elements are not known well for the Moon; such details will determine what their availability will be in any processing.

There are other non-silicate minerals on the Moon as well, including oxides and spinels. Oxides, such as ilmenite ($FeTiO_3$), are a desirable source for oxygen production, as compared to the silicate minerals. Chromite ($FeCr_2O_4$) is a spinel, and is a chromium ore when concentrations are high enough. It is also a desirable source for oxygen production, as compared to the silicate minerals. As with ilmenite, it occurs as a common trace mineral and is extremely abundant in very distinct regions on the lunar surface. Spinel (senso stricto, $MgAl_2O_4$) and spinel group minerals are common trace minerals in the lunar regolith. They may be of special interest when considering abrasion as they can be extremely hard, tough, and break with sharp edges.

Compared to Earth, the types and number of minerals that occur on the Moon are limited. For example, few of the hydrated minerals, minerals containing either H_2O or OH in their structure, are known from the Moon. Such minerals are very common on Earth. Clay minerals and other minerals associated with terrestrial weathering are absent on the Moon. Minerals which form by precipitation from, or reaction in the presence of, water are absent. Thus sulfates, carbonates, and similar species do not exist on the Moon. Ultimately, the absence of such minerals will preclude using many of the resource recovery technologies on the Moon that are used on Earth. For example aluminum production requires a significant source of fluorine. On Earth, fluorine comes from the mineral fluorite, the deposits of which were formed by hot water. Similarly, commercial iron production requires the use of limestone, which is calcium carbonate.

1.5.5 Lithology

The word "lithology" literally means the study of rocks, and is a combination of the Greek words "lithos" for stone and "logia" for logic. The names of rocks place them in a conceptual framework – that is, the name of a rock can give one an idea of its chemical and physical origin, mineralogy, particle size, and closely associated rocks. There are different types of rocks on the lunar surface; lunar geologists most commonly use igneous and impact rock type descriptors for lunar rocks.

Igneous rocks can be divided into two types – extrusive (i.e. lava, which has reached the surface) and intrusive (i.e. magma, which has not reached the surface). Extrusive rocks cool relatively quickly compared to intrusive rocks, which are insulated by surrounding rocks. The size and composition of mineral crystals in the rock is a record of both depth of crystal formation and residence time of the magma as it ascends. Volcanic glass beads are one example of extrusive rocks on the Moon; they cooled exceedingly quickly. The mare basalts are also examples of extrusive rocks and are mixtures of crystals and glass with varying degrees of devitrification. Intrusive rocks are completely crystalline. Examples of intrusive rocks include the lunar magnesian suite. Table 1.2 provides a short list of the common rock types found on the Moon.

Table 1.2 Common igneous rock types on the lunar surface and their major minerals

Rock type	Major mineral(s)	Location
Anorthosite	Plagioclase (anorthite), some olivine and/or pyroxene	Highlands
Norite	Plagioclase, pyroxene	Highlands
Troctolite	Plagioclase, olivine	Highlands
Basalt	Pyroxene, olivine, some plagioclase and ilmenite	Mare

When rocks are shocked or heated by impact, new rocks can form and old rocks can be modified. These are known as impactites, a sub-class of metamorphic rocks. Stöffler and Grieve (2007) provided a classification scheme for impactites; Figure 1.8 represents the types of impactites.

Metamorphic rocks, such as granulites, are also present on the Moon. The heat and pressure required to metamorphose the samples could be the result of heating from volcanic activity or other lunar interior heat sources (such as radioactivity or proximity to magma), or from the heat of impact.

1.6 Locations of High Concentrations of Specific Minerals or Rocks

It is becoming easier to identify locations with potential in situ resource utilization capability with increasing resolution of remote sensing techniques, and further study of "ground truth" samples collected by the Apollo and Luna missions.

1.6.1 Ilmenite

Ilmenite is in trace abundance in the highlands; it is found in much greater abundance in some of the high-Ti mare basalts (e.g. McKay and Williams 1979; Heiken and Vaniman, 1990). There are no good terrestrial analogs of these rocks, as ilmenite in the oxidized terrestrial environment has exsolutions of magnetite unlike the highly reduced pure ilmenite on the Moon.

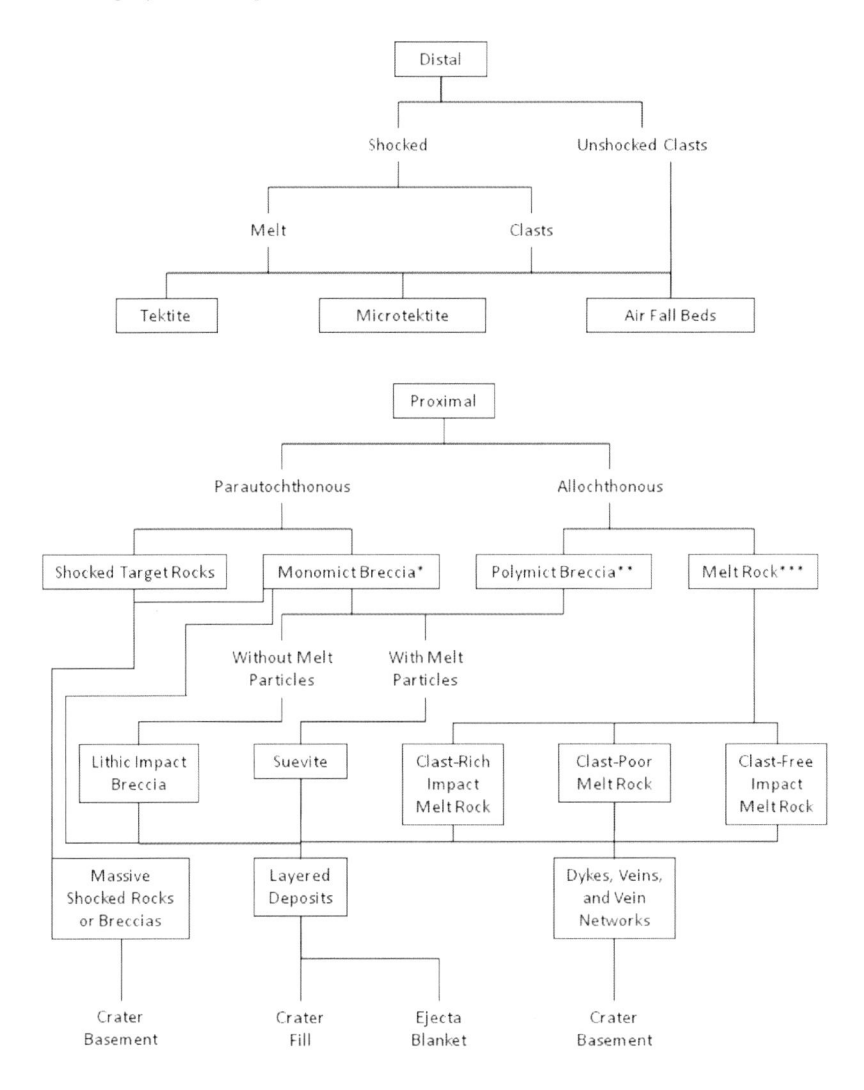

Fig. 1.8 Impactite classification after Stöffler and Grieve (2007), as presented by Stoeser et al. (2011). *Typically monomict. ** Generally polymict but can be monomict (e.g., in a single lithology target). ***Includes glassy, hypocrystalline, and holocrystalline varieties.

Ilmenite is important, because it is readily capable of being reduced in order to extract oxygen from the regolith (e.g. Chambers et al. 1995). Other materials such as agglutinates, volcanic glass beads, olivine, and pyroxene can also be reduced to provide oxygen for a lunar habitat (e.g. Brecher et al. 1975; Allen et al. 1996) but ilmenite is a preferred source for reasons of energy requirements (e.g. Taylor and Carrier 1993).

1.6.2 Chromite

Sunshine et al. (2010) determined that the spectral signature of two portions of Sinus Aestuum, measured by the Moon Mineralogy Mapper (M^3), match the mineral chromite. The chromite mineral signature is dominant in the spectra, although there may be minor amounts of other minerals such as plagioclase, pyroxene, and olivine associated with these deposits, thus physical sampling is required to completely characterize the deposits. There are very large ore grade deposits of chromite on Earth, and they are in rocks with compositions and origins analogous to those proposed by models of the lunar crust's origin.

1.6.3 Spinel

Large concentrations of the mineral spinel have been reported in the inner ring of the Moscoviense Basin by Pieters et al. (2010). It was detected using the M^3 instrument aboard Chandrayaan-1. There are no similar rocks on Earth.

1.6.4 Anorthosite

Anorthosite, senso stricto, was reported in lunar craters by Ohtake et al. (2010). The Multiband Imager on the Selenological and Engineering Explorer (SELENE) spacecraft detected areas of high abundance, nearly 100%, of plagioclase. Sixty-nine areas in crater peaks, rims, rings, and ejecta were detected. These areas were chosen because of their high albedo and low (to no) regolith cover. A map of each detected anorthosite location is provided in the Ohtake et al. (2010) paper.

1.6.5 Regolith

Regolith can be used for many things, including sintering (e.g. Taylor and Meek 2005), radiation shielding (e.g. Nealy et al. 1988; Miller et al. 2009), agriculture (e.g. Drees and Wilding 1989) and berm or habitat construction (e.g. Faierson et al. 2010). It is a resource found all over the lunar surface.

1.7 Topographic Features

The lunar surface is covered in interesting geologic features, including craters, volcanoes, rilles, and lava flows. These features in themselves provide a natural resource for lunar habitation and exploration.

Craters are the most abundant topographic feature on the Moon. They are essentially circular valleys with a surrounding rim mountain chain, some with ejecta rays. One could use the valley floors to provide a fairly flat surface for landing pads, road networks, and foundations for habitats. Crater rims can be used as a strategic position or communication base, provide the best views of the lunar landscape, and absorb the greatest number of solar rays.

Because of the low tilt ($1.5°$) of the Moon's spin axis relative to the ecliptic plane, craters that are in the polar regions provide areas of permanent shadow (valleys that receive no sunlight because it is blocked by the surrounding rim) and permanent sunlight. Regions of permanent darkness are exceedingly cold, as shown by the Lunar Reconnaissance Orbiter Diviner instrument results (Paige et al. 2010). The temperature of one permanently shadowed region was measured to be 29 Kelvin or -244°C. In stark contrast, nearby areas of constant sunlight can have an average temperature around 200 K, and peak temperatures exceeding 300 K (27°C). Bussey et al. (2005, 2010) estimated the locations of near-permanent illumination at the lunar poles by examining lunar topography and the astronomical properties of the Moon.

Other topographic features of the Moon are related to volcanic activity. The relatively smooth lunar mare surfaces formed as a massive body of lava cooled. Contraction of the lava during cooling created some rilles, which are long, narrow valleys. Smaller and more sinuous rilles are thought to be collapsed lava tubes (tunnels in which lava flowed, which remained after the lava stopped flowing). The sinuous rilles may be the result of completely or only partially-collapsed lava tubes. It may be possible to place a lunar habitat within a lava tube at the end of a sinuous rille, which would greatly protect the habitat from space weathering.

One other benefit to living within a rille or other embedded habitat is the relatively constant temperature. The surface of the Moon at the equator can heat up to 390 K and cool to 104 K during the Moon's day-night cycle (Glasstone 1965). This daily change in temperature is only observed to a depth of approximately 1 m. Below 1 m, the temperature was measured during the Apollo missions to be relatively stable between 250 K and 260 K (Vaniman et al. 1991). The temperature below this depth depends on the heat flow from the lunar core (Weber et al. 2011) and the location and concentration of heat-producing radioactive elements. However, given the observed lack of temperature changes below 1 m, man-made structures below this depth could anticipate constant thermal conditions over the lifetime of the structure.

1.8 Summary

This chapter describes most of the geologic resources available on the Moon, and complements chapters involving water on the Moon and such energy sources as helium-3. The regolith, geologic concentrations of minerals, and physical features of the Moon enable abundant in situ resource utilization and habitation opportunities. With the ever-increasing data acquired from and about the Moon, at higher resolutions, the pathway to self-sustaining lunar habitation and resource use will become clearer.

References

Allen, C.C., Morris, R.V., McKay, D.S.: Oxygen extraction from lunar soils and pyroclastic glass. Journal of Geophysical Research 101, 26085–26095 (1996)

Bowen, N.L.: Melting phenomena in plagioclase feldspars. American Journal of Science 35, 577–599 (1913)

Brecher, A., Menke, W.H., Adams, J.B., Gaffey, M.J.: The effects of heating and subsolidus reduction on lunar materials: An analysis by magnetic methods, optical, Mössbauer, and X-ray diffraction spectroscopy. In: Proceedings of the 6th Lunar Science Conference, pp. 3091–3109 (1975)

Bussey, D.B.J., Fristad, K.E., Schenk, P.M., Robinson, M.S., Spudis, P.D.: Constant illumination at the lunar north pole. Nature 434, 842 (2005)

Bussey, D.B.J., McGovern, J.A., Spudis, P.D., Neish, C.D., Noda, H., Ishihara, Y., Sørensen, S.A.: Illumination conditions of the south pole of the Moon derived using Kaguya topography. Icarus 208, 558–564 (2010)

Canup, R.M., Asphaug, E.: Origin of the Moon in a giant impact near the end of the Earth's formation. Nature 412, 708–712 (2001)

Carrier III, W.D., Olhoeft, G.R., Mendell, W.: Physical properties of the lunar suface. In: Heiken, G.H., Vaniman, D.T., French, B.M. (eds.) Lunar Sourcebook: A User's Guide to the Moon, pp. 475–594 (1991)

Carrier III, W.D.: Particle size distribution of lunar soil. Journal of Geotechnical and Geoenvironmental Engineering 129, 956–959 (2003)

Chambers, J.G., Taylor, L.A., Patchen, A., McKay, D.S.: Quantitative mineralogical characterization of lunar high-Ti mare basalts and soils for oxygen production. Journal of Geophysical Research 100, 14391–14401 (1995)

Cintala, M.J.: Impact-induced thermal effects in the lunar and mercurian regoliths. Journal of Geophysical Research 97, 947–973 (1992)

Cole, D.M., Taylor, L.A., Liu, Y., Hopkins, M.A.: Grain-scale mechanical properties. In: Engineering Science, Construction and Operations in Challenging Environments (12th), Earth and Space 2010, March 14-17, American Society of Civil Engineers, Honolulu HI (2010)

Deer, W.A., Howie, R.A., Zussman, J.: An Introduction to the rock-forming minerals, 2nd edn., p. 696. Longman Scientific and Technical, Harlow (1992)

Dence, M.R.: Shock zoning at Canadian craters: Petrography and structural implications. In: French, B.M., Short, N.M. (eds.) Shock Metamorphism of Natural Materials, pp. 169–184. Mono Book Corp., Baltimore (1968)

Drees, L.R., Wilding, L.P.: Pedology, pedogenesis, and the lunar surface. In: Ming, D.W., Henninger, D.L. (eds.) Lunar Base Agriculture: Soils for Plant Growth, pp. 69–83 (1989)

Elkins-Tanton, L.T., Burgess, S., Yin, Q.Z.: The lunar magma ocean: Reconciling the solidification process with lunar petrology and geochronology. Earth and Planetary Science Letters 304, 326–336 (2011)

Faierson, E.J., Logan, K.V., Stewart, B.K., Hunt, M.P.: Demonstration of concept for fabrication of lunar physical assets utilizing lunar regolith simulant and a geothermite reaction. Acta Astronautica 67, 38–45 (2010)

French, B.M.: Traces of Catastrophe: A handbook of shock-metamorphic effects in terrestrial meteorite impact structures. LPI Contribution No. 954, p. 120. Lunar and Planetary Institute, Houston (1998)

Glasstone, S.: Sourcebook on the Space Sciences, p. 937. Van Nostrand, Princeton (1965)

Hapke, B.: Space weathering from Mercury to the asteroid belt. Journal of Geophysical Research 106, 10039–10073 (2001)

Hapke, B., Cassidy, W., Wells, E.: Effects of vapor-phase deposition processes on the optical, chemical, and magnetic properties of the lunar regolith. The Moon 13, 339–353 (1975)

Hartmann, W.K., Davis, D.R.: Satellite-sized planetesimals and lunar origin. Icarus 24, 504–515 (1975)

Heiken, G.H., Vaniman, D.T.: Characterization of lunar ilmenite resources. In: Proceedings of the 20th Lunar and Planetary Science Conference, pp. 239–247 (1990)

Heiken, G.H., McKay, D.S., Brown, R.W.: Lunar deposits of possible pyroclastic origin. Geochimica et Cosmochimica Acta 38, 1703–1718 (1974)

Heiken, G., Vaniman, D., French, B.M.: Lunar Sourcebook: A User's Guide to the Moon, p. 736. Cambridge University Press, New York (1991)

Hess, P.C.: Origins of igneous rocks, p. 336. Harvard University Press, Cambridge (1989)

Hörz, F., Grieve, R., Heiken, G., Spudis, P., Binder, A.: Lunar surface processes. In: Heiken, G.H., Vaniman, D.T., French, B.M. (eds.) Lunar Sourcebook: A User's Guide to the Moon, pp. 61–120. Cambridge University Press, Cambridge (1991)

James, C., Letsinger, S., Basu, A., Wentworth, S.J., McKay, D.S.: Size distribution of Fe^0 globules in lunar agglutinitic glass. In: 33rd Lunar and Planetary Science Conference. Abstract #1827 (2002)

Karner, J., Papike, J.J., Shearer, C.K.: Plagioclase from planetary basalts: Chemical signatures that reflect planetary volatile budgets, oxygen fugacity, and styles of igneous differentiation. American Mineralogist 89, 1101–1109 (2004)

Keller, L.P., McKay, D.S.: Discovery of vapor deposits in the lunar regolith. Science 261, 1305–1307 (1993)

Keller, L.P., McKay, D.S.: The nature and origin of rims on lunar soil grains. Geochimica et Cosmochimica Acta 61, 2331–2341 (1997)

Klein, C., Hurlbut Jr., C.S.: Manual of Mineralogy, p. 681. John Wiley & Sons, Inc., New York (1993)

Lindsay, J.F.: A steady-state model for the lunar soil. Geological Society of America Bulletin 86, 1661–1670 (1975)

Lingenfelter, R.E., Canfield, E.H., Hampel, V.E.: The lunar neutron flux revisited. Earth and Planetary Science Letters 16, 355–369 (1972)

Liu, Y., Park, J., Schnare, D., Hill, E., Taylor, L.A.: Characterization of lunar dust for toxicological studies II: Texture and shape characteristics. Journal of Aerospace Engineering 21, 272–279 (2008)

Longhi, J.: A new view of lunar ferroan anorthosites: Postmagma ocean petrogenesis. Journal of Geophysical Research 108 (E8), 5083 (2003), doi:10.1029/2002JE001941

Lucey, P.G., Blewett, D.T., Taylor, G.J., Hawke, B.R.: Imaging of lunar surface maturity. Journal of Geophysical Research 105, 20377–20386 (2000)

Lucey, P., Korotev, R.L., Gillis, J.J., Taylor, L.A., Lawrence, D., Campbell, B.A., Elphic, R., Feldman, W., Hood, L.L., Hunten, D., Mendillo, M., Noble, S., Papike, J.J., Reedy, R.C., Lawson, S., Prettyman, T., Gasnault, O., Maurice, S.: Understanding the lunar surface and space-Moon interactions. In: Jolliff, B.L., Wieczorek, M.A., Shearer, C.K., Neal, C.R. (eds.) Reviews in Mineralogy and Geochemistry, 60, New Views of the Moon, pp. 83–219. Mineralogical Society of America, Chantilly (2006)

Mathew, K.J., Marti, K.: Lunar nitrogen: Indigenous signature and cosmic-ray production rate. Earth and Planetary Science Letters 184, 659–669 (2001)

McCubbin, F.M., Steele, A., Hauri, E.H., Nekvasil, H., Yamashita, S., Hemley, R.J.: Nominally hydrous magmatism on the Moon. Proceedings of the National Academy of Sciences 107, 11223–11228 (2010)

McKay, D.S., Williams, R.J.: A geologic assessment of potential lunar ores. In: Billingham, J., Gilbreath, W., O'Leary, B. (eds.) Space Resources and Space Settlements, pp. 243–255 (1979)

McKay, D.S., Heiken, G., Basu, A., Blanford, G., Simon, S., Reedy, R., French, B.M., Papike, J.: The lunar regolith. In: Heiken, G.H., Vaniman, D.T., French, B.M. (eds.) Lunar Sourcebook: A User's Guide to the Moon, pp. 285–356. Cambridge University Press, Cambridge (1991)

McKay, D.S., Fruland, R.M., Heiken, G.H.: Grain size and the evolution of lunar soils. In: Proceedings of the 5th Lunar Science Conference, pp. 887–906 (1974)

Miller, J., Taylor, L., Zeitlin, C., Heilbronn, L., Guetersloh, S., DiGiuseppe, M., Iwata, Y., Murakami, T.: Lunar soil as shielding against space radiation. Radiation Measurements 44, 163–167 (2009)

Morimoto, N.: Nomenclature of pyroxenes. Mineralogical Magazine 52, 535–550 (1988)

Morris, R.V.: The surface exposure (maturity) of lunar soils: Some concepts and I_s/FeO compilation. In: Proceedings of the 9th Lunar and Planetary Science Conference, pp. 2287–2297 (1978)

Nealy, J.E., Wilson, J.W., Townsend, L.W.: Solar-flare shielding with regolith at a lunar base site. NASA Technical Paper 2869, p. 18 (1988)

Ohtake, M., Matsunaga, T., Haruyama, J., Yokota, Y., Morota, T., Honda, C., Ogawa, Y., Torii, M., Miyamoto, H., Arai, T., Hirata, N., Iwasaki, A., Nakamura, R., Hiroi, T., Sugihara, T., Takeda, H., Otake, H., Pieters, C.M., Saiki, K., Kitazo, K., Abe, M., Asada, N., Demura, H., Yamaguchi, Y., Sasaki, S., Kodama, S., Terazono, J., Shirao, M., Yamaji, A., Minami, S., Akiyama, H., Josset, J.L.: The global distribution of pure anorthosite on the Moon. Nature 461, 236–240 (2010)

Pahlevan, K., Stevenson, D.J.: Equilibration in the aftermath of the lunar-forming giant impact. Earth and Planetary Science Letters 262, 438–449 (2007)

Paige, D.A., Siegler, M.A., Zhang, J.A., Hayne, P.O., Foote, E.J., Bennett, K.A., Vasavada, A.R., Greenhagen, B.T., Schofield, J.T., McCleese, D.J., Foote, M.C., DeJong, E., Bills, B.G., Hartford, W., Murray, B.C., Allen, C.C., Snook, K., Soderblom, L.A., Calcutt, S., Taylor, F.W., Bowles, N.E., Banfield, J.L., Elphic, R., Ghent, R., Glotch, T.D., Wyatt, M.B., Lucey, P.G.: Diviner lunar radiometer observations of cold traps in the Moon's south polar region. Science 330, 479–483 (2010)

Papike, J.J., Hodges, F.N., Bence, A.E., Cameron, M., Rhodes, J.M.: Mare basalts: Crystal chemistry, mineralogy and petrology. Reviews in Geophysics and Space Physics 14, 475–540 (1976)

Papike, J.J., Simon, S.B., Laul, J.C.: The lunar regolith: Chemistry, mineralogy, and petrology. Reviews of Geophysics and Space Physics 20, 761–826 (1982)

Papike, J.J., Ryder, G., Shearer, C.K.: Lunar samples. In: Papike, J.J. (ed.) Planetary Materials, p. 234. Mineralogical Society of America, Washington DC (1998)

Papike, J.J., Karner, J.M., Shearer, C.K.: Comparative planetary mineralogy: Valence state partitioning of Cr, Fe, Ti, and V among crystallographic sites in olivine, pyroxene, and spinel from planetary basalts. American Mineralogist 90, 277–290 (2005)

Parmentier, E.M., Zhong, S., Zuber, M.T.: Gravitational differentiation due to initial chemical stratification: Origin of lunar asymmetry by the creep of dense KREEP? Earth and Planetary Science Letters 201, 473–480 (2002)

Pieters, C.M., Fischer, E.M., Rode, O., Basu, A.: Optical effects of space weathering: The role of the finest fraction. Journal of Geophysical Research 98, 20817–20824 (1993)

Pieters, C.M., Boardman, J., Buratti, B., Clark, R., Combe, J.P., Green, R., Goswami, J.N., Head III, J., Hicks, M., Isaacson, P., Klima, R., Kramer, G., Kumar, K., Lundeen, S., Malaret, E., McCord, T., Mustard, J., Nettles, J., Petro, N., Runyon, C., Staid, M., Sunshine, J., Taylor, L.A., Thaisen, K., Tompkins, S., Varanasi, P.: Identification of a new spinel-rich lunar rock type by the Moon Mineralogy Mapper (M^3). In: 41st Lunar and Planetary Science Conference, Abstract #1854 (2010)

Sasaki, S., Nakamura, K., Hamabe, Y., Kurahashi, E., Hiroi, T.: Production of iron nanoparticles by laser irradiation in a simulation of lunar-like space weathering. Nature 410, 555–557 (2001)

Sharp, Z.D., Shearer, C.K., McKeegan, K.D., Barnes, J.D., Wang, Y.Q.: The chlorine isotope composition of the Moon and implications for an anhydrous mantle. Science (2010), doi:10.1126/science.1192606

Shkuratov, Y.G., Bondarenko, N.V.: Regolith layer thickness mapping of the Moon by radar and optical data. Icarus 149, 329–338 (2001)

Smith, J.V., Anderson, A.T., Newton, R.C., Olsen, E.J., Wyllie, P.J., Crewe, A.V., Isaacson, M.S., Johnson, D.: Petrologic history of the Moon inferred from petrography, mineralogy, and petrogenesis of Apollo 11 rocks. In: Proceedings of the Apollo 11 Lunar Science Conference, pp. 897–925 (1970)

Stoeser, D.B., Benzel, W.M., Schrader, C.M., Edmunson, J.E., Rickman, D.L.: Notes on lithology, mineralogy, and production for lunar simulants. NASA technical memorandum 2011-216454, 49 p (2011)

Stöffler, D., Grieve, R.A.F.: Impactites. Recommendations by the IUGS Subcommission on the Systematics of Metamorphic Rocks. IUGS Subcommission on the Systematics of Metamorphic Rocks 15 (2007)

Sunshine, J.M., Besse, S., Petro, N.E., Pieters, C.M., Head, J.W., Taylor, L.A., Klima, R.L., Isaacson, P.J., Boardman, J.W., Clark, R.C., M^3 Team.: Hidden in plain sight: Spinel-rich deposits on the nearside of the Moon as revealed by Moon Mineralogy Mapper (M^3). In: 41st Lunar and Planetary Science Conference, Abstract #1508 (2010)

Taylor, L.A., Carrier III, W.D.: Oxygen production on the Moon: An overview and evaluation. In: Lewis, J., Matthews, M.S., Guerrieri, M.L. (eds.) Resources of Near-Earth Space, pp. 69–108 (1993)

Taylor, L.A., Meek, T.T.: Microwave sintering of lunar soil: properties, theory, and practice. Journal of Aerospace Engineering 18, 188–196 (2005)

Taylor, L.A., Pieters, C.M., Keller, L.P., Morris, R.V., McKay, D.S.: Lunar mare soils: Space weathering and the major effects of surface-correlated nanophase Fe. Journal of Geophysical Research 106, 27985–27999 (2001)

Taylor, L.A., Pieters, C., Patchen, A., Taylor, D.-H.S., Morris, R.V., Keller, L.P., McKay, D.S.: Mineralogical and chemical characterization of lunar highlands soils: Insights into the space weathering of soils on airless bodies. Journal of Geophysical Research 115, E02002 (2010), doi:10.1029/2009JE003427

Vaniman, D., Reedy, R., Heiken, G., Olhoeft, G., Mendell, W.: The lunar environment. In: Heiken, G.H., Vaniman, D.T., French, B.M. (eds.) Lunar Sourcebook: A User's Guide to the Moon, pp. 27–60. Cambridge University Press, Cambridge (1991)

Weber, R.C., Lin, P.Y., Garnero, E.J., Williams, Q., Lognonné, P.: Seismic detection of the lunar core. Science 331, 309–312 (2011)

Wilhelms, D.E.: The geologic history of the Moon. United States Geological Survey Professional Paper 1348 (1987)

Wong, N., Santarius, J., Taylor, L.A., Garrison, D.H., James, J.T., McKay, D.S., Kuhlman, K.R.: Decay of reactivity in forsterite due to UV radiation. Earth and Space, 29–35 (2010)

Wood, J.A., Dickey, J.S., Marvin, U.B., Powell, B.N.: Lunar anorthosites and a geophysical model of the Moon. In: Proceedings of the Apollo 11 Lunar Science Conference, pp. 965–988 (1970)

Yoder, H.S., Stewart, D.B., Smith, J.R.: Ternary feldspars. Carnegie Institution of Washington Yearbook 56, 206–216 (1957)

2 Helium Isotopes in the Lunar Regolith – Measuring Helium Isotope Diffusivity in Lunar Analogs

Kimberly R. Kuhlman[1] and Gerald L. Kulcinski[2]

[1] Planetary Science Institute, Tucson
[2] University of Wisconsin, Madison, USA

2.1 Introduction

The purpose of the present chapter is to provide information about the diffusivities and activation energies of ^4He and ^3He in analogs of lunar ilmenite, which has been theorized to retain helium isotopes better than other lunar minerals. The presence of noble gases in the surfaces of lunar fines was discovered in early Apollo samples by several investigations (Bauer et al. 1972, Ebergart et al. 1970, Hintenberger et al. 1970). The correlation between lunar helium and the mineral ilmenite was discovered by Eberhart et al. (1970) and subsequently others (Muller et al. 1976, Signer et al. 1977). This correlation was rediscovered by researchers at the University of Wisconsin in the early 1980's while researching sources of ^3He for fuel for future fusion reactors (Cameron 1988). The data presented here raise questions about the diffusivities of helium in other lunar minerals and will be useful in the design of future lunar miners. This chapter also provides insight into the mechanisms of space weathering on airless bodies.

2.1.1 Why Is Lunar ^3He Important?

Helium-3 is rare isotope of helium that has many important uses, including low-temperature physics research, medical lung imaging, neutron detection and potentially nuclear fusion (Fetter 2010). In its function as a neutron detector, ^3He is used in basic research, nuclear safeguards, homeland security and oil and gas exploration. The current U.S. reserve of ^3He is less than 10 kg, and the U.S. Dept. of Energy Isotope Program is only capable of generating approximately 8,000 liters at standard temperature and pressure (STP) per year from the decay of tritium or 1.07 kg per year (Fetter 2010). The terrestrial demand for ^3He will grow substantially in 2018 when the Spallation Neutron Source comes online.

Natural terrestrial stores of ^{3}He include the atmosphere (7.7 x 10^{-6} ppmv) and ^{3}He collected from natural gas (0.2 ppmv) (Fetter 2010). Helium-3 is also produced from tritium produced for nuclear weapons, which decays to ^{3}He with a 12.4 year half-life (Fetter 2010). These sources are not enough to meet the demand, particularly if ^{3}He fusion becomes a reality.

The impending global energy crisis that looms over the 21st century will require massive action and innovative energy solutions (Kulcinski and Schmitt 2000). Nuclear fusion is potentially one of those solutions. Traditional fusion research utilizes the first generation deuterium (D) and tritium (T) fuel cycle, which results in nuclear waste that must be stored. The second and third generation deuterium (D)-^{3}He and ^{3}He-^{3}He fuel cycles, respectively are more desirable because they produce far less radioactivity. Comparison with wind and solar power sources are difficult to make since there are no ^{3}He-based fusion plants. There have been D^{3}He studies done and the cost of electricity (COE) was determined to be 7.7 cents/kWh in 1990 (Kulcinski et al. 1992). This would be approximately 12 cents/kWh in today's currency. That is on the high end of current electrical energy sources. The equilibrium cost (after ≈ 20 years of successful mining) in 1990 dollars was ≈ \$300 per gram (≈ \$500/g in 2010 dollars) and that contributes ≈ 0.5 cent per kWh in today's dollars (Wittenberg et al. 1992). The ^{3}He-^{3}He cycle is of particular interest because it is aneutronic, i.e. it produces no neutrons and thus, no nuclear waste (Kulcinski and Schmitt 2000). However, the second and third generation cycles require more challenging physics than the first generation deuterium (D) and tritium (T) fuel cycle. They also suffer from a lack of practically extractable terrestrial stores of ^{3}He, which was recognized in 1986 by scientists at the University of Wisconsin's Fusion Technology Institute (Wittenberg et al. 1986). Faced with a future need for ^{3}He, scientists and engineers at the University of Wisconsin's Fusion Technology Center started looking for extraterrestrial sources of ^{3}He, starting with the Moon. Wittenberg et al. (1986) realized that the Moon contains as much as one million tonnes of ^{3}He implanted into the lunar regolith by the solar wind. As an example, as little as 150 tonnes of ^{3}He would be enough to fuel the electrical needs of the entire world in the year 2000 (Kulcinski and Schmitt 2000).

2.1.2 Where Is Lunar ^{3}He Located?

The correlation between helium and ilmenite of Moon's regolith. allows a map of titanium (Fig. 2.1) to approximate a map of ^{3}He concentration. However, Cameron (1988) stressed the fact that we only have minute fractions of a few maria upon which to base this correlation. Ilmenite (FeTiO$_3$) is the major titanium-containing mineral on the Moon, and has been shown to retain ^{3}He differently than other major minerals (Signer et al. 1977). The finest fraction of the regolith contains the greatest proportion of ^{3}He because implantation by the solar wind is a surface process. Helium is implanted a few 10's of nanometers into exposed surfaces of regolith particles. Early experiments showed that 80-90% of the ^{3}He is removed from the regolith by simply heating it to approximately 700°C (Gibson and Johnson 1971).

In addition to the Ti correlation, other scientists have discovered that the amount of ^{3}He present is correlated with the maturity of the regolith, i.e. how long it has been exposed. Two comparable measurements of maturity are the ferromagnetic resonance (FMR) surface exposure index, I_s/FeO (Morris 1978), and optical maturity parameter (OMAT) (Lucey et al. 2000b). I_s/FeO is the ratio of the saturation magnetization, I_s, (value of the intensity of the FMR resonance at g ~2.1) to the value of FeO concentration (Morris 1978). As such, it must be measured in the laboratory or *in situ*. The OMAT parameter is defined as the Euclidean distance from the dark red origin (hypermature end-member) to each point (Grier et al. 1999). It is measured using Clementine data and is defined as Eq. (2.1):

$$OMAT = \left[\left(R_{750} - x_0 \right)^2 + \left(\frac{R_{950}}{R_{750}} - y_0 \right)^2 \right]^{1/2} \tag{2.1}$$

where x_0 is the reflectance of the origin, y_0 is the ratio value of the origin, R_{750} is the reflectance at 750 nm of a pixel or spectrum, and R_{950} is the reflectance at 950 nm of a pixel or spectrum (Lucey et al. 2000b). More recently, a new lunar optical maturity index has been developed based on Moon Mineralogy Mapper (M^3) data using three spectral parameters instead of two (Nettles et al. 2010). When superposed upon Ti maps measured either by Clementine or M^3, respectively, these maturity indices help to define the richest deposits of ^{3}He for mining.

Highly mature highland soils are also shown to contain relatively large amounts of helium in agglutinates (Signer et al. 1977). Only ilmenite was shown to be the exception from this observation. These trends can be general guides in the selection of favorable areas for mining. Understanding of the He diffusion and trapping properties of the minerals involved will help refine these guides. For instance, if magnesian ilmenite ($Mg_xFe_{1-x}TiO_3$) is found to retain He better than stoichiometric ilmenite ($FeTiO_3$), remote sensing data for Mg could be considered in addition to Ti and maturity data as has been the practice (Swindle et al. 1990).

Interest in lunar ^{3}He has recently been rekindled by Chinese investigators and the wealth of data from the Chang'E-1 mission (Fa and Jin 2007, Li et al. 2010, Wang et al. 2010). New global distributions of ^{3}He have recently been estimated using data about regolith depth provided by the ChangE-1 Lunar Microwave Sounder. These new results are compared to previous estimations by Starukhina (2006), Li et al. (2010) and Wang et al. (2010). The general consensus is that the lunar inventory of ^{3}He is approximately 1.3×10^6 tonnes (Li et al. 2010). However, the relation between ^{3}He abundance and depth of lunar regolith remains poorly understood below the Apollo core depths (Li et al. 2010).

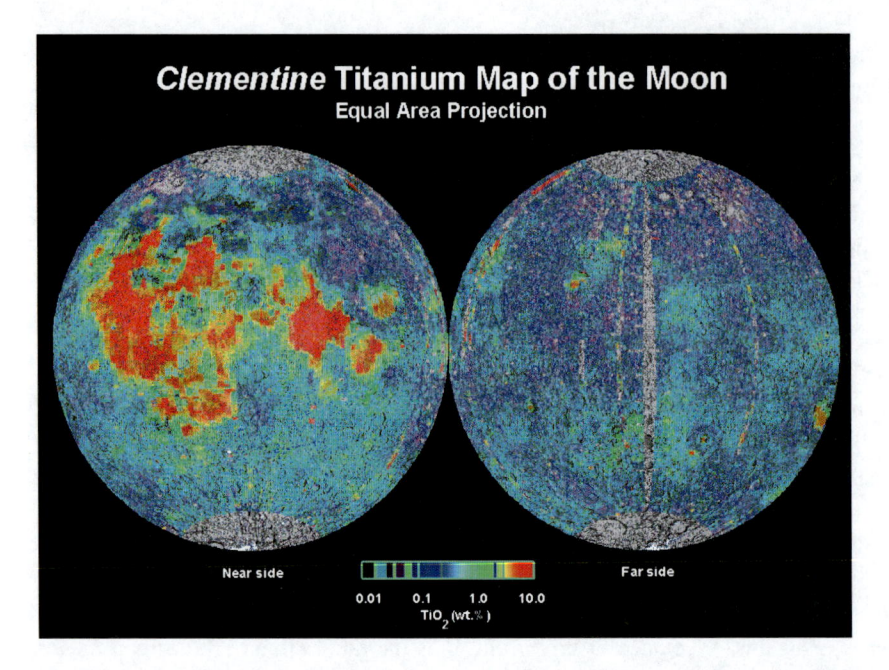

Fig. 2.1 Image derived from the Clementine global color data (in 415 and 750 nm wavelengths) showing the concentration of titanium in the soils of the lunar surface (Lucey 2011, Lucey et al. 2000a).

2.1.3 ^{3}He Mining

To recover this valuable resource, scientists in the UW Fusion Technology Institute have designed a solar-powered autonomous Mark II lunar miner, which is capable of separating out the large aggregates that do not contain much ^{3}He (Fig. 2.2) (Sviatoslavsky 1992). The miner heats the regolith to extract all volatile elements, cools the regolith to recover the energy and finally returns it to the lunar surface. The evolved gases are compressed into cylinders, which are separated at a central facility. Helium-3 is then returned to Earth as a liquefied gas. Schmitt (2006) discusses ^{3}He power economics, the geology of lunar ^{3}He and human settlement to support ^{3}He production in the context of a return of humans to the Moon in great detail. Dr. Schmitt also delves into the many financial and managerial approaches that could be used to develop a permanent human presence on the Moon based upon a ^{3}He economy.

2.1.4 Issues to Consider for Helium Trapping in Minerals

Previous studies concerning lunar helium have involved stepwise heating e.g. (Futagami et al. 1990,1993, Pepin et al. 1970b, Srinivasan et al. 1972), stepwise etching (Signer et al. 1991) and the rare gas ion probe (Muller et al. 1976). Each of these methods has produced useful information concerning the ratios of rare gas isotopes, gas release as a function of temperature and in the latter case, helium profiles as a function of depth. However, none of these studies have generated consistent values of the diffusivity or the activation energy of helium in ilmenite. Nor has the trapping mechanism been identified.

It has been postulated that either an amorphous coating, or petina, or possible radiation damage resistance of ilmenite is responsible for the relative enrichment of He in high-Ti regoliths. Helium diffuses more quickly through an amorphous coating or damaged surface layer than through the crystal matrix (Frick et al. 1988). Therefore, the diffusivity of He in crystalline ilmenite (volume diffusion) is thought to be the limiting factor.

The resistance of ilmenite to radiation damage has been investigated using 400 keV Xe^{2+} ions (Mitchell 1996; Mitchell et al. 1996). Terrestrial ilmenite was shown to amorphize at temperatures between 173 and 473 K under relatively low ion fluences - approximately 1×10^{15} ions/cm^2. Concurrent ion-electron experiments demonstrated that electron bombardment significantly delayed amorphization.

The structure of a mineral may also contribute to its ability to retain noble gases. Ilmenite has the same crystal structure as hematite (Fe_2O_3), based on the hexagonal closest packing of the oxygen lattice with cations in octahedral positions between them. This configuration is less tightly packed than a face-centered cubic lattice such as that found in Fe-Ni alloys. Silicates, such as olivine, are significantly less tightly packed than ilmenite. The degree of packing may explain the observation that helium retentivity follows the sequence: Fe-Ni > ilmenite > olivine-pyroxene > plagioclase (Signer et al. 1977).

Fig. 2.2 The University of Wisconsin's Mark II Lunar Miner (Courtesy of the University of Wisconsin – Madison Fusion Technology Institute).

2.1.5 Implantation Studies of Helium Diffusion in Terrestrial Minerals

The diffusion coefficients of He in ilmenite and olivine have been reported by Futagami et al. (1993). These measurements were performed on single crystal synthetic olivine (Mg_2SiO_4) and natural ilmenite ($Fe_{1.0}Ti_{1.1}Mn_{0.2}O_{3.4}$, as quoted by Futagami et al. (1993). Helium was implanted using an ion implanter at energies of 5-50 keV/amu along the (001) direction in olivine and the (113) direction in ilmenite. The only surface treatment prior to implantation involved polishing with 0.06 micron alumina powder. Thermal release patterns were obtained, and depth profiles prior to heating were calculated using the Stopping and Range of Ions in Matter (SRIM) code (Ziegler 1996). The diffusion coefficients of helium, neon and argon in ilmenite and olivine as a function of inverse temperature were obtained (Fig. 2.3). They obtained activation energies of 0.48 eV and 1.34 eV for ilmenite and olivine, respectively. These activation energies imply that olivine should retain more ^{4}He at lunar temperatures than ilmenite, contradicting the observation that Ti correlates with He in Apollo samples.

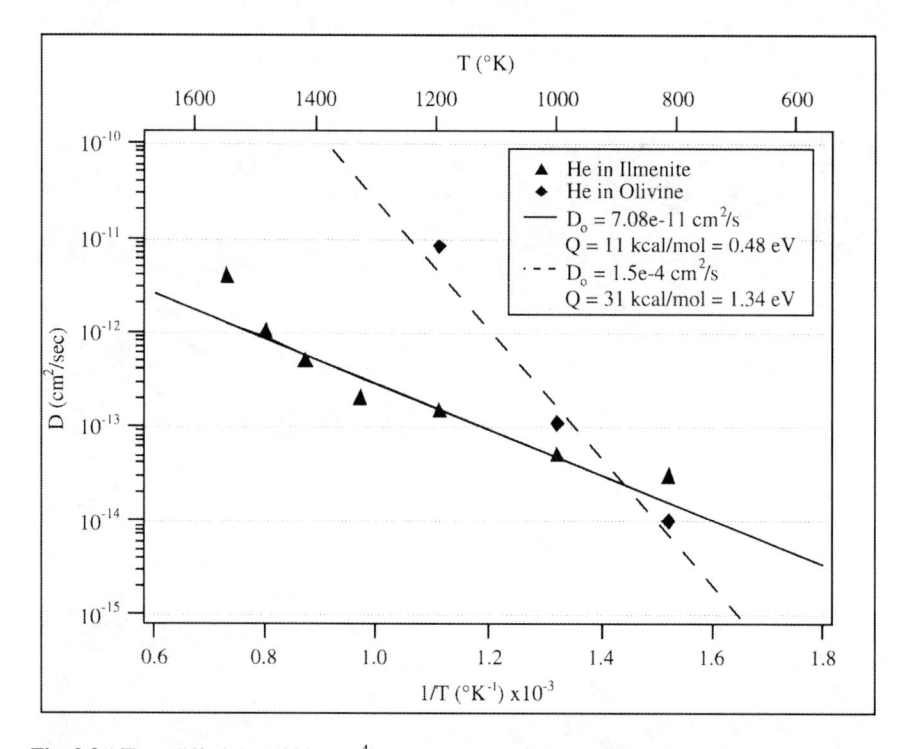

Fig. 2.3 1/T vs. diffusivity of 20 keV ^{4}He implanted in ilmenite and olivine [After Futagami et al. 1993].

2.2 Experimental Approach

2.2.1 Analogs of Lunar Ilmenite

Samples of several terrestrial ilmenites were analyzed using electron probe micro-analysis (EPMA) to determine the closest analog of lunar ilmenites (Table 2.1 and Table 2.2). The most suitable analogs were ilmenite from Quebec and New York. Ilmenite is normally found in solid solution with hematite on Earth, but even deep mantle ilmenites were found to contain a high number density of chromite crystallites. There are major differences between lunar ilmenite and terrestrial ilmenites. The main difference involves the fact that minerals on the Earth solidified in more oxygen-rich environments than did those on the Moon. Terrestrial minerals have also been formed and preserved in the presence of water. Both of these differences resulted in mineral phases present in ilmenite on Earth that do not exist on the Moon.

Table 2.1 The compositions of ilmenite xenocrysts from Sierra Leone and Liberia and ilmenite from anorthites at Sanford Lake, New York and St. Urbaine, Quebec analyzed with EPMA.

	Terrestrial Samples			
Sample Number	Liberia	Sierra Leone	New York	Quebec
TiO_2	50.53	50.12	49.14	51.06
SiO_2	0.09	0.02	0.33	0.02
Cr_2O_3	0.23	0.23	0.02	0.26
Al_2O_3	0.40	0.34	0.00	0.00
CaO	N/A	N/A	N/A	N/A
MnO	0.26	0.25	0.30	0.00
MgO	9.43	9.32	1.99	3.28
FeO	28.68	28.58	40.69	40.06
Fe_2O_3	8.11	9.77	8.03	3.48
Total Oxides	97.73	98.62	100.50	100.37

Terrestrial minerals in the crust formed in oxygen partial pressure (P_{O2}) regions above the magnetite/wustite boundary (above about $logP_{O2} = -9$) while the lunar minerals formed in the region below the wustite/iron boundary (below about $logP_{O2} = -12$ at 1200°C) (Papike et al. 1991; Webster and Bright, 1961). Consequently, the minerals hematite and magnetite are not found on the Moon, while most terrestrial ilmenites are in solid solution with hematite. We recognize that much of the noble gas resides in agglutinitic glasses, even in ilmenite-rich regoliths, however, the gases are not surface correlated (Etique et al. 1978). However, we have attempted to simplify the experimental conditions as much as possible in this study using a single mineral. Gases in agglutinitic glasses should also be released at lower temperatures than those in mineral grains during mining as one would expect higher diffusivity.

Sample selection may not be as critical as one might think since modal analyses of the Apollo and Luna samples show relatively small percentages (0-20% by volume) of ilmenite (Papike et al. 1991). It will be assumed that a terrestrial sample that is mainly ilmenite will display release behavior similar to that of the lunar samples. This hypothesis is strengthened by the empirical correlation between helium and titanium (Baur et al. 1972; Cameron, 1988; Eberhardt et al. 1970; Fegley and Swindle 1993; Hintenberger et al. 1970). Other phases presumably do not trap helium nearly as well and thus, will release their small helium inventory at lower temperatures than will ilmenite. Care has been taken to use samples with the highest proportions of ilmenite possible.

Table 2.2 Compositions (weight percent) of ilmenite grains from Apollo 11 samples 10046 (average of first 6 points) (Lovering and Ware 1970), 10019 (Keil et al. 1970), 10020 and 10071 (Haggerty et al. 1970), 10085, 10045 (average of 6 points), and 10017 (average of 9 points) (Agrell et al. 1970), rock 10017 (Brown et al. 1970).

	Apollo 11 Samples						
Sample Number	10046-5	10019-22	10020-40	10071-28	10085- 4-10	10045-35	10017-12
TiO_2	53.13	51.8	53.9	52.8	56.3	53.65	53.95
SiO_2	0.08	N/A	N/A	N/A	N/A	0.08	N/A
Cr_2O_3	0.46	0.59	1.3	0.59	0.34	0.66	0.66
Al_2O_3	0.07	0.15	0.11	0.15	1.64	0.42	N/A
CaO	0.27	0.06	0.07	0.01	0.44	0.17	N/A
MnO	0.23	0.51	0.47	0.32	0.34	0.38	0.40
MgO	2.87	2.18	2.59	2.06	9.63	3.38	0.97
FeO	42.76	45.3	41.9	44.2	32.39	41.73	43.80
Fe_2O_3							
Total Oxides	99.87	100.59	100.34	100.13	101.08	100.30	99.78

Natural polycrystalline ilmenites from Sierra Leone, Liberia, New York and Quebec have been considered in this study. The samples were examined using secondary electron microscopy (SEM), high-resolution transmission electron microscopy (HRTEM), energy dispersive x-ray analysis (EDX) and electron probe microanalysis (EPMA). These analyses were complicated due to sub-micron diameter particles in some of the samples.

Ilmenite Xenocrysts

Samples of natural ilmenite were obtained from Professor Haggerty of the University of Massachusetts. These ilmenite xenocrysts were formed in the deep mantle of the Earth at higher pressure than lunar ilmenites, but they are known to be more reduced than most terrestrial ilmenites (Haggerty 1996). The samples were cross-sectioned, polished and analyzed using EPMA and the standards mentioned above.

The ilmenite in these samples is similar in bulk composition to the lunar regoliths from the Apollo 11 site (Table 2.1 and Fig. 2.4). The EPMA measurements for these xenocrysts are only qualitative since a standard of magnesian ilmenite was not available. The amount of Fe_2O_3 in the xenocrysts indicates that they are not as reduced as the lunar regolith or the lunar rocks. The xenocrysts all have amounts of $MgTiO_3$ (geikielite) similar to a geikielo-ilmenite fragment from lunar fines 10085. An even distribution of oriented precipitates have also been found in these xenocrysts (Fig. 2.5). Quantitative analysis of these precipitates has proven difficult due to their small dimensions. Energy dispersive x-ray analysis of the precipitates shows elevated levels of chromium and aluminum.

Marcy Anorthosite Massif, Adirondack Mountains, New York

A granular ilmenite from Sanford Lake in the Adirondack Mountains was provided by Professor John W. Valley of the University of Wisconsin-Madison. This ilmenite probably formed under similar conditions as those in the crust of the Moon, although dissimilar to those formed in the lunar basalts that are the primary contributor to the regolith. This sample was polished and examined using EPMA. The composition of this ilmenite is very similar to that of most lunar ilmenites with a small amount of hematite present (Table 2.1, Table 2.2 and Fig. 2.4). This ilmenite also contains much less geikielite than the xenocrysts. This sample was useful in the current study since the small volume of the hematite lamellae and other inclusions ($\approx 10\%$) will probably contribute little to the total evolved helium and/or will evolve helium at lower temperatures.

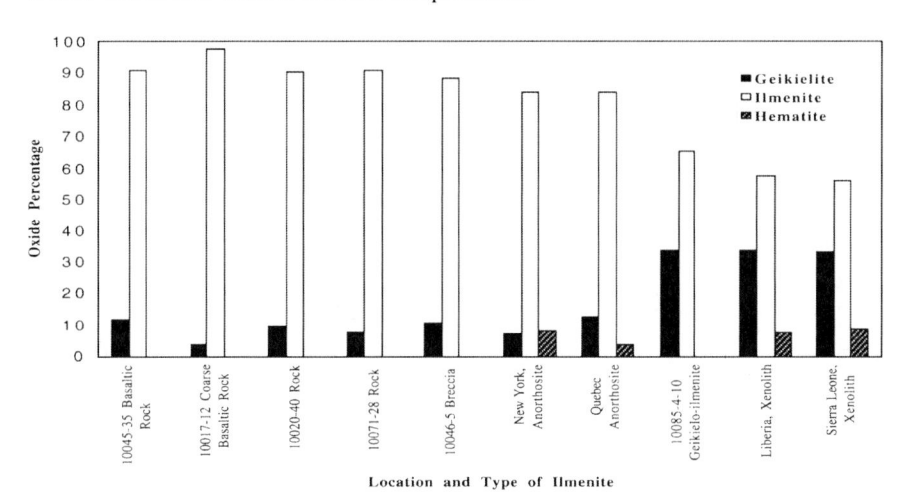

Fig. 2.4 Mineral content of samples in Table 2.2 and Table 2.1. The terrestrial ilmenites from anorthosites are very similar to the Apollo 11 (Apollo 11 samples have the form 10XXX-XX) ilmenites except for the small amount of hematite present. The results for the terrestrial samples are qualitative only. The analyses are of the bulk ilmenite.

St. Urbain, Quebec, Canada

A sample of ilmenite from a group of anorthosite dikes and lenses in St. Urbain township of Quebec were provided from the collection of Professor Eugene N. Cameron at the University of Wisconsin - Madison. The sample was polished and examined using EPMA. This ilmenite is the most lamellae-rich sample of those examined (Fig. 2.6). Note that the lamellae exist on different dimensional scales - small hematite lamellae within the bulk ilmenite and small lamellae of ilmenite within the hematite lamellae. The oxide composition of the ilmenite in this sample is given in Table 2.1, and its mineral composition is shown Fig. 2.4. The ilmenite in this sample appears to be somewhat more reduced than that of the previous samples since the amount of dissolved hematite is less than 4 percent as determined by EPMA. It also contains more geikielite than the Apollo 11 ilmenites.

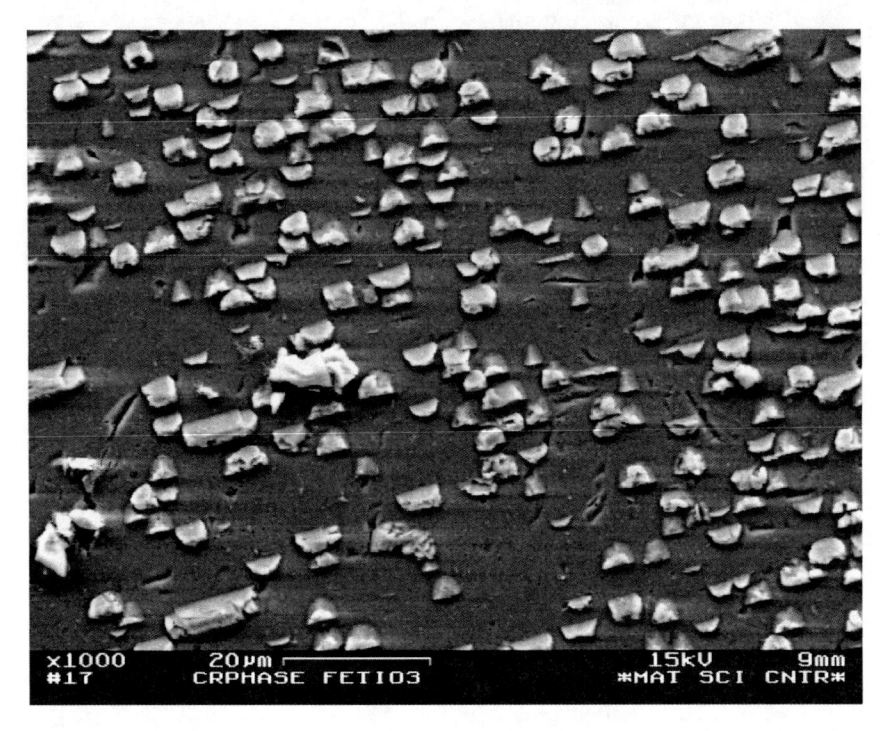

Fig. 2.5 Secondary electron micrograph of polished section of ilmenite xenocryst from Sierra Leone etched in 25% HF for 5 min. (1000X). The precipitates appear to be Cr and Al-rich compared to the bulk.

2.2.2 Plasma Source Ion Implantation

General Principles. Plasma source ion implantation (PSII) is a non-line of sight technique for the surface modification of materials (Conrad 1988; Conrad et al. 1990). The target - in this case a silicon wafer with wafered samples lying on top - is placed in a 1 m^3 chamber which is evacuated to a base pressure of about 10^{-6} torr. Gas of the species to be implanted is allowed to flow through the chamber at a pressure of several millitorr.

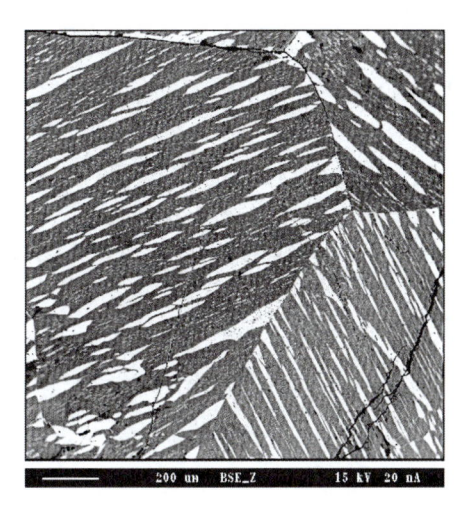

Fig. 2.6 Backscattered electron image of a granular ilmenite from St. Urbain, Quebec. Medium gray areas are bulk ilmenite and white areas are hematite lamellae.

A plasma is generated using tungsten filaments to ionize the gas by energetic primary electron impact. Other ways of generating a plasma for higher energy implantations are by radiofrequency (RF) or glow discharge methods.

A series of negative high voltage pulses are applied to the target, and the resulting electric field accelerates the ions in the plasma to high energies normal to the surface of the target (Fig. 2.7).

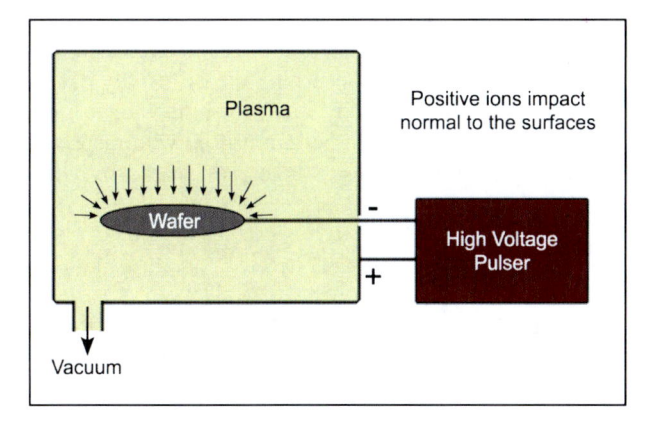

Fig. 2.7 Schematic diagram of the Plasma Source Ion Implantation (PSII) process (Malik 1998)

In the case of a wafer, significant asymmetries can occur at the edges where the electric field is changing rapidly. Focusing of the ions at the center of the wafer can also occur (Fetherston 1997). Therefore, the samples of ilmenite are placed off-center and away from the edges to maintain a uniform implantation.

The time needed to implant a desired fluence is given by Eq. (2.2):

$$time = \frac{Fluence \cdot Area \cdot e \cdot (\gamma + 1)}{I_p \cdot RR \cdot N} \qquad (2.2)$$

where *time* is the total time of the implantation (sec), *Fluence* is the desired final dose (ions/cm^2), *e* is the charge on an electron (Coulomb per electron), γ is the secondary electron emission coefficient (SEEC) (electrons emitted per incident ion), I_p is the average current per "on-time" of the pulse (Ampere), *RR* is the repetition rate (Hz) and *N* is the number of atoms per molecule of the precursor gas (i.e. $N=1$ for He, $N=2$ for H_2). The SEEC's used were taken from a study of electron emission from clean metal surfaces (Baragiola et al. 1979). These values do not change drastically for other elements ($\pm$ at most 0.2) at these energies and the choice of the value to use will not affect the calculation of time needed for the dose. The values of the SEEC were generalized to 0.25 for H and 0.5 for He in the current implantations.

2.2.3   Stepwise Heating with Mass Spectroscopy

One of the methods used to analyze rare gases in minerals is linear or stepwise heating. However, these heating methods allow for the activation of diffusion mechanisms during the analysis, possibly increasing or decreasing the concentrations found at various depths, effectively "smearing" the results. Bubble formation and phase changes can overshadow the diffusion mechanisms. A further complication results from the temperature extremes on the Moon ranging from about 100 K at night to a maximum on the surface of 380 K during the day (Vaniman et al. 1991). It is useful to look at the release of helium as a function of temperature, but this information is only one parameter. Information concerning the diffusivity can be only be obtained using data from an anneal with different time steps or a linear temperature ramp or in combination with an initial helium profile with respect to depth. Note that about 50% of the helium in these studies is released by about 500°C and 100% of the helium is released at 1000°C.

Long Heating Time Furnace. One sample, 050997 (New York ilmenite, implanted with 4 keV ^{4}He, dose = 1×10^{16} ^{4}He/cm^2), was annealed using 30 minute temperature steps of approximately 50°C each. The sample was analyzed (Becker 1997) using the experimental set up at the University of Minnesota described in detail by Frick et al. (1988) and Frick and Pepin (1981ab). The sample was wrapped in platinum foil for analysis and heated by an external resistance heater capable of at most 1200 - 1260°C. The system is evacuated to an equilibrium pressure of $\ll 1 \times 10^{-10}$ torr. A ZrTi bulk getter pump was used to initially "clean-up" the noble gases. Any heavier noble gases were frozen out on a metal frit. Further cleaning of the helium gas released was accomplished with ion getter pumps and noble gas separation on charcoal. Hydrogen was removed using a cold ZrTi alloy getter and N_2 was frozen out in another metal frit. The helium was then analyzed by a 6-inch double focusing Nier-type mass spectrometer with high mass resolution ($M/\Delta M$''800) (Frick et al. 1988). The temperature of each step was measured using a thermocouple mounted on the quartz furnace.

Blank measurements were taken at 300°C, 600°C and 950°C. These measurements basically measure the amount of helium that diffuses into the quartz sample furnace from the outside atmosphere. All measurements and errors for this method were calculated using a spreadsheet of measurement and error analysis formulas developed by Becker (1997). The average of the low temperature blanks was used for analysis of the measurements for temperatures below 600°C and the average of the blanks taken at 950°C was used for analysis of the remaining data points. The helium measurements for sample 050997 will be given below.

Small Particle Furnace. All other samples were analyzed (Schlutter 1997) using a pulse heating furnace developed for accurately measuring the noble gases released from interplanetary dust particles at the University of Minnesota (Nier and Schlutter 1990, 1992; Nier and Schlutter, 1993). This furnace was used for its capability to analyze for both ^3He and ^4He. The main goal of this experimental setup is to reduce the mass of the furnace as much as possible and thus reduce the blank measurements since the amount of gas contained in these samples is very small (Nier and Schlutter 1992).

The samples were wrapped in a tantalum foil, which comprises the oven. The oven is then attached by spring tension to tungsten wires which are spot-welded to heavy nickel leads (Nier and Schlutter 1992). The oven material and tungsten springs are thoroughly outgassed prior to use. However, these materials still release small amounts of gas during sample analysis. Blank runs are interspersed with the analysis runs. The gases are purified using liquid nitrogen-cooled charcoal traps and hydrogen was gettered by a liquid nitrogen-cooled volume freshly coated with titanium. The mass spectrometer used is same one used in work reported by (Nier and Schlutter 1990, 1992; Nier and Schlutter, 1993) and is similar to the one described earlier by Nier and Schlutter (1985). The sensitivity of the mass spectrometer in static operational mode is such that approximately 3×10^6 atoms of He produces an ion current of one ion per second (Nier and Schlutter 1990). Ion counting is used with a background counting rate of less than 1 count per 200 seconds (Nier and Schlutter 1990).

Heating in this apparatus is accomplished by resistive heating of the Ta foil furnace (Nier and Schlutter 1990). The nickel leads and tungsten springs are used to apply an electric current directly to the foil. Nier and Schlutter (1992) devised a circuit that multiplies the instantaneous current and voltage with this product determining the voltage needed to keep the power constant during a step (Nier and Schlutter 1992). Temperatures above 770°C are typically measured using an optical pyrometer and lower temperatures were extrapolated using an empirical formula which assumes two modes of power loss: 1) conduction and 2) radiation (Nier and Schlutter 1990). Raw data from these measurements are available elsewhere (Harris-Kuhlman 1998).

2.2.4 Numerical Analysis of Stepwise Heating Profiles

The three-dimensional ANNEAL code was developed for this research for calculating diffusivity based on the fraction of solar wind implanted gas released during each annealing step. It uses the alternating direction method to solve the parabolic differential equation that describes diffusion as well as heat transfer. This equation is of the general form in three-dimensions:

$$\frac{\partial C}{\partial t} = D\nabla^2 C \qquad (2.3)$$

where D is the diffusivity (cm^2/s), C is the concentration (atoms/cm^3).

The two inputs to ANNEAL are the measured fractional release of helium at each temperature step and the depth profile of ^{3}He or ^{4}He calculated using the program SRIM mentioned above. The details of the ANNEAL code will be discussed in detail in elsewhere (Kuhlman 2011). The diffusion coefficient and activation energies of helium are then calculated by fitting the resulting diffusivities to Fick's Law. It must be recognized that SRIM lacks information about crystalline structure and measurements of stopping powers at low energies, thus making the results qualitative.

2.3 Results

2.3.1 Initial Solar Wind Simulation

The first experiment involved simulating lunar ilmenite samples using Plasma Source Ion Implantation (PSII) (Fig. 2.7) to generate a solar-wind-like flux on samples of terrestrial ilmenite. To determine how well the ilmenite from anorthite in New York matches the behavior of lunar ilmenites under irradiation by solar-wind helium, PSII was used to simulate the solar wind. The samples were polished with 0.05 micron colloidal silica and mounted on a 6 inch (15.24 cm) silicon wafer to maintain normal ion incidence during PSII. After the PSII chamber was evacuated to a base pressure of about 10^{-6} torr, the samples were sputter-cleaned with 600 eV argon, and implanted with 4 keV ^{4}He to a dose of 1×10^{16} ions/cm^2. This dose was chosen to be below the amount of ^{4}He in grains of ilmenite from the mature lunar regolith sample 76501,44 (Ducati et al. 1973; Müller et al. 1976) in order to avoid saturation of the sample. The samples were then brought up to air before being put into the long time heating high-vacuum furnace/mass spectrometer system (Frick et al. 1988). Similar procedures were used for the original Apollo samples. The implanted helium was released using time steps of 30 minutes and temperature steps of 50°C to approximate the helium release data from the Apollo 11 (Pepin et al. 1970a) and Apollo 17 samples (Frick et al. 1988). The helium release behavior of the implanted New York ilmenite is compared to the release from Apollo 11 rock 10069-21, regolith 10084-48, and two single grains of Apollo 17 ilmenite - from sample 71501-38 (a sub-mature soil) and sample 79035-24 (a mature breccia). The integrated release of helium from the implanted sample is quite similar to that of the Apollo samples (Fig. 2.8).

Several observations can be made with respect to Fig. 2.8. First, two slopes appear in the release from the simulant and are similar to the release of 10069-21, possibly indicating two different release mechanisms. The slopes of each of these curves change at about 590°C. Second, the slopes of these two curves are somewhat different from those of 71501-38 and 79035-24 below approximately 80% of total release. Finally, the release from the simulant begins at a lower temperature than for any of the lunar samples. This may be due to implantation rate effects or the fact that ilmenite is a semiconductor and may have been somewhat shielded during implantation. Longer implantation times are not expected to have much effect since the magnitude of the difference in exposure time is so great. This early release may also be a result of the "contamination" of the sample by hematite and/or other minerals. The lack of early loss from the Apollo samples may be due to loss of the least tightly bound fraction during collection and curation.

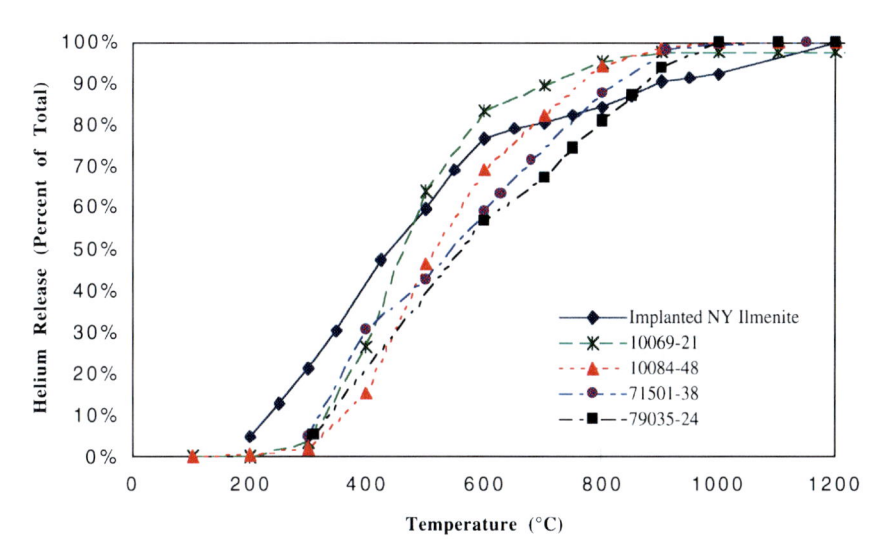

Fig. 2.8 Release data of helium from implanted New York ilmenite and samples of rocks and regolith from Apollo 11 (Pepin et al. 1970) and Apollo 17 (Frick et al. 1988). Annealing steps of 50°C and 30 minutes were used for the implanted terrestrial sample while steps of approximately 100°C and 1 hour were used for the Apollo samples.

It was concluded that PSII is capable of simulating the solar-wind implantation of volatile elements into minerals. The experiment also indicates that the release of helium by regoliths from high-titanium locations on the Moon may be a function of their ilmenite content.

2.3.2   Parametric Studies of Hydrogen Implantation

Three sets of samples were implanted with ^{4}He to 10^{16} ions/cm^2, ^{3}He to 10^{13} ions/cm^2 and three different fluences of H from 10^{14} ions/cm^2 to 10^{17} ions/cm^2 at solar-wind energies using PSII. Immediately after implantation, the samples were brought up to air, and placed in a dessicator. They were stored in the dessicator for various periods before annealing. Prior to analysis by step-wise heating, the samples were broken to make particles small enough to fit into the small particle furnace. These furnaces require that the samples be small so that the mass spectrometers were not overwhelmed with gas. Optical micrographs were taken of the polished surface areas of the individual grains produced by crushing so that the areas could be quantified using the computer program NIH Image (National Institute of Health 1996).

2.3.4   High-Resolution Transmission Electron Microscopy

In parallel, high-resolution transmission electron microscopy images of an ilmenite implanted with approx. 3×10^{16} atoms/cm^2 ^{4}He and 1×10^{17} atoms/cm^2 H, at 1keV/amu respectively. The samples show evidence of two damage regimes, one due to He implantation and one due to H implantation (Fig. 2.9).

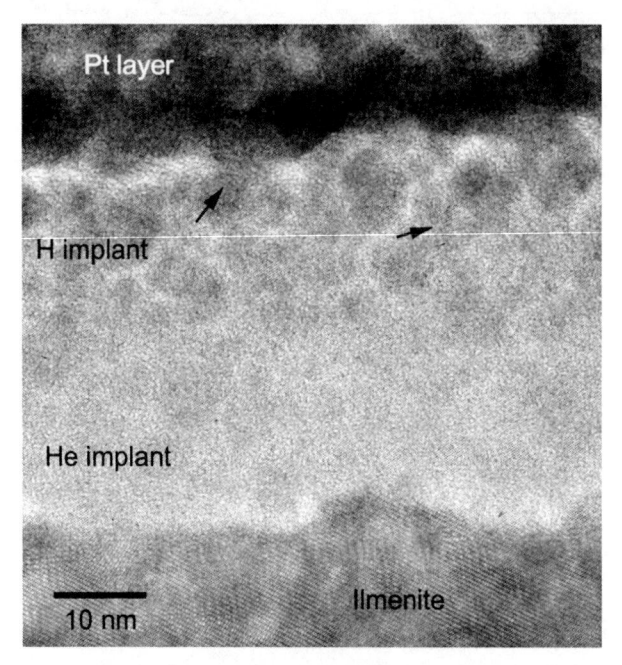

Fig. 2.9 Bright field HRTEM image of ilmenite implanted (from top) with 1keV/amu 1×10^{17} atoms/cm^2 H and 3×10^{16} atoms/cm^2 ^{4}He to simulate lunar space weathering. The arrows indicate regions of weak crystallinity. The image was taken on the zone axis of the ilmenite source material.

The depths of these regimes correspond quite well to the implantation ranges of the H and ^{4}He at 1 keV/amu (i.e. solar wind energy) as determined using SRIM. The unimplanted ilmenite shows strong crystallinity, while the depth implanted with ^{4}He is quite amorphous. Weak crystallinity appears in the hydrogen implanted surface layer approximately 10 nm deep. This crystallinity appears to be quite different from the substrate material (diffraction patterns not shown here), but further examination is needed to characterize this second phase. Scanning transmission electron microscopy (STEM) with EDS is needed to measure the elemental compositions because of the small size of the phases involved. It is not clear that this phase is the beginning of the formation of nanophase iron, but this possibility cannot be ruled out. It is also quite notable that the interface between the amorphous layer generated by implantation and the host mineral is irregular but very abrupt on the nanometer scale. This was noted for rims on lunar grains by Keller and McKay (1997) who also noted that these irregular sharp interfaces had been suggested to be associated with radiation effects by Bradley et al. (1996).

2.4 Experimental Results

One experiment was run using the long duration furnace. The sample, 050997, was only implanted with ^{4}He. This experiment is discussed in Section 2.4.1. A set of experiments was run with varying hydrogen fluences to investigate the effect that implanted hydrogen might have on helium diffusion. The first experiment was a control experiment with no hydrogen implanted. The next three experiments with a helium fluence of 1 keV/amu 1×10^{16} atoms/cm^2 and varying H fluences between 1×10^{14} /cm^2 and 1×10^{17} /cm^2 were run. These runs are summarized in Table 2.3.

Table 2.3 Summary of stepwise heating experiments

Sample Number	H Fluence	^{3}He Fluence	^{4}He Fluence
050997			1×10^{16} /cm^2
041098		1×10^{13} /cm^2	1×10^{16} /cm^2
041398	1×10^{17} /cm^2	1×10^{13} /cm^2	1×10^{16} /cm^2
040998	1×10^{14} /cm^2	1×10^{13} /cm^2	1×10^{16} /cm^2
041698	5×10^{15} /cm^2	1×10^{13} /cm^2	1×10^{16} /cm^2

2.4.1 Sample 050997.1

The result of applying the ANNEAL code to the ^{4}He release data of sample 050997.1 is shown in Fig. 2.10. This sample was implanted with 1×10^{16} ^{4}He/cm^2. The error is that due to the assumption of a monoenergetic implantation. The dip in the curve at approximately 675°C is likely a phase change due to the long heating times (30 minutes) and high vacuum (i.e. very low P_{O2}) (Muan and Osborn 1965). Three regions of diffusion are identified and summarized in Table 2.4 and Fig. 2.10.

Table 2.4 Summary of diffusion characteristics of ^{4}He release from sample 050997.1.

Temperature Range (°C)	Diffusion Coefficient (cm^2/s)	Activation Energy (eV)
200-350	1×10^{-13}	0.26
350-675	1×10^{-11}	0.54
700-900	4×10^{-08}	1.38

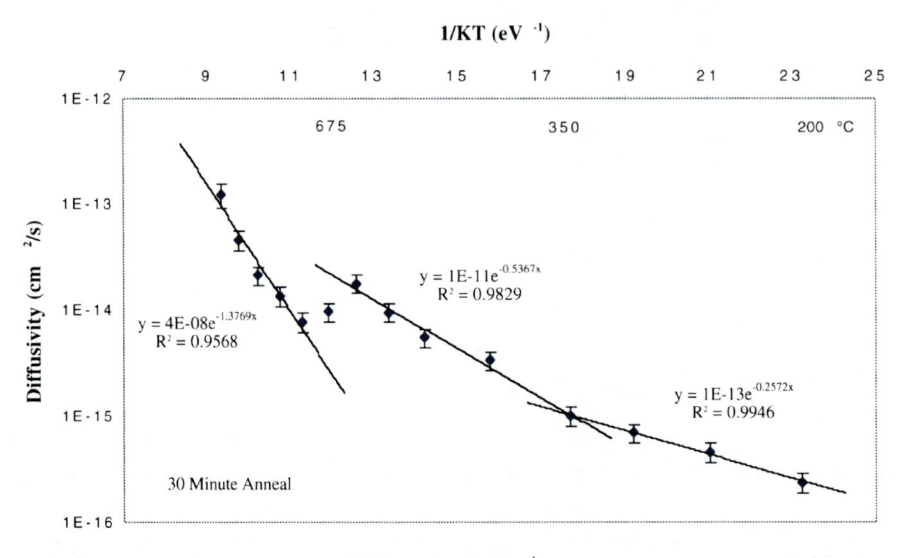

Fig. 2.10 Results of applying the ANNEAL code to the ^{4}He release data from sample 050997.1. This sample was only implanted with 1×10^{16} ^{4}He/cm^2. The error is that due to the assumption of a monoenergetic implantation.

2.4.2 Sample 041098.4

The result of applying the ANNEAL code to the ^{4}He and ^{3}He release data of sample 041098.4 is shown in Fig. 2.11.

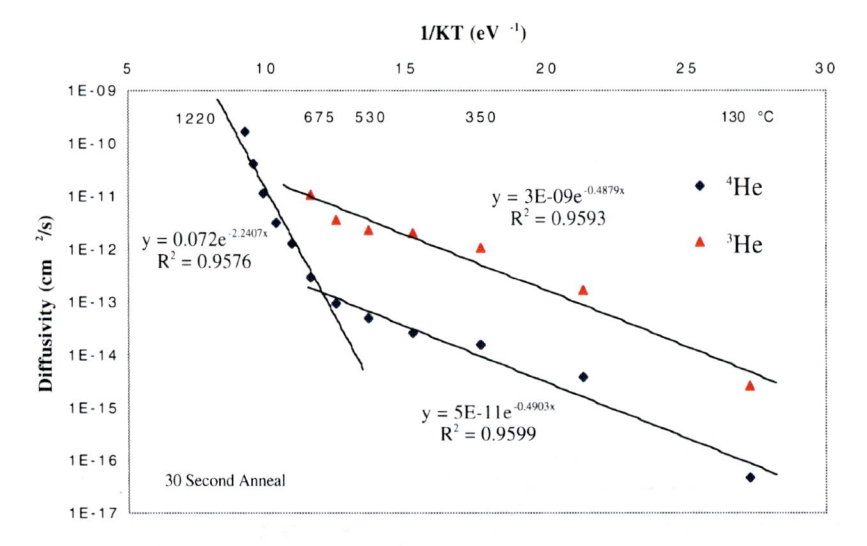

Fig. 2.11 Results of applying the ANNEAL code to the ^{4}He and ^{3}He release data from sample 041098.4. This sample had no hydrogen preimplantation and was implanted with ^{4}He to 1×10^{16} ions/cm^2 and ^{3}He to 1×10^{13} ions/cm^2.

This sample had no hydrogen preimplantation and was implanted with ^{4}He to 1×10^{16} ions/cm^2 and ^{3}He to 1×10^{13} ions/cm^2. Two regions of diffusion are identified in the release of both ^{4}He and ^{3}He. The diffusion characteristics are summarized in Table 2.5. Note the lack of a dip in the data. The short heating times (30 seconds) prevent any phase change from occurring.

Table 2.5 Summary of diffusion characteristics of ^{4}He and ^{3}He release from sample 041098.4.

	^{4}He		^{3}He	
Temperature Range (°C)	Diffusion Coefficient (cm^2/s)	Activation Energy (eV)	Diffusion Coefficient (cm^2/s)	Activation Energy (eV)
130-675	5×10^{-11}	0.49	3×10^{-09}	0.49
675-1000	7.2×10^{-02}	2.24		

2.4.3 Sample 041398.5

The result of applying the ANNEAL code to the ^{4}He and ^{3}He release data of sample 041398.5 is shown in Fig. 2.12. This sample was implanted with H to 1×10^{17} ions/cm^2, ^{4}He to 1×10^{16} ions/cm^2, and ^{3}He to 1×10^{13} ions/cm^2. Two regions of diffusion are identified in the release of both ^{4}He and ^{3}He. However, in this case, not enough data was analyzed to make a good fit of the ^{3}He data in the higher temperature range. The fit of ^{3}He in this range was done with two points simply to show that the activation energy is increasing. The diffusion characteristics are summarized in Table 2.6. Again, note the lack of a dip in the data.

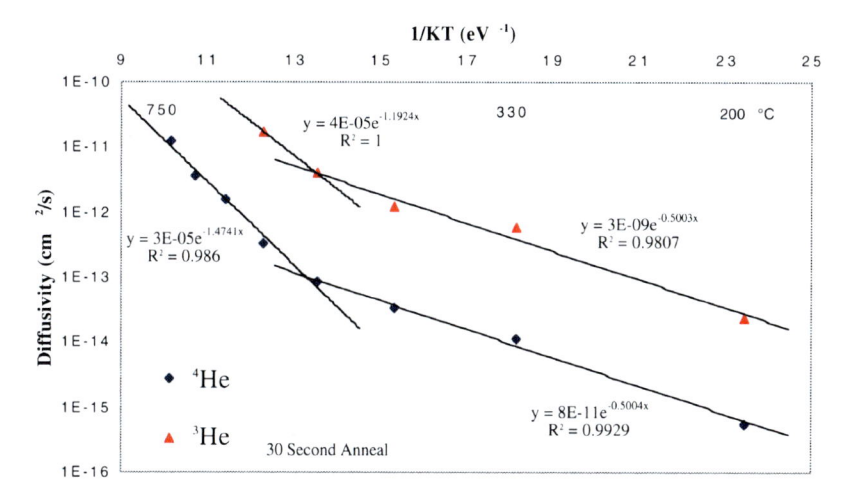

Fig. 2.12 Results of applying the ANNEAL code to the ^{4}He and ^{3}He release data from sample 041398.5. This sample was implanted with H to 1×10^{17} ions/cm^2, ^{4}He to 1×10^{16} ions/cm^2, and ^{3}He to 1×10^{13} ions/cm^2.

Table 2.6 Summary of diffusion characteristics of ^{4}He and ^{3}He release from sample 041398.5.

	^{4}He		^{3}He	
Temperature Range (°C)	Diffusion Coefficient (cm^2/s)	Activation Energy (eV)	Diffusion Coefficient (cm^2/s)	Activation Energy (eV)
200-675	8×10^{-11}	0.50	9×10^{-09}	0.55
675-750	3×10^{-05}	1.47	4×10^{-05}	1.19

2.5 Discussion

The interpretation of the experimental results presented here requires an understanding of several properties of ionic crystalline solids. The "free" space available between atoms in the structure, the mobility of the constituent ions within that structure, the number and type of point, line and plane defects present, and the solubility of helium all play a role in determining the trapping mechanism of helium in ilmenite. Several major observations emerge from the experiments documented here regarding possible trapping and diffusion mechanisms:

- Helium release from implanted terrestrial ilmenites is diffusion-limited as opposed to reaction rate-limited.
- Helium diffusion in terrestrial ilmenite implanted with ^{4}He and ^{3}He is characterized by four distinct activation energies, $E_1 = 0.26$ eV, $E_2 \cong 0.5$ eV, $E_3 \cong 1.4$ eV and $E_4 > 2.2$ eV (Harris-Kuhlman 1998).
- The diffusivity of ^{3}He in terrestrial ilmenite implanted with ^{4}He and ^{3}He is greater than the diffusivity of ^{4}He.
- Moderate fluences ($<1 \times 10^{15}$ cm^{-2}) of hydrogen preimplantation retard ^{3}He diffusion in terrestrial ilmenite implanted with ^{4}He and ^{3}He. High fluence H preimplantation (10^{17} cm^{-2}) has virtually no effect on ^{3}He diffusion.

2.5.1 A Model for the Diffusion of Helium Implanted into Ilmenite

The results of the ANNEAL code applied to the fractional release of the implanted helium indicate four activation energies: $E_1 = 0.26$ eV, $E_2 \cong 0.5$ eV, $E_3 \cong 1.4$ eV and $E_4 > 2.2$ eV. It is clear that the diffusivity of the implanted helium is several orders of magnitude greater than the self-diffusivities of component ions observed for other rhombohedral oxides or other iron oxide. Helium appears to be diffusing through a quasi-stationary structure, i.e. ionic diffusion processes and related point defect relaxations are negligible.

The implantation of He and H (as well as Ar during sputtering cleaning) has modified the surface of the ilmenite to a depth of approximately 50 nm (Fig. 2.9). In addition to raising the ilmenite to a higher energy state by the introduction of He and H, the crystalline structure of the ilmenite is partially destroyed by the irradiation. The activation energy, $E_1 = 0.26$ eV, could be explained by radiation damage caused by the Ar sputter-cleaning which damages the upper 5 nm of the surface. The activation energy, $E_2 \cong 0.5$ eV, might be explained by radiation damage caused by the implantation of ^{4}He or the numerous exsolution lamellae shown to exist in terrestrial ilmenites (Fig. 2.6). The activation energy, $E_3 \cong 1.4$ eV, could correspond to the binding energy for helium trapping in oxygen vacancies as calculated by Welch et al. (1976) and observed by Allen (1991) for α-Al_2O_3. Finally, the activation energy, $E_4 > 2.2$ eV, may be an indication of helium trapping in constitutional vacancies (naturally vacant octahedral sites) in the corundum-like structure of ilmenite, also calculated by Welch et al. (1976) and observed by Allen (1991) for α-Al_2O_3. These latter two mechanisms are likely to be high-energy states for He and are likely obtained only by the energetic implantation of He into the crystalline structure. Once the sample is heated enough to overcome the binding energy of He to either type of trap, the He appears to diffuse to the surface and escape. The various influences on He diffusion in this model are discussed in more detail below.

The existence of two or more activation energies in solar wind implanted materials has recently been observed by Heber et al. (2009) in diamond on silicon from the Genesis mission. The lower activation energy release was attributed to strong radiation damage at the surface of the material, similar to the radiation damage demonstrated in artificially implanted ilmenite in Fig. 2.9.

2.5.2 Solubility of Helium in Minerals

In order to be able to discuss diffusion of ^{4}He in ilmenite, one must first have an idea of the amount of helium that is naturally soluble in the lattice. While solubility data for helium in ilmenite has not been found, an estimate can be made using data for magnetite and basaltic glass. The solubility limit of helium in magnetite has been measured to be 6.1×10^{16} atoms/g or 23 ppm at 723°C (Lancet and Anders 1973). Calculations using SRIM indicate that the maximum concentration of implanted ^{4}He is about 4.2×10^{20} atoms/g for an implantation of 1×10^{16} ions/cm^2. Thus, the solubility limit is about four orders of magnitude lower than peak implantation concentration. The profile of implanted helium provides the driving force for diffusion because of the local supersaturation of helium caused by the implantation.

2.5.3 Lattice Disorder and Radiation Damage

The diffusion constant, D_0, is a convolution of the distance between equilibrium positions of the diffusing species and an entropy term, which is given by (Jost 1952):

$$D = D_0 \exp\left(\frac{\Delta H}{RT}\right) = a^2\left(\frac{kT}{h}\right)\exp\left(\frac{\Delta S}{R}\right)\exp\left(\frac{\Delta H}{RT}\right) \qquad (2.4)$$

In Eq. (2.4), a is the distance between equilibrium positions (effective jump distance), k is Boltzmann's constant, T is the temperature in Kelvin, h is Planck's constant, R is the gas constant, ΔS is the change in entropy and ΔH is the effective activation energy between the activated and normal state. For diffusion in a crystalline solid, the entropy term is typically the order of unity. However, since the helium is being placed in the lattice by irradiation, it is instructive to estimate the magnitude of the energy of disorder, $T\Delta S$, for each of the temperature regimes observed (Table 2.7). In this estimation, the effective jump distance has been assumed to be 0.5 nm, the lattice parameter in the a-direction in ilmenite. This energy of disorder can then be compared to the energy of fusion for ilmenite, $\Delta H = 0.94$ eV (91 kJ/mol).

Table 2.7 Calculated entropy contribution to the diffusion constant

Possible Diffusion/ Trapping Mechanism	Temperature Range (°C)	Experimentally Measured Activation Energy (eV)	Diffusion Constant, D_0 (cm^2/s)	Energy of Disorder, $T\Delta S$ (eV)
Radiation Damage Planar Defects	130-300	0.26	1×10^{-13}	-0.90 (130°C)
Radiation Damage or Planar Defects	130-675	0.49	5×10^{-11}	-0.69 (130°C)
Oxygen Vacancies	675-750	1.47	3×10^{-05}	-0.62 (700°C)
Constitutional Vacancies	750-1000	2.24	7×10^{-02}	0.006 (1000°C)

Table 2.7 reveals that the energy of disorder for the temperature range 130 - 675°C is considerably larger than the activation energy measured for helium diffusion and approaches the enthalpy of fusion for ilmenite. The entropy term is negative as opposed to the positive activation energy. The large negative entropy term suggests that the highly damaged lattice is relaxing to a more ordered, lower energy state as the sample is heated. This entropy contribution also suggests that the ilmenite has been at least partially amorphized by the irradiation. The metamict state may be a good approximation of the state attained by the implantation (Fig. 2.9). This layer may contain very small regions of crystalline order among basically amorphous material. These tiny crystalline domains may exhibit behavior similar to that seen in Ar dating studies of systems with multiple diffusion domains (MDD's) (Lovera et al. 1989). This phenomenon could be explored using the current sample set with much smaller temperature steps at low temperatures. It should be noted that dating studies assume uniform distribution of the noble gas within the material and the models derived typically do not apply to irradiation-damaged material.

Amorphization is further corroborated here since the diffusion of helium in this temperature range with activation energies of 0.26 eV and 0.5 eV is similar to the diffusion of helium in glass. At higher temperatures, the activation energy becomes significantly larger than the entropy term, indicating that most of the energy of disorder has been released. Finally, the entropy term becomes positive and very small. At this point, the entropy term becomes negligible and the effective jump distance for diffusion is approximately 0.5 nm.

Another possible explanation for the activation energies, E_1 and E_2, and the corresponding diffusion constants is the detrapping from high concentrations of planar defects associated with the exsolved phases within the terrestrial ilmenite. The helium could also be interacting with the stresses around those planar defects induced by thermal expansion and phase changes as the temperature increases. Further work is required using high resolution TEM to justify this explanation since it was not possible to quantify the spatial defect concentrations in the current study. The methods of Lovera et al. (1989) may again be applicable since there may be more than one type of defect, but on a different spatial scale than discussed earlier.

2.5.4 Point Defect Concentrations in Ilmenite

The point defect thermodynamics for hematite developed by Dieckmann (1993) can be used as a model for ilmenite. If a closure temperature (the temperature at which the material "closes" to diffusion of the constituent species) of 600°C is assumed for terrestrial ilmenite and 750°C for lunar ilmenite, the oxygen vacancy concentrations are found to be 13 ppm and 66 ppm, respectively. This concentration of oxygen vacancies is smaller than the local implanted ^{4}He concentration, but provides a uniform concentration of traps as the helium is implanted further into the undamaged lattice. Additional evidence that helium is bound to oxygen vacancies in ilmenite is provided by the observations of Dieckmann (1996) that oxygen vacancies in hematite are capable of storing He. These vacancies are naturally present in the undamaged lattice and may provide the trapping mechanism for helium implanted beyond the "amorphous" layer.

2.5.5 Traps for Helium in Implanted Ilmenite

The calculations by Welch et al. (1976) and the experimental observations of Allen (1991) and Allen and Zinkle (1992) for He in α-Al_2O_3 can also provide some guidance in identifying the trapping mechanisms above. It must be kept in mind that the bonding in α-Al_2O_3 and in $FeTiO_3$ will be quite different despite the fact that they have the same crystal structure. Welch et al. (1976) calculated that the energy difference between a He atom in an oxygen vacancy and a He atom in an interstitial site in α-Al_2O_3 is 1.7 eV. An analogous situation in ilmenite could be indicated by the measured activation energy, $E_2 \cong 1.4$ eV. A separate calculation by Welch et al. (1976) for He in an intrinsically vacant octahedral site relative to a perfect crystal resulted in an energy difference of 2.7 eV. In the current study, this trapping mechanism may be associated with the increase in activation energy measured at higher temperatures, $E_4 > 2.2$ eV. The lower population in these high-energy traps may be related to the energy barrier required for He to get into these traps in the first place.

2.5.6 Concentrations Calculated by SRIM

The number of oxygen vacancies caused by the 4 keV ^{4}He irradiation to 1×10^{16} ions/cm^2 can be estimated using SRIM. It has been established that oxygen vacancies are larger than helium atoms while the cation vacancies are smaller than helium atoms in ilmenite (Harris-Kuhlman 1998). The results for oxygen vacancies are shown in terms of number per lattice atom in Fig. 2.13 along with the concentrations of ^{4}He and ^{3}He atoms in terms of number per lattice atom (9.5×10^{22} atoms/cm^3 for FeTiO$_3$). The calculation of the vacancy concentrations has been reduced by a factor of 10 based on the observations of Biersack (1987) which indicate that the residual number of point defects is about an order of magnitude less than that calculated by SRIM. The peak number of vacancies occurs closer to the surface, at about 17 nm, whereas the peak ^{4}He concentration occurs at 33 nm from the surface. Since the ^{4}He must diffuse through this region of a large concentration of oxygen vacancies before escaping from the solid, this calculation supports the observation of trapping in oxygen vacancies with $E_3 = 1.4$ eV and demonstrated by Allen (1991).

The ratio of the diffusivities of ^{3}He and ^{4}He, D_{3He}/D_{4He}, for sample 041098.4 in the temperature range 130-675°C is about 60 (Table 2.5). This anomalous diffusion of ^{3}He is much greater than expected for the isotope effect:

$$\frac{D_{^3He}}{D_{^4He}} \approx \left(\frac{m_{^4He}}{m_{^3He}}\right)^{-1/2} = 1.15 \tag{2.5}$$

where m is the mass of the isotope (Jambon and Shelby 1980). The only explanation for the increased diffusivity of ^{3}He is that since it is implanted last, the ^{4}He have saturated the available traps. The majority of the implanted ^{3}He must also diffuse out through the damaged lattice with activation energy $E_2 = 0.5$ eV since the ^{3}He is exposed to few traps. The fraction of ^{3}He implanted into the relatively undamaged portion of the lattice may be responsible for the appearance of higher activation energies at higher temperatures (Fig. 2.12). Figure 2.14 illustrates the concentrations of the various implanted species and the oxygen vacancies produced. The calculations displayed in these figures were performed using SRIM. The results for oxygen vacancy concentrations have again been reduced by a factor of 10 to reflect observations that the residual number of point defects is about an order of magnitude less than the number calculated by SRIM (Biersack 1987). Again, it is likely that ^{4}He is saturating the O vacancies produced. The vacancies are not available to trap the ^{3}He except near the surface of the amorphous region and in the relatively undamaged bulk. Therefore, the activation energy, $E_2 \cong 0.5$ eV, is observed for the ^{3}He diffusing through the amorphous region.

The indication of rising activation energy above 675°C (Figs. 2.11 and 2.12) is likely due to the tail of the ^{3}He (or ^{4}He) distribution deeply implanted into the crystalline bulk. A similar situation exists for sample 040998. The retardation of the release of ^{3}He from this sample (040998) relative to sample 041098 may be a result of additional vacancies produced by the implantation of hydrogen (1×10^{14} cm^{-2}). The decrease in diffusivity of ^{3}He may also be due to the implantation of hydrogen that subsequently reacts with oxygen, creating a small energy barrier for ^{3}He diffusion. The hydrogen may also react with the constituent atoms, forming new traps or making it possible for oxygen vacancies to be available to trap the ^{3}He. The effect of retarding ^{3}He release increases as the fluence of implanted H increases to 5×10^{15} cm^{-2}. This delay in ^{3}He release may again be the result of H reacting with the constituent atoms, forming new traps or leaving oxygen vacancies available to the ^{3}He.

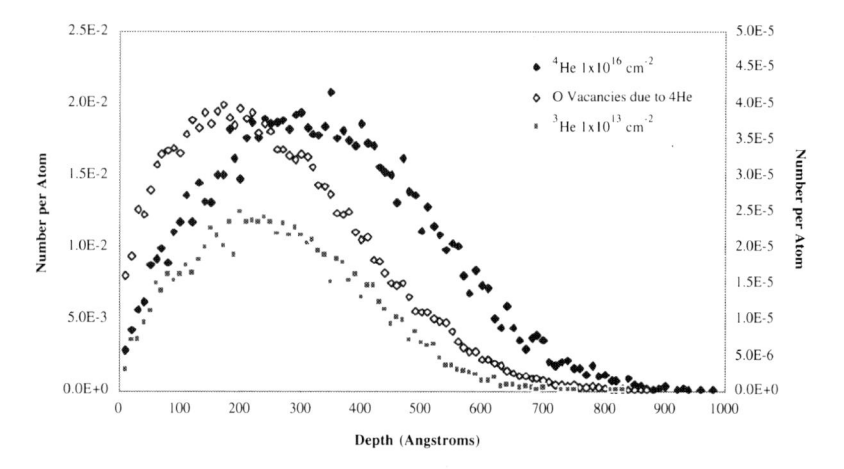

Fig. 2.13 Calculations of the number of oxygen vacancies and helium atoms per lattice atom performed using SRIM, ^{4}He implanted to 1×10^{16} ions/cm^2 and ^{3}He implanted to 1×10^{13} ions/cm^2 (Sample 041098). Note the difference in scale for the ^{3}He concentrations. The number of oxygen vacancies reflects the observations of Biersack (1987) that the residual number of point defects is about an order of magnitude less than that calculated by SRIM.

Finally, a fluence of hydrogen approaching the solar wind ratio compared with the implanted helium fluences results in a release profile similar to the case with no hydrogen implantation. Figure 2.14 illustrates the fact that the number of oxygen vacancies produced by the implantation of H has approached the number of oxygen vacancies produced by the ^{4}He implantation to a depth of about 10 nm. The lack of a difference between no H implantation and the implantation of H to 1×10^{17} cm^{-2} may be a result of H simply saturating all of the oxygen vacancies produced. In this case, the only oxygen vacancies remaining will be in the relatively undamaged region below about 50 nm. With no traps available to the ^{3}He implanted last, the activation energy, E_2, will again be observed.

2.5.7 Effects of Hydrogen Implantation

None of the studies in the literature have explored the effect of hydrogen on the diffusion of solar wind implanted He isotopes. Since protons make up the majority of the solar wind and are very reactive, it seems plausible that they could control the diffusion of He in lunar minerals by modifying existing traps. Lord (1968) suggested that the implanted H could bond to oxygen atoms in the lattice. The implantation of 1 keV H also creates additional damage in the lattice (Fig. 2.14) which could generate traps for He. This is seen in the current experiment by the difference in total counts of evolved ^{4}He measured in samples 041098.4 (no H implantation) and 041398.5 (H implanted to 1×10^{17} cm^{-2}).

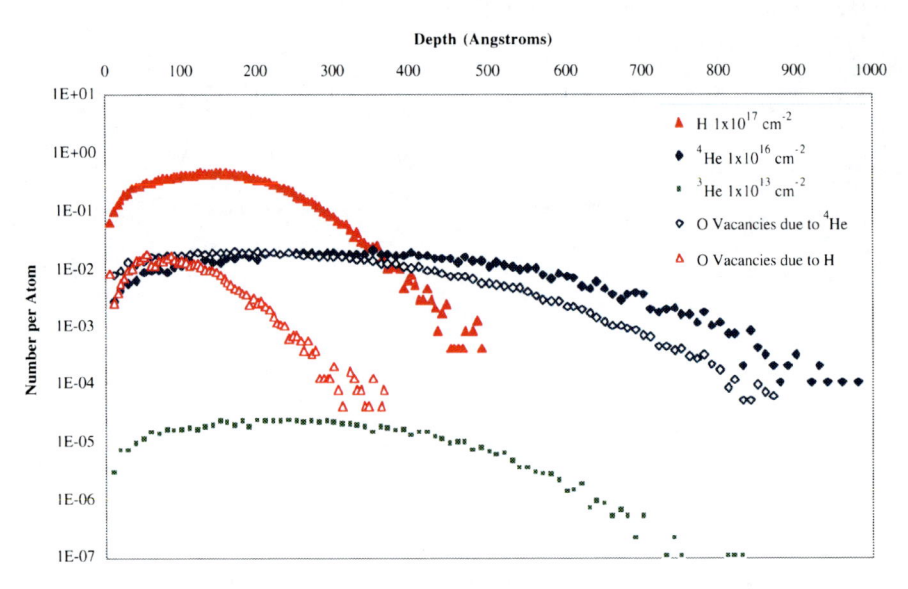

Fig. 2.14 Calculations of the number of oxygen vacancies produced and helium and hydrogen atoms deposited per lattice atom. These calculations were performed using SRIM and the implantation fluences shown for sample 041398.

Assuming that the surface area of the two samples are approximately the same, the fact that sample 041398.5 retains about 3.7 times more ^{4}He than sample 041098.4 could be indicative of a higher trap density caused by the H implantation. Conversely, sample 041098.4 retained 2.5 times more ^{3}He than did 041398.5. This observation supports the theory that the anomalously high diffusivity of ^{3}He is due to the order of implantation in this experiment. The low ^{3}He retentivity of 041398.5 may be due to the occupation of the available traps by the high doses of H and ^{4}He implanted prior to the ^{3}He implantation.

Release of ^{4}He with Hydrogen Implantation

The fractional release of ^{4}He for the 04XX98 series of samples is shown in Fig. 2.15. The release is fairly consistent with varying fluences of implanted H and no systematic correlations with H are apparent. The release of ^{4}He begins at approximately the same temperature, and the shapes of the curves are similar.

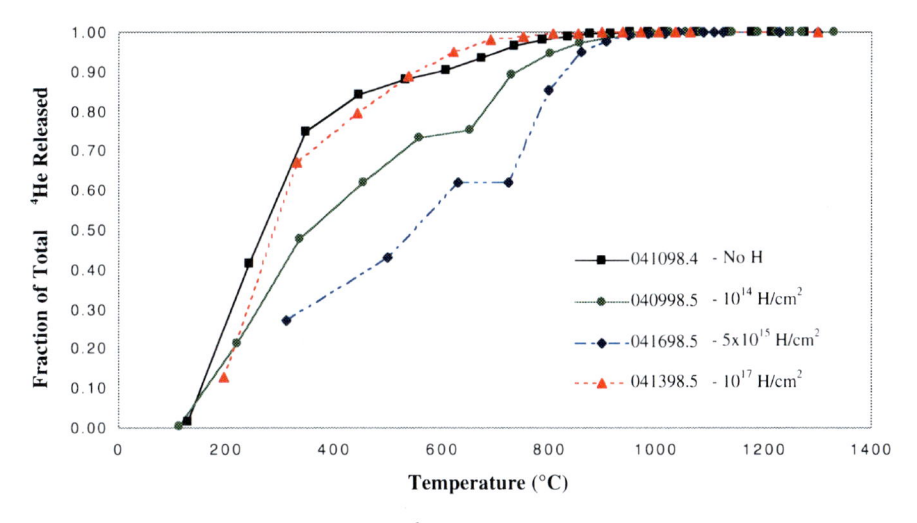

Fig. 2.15 Cumulative fractional release of ^{3}He from the 04XX98 series of samples. The hydrogen initially retards ^{3}He evolution but the highest H fluence has virtually no effect on ^{3}He release. The lines are included to guide the eye.

The 4 keV ^{4}He generates a large number of vacancies at its mean range, about 33 nm, which could be responsible for the ^{4}He trapping and release with characteristic activation energy of 1.38 - 1.47 eV. The activation energy for ^{4}He with no H implantation was measured as 2.24 eV and increases at higher temperature. This release may be due to constitutional vacancies (naturally vacant octahedral sites which account for 1/3 of all cation sites). The data supports a combination of both types of traps since the diffusivity shows a smooth increase in activation energy from 1.38 to 2.24 eV.

Release of ^{3}He with Hydrogen Implantation

The fractional release of ^{3}He for the 04XX98 series of samples is shown in Fig. 2.16. The release is seen to change dramatically with varying fluences of implanted H. The release of ^{3}He occurs very early in the case of no H implantation relative to the ^{4}He release of all these samples. As the fluence of H is increased, the ^{3}He release occurs at progressively higher temperatures. However, the behavior of the sample implanted with solar wind ratios of H, ^{3}He and ^{4}He returns to the release behavior of the sample with no hydrogen implanted.

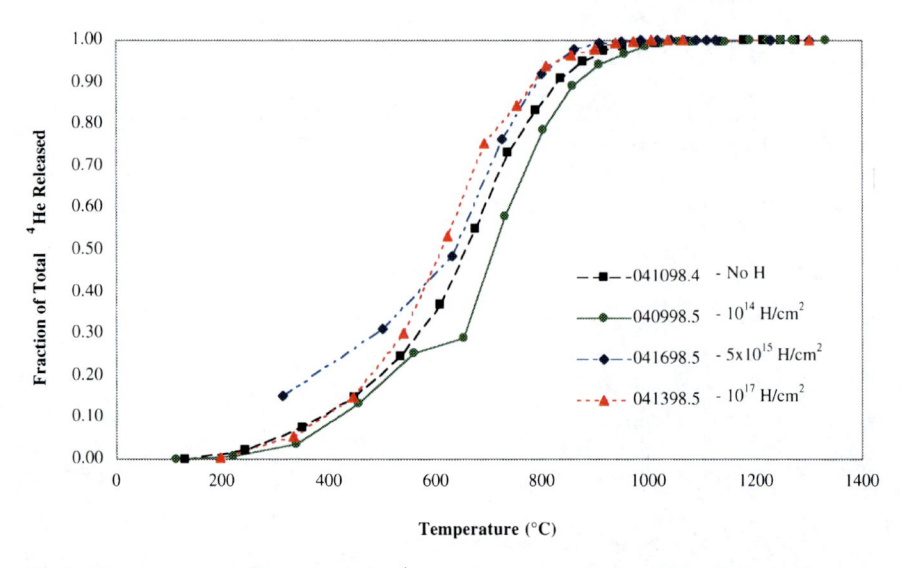

Fig. 2.16 Cumulative fractional release of ^{4}He from the 04XX98 series of samples. Increasing H fluence has virtually no influence on the release of ^{4}He. The lines are included to guide the eye.

2.5.8 Implantation of Helium in Terrestrial Ilmenite

Futagami et al. (1993) performed an experiment similar to the one described here, but used 20 keV ^{4}He rather than the 4 keV ^{4}He and 3 keV ^{3}He characteristic of the solar-wind. In earlier work, Futagami et al. (1990) used 3.6 keV ^{4}He on ilmenite, olivine, magnetite and rutile, but activation energies were not reported. They concluded that the activation energies for the release of ^{4}He from implanted ilmenite and olivine were 0.48 eV and 1.34 eV, respectively. This is counterintuitive to the higher retention of helium by high-Ti regoliths observed by Cameron (1988). The activation energy obtained for the release of He from ilmenite, 0.48 eV, is consistent with the low temperature behavior seen in the current research and is likely due to surface amorphization mentioned earlier. The activation energy for olivine, 1.34 eV, is relatively low compared to other studies of native He diffusion in olivine, which have obtained a value of 5.21 eV (Hart 1984). The activation energy for He release from olivine measured by Futagami et al. (1993) is somewhat higher than the activation energy of about 1.09 eV obtained by Trull et al. (1991) for cosmogenic ^{3}He in olivine. It is likely that the high activation energy measured by Hart (1984) is due to a small population of high-energy traps which have retained their He over a geologic time period while the He in lower energy traps diffused out.

2.6 Conclusions

Several major observations emerge regarding possible trapping and diffusion mechanisms:

- Helium release from implanted terrestrial ilmenites is diffusion limited as opposed to reaction rate-limited.
- Helium diffusion in terrestrial ilmenite implanted with ^{4}He and ^{3}He is characterized by four distinct activation energies, $E_1 = 0.26$ eV, $E_2 \cong 0.5$ eV, $E_3 \cong 1.4$ eV and $E_4 > 2.2$ eV.
- In the temperature range studied here, the diffusivity of helium implanted in ilmenite is much greater than the self-diffusivities observed for anions and cations in rhombohedral oxides or other iron oxides.
- The diffusivity of ^{3}He in terrestrial ilmenite implanted with ^{4}He and ^{3}He is greater than the diffusivity of ^{4}He. In fact, it is greater than can be accounted for by the isotope effect.
- Helium implanted terrestrial ilmenite exhibits release behavior similar to that of the lunar regolith during 30 minute stepwise heating.
- Moderate fluences ($<1 \times 10^{15}$ cm^{-2}) of hydrogen implantation apper to retard ^{3}He diffusion in terrestrial ilmenite implanted with ^{4}He and ^{3}He. High fluence H implantation (10^{17} cm^{-2}) has virtually no effect on ^{3}He diffusion.

Helium (both ^{3}He and ^{4}He) diffusion in these samples is characterized by four distinct activation energies, characteristic of diffusion through two amorphous layers, detrapping from oxygen vacancies and constitutional vacancies. The diffusivity of ^{3}He was seen to be at least a factor of 10 higher than the diffusivity of ^{4}He. A diffusivity of 9×10^{-24} cm^2/s at 400K was determined for diffusion of helium with activation energy of 1.4 eV. This behavior is consistent with helium trapping in oxygen vacancies.

What does this mean for the recovery of helium from lunar regoliths? The data presented here allows for the more efficient design of methods for heating the regolith to extract the gases from relatively ilmenite-rich regolith. Additional research is needed to determine the diffusivities and activation energies of the major minerals in the lunar highlands to more precisely determine the inventory of helium in these areas and determine the yield of helium from these potential "ore" materials. Finally, the diffusivity and activation energy of olivine needs to be reexamined to verify the results of Futagami et al. (1993), which appears to indicate that olivine retains helium more than ilmenite at lunar temperatures.

Acknowledgements. Thanks go to Professor Harrison H. Schmitt for encouraging this research and for many insightful suggestions. Many people have been instrumental to various parts of this research. We want to thank the staff of the Fusion Technology Institute at the University of Wisconsin for all of their support, moral and financial. Our sincere thanks also go to the generous members of the Plasma Source Ion Implantation (PSII) group at the University of Wisconsin: Professor John R. Conrad, Dr. Robert Breun, Dr. Muhammad Shamim, Dr. Kumar Sridharan and Paul Fetherston.

We also want to thank the staff of the Wisconsin Space Grant Consortium for their generous support. This research has also been generously supported by NASA Johnson Space Center and by the Graduate School of the University of Wisconsin. Thanks also to Dr. William Barker and Dr. John Fournelle (University of Wisconsin) for performing the high resolution TEM work and assistance in performing the electron probe microanalysis, respectively.

We want to extend out sincerest appreciation to Professors Donald H. Lindsley (SUNY - Stony Brook), Stephen E. Haggerty (University of Massachusetts), Lawrence A. Taylor (University of Tennessee), Eugene N. Cameron and John W. Valley (University of Wisconsin) for suggesting and providing samples of terrestrial ilmenites as well as much needed advice concerning the geological aspects of this work.

We also want to thank Professor Robert O. Pepin, Dr. Richard H. Becker and Dr. Dennis J. Schlutter of the University of Minnesota for generously performing the stepwise heating with mass spectrometry in this work and for considerable feedback concerning this work. Special thanks to Dr. Becker for the conversation which stimulated the development of the ANNEAL code.

Finally we want to profusely thank Dr. William C. Feldman and Professor Rainer Wieler for thorough and extremely helpful reviews of this chapter.

References

Agrell, S.O., Scoon, J.H., Muir, I.D., Long, J.V.P., McConnell, J.D.C., Peckett, A.: Observations on the chemistry, mineralogy and petrology of some Apollo 11 lunar samples. Paper Presented at Proceedings of the Apollo 11 Lunar Science Conference, Houston, TX (1970)

Allen, W.R.: Lattice Location of Ion Implanted 3He in Sapphire. Nuclear Instruments and Methods in Physics Research B61, 325–336 (1991)

Allen, W.R., Zinkle, S.J.: Lattice Location and Clustering of Helium in Ceramic Oxides. J. Nucl. Mater. 191-194, 625–629 (1992)

Baragiola, R.A., Alonso, E.V., Ferron, J., Oliva Florio, A.: Ion-Induced Electron-Emiission from Clean Metals. Surface Science 90(2), 240–255 (1979)

Baur, H., Frick, U., Funk, H., Schultz, L., Signer, P.: Thermal Release of Helium, Neon, and Argon from Lunar Fines and Minerals. Paper Presented at Proceedings of the Third Lunar Science Conference (1972)

Becker, R.H.: Personal Communication (1997)

Biersack, J.P.: Computer-simulations of Sputtering. Nuclear Instruments & Methods in Physics Research Section B - Beam Interactions with Materials and Atoms 27(1), 21–36 (1987)

Bradley, J.P., Dukes, C., Baragiola, R., McFadden, L., Johnson, R.E., Brownlee, D.E.: Nanometer-scale mineralogy and petrography of fine-grained aggregates in anhydrous interplanetary dust particles. Paper Presented at Lunar and Planetary Science Conference XXVII. Lunar and Planetary Institute, Houston (1996)

Brown, G.M., Emeleus, C.H., Holland, J.G., Phillips, R.: Mineralogical, chemical and petrological features of Apollo 11 rocks and their relationship to igneous processes. Paper Presented at Proceedings of the Apollo 11 Lunar Science Conference, Houston, TX (1970)

Cameron, E.N.: Helium Mining on the Moon: Site Selection and Evaluation. Paper Presented at Second Conference on Lunar Bases and Space Activities of the 21st Century. Lunar and Planetary Institute, Houston (1988)

Conrad, J.R.: Method and Apparatus for Plasma Source Ion Implantation, edited, United States (1988)

Conrad, J.R., Dodd, R.A., Han, S., Madapura, M., Scheuer, J., Sridharan, K., Worzala, F.J.: Ion-Beam Assisted Coating and Surface Modification with Plasma Source Ion-implantation. Journal of Vacuum Science Technology A 8(4), 3146–3151 (1990)

Dieckmann, R.: Point-defects and Transport in Hematite (Fe_2O_3-Epsilon). Philos. Mag. A-Phys. Condens. Matter Struct. Defect. Mech. Prop. 68(4), 725–745 (1993)

Dieckmann, R.: Personal Communication (1996)

Ducati, H., Kalbitzer, S., Kiko, J., Kirsten, T., Müller, H.W.: Rare Gas Diffusion Studies in Individual Lunar Soil Particles and in Artificially Implanted Glasses. The Moon 8, 210–227 (1973)

Eberhardt, P., Geiss, J., Graf, H., Grogler, N., Krahenbuhl, U., Schwaller, H., Schwarzmuller, J., Stettler, A.: Trapped Solar Wind Noble Gases, Exposure Age and K/Ar-age in Apollo 11 Lunar Fine Material. Paper Presented at Proceedings of the Apollo 11 Lunar Science Conference (1970)

Etique, P., Baur, H., Derksen, U., Funk, H., Horn, P., Signer, P., Wieler, R.: Light Noble Gases in Agglutinates: A Record of Their Evolution? Paper Presented at Ninth Lunar and Planetary Science Conference, Houston, TX (1978)

Fa, W.Z., Jin, Y.Q.: Quantitative estimation of helium-3 spatial distribution in the lunar regolith layer. Icarus 190(1), 15–23 (2007)

Fa, W.Z., Jin, Y.Q.: Global inventory of Helium-3 in lunar regoliths estimated by a multi-channel microwave radiometer on the Chang-E 1 lunar satellite. Chinese Science Bulletin 55(35), 4005–4009 (2010)

Fegley Jr., B., Swindle, T.: Lunar Volatiles: Implications for Lunar Resource Utilization. In: Lewis, J.S., Matthews, M.S., Guerrieri, M.L. (eds.) Resources of Near-Earth Space, pp. 367–426. The University of Arizona Press, Tucson (1993)

Fetherston, P.: Personal Communication (1997)

Fetter, S.: Overview of Helium-3 Supply and Demand. Paper Presented at AAAS Workshop on Helium-3, Washington, DC, April 6 (2010)

Frick, U., Pepin, R.O.: On the Distribution of Noble Gases in Allende: A Differential Oxidation Study. Earth and Planetary Science Letters 56, 45–63 (1981)

Frick, U., Pepin, R.O.: Microanalysis of Nitrogen Isotope abundances: Association of Nitrogen with Noble Gas Carriers in Allende. Earth and Planetary Science Letters 56, 64–81 (1981)

Frick, U., Becker, R.H., Pepin, R.O.: Solar Wind Record in the Lunar Regolith: Nitrogen and Noble Gases. Paper Presented at Eighteenth Lunar and Planetary Science Conference. Cambridge University Press (1988)

Futagami, T., Ozima, M., Nakamura, Y.: Helium Ion Implantation into Minerals. Earth and Planetary Science Letters 101, 63–67 (1990)

Futagami, T., Ozima, M., Nagai, S., Aoki, Y.: Experiments on Thermal Release of Implanted Noble-Gases from Minerals and Their Implications for Noble-Gases in Lunar Soil Grains. Geochimica Cosmochimica Acta 57(13), 3177–3194 (1993)

Gibson, E.R., Johnson, S.M.: Paper Presented at Second Lunar Science Conference (1971)

Grier, J.A., McEwen, A.S., Lucey, P.G., Milazzo, M., Strom, R.G.: The optical maturity of ejecta from large rayed craters: Preliminary results and implications. Paper Presented at Workshop on New Views of the Moon II, LPI Contribution no. 980, Lunar and Planetary Institute, Houston (1999)

Haggerty, S.E.: Personal Communication (1996)

Haggerty, S.E., Boyd, F.R., Bell, P.M., Finger, L.W., Bryan, W.B.: Opaque minerals and olivine in lavas and breccias from Mare Tranquillitatis. Paper Presented at Proceedings of the Apollo 11 Lunar Science Conference (1970)

Harris-Kuhlman, K.: Trapping and Diffusion of Helium in Lunar Minerals. Ph.D. thesis. University of Wisconsin - Madison, Madison (1998)

Hart, S.R.: He Diffusion in Olivine. Earth and Planetary Science Letters 70, 297–302 (1984)

Heber, V., Wieler, R., Baur, H., Olinger, C., Friedmann, T.A., Burnett, D.S.: Noble gas composition of the solar wind as collected by the Genesis mission. Geochimica et Cosmochimica Acta 73, 7414–7432 (2009)

Hintenberger, H., Weber, H.W., Voshage, H., Wanke, H., Begemann, F., Wlotzka, F.: Concentrations and Isotopic Abundances of the Rare Gases, Hydrogen and Nitrogen in Apollo 11 Lunar Matter. Paper Presented at Proceedings of the Apollo 11 Lunar Science Conference (1970)

Jambon, A., Shelby, J.E.: Helium Diffusion and Solubility in Obsidians and Basaltic Glass in the Range 200-300°C. Earth and Planetary Science Letters 51, 206–214 (1980)

Jost, W.: Diffusion in Solids, Liquids, Gases. Academic Press, New York (1952)

Keil, K., Bunch, T.E., Prinz, M.: Mineralogy and composition of Apollo 11 lunar samples. Paper Presented at Proceedings of the Apollo 11 Lunar Science Conference, Houston, TX (1970)

Keller, L.P., McKay, D.C.: The nature and origin of rims on lunar soil grains. Geochimica Et Cosmochimica Acta 61(11), 2331–2341 (1997)

Kuhlman, K.R.: Numerical Simulation of Solar Wind Implantation and Diffusion in Regoliths on Airless Bodies. Nuclear Instruments & Methods in Physics Research Section B - Beam Interactions with Materials and Atoms (in Preparation, 2011)

Kulcinski, G., et al.: Summary of Apollo. A D-HE-3 Tokamak Reactor Design. Fusion Technol. 21(4), 2292–2296 (1992)

Kulcinski, G.L., Schmitt, H.H.: Nuclear Power Without Radioactive Waste - The Promise of Lunar Helium-3. Paper Presented at Second Annual Lunar Development Conference, Las Vegas, NV, July 20-21 (2000)

Lancet, M.S., Anders, E.: Solubilities of Noble Gases in Magnetite: Implications for Planetary Gases in Meteorites. Geochimica et Cosmochimica Acta 37, 1371–1388 (1973)

Li, D.H., Liu, H.G., Zhang, W.G., Li, Y., Xu, C.D.: Lunar (3)He estimations and related parameters analyses. Sci. China-Earth Sci. 53(8), 1103–1114 (2010)

Lovera, O.M., Richter, F.M., Harrison, T.M.: The 40Ar/39Ar Thermochronometry for Slowly Cooled Samples Having A Distribution of Diffusion Domain Sizes. Journal of Geophysical Research 94(B12), 17917–17935 (1989)

Lovering, J.F., Ware, N.G.: Electron probe microanalysis of minerals and glasses in Apollo 11 lunar samples. Paper Presented at Proceedings of the Apollo 11 Lunar Science Conference, Houston, TX (1970)

Lucey, P.G.: Clementine Images (2011),
    http://www.lpi.usra.edu/lunar/missions/clementine/images/

Lucey, P.G., Blewett, D.T., Jolliff, B.L.: Lunar Iron and Titanium Abundance Algorithms Based on Final Processing of Clementine UVVIS Images. Geophysical Research 105(E8), 20377–20386 (2000a)

Lucey, P.G., Blewett, D.T., Taylor, G.J., Hawke, B.R.: Imaging of Lunar Surface Maturity. Journal of Geophysical Research 105(E8), 20297–20305 (2000b)

Malik, S.M.: Personal Communication (1998)

Mitchell, J.N.: Personal Communication (1996)

Mitchell, J.N., Devanathan, R., Sickafus, K.E., McClellan, K.J.: The Mineralogy of Radiation-resistant Ceramics and the Suitability of Ilmenite-group. Minerals for Fusion Reactor Applications (1996)

Morris, R.V.: The Surface Exposure (Maturity) of Lunar Soils: Some Concepts and Is/FeO compilation. Paper Presented at Ninth Lunar and Planetary Science Conference, Houston, TX (1978)

Müller, H.W., Jordan, J., Kalbitzer, S., Kiko, J., Kirsten, T.: Rare Gas Ion Probe Analysis of Helium Profiles in Individual Lunar Soil Particles. Paper Presented at Seventh Lunar Science Conference, Houston, TX (1976)

National Institutes of Health, NIH Image, edited (1996)

Nettles, J.W., et al.: Progress Toward a New Lunar Optical Maturity Measure Based on Moon Mineralogy Mapper (M3) Data. Paper Presented at 41st Lunar and Planetary Science Conference, The Woodlands, TX (2010)

Nier, A.O., Schlutter, D.J.: High-Performance Double-Focusing Mass Spectrometer. Rev. Sci. Instrum. 56, 214–219 (1985)

Nier, A.O., Schlutter, D.J.: Helium and Neon Isotopes in Stratospheric Particles. Meteoritics 25(4), 263–267 (1990)

Nier, A.O., Schlutter, D.J.: Extraction of Helium from Individual Idps and Lunar Grains by Pulse Heating. Meteoritics 27(3), 268–269 (1992)

Nier, A.O., Schlutter, D.J.: Extraction of He and Ne from Individual Lunar Ilmentite Grains by Pulse Heating. Meteoritics 28, 412 (1993)

Papike, J., Taylor, L., Simon, S.: Lunar Minerals. In: Heiken, G., Vaniman, D., French, B.M. (eds.) Lunar Sourcebook: A User's Guide to the Moon, p. 736. Cambridge University Press, Cambridge (1991)

Pepin, R.O., Nyquist, L.E., Phinney, D., Black, D.C.: Rare Gases in Apollo 11 Lunar Material. Paper Presented at Proceedings of the Apollo 11 Lunar Science Conference (1970a)

Pepin, R.O., Nyquist, L.E., Phinney, D., Black, D.C.: Rare Gases in Apollo 11 Lunar Material. Paper Presented at Apollo 11 Lunar Science Conference, Houston, TX (1970b)

Schlutter, D.J.: Personal Communication (1997)

Schmitt, H.H.: Return to the Moon: Exploration, Enterprise, and Energy in the Human Settlement of Space, p. 335. Praxis Publishing Ltd., London (2006)

Signer, P., Baur, H., Wieler, R.: Closed System Stepped Etching; An Alternative to Stepped Heating. Paper Presented at Alfred O. Nier Symposium on Inorganic Mass Spectrometry, Durango, CO (1991)

Signer, P., Baur, H., Derksen, U., Etique, P., Funk, H., Horn, P., Wieler, R.: Helium, neon, and argon records of lunar soil evolution. Paper Presented at Eighth Lunar Science Conference, Houston, TX, March 14-18 (1977)

Srinivasan, B., Hennecke, E.W., Sinclair, D.E., Manuel, O.K.: A Comparison of Noble Gases Released From Lunar Fines (#15601,64) with Noble Gases in Meteorites and in the Earth. Paper Presented at Third Lunar Science Conference, Houston, TX (1972)

Starukhina, L.V.: Polar regions of the moon as a potential repository of solar-wind-implanted gases. In: Ehrenfreund, P., Foing, B., Cellino, A. (eds.) Moon and Near-Earth Objects, pp. 50–58. Elsevier, Oxford (2006)

Sviatoslavsky, I.N.: Lunar He-3 Mining: Improvements on the Design of the UW Mark II Lunar Miner Rep. Wisconsin Center for Space Automation and Robotics, Madison, WI, WCSAR-TR-AR3-9201-2 2 (1992)

Swindle, T.D., Glass, C.E., Poulton, M.M.: Mining Lunar Soils for HeRep. NASA Space Engineering Research Center TM-90/1, p. 99. University of Arizona (1990)

Trull, T.W., Kurz, M.D., Jenkins, W.J.: Diffusion of cosmogenic 3He in olivine and quartz: Implications for surface exposure dating. Earth and Planetary Science Letters 103, 241–256 (1991)

Vaniman, D., Reedy, R., Heiken, G., Olhoeft, G., Mendell, W.: The Lunar Environment. In: Heiken, G., Vaniman, D., French, B.M. (eds.) Lunar Sourcebook, pp. 27–60. Cambridge University Press, Cambridge (1991)

Wang, Z.Z., Li, Y., Jiang, J.S., Li, D.H.: Lunar surface dielectric constant, regolith thickness, and (3)He abundance distributions retrieved from the microwave brightness temperatures of CE-1 Lunar Microwave Sounder. Sci. China-Earth Sci. 53(9), 1365–1378 (2010)

Webster, A.H., Bright, N.F.H.: The System Iron-Titanium-Oxygen at 1200°C and Oxygen Partial Pressures Between 1 Atm. and 2 x 10-14 Atm. J. Am. Ceram. Soc. 44(3), 112 (1961)

Welch, D.O., Lazareth, O., Dienes, G.J., Hatcher, R.D.: Theory of Helium Migration and Trapping in α-Al2O3. Radiation Effects 28, 195–198 (1976)

Wittenberg, L.J., Santarius, J.F., Kulcinski, G.L.: Lunar Source Of He-3 For Commercial Fusion Power. Fusion Technol. 10(2), 167–178 (1986)

Wittenberg, L.J., Cameron, E.N., Kulcinski, G.L., Ott, S.H., Santarius, J.F., Sviatoslavsky, G.I., Sviatoslavsky, I.N., Thompson, H.E.: A Review of He-3 Resources and Acquisition for Use as Fusion Fuel. Fusion Technol. 21(4), 2230–2253 (1992)

3 Water on the Moon: What Is Derived from the Observations?

Larissa Starukhina

Astronomical Institute of Kharkov National University, Ukraine

3.1 Introduction

Further exploration and utilization of the Moon crucially depends on the answer to the question: is there water on the Moon and, if so, in which form? Being a differentiated silicate planet without an atmosphere and with high day temperature, the Moon did not inspire many expectations for presence of water or any other volatile with small molecular weight. A hope of finding ice on the lunar surface was resuscitated by the hypothesis of ice delivery to the lunar poles after cometary impacts (Watson et al. 1961, Arnold 1979). Water molecules from a comet nucleus were supposed to migrate in the lunar exosphere until being trapped in cold polar regions. Less important sources of migrating water could be water-bearing meteorites and volcanic gases. Much effort has been devoted to study formation of "polar caps" and ice stability (e.g. Butler 1997, Starukhina 2008).

However, the cometary hypothesis of ice origin has a weak point that is difficult to overcome. High temperatures in the vapor cloud produced by high-velocity impact may result in either destruction or escape of all projectile molecules with small masses from the Moon before substantial cooling of the vapor. The average energy per one projectile atom in high-velocity impact is 20÷40 eV (up to 400 eV), which is more than enough for ionization, dissociation and other destruction process for molecules at the moment of fast energy release. The temperatures of the impact vapor cannot be calculated correctly, because this requires optical constants of the expanding hot vapor in wide rage of pressures and temperatures. These optical constants are unknown, so thermal emission cannot be taken into account, which does not allow for adequate modeling of cooling regimes and recombination processes. In expansion of hot impact vapor, escape of all survived or recombined water molecules from the Moon may occur.

Another hypothesis of the origin of lunar water suggests that H_2O molecules could have formed in bombardment of the lunar surface by solar wind protons (e.g. Housley et al. 1973), hydrogen being a reducing agent for cations, mostly

Fe^{2+}, favoring formation of nano-grains of metallic iron ($npFe^0$) ubiquitous in the rims of lunar soil particles. Presence of $npFe^0$ is the most remarkable distinction of the lunar soils from the terrestrial ones and is typical of the soils exposed to solar wind and micrometeoritic bombardment of the cosmic environment. Soils with highest ratio of the abundance of $npFe^0$ to native iron content are supposed to be subjected to longest cosmic exposure and are called mature (McKay et al. 1991). Maturation degree of a soil can be estimated from its optical spectrum in the near-infrared range and is characterized by the optical maturity index.

However, in the lunar agglutinitic glasses, which contain the largest abundance of $npFe^0$-grains, no water was found (Taylor et al. 1995). This could be expected, because H_2O molecules in silicates (major constituents of the lunar surface) tend to dissociate into OH and H ions, the latter combining with an oxygen atom of silicate. The process was discovered by Moulson and Roberts (1960) who studied penetration of H_2O into silica glass and found that water molecules are split inside a silicate matrix. Therefore, formation of H_2O from constituent atoms in minerals containing SiO_4-tetrahedrons is hardly possible, and proton implantation may be expected to produce OH-groups only.

In studies of Apollo samples with secondary ion mass spectrometry, hydration (up to a few tens ppm) of the interior of Apollo 15 and 17 volcanic glasses was found (Saal et al. 2008). The result is often referred to as discovery of water, though in such experiments only OH$^-$ ions are detected. This revived search of native water in lunar minerals. The most promising object for this was the last portion of the crystallizing melt, apatite $Ca_5(PO_4)_3(F,Cl,OH)$. OH abundance found there was up to about 0.46 wt% (McCubbin et al. 2010), D/H values being suggestive of cometary input, probably incorporated into the ancient Moon (Greenwood et al. 2010). Since minerals like apatite are rare on the lunar surface, the total amount of hydration makes it interesting rather for geologic history of the Moon and modeling of the lunar interior than for practical purposes.

Thus, now we have no indubitable arguments for the possibility of presence of considerable amount of water molecules, either guest or radiation-induced, on the lunar surface. This requires thorough analysis of the observations that are indicative of such a possibility. In the present work we look for answer to the question: are there unambiguous evidences for water molecules on the Moon?

Many works aimed on search of water were focused on the upper limits of water content that can be derived from the observations. Here we consider these results from another side: are the measured data (or even the upper estimates derived from models) consistent with other interpretations of the observations and are there facts that exclude such interpretations?

Till present, three remote sensing methods have been used for search of water. They are neutron spectrometry, infrared (IR) reflectance spectrometry near 3 μm, and radar observations. The methods study the upper layer of the lunar regolith within different depths. Neutrons are detected from 0 to about 1 m depth, IR light is scattered in the upper few millimeters, and the radar signals come from depths up to a few radar wavelengths (from a few cm to tens of cm).

All the remote sensing methods give indirect evidence for water (Table 3.1), and their alternative interpretations were proposed and validated by model calculations (Starukhina 2001). Though at the moment of the publication (Starukhina 2001) 3 μm-absorption was observed for asteroids only and ascribed to water (Rivkin et al. 1995), such an absorption, attributed to solar wind protons, was predicted for the Moon and first model spectra were calculated (Starukhina 1998b, 1999b,c, 2001).

Recent observations provide new material for theoretical analysis and modeling. Thus, 3 μm absorption on the Moon has been reported from three independent sources: the Visual and Infrared Mapping Spectrometer (VIMS) on Cassini (Clark 2009), near-infrared spectrometer on Deep Impact spacecraft (Sunshine et al. 2009), and the Moon Mineralogy Mapper (M^3) on Chandrayaan-1 (Pieters et al. 2009, McCord et al. 2011). Measurements of the band with high spectral and spatial resolution were carried out over large parts of the Moon.

Table 3.1 Three methods for search of water and interpretations of their results

Methods and objects of study	What was found in fact	Possible interpretation
(1) Neutron spectrometry (the Moon)	Hydrogen atoms (H)	Implanted solar wind protons
(2) Reflectance spectroscopy near 3 μm (the Moon, asteroids)	Hydroxyl groups (OH)	Chemically bound solar wind protons
(3) Radar observations (Mercury, the Moon)	Low microwave absorption on Mercury, mainly near the poles	Low dielectric loss, especially in cold silicates

After neutron leakage measurements by Lunar Prospector (Feldman et al. 1998, 2000, 2001), the Lunar Exploration Neutron Detector (LEND) on the Lunar Reconnaissance Orbiter (LRO) provided maps of hydrogen content in polar craters with higher spatial resolution (Mitrofanov et al. 2010, 2011). All these observations give valuable experimental material for further analysis of the lunar water problem.

Another recent experiment aimed to search of water was the impact of the Lunar Crater Observation and Sensing Satellite (LCROSS) into the permanently shadowed part of south polar crater Cabeus (Colaprete et al. 2010). Spectra of the impact ejecta, after their rise to sunlit height, were measured in UV to near-IR range. The problem of decomposition of spectra of multicomponent mixtures, containing both gas phase and solid particles of unknown sizes and compositions, is very complicated and has multiple solutions. A solution with 5.6±2.9 wt.% of water ice in the ejecta was found (Colaprete et al. 2010), but this is one version of wide variety of possibilities. Too much model parameters are involved in the problem, so the LCROSS impact experiment presents a hope, but not a proof of presence of ice or any other substance in permanently shadowed areas of the Moon, so this experiment will not be considered below.

Any alternative interpretation of the observations should account for the parameters measured in the experiments and regional (or other, if any) variations of these parameters over the Moon. Such parameters are: (1) the amount of hydrogen found by neutron spectrometry, (2) 3 μm band depths in the optical observations, (3) high reflectance and depolarization of the reflected signal discovered in radar observations. Each of the first three sections of this chapter is devoted to analysis of one of the water detection methods and proper parameters.

The most important of them is the abundance of H in the lunar regolith. Note that in interpretation of the results of neutron spectrometry, the authors of the experiments present two types of H abundance. First of them is what was actually measured, and the other values are what can be derived from models. In such models the authors try to compensate for the lack of spatial resolution and unknown depth of the hydrogen found. The largest H abundance measured by Lunar Prospector (Feldman et al. 2001) was 145 ppm (in crater Peary near the lunar north pole); in assumption that the hydrogen near the lunar south pole is concentrated in permanently shadowed regions of craters Shoemaker and Faustini, the calculated H abundance was 1700 ppm. In further analysis of the Lunar Prospector results (Lawrence et al. 2006), the average hydrogen abundance near both lunar poles 100-150 ppm was confirmed (H being likely buried by 10 ± 5 cm of dry lunar soil). As for the localized H abundance for small (<20 km) areas of permanently shaded regions, the authors admitted it to be highly uncertain: from 200 ppm up to 4.4 wt% in some isolated regions, depending on the presumed location area and depth.

In LEND experiment (Mitrofanov et al. 2010), the highest H content estimated from the measured parameters of neutron leakage was 470 ppm for hydrogen uniformly distributed with depth. If H was supposed to be buried by a 40-, 50-, or 60-cm thick layer with 100 ppm H, the H content at the depth was estimated as 600, 1200, or 4500 ppm, respectively.

Thus, H abundance provided by neutron spectroscopy is extremely sensitive to moderate change of the area and, particularly, of the depth presumed for hydrogen location. This should be kept in mind, when the values of H abundance reported in literature are compared to the calculated below. In such comparison, we take both measured values and the upper estimates into account.

3.2 Implanted Hydrogen in Lunar Regolith

The surfaces of atmosphereless celestial bodies are exposed to energetic ions – mainly to solar wind protons of 0.2–4 keV range, with the maximum flux near 1 keV. They displace atoms in the rims of lunar soil particles creating vacancies and dangling chemical bonds which are trapping sites for the implanted solar wind ions. The He ions of solar wind (~4%) also contribute to radiation damage. Modeling of solar wind damage to lunar-like silicates with TRIM code (Biersack and Haggmark 1980, Ziegler et al. 1996) yields ~4 displacements per incident ion.

The most damaged parts of the rims are amorphous, their depths being $h \approx$ 1000 Å (Dran et al. 1970, Borg et al. 1971). This value exceeds the average implantation depths $h_a \approx$ 250-500 Å for keV-protons and is a little less than total width of the implanted and damaged zone. In lunar silicates, where oxygen atoms amount to ≈60% of the total, broken chemical bonds of oxygen are most frequent trapping sites. They tend to be saturated by the implanted H-atoms so that OH-groups are formed. They are disordered and belong neither to water molecules nor to hydrated silicate compound.

In the early laboratory simulations of the effect of energetic ions (MeV protons) on the lunar surface (Zeller et al. 1966), proton-induced hydroxyl formation was observed and predicted for the Moon. Hydroxyl formation in irradiation of silicates was also detected by Bibring et al. (1982). In material science, irradiation-induced hydroxyl (or deuteroxyl) groups were observed in a number of laboratory experiments on implantation of H^+ and D^+ ions of energies from 10 keV to 10 MeV into different oxides (see Roth 1983 and references therein). Recently, Zent et al. (2010) reproduced OH in preliminary dehydrated Apollo lunar soil samples irradiating them with protons of solar wind energies. In all these experiments, OH-groups were found in observation of the optical absorption feature near 3 μm, which is due to O-H stretch of hydroxyl groups.

To ascertain that the chemically trapped solar wind protons can be detected and even mistaken for water molecules on the Moon, theoretical simulation of 3 μm absorption band for lunar regolith is required. It should be started with estimates of possible hydrogen content and, in particular, OH-content in lunar regolith (Starukhina 1998a, 1999a, 2000b, 2001).

3.2.1 *Trapping of Hydrogen in Lunar Regolith*

As ion implantation is surface-correlated process, its primary characteristics is the number n_s of solar wind protons trapped per unit area of the exposed particle rims: $n_s = n_v h$, where n_v is the average number of hydrogen atoms per unit volume of the implanted zone of a total width h. Measured depths of hydrogen distribution in the rims of lunar soil particles exceed 4000 Å (Leich et al. 1973), the width of the concentration profile taken at 1/2 of its maximum being 1500-2000Å. To approximate the profile by a rectangle with the maximum n_v, $h = 1000$ Å can be taken. The upper limits of n_v and n_s depend on trapping mechanism.

3.2.1.1 Gas Trapping on Radiation Defects

Consider the first of the trapping mechanisms, trapping in radiation defects, in particular, broken bonds of oxygen. Their concentration is limited by the volume concentration of oxygen atoms in lunar silicates, i.e. by $(4\div5)\times10^{22}$ cm^{-3}. Up to about a half of this amount can be bound to protons in implantation (Mattern et al. 1976). For trapping in OH, this yields n_s up to 2×10^{17} cm^{-2}. This is in agreement with laboratory experiments. In proton and deuterium implantation into oxygen-bearing materials, including natural silicates like olivine and enstatite, all the implanted hydrogen was chemically trapped up to surface concentrations $n_s = (1\div2)\times10^{17}$ cm^{-2}; at $n_s \geq (2\div4)\times10^{17}$ cm^{-2} chemical trapping efficiency gradually decreased and chemical trapping changed to physical trapping, saturation value of n_s being ≈5×10^{17} cm^{-2} (Lord 1968). This amount may include hydrogen trapped in radiation defects other than dangling bonds of oxygen atoms. Further irradiation did not increase concentration of hydrogen because H and D atoms escaped from the samples. These results are typical of implantation – desorption experiments on oxygen-bearing targets.

The importance of trapping on radiation defects can be demonstrated by comparison of the above n_s values to concentrations n_{s0} that could be achieved in absence of trapping, when hydrogen can diffuse out. In keV-proton irradiation with solar wind normal flux $j_0 = 4\times10^8$ cm^{-2}s^{-1}, $n_{s0} \approx j_0 h_a^2/2D = j\tau_D$, where D is hydrogen diffusion coefficient in silicate particle and $\tau_D \approx h_a^2/2D$ is the escape time. For example, diffusion coefficient in quartz extrapolated to 350 K (lunar day temperature at low latitudes) is $D \approx 2\times10^{-10}$ cm^2/s (Lee 1963), the extrapolation being underestimation (see Sect. 3.2.3). Then $n_{s0} < 10^7$ cm^{-2}, i.e., even overestimated n_{s0} is 10 orders of magnitude less than saturation values of n_s. Moreover, without trapping, all hydrogen escapes in time $\tau_D \sim 10^{-2}$ s, i.e., immediately after interruption of irradiation subsequent to sunset or particle burial.

In samples of lunar soil from low latitudes, concentrations of surface-correlated hydrogen $n_s = (1.5\div6)\times10^{16}$ cm^{-2} were found (Des Marais 1974). In individual grains, n_s may reach $\approx10^{17}$ cm^{-2} (Leich et al. 1973). The sample-averaged values of n_s are very much higher than n_{s0} and by an order of magnitude less than the saturation values of n_s. This points, on the one hand, to strong trapping of the implanted hydrogen and, on the other hand, to desorption of a part of the implanted species in hot lunar areas (see Sect. 3.2.3). The values of $n_s = 5\times10^{16}$ cm^{-2} and 2×10^{17} cm^{-2} are used in Sect. 3.3 for simulation of the 3 μm band depth in the equatorial and polar regions, respectively.

Another evidence for strong trapping of hydrogen is the total hydrogen content in the lunar samples (up to $10\div100$ ppm (Haskin and Warren 1991)) that exceeds by orders of magnitude the thermodynamic equilibrium values at the day temperatures in low-latitude lunar regions. Most of the hydrogen is surface correlated (Des Marais 1974). The same applies to the other implanted gases. This means that most gas atoms in lunar soils are retained by chemical forces, i.e., much stronger than just physically. The thin implanted layers around regolith particles are responsible for the retention of gas atoms and control the gas abundance in the lunar surface.

Let us estimate the upper limit for mass fraction [H] of hydrogen trapped in radiation defects in mature lunar soil. Regolith mixing in meteoritic and micrometeoritic bombardment may provide solar wind irradiation for all particles up to meter depths (Shoemaker et al. 1970), including exposure of the entire surface of each grain (Borg et al. 1976). In this case,

$$[H] = S m_p n_s, \tag{3.1}$$

S being specific area of regolith, and m_p being proton mass. Specific surface area depend on the average size l of regolith particles: $S = 6\xi/\rho l$, where $\rho \approx 2.8$ g/cm3 is the density of regolith material and ξ is the average ratio of the surface of a particle of a size l to that of a sphere of the same diameter. Measured S vary from 800 (Grossman et al. 1972) to 5000 cm^2/g (Cadenhead et al. 1977); at typical $l \approx 50$ mm (McKay et al. 1991) this corresponds to x from 2 to 12.

The latter value is too high, so large S are, probably, due to l = 15-30 mm also typical for lunar soils. Substitution of the minimum S and moderate n_s = $1.2´10^{17}$ cm^{-2} to Eq. (3.1) yields [H] = 150 ppm. Thus, at least [H] < 150 ppm, i.e., all the values measured on Lunar Prospector, are consistent with solar wind origin of the observed hydrogen. Even the maximum [H] = 1700 ppm (0.17 wt%) suggested by Feldman et al. (2001) can be achieved at maximum S and n_s = $2.1´10^{17}$ cm^{-2}, i.e., may also be due to hydrogen trapping in OH or other radiation defects. If this hydrogen is interpreted as a constituent of water ice, its weight fraction would be [H_2O] = [H]m_W/2 = 1.5 wt% (m_W = 18 being molecular mass of water).

The largest [H] = 470 ppm measured in the recent higher resolution observations of the lunar polar craters by neutron telescope LEND implies either intermediate values of S or, at minimum S, n_s = $3.7×10^{17}$ cm^{-2}, which is about the upper limit for H-trapping in radiation defects. The highest [H] = 0.45 wt% derived from the interpretation of the LEND data is extremely model-dependent and hence, uncertain; however, even this value of [H] may be still consistent with implantation origin of hydrogen. At low polar temperatures, a trapping mechanism more effective than trapping in radiation defects may dominate and provide higher gas concentrations in lunar regolith of polar regions.

3.2.1.2 Hydrogen Trapping in Pores

Another mechanism of trapping of energetic gas ions in solid materials is radiation blistering. When ion fluence is so high that concentration of the implanted ions exceeds the concentration of trapping sites in the implanted layer and gas desorption is hindered (mainly, because of low temperature), gas atoms gather in bubbles (blisters) located at the maximum of the implantation profile, bubble diameter being of the order of the profile width (see Scherzer 1983 and references therein). Gas pressure in blisters increases until the limit of material strength or yield strength is reached. After this, the pressure is relaxed either by breaking the tops of blisters or by their growth.

Radiation blistering was well studied for metals (Guseva and Martynenko 1981, Scherzer 1983) as constructional materials; natural minerals were not objects of such studies. In investigation of lunar glasses with electron microscopy, numerous vesicles were observed (e.g., Housley et al. 1973) and evidence for their solar wind implantation origin were presented. In detailed study of the rims of lunar soil particles (Keller and McKay 1997), amorphous vesicular rims of typical width ≈1000 Å were found around significant part of them, vesicle diameter being 500 Å. This suggests blistering in such particles.

At lower temperatures of polar regions, vesicular rims may be much more frequent and contain more gas. Volume concentration n_g of hydrogen atoms in a blister can be estimated as n_g = $2p/kT$, where T is temperature, k is Boltzmann constant, the factor 2 is due to hydrogen association in molecules H_2, and gas pressure $p ≈ 0.1E$ (Guseva and Martynenko 1981), E being the elastic modulus. Substituting E = $7×10^{11}$ dyn/cm^2, typical of glasses, obtain n_g = 10^{25} and $2×10^{25}$ cm^{-3} for T = 100 and 50 K, respectively. Supposing volume fraction of blisters $c_b = 0.05$ in the $h_b = 1000$ Å-thick layer, obtain $n_s = c_b h_b n_g = 5 \times 10^{18}$ and 10^{19} cm^{-2} at $T = 100$ and 50 K, respectively. Thus, in case of low-temperature blistering, concentration of the implanted hydrogen on particle surfaces and, consequently, hydrogen content in regolith, may by an order of magnitude exceed those achieved in case of hydrogen trapping on radiation defects. Substituting the minimum $S = 800$ cm^2/g to Eq. (3.1), yields [H] ≈ 0.7 and 1.4wt% (equivalent [H$_2$O] $\approx$ 6 and 12wt%) at $T = 100$ K and 50 K, respectively, if all regolith particles are surrounded by blistered rims. So the retention in blisters may provide even the highest hydrogen content [H] = 0.45wt% supposed in polar crater Cabeus (Mitrofanov et al. 2010). If the largest measured $S = 5000$ cm^2/g are taken, we obtain [H] ≈ 4.2 and 8.5wt% (equivalent [H$_2$O] $\approx$ 38 and 76wt%) at $T = 100$ K and 50 K, respectively, i.e., hydrogen retention in blisters allows also for the highest localized [H] proposed by Lawrence et al. (2006). Note, that the estimates obtained in this section for n_s and [H] characterize hydrogen *retention ability* of lunar soils, i.e., *potential* for hydrogen accumulation. Its realization, i.e., the factual H abundance, depends on two factors: (1) on proton flux and irradiation times, that is on the total proton fluence incident to a permanently shadowed area after the last large impact (and hence, outgassing) event; (2) on outgassing rates. This is considered in the next sections.

3.2.2 Hydrogen Accumulation Times

3.2.2.1 Accumulation in Sunlit Areas

For applications to celestial bodies, the time τ_s of hydrogen accumulation is important. If outgassing during irradiation is negligible (see below), the time to reach concentration n_s in the entire rim around a particle exposed to the flux j of solar wind is $\tau_s \approx 4n_s/j$ (4 stands for the average number of surface exposures to irradiate a particle from all sides). The average flux is $j = (j_0 \sin\theta)/4$, where the normal flux $j_0 = 4 \times 10^8$ cm^{-2}s^{-1}, the factor 1/4 is due to rotation of the Moon, and θ is the angle between the normal to the exposed surface and the ecliptic pole. At low latitudes of the Moon ($\theta \approx 60 \div 90°$) and at the illuminated sites near lunar poles ($\theta \approx 1.5°$), the upper concentrations $n_{s,OH} \approx 2 \times 10^{17}$ cm^{-2} of hydrogen trapped in OH can be achieved in $\tau_{s,OH} \approx 8 \times 10^9$ s ≈ 250 years and in $\tau_{s,OH} \approx 10^4$ years, respectively. Decrease of trapping efficiency before saturation, as well as different processes of hydrogen loss (backscattering, sputtering, diffusion at lunar temperatures), increase saturation times within the same order of magnitude, so all the times are short at a timescale of regolith processes.

The accumulation times should be compared to the total time of exposure of a regolith particle to solar wind. According to computer simulation by Borg et al. (1976), the average time of one surface exposure is $\tau_1 \approx 5000$ years, the average number of the exposures for each particle being ≈ 30, so that the total exposure time is $\tau_{ex} \approx 1.5 \times 10^5$ years. This exceeds both above-estimated times τ_s by more than an order of magnitude, from which two conclusions can be drawn.

(1) A regolith particle have more than enough time for saturation with hydrogen both at the equator and in the sunlit areas near the lunar poles, so that H-accumulation near the poles is no less effective than at the equator. The same applies to helium, so that mapping of H or He abundance proportionally to the incident fluxes is not correct.

(2) At low latitudes, decrease of proton flux by a factor up to 600 cannot prevent saturation of regolith particles with hydrogen, provided that subsurface soil is much older than τ_{ex}. This means that weak magnetic fields of the lunar magnetic anomalies, in particular, in lunar swirls, can hardly shield the regolith in the anomalies from being saturated with hydrogen.

3.2.2.2 The Importance of Regolith Maturity

Thus, accumulation of solar wind protons in the rims of the exposed lunar soil particles is fast enough all over the Moon and does not control hydrogen abundance in mature lunar regolith, where almost all particles were irradiated by solar protons. For immature soils, smaller portions of particles (or smaller portions of their surfaces) could be exposed to solar wind. Though the exposed parts of the particle rims may be saturated with hydrogen, bulk weight fraction of H and bulk OH content in immature soils should be lower.

Short exposure history of the subsurface layers in immature soils may be expected to produce weaker light absorption by irradiation-induced OH groups, i.e., weaker 3 µm bands in reflectance spectra of such soils. The reflected infrared light is formed in many particulate layers in the upper few millimeters of the surface; so the lack of exposure for the particles within the upper millimeters may display itself in the lack of 3 µm absorption.

This may explain lower OH absorption observed in lunar swirls (Kramer et al. 2011) – curl-like patterns distinguished by higher albedo and low optical maturity index. Though Kramer et al. (2011) suppose the weak OH-bands to be due to shielding from solar wind protons by the local magnetic fields, the above estimates show that even 99% shielding efficiency (which is rather an overestimation), cannot prevent saturation of the exposed particles in the equatorial regions. According to the above considerations, the lack of OH in the lunar swirls is also consistent with the mechanisms of swirl formation that imply young exposure age of the subsurface soils of the lunar swirls (see, e.g., Starukhina and Shkuratov (2004) and references therein), for instance, with cometary or meteorite swarm hypotheses.

The effect of the lack of OH on the depth of 3µm absorption may be partially compensated (and even overcompensated) by the effect of surface brightness. The absorption depth is enhanced in bright soils, which is considered in Sect. 3.3.1. The interpretation of weak 3µm absorption in swirls as a result of the lack of OH-groups is especially convincing because weak bands are combined with higher reflectance than that for surrounding soils with deeper OH-feature.

Thus, in immature regolith, bulk H and OH abundance may depend on the time of the exposure; in mature regolith, gas abundance is controlled by outgassing processes considered in Sect. 3.2.3.

3.2.2.3 Hydrogen Accumulation in Permanently Shadowed Areas

In permanently shaded areas, the regolith is irradiated by keV protons when the Moon passes through in the Earth's magnetotail (Arnold 1979). Penetration of keV protons into permanently shadowed craters was analyzed in detail by Feldman et al. (2001). They took solar wind, as well as different parts of the Earth's magnetosphere, into account and pointed out that the proton flux at the crater center depends on crater morphology, namely, on depth to diameter ratio d/D_c. Penetration of the protons to a crater depth is due to the randomly oriented thermal velocities of protons that flow over a crater with bulk velocity v. Thermal velocities can be directed down to the crater center; for average vertical velocity $v_\perp$, penetration probability is $w = \exp[-2(d/D_c)^2(v/v_\perp)^2]$, being larger for shallow craters. For craters with $D_c < 10$ km, $d/D_c = 0.2 \div 0.25$ is usually assumed, for larger craters, d/D_c is lower. Taking the data of Table 2 and $d/D_c = 1.044(D_c/\text{km})^{-0.7}$ from (Feldman et al. 2001), obtain the time-averaged flux of keV protons in the shaded polar craters $j = 4.5 \times 10^6$ cm^{-2}s^{-1}, 10^6 cm^{-2}s^{-1}, and 3×10^5 cm^{-2}s^{-1} for and $D_c = 100$ km, 30 km, and $D_c < 10$ km, respectively.

3.2.2.4 Contribution of Solar Wind Backscattering to Hydrogen Flux in Polar Regions

However, in calculation of these fluxes, the solar wind protons backscattered from the illuminated walls and other topographic details near the shadowed areas were not taken into account. Numerical simulation of keV-proton bombardment of the lunar surface with TRIM code (Biersack and Haggmark 1980, Ziegler et al. 1996) has shown that in average 23% of the incident solar wind protons are backscattered (Starukhina 2003). Global average flux of backscattered and sputtered H was found to be too small $\sim 10^4$ cm$^{-2} \cdot$s^{-1}; however, the local contributions from the neighbor sites are significant.

According to the backscattering indicatrix of a regolith particle calculated in (Starukhina 2003), $\approx 40\%$ of the backscattered ions are directed between 55 and 90° from the irradiation beam. Taking 40% of this amount to be directed from the crater wall to its bottom, obtain that $\approx 4\%$ of the protons incident to crater wall are backscattered to a permanent shade. Further calculations with TRIM code have shown that $\sim 90\%$ of backscattered solar wind hydrogen penetrates into soil particle rims to depths 20- 500 Å. The ratio of the illuminated to permanently shaded area can be assumed ≈ 0.25 at latitudes lower than about 85–87° and ≈ 0.1 for the craters within a few degrees from the lunar poles. Then, for a polar crater wall directed at 55° to the incident solar wind protons, the average backscattered flux that falls to the shade is $j_b \approx 0.04 \cdot 0.9 \cdot 0.25(j_0/4)\cos(55°) \approx 5 \times 10^5$ cm^{-2}s^{-1}, the minimum value being $j_{b,min} \approx 0.04 \cdot 0.9 \cdot 0.1(j_0/4)\cos(55°) \approx 2 \times 10^5$ cm^{-2}s^{-1}.

This estimate shows that backscattered solar wind can make up substantial contribution to proton flux to the permanently shadowed craters with diameters $D_c < 10$ km, especially for the craters at latitudes <85°. Real local topography is complicated, so that some shadowed spots may "see" much more illuminated area than others. The highest H-content can be expected in the permanently shaded sites from which more illuminated area is visible.

The above estimates show that, in small permanently shadowed craters, a typical value for contemporary keV-proton flux $j \approx 8\times10^5$ cm^{-2}s^{-1} and the minimum value $j_{min} \approx 5\times10^5$ cm^{-2}s^{-1} can be taken; in 30 km-craters, the values are $j \approx 1.8\times10^6$ cm^{-2}s^{-1} and $j_{min} \approx 1.5\times10^6$ cm^{-2}s^{-1}. Therefore, the minimum number $n_{s,min}$ of protons accumulated in the entire rim of a regolith particle in the total time τ_{ex} of its exposure in a small ($D_c <$ 10 km) permanently shadowed polar crater is $n_{s,min} \approx j_{min}\tau_{ex}/4 \approx 6\times10^{17}$ cm^{-2}. For 30 km-craters, $n_{s,min} \approx 2\times10^{18}$ cm^{-2}. Both values exceed that allowed for OH-trapping case ($n_s = 2\times10^{17}$ cm^{-2}), but are still lower than the maximum $n_s \approx 5\times10^{18}$ cm^{-2} allowed for blistering case. Substituting $n_{s,min}$ and the total range of specific surface areas $S = $ 800-5000 cm^2/g to Eq. (3.1) yields minimum values for [H] = 770 ppm to 0.48% for small shaded craters and [H] = 0.26-1.6% for 30 km-craters, which covers the highest estimates from LEND observations. However, these values are less than hydrogen retention ability of lunar soils at low temperatures estimated in Sect. 3.2.1 where blistering was considered. The maximum retention ability might be realized in two cases: (1) if the ancient solar wind was much more intensive, (2) if lunar the contemporary polar regions were illuminated in the ancient times. Both cases are probable, especially if the lunar obliquity history was so complicated as described in (Siegler et al. 2011).

Beside the "local" times τ_s of H-accumulation on particle surfaces, the total times τ_v for regolith volume should be evaluated. Their lower estimate is obtained in the assumption of no loss, when hydrogen column density (the number of H-atoms per cm^2 of regolith column) is equal to the total proton fluence. The assumption can be satisfied in polar regions between large impact events when degassing of large areas occurs. In permanent shadows, even gas atoms physically trapped in the interstitials can be "frozen" in the regolith particles, so that all the incident atoms can be retained in regolith (see Sect. 3.2.3).

The times τ_v depend on the thickness H_r of regolith layer considered: $\tau_v = $ [H]$\rho_r H_r/m_p j$, where $\rho_r \approx$ 1.5 g/cm^3 is regolith density. For LEND experiment $H_r \approx$ 50 cm can be taken, then, at the above-estimated minimum proton flux to shaded areas $j = 5\times10^5$ cm^{-2}s^{-1} and the largest measured [H] = 470 ppm, obtain $\tau_v \approx$ 1.4$\times10^9$ years. At maximum [H] = 0.45wt% derived for 60-cm depth, the time τ_v is an order of magnitude longer. The times τ_v can vary in both directions: to increase because of degassing in large impacts and to decrease due to more intensive ancient solar wind or long illumination of the present shadowed areas in the lunar obliquity history. Beside this, high sensitivity of the derived [H] to the model parameters (see the introduction), as well as unknown thickness H_r of hydrogen-enriched layer do not enable us to reject the hypothesis of implantation origin of the lunar polar hydrogen.

3.2.3 Retention of Gases in Lunar Regolith and Regional Variations of Hydrogen Content

The most convincing argument in favor of degassing as the process that controls the abundance of solar wind implanted elements is provided by chemical analysis of lunar soils. The ratios between the elements implanted in the regolith are far from those in the solar wind, the contrast being most pronounced for the pairs with high difference in diffusivity (Starukhina 2006). For example, in samples with low H content ($\approx$25% of all implanted atoms), the fraction of C may reach 65%; at high H content ($\approx$90% of all implanted atoms), the fraction of C is 8% (Haskin and Warren 1991), whereas in solar wind H and C constitute 96 and 0.04%, respectively. This demonstrates that variations of the concentration of solar wind implanted elements are due to difference in gas loss rates and not in accumulation rates. This applies also to regional variations of hydrogen.

There are two types of H and OH regional variations over the Moon: (1) global, temperature-correlated, excess H and OH being concentrated in the polar regions; (2) local anomalies of H and OH abundance, when spots of the excess H and OH are found at low latitudes or in the illuminated polar areas, and lack of H is found in some permanently shaded areas. Such anomalies were observed in neutron spectrometry experiment on Lunar Prospector (Feldman et al. 1998, 2000) and confirmed by more detailed LEND observations of polar craters (Mitrofanov et al. 2011). Many polar craters with permanently shadowed regions did not manifest any detectable signature of the excess hydrogen, whereas large sunlit area close to the North pole showed high H content.

Note that ice hypothesis implies high correlation of H_2O abundance with surface temperature and have problems in explanation of the local anomalies of H and OH content. Random migration of water molecules, either in collisions between them or in ballistic hops in the exosphere, and trapping of the molecules in cold regions only, does not produce the local [H] anomalies. Burial of ice (or H-enriched soils) can account for excess H in some sunlit polar areas, but not at low latitudes. Buried ice deposits were shown to be able to survive up to regolith temperatures $\approx$150 K (Starukhina 2008) due to return of most part of the evaporated molecules back to cold traps. The ice-depletion of many permanent shadows suggests recent outgassing in large impact, which may be the case for young craters only. Thus, in explication of the local [H] anomalies by water ice, many additional hypotheses and assumptions are required.

Both types of regional variations are consistent with the implantation hypothesis and can be accounted for by wide variety of outgassing rates (Starukhina 2000b, 2001; Starukhina and Shkuratov 2000). The variety is provided by exponential dependence of diffusion coefficients $D = D_0\exp(-U/kT)$, the temperature T and the activation energy U being responsible for global concentration trend and for local anomalies, respectively. At typical $U \approx 0.3 \div 3$ eV, $U/k \approx (0.3 \div 3) \times 10^4$ K (compare $U \approx 1$ kJ/mol ≈ 0.01 eV, $U/k \sim 100$ K for physical adsorption of H_2 molecules), gas loss rate is extremely sensitive to temperature, so that the decrease of outgassing at lunar day temperatures from the equatorial to polar regions overcompensates for the geometric decrease of the incident ion flux.

At low temperatures or high activation energies U, escape of the implanted atoms from the regolith particles is suppressed; at small U, gas release may occur even at low temperatures. This, together with difference in regolith maturity, may explain the lack of correlation between temperature and H or OH content.

The most pronounced variance of gas escape rates from different materials is expected in polar regions, especially in the permanently shadowed areas, where $T = 40 \div 100$ K. Very small difference of the activation energies U in similar materials may enable gas release from one of them and stop it in the other. This is consistent with LEND observations of polar craters (Mitrofanov et al. 2011).

A global and local distribution similar to that of hydrogen can be expected for the other solar-wind-implanted elements, e.g., for ^{3}He (Starukhina 2006). Consequently, polar regions of the Moon, in particular, permanently shaded areas, are the most promising place to extract this potential thermonuclear fuel.

3.2.3.1 The Role of Micrometeoritic Bombardment

In addition to surface temperature, there is another outgassing agent for lunar soils, namely, micrometeoritic bombardment. It acts more frequently and homogeneously over the lunar surface than the impacts of large meteorites. The degassed volume at each impact can be estimated as 10 projectile volumes (about that of the impact melt). However, the role of micrometeoritic bombardment is two-fold. Beside outgassing, it favors saturation of the bulk of lunar regolith with solar wind atoms due to mixing of the adjacent soil. Regolith volume excavated at each impact is about half of the crater volume, that is $\approx 10^3$ projectile volumes for small craters in sand-like material (Basilevsky et al. 1983). Hence, the degassed volume is $\approx 1\%$ of that subjected to a new exposure and distributed in the area much greater than the degassed area.

The intervals τ_i between micrometeoritic impacts to the same area of the order of the projectile area are much longer than hydrogen saturation times τ_s. In permanently shadowed areas, where hydrogen accumulation occurs slower, outgassing and mixing in micrometeorite impacts are also expected to be slower, because the micrometeorite flux in the directions facing poles is lower than the ecliptic flux by a factor of ~7 (Hutcheon 1975).

3.2.3.2 Gas Retention Due to Surface Processes

As distinct from the case of noble gases, desorption of hydrogen and other gases with two atoms per molecule can be hindered by processes on the very surface of regolith particles. This provides an additional retention mechanism for such gases (Starukhina 2006).

Hydrogen outgassing from a particle requires an encounter of two OH groups and breaking of two O-H bonds on the particle surface. The potential barrier for this process can be high, so that hydrogen outgassing is hindered in H_2-molecule formation on the boundary. As a result, hydrogen diffusion may occur mostly inward, which prevents hydrogen loss. If a particle with high OH abundance all over the volume is heated for a short time, outgassing of hydrogen from the boundary can be activated and a part of hydrogen removed from the rims. In such case, an inverse OH gradient, such as observed in (Saal et al. 2008), can be formed. If no further irradiation take place, the inverse OH gradient can be preserved after cooling.

3.2.3.3 Characteristics of Diffusion in Solids

The main objection to the implantation origin of the excess of hydrogen near the lunar poles was that diffusion of hydrogen and hydroxyl in silicates can be either too fast or too slow all over the Moon, or at least, within both permanently shaded and sunlit polar craters (Feldman et al. 2001). The statement was based upon extrapolation of diffusion data from elevated temperatures where they were measured (>800 K) to lunar temperatures. However, such extrapolation and quantitative conclusions based on it are not correct. Difficulties in extrapolation of the diffusion coefficients to lunar temperatures were analyzed in (Starukhina 2006).

In spite of simple qualitative description, the temperature dependence of outgassing is difficult to calculate. The reason of it is that both preexponent factor D_0 and activation energy U of diffusion coefficient D are not universal characteristics of solid material in all the temperature range where the solid phase exists. In general, they are not constants of a material but rather characterize a *given sample* with its individual mechanical (plastic deformation, shock, cracking), thermal (cooling from a melt, subsequent temperature jumps, annealing, etc.), and irradiation history, as well as minute variations in chemical composition and surroundings. Each of these factors produces or heals defects (i. e. specific atomic configurations) in the crystalline (or amorphous) structure of a sample. The defects provide, on the one hand, trapping sites for the implanted atoms, and on the other hand, paths of enhanced diffusion, especially in the implanted zones of irradiated materials.

This imposes severe restrictions on extrapolations of diffusion data, especially for natural materials, from high to room and lower temperatures.

At the temperatures of the lunar surface, diffusion coefficients in solid materials are usually very small, which makes them difficult to measure. Diffusion data available in literature were obtained at elevated temperatures when diffusion coefficients are large enough for measurements and diffusion occurs by vacancy mechanism. Diffusion coefficients can be reliably extrapolated only to temperatures high enough for vacancies to be still abundant and provide most of diffusion transport of atoms.

At low temperatures, concentration of vacancies becomes much smaller than the abundance of the other structural defects and diffusion occurs due to other transport paths, either temperature-independent or with lower U (and much lower D_0). Sharp decrease of D with inverse temperature $1/T$ characteristic of vacancy mechanism becomes more and more slow as $1/T$ increases (T decreases). This is why the values of D extrapolated from high to room temperatures are always underestimated (mostly by many orders of magnitude). Individual concentrations of different defects and the activation energies U associated with them provide rather unique temperature dependence of diffusion coefficient D and outgassing rate for each individual sample in low temperature range, especially in the irradiated zones, as explained in detail and illustrated in (Starukhina 2006). This considerations support the plausibility or solar wind retention scenario.

3.3 Optical Effects of Solar-Wind-Induced OH

Irradiation-induced OH-groups show optical absorption near 3 μm. Absorption frequencies due to H_2O molecules are also present in this spectral range (at ~3.05 μm); though they are overtones, the H_2O-absorption feature may be strong and broad (Starukhina 2001). As explained above, occurrence of water molecules in silicates (except interlayer water in phyllosilicates) is hardly probable, because of dissociation of H_2O into OH-groups (Moulson and Roberts 1960). Taking this into account, below we suppose 3 μm absorption to be due to OH, although, in multi-component natural silicates, the absorption can be easily mistaken for signal of water even at high spectral resolution.

The large width of the 3 μm OH-band in natural silicates may be due to different vibration frequencies of OH-groups located near different cations. For example, in mono-oxides, the minima of irradiation-induced bands are 2.73 μm in SiO_2 (Mattern et al. 1976), 2.94 μm in Al_2O_3 (Gruen et al. 1976), and 3.06 μm in TiO_2 (Siskind et al. 1977). Disordering of atomic structure by ion bombardment also contributes to the variety of OH positions and, consequently, vibration frequencies, which can result in additional broadening of OH-absorption bands and their similarity to those of water.

In the M^3 spectral observations of the Moon, two types of light absorption near 3μm were found (McCord et al. 2011) : (1) narrow and shallow bands centered near 2.8μm and present everywhere, and (2) broad deep bands centered beyond 2.9μm that occur preferentially in highland-type material. Though the latter type of absorption is suggestive of water, McCord et al. (2011) pointed out that the spectral features in the M^3 could be created only by OH. For the reasons mentioned above, both types of the absorption features are treated here as OH-bands. The distinction between them can be due to difference, e.g., in surface mineralogy, in concentration of OH groups, and in spectral contrasts for surfaces of different brightness (Starukhina 2011).

3.3.1 Simulation of the Observed 3μm Bands

In early simulations of solar-wind-induced 3μm absorption (Starukhina 1998b, 1999b,c, 2001), the band minima for all OH groups were supposed to be the same, so the band depths were overestimated and band widths underestimated. First attempt to reproduce the real band shape was made for asteroids (Starukhina 2002). After discovery of 3 μm absorption on the Moon, the shape and depth of the feature was simulated in the assumption of its solar wind origin (Starukhina and Shkuratov 2010).

All these simulations were based on the spectral model (Starukhina and Shkuratov 1996, Shkuratov et al. 1999) for multicomponent regolith-like surfaces. The model is invertible, i.e., reflectance spectrum $R(\lambda)$ (λ being wavelength) can be calculated from the spectrum $\tau(\lambda)$ of the optical density of the particles and vice versa. This enables one to predict the changes in reflectance of a surface resulting from modification of the size or composition of the particles. As a first step, spectrum $\tau(\lambda)$ for the particles of the initial featureless material is calculated from its reflectance spectrum, then the absorption due to OH is added, after which a modified reflectance spectrum is obtained.

The contribution of OH into optical density of a regolith particle was calculated as a sum

$$\Delta\tau(\lambda) = \Sigma 2 n_s \chi_i(\lambda) c_i, \tag{3.2}$$

where summation was made over 10 model positions of the OH band minima (from 2.78 to 3.35 μm) ascribed to different OH sites, c_i being the fraction of each site ($\Sigma c_i = 1$). Here the factor 2 is due to double intersection of the irradiated particle rims at each scattering, $\chi(\lambda)$ is the spectrum of the absorption coefficient of light near 3 μm per unit volume concentration of OH in a single-cation material. It was taken from (Mattern et al. 1976) and normalized to the value $\chi = 3{\times}10^{-19}$ cm^2 at the absorption maxima (Moulson and Roberts 1960, Peri and Hannan 1960). For each model OH-site, $\chi(\lambda)$ was shifted to a new position of the local band minimum and the c_i-weighted sum Eq. (3.2) of the overlapping spectra was calculated.

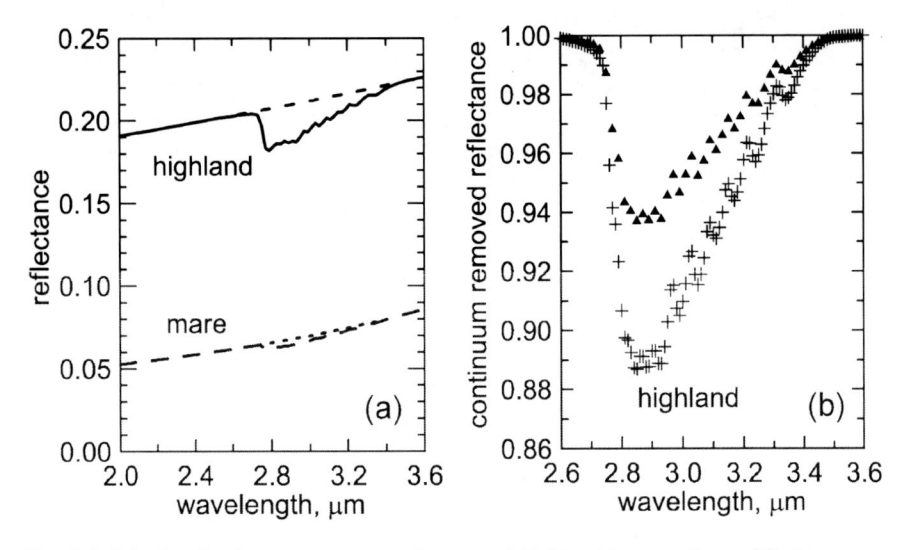

Fig. 3.1 Calculated reflectance spectra of mare and highland lunar soils modified by presence of OH-groups at concentration $n_s = 2 {\cdot} 10^{17}$ cm^{-2} on the surface of each regolith particle of the few upper layers where reflected light is formed (*a*); the shape and depth of 3μm band are shown in continuum removed spectra (*b*)

In Fig.3.1, a fit for the deepest 3μm band observed by Sunshine et al. (2009) in cold lunar regions is presented. The band shape was reproduced at a reasonable set of the spectral parameters (local band positions and corresponding c_i), and the maximum band depth was obtained at a reasonable surface concentration of OH

$n_s = 2 \times 10^{17}$ cm^{-2} (see Sect. 3.2.1). To show the band shape, the continuum removed spectra were calculated (Fig.3.1b).

Beside simulation of highland-type spectrum, 3μm absorption for mare type-spectrum at the same value of n_s is also shown in Fig 3.1. The maximum band depth for cold mare soils shown in (Sunshine et al. 2009) can be achieved at $n_s = 3 \times 10^{17}$ cm^{-2}.

3.3.1.1 Surface Brightness Effect

As shown in Fig.3.1, the bend depth at the band minimum depends on the surface reflectance, being larger for brighter highland soils than for darker mare soils at the same OH content. This applies both to absorption depth ($R_{contimuun}-R$) measured from the background $R_{contimuun}$ of a spectrum (dashed lines in Fig.3.1a) and to relative depth $H = (R_{contimuun}-R)/R_{contimuun}$ shown in Fig.3.1b. Normalization of the spectra (frequently used in astronomy and spectroscopy) does not remove the effect of surface brightness on spectral contrast because the contrast tends to zero both for too bright ($R \rightarrow 1$) and too dark ($R \rightarrow 0$) surfaces. This was observed in laboratory measurements (Adams and Filice 1967) and simulated theoretically (Shkuratov et al. 1999).

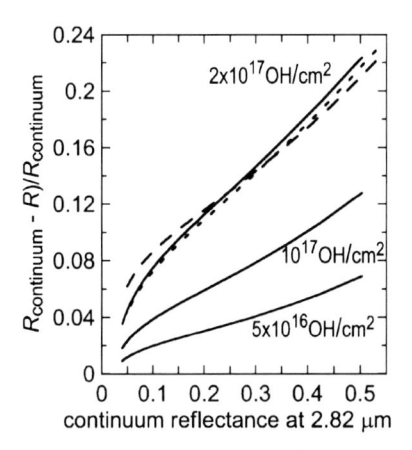

Fig. 3.2 Dependence of OH band depth on surface brightness: 3μm band depth vs. reflectance of regolith at different concentrations of OH-groups on the surface of regolith particles (solid lines), dashed line showing 3μm depth for 2.5wt% of coarse (>10μm) ice particles, dot line showing the depth for 0.006wt% of submicron ice

For OH-absorption, theoretical modeling of the effect is presented in Fig.3.2. Calculated 3μm depths are shown vs. reflectance of regolith at different surface concentrations n_s of OH. The range of OH abundance is taken from 5×10^{16} cm^{-2} typical of lunar regolith particles from low-latitude regions (Des Marais 1974) to 2×10^{17} cm^{-2} estimated above for polar regolith.

In Fig.3.3, OH abundance is varied and 3μm band depths are plotted for different values of continuum reflectance of a powdered surface. The upper (dash-dot) and the lower (solid) lines present highland-like and mare material, respectively.

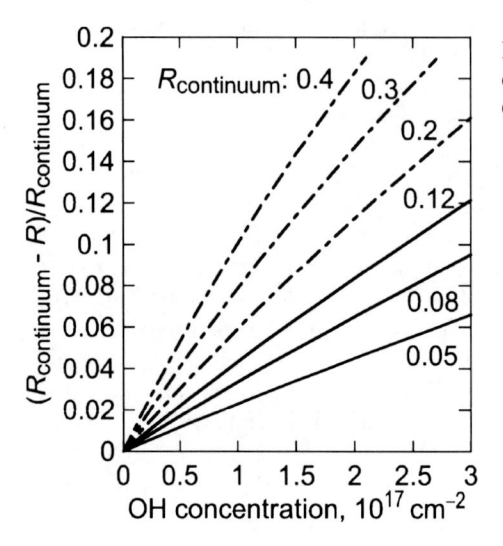

Fig. 3.3 OH band depth vs. surface concentration of OH-groups at different reflectance of regolith

The calculated 3 μm depth ranges are consistent with M^3 observations (McCord et al. 2011). The lower line in Fig.3.2 covers 2.8 μm depth range 0.015-0.045 observed for equatorial regions in reflectance range 0.067-0.32 at $n_s = 5 \cdot 10^{16}$ cm^{-2} typical for such regions. The deepest absorption (0.1-0.2) shifted to longer wavelengths is observed in polar regions. There, highest OH abundance (due to lower temperatures) and bright highland material dominate (see the top right corners of Figs.3.2, 3.3).

Figures 3.2 and 3.3 show that the same relative band depth may be achieved in dark soils at OH-content by a factor 2–4 higher than in bright soils. The two examples of the pairs "dark - bright" most important for the lunar soils are (1) "mare - highland" and (2) "mature - immature". Though immaturity implies lower bulk concentration of OH and weaker OH absorption (see Sect. 3.2.2), brightness of the immature soils may compensate for the lack of OH, so that the bands produced at lower OH content may look even stronger than those for mature soils of similar composition.

Unambiguous interpretation of stronger 3 μm absorption as due to higher OH abundance can be made in a few cases: (1) at the same apparent reflectance of the compared soils, and (2) when deeper 3μm bands are observed for darker soils. The latter is the case for lunar swirls (see Sect. 3.2.2); the first case is shown in (Pieters et al. 2009, supporting online material): fresh craters in a feldspathic region have deeper 3 μm bands than the background mature soil of approximately the same reflectance. Figures 3.2 and 3.3 support the author's interpretation of this fact as due to higher OH content in fresh feldspathic material from the craters. McCord et al. (2011) supposed this to be due to high reactivity of fresh crystalline feldspathic surfaces that favors OH formation. This explanation is consistent with measurements of water absorption and chemisorption for terrestrial analogs of lunar materials (Hibbitts et al. 2011): crystalline feldspars were found to be more absorptive than, at least, basalt glasses.

One more case of unequivocal comparison of OH content in different soils is when the normalized 3 μm band depths are the same for different surface reflectance. In this case, OH abundance should be *higher in the darker material*. An example close to this situation was shown in Fig.11 of (McCord et al. 2011), though the normalized depths were a little lower for darker iron-rich surfaces than for feldspathic material, which does not allow for definite conclusions about OH content without scales on proper spectra. Plots shown in Figs.3.2 and 3.3 may help to distinguish high OH-content from the effect of surface brightness in further analysis of M^3 or other spectral data and provide evaluation of OH abundance in different lunar materials.

3.3.1.2 Comparison to 3 μm Absorption by Ice

For comparison, the spectral effect of small admixtures of ice to lunar regolith have been calculated (Starukhina 1999c, 2001). The depths of the 3μm bands due to ice crucially depend on the range of ice particle sizes. At the absorption wavelengths, coarse ice particles are nontransparent and do not take part in light absorption (they are just "black"), whereas fine ice is semitransparent, which makes it an effective absorber. That is why the spectral effect is much more pronounced for submicron ice grains or ice coatings around regolith particles than for ice grains of sizes much greater than the wavelength of light. This makes it difficult to estimate ice content from 3 μm absorption depth, if attributed to water. For ice accumulated in gradual deposition due to random hops, submicron form is more probable. In Fig.3.2 examples of the dependence of 3 μm band depth H vs. reflectance R are shown for fine and coarse ice. The difference by a factor more than 400 in the ice concentrations giving nearly the same H-R plots (close to that for $n_s = 2 \times 10^{17}$ cm^{-2}) is eloquent.

Thus, simulation of 3 μm absorption for regolith-like surface shows that, if the variance of spectral positions of OH absorption maxima near different cations is taken into account, neither depth nor width, shape and position of 3 μm band can be used to distinguish between H_2O molecules (or true hydrated silicates) and OH-groups of radiation origin, at least at low band depths (which is the case for most observations).

The interpretation of deep 3 μm absorption in the irradiated silicates requires further laboratory studies. Consider the conditions necessary for such experiments.

3.3.1.3 The Importance of Multiple Scattering for Detectability of OH

The essential feature of the theoretical model (Starukhina and Shkuratov 1996, Shkuratov et al. 1999) used for calculation of the lunar 3 μm absorption is that the model operates with particulate surfaces similar to lunar soils. The light reflected by such surfaces is multiply scattered in the particles before emergence. If absorbing species are present in particles, the light beam is absorbed at each scattering (at each passage through a particle), so that the total absorption is enhanced many times. This is the physical mechanism that provides detectability of the absorbers, in particular, OH, in transparent particles of regolith.

All attempts of simulation, either theoretical or experimental, of irradiation-induced 3 μm absorption in reflectance spectra of polished coarse-grain crystals are irrelevant. This is the reason of the failure in detection of significant absorption in the experiments of Burke et al. (2010) (which was supposed also by McCord et al. (2011)). To obtain detectable OH band in reflectance spectra, regolith-like surface (fine powders with particle size ~50 μm and less) should be irradiated many times, each time after mixing of the sample, so that fresh surface could be exposed to irradiation. This would provide formation of the irradiated rims around many particles so that the incident light beam could intersect the zones enriched in OH-groups many times before emergence and the hydroxyl could contribute to light absorption.

The OH band depth may be noticeable also for proton irradiated polycrystal with very fine grains in the surface layer. Quantitative characteristics of OH detectability is the ratio of the maximum addition $\Delta\tau$ (Eq. (3.2)) to the optical density τ. Calculations show that the relative band depth is $H \sim (1/3)\Delta\tau/\tau$ and $H \sim (1/2)\Delta\tau/\tau$ for surface reflectance $R \approx 0.2$ and 0.06, respectively.

The $\Delta\tau/\tau$ criterion is more important for simulation of the lunar OH absorption than the others, e.g., the incident ion energy.

3.3.2 Variations of 3 μm Band Depth

The considerations about spatial variations of the implanted hydrogen given in Sect. 3.2.3 are valid for OH groups. To this, spatial variations of spectral contrasts with surface brightness are added. Both global trend and local anomalies of OH were observed for the Moon (Pieters et al. 2009, McCord et al. 2011).

Note that, in general, H and OH abundances should not correlate with each other. The lack of correlation was actually found by Pieters et al. (2009) when spatial distribution of OH absorption bands near 3μm in sunlit M^3 data were compared to neutron spectrometer hydrogen abundance data. The correlation is destroyed by factors considered above. (1) Hydrogen can be retained in radiation defects other than oxygen dangling bonds, up to retention in blisters where hydrogen is in molecular form (see Sect. 3.2.1). (2) IR and neutron spectrometry study different regolith depths. (3) OH band depths depend on surface brightness as shown in the previous section.

As for diurnal temporal variations of 3 μm absorption, simulation of the lunar 3 μm bands enables us to estimate the difference in OH surface concentration n_s required to obtain the difference in the morning and noon band depths reported by Sunshine et al. (2009). The loss from particle rims is $\Delta n_s \approx 8 \times 10^{16}$ cm^{-2} at least in the upper ~10 regolith layers of a depth $h_r \sim 1$ mm, so the total loss per unit regolith area is $\Delta N_s \approx \Delta n_s S \rho_r h_r > 10^{19}$ cm^{-2}. If such amount of OH was released during the half of lunar day, the number density of temporary atmosphere would be $n \sim \Delta N_s g/v^2$, where the lunar gravity $g = 162$ cm/s^2 and hydroxyl thermal velocity at lunar day temperatures $v \approx 6 \times 10^4$ cm/s. Then $n > 4 \times 10^{11}$ cm^{-3}, which is not the case either for hydrogen, or for OH: the observed density of the lunar day-time exosphere is $n < 10^6$ cm^{-3} (Stern 1999). To restore the morning amount of OH would take at least $t_r \sim \Delta N_s/j \approx 10^{11}$ s $\sim 10^4$ years, i. e., the lost hydrogen could not be restored by next lunar morning.

If the 3 μm feature was due to water molecules, noticeable temporal variations of the band depth could not be obtained either. If the upper millimeter of regolith lost 0.1 wt.% of water, the number of water molecules released to the lunar exosphere would be $\Delta n_{H2O} \approx 5 \cdot 10^{18}$ cm^{-2}. The number density of the atmosphere created in such release would have been $n \sim \Delta n_{H2O} g/v^2 \sim 2 \times 10^{11}$ cm^{-3}, whereas there has never been a confirmed detection of water or its dissociation products in the lunar atmosphere (Stern 1999), the upper estimate for OH being $n < 10^6$ cm^{-3}.

This means that the temporal variations found by Sunshine et al. (2009) are most probably due to thermal emission whose contribution in the daytime is significant. It is difficult to remove the emission from the absorption band in reflectance spectrum without exact information about surface temperature and taking surface reflectivity into account. In removal of thermal emission from the absorption band in reflectance spectrum, emissivity was assumed to equal 1 at all wavelengths, which is not correct. Another potential source of the enhancement of 3 μm absorption in the lunar mornings and evenings is photometric: depth of absorption bands slightly changes with illumination/observation geometry. The contamination of the lunar spectra by thermal emission, as well as photometric effects were considered by McCord et al. (2011).

3.4 High Radar Response of Mercury Polar Craters: Water Ice or Cold Silicates?

Here we shortly consider results concerning Mercury, because it was on Mercury that the first observations suggesting water near the lunar poles was made. Slade et al. (1992) observed high radar brightness of Mercurian polar regions and high depolarization of the reflected signal. All this is typical of radar response from ices in Earth, Mars, and Galilean satellites, so the observations were interpreted as ice on Mercury.

After discovery of the microwave "polar caps" on Mercury, it was natural to look for such "caps" on the Moon. Though the Clementine radar team reported detection of lunar polar brightness (Nozette et al. 1996), this result was not confirmed in reanalysis of the Clementine radar data (Simpson and Tyler 1999). Besides, no polar brightness was found in ground radar measurements of the Moon on Arecibo (Stacy et al. 1996).

The difference in radar brightness between polar craters on Mercury and the Moon is itself a problem for ice hypothesis. Among suggested explanations were the difference in obliquity history or endogenic sources of ice and the difference in crustal chemistry and/or outgassing history of the two bodies. The problems of the water ice hypothesis, their possible solutions and the problems concerning the solutions were considered by Harmon et al. (2001). Suggested solutions require many additional assumptions to be consistent with all the observations.

The most puzzling problems for ice hypothesis on Mercury are of the same type as for H content on the Moon. They are associated with regional variations of radar brightness and polarization ratio: local anomalies such as excess brightness in craters with temperatures too high for ice stability and the lack of radar brightness in some permanently shaded areas. The anomalies can be found by comparison of the high resolution radar data (Harmon et al. 2001) and temperatures calculated for Mercurian craters (Vasavada et al. 1999).

Physical property responsible for radar brightness of dielectric material is low dielectric loss. It enables penetration of the radar signal into the volume of material and multiple scattering from density fluctuations therein, as well as coherent backscattering (Hapke 1990), which yields both high reflectance and high depolarization for the signal scattered in the direction opposite to the incident wave.

Dielectric loss in solid materials in the microwave range is much more sensitive to temperature than the losses in the optical range and is characterized by exponential temperature dependence, similar to that for outgassing rates: $\sim\exp(-U/kT)$. The low dielectric loss is characteristic not only of water ice, but also of cold silicates that are major constituents of the surfaces of the Earth-like planets. Extrapolation of dielectric loss of silicates to low temperatures and radar wavelengths (Starukhina 2000a, 2001) yields loss tangents as low as $\sim10^{-5}$, which may account for high radar response from cold regions and highly depolarized component of the response.

Being temperature-activated, dielectric losses at microwave frequencies depend on activation energies U of the absorption centers: $\sim\exp(-U/kT)$. They vary with mineral type, as well as thermal and irradiation history, which can account for local anomalies of radar brightness and the difference in radar properties between polar craters of Mercury and the Moon.

3.5 Conclusions

Thus, the amount of hydrogen detected near the lunar poles up to present cannot exclude the interpretation of this hydrogen as protons of solar wind trapped in the lunar regolith. A variety of trapping mechanisms is possible: trapping on radiation defects, in particular, chemical trapping in hydroxyl groups, blistering or just physical absorption in permanently shadowed areas. As distinct from the water ice hypothesis, this interpretation may explain both global and local variations of H abundance on the Moon.

Light absorption near 3 μm detected in reflectance spectra of the Moon can be due to OH-groups of solar wind origin. Theoretical simulation of such absorption have shown that neither the depth nor width, shape and position of 3μm band can be used to distinguish between H_2O molecules and OH-groups of radiation origin. For this, intensive laboratory simulation of solar wind effect on silicates are required.

The observed spatial variations of the OH absorption are due not only to dependence of outgassing rates on surface temperature and composition, but also to variation of regolith brightness, brighter soils showing deeper bands than the darker ones at the same OH concentration on the surfaces of regolith particles. Simulated variations of 3μm band depth with surface brightness and OH abundance are consistent with those observed for lunar surface.

Low dielectric loss of surface material that is responsible for high radar reflection and depolarization of the reflected signal can be characteristic of silicates cooled to low temperatures. If low-loss cold silicates are found, the radar brightness of the polar craters of Mercury can be explained without water ice.

Thus neither of the methods for search of water or hydrosilicate compounds on the lunar surface can be used unreservedly. To interpret the results of the remote sensing experiments aimed on detection of water, possible alternative explanations should be taken into account.

3.5.1 Experimental Support Necessary for Remote Sensing Observations of the Lunar Polar Regions

As follows from the above considerations, the key to unambiguous interpretation of the observations aimed on search of water is not in cosmic space, but in the terrestrial laboratories. Each type of remote sensing methods requires substantial support from laboratory experiments on various lunar-like silicates.

(1) To determine the upper limit of hydrogen content in polar craters, study of retention of keV protons and blistering at low temperatures should be carried out. The irradiation should be accompanied by measurements of gas (H_2) desorption and determination of hydrogen concentration profile.

To maintain temperature in the range expected for permanently shaded craters ($T = 40$–100 K), special cooling of the irradiated specimen should be provided to compensate for heating by the energetic ion beam. In irradiation at room temperatures, extra heating is avoided due to low current density ($\sim 1 \mu A \cdot cm^{-2} s^{-1}$); at lower current densities, the time to reach the necessary fluences 10^{18}–10^{19} cm^{-2} is unreasonably long.

(2) Spectral measurements near 3 μm of the irradiated silicates. For reflectance measurements, fine powders should be irradiated to fluences up to 5×10^{17} cm^{-2} many times, each time after mixing of the sample, exposing fresh surface.

These experiments may be carried out on polycrystal, but its grain size should be very small (<50 μm)

(3) To provide more confident interpretation of the radar brightness of cold surfaces, measurements of dielectric losses at temperatures of 40 to 400 K and Arecibo and Goldstone microwave frequencies (wavelength 12.6 and

3.5 cm, respectively) are necessary, lower dielectric losses being more probable for optically bright materials with lower static permittivity.

All these experiments are much less expensive than any spacecraft mission. However, the best proof of water is drinking, just as the best proof of pudding is eating. Incontrovertible evidence for water ice on the Moon would be delivery or *in situ* analysis of a column of lunar regolith from a hydrogen-rich permanently shadowed area.

3.5.2 Implications for Lunar Exploration

3.5.2.1 Water Production

Even if the detected hydrogen is not associated with water, polar regions and, in particular, permanently shaded areas, are, perhaps, most promising in the context of lunar exploration. If the excess hydrogen found by Lunar Prospector is in the form of implanted atoms, this is still a promising discovery, because the hydrogen can be used in water production. For example, at the maximum values of [H] = 0.17 ± 0.09 wt.% suggested by Feldman et al. (2001), the reaction

$$FeO + 2H = Fe + H_2O \qquad (3.3)$$

can yield about 15 ± 8 litres of water per ton of regolith (6 ± 3 wt.% of FeO being required for the process). This is of the same order of magnitude as the projected yield of a device for water production with hydrogen brought from the Earth (Eguchi et al. 1999). The water output may be greater by several fold for maximum [H] = 0.45wt.% suggested by Mitrofanov et al. (2010). Then supply of hydrogen from the Earth may be not necessary for water production on the Moon, which may be important for future exploration.

3.5.2.2 Extraction of [3]He

Since sublimation energies of volatile substances are typically lower than trapping energies of the implanted atoms, the conditions in polar craters are more favorable for long retention of solar wind implanted elements than for retention of volatiles. If polar regions are rich in volatiles, the implanted gases are also abundant there. The opposite statements is, in general, not correct.

Polar regions of the Moon, in particular, permanently shaded areas, are, perhaps, the most promising sites for lunar utilization, including volatile extraction or water production. Accumulation of solar wind [3]He makes the high latitudes of the Moon the most likely candidate for "the Persian Gulf of the XXI century".

3.5.2.3 General Prospects

Is the absence of water on the Moon a pessimistic perspective? Certainly, not. Whatever the result of "water race", probably the most valuable of the lunar resources is high vacuum in large scale. In this connection, the lack of volatiles may be more helpful than their presence.

Moreover, the "dry" Moon would be less attractive object for human colonization, which may help to preserve its present exosphere from contamination. The Moon is unique near-Earth object that can provide high purity of the surfaces unachievable in terrestrial laboratories. This may determine further stage of the progress of human technologies.

Acknowledgments. I am grateful to Mikhail Kreslavsky for providing me with new information, for discussions and captious check of calculations. I appreciate encouraging and helpful comments from Thomas McCord. I thank Yurij Shkuratov for discussions and CRDF, grant UKP2-2897-KK-07, for support.

References

Adams, J., Filice, A.: Spectral reflectance 0.4 to 2.0 microns of silicate rock powders. J. Geophys. Res. 72, 5705–5715 (1967)

Arnold, J.R.: Ice in the lunar polar regions. J. Geophys. Res. 84, 5659–5668 (1979)

Basilevsky, A.T., Ivanov, B.A., Florensky, K.P., Yakovlev, O.I., Feldman, B.I., Granovsky, L.V.: Impact craters on the Moon and planets Nauka, Moscow (1983) (in Russian)

Bibring, J.-P., Langevin, Y., Rocard, F.: Synthesis of molecules by irradiation in silicates. In: Proc. Lunar. Sci. Conf.13, pp. A446–A450. LPI, Houston (1982)

Biersack, J.P., Haggmark, L.G.: A Monte Carlo computer program for the transport of energetic ions in amorphous target. Nucl. Instr. Methods 174, 257–269 (1980)

Borg, J., Maurette, M., Durrieu, L., Jouret, C.: Ultramicroscopic features in micron-sized lunar dust grains and cosmophysics. In: Proc. Lunar. Sci. Conf. 2nd, vol. 3, pp. 2027–2040. LPI, Houston (1971)

Borg, J., Comstick, G.M., Langevin, Y., Maurette, M., Jouffrey, B., Jouret, C.: A Monte-Carlo model for the exposure history of lunar dust grains in the ancient solar wind. Earth Planet Sci. Lett. 29, 161–174 (1976)

Burke, D.J., Dukes, C.A., Kim, J.H., Shi, J., Famá, M., Baragiola, R.A.: Solar wind contribution to surficial lunar water: laboratory investigations. Icarus 211, 1082–1088 (2011)

Butler, B.J.: The migration of volatiles on the surfaces of Mercury and the Moon. J. Geophys. Res. 102, 19283–19291 (1997)

Cadenhead, D., Brown, M., Rice, D., Stetter, J.: Some surface area and porosity characterization of lunar soils. In: Proc. Lunar. Sci. Conf. 8, pp. 1291–1303. LPI, Houston (1977)

Clark, R.N.: Detection of adsorbed water and hydroxyl on the Moon. Science 326, 562–564 (2009)

Colaprete, A., 16 co-authors: Detection of water in the LCROSS ejecta plume. Science 330, 463–468 (2010)

Des Marais, D.J., Hayes, J.M., Meinschein, W.G.: The distribution in lunar soil of hydrogen released by pyrolysis. In: Proc. Lunar. Sci. Conf. 5, pp. 1811–1822. LPI, Houston (1974)

Dran, J.C., Durrieu, L., Jouret, C., Maurette, M.: Habit and texture studies of lunar and meteoritic material with a 1 MeV electron microscope. Earth Planet Sci. Lett. 9, 391–400 (1970)

Eguchi, K., Ogiwara, S., Oguchi, M., et al.: A design concept of water production experiment mission for lunar resource utilization. Solar System Res. 33, 376–381 (1999)

Feldman, W.C., Maurice, S., Binder, A.B., Barraclough, B.L., Elphic, R.C., Lawrence, D.J.: Fluxes of fast and epithermal neutrons from Lunar Prospector: Evidence for water ice at the lunar poles. Science 281, 1496–1500 (1998)

Feldman, W.C., Lawrence, D.J., Elphic, R.C., Barraclough, B.L., Maurice, S., Genetay, I., Binder, A.B.: Polar hydrogen deposits on the Moon. J. Geophys Res. 105, 4175–4195 (2000)

Feldman, W.C., 11 colleagues: Evidence for water ice near the lunar poles. J. Geophys Res.-Planets 106, 23231–23251 (2001)

Greenwood, J.P., Itoh, S., Sakamoto, N., Taylor, L.A., Warren, P.H., Yurimoto, H.: Water in Apollo rock samples and the D/H of lunar apatite. Lunar Planet Sci. 41, LPI, Houston, abstract # 2439 (2010)

Grossman, J.J., Mukherjee, N.R., Ryan, J.A.: Microphysical, microchemical and adhesive properties of lunar material III: Gas interaction with lunar material. In: Proc. Lunar Sci. Conf., vol. 3, pp. 2259–2269. LPI, Houston (1972)

Gruen, D.M., Siskind, B., Wright, R.B.: Chemical implantation, isotopic trapping effects, and induced hydroscopicity resulting from 15 keV ion bombardment of sapphire. J. Chem. Phys. 65, 363–378 (1976)

Guseva, M.I., Martynenko, Y.V.: Radiation blistering. Physics-Uspekhi (Advances in Physical Sciences) 24, 996–1007 (1981)

Hapke, B.: Coherent backscatter and the radar characteristics of outer planet satellites. Icarus 88, 407–417 (1990)

Harmon, J.K., Perillat, P.J., Slade, M.A.: High-resolution radar imaging of Mercury's north pole. Icarus 149, 1–15 (2001)

Haskin, L., Warren, P.: Lunar chemistry. In: Heiken, G.H., Vaniman, D.T., French, B.M. (eds.) Lunar Sourcebook, pp. 357–474. Cambridge University Press, New York (1991)

Hibbitts, C.A., Grieves, G.A., Poston, M.J., Dyar, M.D., Alexandrov, A.B., Johnson, M.A., Orlando, T.M.: Thermal stability of water and hydroxyl on the surface of the Moon from temperature-programmed desorption measurements of lunar analog materials. Icarus 213, 64–72 (2011)

Housley, R.M., Grant, R.W., Paton, N.E.: Origin and characteristics of excess Fe metal in lunar glass welded aggregates. In: Proc Lunar Sci Conf. 4th, pp. 2737–2749. LPI, Houston (1973)

Hutcheon, I.D.: Micrometeorites and solar flare particles in and out of the ecliptic. J. Geophys Res. 80, 4471–4483 (1975)

Keller, L.P., McKay, D.S.: The nature and origin of rims on lunar soil grains. Geochim Cosmochim Acta 61, 2331–2340 (1997)

Kramer, G.Y., Besse, S., Dhingra, D., Nettles, J., Klima, R., Garrick-Bethell, I., Clark, R.N., Combe, J.-P., Head, J.W., Taylor, L.A., Pieters, C.M., Boardman, J., McCord, T.B.: M^3 spectral analysis of lunar swirls and the link between optical maturation and surface hydroxyl formation at magnetic anomalies. J. Geophys Res. 116 (2011) (in Press), doi:10.1029/2010JE00

Lawrence, D.J., Feldman, W.C., Elphic, R.C., Hagerty, J.J., Maurice, S., McKinney, G.W., Prettyman, T.H.: Improved modeling of Lunar Prospector neutron spectrometer data: Implications for hydrogen deposits at the lunar poles. J. Geophys Res. 111, E08001 (2006)

Lee, R.W.: Diffusion of hydrogen in natural and synthetic fused qurtz. J. Chem. Phys. 38, 448–455 (1963)

Leich, D.A., Tombrello, T.A., Burnett, D.S.: The depth of distribution of hydrogen in lunar material. Lunar Sci. 4, 463–465 (1973) (abstract)

Lord, H.C.: Hydrogen and helium implantation into olivine and enstatite: Retention coefficients, saturation concentrations, and temperature-release profiles. J. Geoph. Res. 73, 5271–5280 (1968)

Mattern, P.L., Thomas, G.J., Bauer, W.: Hydrogen and helium implantation in vitreous silica. J. Vac. Sci. Technol. 13, 430–436 (1976)

McCord, T.B., Taylor, L.A., Combe, J.P., Kramer, G., Pieters, C.M., Sunshine, J.M., Clark, R.N.: Sources and physical processes responsible for OH/H_2O in the lunar soil as revealed by the Moon Mineralogy Mapper (M^3). J. Geophys. Res. 116 (2011), doi:10.1029/2010JE003711

McCubbin, F.M., Steele, A., Nekvasil, H., Schnieders, A., Rose, T., Fries, M.: Detection of structurally bound hydroxyl in apatite from Apollo mare basalt 15058, 128. Lunar Planet. Sci 42, LPI, Houston, abstract #2468 (2010)

McKay, D.S., Heiken, G., Basu, A., Blanford, G., Simon, S., Reedy, R., French, B.M., Papike, J.: The lunar regolith. In: Heiken, G.H., Vaniman, D.T., French, B.M. (eds.) Lunar Sourcebook, pp. 285–356. Cambridge University Press, New York (1991)

Mitrofanov, I.G., 28 co-authors: Neutron mapping of the lunar south pole using the LRO neutron detector experiment LEND. Science 330, 483–486 (2010)

Mitrofanov, I.G., 15 co-athours: Neutron suppression regions at lunar poles as local areas of water-rich permafrost. Lunar Planet. Sci. 42, LPI, Houston, abstract #1787 (2011)

Moulson, J., Roberts, J.P.: Water in silica glass. Trans. Brit. Ceramic Soc. 59, 388–394 (1960)

Nozette, S., Lichtenberg, C., Spudis, P., Bonner, R., Ort, W., Malaret, E., Robinson, M., Shoemaker, E.: The Clementine bistatic radar experiment. Science 274, 1495–1498 (1996)

Peri, J.B., Hannan, R.B.: Surface hydroxyl groups on γ-alumina. J. Phys. Chem. 64, 1526–1530 (1960)

Pieters, C.M., 28 colleagues: Character and spatial distribution of OH/H_2O on the surface of the Moon seen by M^3 on Chandrayaan-1. Science 326, 568–572 (2009)

Rivkin, A.S., Howell, E.S., Britt, D.T., Lebovsky, L.A., Nolan, M.C., Branston, D.D.: 3µm photometric survey of M- and E-class asteroids. Icarus 117, 90–100 (1995)

Roth, J.: Chemical sputtering. In: Behrisch, R. (ed.) Sputtering by Particle Bombardment II, pp. 91–146. Springer, New York (1983)

Saal, A.E., Hauri, E.H., Cascio, M.L., Van Orman, J.A., Rutherford, M.C., Cooper, R.F.: Volatile content of lunar volcanic glasses and the presence of water in the Moon's interior. Nature 454, 192–195 (2008)

Scherzer, B.M.U.: Development of surface topography due to gas ion implantation. In: Behrisch, R. (ed.) Sputtering by Particle Bombardment II, pp. 271–356. Springer, New York (1983)

Shkuratov, Y.G., Starukhina, L.V., Hoffmann, H., Arnold, G.: A model of spectral albedo of particulate surfaces: Implications for optical properties of the Moon. Icarus 137, 235–246 (1999)

Shoemaker, E.M., Hait, M.H., Swann, G.A., Schleicher, D.L., Schaber, G.G., Sutton, R.L., Dahlem, D.H., Goddard, E.N., Waters, A.C.: Origin of the lunar regolith at Tranquility Base. In: Proc. Apollo 11 Lunar Sci. Conf., LPI, Houston, pp. 2399–2412 (1970)

Siegler, M.A., Bills, B.G., Paige, D.A.: Effects of orbital evolution on lunar ice stability. J. Geophys. Res. 116, E03010 (2011), doi:10.1029/2010JE003652

Simpson, R.A., Tyler, G.L.: Reanalysis of Clementine bistatic radar data from the lunar South Pole. J. Geoph. Res. 104, 3845–3862 (1999)

Siskind, B., Gruen, D.M., Varma, K.: Chemical implantation of 10 keV H^+ and D^+ in rutile. J. Vac. Sci. Technol. 14, 537–542 (1977)

Slade, M.A., Butler, B.J., Muhleman, D.O.: Mercury radar imaging: Evidence for polar ice. Science 258, 635–640 (1992)

Stacy, N.J.S., Campbell, D.B., Ford, P.G.: Arecibo radar mapping of the lunar poles: A search for ice deposits. Science 276, 1527–1530 (1997)

Starukhina, L.V.: Excess hydrogen on the lunar poles: water ice or solar wind induced OH? (abstract). In: The 3rd International Conference on the Exploration and Utilization of the Moon, Moscow. Russian Acad. Sci., p. 38 (1998a)

Starukhina, L.V.: Estimation of solar wind induced 2.9 μm absorption in lunar regolith: implication for the problem of water detection on the lunar poles (abstract). In: The 3rd International Conference on the Exploration and Utilization of the Moon, Moscow. Russian Acad. Sci., p. 39 (1998b)

Starukhina, L.V.: The excess hydrogen on the lunar poles: water ice or chemically trapped solar wind? Lunar Planet. Sci. 30, abstract #1093 (1999a)

Starukhina, L.V.: Estimation of 3μm light absorption by hydroxyl of solar wind origin: implication for the problem of water detection on the surfaces of atmosphereless celestial bodies. Lunar Planet. Sci. 30, LPI, Houston, abstract #1094 (1999b)

Starukhina, L.V.: Light absorption by radiation-induced hydroxyl ions and the problem of finding water on atmosphereless celestial bodies. Solar System Research 33, 291–295 (1999c)

Starukhina, L.V.: High radar response of Mercury polar regions: water ice or cold silicates? Lunar Planet. Sci. 31, LPI, Houston, abstract # 1301 (2000a)

Starukhina, L.V.: On the origin of excess hydrogen at the lunar poles. Solar System Res. 34, 215–219 (2000b)

Starukhina, L.V.: Water detection on atmosphereless celestial bodies: Alternative explanations of the observations. J. Geophys. Res.-Planets 106, 14701–14710 (2001)

Starukhina, L.V.: 3μm light absorption by hydroxyl of solar wind origin and the prospects of water detection on asteroids with NIR spectroscopy. In: Proc. ACM 2002, Berlin, pp. 513–516 (2002)

Starukhina, L.V.: Computer Simulation of Sputtering of Lunar Regolith by Solar Wind Protons: Contribution to Alteration of Surface Composition and to Hydrogen Flux at the Lunar Poles. Solar System Research 37, 36–50 (2003)

Starukhina, L.V.: Polar regions of the Moon as a potential repository of solar-wind-implanted gases. Adv. Space Res. 37, 50–58 (2006)

Starukhina, L.V.: Ice on the moon and Mercury: reanalysis of the origin and survival conditions. Lunar and Planetary Science 39, LPI, Houston, abstract #1141 (2008)

Starukhina, L.V.: Depth of 3μm lunar absorption bands: the effect of surface brightness. A Wet vs. Dry Moon: Exploring Volatile Reservoirs and Implications for the Evolution of the Moon and Future Exploration, LPI, Houston, Abstr. #6002 (2011)

Starukhina, L.V., Shkuratov, Y.G.: A model for spectral dependence of albedo for multicomponent regolith-like surfaces. Solar System Res. 30, 258–264 (1996)

Starukhina, L.V., Shkuratov, Y.G.: The lunar poles: water ice or chemically trapped hydrogen? Icarus 147, 585–587 (2000)

Starukhina, L.V., Shkuratov, Y.G.: Swirls on the Moon and Mercury: meteoroid swarm encounters as a formation mechanism. Icarus 167, 136–147 (2004)

Starukhina, L.V., Shkuratov, Y.G.: Simulation of 3μm absorption band in lunar spectra: water or solar wind induced hydroxyl? Lunar and Planetary Science 41, LPI, Houston, abstract #1385 (2010)

Stern, S.A.: The lunar atmosphere: History, status, current problems, and context. Rev. Geophys. 37, 453–491 (1999)

Sunshine, J.M., Farnham, T.L., Feaga, L.M., Groussin, O., Merlin, F., Milliken, R.E., A'Hearn, M.F.: Temporal and spatial variability of lunar hydration as observed by the deep impact spacecraft. Science 326, 565–568 (2009)

Taylor, L.A., Rossman, G.R., Qi, Q.: Where has all the lunar water gone? Lunar Planet. Sci. 26, LPI, Houston, 1399–1400 (1995)

Vasavada, A.R., Paige, D.A., Wood, E.S.: Near-surface temperatures on Mercury and the Moon and the stability of polar ice deposits. Icarus 141, 179–193 (1999)

Watson, K., Murray, B.C., Brown, H.: The behavior of volatiles on the lunar surface. J. Geophys. Res. 66, 3033–3045 (1961)

Zeller, E.J., Ronca, L.B., Levy, P.W.: Proton-induced hydroxyl formation on the lunar surface. J. Geophys. Res. 71, 4855–4860 (1966)

Zent, A.P., Ichimura, A.I., McCord, T.B., Taylor, L.A.: Production of OH/H2O in lunar samples via proton bombardment. Lunar Planet. Sci. 41, LPI, Houston, abstract # 2665 (2010)

Ziegler, J.F., Biersack, J.P., Littmark, U.: The Stopping and Range of Ions in Solids. Pergamon Press, New York (1996)

4 Theoretical Modeling, Numerical Simulation, and Retrievals from Chang'E-1 Data for Microwave Exploration of Lunar Surface/Subsurface

Ya-Qiu Jin

Fudan University, Shanghai, China

4.1 Introduction

China had successfully launched its first lunar exploration satellite Chang'E-1 (CE-1) on 24 October 2007 at lunar circle orbit ~200 km. A duplicate CE-2 at lower orbit ~100 km was also launched in 1 October 2010. A multi-channel microwave radiometer, for the first time, was aboard the CE-1 (and CE-2) satellite with the purpose of measuring the microwave thermal emission from the lunar surface layer (Jiang and Jin 2011). There are four frequency channels for CE-1 microwave radiometer: 3.0, 7.8, 19.35 and 37.0 GHz. The observation angle is $0°$, the spatial resolution is about 35 km (for the channels 7, 19, 37 GHz) and 50 km (for 3 GHz), and the radiometric sensitivity about 0.5 K. The measurements of the multi-channel brightness temperature, T_B, are applied to invert the global distribution of the regolith layer thickness, from which the total inventory of ^{3}He (Helium-3) stored in the lunar regolith layer can be estimated quantitatively (Jiang and Jin 2011; Fa and Jin 2007a,b; 2010a,b; Jin and Fa 2009, 2010).

In this Chapter, an overall work by author's Laboratory for this mission is reported, including the microwave emission modeling and numerical simulation of T_B from lunar surface media (Fa and Jin 2007a,b; Jin and Fa 2009), data validation and retrieval of lunar regolith layer thickness from multi-channel CE-1 T_B data (Fa and Jin 2010a), and evaluation of global inventory of ^{3}He in lunar regolith (Fa and Jin 2010b), and diurnal temperature change of the cratered lunar surfaces (Gong and Jin 2011a,b).

On the other hand, due to the advancement of spaceborne radar technique, spaceborne radar sounder and synthetic aperture radar (SAR) have been utilized for lunar, Martian and other planetary explorations. Based on statistics of lunar crater population and morphology, the inhomogeneous undulated cratered lunar surface is numerically generated. Radar echoes from the cratered lunar surface are numerically calculated. Employing a back projection (BP) algorithm, SAR image of lunar surface is numerically simulated (Jin et al. 2007, Fa et al. 2009, Fa and Jin 2010).

Lunar subsurface features preserve the history information of lunar geology. Knowledge of lunar subsurface features can provide rich information about physical status, conformation, origin and evolution of the Moon. High frequency (HF) radar echoes from the cratered lunar surface/subsurface is studied. On the basis of Kirchhoff rough surface scattering and ray tracing of geometric optics of layered structures, an effective simulation of radar echoes and imaging is developed. Then, radar echoes and range images are simulated, sequentially, and their dependences on the physical properties of lunar layering structures are analyzed (Fa et al. 2009, Fa and Jin 2010).

4.2 Model of Microwave Emission from Lunar Surface Media

Brightness temperature of the lunar surface mainly correlates with the physical properties of the regolith layer, e.g. regolith thickness, physical temperature, dielectric permittivity of regolith, etc. According to the variation of physical temperature and bulk density with depth, a three-layer model consisting of an upper dust layer with temperature T_1 and thickness d_1, a regolith layer with temperature T_2 and thickness d_2, and an underlying bedrock with temperature T_3 is developed for numerical simulation of lunar surface T_B, as shown in Fig. 4.1.

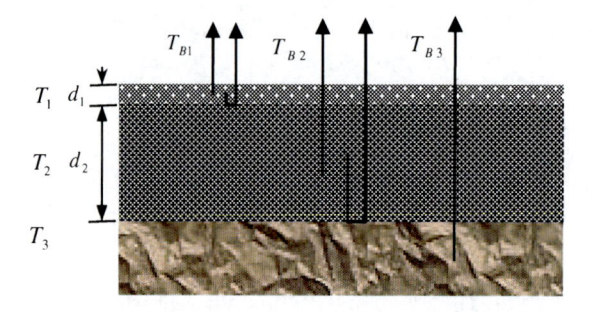

Fig. 4.1 A three-layer model for CE-1 T_B observation

Using the fluctuation dissipation theorem and radiative transfer of stratified media (Jin 1994, 2005), numerical simulations of multi-channel T_B from global lunar surface are obtained. The brightness temperature at $0°$ observation angle is finally derived as follows (Fa and Jin 2007a):

$$
\begin{aligned}
T_B = & \frac{k_0 \varepsilon''_{g1}}{2\varepsilon_0 k''_1} \mid X_{01} \mid^2 (1 - e^{-2k''_1 d_1})(1 + \mid R_{12} \mid^2 e^{-2k''_1 d_1}) T_1 \\
& + \frac{k_0 \varepsilon''_{g2}}{2\varepsilon_0 k''_2} \mid X_{01} X_{12} \mid^2 (1 - e^{-2k''_2 d_2})(1 + \mid R_{23} \mid^2 e^{-2k''_2 d_2}) e^{-2k''_1 d_1} T_2 \\
& + \frac{k_0 \varepsilon''_3}{2\varepsilon_0 k''_3} \mid X_{01} X_{12} X_{23} \mid^2 e^{-2k''_1 d_1} e^{-2k''_2 d_2} T_3
\end{aligned}
\tag{4.1}
$$

where three terms on the right hand side are contributed, respectively, by T_{B1}, T_{B2}, T_{B3} as described in Fig. 4.1, and ε_{g1}, ε_g, ε_3 are the effective dielectric constants of the media 1, 2 and 3, respectively. $k_n = k'_n + k''_n$ is the wave number of the media n ($n=0,1,2,3$), respectively. R_{nm} and X_{nm} are the reflection and transmission coefficients between the media n and m (n, $m=0,1,2,3$), respectively.

To numerically simulate T_B of Eq. (4.1), the parameters, ε_{g1}, d_1 and T_1 of the top dust layer are pre-determined. Other parameters, i.e. ε_{g2}, d_2 and T_2 of the regolith layer and ε_3, T_3 of bedrock media, are to be determined for simulation.

Correspondence of the lunar regolith layer thickness to the lunar digital elevation mapping (DEM) is proposed to tentatively construct the global distribution of lunar regolith layer thickness d_2 (Fa and Jin 2007a, Jin and Fa 2009). Using Clementine UVVIS multispectral data, the global spatial distribution of $FeO+TiO_2$ content on the lunar regolith media is calculated for the dielectric permittivities of lunar media, ε_{g2} and ε_3.

Based on some measurements of physical temperature of the lunar surface, an empirical formula of physical temperature distribution, T_2, T_3 is presented (Fa and Jin 2007a, Jin and Fa 2009). These parameters are applied to simulate the brightness temperature from lunar layered media (Fa and Jin 2007a, Jin and Fa 2009), as shown in Eq. (4.1), and further to analyze and retrieve the regolith layer thickness, and finally to evaluate global distribution of ^{3}He content in regolith media derived from global distribution of solar flux and lunar surface optical maturity (Fa and Jin 2007b).

4.3 CE-1 T_B Data

The primary 621 tracks of swath data measured by CE-1 microwave 4- channel radiometer from November 2007 to February 2008 are collected and analyzed. Using nearest neighbor interpolation based on the sun incidence angle in observations, global distributions of T_B from the lunar surface at lunar daytime and nighttime are constructed. On the basis of the observed relation between brightness temperature and solar incidence angle, 264 orbital tracks with solar incidence angles of $0°-14°$ and $166°-180°$ at the lunar equator were selected to construct global brightness temperature maps near lunar noon and midnight, respectively. As an example, Fig. 4.2 shows T_B observed by CE-1 near lunar noon.

In the global T_B distribution over lunar surface, brightness temperature is high at lunar equatorial area and decreases as latitude increases, and finally reaches the minimum at the pole area. This variation trend is dominated by the physical temperature decrease toward the poles.

The significant difference in brightness temperature between daytime and nighttime (e.g. about 50 K at 37 GHz channel) is due to the extremely high variation of the physical temperature of lunar media during a whole lunar day. Figure 4.3 shows one swath T_B observation (orbit number 0247 of CE-1) on 28 November 2007.

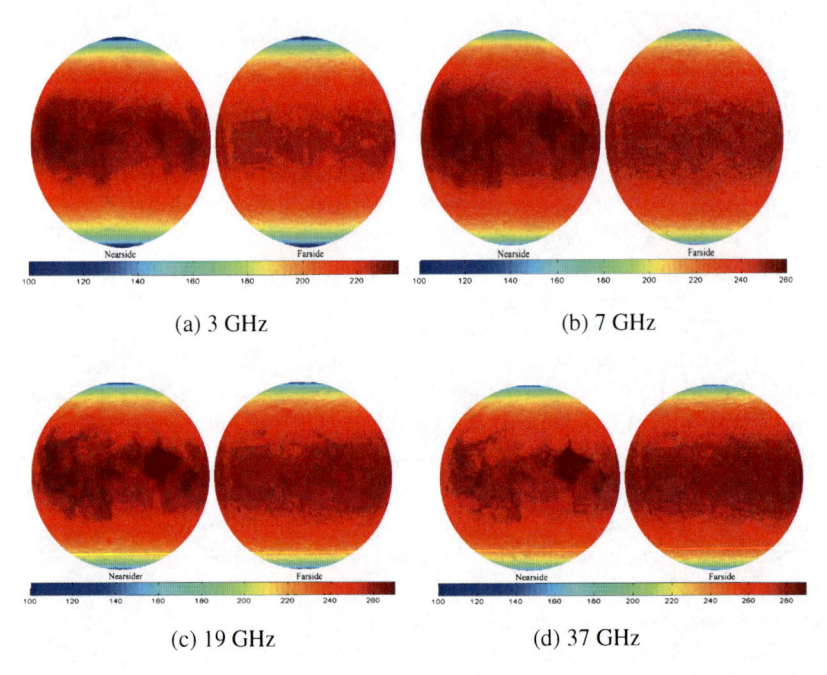

Fig. 4.2 T_B data collected from CE-1 observations at lunar local noon (solar incidence angles between $0°$ and $14°$)

Comparing the T_B distribution in lunar daytime and FeO+TiO$_2$ content, brightness temperature at 7.8, 19.35 and 37 GHz in lunar daytime look similar to the distribution of FeO+TiO$_2$ content, i.e. high FeO+TiO$_2$ content corresponds to high brightness temperature.

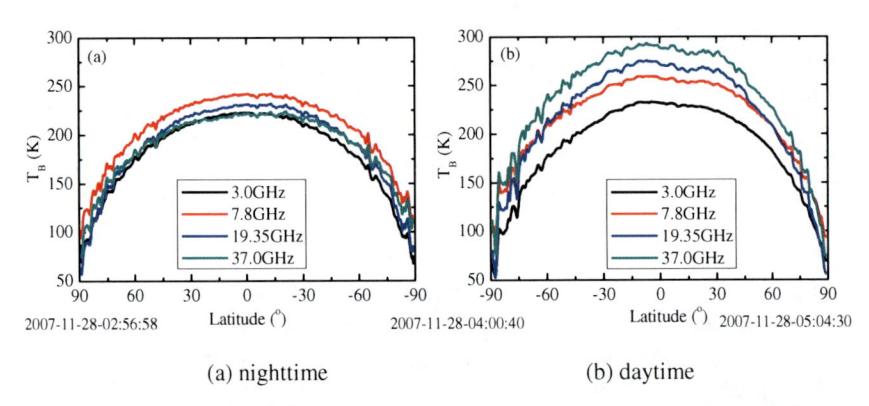

Fig. 4.3 CE-1 T_B data on 28 November 2007, where the radiometer flew from the north pole (UTC 2007-11-28 02:56:58) to the south pole (UTC 2007-11-28 04:00:40) at lunar nighttime and then from the lunar south pole to the north pole (UTC 2007-11-28 05:04:30) at lunar daytime (corresponds to the other side of the Moon)

The maria with high FeO+TiO$_2$ content have a large T_B value, while the highlands with low FeO+TiO$_2$ content have a small value of T_B, and the difference between maria and highlands is large.

4.4 Inversion of Regolith Layer Thickness

On the basis of the ground measurements at Apollo landing sites (Shkuratov and Bondarenko 2001, Oberbeck and Quaide 1968), the observed T_B at these locations are validated and calibrated by numerical three-layer modeling. Lunar surface temperature is first determined by high frequency channels T_B, such as 19 and 37 GHz. Using the empirical dependence of physical temperature upon the latitude verified by the measurements at Apollo landing sites, the global distribution of physical temperature over the global lunar surface is proposed, and some problems to invert the global physical temperature are avoided.

When the physical temperature and the dielectric constant of each layer in our model is known a priori, the 3 GHz channel observations, which possess the largest penetration depth, can thus be used to invert for the regolith layer thickness. Figure 4.4 shows our inversion results for the thickness of the regolith layer using the 3 GHz noontime data (Fa and Jin 2010a), which is the first map of global regolith layer thickness inverted from CE-1 data.

In Fig. 4.4, the low regolith thicknesses on the lunar nearside correspond to the mare, whereas the higher values correspond to the highlands. The systematic difference in regolith thickness between the mare and highlands is a natural consequence of the difference in age of the surfaces. Results show that the average regolith thickness of the maria is 4.5 m, and that of the highlands at intermediate latitude area (60°S–60°N) is 7.6 m.

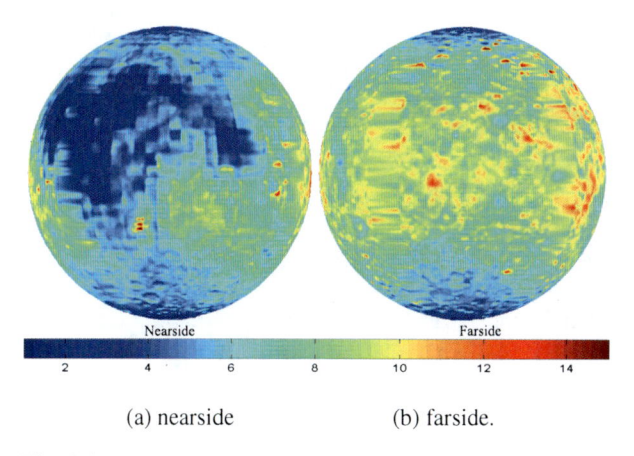

(a) nearside   (b) farside.

Fig. 4.4 Lunar regolith layer thickness (in meters) inverted from CE-1 daytime T_B data

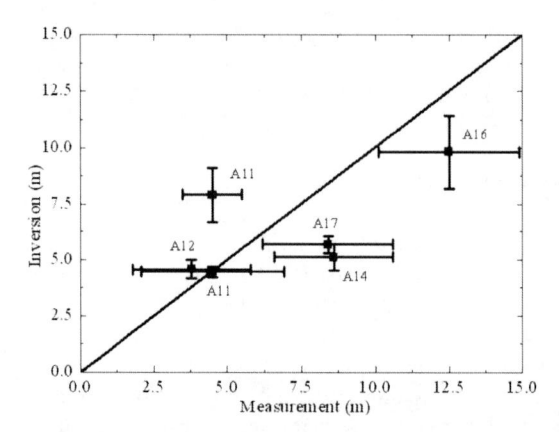

Fig. 4.5 Comparison of the regolith thickness measurements obtained from the inversion technique and from the Apollo landing sites. The horizontal error bars show the error range of the measurement, the vertical error bars are the inversion error due to the uncertainty of 0.5 K brightness temperature at 3 GHz

Figure 4.5 shows a comparison of the inverted regolith thickness with *in situ* point measurements at the Apollo landing sites. Because of the coarse resolution of the CE-1 radiometer, 30–50 km, the inversion results around the landing site are actually the averaged value over a larger area (a resolution cell), where any amount of variation is possible. However, a good comparison, (as shown in Fig. 4.5), between our inversion approach and the measurement at the Apollo 12 landing site, which was a fairly homogeneous surface with rather small variation in regolith thickness in the southeastern Oceanus Precellarum, demonstrates that our inversion technique works well.

The inverted regolith thickness (Fa and Jin 2010a) is also well compared with some results from the Earth-based radar observations (Shkuratov and Bondarenko 2001, Oberbeck and Quaide 1968, Keihm and Cutts 1981, Keihm and Langseth 1975). Our results also confirm that the regolith thickness is correlated with the age of the local regions.

4.5   Estimation of ^{3}He

The abundance of ^{3}He in the lunar regolith is mainly governed by two processes: implantation by the solar winds and outgassing of the lunar regolith. If not all grain surfaces are saturated with solar wind particles, then ^{3}He abundance should be dependent on the solar wind flux over the lunar surface. The ^{3}He abundance in the regolith is governed by the efficiency with which implanted ^{3}He is retained, i.e., the process of regolith outgassing, which is determined by the structure and chemical composition, e.g. lunar soil maturity and TiO_2 content, of the regolith itself (Grier et al. 1999, Lucey et al. 2000, Shkuratov et al. 1999, Heiken et al. 1991).

Because the solar wind is the only source of ^{3}He in the lunar regolith, its concentration should exhibit a global latitudinal variation as the lunar surface is tilted to the solar rays they receive a smaller flux of solar wind particles. As the Moon moves in the tail of the Earth's magnetotail and deflects the solar wind, the lunar nearside receives less solar wind exposure than the farside, which further causes a longitudinal variation in ^{3}He. In this study, the solar wind flux distribution model is used to calculate the normalized solar wind flux over the lunar surface.

The second factor affecting the ^{3}He abundance is the maturity of the lunar soil, i.e. OMAT (Optical MATurity), which is actually the amount of time that the lunar soil has been exposed to the environment. In this study, the optical maturity is used to quantify the maturation of the lunar surface.

The third factor is the TiO_2 content. Comparisons of lunar ilmenite, olivine, pyroxene and plagioclase show that ilmenites in the same grain-size ranged from the same soil may contain 10 to 100 times as much ^{3}He as. Since most lunar TiO_2 is found in ilmenites, the TiO_2 content serves as a good tracer of ilmenite abundance, and hence ^{3}He retentivity.

Considering these three factors that affect the abundance of ^{3}He in the lunar regolith and using the measurement of Apollo regolith samples, we presented a linear relation between ^{3}He abundance C_0 of the lunar surface (in ppb, part per billion), the normalized solar wind flux F, the TiO_2 content S_{Ti} and the OMAT as

$$C_0 = 0.56 \times [S_{Ti} \times F/OMAT] + 1.62 \tag{4.2}$$

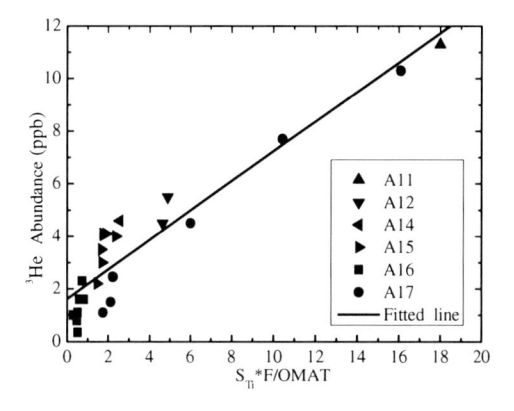

Fig. 4.6 Regression analysis for ^{3}He abundance, solar wind flux, TiO_2 content, and OMAT for the Apollo regolith sample

Figure 4.6 shows a regression function between ^{3}He abundance, normalized solar wind flux, TiO_2 content, and OMAT that was derived from the Apollo regolith samples.

As the ^{3}He is first implanted on micron depth of the lunar superficial media, it then spreads to depths of several meters as the surface material is broken down, stirred even melted by the continuous impacting of large and small meteoroids over the long geologic history of the Moon. This stochastic process of impact gardening makes ³He distributed in whole regolith media.

As the mixing of regolith grains becomes weaker in the deeper depth and yields the ³He profile with depth z, it is supposed as an exponential form as:

$$C(z) = C_0 e^{-\alpha z} \tag{4.3}$$

³He stored at a given location can be finally calculated as:

$$M = \int_0^d \rho(z) \cdot C(z) dz \tag{4.4}$$

where d is the regolith thickness, ρ is the bulk density of the regolith usually assumed as $\rho=1.8$ g/cm³, and $\alpha = 3$/m since most of ³He are concentrated on the top~3 m layer of the regolith media.

By adding all the ³He content in each pixel for the lunar nearside and farside, the global inventory of ³He for the whole lunar regolith can be obtained. To take account of the spatial resolution of the CE-1 radiometers in 30 to 50 km², it is advised to keep two effective numbers for the ³He content estimation. Doing this yields a total of 6.6×10^8 kg ³He in the entire lunar regolith, with 3.7×10^8 kg on lunar nearside and 2.9×10^8 kg on lunar farside (Fa and Jin 2010b)(Fig. 4.7).

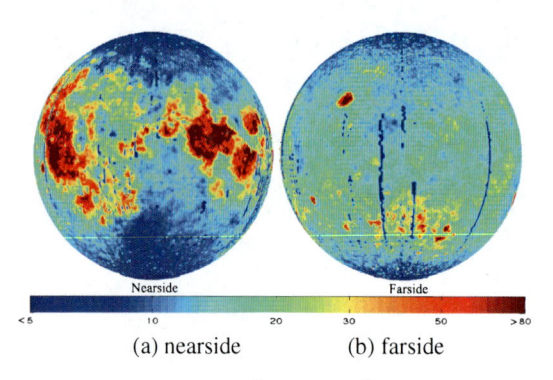

(a) nearside   (b) farside

Fig. 4.7 Total amount of ³He per 1m² of the regolith (ppb/m²) over lunar surface retrieved from CE-1 T_B data

4.6 Some Issues of the Modeling

Some issues about scattering from layering model, particulate lunar media and cratered rough surface, temperature profile in regolith media with dependence on depth, and influences on T_B from lunar surface topography etc. are also discussed (Jin and Fa 2010, Gong and Jin 2011a,b).

It is found that under the spatial resolution of 30 km×30 km of CE-1 radiometer observation, the lunar surface can be well modeled as a flat surface of layering media. It makes the predominance of the parameters for information retrieval, such as the regolith layer thickness and stratified structures.

Using the radiative transfer equation of stratified media with dense scatterers, the scattering coefficient of the regolith layer is found negligible, and the emission is mainly governed by the absorptive property, which is related with the local content of $FeO+TiO_2$ in regolith media.

It can be seen that high frequency channels, i.e. 19 and 37 GHz, are sensitive to the temperature profile of the top media, and low frequency channels, i.e. 7 GHz and especially 3 GHz, are sensitive to the whole regolith media, and the emissions are mainly contributed by the regolith and underlying rock media.

4.7 Diurnal Change of Cratered Surface and Hot-Cold Spots Retrieved from CE-1 Data

CE-1 multi-channel T_B data at the Sinus Iridum (i.e. Bay of Rainbow) area, which might be the CE-3 landing site, are especially collected at different lunar local time from the transformation between the principal coordinates and local coordinates at the observed site. Physical temperatures of both the dust layer and the regolith layer at different lunar local time are inverted. It shows the diurnal temperature changes (Gong and Jin 2011a) in the Sinus Iridum area. Figure 4.8 shows a result of physical temperature of top soil layer, T_1, of lunar surface inverted from CE-1 T_B data during daytime.

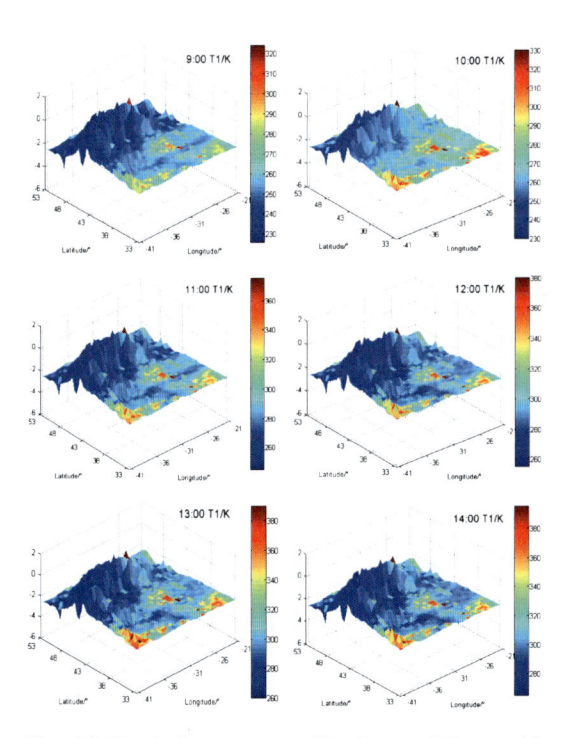

Fig. 4.8 Physical temperature T_1 of top soil layer of lunar surface inverted from CE-1 T_B data during daytime

Hundreds of abnormally "hot spots" on lunar surface had been revealed by infrared scanning image during totality using grounded radar in early 1960s (Shorthil and Saari 1965, Starukhina 2006), which turned out to be "cold spots" on the microwave brightness temperature distribution at night according to CE-1 radiometer observations.

This abnormal phenomenon of diurnal changes can be illustrated by CE-1 T_B observations. Two craters, the Tycho and Maurolycus Craters, typically representing the abnormal fresh crater with a rich rock abundance and a normal old one with rock free, respectively, located on similar latitudes are chosen for comparison of diurnal change. The CE-1 T_B diurnal change and inverted physical temperatures of regolith media, based on the three-layer radiative transfer model, can show the correlation between the abnormal "hot and cold spots" and the rock abundance of cratered surfaces. Figures 4.9(a,b) show a comparison of the diurnal change of physical temperature T_1 at the Tycho and Maurolycus Craters (Gong and Jin 2011b).

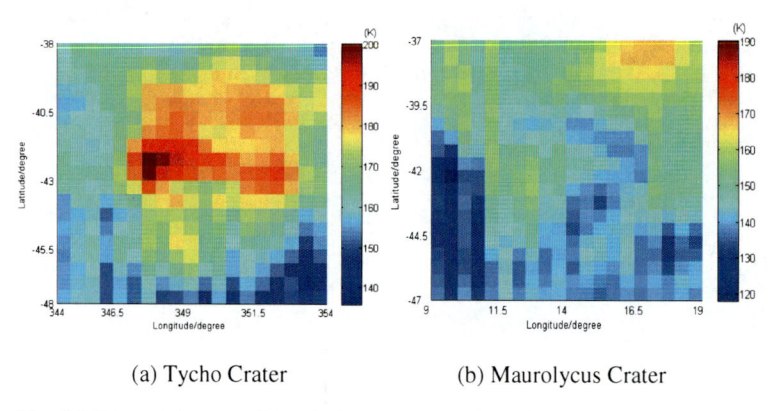

(a) Tycho Crater   (b) Maurolycus Crater

Fig. 4.9 Diurnal change of physical temperature T_1 at Tycho and Maurolycus Craters

4.8 SAR Imaging of Lunar Surface

Numerical topography of the lunar surface based on some statistics is constructed for SAR image simulation. The statistical properties of the cratered terrain, e.g. crater population, dimension and shape are controlled by given parameters to numerically generate an inhomogeneous undulated lunar surface. The position of an impact point is randomly chosen using Monte Carle (MC) method, and its dimension is also chosen randomly under the statistics of cumulative number of craters, as shown in Fig. 4.10(a).

To calculate electromagnetic scattering and simulate SAR imaging, the lunar surface is first divided into discrete meshes. The triangulated irregular network (TIN) is employed for inhomogeneous surface subdivision, as shown in Fig. 4.10(b). The Kirchhoff approach (KA, i.e. the tangent plane approximation) (Jin 1994, 2005) is applied to calculation of scattering from the lunar surface.

Using KA, the backscattering field at frequency f from a triangle mesh is written as:

$$\mathbf{E}_S^n(f,\mathbf{R}) = \frac{-jk}{4\pi R} E_0 \left(\bar{\mathbf{I}} - \hat{\mathbf{k}}_s \hat{\mathbf{k}}_s\right) \cdot \mathbf{F}\left(\hat{\mathbf{k}}_i,\hat{\mathbf{e}}_i,\hat{\mathbf{n}}\right) I_A \tag{4.5}$$

where $\mathbf{R}$ is the observation point, $\bar{\mathbf{I}}$ is the unit dyadic, k is the wave number, $\hat{\mathbf{k}}_i$ and $\hat{\mathbf{k}}_s$ are the incident and scattering wave vectors, respectively. I_A is the phase integral and can be analytically calculated. The function $\mathbf{F}\left(\hat{\mathbf{k}}_i,\hat{\mathbf{e}}_i,\hat{\mathbf{n}}\right)$ can be found in Jin (1994, 2005). Total echoes from all triangles illuminated by incident radar wave in the frequency-domain are written as

$$\mathbf{E}_S(f) = \sum_{n=1}^{N} \mathbf{E}_S^n(f) \tag{4.6}$$

where N is total triangles number illuminated by radar.

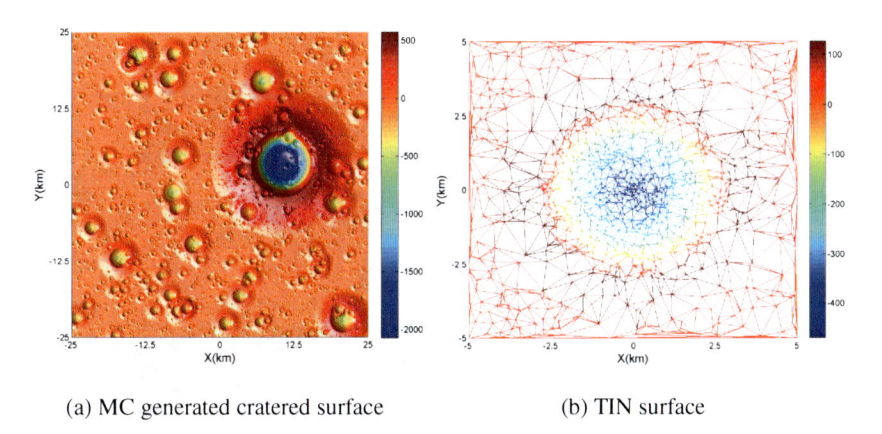

(a) MC generated cratered surface (b) TIN surface

Fig. 4.10 Simulated cratered lunar surface terrain of highland region (m), and an example of TIN for a bowl crater. The spatial extent of the area is 50 km×50 km. There are total 10000 craters in this area with the minimum crater radius 50m and no upper limit of crater radius is introduced

Taking the inverse Fourier transform of Eq. (4.6), the radar echoes in time domain is written as

$$\mathbf{E}_S(t) = \frac{1}{2\pi} \int_{-\infty}^{+\infty} \mathbf{E}_S(f) e^{-i2\pi ft} df \tag{4.7}$$

Back projection (BP) SAR imaging algorithm is directly applied to the received raw echo for azimuth compression and imaging. The principle of BP is to choose the sampling signal at the focusing position from two-dimensional compressed radar signal, then to compensate the Doppler phase, and finally to coherently add those echoes. Generally, SAR imaging algorithm consists of range and azimuth compressions.

In the simulation, the SAR raw signal is calculated in frequency domain, and then converted to time domain using inverse Fourier transform. Radar echo in space domain, i.e., radar echo variation with range is obtained.

Azimuth compression is modelled by taking a cross correlation of the reference signal and the received echo. For the point at (x,r) in the scene denoted by time variable (t_1,t), the compressed signal is

$$\mathbf{E}_{\text{image}}(t_1,t) = \int_{-\infty}^{+\infty} F_{\text{ref}}^*(t_1,t|\tau)\mathbf{E}_S(\tau|t_1,t)d\tau \qquad (4.8)$$

where $F_{\text{ref}}(t_1,t|\tau)$ is the reference signal.

At Apollo 15 landing site given in Fig. 4.11, an area of 12 km × 12 km (3.4~3.8°E, 25.9~26.3°N) is chosen for SAR image simulation.

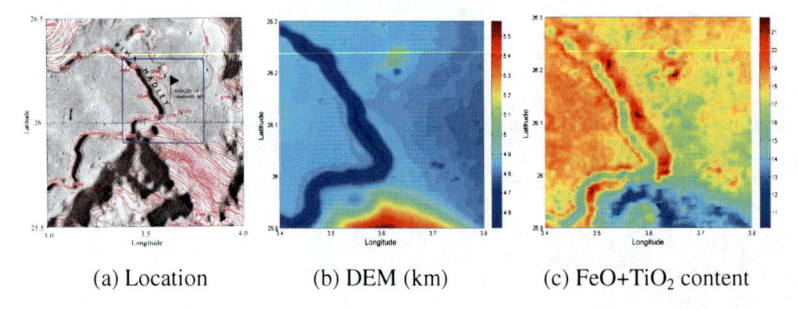

    (a) Location   (b) DEM (km)   (c) FeO+TiO$_2$ content

Fig. 4.11 Geographic map of Apollo 15 landing site area

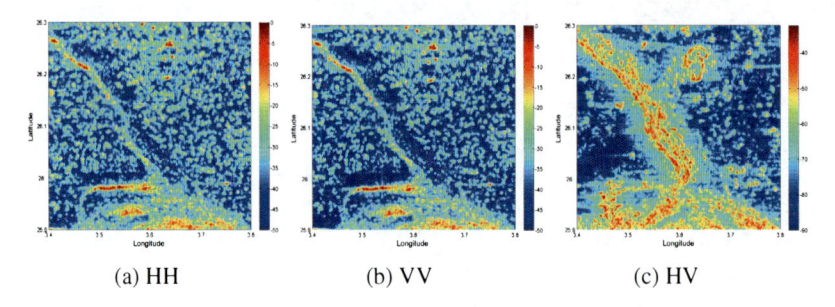

        (a) HH   (b) VV   (c) HV

Fig. 4.12 Simulated SAR Images at Apollo 15 landing site

Figure 4.12 shows the simulated SAR images with HH, VV and HV polarizations at Apollo 15 landing site.

Our SAR imaging simulation might provide a forward tool for data and image evaluation, feature identification and information extraction in lunar exploration, such as Mini-SAR in Chandrayaan-1 and LRO.

4.9 Radar Sounder Echoes from Lunar Surface/ Subsurface

Figure 4.13 shows the principle of radar sounder detection in a two layer model. As a radar sounder transmits a short pulse with the beam width θ upon the top surface, a first echo from the surface nadir point A to the antenna takes the time delay $2H/c$. A portion of the incident energy is transmitted through the surface to reach the lunar subsurface with attenuation. Generally, if the wavelength is large enough, a significant fraction of the energy can penetrate into the lunar crust. For example, reflection from the subsurface nadir point B can be generated and causes more radar echo upwards finally to the radar antenna. It is generally weaker and takes more time delay $2D(\varepsilon'_1)^{1/2}/c$ than the surface echo. If the dielectric constant ε_1 of the top layer 1 is known, the penetration depth can be inverted from the echoes time delays. The echoes intensities can be also analyzed for estimating the interface reflectivity and attenuation of the intervening layers (Starukhina 2000, 2001).

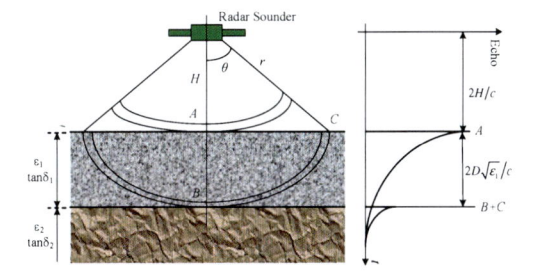

Fig. 4.13 Radar detection of lunar subsurface structures

Since the real lunar surface is randomly undulated, more echoes from off-nadir regions (e.g. the point C) with longer time delays can also reach the radar antenna, even they are generally weaker than the nadir surface echo. If the time delay of subsurface nadir echo happens to equal that of the surface off-nadir echo, the radar sounder receive these two echoes simultaneously. Thus, the stronger surface off-nadir echo might mask the weak subsurface nadir echo.

Suppose that the dipole antenna parallels to the lunar surface as radar flying along the satellite orbit. The radiation field of the dipole antenna with length L is written as

$$\mathbf{E}(r,\theta_i) = -i\frac{kIL}{4\pi}\sqrt{\frac{\mu_0}{\varepsilon_0}}\frac{e^{ikr}}{r}\sin\theta_i\hat{\theta}_i \qquad (4.9)$$

Using the ray tracing, the wave scattering and propagation can be described as shown in Fig. 4.14.

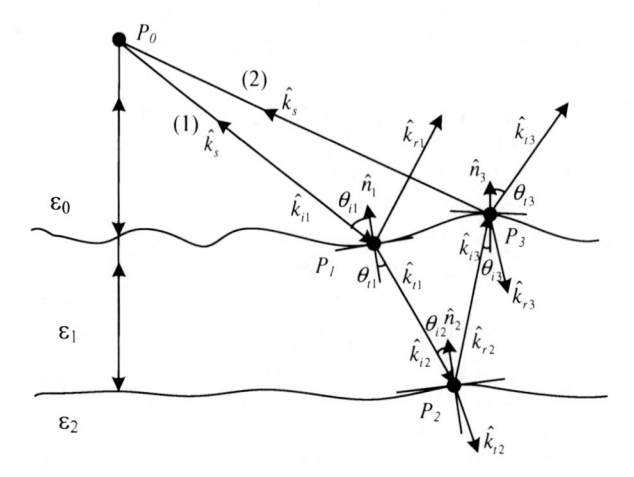

Fig. 4.14 Radar echoes from lunar surface and subsurface

As the lunar surface is numerically divided into discrete triangular meshes, the scattering from each mesh is calculated using Stratton-Chu integral formula in KA. Summing up all scattering fields from those meshes, whose ranges fall in the range bin r_n (the n-th range bin), the scattering field from lunar surface (*sur*) $E_{sur}(r_n)$ and subsurface (*sub*) $E_{sub}(r_n)$ received by radar sounder are obtained, respectively, as

$$\mathbf{E}_{sur}(\mathbf{r}) = -\int_S dS[(\hat{n}_1 \cdot \mathbf{E}_1)\nabla G_{10} + (\hat{n}_1 \times \mathbf{E}_1) \times \nabla G_{10} + (\hat{n}_1 \times \nabla \times \mathbf{E}_1)G_{10}] \quad (4.10a)$$

$$\mathbf{E}_{sub}(r_n) = ikE_0 \exp\left[ik_1(r_{12} + r_{23})\right] \sum_{range \subseteq r_n} \mathbf{F} \cdot G_{30}\Delta S \quad (4.10b)$$

where G_{ij} is the Green's function of layered media ij. Thus, the total field received by radar sounder is

$$\mathbf{E}(r_n) = \mathbf{E}_{sur}(r_n) + \mathbf{E}_{sub}(r_n) \quad (4.11)$$

To make enough range resolution and transmitted energy, linear frequency modulation (LFM) pulse, $T_r(t)$, is employed as the transmission signal. The electric field $\bar{E}(t)$ received by radar sounder is synthesized as

$$\mathbf{E}(t) = \sum_{n=1}^{N} \mathbf{E}(r_n)T_r(t, \tau_n) \quad (4.12)$$

where $\tau_n = 2r_n/c$ is the time delay.

Taking a Fourier transform of the mixed signal $\mathbf{E}_{mix}(t)$, it gives

$$\mathbf{E}(f) = \frac{1}{2\pi}\int_{-\infty}^{+\infty} \mathbf{E}_{mix}(t)e^{i2\pi t}dt \quad (4.13)$$

Suppose the altitude of radar orbit 50km, the center frequency f = 5 MHz, the length of dipole antenna L = 30 m, the antenna impedance 50 Ω, and the transmitted power of radar sounder 800W, it yields the applied electric current intensity on the dipole antenna I = 4 A/m^2.

The mare surface and subsurface are both assumed as the Gaussian random surfaces with the roughness $k\delta_1$ and $k\delta_2$, where δ_1 and δ_2 are the root mean square (RMS) of the surface heights, respectively. The length of the transmission pulse is $\tau = 2.67 \times 10^{-5}$ s.

Using Monte Carlo method and lunar surface statistics, the cratered lunar surface is realized as shown in Fig. 4.15(a), which is 40 km $\times$ 40 km with 1000 craters.

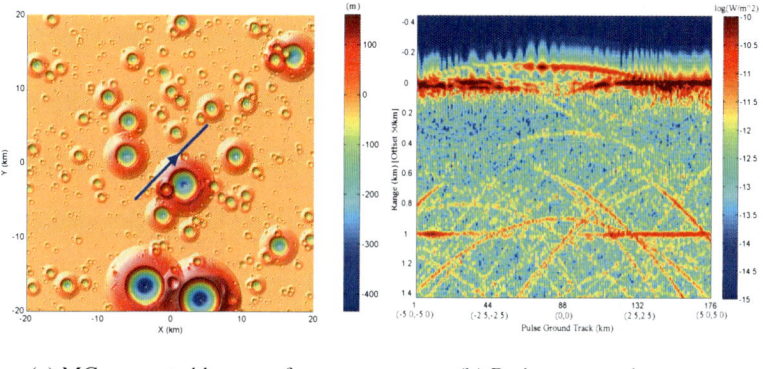

(a) MC-generated lunar surface   (b) Radar range echoes

Fig. 4.15 A cratered lunar surface with 1000 craters, and radar echo range image of a cratered lunar surface with B_w=8 MHz, $k\delta_2$=0

Figure 4.15(b) shows the radar range image as radar moves, where the vertical ordinate is the range from radar to target (crater), and the horizontal ordinate indicates the observation number with the interval 80 m. As the radar moves, the range from radar to each crater changes, and radar echo from a single large crater in radar image looks an arc. In this example, the subsurface is assumed flat. It makes the radar range echoes as a straight line, which can be identified in Fig. 4.15(b). The intensity, its fluctuation and arc shape of the radar range echoes are dependent upon the surface/subsurface angular scattering pattern and other physical properties. If there are strong echoes from the subsurface, it might mean water existence.

The SELENE LRS has made the initial observation for lunar Poisson area (30.4°S, 10.6°E) on 20 November, 2007 (http://www.jaxa.jp/ press/ 2008/01/ 20080110_kaguya_e.html). Figure 4.15(b) looks similar to the released LRS observation. It is believed that this simulation is feasible for future analysis of the observations.

Acknowledgments. This work was supported in part by the National Natural Science Foundation of China NSFC 41071219 and 60971091.

References

Fa, W., Jin, Y.Q.: Simulation of brightness temperature from lunar surface and inversion of regolith-layer thickness. J. Geophys Research 112, E05003 (2007a)

Fa, W., Jin, Y.Q.: Quantitative estimation of helium-3 spatial distribution in the lunar regolith layer. Icarus 190, 15–23 (2007b)

Fa, W., Jin, Y.Q.: Analysis of microwave brightness temperature of lunar surface and inversion of regolith layer thickness: primary results from Chang-E 1 multi-channel radiometer observation. Icarus 207, 605–615 (2010a)

Fa, W., Jin, Y.Q.: Global inventory of Helium-3 in lunar regolith estimated by multi-channel microwave radiometer on Chang'E-1. Chinese Science Bulletin 55, 4005–4009 (2010b)

Fa, W., Xu, F., Jin, Y.Q.: Image simulation of SAR remote sensing over inhomogeneously undulated lunar surface. Science in China (F) 52, 559–574 (2009)

Fa, W., Jin, Y.Q.: Simulation of radar sounder echo from lunar surface and subsurface structure. Science in China (D) 53, 1043–1055 (2010)

Gong, X., Jin, Y.Q.: Diurnal Physical Temperatureio at Sinus Iridum Area Retrieved from Observations of CE-1 Microwave Radiometer. Icarus (in press, 2011a)

Gong, X., Jin, Y.Q.: Diurnal Change of Thermal Emission from Lunar Craters with Relevance to Rock Abundance. Earth and Planetary Science Letters (in press, 2011b)

Grier, J.A., McEwen, A.S., Lucey, P.G., Milazzo, M., Strom, R.G.: The optical maturity of ejecta from large rayed craters: preliminary results and implications. In: Workshop on New Views of the Moon II, LPI Contribution no. 980, Lunar and Planetary Institute, Houston, pp. 19–20 (1999)

Heiken, G.H., Vaniman, D.T., French, B.M.: Lunar Source-Book: A User's Guide to the Moon. Cambridge University Press, London (1991)

Jiang, J.S., Jin, Y.Q. (eds.): Selected paper on Microwave Exploration of Lunar Surface in Chinese Chang'E-1 Project. Science Press, Beijing (2011)

Jin, Y.Q.: Electromagnetic Scattering Modelling for Quantitative Remote Sensing. World Scientific, Singapore (1994)

Jin, Y.Q.: Theory and Approach of Information Retrievals from Electro- magnetic Scattering and Remote Sensing. Springer, Netherlands (2005)

Jin, Y.Q., Fa, W.: An inversion approach for lunar regolith layer thickness using optical albedo data and microwave emission simulation. Acta Astronautica 65, 1409–1423 (2009)

Jin, Y.Q., Fa, W.: The modeling analysis for microwave emission from stratified media of non-uniform lunar cratered terrain surface in Chinese Chang'E-1 observation. IEEE Geoscience and Remote Sensing Letters (3), 530–534 (2010)

Jin, Y.Q., Xu, F., Fa, W.: Numerical simulation of polarimetric radar pulse echoes from lunar regolith layer with scatter inhomogeneity and rough interfaces. Radio Science 42, RS3007 (2007), doi:10.1029/RS2006003523:1-10

Keihm, S.J., Cutts, J.A.: Vertical-structure effects on planetary microwave brightness temperature measurements: applications to the lunar regolith. Icarus 48, 201–229 (1981)

Keihm, S.J., Langseth, M.G.: Lunar microwave brightness temperature observations reevaluated in the light of Apollo program findings. Icarus 24, 211–230 (1975)

Lucey, P.G., Blewett, D.T., Taylor, G.J., Hawke, B.R.: Imaging of lunar surface maturity. Journal of Geophys. Research 105(E8), 20377–20386 (2000)

Oberbeck, V.R., Quaide, W.L.: Genetic implication of lunar regolith thickness variations. Icarus 9, 446–465 (1968)

Shkuratov, Y.G., Bondarenko, N.V.: Regolith layer thickness mapping of the Moon by radar and optical data. Icarus 149, 329–338 (2001)

Shkuratov, Y.G., Starukhina, L.V., Kaydash, V.G., Bondarenko, N.B.: Distribution of 3He abundance over the lunar nearside. Solar System Research 33, 409–420 (1999)

Shorthill, R.W., Saari, J.M.: Nonuniform cooling of the eclipsed Moon: a listing of thirty prominent anomalies. Science 150, 210–212 (1965)

Starukhina, L.V.: High radar response of Mercury polar regions: water ice or cold silicates? Lunar and Planetary Science XXXI, Abstr.#1301 and sections 5.2–5.4 (2000)

Starukhina, L.V.: Water detection on atmosphereless celestial bodies: alternative explanations of the observations. J. Geophys. Res.-Planets 106, 14701–14710 (2001)

Starukhina, L.V.: Polar regions of the Moon as a potential repository of solar- wind-implanted gases. Advanced Space Research 37, 50–58 (2006)

5 Lunar Minerals and Their Resource Utilization with Particular Reference to Solar Power Satellites and Potential Roles for Humic Substances for Lunar Agriculture

Yuuki Yazawa[1], Akira Yamaguchi[2], and Hiroshi Takeda[1,3]

[1] Chiba Institute of Technology
[2] National Institute of Polar Research
[3] University of Tokyo, Japan

5.1 Introduction

Plans for manned bases on the Moon have been conducted by engineers and scientists for many years after the Apollo missions. Reports on lunar bases and space activities of the 21st century have been published in NASA Conference Publication 3166 (Mendell 1992). Taking human-kind to another world to stay is after all a major philosophical question. So, we should ask the rhetorical question: "What is required for humans to survive on the Moon"? The "basic needs" to be met in comparison with earth, include food, shelter, and an energy source. Water and air are so abundant that often they are overlooked. On the moon humans will need to have oxygen and water supplied in addition to food, shelter and energy to survive. Since some of these topics are explained in other chapters, after looking at the usefulness and global distribution of minerals, we will introduce utilization of plagioclase, the most abundant minerals of the highland. This topic is also related to our science themes of why we want to be on the moon. The formation of plagioclase highland crust from the lunar magma ocean is the most important questions to be answered. For such lunar exploration, we will not establish multi purpose permanent bases. Instead, we are assuming a small base for construction of a solar power station to solve energy crises on Earth, and small agricultural dome to glow green vegetables.

On the Moon, the raw materials required to meet basic needs are distinctly limited by terrestrial standards, being essentially tied to lunar minerals. That fact should be a guiding principle for this chapter, and provides the rationale for considering a wide range of lunar minerals as is done here. Which mineral can most effectively be tied to which basic needs, has been written in some chapters in NASA

Conference Publication 3166 (Taylor 1992). For oxygen, ilmenite; for water, regolith via solar wind hydrogen, again with an emphasis on ilmenite in the regolith (Haskin 1992); for shelter, anorthosite for Al for building materials; for energy, anorthosite again for Si solar cells. Food remains a primary need for which a single mineral source is not immediately obvious. We consider Ca-plagioclase (anorthosite) as the most important mineral for soil development at the highland base, but it is obvious that other minerals in the mare regions should not be ignored. It seems possible, for example, that areas of the moon that are much more K, P, and trace-element (such as rare earth element, REE) rich should be considered for lunar agriculture. This area is within the Procellarum KREEP Terrane (PKT) defined by Jolliff et al. (2000), but is different from Feldspathic Highlands Terrane (FHT), which we are going to emphasize in this chapter. At any rate, this chapter considers more thoroughly the advantages of plagioclase, and not each type of mineral/area, and the trade-offs among them.

In this chapter we review the information about materials present on the surface of the Moon and then discuss available resources and energy sources that can be utilized to support human activities on the Moon and to solve some energy problems on the Earth too. However, we will not cover the entire view of such activities, because many experts on the specific activities wrote them in other chapters. After overview of the lunar minerals in general, we will cover utilization of *anorthosites*, which are most abundant materials of the lunar surface, especially of the highland. There is a predominance of extreme calcic plagioclase (anorthite $CaAl_2Si_2O_8$) in comparison with chemically equivalent terrestrial plagioclase. Calcic plagioclase is the most important lunar mineral not only because of its quantity but also because of the information that contributes to our understanding of formation of the primordial lunar crust from the magma ocean. We will focus on the utilization of Si, Al and Ca in this mineral. Lunar resource utilization studies in Japan have developed a method to use lunar anorthosite as an important resource.

During the preparation of this chapter, we encountered the disaster at the Fukushima Daiichi nuclear power station induced by the Great East Japan Earthquake and Tsunami on March 11, 2011. A loud chorus in the U.S. and Europe is calling for a rethink of atomic energy. The U.S. Nuclear Regulatory Commission is taking a fresh look at safety in light of the problems at the Fukushima plant. If it may cost over several times more to construct a nuclear plant than a natural gas-fired thermal one, construction of a Solar Power Satellite station may not be just a far future dream. Late Prof. Makoto Nagatomo in 2000 formed the ISAS Solar Power Satellite (SPS) Working Group (Sasaki 2006) and developed a concept, which helped for their research to have firm bases on realistic demand and practical technologies to be presented. Susumu Sasaki (2006) of ISAS/JAXA proposed Tethered SPS consisting of a large panel with a capability of power generation/transmission and a bus system. We presented a plan to use silicon extracted from the lunar calcic plagioclase to produce SPS. We will show a more developed concept in this chapter.

Mineralogical and petrologic characteristics of the materials on the surface of the Moon, are less advanced than Mars in terms of our concept of "*Evolution of Planetary Materials*" (Takeda 2009). Because of the absence of air and water on the surface of the Moon, weathered materials as are common on the Earth and Mars are absent. We have to produce soils from lunar materials for terrestrial vegetables keeping microorganisms as small as possible in a small dome, because evolved microorganisms brought from the Earth will be mistaken as a lunar one, when future space biologists look for lunar microorganisms. Dissolution of main rock-forming silicates by inorganic and organic acids is the basic process for producing clay minerals on the earth or Mars. We investigated a process of soil production from Ca-rich plagioclase similar to lunar ones by geochemical weathering with fulvic acid in comparison with normal acid (hydrochloric, oxalic, lactic and garlic acids). Fulvic acid is a complex natural organic acid produced in humified soils. The Ca ion released from plagioclase can be used to fix carbon dioxide as calcite. This information is useful for reducing carbon dioxide from the atmosphere on the Earth.

5.2 Lunar Materials

Our knowledge on lunar materials largely depends on the samples returned by the Apollo missions (Frondel 1975). Since the numbers of mineralogists and petrologists who extensively worked on such samples, are getting smaller and smaller, real entire pictures of the lunar materials are difficult to envision now. Many data of the lunar materials covered by literatures and the review books (e.g., Lucey et al. 2006) are on the pristine crystalline rocks directly crystallized form the magma ocean and basaltic magma, but most abundant materials available on the surface of the Moon are mixed rocks so-called breccias. The texture of a good example of such sample is shown by the cut surface of highland regolith breccia, 60019, the largest rock sample (Fig. 5.1) recovered by the Apollo mission. More data of such rocks should be investigated before we talk about utilization of lunar materials in general.

Fig. 5.1 Photograph of a cut surface of a slice of one of the largest breccia samples of the Apollo samples (Photo from NASA)

In addition to the Apollo samples, we have two new sources about samples of the Moon. There are mainly two available sources to obtain information about rocks and soils on the Moon in addition to the Apollo samples. One is from direct analyses of meteorites originating from the Moon, ejected by heavy impacts that eventually reached the Earth. About 144 lunar meteorites recovered at present (Meteoritical Society database) from Antarctica and hot deserts on the Earth. The other is from data collected by spacecraft exploration of the Moon. New spectroscopic information of minerals was collected by spacecraft exploration to the Moon, for example by Kaguya (e.g., Ohtake et al. 2009). Both sources have provided us with valuable insight into the geochemistry and mineralogy of the lunar surface materials.

Half of the lunar meteorites were likely derived from the farside. Two lunar meteorite groups of possible farside origin have been reported (Korotev et al. 2006; Takeda et al. 2006; Yamaguchi et al. 2010). Although low Th (Kobayashi et al. 2011) and FeO concentrations are common features, the types of mineral fragments and clasts are different. The Yamato (Y-) 86032 group (Yamaguchi et al. 2010) has been reported to have some similarity to the Apollo samples, but the Dhofar 489 group has some features unique to the farside. The mineralogical data of the Apollo samples will be summarized before we discuss the characteristics of the farside obtained by the new lunar meteorites and the Kaguya mission. Then, we will try to find differences in mineralogy and clast types between the farside and the nearside.

Lunar materials can be classified on the basis of texture and composition into four distinct groups: (1) anorthosites, (2) KREEP-rich rocks, (3) basaltic volcanic rocks, (4) the lunar regolith, and (5) polymict clastic breccias and related materials. Details of the classification are given in review textbooks. We summarize each group in the following sections.

5.2.1 Anorthositic Rocks of the Highland Regions

Pristine highland rocks are primordial igneous rocks, uncontaminated by impact mixing. The lunar highland crust composed of such rocks is proposed to have been formed by the crystallization and flotation of plagioclase from a global magma ocean of the Moon. The compositions of the lunar highland rocks are therefore critical to the investigation of a magma ocean and the subsequent evolution of the Moon. The major rocks can be subdivided into two chemical groups based on their molar Na/(Na+Ca+K) content versus the molar Mg/(Mg+Fe) content of their bulk rock compositions (e.g. Warner et al. 1976; Papike et al. 1998). One trend represents anorthosites or ferroan anorthosites (FANs), except for minor varieties. Such rocks should yield ages close to 4.5 Ga, but the Ar-Ar ages of most of such FAN (Fig. 5.2) are reset in the range 4.4–4.2 Ga; the magnesian-suite rocks (high Mg/Fe) are, as a group, somewhat younger (4.43–4.17 Ga) (Taylor et al. 1992) and contain dunites, troctolites, norites, and gabbronorites. A less abundant alkali suite contains similar rock types, but enriched in alkali and other trace elements, and extending to somewhat younger ages. This rather simple picture has been complicated by more recent work (e.g., see Shearer and Newsom 2000). The magnesian suite may mark the transition between magmatism associated with the magma ocean and serial magmatism. This transition period may have occurred as early as 30 Ma (Shearer and Newsom 2000) or as late as 200 Ma after the formation of the magma ocean (Solomon and Longhi 1977).

Fig. 5.2 Photomicrograph of a part of a thin section (PTS) of the Apollo16 pristine FAN, 60025. Note the twin texture in a plagioclase grain, but we can recognize the monomict breccia texture in a part of the PTS, in spite of its proposed pristine nature

On the basis of mineralogy obtained from the new lunar meteorites and the data from the Kaguya mission, we examined how to utilize these materials for sustaining human activities on the Moon. We developed methods to produce silicon, aluminum, calcium and oxygen by dissolution of plagioclase. Silica powders can be used to produce silicon solar cells. Al metal can be obtained by the electric fusion of aluminum oxide with cryolite and calcium fluoride. Carbon dioxide generated in a dome of the base can be fixed by Ca ion as calcite.

The Multiband Imager (MI) on the Selenological and Engineering Explorer (SELENE or Kaguya) has a high spatial resolution of optimized spectral coverage, which enables us to acquire a clear view of the composition of the lunar crust. Ohtake et al. (2009) reported the global distribution of rocks of extremely high plagioclase abundance (nearly 100 vol.%) using an unambiguous plagioclase absorption band found by the MI. The plagioclase abundance in the lunar upper crust is possibly much higher than previously estimated (82 to 92 vol.%), suggesting the high modal abundance of plagioclase of the upper crust calls for mechanisms to generate a stratum consisting of pure plagioclase.

5.2.2 Lunar Meteorites

The astronauts of the Apollo missions were limited to the near side, equatorial region of the Moon, and they sampled a relatively small area of the Moon. All of the landed sites outline only 4 to 5 % of the lunar surface. After the Apollo era, it was realized that impact events on the lunar surface eject material into space that falls to Earth as meteorites. More than 140 lunar meteorites have now been recovered from Antarctica and hot deserts. These are invaluable samples, because they represent a much broader region of the lunar surface, including the farside of the Moon. Good summaries of lunar meteorites have been assembled by Prof. Randy Korotev at Washington University (in St. Louis) (Korotev et al. 2003) and by Dr. Kevin Righter at NASA's Johnson Space Center.

5.2.2.1 Lunar Meteorites from the Farside

The Dhofar 489 and Y-86032 group, with very low Th and FeO may have originated from the farside, where the remote sensing data suggest their concentrations
are very low (Kobayashi et al. 2010; Lawrence et al. 1998). Pyroxene fragments
chemically similar to those of so-called Mg-suite rocks are present in the matrices
of the Y-86032 group (Yamaguchi et al. 2010), but more olivine fragments are
found in the matrices of the Dhofar 489 group (Takeda et al. 2006, 2010) (e.g.,
Dhofar 307, 309, 489, 908 etc.). Nearly pure anorthositic clasts (Ohtake et al.
2009) are major clasts in both groups, but more magnesian anorthosites (Fig. 5.3)
are dominant in the Dhofar 489 group (Fig. 5.4).

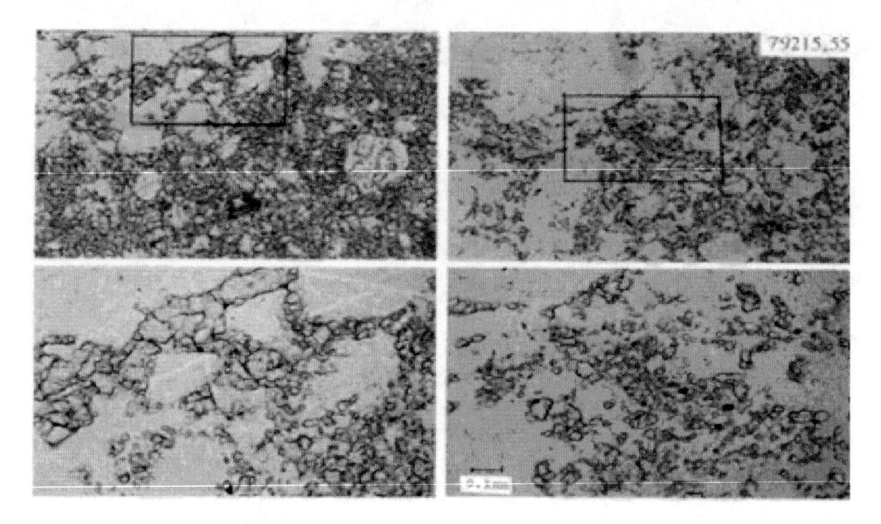

Fig. 5.3 Photomicrograph of the textures of granulitic breccias, 79215. A part of the PTS
with an apparent breccia texture (square) is enlarged below to show a crystalline granoblastic texture.

Y-86032 (Nyquist et al. 2006; Yamaguchi et al. 2010) contains a light-gray
feldspathic (LG) breccia, which mainly consists of fragments of anorthosites
(An93 anorthosite) more sodic than nearside FANs. The LG clast also contains an
augite-plagioclase clast. The mineralogical and geochemical differences between
the two groups have been attributed to the locations of the ejection sites. Dhofar
489 may have originated from one of the farside basins in the earliest crust of the
Moon with the lowest Th concentration (Kobayashi et al. 2010). Likely candidates
are old basins with heavily cratered floors. The lunar magma ocean (LMO) model
initially proposed by Warren (1985) is based on the mineralogy of the FANs. The
lunar crustal asymmetries we found may be explained by "convective processes"
acting early in the Moon's history, during the LMO phase, and after synchronous
rotation was established, as was proposed by Loper and Werner (2002).

Fig. 5.4 Top: Photomicrograph of a magnesian anorthosite clast in Dhofar 489 lunar meteorite, with a small olivine grain in the left side. Bottom: The cross polarized view of the top view, showing twinning texture of a plagioclase grain (Takeda et al. 2006)

5.2.2.2 Differentiation Trends of the Anorthositic Clasts

Anorthositic clasts are major clasts both in the Dhofar 489 and Y-86032 groups, but magnesian anorthosites (Mg An) are dominant in the Dhofar 489 group. The distributions of An values of the Dhofar 489 and Y-86032 groups are within a limited range, except for one light gray ("LG") sodic clast of Y-86032. Histograms of the olivine and pyroxene compositions are more complex. The Mg# vs. An diagram revealed two distinct trends (Warren 1993). One trend is called the Mg-suite trend, and was originally thought to be compound of the crustal rocks, but now many investigators proposed that these rocks were deep seated rocks intruded into the lower crust of the PKT region. The other trend is called the "FAN" trend of the anorthositic crust. It is difficult to identify such pristine crustal rocks by the present remote sensing data, because information on their textures is not sensitive to the spectroscopic data.

The differentiation trends of the Dhofar 489 clasts extend more towards the magnesian side of the FAN trend (Fig. 5.5). There are many previously reported clasts, which are plotted between the Mg suite and FAN in Fig. 5.5 (dotted line), but they are granulites and not a crystalline material as was found in a Mg An clast in Dhofar 489. The spinel troctolite in Dhofar 489 is more magnesian than the Mg An trend. This trend is in line with a proposal that spinel troctolites are present ca. 15 km beneath the magnesian anorthosites. We suggested that the Mg An crust constitute major parts of the northern farside highland. Differentiation tends of Y-86032 are more complex and the light gray (LG) anorthosite clast distribute in the range not common in the Apollo samples. Pyroxene fragments compositionally similar to those of so-called Mg-suite rocks are present in the matrices of the Y-86032 group, but more olivine fragments are found in the matrices of the Dhofar 489 group. The differences have been attributed to the locations of the ejection sites. The presence of the light gray (LG) clast with An93 and the Aug-Plag clast

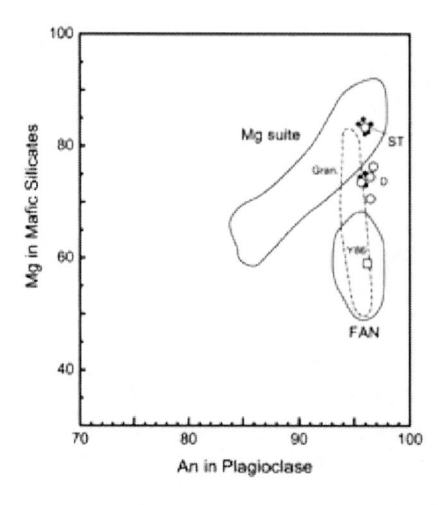

Fig. 5.5 Differentiation trends of magnesian anorthosites, spinel trocktolite (ST), and granulitic breccias in the Dhofar 489 and Y-86032 lunar meteorites, plotted in the An vs. Mg number diagram. The dotted line (Gran.) is for lunar granulitic breccias (Takeda et al. 2006)

suggests that the farside feldspathic highland has some variations of the anorthositic lithologies in addition to PAN (Yamaguchi et al. 2010).

There are some common clasts in Dhofar 489, 309 and 307. The Dhofar 307 polished thin section (PTS) contains magnesian granulitic (GR) clasts similar to the spinel troctolite (ST) clast in Dhofar 489 in mineral chemistry. The grain sizes and mineral chemistry of the GR clasts are the same as those of impact melt (IM) clasts of the Dhofar 309 clasts. Moderately rapid growth features of the IM clasts indicate that these clasts formed in an impact melt pool or a thick sheet. The range of the modal abundances of the minerals and their compositions of three clast types (ST, IM, and GR) obtained from the mineral distribution maps are practically the same and suggest that these breccias were produced at a floor of a large basin, and in subsequent cratering events (Takeda et al. 2010). The plagioclase-rich terrane covers most of the two lunar highland terranes defined by Jolliff et al. (2000), but is different from the PKT Terrane.

5.2.3 Redistributed KREEP-Rich Rocks of the PKT

KREEP basalts, highly enriched in incompatible elements, differ fundamentally from mare basalts and are thought to have formed as the final dregs of the magma-ocean (Warren 1985). KREEP rocks have been found as small basaltic fragments in Apollo 14, 15 and 17 breccia samples. There are a few lunar meteorites that are enriched in KREEP components. Volumetrically, KREEP rocks are far less important than ferroan anorthosites and Mg-rich rocks. The KREEP rocks are named for their characteristically high contents of incompatible elements, especially K, rare earth elements (REE), and P. In practice, because of their generally fine-grained crystalline textures, KREEP rocks are also considered to be basalts, crystallized on or just below the lunar surface. KREEP rocks seem to be widely distributed in the PKT. KREEP-rich rocks may make fertile agricultural soils.

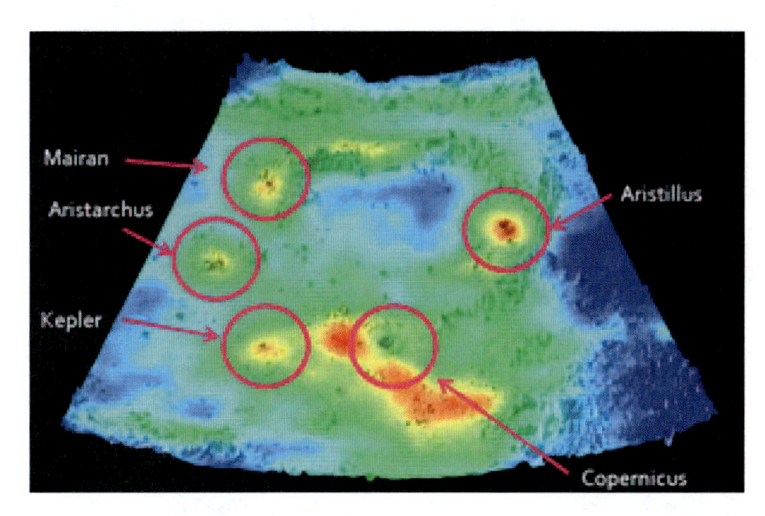

Fig. 5.6 The elemental distribution of Th measured by the gamma-ray spectrometers (GRS) onboard the Kaguya orbiters (Kobayashi et al. 2011). The concentration is high for the orange-yellow regions. K, Th and U are considerably concentrated in the Procellarum, Imbrium and Fra Mauro region (Procellarum KREEP Terrain, PKT)

The elemental composition of the lunar surface has been measured by several gamma-ray spectrometers (GRS) onboard lunar orbiters (Lawrence et al. 1998; Kobayashi et al. 2010). One of the most important discoveries accomplished by GRSs is the dichotomic distribution of K, Th and U on the global lunar surface. GRSs have found that K, Th and U are considerably concentrated in Procellarum, Imbrium and Fra Mauro region (Procellarum KREEP Terrain, PKT as shown in Fig. 5.6. The most important question, however, is whether the bulk of the underlying crust in PKT also contains much of K and Th or not. The total abundance of K, Th and U in bulk PKT is a key parameter to understand several important lunar issues: the bulk composition of refractory elements, the thermal history of the Moon and the nearside unique volcanic activity (Kobayashi et al. 2010), the origin of the PKT itself and the above issues are related to Imbrium basin. Therefore, the concentration of K, Th and U in bulk PKT is closely related to the origin and evolution of the Moon. KREEP-rich rocks may also make fertile agricultural soils.

5.2.4 Basaltic Volcanic Rocks of the Mare Regions

Basaltic rocks including lava flows and pyroclastic deposits cover the surface of the mare regions of the Moon. Basaltic lava flows are commonly found in the terrestrial planets, such as Mars, Venus and Mercury. The area of the mare regions of the Moon is less than the basaltic regions of other planets. Mare basalts are variably enriched in FeO and TiO_2, depleted in Al_2O_3, and have higher CaO/ Al_2O_3 ratios than highland rocks (Taylor et al. 1992). Utilization of high Ti basalts for production of oxygen has been described in detail by previous workers (Gibson and Knudsen 1988). Mineralogically, mare basalts are enriched in olivine and/or pyroxene, especially clinopyroxene (Fig. 5.7), and depleted in plagioclase compared

Fig. 5.7 Photograph of clinopyroxene crystals (pigeonite with rims of augite) separated from vesicles of the Apollo 15 lunar basalt, 12052 (Takeda 1972)

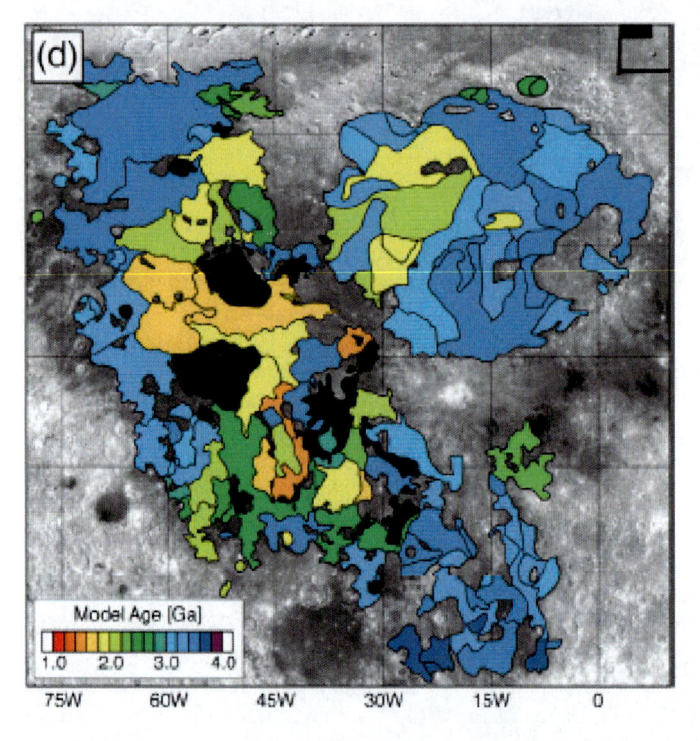

Fig. 5.8 Distribution of the youngest basalt flows around the Aristarchus plateau. Diagram after Morota et al. (2011)

to highland rocks. It is widely accepted that mare basalts originated from remelting of mantle cumulates produced in the early differentiation of the Moon. Several classification schemes exist for lunar mare basalts based on their petrography, mineralogy, and chemistry (e.g. Papike et al. 1998).

Unraveling the timing and duration of mare volcanism on the Moon is essential for understanding its thermal evolution. The end of mare volcanism is poorly constrained, because mare basalts are incompletely sampled. Employing SELENE (Kaguya) high-resolution images, Morota et al. (2011) performed new crater size-frequency measurements for 49 young mare units (less than ~3.0 Ga) in the PKT, where the latest magma eruption of the Moon occurred. They suggested that volcanic activity in this region ceased ~1.0 Ga after the magma eruption had globally ceased at 2.5–3.0 Ga. Volcanic activity may have peaked 1.8–2.2 Ga ago (Morota et al. 2011). The youngest basalts occur around the Aristarchus plateau and the Kepler crater (Fig. 5.8), which are located in the central region of the PKT. In contrast with previous basalt dating in this region, the recent results indicate a higher correlation between ages and spectral types of mare basalts. The young mare units in the PKT tend to have spectral types corresponding to high titanium contents, while low titanium basalts occur mainly in the early stage (Morota et al. 2011). If one wants proximity to young basalts in the PKT, one could pick a site in the south-central KREEP rich areas near Fra Mauro. An Aristillus site would be better if one wanted proximity to the backside highlands.

5.2.5 Lunar Regolith

"Lunar regolith" usually refers to the fine-grained fraction (mostly <1 cm) of unconsolidated surface material. The term "soil" has often been misused in the literature to describe this material. Regolith materials are the ones on which human activities will take place. Lunar regolith ranges in composition from basaltic to anorthositic (with a small meteoritic component usually <2%), has an average grain size of ~60–80 μm, and consists of mainly five particle types: mineral fragments, pristine crystalline rock fragments, breccia fragments, glasses, and agglutinates (McKay et al. 1991). Agglutinates are aggregates of smaller particles welded together by glasses produced in micrometeorite impacts, which also produces an auto-reduction of FeO to metallic Fe, increasing the abundance in nanophase iron in the more "mature" regolith. Lunar regolith exhibits variations in maturity, a quality which is roughly proportional to the time the soil is exposed to micrometeorite bombardment and the agglutinate abundances. With the exception of a few small-scale exposures of bedrock, lunar regolith covers more or less the entire lunar surface. Thus, understanding the optical properties, for example the effects of grain size and maturity of the lunar regolith, is extremely important for optical remote-sensing techniques such as the mapping of TiO_2 and FeO abundances. Solar thermal power system, for oxygen production from lunar regolith were given by Nakamura et al. (2008).

5.2.6 Polymict Clastic Breccias and Related Materials

These materials include impact melt rocks, and thermally metamorphosed granulitic breccias. Lunar impact breccias are produced by single or multiple impacts and are a mixture of materials derived from different locations and different kinds of bedrock. They contain various proportions of clastic rock fragments and impact

melts, and show a wide variety of textures, grain sizes, and chemical compositions. A widely adopted classification of breccias was presented by Stöffler et al.
(1980), who discriminated among fragmental, glassy melt, crystalline melt, clast-
poor impact melt, granulitic, dimict, and regolith breccias. A detailed discussion
of these breccias is provided in Taylor et al. (1992) and Papike et al. (1998). There
is not something about granulitic breccias, for example, that affects their usefulness as resources. It should be mentioned that granulites are the most likely highland materials to be encountered at a highland site. Because these breccias contain
more mafic silicates than anorthosites, such materials may find them useful for
some agricultural soils.

The textures of granulitic breccias and granulites are extremely varied, ranging
from granoblastic to poikiloblastic, from homogeneous to heterogeneous, and
from coarse- to fine-grained (Fig. 5.3). In this photograph, we can recognize the
original breccia textures. Granulitic breccias tend to have fine-grained matrices,
whereas poikiloblastic breccias tend to be coarse-grained throughout. Generally,
the clasts are as recrystallized as the matrices. The distinction between poikiloblastic textures (metamorphic, recrystallized) and poikilitic textures (primary igneous) on the basis of texture alone is all but impossible. The granoblastic rocks are
dominated by anhedral and equant plagioclase (70-80%); smaller pyroxenes (15-
25%) and small anhedral olivines (absent to a few percent) occur within and between the larger plagioclase crystals. The poikiloblastic rocks tend to have blocky
plagioclase grains, partially enclosed in larger pyroxene crystals (*oikocrysts*). Ilmenite is rare and generally subhedral. Olivine inclusions commonly form strings
or "necklaces" inside large plagioclase crystals; the strings possibly indicate the
location of premetamorphic grain boundaries. In most samples, pyroxene and olivine grains are of nearly constant composition and are unzoned.

Most of the granulitic breccias and granulites, including the poikiloblastic varieties, have major element chemical compositions similar to anorthositic norites.
They have low abundances of the incompatible elements, and are contaminated
with meteoroid siderophile elements. There are small but important differences in
the major elements, as well as larger variations in the trace elements. Lindstrom
and Lindstrom (1986) divided the range of compositions into two groups: ferroan
[Mg/(Mg + Fe) <0.7 molar] and magnesian [Mg/(Mg + Fe) >0.7]. Although the
major-element data do not show a clear break between the two groups, the element
Sc appears to define the groups more precisely. Ferroan granulites have about
twice as much Sc as do magnesian granulites. Ferroan granulites also tend to have
lower abundances of the incompatible elements. There is no correlation of microscopic texture with composition.

5.3 Lunar Minerals

About 60 valid minerals are known in the Apollo lunar samples (Frondel 1975).
This number is quite small in comparison with more than 4300 known terrestrial
minerals. Olivine, pyroxenes, plagioclase, ilmenite and spinels are five common
rock-forming minerals known in the lunar materials. Some topics related only to

those minerals are given below with reference to their utilization. Taylor (1992) introduced rocks, minerals, and soil of the Moon for resources for a lunar base.

5.3.1 Plagioclase

Plagioclase is the most abundant mineral in the lunar crust, because the lunar highland crust was formed by accumulation of plagioclase. The chemical composition is mostly calcic and their range is rather small. This calcic composition with very low incompatible trace elements makes it one of the most useful minerals for resources. The feldspar minerals are known to show very complex micro-textures, but crystallographic study of lunar plagioclase has not been performed too well. In addition, shock and annealing textures introduced by the impact processes obliterated interpretation of the original textures.

Within this three-dimensional framework of tetrahedra containing Si and Al in plagioclase, much larger sites with 8 to 12 coordination occur that accommodate large cations (Ca, Na, K, Fe, Mg, Ba). Most lunar feldspars belong to the plagioclase series, which consists of solid solutions between albite ($NaAlSi_3O_8$) and anorthite ($CaAl_2Si_2O_8$). The maximum chemical variation involves solid solution between albite (Ab) and anorthite (An), a variability that can also be described as the coupled substitution between NaSi and CaAl, in which the CaAl component represents anorthite. Lunar plagioclases are depleted in Na (the Ab component) relative to terrestrial plagioclases indicating the alkali depleted nature of the Moon. The Ca abundance in the plagioclase, and therefore mol% anorthite (An) correlate positively with the Ca/Na ratio in the host magma.

5.3.2 Pyroxenes

Pyroxenes should be classified by chemical compositions and crystal structures. Here, we just mention four pyroxene polymorphs: (a) Protoenstatite; (b) orthopyroxene (Opx), orthorhombic pyroxene with Space group, *Pbca*; (c) Pigeonite (Pig), Low-Ca clinopyroxene (LCP) with monoclinic structure and $P2_1/c$ space group; (d) Inverted Pigeonite (Opx+Aug), Opx inverted from pigeonite by slow cooling and exsolved augite; (e) Augite (Aug), high-Ca clinopyroxene, (HCP) with space group *C2/c*. Information of the space groups is related to the symmetry of the oxygen coordination polyhedra around the M2 cations, and is important to interpret their reflectance spectra. Although pyroxene may not be suited for resource utilization, strong absorption bands of Fe in the M1 and M2 sites of the structure are useful for remote sensing studies.

Although low-Ca pyroxene fragments are common in many breccias of the Apollo samples and the lunar meteorites, some of them in the Mg-suite rocks are orthopyroxene (Opx) inverted from pigeonite (Pig). A meteorite, which is entirely composed of primary Opx up to a few cm in size, in the differentiated meteorite group, is possibly lower crust or upper mantle of a differentiated asteroid (Takeda 1997). Fragments of such Opx are also common in the breccias developed on the same body. Such Opxs are not known in the lunar samples. Polymorphism of pyroxene with respect to differentiation of magma is the most useful information for the evolution of lunar magma, and magma ocean. Hiesinger and Head (2006) gave conceptual cross section of lunar crust and upper mantle, in which norite layer is illustrated between anorthosite and olivine-rich mantle. However, we do not have fragments of such Opx in the Apollo samples.

The first Opx found in the Apollo studies is in the low-K Fra Mauro rock, impact melt rock, 14310 (Takeda and Ridley 1972). Opx grown from an impact melt, is also found in the poikilitic clast in lunar highland breccia, 60019 (Takeda et al. 1988). Some Opx fragments common in the Apollo samples are from the Mg-suite rocks, such as norites. Opx from the farside highland is found in devitrified impact melt matrix in Dhofar 489.

Opx detected by spectroscopists is practically low-Ca pyroxene including true Opx and Pig. These pyroxenes should be called LCP. The oxygen coordination polyhedra around the mafic cations are similar, but the detailed configurations of oxygens are distinct. Cpx by spectroscopists is practically high-Ca pyroxenes including true Augite (Aug). These pyroxenes should be called high-Ca clinopyroxene (HCP).

5.3.3 Oxide Minerals

The oxide minerals are of great significance because they retain signatures of critical conditions of formation. Several oxide minerals are important constituents of lunar samples (Lucey et al. 2006): ilmenite, $FeTiO_3$; spinels with extensive chemical variations: $(Fe,Mg)(Cr,Al,Fe,Ti)_2O_4$; and armalcolite, $(Fe,Mg)Ti_2O_5$. Because their oxygen is more weakly bonded than that in silicate minerals, oxide minerals are obvious and important potential feedstocks for future production of lunar oxygen and metals. Lunar oxygen production from ilmenite is one of the most well known utilization of lunar minerals (Gibson and Knudsen 1988). Chromite is an ore mineral in some terrestrial ultramafic bodies. One possible conclusion on importance among different minerals, especially between ilmenite and plagioclase might be that a lunar colony would want to have more than one base of operations. Oxygen and food production for life support might favor ilmenite-rich and K-rich sites, for example, whereas Si and Al production for energy and construction might favor a plagioclase-rich highland site.

With the ideal formula $FeTiO_3$, ilmenite is the most abundant oxide mineral in lunar rocks. The amount of ilmenite in a rock is a function of the bulk Ti composition of the magma from which the rock crystallized. The volume percentages of ilmenite in mare basalts vary widely across the Moon, as indicated by the range of TiO_2 contents in samples from different lunar missions and from remotely sensed data.

The ilmenite crystal structure is hexagonal and consists of the closest hexagonal packing of oxygen as in the corundum structure, and of alternating layers of Ti- and Fe-containing octahedra. Most lunar ilmenite contains some Mg substituting for Fe, which arises from the solid solution that exists between ilmenite ($FeTiO_3$) and $MgTiO_3$, the mineral geikielite. Other elements are present only in minor to trace amounts (i.e., <1 wt%); these include Cr, Mn, Al, V, and Zr. Ilmenite commonly occurs in mare basalts as bladed crystals up to a few millimeters long. In

Apollo 17 rocks, ilmenite is frequently associated with armalcolite and occurs as mantles on armalcolite crystals. The composition of lunar ilmenite plots along the $FeTiO_3$-$MgTiO_3$ join; variation from $FeTiO_3$ is often expressed in wt% of MgO. In general, the ilmenite with the highest Mg contents tends to come from relatively high-Mg rocks; ilmenite composition correlates with the bulk composition of the rock and therefore reflects magmatic chemistry rather than pressure.

Spinel is the name for a group of oxide minerals, all with cubic crystal symmetry, that have extensive solid solution within the group. The basic spinel structure is a cubic array of oxygen atoms. Oxygen atoms stack along the diagonal axis of the cube as the cubic closest packing sequence, A-B-C. Within the array of oxygen, the tetrahedral A-sites are occupied by one-third of the cations, and the octahedral B-sites are occupied by the remaining two-thirds of the cations. In a normal spinel structure, the divalent cation, such as Fe^{2+}, Mg^{2+}, occupies only the tetrahedral sites, and the two different sites each contain only one type of cation (e.g., $FeCr_2O_4$). If the divalent cation occurs in one-half of the B-sites, the mineral is referred to as an inverse spinel [e.g., $Fe(Fe,Ti)_2O_4$].

Spinels are the second most abundant opaque mineral on the Moon, second only to ilmenite, and they can make up as much as 10 vol% of certain basalts, most notably those from the Apollo 12 and 15 sites. The general structural formula for these minerals is $IVAVIB_2O_4$, where IV and VI refer to cations with tetrahedral and octahedral coordination, respectively. In lunar spinels, the divalent cations (usually Fe^{2+} or Mg^{2+}) occupy either the A- or both A- and B-sites (i.e., there are both normal and inverse lunar spinels), and higher-charge cations (such as Cr^{3+}, Al^{3+}, Ti^{4+}) are restricted to the B-sites.

The relations of the various members of the spinel group can be displayed in a diagram with the end-members represented including chromite, $FeCr_2O_4$; ulvöspinel, $FeFeTiO_4$ (commonly written as Fe_2TiO_4;, but this is an inverse spinel with Fe^{2+} in both A- and B-sites); hercynite, $FeAl_2O_4$; and spinel (*sensu stricto*), $MgAl_2O_4$. Intermediate compositions among these end-members are designated by using appropriate modifiers (e.g., chromian ulvöpsinel or titanian chromite). Most lunar spinels have compositions generally represented within the three-component system: $FeCr_2O_4$ - $FeFeTiO_4$ - $FeAl_2O_4$, and their compositions can be represented on a simple triangular plot. The addition of Mg as another major component provides a third dimension to this system; the compositions are then represented as points within a limited Johnston compositional prism in which the Mg-rich half (Mg>Fe) is deleted. Most lunar spinel compositions fall between chromite and ulvöspinel. The principal cation substitutions in these lunar spinels can be represented by $Fe^{2+} + Ti^{4+} = 2(Cr,Al)^{3+}$. Other cations commonly present include V, Mn, and Zr. Spinels are ubiquitous in mare basalts, where they occur in various textures and associations. The spinels are invariably zoned chemically. Such zoning occurs particularly in Apollo 12 and 15 rocks, in which chromite is typically the first formed mineral.

Certain highland rocks, notably the olivine-feldspar types (troctolites), contain pleonaste spinel. The composition of this spinel is slightly more Fe- and Cr-rich than an ideal composition precisely between the end members $MgAl_2O_4$ and $FeAl_2O_4$. Such spinel pleonastes are not opaque; viewed with a petrographic microscope, and it stands out because of its pink color (pink spinel), high index of refraction, and isotropic character in cross-polarized light.

5.3.4 Olivine and Some Other Lunar Minerals

Olivine is quite common in many mare basalts, but we have not found lunar mantle olivines in the Apollo samples and lunar meteorites. Famous olivine crystals in dunite 72415 show chemical zoning of Fe/Mg, indicating the shallower origin. Spinel troctolite in the Dhofar 489 lunar meteorite form the farside contains spinel grains and olivine, but not of mantle origin. In the matrices of the Dhofar 489 group meteorites, there are more fragments of olivine than those of pyroxenes (Takeda et al. 2006). Because such lunar olivines do not occur in a large mass and it is difficult to decompose them, we do not find them useful for a specific purpose. Olivine is susceptible to terrestrial weathering and other types of chemical attack. Perhaps this susceptibility would be an exploitable advantage for generation of agricultural soils.

Since the Moon is enriched in refractory elements, some minerals unique to the Moon also contain such elements. Minerals containing involatile elements are Baddeleyite ZrO_2, Dysanalyte $(Ca,Fe)(Ti,REE)O_3$, Rutile TiO_2, Thorite $ThSiO_4$, Titanite $CaTiSiO_5$, Tranquillityite $Fe_8(Zr,Y)_2Ti_3Si_3O_{24}$, Yttrobetafite $(Ca,Y,U,Th,Pb,REE...)_2(Ti,Nb...)_2O_7$, Zircon $ZrSiO_4$, Zirconolite $(Ca,Ce)Zr(Ti,Nb)_2O_7$, Zirkelite $CaZrTiO_5$ (Frondel 1975). Tranquillityite is one of the new minerals found only on the Moon. It contains moderate amounts of Rare Earth Elements (REE). Dysanalyte and Zirconolite contains REE. Yittrobetafite found in 14321, 1494 contains Nb_2O_3 17.67 wt%, UO_2 4.33, Y_2O_3 12.00, and REE 12.81 (Meyer and Yang 1988). Nd_2O_3 concentration estimated from REE pattern gives 2.38 wt%. Other minerals containing REE of the KREEP type rocks include whitlockite. A synthetic mineral has chemical formula $Ca_{18}Mg_2H_2P_{14}O_{56}$. Because of the absence of water in the Moon, lunar whitlockite contains REE substituting for Ca and H is replaced by Na and K. These minerals seem to be unique to PKT. The chemical formulae of whitlockite in 14310,123 is

$$Ca_{17.3}Y_{0.4}La{\rightarrow}Lu_{0.88}(Mg,Mn,Fe)_{2.4}(Na,K)_{0.07}(P,Al,Si)_{13.9}O_{56} \quad (5.1)$$
$$(14310,123).$$

Modal abundances of these minerals are so small even for the KREEP rocks that it is difficult to concentrate these minerals for utilization. Whitlockite is one of abundant such minerals, but it makes up to 3% of the rock's volume (Frondel 1975).

Some minerals containing useful metals and sulfides include Chalcopyrite $CuFeS_2$, Copper Cu, Nickel-iron (Ni,Fe), Troilite FeS, Mackinawite FeS or $(Fe,Ni)_{1+x}S$, Sphalerite ZnS, Tin Sn, Pentlandite $(Fe,Ni)_9S_8$. These minerals are called ore minerals, but their abundance is not large enough to be called ores. Although tranquillityite contains REE, this silicate is difficult to dissolve to extract rare metals. Considerable amounts of REE in the KREEP rocks are concentrated in whitlockite as explained above, but amounts of such minerals are also too small to extract rare metal such as Nd as resources of the Moon. One possible process to be explored is repeated dissolution of β-tricalcium phosphate by solution (Takeda et al. 2010). Then, these rare minerals could be a source of trace elements, such as rare metals, if those were needed in colony, or return to Earth. The P in whitlockite could be useful for lunar farming.

5.4 Scientific Goals of Lunar Exploration

The best rocks which will tell us the origin and evolution of the Moon are called "pristine nonmare rocks" (Warren 1993). The rock fragments of the pristine nonmare rocks were collected as a size of nuts, called, "Rake samples". The pristine nonmare rocks should preserve the original igneous crystallization textures from a magma or magma ocean, but some are brecciated rocks (e.g., 60025), but are chemically prisitine (i.e., low Ir etc.). The original crystallization textures are disturbed by a shock event or thermal metamorphism, and it is difficult to identify their pristinity. Twinning textures preserve the grain size of the original crystallization and such rocks contain textures of slow cooling in a deep crust. Such rocks should not be contaminated by elements introduced by a meteorite impact. It should be noted that they have very old crystallization age.

Our scientific goals of lunar exploration are to obtain the earliest crustal rocks directly crystallized from magma ocean by future sample return missions. The landing site for such sample return will be selected by future remote sensing and landing missions.

5.4.1 Lunar Surface Materials Identified by Remote Sensing

In order to identify minerals on the surface of the Moon, we use information obtained by the Spectroprofiler (SP) and the Multiband Imager (MI) on board the Kaguya mission (Ohtake et al. 2009). Identification of geological terrain units done by MI is followed by identification of rock-types by SP. However, it is difficult to identify terrain units, because lunar surface materials are regoliths including various breccias from wide regions of the lunar surface and include shocked minerals, impact melts and fragments of many rocks and minerals. In this sense, polymict breccias provide us with more unbiased information of the lunar crust.

Highland rocks also include many metamorphic rocks, granulites with compositions between FAN and Mg-suite rocks in the An-Mg# diagram. Many investigators talked about "so-called" magnesian anorthosite (MAN) with reference to the pristine Magnesian Anorthosite (Mg An) clast found in the Dhofar 489 group, but some clasts identified as MAN include magnesian granulites. Pristine crystalline clasts of MAN are rather rare and it is difficult to prove its pristine nature, because of their small sizes. The photomicrographs of the PTS (Fig. 5.4) of the magnesian anorthosite clast in Dhofar 489 shows the twinning lamellae extended to the entire original crystal. This kind of texture is obesrved in 60025 FAN. The olivine grains in this pristine clast show granulitic subrounded shape. MAN includes granulites and mixtures of spinel troctolite and anorthosite. The purest anorthosite (PAN) has been recognized by the Kaguya mission (Ohtake et al. 2009). We looked for such anorthosite in the farside lunar meteorites. Pure clear plagioclase crystals were identified in Dhofar 307 and others as a large fragments or a crystal. Mafic silicates are not fond within this large crystal, but small wormy mafic silicates are found together with fine grained anorthite crystals at grain boundaries.

The bigger question is why one would want to have a colony on the Moon. Science can never be the entire answer, but it should always be part of the answer. Our science rationale has been presented at the beginning of this section. Again, we are not proposing resource utilization to establish a colony. We will establish a small base for scientific exploration and for production of the SPS and small dome for agriculture to support such activities.

5.5 Energy Resources of the Moon

In this section, we now address how to utilize lunar materials for sustaining small scale human activities on the Moon, on the bases of the contribution that came from the Kaguya mission. What kinds of materials can be used as resources may depend on the nature of human activities on the Moon. Plagioclase is the most abundant mineral in the lunar highlands and some part of the lunar crust is nearly pure anorthosite (Ohtake et al. 2009). Such rocks can be found close to the lunar base. An important material for energy resources of the Moon is silicon to be used for solar power stations (SPS). If such a silicon panel factory is difficult to be brought to the SPS orbit, silicon panel can be produced on the Moon from pure anorthosites. It is difficult to separate resource minerals by some of the mechanical processes used on the earth, because the equipment is too heavy to be carried to other planets. Pure anorthosite can be used on the Moon without separation processes.

Utsunomiya and Takeda (1996) used plagioclase ($An_{78}Ab_{22}$) to produce silicon, aluminum, calcium and oxygen by dissolution. Plagioclase is dissolved in hydrochloric acid and after filtration of residual plagioclase, the solution is evaporated to dryness and leached by hydrochloric acid to dissolve aluminum chloride. Silica powders are obtained by filtration (Fig. 5.9). This material can be used to produce silicon solar cells. Pure anorthosite can be used on the Moon without separation processes. Plagioclase is dissolved in hydrochloric acid or also some organic acids such as fulvic acid. The dissolution rate of calcium for oxalic acid is very low in comparison with organic acids with high aromatic components because calcium oxalate is not a soluble compound. Fulvic acid may decompose calcic plagioclase for soil production.

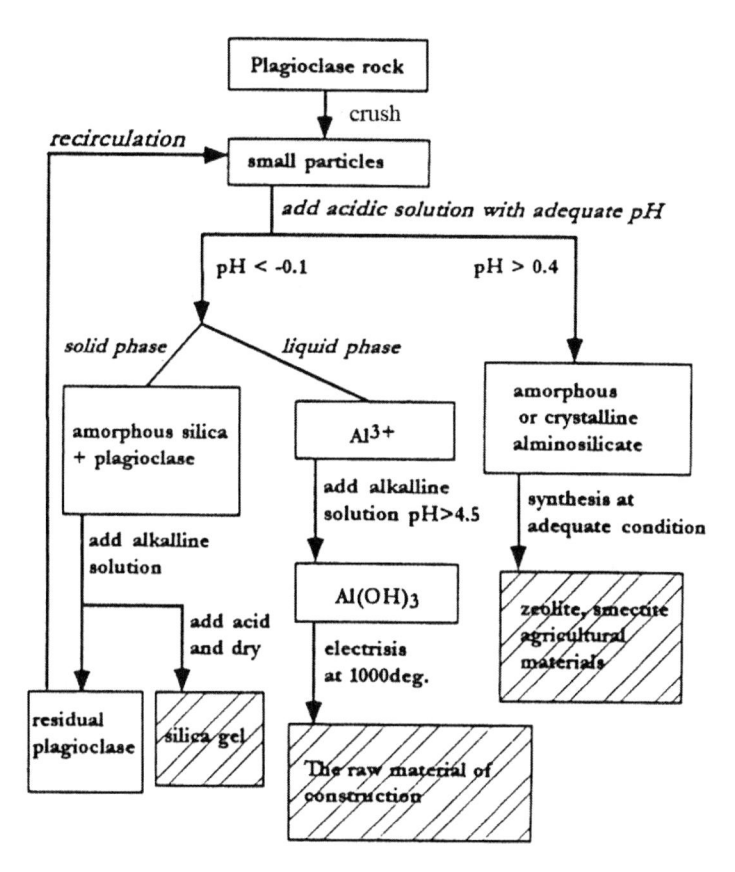

Fig. 5.9 A diagram showing separation processes of silicon, aluminum and calcium from plagioclase (Utsunomiya and Takeda 1996)

In this report, we tried to correlate data obtained by the Kaguya mission and the lunar meteorites from the farside, but we found so much evidence that our image of the Moon developed from the Apollo samples should be revised. We will be able to obtain more information from the Kaguya data in the near future, but future sample return from the farside of the Moon will be an important source of information to advance lunar science.

5.5.1 Production of Silicon Solar Cells for Lunar SPS

We presented a plan to use silicon extracted from the lunar calcic plagioclase to produce SPS. We will show more developed concept in this chapter. One problem in production of silicon solar cells is that construction of semiconductor factories on the Moon seems to be difficult. We proposed a method to extract silicon from lunar plagioclase (Utsunomiya and Takeda 1996). Recent developments of producing silicon solar power cell will give us a simpler process. Conventional vacuum processes and vapour-phase deposition—for the fabrication of electronic devices are not practical on the Moon. The use of solution processes (Iwasawa et al. 2006) has received considerable attention for a wide range of applications. The ability to print semiconductor devices using liquid-phase materials could prove essential for the lunar applications, such as large-area flexible displays. Recent research in this area has largely been focused on organic semiconductors, some of which have mobilities comparable to that of amorphous silicon (a-Si). Solution processing of metal chalcogenide semiconductors to fabricate stable and high-performance transistors has also been reported. If high-quality silicon films could be prepared by a solution process, this situation might change drastically. The solution processing of silicon thin-film transistors (TFTs) using a silane-based liquid precursor has been demonstrated. Using this precursor, we have prepared polycrystalline silicon (poly-Si) films by both spin-coating and ink-jet printing, from which we fabricate TFTs. Although the processing conditions have yet to be optimized, these mobilities are already greater than those that have been achieved in solution-processed organic TFTs.

5.5.2 Solar Power Station by Utilization of Lunar Materials

A concept of the ISAS Solar Power Satellite (SPS) proposed by Late Prof. Makoto Nagatomo in 2000 has been developed by S. Sasaki (2006) of ISAS/JAXA. This proposed Tethered SPS consists of a large panel with a capability of power generation/transmission and a bus system, which are connected by multi-wires. This power generation/transmission panel is composed of a huge number of perfectly equivalent power modules. The electric power generated by the solar cells at the surface of each module is converted to the microwave power in the same module. No wired signal/power interfaces are required between the modules, because the modules are controlled by the bus system using wireless LAN. The attitude in which the microwave transmission antenna is directed to the ground is maintained by the gravity gradient force. The SPS planned by the JAXA group is placed as a satellite at the Geostationary Earth orbit, 36,000 km high (Asahi Newspaper, May 17, 2011). The SPS can receive sun light all the time, and produce electricity comparable to that of one nuclear power unit. The tethered panel is composed of individual tethered subpanels which are loosely connected to each other. This configuration enables an evolutional construction in which the function of the SPS grows as the construction proceeds (Sasaki et al. 2006).

The final lunar SPS proposed here will be in an orbit at Lagrangian points in the Moon-Earth system, the antenna is directed to a satellite at the Geostationary Earth orbit. The electric power generated at SPS is planned to send to a satellite at the Geostationary Earth orbit by microwave transmission system, the same as the above Earth orbit SPS. Serious problems pointed out in the past include environmental damages to the Earth system. However, if the microwave is received by a satellite at the Geostationary Earth orbit, the electric power can be sent to a small island in the ocean by a Space elevator with carbon nano-tube wire to the ground, from where we can use the normal electric transport system. The cost of lift off from the Earth is estimated to be more than that from the lunar surface, and many space debris will be generated

5.5.3 Utilization of Lunar-Produced Materials Other Than Silicon

Utsunomiya and Takeda (1996) used plagioclase ($An_{78}Ab_{22}$) to produce silicon as described above. After removal of silicon, aluminum, oxygen and calcium can be extracted by the following methods. Aluminum hydro-oxide can be precipitated by adding sodium hydroxide solution to the above solution in the experimental run. Aluminum metal can be obtained by the electric fusion of aluminum oxide with catalytic compound, cryolite (Na_3AlF_6) and CaF_2. Aluminum is a good construction material for low gravity on the Moon. Electric power to reduce aluminum oxides is available on the Moon by silicon solar panels. We use graphite (carbon) electrodes to reduce fused materials on the earth, but carbon dioxide bubbles are generated by combining C with O_2. If we replace graphite by platinum electrodes on the Moon, we can generate oxygen bubbles to be used for breathing in the base. If we extract hydrogen from the regolith, which was brought by the solar wind, we can produce water or use it for hydrogen batteries. The solar implanted hydrogen abundance in the regolith is given by Haskin (1992) together with C and N on the Moon. The water equivalent of H in the upper 2 m of the regolith averages at least 1.3 million litters per square kilometer. Other investigator used fluoride gas to produce oxygen from lunar minerals by making mixtures of CaF_2 (fluorite) and aluminum fluoride.

Calcium ion left in the solution after precipitation of aluminum, can be used to fix carbon dioxide generated in a dome of the base. Carbon dioxide is in the atmosphere of Mars, but it is not permitted to release it on the Moon. Calcium carbonate thus produced can be used as construction materials of the base such as limestone or marble, or used as cement-like materials to join bricks to build extension buildings of the base. This technique to produce calcium carbonate from plagioclase should also be developed on the earth to fix carbon oxide to reduce the green house effects.

Fixation of carbon dioxide by calcium disolved from calcium silicate minerals such as plagioclase, after removal of aluminum as discussed above, will be applied for decreasing carbon dioxide concentration in the atmosphere of the lunar base. If carbon dioxide generated by human activities in the lunar base, is released outside the lunar dome, the atmosphere close to the lunar surface is no longer vacuum, and carbon dioxide will be adsorbed into the fine-grained regolith, resulting in chemical contamination of the Moon.

This technique will also be applied to remove carbon dioxide from the terrestrial atmosphere to minimize the "global warming of the earth". Silicate minerals which are present on the surface of the terrestrial continents do not contain as much calcium as those in plagioclase. Augite (augite - diopside) also contains a lot of calcium, but decomposition of this mineral is not as easy as plagioclase decomposition. If we pay attention to the regional distribution of plagioclase on the Earth, plagioclase enrichment is found in some regions such as north of the Great Lakes in Canada. There is enough plagioclase on the Earth, so as to fix carbon dioxide in the atmosphere as calcium carbonate. A few thousand million years ago in the Earth history, all the region on the Earth down to the true equator was frozen. Our planet was called "Snowball Earth" for this period. One of the reasons for this cool down of the Earth, is that the total surface area of the continents was larger than the present one, because plate movements were not extensive at this time. The larger surface areas produced more Ca ions, which fixed more carbon dioxide in the atmosphere, resulting in negative "global warming"

Pesent simulation of the "global warming" of the Earth have not been taken into account, in reactions of the surface materials with carbon-dioxide enriched atmosphere, since those geophysisists performing such simulation are not aware of the dissolution rate of the surface minerals by organic acids produced in the forests. The amounts of plagioclase available on the continents and the amount of carbon dioxide in the air and the reaction between them should be predicted for the future simulation of the global warming. The study of lunar plagioclase disolution by fulvic acids and other organic and inorganic acids will be useful for such simulation in addition to the increase of the amount of carbon dioxide. Human activities on the Moon to ensure enough food will encounter a serious problem for balancing to produce calcium carbonate and to use carbon dioxide for growing plants and generating oxygen by plants. It is a problem of space agriculture in general. The soils for space agriculture can be manufactured in the following ways as in the next section. The calcium ion can be combined with phosphate as apatite to help eliminate the extra phosphorus.

5.6 Problems of Soils for Agriculture on the Moon

Wada et al. (2009) examined the possibility of Martian agriculture, for which air, water, and regolith are brought into a pressurized dome at the first stage. Agriculture on the Moon can be undertaken in a manner similar to that of Martian agriculture. Recycled use of materials is also a major function of agriculture on the Moon. Lunar agriculture is also characterized by limited species in the ecosystem, contained inside a small dome. Because no clay minerals have been reported on the surface of the Moon, we have to produce clay minerals from the lunar surface materials.

Clay minerals on the earth have crystal structures of layer silicates or phyllosilicates. Clay minerals are main components of terrestrial soils. Since these clays are not available for lunar agriculture, we recommend producing clays from common rock-forming silicates abundant on the lunar surfaces. Considering that the materials around a lunar base are regoliths, basalts and anorthosites, we have to design the simplest assemblage of clay minerals and organic compounds to grow special spices of vesitables.

5.6.1 Alteration of Rock-Forming Silicates

The goal of space agriculture is to create and maintain optimum living environment on extraterrestrial planet for human life as was discussed for Mars (Yazawa et al. 2010). Although some scientists promoting an idea to utilize ecological system of the hyperthermophilic aerobic composting bacteria for producing soil from human and animal waste and inedible biomass, future astrobiological exploration of the Moon may encounter problems for detecting signs of life on the Moon if living terrestrial bacteria were introduced to the Moon. We are trying to introduce very limited fresh vegetables to support our life on the Moon. We have to produce soils from lunar rocks for terrestrial vegetables keeping microorganisms as small as possible.

Table 5.1 Weathering reactions of Ca- or Na-plagioclase and its equilibrium constants at 298 K and 1 atm (Ichikuni 1989)

Chemical equations	Equilibrium constants
(1) Ca-plagioclase - Gibbsite	
$CaAl_2Si_2O_8 + 2H^+ + 6H_2O \rightarrow Ca^{2+} + 2Si(OH)_4 + 2Al(OH)_3$	$Log\ (\alpha_{Ca}/\alpha_{2H}) = 4.41 -$
(2) Na-plagioclase - Gibbsite	$2Log\ \alpha_{Si(OH)4}$
$NaAlSi_3O_8 + H^+ + 7H_2O \rightarrow Na^+ + 3Si(OH)_4 + Al(OH)_3$	
	$Log\ (\alpha_{Na}/\alpha_H) = -4.65 -$
	$3Log\ \alpha_{Si(OH)4}$
(3) Ca-plagioclase - Kaolinite	
$CaAl_2Si_2O_8 + 2H^+ + H_2O \rightarrow Ca^{2+} + Al_2Si_2O_5(OH)_4$	$Log\ (\alpha_{Ca}/\alpha_{2H}) = 14.5$
(4) Na-plagioclase - Kaolinite	
$2NaAlSi_3O_8 + 2H^+ + 9H_2O \rightarrow 2Na^+ + 4Si(OH)_4 + Al_2Si_2O_5(OH)_4$	$Log\ (\alpha_{Na}/\alpha_H) = 0.39 -$ $2Log\ \alpha_{Si(OH)4}$
(5) Ca-plagioclase – Ca-montmorillonite	
$7CaAl_2Si_2O_8 + 12H^+ + 8Si(OH)_4 \rightarrow 6Ca^{2+} + 16H_2O +$ $3Ca_{0.33}Al_4(Si_{3.67}Al_{0.33}O_{10})_2(OH)_4$	$Log\ (\alpha_{Ca}/\alpha_{2H}) = 19.87 +$ $1.33Log\ \alpha_{Si(OH)4}$
(6) Na-plagioclase – Na-montmorillonite	
$7NaAlSi_3O_8 + 6H^+ + 20H_2O \rightarrow 6Na^+ + 10Si(OH)_4 +$ $3Na_{0.33}Al_2(Si_{3.67}Al_{0.33}O_{10})(OH)_2$	$Log\ (\alpha_{Na}/\alpha_H) = 1.84 -$ $1.67Log\ \alpha_{Si(OH)4}$
(7) Ca-montmorillonite - Kaolinite	
$3Ca_{0.33}Al_4(Si_{3.67}Al_{0.33}O_{10})_2(OH)_4 + 2H^+ + 23H_2O \rightarrow Ca^{2+} + 8Si(OH)_4 +$ $7Al_2Si_2O_5(OH)_4$	$Log\ (\alpha_{Ca}/\alpha_{2H}) = -17.79 -$ $8Log\ \alpha_{Si(OH)4}$
(8) Na-montmorillonite - Kaolinite	
$6Na_{0.33}Al_2(Si_{3.67}Al_{0.33}O_{10})(OH)_2 + 2H^+ + 23H_2O \rightarrow 2Na^+ + 8Si(OH)_4 +$ $7Al_2Si_2O_5(OH)_4$	$Log\ (\alpha_{Na}/\alpha_H) = -8.27 -$ $4Log\ \alpha_{Si(OH)4}$
(9) Kaolinite - Gibbsite	
$Al2Si_2O_5(OH)_4 + 5H_2O \rightarrow 2Si(OH)_4 + 2Al(OH)_3$	$Log\ (\alpha_{Ca}/\alpha_{2H}) = -5.04$

Global scale soil formation on the earth, depends on latitude, but at a local scale, the nature of the country rock is a predominant factor that governs soil formation at the local scale. The primary lunar rocks are essentially basaltic for the mare regions and anorthosites for the highlands. Many studies emphasize the importance of the primary mineralogy in addition to the physicochemical conditions of the alteration processes. Dissolution of main rock-forming silicates by inorganic and organic acids is the basic process for producing clay minerals either on the earth or the Moon. Organic acids produced by decomposition or fermentation of plant leaves and roots are important reagents for natural weathering in fields or forests on the earth (Yazawa and Takeda 2008). The dissolution rates of main rock-forming silicates by organic acids have been discussed by Yazawa et al. (2007).

Dissolution rates of major minerals for soil production under relatively equilibrium conditions are listed in Table 5.1 (Lasaga et al. 1994). Olivines (forsterite) and pyroxenes (enstatite and diopside) are more easily dissolved, but they do not contain much silicon and aluminum to synthesize layer silicates. Mg and Fe in these mafic silicates can be included in some clay minerals, which will be good for some plants. Nevertheless feldspars are good candidates for clay mineral formation. Ullman and Welch (2002) studied organic ligands and feldspar dissolution and pointed out that organic compounds can both enhance and inhibit the rates of feldspar dissolution in the laboratory and presumably in field settings where organic matter is abundant. Organic acids can directly enhance dissolution by either a proton- or a ligand-mediated mechanism. Organic polymers can either inhibit dissolution by irreversibly binding to the mineral surface and reducing the effective reactive surface area (i.e. the number of reactive sites) of the mineral or enhance dissolution by a ligand-mediated mechanism. The experimental determinations of the impact of a wide variety of organic ligands on feldspar dissolution rates provide evidence for multiple mechanisms of mineral ligand interaction that can impact the dissolution process.

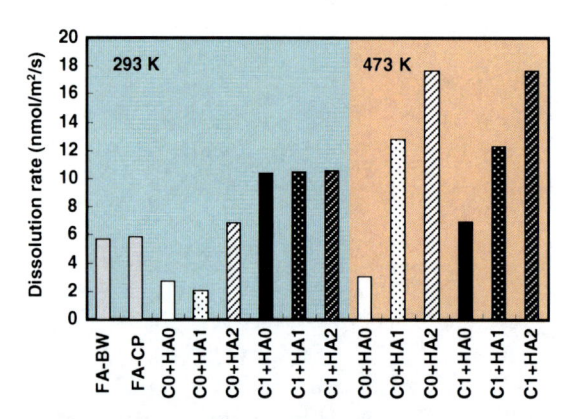

Fig. 5.10 Comparison in dissolution rate of Ca-plagioclase between fulvic acid and combinative acid of humic-CO_2 under 293 or 473 K. FA: fulvic acid, BW: brine water, CP: Canadian peat moss, C0: without carbonated water (deionized water), C1: with carbonated water (320 mmol-CO_2/L), HA0~2: 0.0~0.2 mg/20 mL of humic acid extracted from Canadian peat moss

Dissolution experiments done by these authors are for a few organic acids and limited compositional ranges of feldspars. For example, Huang and Kiang (1972) dissolved anorthite and five feldspars by acetate, salicicylate, aspartate, and citrate, and found that the net dissolution of feldspar into solutions of organic acids is proportional to the anorthite (An) content. Yazawa et al. (2007) performed more systematic studies comparing lactic acid, oxalic acid, garlic acid, and fulvic acid for dissolving Ca-rich plagioclase (Fig. 5.10). They found that fulvic acid could dissolve Ca-rich plagioclase by a process similar to natural weathering involving plants. Because dissolution rates of more Ca-rich plagioclases found in lunar rocks are higher than the Na-rich members of the Earth, clay formation of Ca-rich members will be faster than those of the Na-rich ones (Blum and Stillings 1995). Concentrations of ions in the solutions by these dissolution experiments will provide us with useful information for plant fertilization, described in detail in the next section.

5.6.2 Production of Soils by Humic Substances and CO_2 for Lunar Agriculture

This section has been designed to determine if it is possible to grow green vegetables for human nutritional requirements during short say utilizing lunar regoliths as the growing medium. Photosynthetic plants require N, P, K, Ca, Mg and S in considerable quantities and Fe, Mn, Zn, Cu, B and Mo in trace quantities. The lunar regolith contains Ca, Mg, Fe, Mn in sufficient quantities, together with Si, Al, Ti, Na, Cr, K. Nitrogen is available in regoliths supplied by solar wind (Haskin 1992), but it is not expected to provide the full needs of plant growth. P, K and S are available in small amounts in the regolith, specially in the KREEP-rich region, but not Zn, Cu and Mo – they will also need to be supplemented. There is no evidence of B on the Moon. Most of the elements that are required are available, but the specific detailed mineralogy and it is unknown that availability of the elements for plants to be able to use them in their present form. Fertilization will be needed to provide some essential elements.

In this chapter, we emphasized utilization of plagioclase, because our scientific investigation is going to be done in the highland. One might think of soils formed on some terrestrial volcanic areas, i.e., from basalts, for example, as being better primary materials for the production of agricultural soils than pure Ca-plagioclase. We make some acknowledgement of such factors, but admit to make obvious conclusion that perhaps KREEP-rich regions of the Moon would be favorable sites for lunar farming only. MacElroy et al. (1985) and Salisbury and Bugbee (1985) introduced farming in a lunar base.

Some other elements induce special problems because of the specific nature of the lunar regolith. Aluminum is present in a larger concentration in the lunar highlanads than is common in earth soils, if we make soils from anorthosites. In soil solution culture experiments, Yazawa et al. (2000) found that addition of humified natural organic matter reduced the amount of monomeric Al present in solution and alleviated the toxic effect of Al on root growth of wheat under acidic condition. Consequently, terrestrial humic matters including fulvic acid can not only buffer to control for nutrient and toxic element but also accelerate to dissolute elements and synthesis clay from lunar regolith with chemical weathering. Because humified natural organic matter is common organic materials in the terrestrial soils, we selected fulvic acid to make soils from the lunar surface materials.

Fulvic acid is a complex organic acid with high molecular weight from about 680 to 920. Fulvic acid has a water bridging ability. It attracts water molecules, helping the soil to remain moist and aiding the movement of nutrients into plant roots. The small molecular size of fulvic acid allows it to enter the vascular system of plants. Fulvic acid easily binds to minerals such as iron, calcium, copper, zinc and magnesium. As it moves through the roots, stem and leaves, it takes these minerals to metabolic sites within the plant. The ability of fulvic acid to bind to trace minerals makes it an effective organic chelating compound. Sprayed on plants, fulvic acid is a non-toxic mineral additive and water binder that maximizes the plant's productivity.

We investigated a process of soil production from Ca-plagioclase and some other minerals by geochemical weathering with fulvic acid in comparison with normal acid (hydrochloric, oxalic, lactic and garlic acids). Fulvic acid is a natural organic matter produced in humified soils. The fulvic acid can be extracted from many natural materials, such as weathered coal, Canadian peat, Andsol, and forest soils. Our recent studies of fossil brine water in the Kazusa formation in the Mobara area in the Southern Kanto gas field, Chiba, Japan, showed high concentration of fulvic acid (Yazawa et al. 2005). The Kazusa formation is a sedimentary formation of interstratified mudstone and sand stone layers of the Quaternary, Pleistocene to Tertiary, Pliocene age (between 3 million and 400,000 years ago) distributed at the Boso peninsula, Chiba, Japan. The brine water in this formation includes soluble pure-methane, iodine (100 ppm), fulvic acid (40~60 ppm) and a similar chemical composition to seawater.

The fulvic acid can be extracted by the procedures by the standard method suggested by the International Humic Substance Society (IHSS). The molecular structure of fulvic acid, extracted from waste brine water, was determined by Yazawa et al. (2005). The average chemical structure of fulvic acid had practically an aliphatic composition (C 48.5, H 5.2, O 43.5) and high content of nitrogen (N 2.8). Now, annually 50 million tons of the brine water after production of methane and iodine around Mobara, Chiba is being discharged to the sea or returned into the gas field as drainage. Large amounts of fulvic acid will be available as industrial product in Chiba. We will be able to make use of this acid for lunar agriculture in future, but we have to bring fulvic acid from the Earth.

We employed Ca-plagioclase from Crystal bay, Minnesota, with An_{76} ($Ca_{0.78}Na_{0.23}Al_{1.73}Si_{2.23}O_8$), which is in between lunar mineral regoliths and Martian ones (Yazawa et al. 2011).After crushing, the powdered Ca-plagioclase was passed through a sieve of pore size less than 75 μm in diameter, and its relative surface area was 1.75 m^2/g. The used humic acid in this experiment was extracted from Canadian peat moss in accordance with the standard method of International Humic Substance Society. This humic material had average molecular weight of 60,000 g/mol, and $C_{2450}H_{3000}N_{86}O_{1650}$ as chemical formula. The nano-bubble aqueous solution of CO_2 gas was prepared by nano/micro bubble generator with

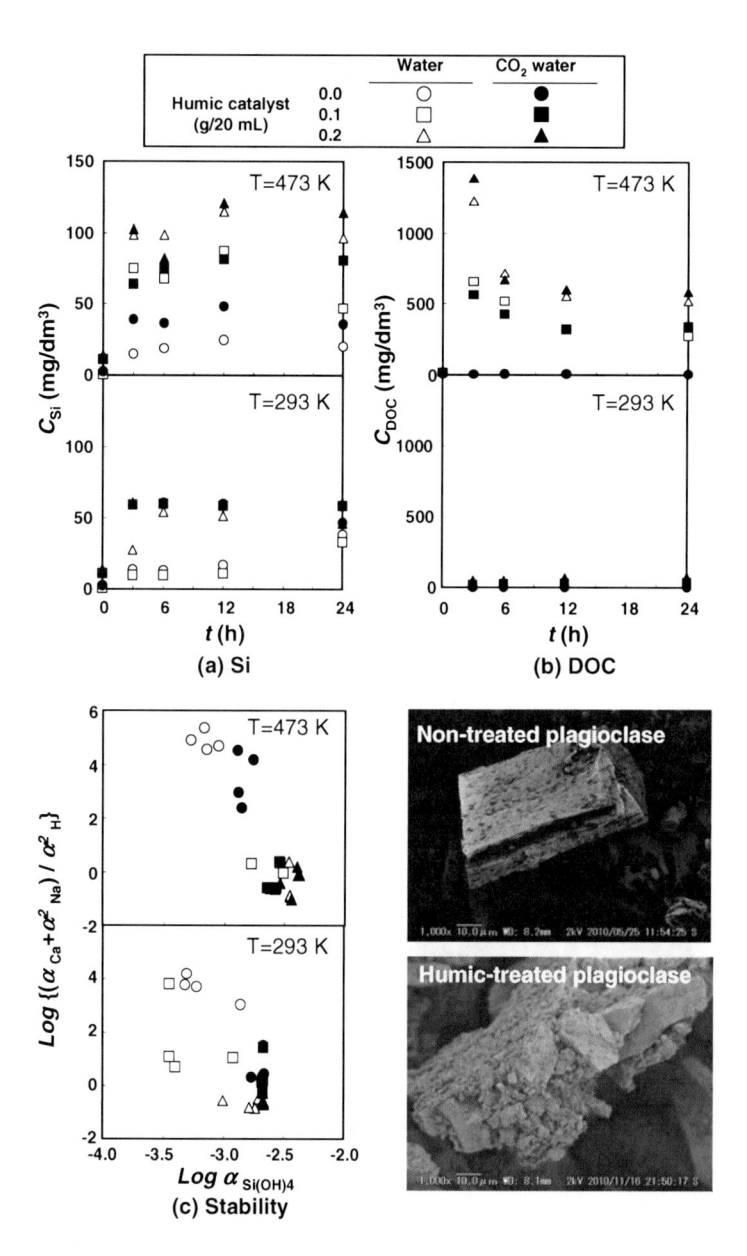

Fig. 5.11 Time dependence of Si dissolved from Ca-plagioclase and DOC by humic acid and carbonated bubble water under 293 or 473 K. Estimation of stability in mineral solutes on a basis of dissolution equilibrium with acid

pressure type (Aura Tec Co., Japan). The formed solution was pH 3.4, which is equivalent to 320 mmol-CO_2/L, with void ratio of 0.15% and bubble diameter of 340 nm. The 0.1 g of Ca-plagioclase and 20 mL of water with/without CO_2 bubble was added into a high-pressure gas vessel with 25 mL. 0.0, 0.1, and 0.2 g of humic acid were added to those reaction systems, and reacted at 293 or 473 K for 24 h. Each calculated pressures in vessel were comparable to 0.02 or 1.62 MPa. After a predetermined time, these reaction solutions were refreshed immediately to 293 K and separated into solid-liquid by centrifuge. Each supernatant liquids were measured for pH, metal ions (Na, Ca, Al, and Si) by atomic absorption spectrophotometer, and dissolved organic C (DOC) by total organic carbon analyzer.

The concentration changes of Si and DOC with reaction time under 293 and 473 K are shown in Fig. 5.11 (a) and (b), respectively. As estimation of process of Si concentration in reaction solution, the carbonated bubble water was accelerated to dissolve Ca-plagioclase under 293 K in comparison with water (reaction route I in Fig. 5.12). Moreover it was confirmed to dissolve Ca-plagioclase by H^+ ion dissociated from humic acid (route II). With rise in temperature and pressure, dissolution rates of all reaction systems were accelerated, the combined effect of humic acid and carbonated bubble water became prominent (route IV). The process of DOC concentration in Fig. 5.11 (b) leads to a further understanding of this advantageous effect, humic acid was slightly soluble under 293 K but freely soluble to 1/4 ~ 1/3 of DOC in initial added humic acid under 473 K.

So, it has gases that water-solubilization of humic acid was accelerated by hydrothermal decomposition and carboxylation with CO_2 gas (route III). As mentioned above, the H^+ ion was sequentially generated in reaction system with exchanging between H^+ in acidic functional groups of water-soluble humic acid and metal cation (M^+) eluviated from Ca-plagioclase, and the chemical weathering made further progress to create soil (organo-clay complex, aggregation, route V).

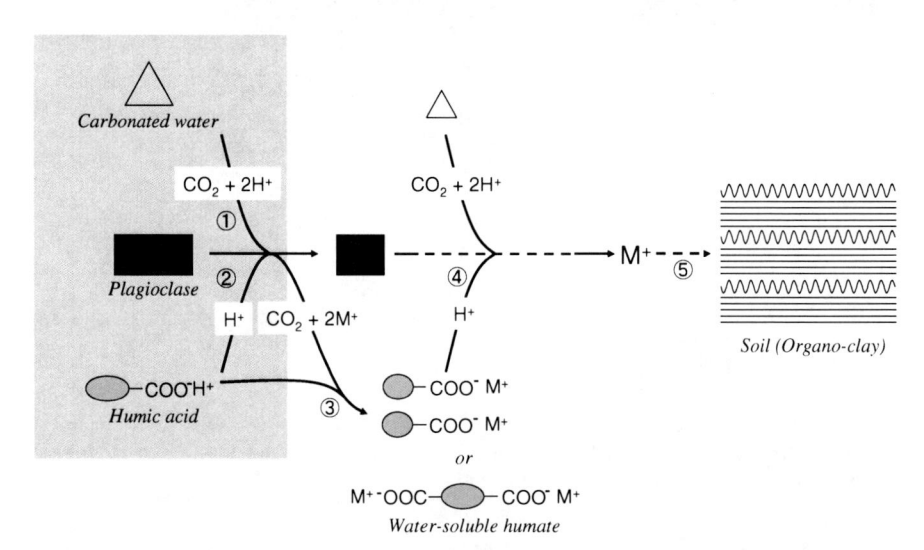

Fig. 5.12 Reaction mechanism of Ca-plagioclase, humic acid, and carbonate water to soil. ①②③④⑤ stand for route I, II, III, IV, V, respectively.

At a carboxylic concentration of 0.01 mol-COOH/L in acid solution under 293 K, the relative effectiveness of these ligands on promoting dissolution rate of Ca-plagioclase was humic acid (2.0 nmol/m^2/s) with the standard of Si element (Fig. 5.10). This value was only little lower than oxalic (14 nmol/m^2/s), lactic (9.0), garlic (7.8) and fulvic (5.7) acid however dramatically increased to 10.5 by a combination of humic acid and carbonate bubble solution. The strengths of organic acids to promote dissolution is related to the initial pK_a (acid dissociation constant) 2.58 (oxalic), 3.52 (lactic), 4.37 (garlic) and 4.16 (fulvic or water-soluble humic), respectively, but it is related to strength of Ca-ligand complex formed in solution (formation of secondary mineral; clay) or on the mineral surface (development of soil aggregation). Formation of clay minerals is expected after adjusting condition of pH and Al/Si ratios to bring them in the stability field of clay minerals. As estimation about temperature dependence of dissolution rate of Ca-plagioclase by the Arrhenius equation, the addition of humic acid caused the frequency factor and activation energy to increase. In contrast, the carbonate bubble solution substantially reduced the activation energy of reaction between Ca-plagioclase and humic acid, therefore acted as catalyst for this reaction.

In the 1950's, soil scientists first clearly described the mineral succession: gibbsite-kaolinite-montmorillonite, in terrestrial soils, as function of climate, topography and   nature of the parent rock. Gibbsite appears in leached conditions in the upper parts of the profiles, in the topographic hights and under humid tropical climates, when the percolating solutions are diluted. Montmorillonite appears in confined conditions, in the lower parts of the profiles, downstream in landscapes and under semi-arid climates, when the percolating solutions are concentrated by evaporation or by a long advance through the weathering zone. At the same time, thermodynamic models and calculations based on natural observations or experiments in the laboratory were undertaken. In weathering profiles, anorthite, albite, gibbsite, kaolinite and Ca- or Na-montmorillonite may form successively from humid to semi arid tropical climates. The solubility products and equilibrium reactions concerning the aluminous silicates at 298 K and 1 atm are written in Table 5.1. Table 5.1 is derived from information presented in the stability diagrams for calcium and sodium silicates given in Lasaga et al. (1994). Figure 5.11(c) are calculated from mineral leachates from Ca-plagioclase with acid. This means the generation of gibbsite has an advantage in the event that log $\alpha_{Si(OH)4}$ value lower, and the generation of clay with layer silicate (i.e. montmorillonite, kaolinite) in the event that both values of log $\alpha_{Si(OH)4}$ and log $\{(\alpha_{Ca}+\alpha^2_{Na})/ \alpha^2_H\}$ are higher. From the result, the addition of humic acid significantly increased the Log $\alpha_{Si(OH)4}$ value but decreased the log $\{(\alpha_{Ca}+\alpha_{2Na})/ \alpha_{2H}\}$ value. Both values were increased by a combination of humic acid and carbonate bubble solution, especially the stabilization of clay formation was expected under 473 K.

Iron is an essential element for plant growth, but iron is converted into insoluble form because of pH change and high oxygen partial pressure on the Earth. However, most of the dissolved iron has been found to form metal-organic complexes. Fulvic acid and low molecular weight organic acids produced in forest and marsh lands on the Earth are responsible for dissolving Fe from minerals in rocks or parent material in soil, and for helping to keep its solubility in the water. Rivers play an important role in transport of the complex to the sea because their source regions are in the mountains and they run through various lands till they reach the sea. To gain a better understanding of the transport mechanism of fulvic acid from the forest to the sea, Fujiyama et al. (2010) analyzed water and soils sampled along sites in the Obitsu River, Chiba, Japan by fluorescence spectrophotometer. Although there is no river or sea on the Moon, Fe-fulvic acid complexes will play an important role to keep iron in water-containing lunar soil for lunar agriculture.

Fulvic acid and its complexes with iron had a strong accelerative effect for physiological activities of root growth in rice and chlorophyll production in phytoplankton. Confirmation of dissolution of the main rock-forming silicates by fulvic acid will give us better understanding of the basic process of weathering in nature on the Earth, as well as of application for lunar agriculture (Yazawa and Takeda 2008). Pandeya et al. (1998) performed experiments to study the influence of fulvic acid on transport of iron in soils and uptake by paddy seedlings, and found that the uptake of Fe by the crop and the percentage of the tissue iron content derived from fertilizer were higher in the case of Fe fulvic acid complex in comparison with $FeCl_3$, indicating the superiority of organically complex Fe fertilizers over inorganic salts.

Some nutrients may be supplied to the mixtures of rock powders, clay minerals and fulvic acid complex. Most of the plant nutrients, except N and P, are expected to be available on the Moon, because they may take forms accessible to plant roots after they are released from minerals by weathering. Application of some N fertilizer (NH_3, NO_3^-, urea, etc.) or cultivation of some green manure crop plants capable of fixation of N derived from regoliths is inevitable. P is actually contained in some minerals of the lunar basalts such as Ca phosphates, and may be released by weathering in a similar way as other nutrients. Again, the KREEP-rich region is favorable for obtaining such materials. Wada et al. (2009) suggest that the released P is considered to be unavailable for plant roots, because of its strong adsorption (P-fixation) on $CaSO_4$, $CaCO_3$ and iron oxides.

Organic wastes accumulated by human activities and agriculture products in the dome, have been proposed to be used as composting. However, recycling of composted materials in lunar agriculture directly applied to soils may have a chance of releasing microorganisms into the lunar ecosystem. Composting should be done in a closed vessel and the products should be sterilized by high temperature steam before being applied to the farming soils in the agricultural dome in a lunar base. By such process recycling of fulvic acid may not be required if it is brought at the first time.

5.9 Environmental Problems for the Moon

Any time soils and rocks are mined for resources (*e.g.*, H, P, Si, etc), attention should be paid as to what will happen to the by-products of that extraction procedure. The use of HCl on the Moon will likely have significant environmental influences, because of toxic nature of chlorine. This is the reason why we select organic acids, especially fulvic acid. Waste generated from the use of fulvic acid can be added to soils without any environmental problems. On the earth, mining operations just pile the rock off to the side in "tailing piles". The implications of doing this on the Moon are different from the surface of the earth. Piles of the rock can be use as protection materials for high energy particles from space. Mixtures of fulvic acid and clay minerals are good garden soils for agriculture. Also, every time the soil and rock is disturbed for resource utilization, crucial geological and maybe biological information is potentially lost. Care should be taken to avoid this, for example, by limiting mining actions only within one crater.

We did not discuss the potential soil fertility on the Moon, since we assumed soils produced from some lunar rocks by fulvic acid are essentially the same as those on the earth. Many inorganic nutrients are likely to be easily available by dissolving rock-forming minerals. What nutrients will need to be brought from the earth except for fulvic acid may depend on where we establish our lunar base. They are probably only nitrogen and a few others. We may have to perform more experiments to see what nutrients are potentially limiting and how they can be more easily extracted. Our dissolution experiments by organic acids, showed that especially fulvic acid will help with Fe and other micronutrients, by forming organo-metallic complex. The point in discussing chemical weathering and phyllosilicate formation from lunar rocks is that the chemical processes proposed here are the same as producing soils in forests on the earth.

Acknowledgments. We thank Drs. Junichi Haruyama, Tsuneo Matsunaga, Tomokatsu Morota, Singo Kobayashi, Prof. Manabu Kato and the SELENE project team members, and Prof. N. Hasebe and the KGRS group, for support and discussions. This work was supported in part by NIPR, Research Project Fund, and by the Research Forum of the Chiba Inst. of Technology, and in part by funds from the cooperative program (No. 005, 2010) provided by ORI, the University of Tokyo. The authors thank Drs. L. E. Nyquist and Makiko Ohtake, whose comments and discussion improved our studies very much and helped clarify the descriptions of this chapter.

References

Blum, A.E., Stillings, L.L.: Feldspar dissolution kinetics. In: White, A., Brantley, S.L. (eds.) Reviews in Mineralogy 31, Chemical weathering rates of silicate minerals, Mineral Soc. America, pp. 291–352 (1995)

Bohner, M., Lemaître, J., Ring, T.A.: Kinetics of dissolution of β-tricalcium phosphate. J. Colloid Interface Science 190, 37–48 (1997)

Frondel, J.W.: Lunar Mineralogy, pp. 84–87. Wiley (1975)

Fuziyama, R., Takeda, H., Yazawa, Y.: The chemical link of forest and sea by river: Materials supply from land-used soil and transport by river with reference to fulvic-Fe complex. In: 19th World Congress of Soil Science, Soil Solutions for a Changing World, Brisbane, Australia (2010)

Gibson, M.A., Knudsen, C.W.: Lunar oxygen production from ilmenite (abstract). Papers Presented to the 1988 Symposium on Lunar Bases and Space Activities of the 21st Century, p. 94. Lunar and Planetary Institute, Houston (1988)

Haskin, L.A.: Water and cheese from the lunar desert: Abundances and accessibility of H, C, and N on the Moon. In: The Second Conference on Lunar Bases and Space Activities of the 21st Century, NASA Conference Publication 3166, vol. 2, pp. 393–396 (1992)

Hiesinger, H., Head III, J.W.: New Views of the Moon. In: Jolliff, B.L., Wieczorek, M.A., Shearer, C.K., et al. (eds.) Reviews in Mineral & Geochem, The Mineral. Society of America, vol. 60, pp. 1–81 (2006)

Huang, W.H., Kiang, W.C.: Laboratory dissolution of plagioclase in water and organic acids at room temperature. Amer. Mineral 57, 1849–1859 (1972)

Ichikuni, M.: Weathering of silicates and its products. In: Soil Chemistry, Chemical Society of Japan, pp. 6–18. Japan Scientific Soc. Press (1989) (in Japanese)

Iwasawa, H., Wang, D., Miyasaka, M., Takeuchi, Y.: Solution-processed silicon films and transistors. Nature 440, 783–786 (2006)

Jolliff, B.L., Gillis, J.J., Haskin, L.A., Korotev, R.L., Wieczorek, M.A.: Major lunar crustal terranes: Surface expressions and crust-mantle origins. J. Geophys. Res. 105(E2), 4197–4216 (2000)

Kobayashi, S., Kobayashi, M., Hareyama, M., Hasebe, N., Shibamura, E., Yamashita, N., Karouji, Y., Okada, T., d'Uston, C., Gasnault, O., Forni, O., Reedy, R.C., Kim, K.J., Takeda, H., Arai, T., Sugihara, T., Dohm, J.M., Kaguya Gamma-Ray Spectrometer Team: The lowest Thorium region on the lunar surface imaged by Kaguya Gamma-Ray Spectrometer. Lunar and Planetary Sci. 41, #1795 (2010)

Kobayashi, S., Mitani, T., Takashima, T., Karouji, Y., Hasebe, N.: The lunar geochemical analysis by a Gamma-Ray spectrometer for next lunar explorations. Lunar and Planetary Science 42, 1721.pdf (2011)

Korotev, R.L., Jolliff, B.L., Zeigler, R.A., Gillis, J.J., Haskin, L.A.: Feldspathic lunar meteorites and their implications for compositional remote sensing of the lunar surface and the composition of the lunar crust. Geochim Cosmochim Acta 67, 4895–4923 (2003)

Korotev, R.L., Zeigler, R.A., Jolliff, B.L.: Feldspathic lunar meteorites Pecora Escarpment 02007 and Dhofar 489: Contamination of the surface of the lunar highlands by post-basin impacts. Geochim Cosmochim Acta 70, 5935–5956 (2006)

Lasaga, A.C., Soler, J.M., Ganor, J., Burch, T.E., Nagy, K.L.: Chemical weathering rate laws and global geochemical cycles. Geochimica et Cosmochimica Acta 58, 2361–2386 (1994)

Lawrence, D.J., Feldman, W.C., Barraclough, B.L., Binder, A.B., Elphic, R.C., Maurice, S., Thomsen, D.R.: Global elemental maps of the Moon: the Lunar Prospector gamma-ray spectrometer. Science 281, 1484–1489 (1998)

Lindstrom, M.M., Lindstrom, D.J.: Lunar granulites and their precursor anorthositic norites of the early lunar crust. In: Proc. Lunar Planet. Sci. Conf. 16th in J. Geophys. Res., vol. 91, pp. D263–D276 (1986)

Loper, D.E., Werner, C.L.: On lunar asymmetries: 1. Tilted convection and crustal asymmetry. J. Geophys. Res. 107, 13-1–13-7 (2002)

Lucey, P., Korotev, R.L., Gillis, J.J., Taylor, L.A., Lawrence, D., Campbell, B.A., Elphic, R., Feldman Bill, L., Hood, L., Hunten, D., Mendillo, M., Noble, S., Papike, J.J., Reedy, R.C., Lawson, S., Prettyman, T., Gasnault, O., Maurice, S.: Understanding the lunar surface and space-Moon interactions. In New Views of the Moon. In: Jolliff, et al. (eds.) Reviews in Mineralogy & Geochemistry, vol. 60, pp. 83–219. Mineralogical Society of America (2006)

MacElroy, R.D., Klein, H.P., Averner, M.M.: The evolution of CELSS for lunar bases. In: Mendell, W.W. (ed.) Lunar Bases and Space Activities of the 21st Century, p. 623. Lunar and Planetary Institute, Houston (1985)

McKay, D.S., Heiken, G., Basu, A., Blanford, G., Simon, S., Reedy, R., French, B.M., Papike, J.: The Lunar Regolith. In: Heiken, et al. (eds.) Lunar Sourcebook: a Users Guide to the Moon. Cambridge Univ. Press (1991)

Mendell, W.W. (ed.) The Second Conference on Lunar Bases and Space Activities of the 21st Century, NASA Conference Publication 3166, vol. 2 (1992)

Meyer, C., Yang, S.V.: Tungsten-bearing yttrobetafite in lunar granophyre. Am. Mineral 73, 1420–1425 (1988)

Morota, T., Haruyama, J., Ohtake, M., Matsunaga, T., Honda, C., Yokota, Y., Kimura, J., Ogawa, Y., Hirata, N., Demura, H., Iwasaki, A., Sugihara, T., Saiki, K., Nakamura, R., Kobayashi, S., Ishihara, Y., Takeda, H., Hiesinger, H.: Timing and characteristics of the latest mare eruption on the Moon. Earth and Planet Sci. Lett. 302, 255–266 (2011)

Nakamura, T., Smith, B.K., Gustafson, R.J.: Solar thermal power system for production from lunar regolith: Engineering system development. In: Joint Annual Meeting of LEAG, ICEUM, and SRR, Florida (2008)

Nyquist, L.E., Bogard, D.D., Yamaguchi, A., Shih, C.-Y., Karouji, Y., Ebihara, M., Reese, Y., Garrison, D., McKay, G., Takeda, H.: Feldspathic clasts in Yamato-86032: Remnants of the lunar crust with implications for its formation and impact history. Geochim Cosmochim Acta 70, 5990–6015 (2006)

Ohtake, M., Matsunaga, T., Haruyama, J., Yokota, Y., Morota, T., Honda, C., Ogawa, Y., Torii, M., Miyamoto, H., Arai, T., Hirata, N., Iwasaki, A., Nakamura, R., Hiroi, T., Sugihara, T., Takeda, H., Otake, H., Pieters, C.M., Saiki, K., Kitazato, K., Abe, M., Asada, N., Demura, H., Yamaguchi, Y., Sasaki, S., Kodama, S., Terazono, J., Shirao, M., Yamaji, A., Minami, S., Akiyama, H., Josset, J.-L.: The global distribution of pure anorthosite on the Moon. Nature 461(7261), 236 (2009)

Pandeya, S.B., Singh, A.K., Dhar, P.: Influence of fulvic acid on transport of iron in soils and uptake by paddy seedlings. Plant Soil 198, 117–125 (1998)

Papike, J.J.: Review Mineralogy, vol. 36. Mineralogical Society of America, Washington DC (1998)

Salisbury, F.B., Bugbee, B.G.: Wheat farming in a lunar base. In: Mendell, W.W. (ed.) Lunar Bases and Space Activities of the 21st Century, p. 635. Lunar and Planetary Institute, Houston (1985)

Sasaki, S., Tanaka, K., Higuchi, K., Okuizumi, N., Kawasaki, S., Shinohara, N., Senda, K., Ishimura, K.: A new concept of solar power satellite: Tethered-SPS. Acta Astronautica 60, 153–165 (2006)

Shearer, C.K., Newsom, H.E.: W-Hf isotope abundances and the early origin and evolution of the Earth-Moon system. Geochim Cosmochim Acta 64, 3599–3613 (2000)

Soloman, S.C., Longhi, J.: Magma oceanography: Thermal evolution. In: Proc. Lunar Sci. Conf. 8th, pp. 583–599 (1977)

Stöffler, D., Knoll, H.-D., Marvin, U.B., Simonds, C.H., Warren, P.H.: Recommended classification and nomenclature of lunar highland rocks – a committee report. In: Merrill, Papike (eds.) Proc. Conf. on The Lunar Highlands Crust. Geochim Cosmochim Acta, vol. (suppl. 12), pp. 51–70. Pergamon Press (1980)

Takeda, H.: Structural studies of rim augite and core pigeonite from lunar rock 12052. Earth Planet Sci. Lett. 15, 65–71 (1972)

Takeda, H.: Mineralogical records of early planetary processes on the HED parent body with reference to Vesta. Meteoritics and Planet Sci. 32, 841–853 (1997)

Takeda, H.: Evolution of Planetary Materials (in Japanese), Gendaitosyo, Sagamihara, pp. 69–110 (2009)

Takeda, H., Ridley, W.I.: Crystallography and chemical trends of orthopyroxene- pigeonoite from rock 14310 and coarse fine 12033. In: Proc. Lunar Sci. Conf. 3rd, pp. 423–430 (1972)

Takeda, H., Miyamoto, M., Mori, H., Tagai, T.: Mineralogical studies of clasts in lunar highland regolith breccia 60019 and in lunar meteorite Y82129. In: Proc. Lunar Planet Sci. Conf. 18th, pp. 33–43. Lunar Planet Institute, Houston (1988)

Takeda, H., Yamaguchi, A., Bogard, D.D., Karouji, Y., Ebihara, M., Ohtake, M., Saiki, K., Ari, T.: Magnesian anorthosites and a deep crustal rock from the farside crust of the moon. Earth Planet Sci. Lett. 247, 171–184 (2006)

Takeda, H., Kobayashi, S., Yamaguchi, A., Otsuki, M., Ohtake, M., Haruyama, J., Morota, T., Karouji, Y., Hasebe, N., Nakamura, R., Ogawa, Y., Matsunaga, T.: Olivine fragments in Dhofar 307 lunar meteorite and surface materials of the farside large basins. Lunar and Planetary Sci. 41, #1572 (2010)

Taylor, L.A.: Resources for a lunar base: Rocks, minerals, and soil of the Moon. In: The Second Conference on Lunar Bases and Space Activities of the 21st Century, NASA Conference Publication 3166, vol. 2, pp. 361–377 (1992)

Taylor, S.R.: Lunar Science: A Post Apollo View. Pergamon Press, Oxford (1975)

Ullman, W.J., Welch, S.A.: Organic ligands and feldspar dissolution, water-rock interactions, ore deposits, and environmental geochemistry: A tribute to David A. Crerar. The Geochemical Society, Special Publication No. 7 (2002)

Utsunomiya, S., Takeda, H.: Hydrothermal treatments of powders of plagioclase –Lunar resource utilization. In: Proc. 29th ISAS Lunar Planet Symp., Inst. Space Astro Sci., pp. 44–47 (1996)

Wada, H., Yamashita, M., Katayama, N., Mitsuhashi, J., Takeda, H., Hashimoto, H.: Agriculture on Earth and on Mars. In: Denis, J.H., Aldridge, P.D. (eds.) Space Exploration Research. Nova Science Publisher (2009) (in press)

Warner, J.L., Simonds, C.H., Phinney, W.C.: Apollo 17, Station 6 boulder sample 76255: Absolute petrology of breccia matrix and igneous clasts. In: Proc. Lunar Sci. Conf. 7th, pp. 2233–2250 (1976)

Warren, P.H.: The magma ocean concept and lunar evolution. Annu. Rev. Earth Planet Sci. 13, 201–240 (1985)

Warren, P.H.: A concise compilation of petrologic information on possibly pristine nonmare Moon rocks. Amer. Mineral 78, 360–376 (1993)

Yamaguchi, A., Karouji, K., Takeda, H., Nyquist, L., Bogard, D., Ebihara, M., Shih, C.-Y., Reese, Y., Garrison, D., Park, J., McKay, G.: The variety of lithologies in the Yamato-86032 lunar meteorite: Implications for formation processes of the lunar crust. Geochim Cosmochim Acta 74, 4507–4530 (2010)

Yazawa, Y., Mikouchi, T., Takeda, H.: Mars: Prospective Energy and Material Resources. In: Badescu, V. (ed.) ch. 19, pp. 483–516. Springer, Berlin (2010)

Yazawa, Y., Wong, M.T.F., Gilkes, R.J., Yamaguchi, T.: Effect of additions of brown coal and peat on soil solution composition and root growth in acid soil from wheatbelt of Western Australia. Commun. in Soil Sci. Plant Anal. 31, 743–758 (2000)

Yazawa, Y., Takarada, T., Irisawa, A., Yamaguchi, T.: An "average" structure proposed for fulvic acid isolated from iodine-brine water in Chiba (in Japanese). Rep. China Inst. Tech. 52, 29–36 (2005)

Yazawa, Y., Saito, M., Takeda, H.: Variation between terrestrial and lunar or martian desert-Surface materials on Mars and its possible use for agriculture (in Japanese). Biol. Sci. Space 21, 129–134 (2007)

Yazawa, Y., Takeda, H.: Roles of terrestrial fulvic acid in producing agricultural soils from regoliths of habitable planets for space agriculture. In: Soils 2008 -The Living Skin of Planet Earth for the Joint Conference of the Australia and New Zealand Societies of Soil Science in conjunction with the International Year of Planet Earth, Abstract, Massey University, NZ, p. 113 (2008)

Yazawa, Y., Asami, S., Yamaguchi, K., Ito, Y., Takeda, H.: Creation of soils by humic substances and CO_2 for Space Agriculture. In: 1st International Conference on Arid Land, Narita Japan (in press, 2011)

6 Lunar Holes and Lava Tubes as Resources for Lunar Science and Exploration

Junichi Haruyama[1], Tomokatsu Morota[2], Shingo Kobayashi[3], Shujiro Sawai[1], Paul G. Lucey[4], Motomaro Shirao[5], and Masaki N. Nishino[1]

[1] Japan Aerospace Exploration Agency (JAXA), Japan
[2] Nagoya University, Japan
[3] National Institute of Radiological Sciences (NIRS), Japan
[4] University of Hawaii at Manoa, USA
[5] Institute of Planetary Geology, Japan

6.1 Introduction

The Moon is the nearest celestial body to the Earth. As such, it has long been investigated to understand its formation and evolution, as a paradigm for better understanding the terrestrial planets, as well as all airless bodies in our solar system (e.g., Vesta, Phobos). The Moon's proximity to the Earth—more than one hundred times closer than any planet — makes it a convenient target for exploration by spacecraft. Since the dawn of the space age in the previous century, we have explored the Moon with several spacecraft and even succeeded in sending astronauts there. One of the lessons of those explorations that hinders any future lunar expeditions is the severe conditions on the lunar surface. The lack of an atmosphere (10-12 torr) means that cosmic/galactic/solar rays, as well as the many micrometeorites directly striking the surface; in addition, surface temperatures vary widely, over a day-night range of more than 300 K.

Long-term activities on the Moon require artificial shelters protected by thick roofs and walls, primarily to protect against the radiation hazards of extra-lunar rays. However, such a construction project would be technically difficult and costly. Natural lunar caves have therefore been considered as alternate candidates for early shelters (e.g., Hörz 1985). Such caves are likely on the Moon, in the form of lava tubes (buried lava conduits), commonly seen in low-viscosity lava-flow areas on the Earth, which are often popular tourist attractions. The expected thick roofs would protect people and instruments inside from solar rays, meteorites, and wide temperature variations. However, lava tubes lie beneath the surface, so it is

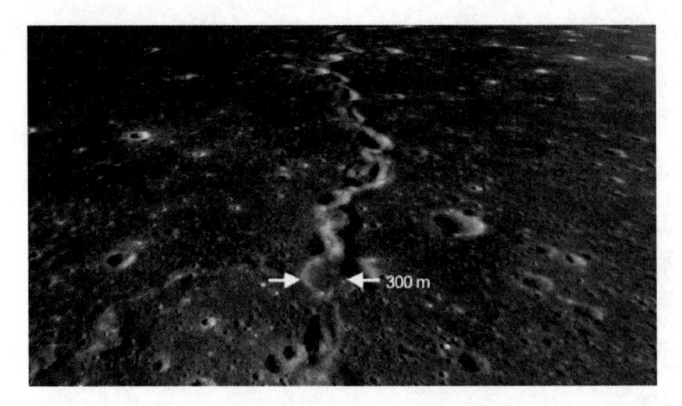

Fig. 6.1 Segmental and elongated depressions in a rille (45°E, 35°N) based on SELENE/TC data (Product ID: TCO_MAP_N37E315N32E318). Previous works inferred that other parts of these depressions may be intact lava tubes. However, no entrance was identified in the Terrain Camera data (10 m/pixel-resolution).

difficult to find them by remote sensing from orbit. Some "sinuous rilles," relics of meandering lava flows, may in fact be collapsed lava tubes (e.g., Oberbeck et al. 1969, Greeley 1971, Hulme 1973, Cruikshank and Wood 1972). Segmented and elongated depressions are seen in some of these sinuous rilles (Fig. 6.1).

It has been inferred that these depressions may be intact lava tubes. Previous researchers (e.g., Coombs and Hawke 1992) have searched for lava tube entrances (i.e., evidence of underlying lava tubes) using Apollo and Lunar Orbiter data, but without success. Recently, however, the discovery of the large vertical shafts on the Moon has rekindled investigations of lunar lava tubes. In this chapter, we introduce three giant lunar holes discovered by SELENE (nicknamed Kaguya), the Japanese lunar orbiter. In the following sections, we outline the significance of such lunar holes and lava tubes probably extending from the bottom the holes for scientific research and their potential for future in-situ resource utilization (ISRU).

6.2 Three Giant Lunar Holes

6.2.1 Discovery and Detailed Observation of Lunar Holes

The 10 m/pixel-resolution Terrain Camera (TC) aboard the Japanese lunar orbiter Selenological and Engineering Explorer (SELENE) detected a possible entrance to an underlying lava tube in 2009 (Haruyama et al. 2009b). The feature is a vertical hole in the middle part of a rille (303.3°E, 14.2°N, -1.65 km elevation from the lunar mean radius of 1737.4 km) in the Marius Hills region (Fig. 6.2, Table 6.1), a region rich in prominent volcanic features (Greely et al. 1971).

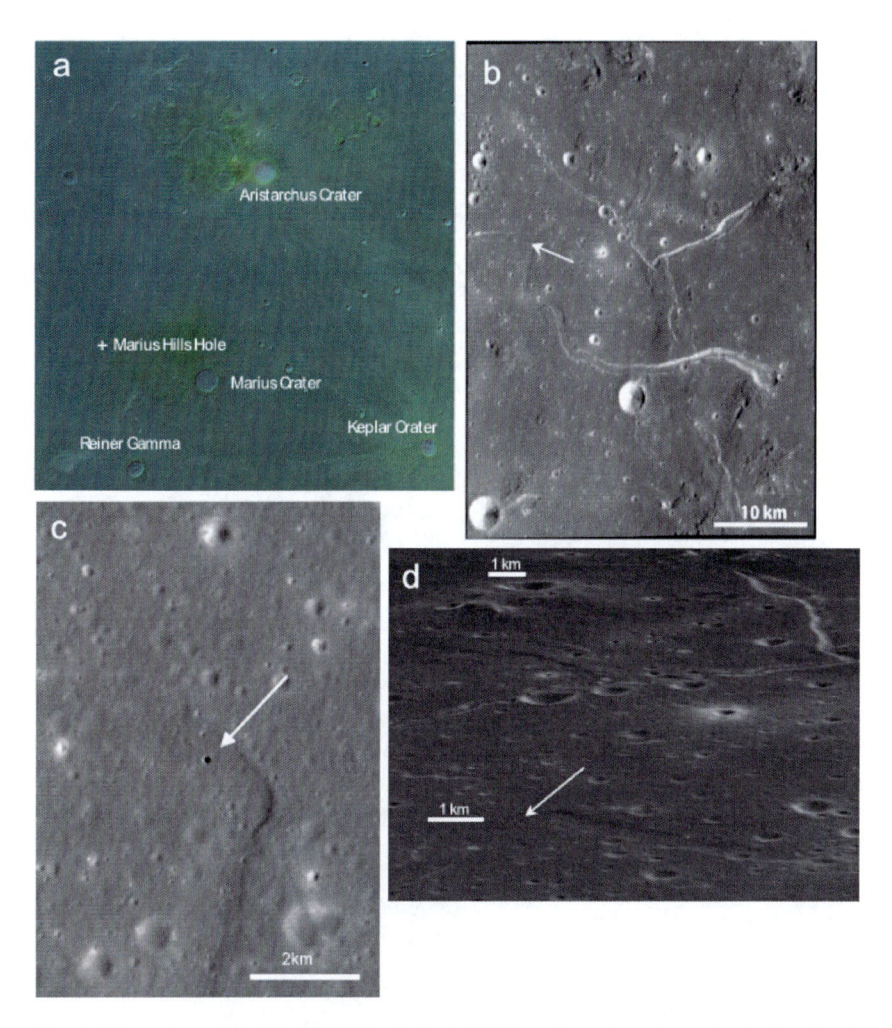

Fig. 6.2 Location of the Marius Hills Hole (MHH). a: Overview image including the location of the hole (250 km × 250 km) which is a composite of ortho image and colored elevation data based on SELENE Terrain Camera stereo-pair data. b: Enlarged image of the region including MHH which is located in a rille. c: MHH at (303.3°E, 14.2°N). The hole is apparently dark and different from normal craters in the SELENE TC data [Product ID: DTMTCO_02742N136E3036, solar elevation angle 48.0°]. d: Perspective view around MHH (from around west) based on [Product ID: DTMTCO_02742N136E3036].

The hole is 65 m in diameter with an estimated depth of 90 m (Haruyama et al. 2009b; more accurate values were subsequently obtained, as described later in this section). The hole thus differs from normal impact craters in the vicinity and is very similar to openings called "skylights" in lava tubes on the Earth and Mars (Okubo and Martel 1998, Wyrick et al. 2004, Cushing et al. 2007, Cushing 2011).

Table 6.1 Three gigantic lunar holes discovered by SELENE (Kaguya)

	latitude, longitude	elevation[a]	long×short-axis lengths[b]	depth[b,c]	long-axis angle[b,d]	surrounding surface age
Marius Hills Hole (MHH)	303.3°E, 14.2°N	-1.65 km	59 m×50 m	48 m	53°	3.56 Gyrs
Mare Tranquillitatis Hole (MTH)	33.2°E, 8.3°N	-0.77 km	98 m×84 m	107 m	165°	3.62 Gyrs
Mare Igenii Hole (MIH)	166.0°E, 35.6°S	-3.62 km	118 m×68 m	45 m	36°	3.17 Gyrs

[a]Elevation from lunar mean radius 1737.4 km
[b]Based on LRO NAC data: M122584310L for MHH, M12671087R for MTH, M123485893R for MIH
[c]Estimated from shadow lengths and solar incidence angles (see Ashley et al. (2011) for various observation conditions)
[d]Measured clock-wise from the north

After the discovery of the Marius Hills Hole (MHH), Haruyama et al. (2010) surveyed the TC data - which cover nearly 100% of the entire lunar surface - for similar lunar hole structures. They took advantage of the extreme darkness of shadows expected in a deep pit when illuminated at about 45 degrees. At these sun angles, the shadowed portions of the vertical holes would be darker than shadows in typical craters would be that more illuminated by sunlight reflected from the crater walls (Haruyama et al. 2008a,b).

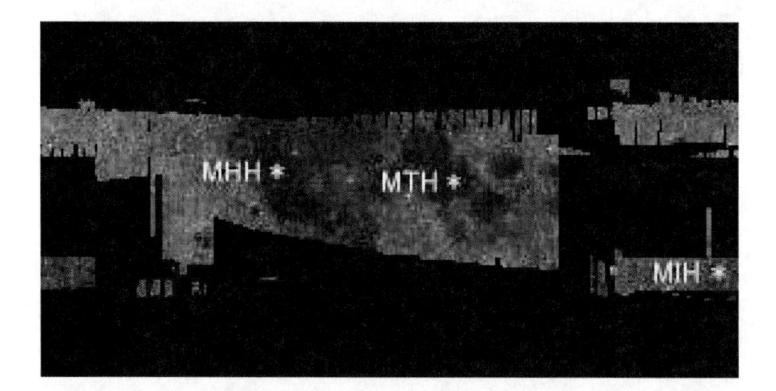

Fig. 6.3 Three lunar hole locations, found in SELENE Terrain Camera (TC) data with solar-elevation angles exceeding 40°, representing a search covering more than 95% of lunar maria. (*1) Marius Hills Hole (303.3°E, 14.2°N). (*2) Mare Tranquillitatis Hole (33.2°E, 8.3°N). (*3) Mare Ingenii Hole (166.0°E, 35.6°S).

The extremely sensitive Terrain Camera is able to detect this difference in brightness, and Haruyama et al. (2010) extracted a data set, consisting of TC data acquired at solar elevation angles of 40° to 60° (Fig. 6.3). As a result, two additional hole structures were discovered, in Mare Tranquillitatis and in Mare Ingenii (Haruyama et al. 2010).

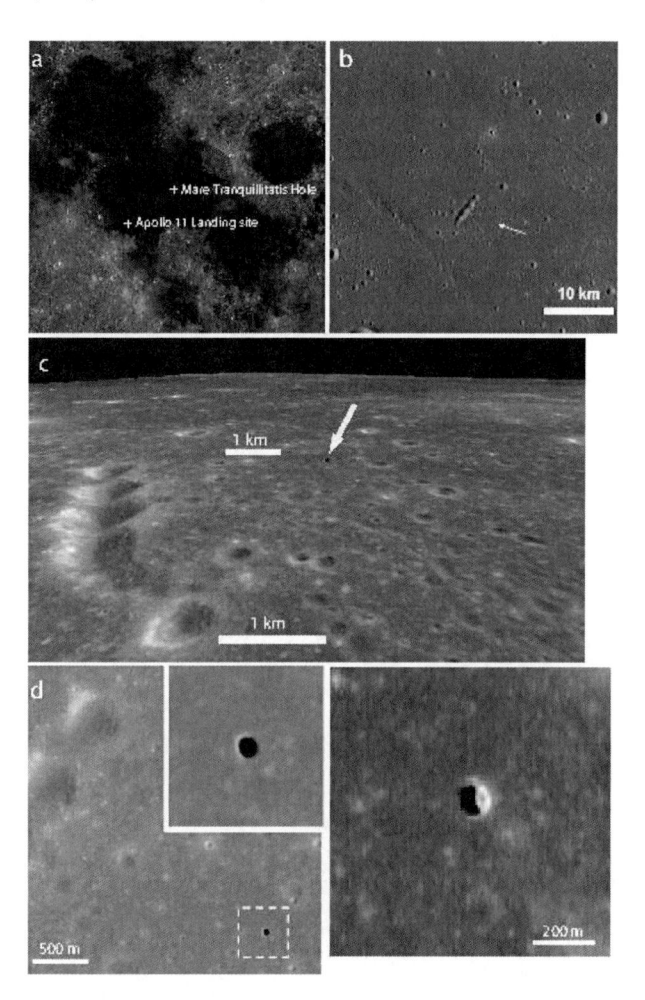

Fig. 6.4 Mare Tranquillitatis Hole (MTH) a: Overview of Mare Tranquillitatis with locations of MTH and the Apollo 11 landing site, which is about 350 km from MTH. b: Enlarged image of the region including MTH [SELENE/TC Product ID: TC_EVE_MAP, solar elevation angle 16.2°]. No rilles are associated with the hole. c: Perspective view around MTH (from South-West) based on [Product ID: TCO_MAP_N12E031N05E036, solar elevation angle 49.6°]. d: MTH at (33.2°E, 8.3°N) [SELENE/TC Product ID: DTMTCO_MAPN12E031N05E036] e: Illuminated MTH floor [Multiband Imager data: SELENE revolution number of 6358, solar elevation angle 74.6°].

The Mare Tranquillitatis Hole (MTH) is located at 33.2°E, 8.3°N, -0.77 km elevation (Table 6.1, Fig. 6.4), 350 km from the Apollo 11 landing site (Fig. 6.4a). It is nearly twice the diameter of Marius Hills feature and is roughly circular with a short-axis length of 110 m (east-west) and a long-axis length of 120 m (north-south) in the TC data (Fig. 6.4d). Floor data were acquired by a Multiband Imager (MI) (Fig. 6.4e). The hole was estimated to be over 100 m deep from shadow measurements. Unlike the Marius Hills hole, the Tranquillitatis hole is not in or near any sinuous rilles (Fig. 6.4b) and no exposed surface bulges, which are often seen with terrestrial lava tubes, are seen around the hole, even in a TC image taken at a very-low solar-elevation angle of 16.2° (Fig. 6.4b). Furthermore, there are numerous craters near MTH, but no other holes are observed in the area at the resolution of TC (10 m/pixel),

The Mare Ingenii Hole (MIH) is located at 166.0 °E, 35.6 °S, -3.62 km elevation, (Table 6.1, Fig. 6.5). It has a slightly irregular shape (rounded triangle) with a long-axis length of 140 m (east-west) and a short-axis length of 110 m (north-south) in the TC data (Fig. 6.5d). The SELENE TC and MI did not observe the bottom of MIH, even at the highest solar elevation angle (47.4° with TC), and thus, it is only possible to place an upper limit on its depth. Like MTH, this hole is not located in a rille (Fig. 6.5b), and no rilles, pits, or bulges are seen near it.

The TC data acquired at solar angles of 40 to 60° covered more than 95% of lunar Maria (see Fig. 6.3). Therefore, MHH, MTH, and MIH are probably the only holes at the scale of one hundred meters on the Moon's surface.

The U.S. lunar satellite Lunar Reconnaissance Orbiter (LRO) that had been launched in 2008 features an extremely high-resolution Narrow Angle Camera (NAC) that images the lunar surface with 0.5 to 1 m/pixel-resolution. Several observations under different lighting conditions of the holes have been made by the LRO NAC and have provided new details regarding the morphologies of the holes.

LRO NAC images confirmed Marius feature to be nearly circular. Additionally, very clear, detailed floor images were acquired at high solar-elevation angles (Fig. 6.6a). From these data, we estimate the floor axis lengths to be 59 m x 50 m with a depth of 47 m, based on a bend in the slope of the inner wall that forms an "inner edge" of the hole at some depth. Ashley et al. (2011) estimated the long-axis length to be 57 m to 62 m, the short-axis length to be 46 m to 61 m, and the depth to be 36 m, based on several observations. SELENE cameras overestimated the depth (90 m) due to their lower spatial resolutions. No ejecta are seen surrounding the hole, confirming that it is neither a normal impact crater nor a volcanic vent.

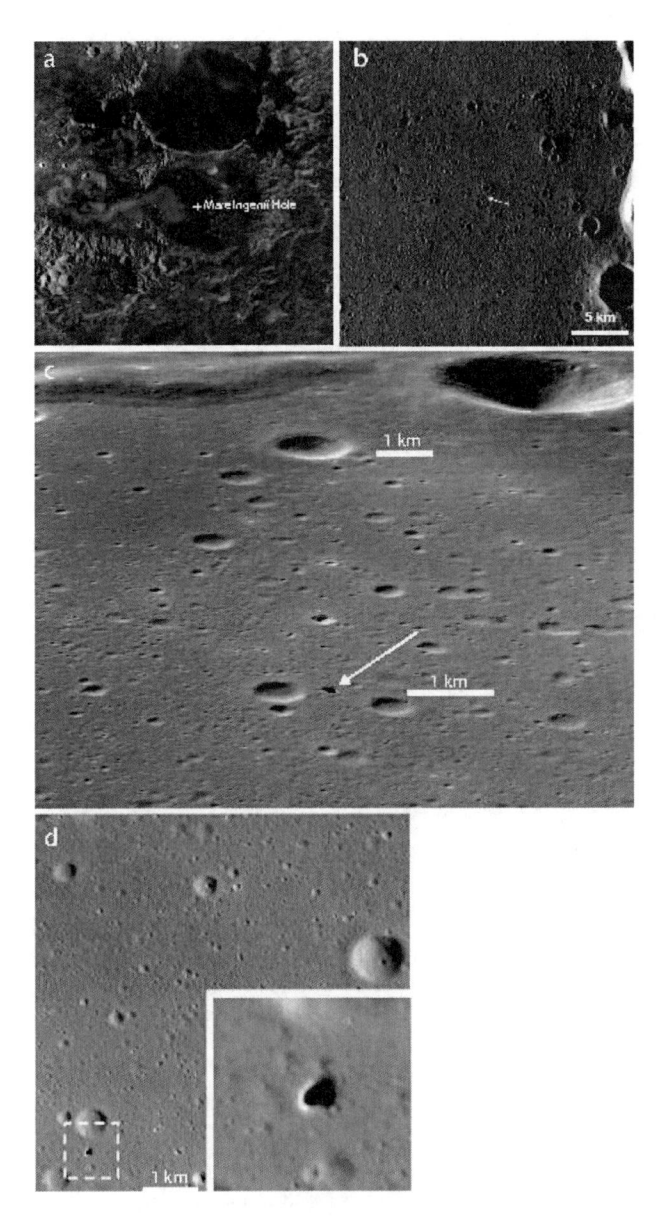

Fig. 6.5 Mare Ingenii Hole (MIH). a: Overview of Mare Ingenii on the far side of the Moon. MIH is located at an edge of a prominent swirl feature inside the south–east crater of the Mare. b: Enlarged image of the location around the MIH attained at a lower solar elevation angle of 7.3°. As with MTH, there is no rille near MIH. c: Perspective view around MIH (from North-West) based on [Product ID: TCO/DTM_MAP_S35E165S38E168, solar elevation angle 25.0°]. d: MIH at (166.0°E, 35.6°S) [Product ID: TCO_MAP_S35E165S38E168].

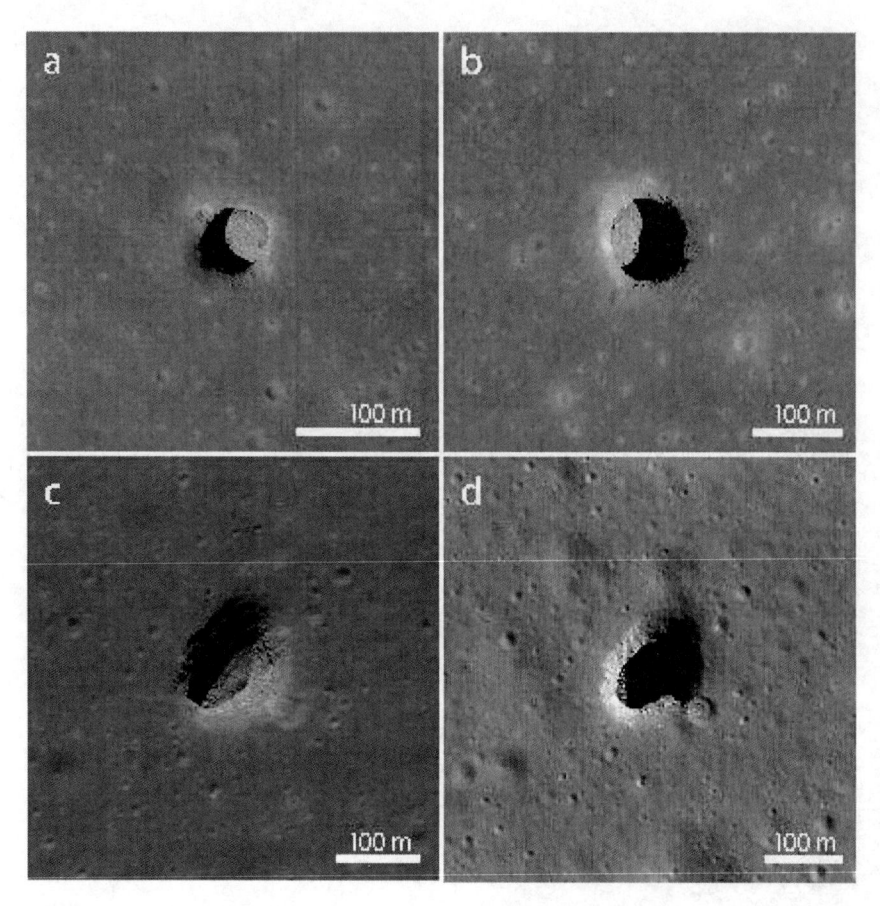

Fig. 6.6 LRO NAC 0.5-1 m/pixel resolution images. a: MHH (Product ID: M122584310L). b: MTH (Product ID: M12671087R). c, d: MIH (Product IDs:M123485893R, M128202846L). All the floors of these holes seem to be covered by fine soil with many boulders of a meter scale. The lowest portions of the inner walls do not continuously conjugate with the floors for all the holes, which strongly indicates the existence of extended voids (possibly cravens of lava tubes) at the bottoms of all the holes.

Notably, the lowest portion of the inner wall of MHH does not continuously follow the floor (Fig. 6.6a), which strongly suggests the existence of extended voids (possibly lava-tube caverns) at the bottom of the hole. The bottom of MHH seems to be covered by fine regolith with many boulders of a meter scale. The geometry of the shadow on the floor is quite similar to the outline of the inner edge of the hole, suggesting that the floor is flat.

The areas surrounding MHH are 3.56 Gyrs old (Haruyama et al. 2009a). A crater-size distribution analysis revealed at least one layer under the surface that was formed 0.1 Gyrs earlier than the surface (Haruyama et al. 2009b). The possibility exists of a lava tube formed from the lower flow, at this earlier age, which has caved in. The possible lava tube formation age of 3.6 Gyrs corresponds to a time of high lunar volcanic activity (Morota et al. 2011). While the age of formation of the hole structure is unclear to date, it could be inferred to be so old that the pile of rubble from a possibly collapsed roof at the opening of the hole has been crushed and flattened by meteorite bombardment and covered by regolith and soil from the inner walls. The meter-scale boulders on the floor may have fallen from the inner wall in recent times.

The LRO images of Mare Tranquillitatis hole also clarified the details of this feature (Fig. 6.6b, LRO product ID is M126710873R). The long-short axes of MTH are measured to be 98 m and 84 m, respectively; and it is up to 107 m deep. The estimates obtained by Ashley et al. (2011) based on data acquired under various imagery conditions are 53 to 80 m for the longer axis and 81 to 88 m for the shorter axis, with a 92 to 106 m depth. The MTH features, as clarified by LRO NAC, are similar to those of MHH. No boulders are seen around the hole, and the surface surrounding MTH formed 3.62 Gyrs ago, corresponding to a time of large amounts of lunar mare volcanism. The upper portion of the MTH wall seems to be sloped, somewhat like a cone, and the lower portion seems to be vertical. The floor is littered with numerous meter-scale boulders and is probably flat, as indicated by shadow analysis, similar to that for MHH. The inner wall apparently does not follow the floor (Fig. 6.6b), similar to the case of MHH. Therefore, there must again be a space between the floor and the inner wall; MTH is thus also suggested to be a skylight of a subsurface cavern (probably a lava tube).

MIH somewhat differs from MHH and MTH in that the LRO NAC images of MIH reveal an eccentric feature (Fig. 6.6c,d). While MIH's rim is shaped as a rounded triangle at the surface (observed by SELENE TC and MI), the inner slope from the rim is more gradual than that of the other holes, and the bottom is almond shaped. The bottom of MIH has a small mound of fine regolith on the eastern side (Fig. 6.6c), while numerous boulders occupy the western side (Fig. 6.6d). The long-short axes are 118 m and 68 m, respectively. MIH is estimated to be 45 m deep from shadow measurements based on an LRO NAC image [product ID: M123485893R]. However, the depth appears variable depending on the observation conditions, probably because the depth of the inner edge from the surface differs and/or the floor is not flat. Ashley et al. (2011) reported a depth of 47 m to 76 m from the inner edge. Based on the crater-size frequency distribution, two formation ages of the surrounding area are deduced, 3.17 Gyrs and 3.59 Gyrs. There are no boulders around the hole, similar to the areas around the other holes, implying that the hole is not a volcanic vent. Like MTH, despite the presence of the hole, there are no additional surface features such as deep holes, pits, or extended surface bulges, which might suggest the existence of subsurface voids.

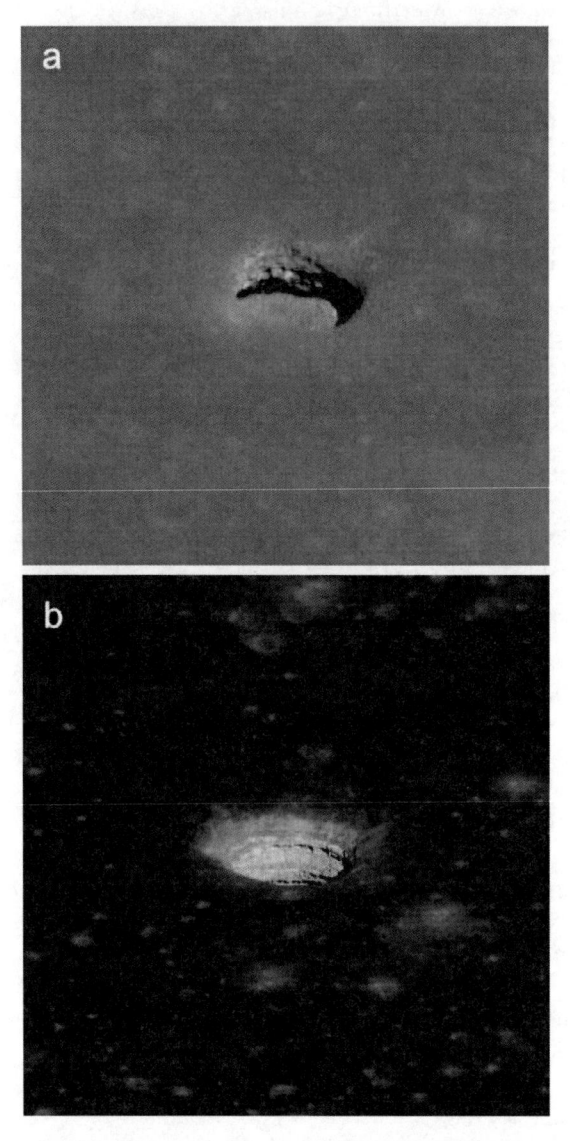

Fig. 6.7 LRO NAC oblique observation images. a: MHH [Product ID: M137929856R]. At the bottom of MHH, an overhanging layer projects its shadow onto the floor, suggesting that entrances to "sublunarean" voids should open at the bottom of the hole. b: MTH [Product ID: M144395745L]. Multiple lava layers of a few meters thickness are observed in the inner wall of MTH.

LRO acquired further valuable information through side-viewing observations (Ashley et al. 2011). In oblique images of MHH (Fig. 6.7a) and MTH (Fig. 6.7b), multi-layer outcrops of meter-scale thickness were observed at the inner walls of these holes. Some layers seem to form overhanging structures. At the bottom of MHH, an overhung layer projected its shadow onto the floor, suggesting that there is no large mound of regolith/soil from the inner wall and that entrances to "sublunar" voids should open at the bottom of the hole, though Ashley et al. (2011) noted that this is speculative because the voids are dark and not well-imaged. The absence of a mound of regolith/soil in the oblique images is consistent with the feature in the nadir images of MHH that indicate a clear boundary of the inner edge of the rim in its depth.

6.2.2 Resources Supporting Geological and Mineralogical Study for Lunar Holes

Various important geological and mineralogical processes could be uniquely and effectively observed inside these holes. The holes therefore have high potential as resources for scientific study.

Outcrops and Dust Environment Study Field

As has been seen in the oblique images acquired by LRO, the inner walls of lunar holes are excellent outcrops exhibiting lava strata (see Sect. 6.2.1). Exposed lava layers could be observed on the inner walls of craters and rilles, but these are largely covered by regolith from upper portions. The nearly vertical walls of the holes exhibit fresh surfaces with a much larger section, thereby preserving more lunar volcanic history, and would be invaluable outcrop for study of successive lunar volcanic activities, buried paleosols, etc.

Lunar mare volcanism (magma eruptions) occurred over a period of 2.3 Gyrs, from about 3.8 Gyrs (e.g., Head 1976) to about 1.5 Gyrs (Morota et al. 2011), more than a half of the age of the Moon. The materials supplied by the volcanism resurfaced the Moon and formed a "secondary" crust (Head and Wilson 1992). Investigation of the lunar mare volcanic history is essential to understanding the lunar evolution complexity and could lead to insights into the evolution of other planets resurfaced by volcanic activities such as Mars, Mercury, and the Earth. A central question for the lunar mare volcanism is what mechanism caused magma eruptions, particularly in the late stage such as the Eratosthenian period (2.5 to 3.0 Gyrs ago). A thick surface crust must have solidified in the Eratosthenian period. This overlaid crust and previously deposited lava terraces would stop internal magma from ascending. For the deep magma to reach the surface, a sufficient ascent driving mechanism must be provided, such as (1) increasing pressure of bubbled volatiles at a depth that might be due to a temperature increase and re-melting caused by the condensation of radiogenic elements in the late stage of internal thermal evolution or (2) dike formation due to cooling and shrinking of previously deposited lava terraces. The former mechanism would cause explosive eruptions with pyroclastic materials piling up as scoria on the lunar surface. The latter would possibly drive magma ascent through the dikes and provide flood basalt lavas on the surface. While the model ages of the surfaces around the holes are estimated to be nearly 3.6 Gyrs old (see Sect. 6.2.1), the thin layers observed at the surface of the inner walls of the holes may be the results of much younger eruptions. Detailed information of well-preserved materials and conditions (lava or scoria) of layers exposed at the inner walls of the holes with correct temporal orders would provide an explicit key to understanding the eruption mechanism and its possible time-dependent transition.

In the paleosol regolith layers sandwiched by subsurface solidified layers, solar-wind elements implanted since over 3 billion years ago may be trapped and preserved. Such elements could provide key information for understanding past solar activity. Measuring magnetic fields remaining in lava layers will be crucial for determining whether or not an intrinsic lunar dynamo ever occurred in the Moon. A regolith layer with no trapped solar-wind particles, if found, might also support the existence of a past lunar dynamo.

Other possibilities for resources provided by the holes include a unique study of dust characteristics on the Moon. An electric potential difference is suggested at the boundary of sunlight and shadow on the Moon (Stubbs et al. 2006), because the sunlit areas would be positively charged by implanted solar-wind protons. This potential difference would induce an electric field along which fine, negatively charged dust particles may migrate from shadows to sunlit areas. Whether or not such a phenomenon occurs is interesting from the viewpoint of understanding the details of the lunar environment, as well as in the study of possible impediments to long-term unmanned and manned activities on the lunar surface. While the shadow-sunlit boundary at sunrise and sunset on the Moon comes around about every two weeks and passes any given site in a single second, the boundaries on the hole floors exist for a longer time during the daytime. The hole bottoms will thus provide a good basis for research into electromagnetic phenomena and dust currents.

Possible Water Resource

The possibility of water storage in lunar holes was explored by (Haruyama et al. 2010; 2011). Clark (2009), Sunshine et al. (2009), and Pieters et al. (2009) reported a 3 μm absorption feature from the lunar surface, demonstrating the existence of hydroxyl and/or water molecules even at higher temperatures than 270 K on the lunar surface. These compounds can be produced by reactions of oxygen bearing lunar materials and solar-wind protons (McCord et al. 2011, Ichimura et al. 2011). If solar-wind protons are implanted into the surface, some of them must have entered into the lunar holes and remained there. Assuming that the lunar holes opened simultaneously with or just after the formation of the surrounding surface and that the solar-wind hydrogen flux was similar to the present flux of 4 x $108/(cm^2sec)$, the maximum column density of the protons inside MHH, MTH, and MIH would be 0.24, 0.24, and 0.19 tons/m^2, possibly corresponding to water-molecule densities of nearly 2.1, 2.2, and 1.8 tons/m^2, very speculative indeed.

While this abundance is not protected from loss, solar-wind protons and/or water molecules should be more effectively retained at the bottom of a lunar hole than at the surface for several reasons. First, the angles through which trapped molecules on the bottoms of the holes can escape to space are smaller than the 2π (hemispherical) angle available at the surface. For MHH and MTH, which have a diameter to depth ratio of 1, the cone angle that corresponds to the escape ratio of molecules from the bottom of the hole is only 10% of the hemispherical angle at the surface.

Second, some areas on the floors of holes are accessible for implantation of solar wind protons but not for solar ultra-violet (UV) radiation. On the surface, UV radiation may dissociate implanted hydrogen molecule hydroxyls and/or water molecules into lighter-ion molecules, which can more easily escape from the surface into space. Some implanted protons and their compounds on the hole bottoms may have avoided the UV dissociation process and have remained there. It may seem that implanted solar-wind protons would be eroded by solar UV, but photon and protons do not follow the same trajectories. The flight direction of solar-wind protons has a dispersion of up to $10°$ (Marsch et al. 1982) around the radial direction of the Sun due to Lamor motion, while UV photons pierce straight through space from the Sun. In addition, solar-wind protons reach the lunar surface with an aberration of $4°$ due to the perpendicular geometrical relationship between the flight direction of the solar wind protons, at 400 km/sec, and the Moon-Earth rotation direction around the Sun, at 30 km/sec.

Third, dust particles produced from the inner walls of holes by micrometeorite impacts cover the floor, protecting implanted protons, hydroxyls, or water molecules and preventing them from being dissociated by UV and escaping into space. Fourth, milder temperature profiles are expected in the shadows on the bottoms of the holes than at the surface (Haruyama et al. 2011; also described in Sect. 6.2.3) aiding retention of implanted protons, hydroxyls, and/or water molecules. Fifth, if lava tubes indeed extend from the hole bottoms, the implanted protons, hydroxyls, and/or water molecules retained there may have diffused and semi-permanently settled inside the tubes because temperatures of the tubes must be lower and much more stable than those of the hole floors.

Thus, some implanted protons, hydroxyls, and/or water molecules might be retained on the hole floors relative to surface abundances, which are now known to be detectable. In addition to solar-wind source, water may have originated from cometary water that migrated from impact locations to the holes and became trapped. Higher deuterium-to-hydrogen ratios (D/H) have been observed in comet water than in the Earth's water (e.g., Villanueva et al. 2009, Greenwood et al. 2011 and references cited in this paper), so we could determine the origin of the water molecules by investigating the D/H of water at the hole bottoms.

Current data prevents an unequivocal conclusion that the amount of hydrogen-related molecules in the holes is sufficiently large to be used for future unmanned and manned activities on the Moon. However, their potentialities may make the holes are aid in understanding the origin of some of the water on the Moon.

Finally, we note here that the stored extra-lunar materials probably contain not only protons and water but also other volatiles, such as carbon, helium, and nitrogen atoms. These molecules are also important materials for space science.

Furthermore, considering that the surface around the holes is 3.6 Gyrs old, terrestrial life that had appeared on the Earth by that time could have been ejected by the more common large impacts of the era on the Earth and might have reached the Moon; traces of ancient life could have settled into the holes and have been preserved there. Though finding ancient terrestrial life preserved intact may be extremely unlikely, certain clues about molecular evolution that have already been lost from the Earth may be found in holes on the Moon.

6.2.3 Potential of Lunar Holes for Site Lunar Bases

In addition to scientific interest in the geological and mineralogical studies described in the previous section, lunar holes are excellent candidates for the placement of future lunar bases. Here we describe the relevant advantages of lunar holes.

Radiation Shielding

One of the most serious concerns for unmanned and manned activities on the Moon is radiation damage due to solar and galactic cosmic rays (SCR and GCR). Denisov et al. (2010) calculated the total annual dose of the SCR protons (p-SCR) to be a few sieverts at the solar maximum on the lunar surface. Hayatsu et al. (2009) estimated that an SCR event on October 28, 2003, exceeded 2 Sv on the lunar surface. These doses of radiation exceed the allowance limit of 1.2 Sv per life long (postulating low dose rate such as 1 to 2 mSv/day) for JAXA male astronaut with his first space flight after age 40 ("JAXA radiation control guideline for astronauts of the International Space Station" 2002), and far exceed of the annual dose recommended by the United States Nuclear Regulatory Committee to protect "individual members of the public from the licensed operation" (1 mSv/yr) ("U.S. Nuclear Regulatory Commission Regulations: Title 10, Code of Federal Regulations, Subpart D" 1991). Furthermore, the values are two orders of magnitude higher than the value published by the Japanese Ministry of Education, Culture, Sports, Science and Technology (MEXT) as the Radiation Safety Standard for schools in Fukushima Prefecture after the nuclear plant accident there (20 mSv/yr based on eight hours outdoor activities and sixteen hours indoor with shielding factor of 0.4 and the air dose of 3.8 μSv/hour; in "Notification of interim policy regarding decisions on whether to utilize school buildings and outdoor areas within Fukushima Prefecture", MEXT No. 134 dated April 19, 2011). A 10.0 g/cm^2 aluminum shield (corresponding to 3.7 cm thickness) could reduce the dose from SCR events by nearly three orders of magnitude (Hayatsu et al. 2009).

The SCR damage could be effectively minimized by shields, but GCR damage is more difficult to shield against. Denisov et al. (2010) evaluated the effects of the GCR. Secondary neutrons (n-GCR) induced by p-GCR (protons and heavier ions, ^{4}He, ^{16}O, ^{28}Si, and ^{56}Fe) produce an annual dose of a few hundred millisieverts, one order of magnitude higher than that o secondary protons due to GCR (p-GCR). p- and n-GCR attenuate much more slowly with lunar soil depth than dose of p-SCR. For example, a 5-m-thick layer of lunar soil is required to reduce the annual total dose of n-GCR to 10 mSv at the solar minimum, and the peak total dose for n-GCR (several hundreds of millisieverts) is found not at the surface, but at a depth of 0.5 to 1.0 m. We note that the lesser amount of lunar soil required to effectively shield against p-SCR may be in fact be counter-productive in reducing the n-GCR. In addition, the contribution of GCR heavy ions is significant at a depth of 0 to 0.5 m because they produce a high linear energy transfer (LET) in the materials. Hayatsu et al. (2009) estimated that the dose due to GCR ions is about five times higher than that of GCR protons on the lunar surface. Adequate shielding of GCR is necessary to conduct lunar activities.

Holes can definitely reduce the effects of cosmic rays because of the limited field of view (FOV) from the hole bottom. For a hole of the same scale of depth and diameter as MHH and MTH, the tangent of FOV/2 from the bottom to space is 0.5, indicating that the amount of GCR including protons and heavy ions entering the hole is less than about 10% of that striking the surface. Thus the total dose rate of p-GCR at the floor would be less than 10 mSv/yr, and the rate of n-GCR induced by p-GCR (protons and heavier ions) and heavy ions would be a maximum of several tens of millisieverts per year during the worst days at solar minimum.

Avoiding Impact Events

Meteorite impacts, in addition to irradiation by cosmic rays, pose a threat to people and instruments operating on the lunar surface. In that regard, Pegasus satellites I and II, which were placed in a 500-km Low Earth Orbit (LEO) from May 25 to July 24, 1965, with 200 m^2 meteoroid detectors deployed may provide the first data on the meteoroid environment around the Earth (Naumann 1965). Based on the Pegasus experiment data on the penetration of 200- and 400-μm thick Al, a fitting function for frequency (F /m2 s) of hole penetration versus the penetration thickness d_p (μm) can be derived as

$$\log_{10} F = -2.91 \log_{10} d_p + 0.151 \tag{6.1}$$

Extrapolating the relation expressed in Eq. (6.1), the frequency of penetration of a 1-mm thick Al would be $10^{-8.6}$ /(m^2 s) or 0.08 /(m^2 yr). The probability of penetrating an astronaut's spacesuit with an equivalent thickness of 1 mm Al and an area of 0.1 m^2 would be 0.8% for an accumulated total exposure time of one year.

Data on the space dust environment have been accumulated after the Pegasus missions. A reliable experiment data set was acquired from the Long-Duration Exposure Facility (LDEF), which was deployed in LEO in 1984 and retrieved in 1990. Based on the LDEF north row (facing space) data (McBride et al. 1996), the penetration frequency F (m^2 s) against the penetration thickness d_p (μm) can be derived as

$$\log_{10} F = -0.175 \left(\log_{10} d_p\right)^2 - 1.03 \left(\log_{10} d_p\right) - 3.06. \tag{6.2}$$

The data from LDEF is consistent with a flux model proposed by Grun et al. (1985), McBride et al. (1996) and also with data from recent similar experiments (Mandeville and Bariteau 2001): the Aragatz experiment on MIR (1988-1990), the ESEF experiment during the Euromir mission (1995), the Particle Impact Experiment on MIR (1996-1997), and the MEEP/PRMD experiment on the Shuttle-MIR docking module (1996-1997). However, the result from LDEF data is inconsistent with that from Pegasus. The LDEF data indicate the frequency of penetration (d_p) of 1-mm thick Al to be 2×10^{-8} /(m^2 sec), which exceeds by several times the result from Eq. (2.1). The probability of spacesuit penetration increased to 8% for a one-year exposure in space. Extrapolating from Eq. (2.2), the flux of particles penetrating 1 cm thick Al in space is 10^{-10} /(m^2 sec) and 3×10^{-3} /(m^2 yr). A space facility with a 10 m×10 m surface area of 1 cm thick Al would have a 30% probability of being penetrated by particles in a year.

The velocity V dependence of the penetration depth of the holes would be $V^{2.37}$ for 10 m, based on considerations of impact cratering (Melosh 1986, Gaults 1974). The escape velocity of the Moon is 2.38 km/sec, and that of the Earth is 11.2 km/s. Therefore, if the Pegasus experiment panels were penetrated by particles drifting around the Earth-Moon system and the impact velocity was dominated by their escape velocities, then the particle impact danger to astronauts on the Moon would decrease to 60% of that in Earth orbit. In contrast, if the dominant particles and dust of cometary origin have high speeds (exceeding 20 km/sec), then the velocity difference between Earth orbit and the Moon is smaller and the flux of dust particles on the Moon is similar to that encountered in Earth orbit.

As noted in the previous section, lunar holes provide good protection from space particles. Specifically, the damage probability at the bottom of the hole is an order of magnitude less than that on the surface. The holes will also guard astronauts and facilities from the danger of hyper-velocity secondary impacts of ejecta that are spattered by primary impacts.

Mild Temperatures Condition / Focusing Sunlight

The lunar surface temperature varies widely, for instance, 102 K to 384 K at the Apollo 15 landing site in the Hadley rille (26°N, 3.6°E) (Keihm and Langseth 1973) and 92 K to 374 K at the Apollo 17 landing site in the Taurus Littrow valley (20°N, 30°E) (Langseth et al. 1976). Wide temperature fluctuations will hamper future manned and unmanned activities on the Moon.

However, the bottoms of lunar holes would provide the milder temperature conditions to be good candidates for lunar base construction. Figure 6.8 plots the surface-temperature profile as a function of local time for a shadowed area (blue) and for an illuminated area (green) at the bottom of a hole 50 meters in diameter and depth, located at 15°N (corresponding to MHH), and also for the surface near the hole (red). The bottom of the hole and the nearby surface are assumed to be covered by a thin (2 cm) regolith layer with an albedo of 0.1 and a thermal emissivity of 0.9.

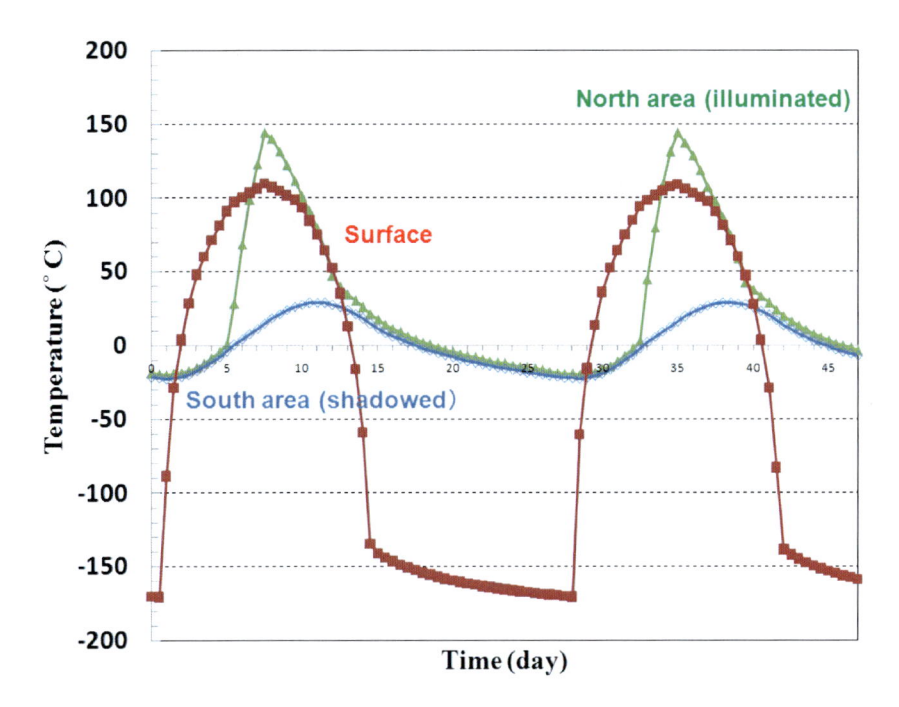

Fig. 6.8 Perspective views of one of the higher Illumination rate Locations found by SELENE (HILS) at 225°E, 89.4°S [based on TC mosaic map produced by extracting illuminated portions]. An arrow indicates a higher topo surrounded by steep down-slopes. Although higher illumination rate locations are often considered as appropriate for future lunar base construction, surrounding surface morphologies may be a negative factor for future exploration and activities at these locations.

The thermal conductivity and the heat capacity are treated as functions of temperature (Vasavada et al. 1999). The temperature in a shadowed portion of the hole bottom ranges from 253 K to 303 K (-20 °C to 30 °C) in one lunar day, but the temperature ranges from -170 °C to 110°C at the surface. If the floors permit mobility, one can secure even smaller temperature variations. Temperatures are mild in the permanently shadowed regions at the bottoms of lunar holes, but could reach 403 K (140°C) in the daytime, hotter than the highest temperature around the holes, due to additional light reflected from the inner walls. This could facilitate effective solar energy collection at the bottom of deep holes. A more accurate quantitative evaluation should be performed, but the general trend of the temperature profiles should not be so different.

6.3 Possible Lunar Lava Tubes

6.3.1 Scientific Interest in Lava Tubes

The holes described in Sect. 6.2 may be lava tube skylights. A lava tube is a subsurface conduit formed as a result of low-viscosity lava flow, as in the case of basalt on the Earth. Most natural tunnels seen in basaltic volcano regions, such as on the flanks of Mt. Kilauea on Hawaii's big island and near Mt. Fuji in Japan and Mt. Etna in Italy, are lava tubes. The surface of lava erupting from a volcano cools and solidifies to form a crust. The lava under the crust does not cool so rapidly and erodes the subsurface, resulting in conduits for lava under the surface. After the lava eruptions cease, lava can flow out of the conduits, leaving tunnels (e.g., Greeley 1971, Cruikshank and Wood 1972, Keszthelyi 1995, Sakimoto et al. 1997, Valerio et al. 2008). Movies about volcanic activities (e.g., "Lava Flows and Lava Tubes," produced by Volcano Video Productions, 2004) are helpful in understanding lava-tube formation.

The maria of the Moon are vast basaltic lava fields and the existence of lava tubes can be assumed. The flat parts seen in places on the rilles of sinuous rilles were considered to be collapsed lava tubes (Coombs and Hawke 1992, Cruikshank and Wood 1972, Greeley 1971, Gornitz 1973) (see Fig. 6.1). However, these morphological characteristics are not a sufficient basis for confirming the existence of intact lava tubes on the Moon. Many researchers have attempted unsuccessfully to locate the entrances of intact lava tubes (e.g., Coombs and Hawke 1992). The lunar holes recently discovered in the SELENE image data (Haruyama et al. 2009b) are probably the most reliable evidence of underlying lunar lava tubes.

Lava-tube formation is a significant factor in lava flows and lava-field formation (Calvari and Pinkerton 1999, Sakimoto et al. 1997). For example, tube structures enable lava to be transported for long distances, which might be important in forming vast lunar mare regions covered by basaltic lavas. Waterlines are often observed on the inner walls of lava tubes and provide information concerning the history of lava eruptions. Whether lunar lava tubes formed or not is an interesting subject in lunar and planetary science and evolution.

6.3.2 Potential of Lunar Lava Tubes as Locations for Constructing Lunar Bases

In addition to scientific interest, lunar lava tubes have been attractive for their potential as future lunar bases. First, as is easily expected, intact lava tubes in a subsurface area of the Moon are potential shelters from severe cosmic radiation and numerous micrometeorite bombardments (e.g., Hörz 1985).

Second, the temperatures inside lava tubes will be very stable. The lunar surface is covered by regolith layers of very low thermal conductivity. Therefore, the diurnal temperature variation decreases rapidly with depth. The skin depth where the thermal variation becomes 1/e is given as $d_s = (2\kappa/\omega)^{1/2}$, where κ is the thermal conductivity of the lunar surface and ω is the angular velocity of the Moon (2.66×10^{-6} rad/sec). Assuming $\kappa = 1$ mm^2/sec, then $d_s = 87$ cm. Thus, temperatures at depths below 1 m should be stable, with a few degrees. The radiation equilibrium temperature of the surface at a latitude (φ) reflecting solar light with an albedo of A and an internal heat flow of q is given in

$$TE = [\{ L_{sol} (1-A) \cos \varphi / (4\pi 2R^2_{S-M}) + q \} / \sigma]^{1/4} \qquad (6.3)$$

where L_{sol} is the solar constant, R_{S-M} is the distance between the Sun and Moon, and σ is the Stefan-Boltzmann constant, Figure 6.9 plots the profile of the thermal equilibrium temperature at various latitudes given in Eq. (2.3), assuming an internal heat flow of 0.02 W/m^2 based on the values estimated by Langseth et al. (1976) and a 10% albedo.

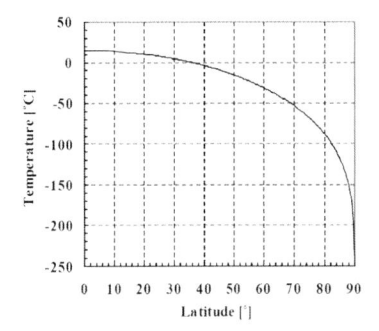

Fig. 6.9 Averaged surface temperature at various latitudes of the Moon, estimated from thermal equilibrium assuming a 10% albedo and an internal heat flow of 0.02 W/m^2. The averaged surface temperature in regions where the three holes are located is around 0°C. We note that the calculated thermal equilibrium temperature at the surface probably differs from the subsurface temperature, even at several tens of meters in depth where lava tubes exist, because solar energy striking the surface may not fully reach the subsurface.

We note that the calculated TE probably differs from the subsurface temperature even at several tens of meters in depth where lava tubes exist because solar energy striking the surface may not fully reach the subsurface. The measured subsurface mean temperatures were -17°C to -16°C at the Apollo 15 and 17 landing sites at 130 cm to 230 cm in depth (Langseth et al. 1976), actually below TE (7.3°C at the Apollo 15 landing site 25.9°N; 10.0°C at the Apollo 17 landing site 20.3°N). Various parameters, such as the surface albedo and the distribution of thermal conductivity, heat capacity, and internal heat flows in deeper levels, must be known to very accurately estimate the subsurface temperature, however, it is certain that temperatures in lava tube will fluctuate very little from an engineering standpoint.

The third benefit of lava tubes as lunar base locations is their expected large size. Previously, the size of a lava tube was inferred to be up to a few hundreds of meters, based on the width of rilles regarded as collapsed lava tubes (e.g., Hörz 1985). Oberbeck et al. (1969) estimated the maximum width of the tube based on simple beam theory to be $(4Sd/3\rho g)^{1/2}$ at lunar gravity (g), where S and ρ are the tensile strength and the density of the ceiling.

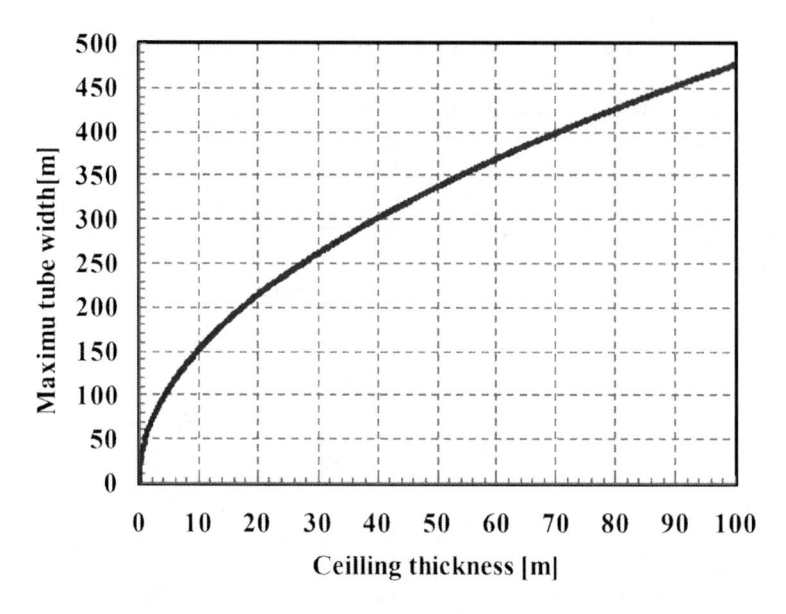

Fig. 6.10 Maximum tube width of lunar lava tubes depending on ceiling thickness, estimated from simple beam theory (Oberbeck et al. 1969) where 6.9 MPa tensile strength and 2500 kg/m^3 density of the ceiling are assumed. A lunar lava tube forming a 50 m thick ceiling could exceed 300 m in width. Lava tubes forming arched roofs could be wider than the width estimated from simple beam theory.

Figure 6.10 plots the maximum width vs. ceiling thickness of the tube, assuming S = 6.9 MPa and ρ = 2500 kg/m^3, which are typical for basaltic materials (Oberbeck et al. 1969). When the ceiling thickness is 50 to 100 m, the width could be a few hundred meters. However, typical lava tubes form arched roofs that would be stronger than a simple beam structure. Lava tubes could thus be wider than estimated from simple beam theory.

There are additional benefits of lava tubes as locations for lunar base construction. For one, the insides of lava tubes can be expected to be free of dust that would interfere with instruments or astronauts. The solid, flat floors and crystallized inner walls that can be supposed by analogy from terrestrial lava tubes are additional advantages of lava tubes as favorable locations for constructing lunar bases.

There are many ideas for utilizing lunar bases inside lunar holes and lava tubes. One involves storing and analyzing extraterrestrial samples, not just lunar samples. In this century, many molecules, dust particles, and rocks from other worlds (Mars, asteroids, and so on) will be sampled. One of the most important research objectives for these samples is to find extraterrestrial life or signs of life. However, there is concern that such samples may be biologically dangerous. Alternatively, study of extraterrestrial biologic samples must preclude contamination. Bases on the Moon would enable the samples to be analyzed first before being returned to Earth.

6.5 Summary and Conclusion

This chapter described lunar holes and lava tubes as potentially advantageous for scientific research and as locations for constructing lunar bases.

In 2009, scientists using data from the 10 m/pixel-resolution Terrain Camera (TC) on board the Japanese lunar orbiter SELENE (Kaguya) discovered three gigantic holes exceeding several tens of meters in both diameter and depth. These holes were located in Marius Hills, Mare Tranquillitatis, and Mare Ingenii. They may be skylights of subsurface voids that are probably lava-tube caverns. The survey for such holes covers more than 95% of the lunar Maria; thus, such large, deep holes are rare and are probably limited to these three. The narrow-angle camera (NAC) on board the Lunar Reconnaissance orbiter (LRO) succeeded in taking higher resolution (0.5 to 1 m/pixel) images of the rims, floors, and vicinities of the holes, providing more-detailed information, including indications that open lava tubes may be present near the bottoms of the holes.

The walls of the lunar holes provide prominent geological outcrops that reveal the complex volcanic history of the moon in detail. Volatile elements such as solar-wind protons and/or water molecules are perhaps trapped in regolith layers sandwiched by multiple lava layers. These volatile elements can provide clues to understanding the solar-wind history and exploring the possibility of a past lunar dynamo. Boundaries of the shadowed and illuminated areas on the surface of the Moon may induce electric currents and/or dust transportation and will be subjects for future considerations regarding overnight stays on the lunar surface. The bottom of a hole where the boundary between the shadowed and illuminated areas passes during the daytime would provide an appropriate environment for studying these phenomena. Extra-lunar materials such as solar-wind protons, hydroxyls, and/or water molecules could be concentrated at the bottom of the holes as well.

The floors of the holes have high potential as locations for constructing future lunar bases. Because of the high vertical walls, fewer extra-lunar rays/particles and micrometeorites can reach the bottoms of the holes than impact the surface. Permanently shadowed areas on the floors of the holes provide much milder temperature variations of a few tens of degrees centigrade than at the surface where the temperatures at the equator can range from -150°C to 120°C. Areas of the holes several tens of meters in diameter would be sufficient for constructing an effective infrastructure for human settlement.

It has been obvious since Apollo that lunar lava tubes occur and may lend important information to the evolution of the Moon. The discovery of these holes in the roofs of lava tubes increases the likelihood of more extensive buried, yet intact, intact lava tubes on the Moon, the insides of which will also provide great amounts of scientifically significant information.

Hole structures that are possible skylights of lava tubes have been discovered not only on the Moon but also on Mars (e.g. Cushing et al. 2007, Cushing 2011). Martian holes and lava tubes are interesting both geologically and biologically as they may contain relics of primitive and/or developed life forms, or even current life forms. Lunar-hole and lava-tube exploration could provide training for Martian hole and lava-tube exploration.

Lunar lava tubes are among the best locations for constructing future lunar settlements, as many earlier writers have suggested. Instruments and people in these tubes will be protected from extra-lunar rays and micrometeorites, and the inside temperature ranges will be considerably reduced. The volume of the tubes leading from the holes may reach 100 meters in width and a few tens of meters in height. Furthermore, there are numerous additional potential advantages: flat floors, sealed inner walls, dust-free environment, and so on. In addition, the ceilings, walls, and floors of the tubes were formed by fresh lava, which perhaps contained mantle-origin rocks and/or volatile materials of scientific interest. Lastly, the unique characteristics and consequent microenvironments of lunar holes and possible related lava tubes represent outstanding resources for exploration and science on the Moon.

Acknowledgement. We thank all the contributors to the SELENE (KAGUYA) and LRO projects for their efforts on the developments, operations, and data processing of their spacecraft and the installed instruments. The SELENE TC/MI data and LRO NAC data used in this chapter are archived at JAXA SELENE data archive website (https://www.soac.selene.isas.jaxa.jp/archive/index. html.en) and the NASA Planetary Data System website (http://pds.nasa. gov), respectively. We acknowledge Dr. M. Yamashita and Dr. B. Foing for their careful reviews and helpful comments.

References

Ashley, J.W., Boyd, A.K., Hiesinger, H., Robinson, M.S., Tran, T., van der Bogert, C.H., Wagner, R.V., LROC Science Team: Lunar pits: Sublunarean voids and the nature of Mare emplacement. In: 42nd Lunar and Planetary Science Conference, #2771. Lunar and Planetary Science Institute, Houston (2011)

Banerdt, W.B., Golombek, M.P., Tanaka, K.L.: Stress and tectonics on Mars. In: Kieffer, H., Jakosky, B.M., Snyder, C.W., Matthews, M. (eds.) Mars, pp. 249–297. Univ of Arizona Press, Tuscon (1992)

Calvari, S., Pinkerton, H.: Lava tube morphology on Etna and evidence for lava flow emplacement mechanisms. J. Volcanol. Geotherm. Res. 90, 263–280 (1999)

Clark, R.N.: Detection of Adsorbed Water and Hydroxyl on the Moon. Science 326, 562–564 (2009)

Coombs, C.R., Hawke, B.R.: A search for intact lava tubes on the Moon Possible lunar base habitats. In: The Second Conference on Lunar Bases and Space Activities of the 21st Century l, pp. 219–229 (1992)

Cruikshank, D.P., Wood, C.A.: Lunar rilles and Hawaiian volcanic features; possible analogues. The Moon 3, 412–447 (1971)

Cushing, G.E., Titus, T.N., Wynne, J.J., Christensen, P.R.: THEMIS observes possible cave skylights on Mars. Geophys. Res. Let. 34, L17201 (2007)

Cushing, G.E.: Visible Evidence of Cave-Entrance Candidates in Martian Fresh-Looking Pit Craters. In: 42nd Lunar and Planetary Science Conference, #2494 Lunar and Planetary Science Institute, Houston (2011)

Denisov, A.N., Kuznetsov, N.V., Nymmik, R.A., Panasyuk, M.I., Sobolevskii, N.M.: On the problem of lunar radiation environment. Cosmic Research 48(6), 509–516 (2010)

Feldman, W.C., Prettyman, T.H., Maurice, S., Nelli, S., Elphic, R., Funsten, H.O., Gasnault, O., Lawrence, D.J., Murphy, J.R., Tokar, R.L., Vaniman, D.T.: Topographic control of hydrogen deposits at low latitudes to midlatitudes of Mars. J. Geophys. Res. 110, E11009 (2005), doi:10.1029/2005JE002452

Gault, D.E., Hörz, F., Brownlee, D.E., Hartung, J.B.: Mixing of the lunar regolith. In: Proc. Lunar Science Conference V, pp. 2365–2386 (1974)

Gornitz, V.: The origin of sinuous rilles. Moon 6, 337–356 (1973), doi:10.1007/BF00562210:

Greeley, R.: Lava tubes and channels in the Lunar Marius Hills. The Moon 3, 289–314 (1971)

Grun, E., Zook, H.A., Fetctig, H., Gese, R.H.: Collisional Balance of the Meteoritic Complex. Icarus 62, 244–272 (1985)

Haruyama, J., Matsunaga, T., Ohtake, M., Morota, T., Honda, C., Yokota, Y., Torii, M., Ogawa, Y.: LISM working group Global lunar-surface mapping experiment using the Lunar Imager/Spectrometer on SELENE. Earth Planets Space 60, 243–256 (2008a)

Haruyama, J., Ohtake, M., Matsunaga, T., Morota, T., Honda, C., Yokota, Y., Pieters, C.M., Hara, S., Hioki, K., Saiki, K., Miyamoto, H., Iwasaki, A., Abe, M., Ogawa, Y., Takeda, H., Shirao, M., Yamaji, A., Josset, J.-L.: Lack of exposed ice inside lunar South Pole Shackleton crater. Science 322, 938–939 (2008b)

Haruyama, J., Ohtake, M., Matsunaga, T., Morota, T., Honda, C., Yokota, Y., Abe, M., Ogawa, Y., Miyamoto, H., Iwasaki, A., Pieters, C.M., Asada, N., Demura, H., Hirata, N., Terazono, J., Sasaki, S., Saiki, K., Yamaji, A., Torii, M., Josset, J.-L.: Long-lived volcanism on the lunar farside revealed by SELENE Terrain Camera. Science 323, 905–908 (2009a)

Haruyama, J., Hioki, K., Shirao, M., Morota, T., Hiesinger, H., van der Bogert, C.H., Miyamoto, H., Iwasaki, A., Yokota, Y., Ohtake, M., Matsunaga, T., Hara, S., Nakanotani, S., Pieters, C.M.: Possible lunar lava tube skylight observed by SELENE cameras. Geophys. Res. Let. 36, L21206 (2009b), doi:10.1029/2009GL040635

Haruyama, J., Hara, S., Hioki, K., Morota, T., Yokota, Y., Shirao, M., Hiesinger, H., van der Bogert, C.H., Miyamoto, H., Iwasaki, A., Ohtake, M., Saito, Y., Matsunaga, T., Nakanotani, S., Pieters, C.M., Lucey, P.G.: New discoveries of lunar holes in Mare Tranquillitatis and Mare Ingenii. In: 41st Lunar and Planetary Science Conference, #1285. Lunar and Planetary Science Institute, Houston (2010)

Haruyama, J., Morota, T., Shirao, M., Hiesinger, H., van der Bogert, C.H., Pieters, C.M., Lucey, P.G., Ohtake, M., Nishino, M., Matsunaga, T., Yokota, Y., Miyamoto, H., Iwasaki, A.: Water in lunar holes? In: 42nd Lunar and Planetary Science Conference, #1134. Lunar and Planetary Science Institute, Houston (2011)

Hayatsu, K., Hareyama, M., Kobayashi, S., Yamashita, N., Sakurai, K., Hasebe, N.: HZE Particle and Neutron Dosages from Cosmic Rays on the Lunar Surface. In: Proc. Int Workshop Advances in Cosmic Ray Science. J. Phys. Soc. Jpn, vol. 78 (suppl. A), pp. 149–152 (2009)

Head, J.W.: Lunar Volcanism in Space and Time. Rev. Geophys. Space Phys. 14(2), 265–300 (1976)

Head, J.W., Wilson, L.: Lunar mare volcanism: Stratigraphy, eruption conditions, and the evolution of secondary crusts. Geochim Cosmochim Acta 56, 2155–2175 (1992)

Hörz, F.: Lava tubes; Potential shelters for habitats. In: Mendell, W.W. (ed.) Lunar Bases and Space Activities of the 21st Century, pp. 405–412 (1985)

Hulme, G.: Turbulent lava flow and the formation of lunar sinuous rilles. Mod. Geol 4, 107–117 (1973)

Ichimura, A.S., Zent, A.P., Quinn, R.C., Taylor, L.A.: Formation and Detection of OH/OD in Lunar Soils After 1H2+/D2+ Bombardment. In: 42nd Lunar and Planetary Science Conference, # 2724. Lunar and Planetary Science Institute, Houston (2011)

Keihm, S.J., Langseth, M.G.: Surface brightness temperatures at the Apollo 17 heat flow site: Thermal conductivity of the upper 15 cm of regolith. In: Proc. Lunar Science Conference IV, pp. 2503–2513 (1973)

Keszthelyi, L.: A preliminary thermal budget for lava tubes. J. Geophys. Res. 100(B10), 20411–20420 (1995)

Langseth, M.G., Keihm, S.J., Peters, K.: Revised lunar heat-flow values. In: Proc. Lunar Science Conference VII, pp. 3143–3171 (1976)

Mandeville, J.C., Bariteau, M.: Cosmic dust and micro-debris measurements on the MIR space station. Adv. Space Res. 28(9), 1317–1324 (2001)

Marsch, E., Mühlhäuser, K.–. H., Schwenn, R., Rosenbauer, H., Pilipp, W., Neubauer, F.M.: Solar Wind Protons - Three-Dimensional Velocity Distributions and Derived Plasma Parameters Measured Between 0.3 and 1 AU. J. Geophys. Res. 87(A1), 52–72 (1982)

McBride, N., McDonnell, J.A.M., Gardner, D.J., Griffiths, A.D.: Meteoroids at 1 AU: Dynamic and Properties. In: Burke, W., Guyenne, T.-D. (eds.) Environment Modeling for Space-based Applications, Symposium Proceedings (ESA SP-392): ESTEC Noordwijk, pp. 335–342 (1996)

McCord, T.B., Taylor, L.A., Combe, J.-P., Kramer, G., Pieters, C.M., Sunshine, J.M., Clark, R.N.: Sources and physical processes responsible for OH/H2O in the lunar soil as revealed by the Moon Mineralogy Mapper (M3). J. Geophys. Res. 116, E00G05 (2010), doi:10.1029/2010JE003711

Melosh, H.J.: Impact Cratering. A Geologic Process. Oxford Univ. Press, New York (1989)

Morota, T., Haruyama, J., Ohtake, M., Matsunaga, T., Honda, C., Yokota, Y., Kimura, J., Ogawa, Y., Hirata, N., Demura, H., Iwasaki, A., Sugihara, T., Saiki, K., Nakamura, R., Kobayashi, S., Ishihara, I., Takeda, H., Hiesinger, H.: Timing and characteristics of the latest mare eruption on the Moon. Earth and Planetary Science Letters 302, 255–266 (2011)

Naumann, R.J.: Pegasus satellite measurements of meteoroid penetration. NASA TM X-1192 (1985)

Oberbeck, V.R., Willam, L.Q., Greeley, R.: On the origin of lunar sinuous rilles. Mod. Geol. 1, 75–80 (1969)

Okubo, C.H., Martel, S.J.: Pit crater formation on Kilauea volcano. Hawaii J. Volcanol. Geotherm. Res. 86, 1–18 (1998), doi:10.1016/S0377-0273(98)00070-5

Pieters, C.M., Goswami, J.N., Clark, R.N., Annadurai, M., Boardman, J., Buratti, B., Combe, J.-P., Dyar, M.D., Green, R., Head, J.W., Hibbitts, C., Hicks, M., Isaacson, P., Klima, R., Kramer, G., Kumar, S., Livo, E., Lundeen, S., Malaret, E., McCord, T., Mustard, J., Nettles, J., Petro, N., Runyon, C., Staid, M., Sunshine, J., Taylor, L.A., Tompkins, S., Varanasi, P.: Character and Spatial Distribution of OH/H2O on the Surface of the Moon Seen by M3 on Chandrayaan-1. Science 326, 568–572 (2009)

Sakimoto, S.E., Crisp, H.J., Baloga, S.M.: Eruption constraints on tube-fed planetary lava flows. J. Geophys. Res. 102, 6597–6613 (1997)

Stubbs, T.J., Vondrak, R.R., Farell, W.M.: A dynamic fountain model for lunar dust. Ad. Spac. Res. 37, 59–66 (2006)

Sunshine, J.M., Farnham, T.L., Feaga, L.M., Groussin, O., Merlin, F., Milliken, R.E., A'Hearn, M.F.: Temporal and Spatial Variability of Lunar Hydration as Observed by the Deep Impact Spacecraft. Science 326, 565–568 (2009)

Valerio, A., Tallarico, A., Dragoni, M.: Mechanisms of formation of lava tubes. J. Geophys. Res. 113, B08209 (2008), doi:10.1029/2007JB005435

Villanueva, G.L., Mumma, M.J., Bonev, B.P., DiSanti, M.A., Gibb, E.L., Böhnhardt, H., Lippi, M.: A sensitive search for deuterated water in Comets 8P/Tuttle. Astrophys J. 690, L5-L9 (2009)

Vasavada, A.R., Paige, D.A., Wood, S.E.: Near-Surface Temperatures on Mercury and the Moon and the Stability of Polar Ice Deposits. Icarus 141, 179–193 (1999)

Wyrick, D., Ferrill, D.A., Morris, A.P., Colton, S.L., Sims, D.W.: Distribution, morphology, and origins of Martian pit crater chains. J. Geophys. Res. 109, E06005 (2004), doi:10.1029/2004JE002240

Zent, A.P., Ichimura, A.I., McCord, T.B., Taylor, L.A.: Production of OH/H2O in Lunar Samples via Proton Bombardment. In: 41st Lunar and Planetary Science Conference, #2665. Lunar and Planetary Science Institute, Houston (2010)

7 Oxygen from Lunar Regolith

Carsten Schwandt[1,2], James A. Hamilton[2],
Derek J. Fray[1], and Ian A. Crawford[3]

[1] University of Cambridge, UK
[2] Green Metals Ltd, London, UK
[3] University of London, UK

7.1 Background and Introduction

In the year 2004 NASA declared its mission to prepare for a return of man to the moon as early as 2015 but no later than 2020, while continuing with robotic missions to Mars (NASA 2004). As a long-term goal, it was intended to establish permanent human presence on the moon and eventually send human missions to Mars. Although the future of US space exploration policy is now more uncertain, following a recent review (Augustine Commission 2009) and the cancellation of the Constellation Program (NASA 2010a), it remains true that an extended human presence on the moon is desirable for scientific and economic reasons (e.g., Crawford 2004; Spudis 2005). For this to become possible, significant progress is needed in the field of 'living off the land', or in situ resource utilisation (ISRU). In 2007 NASA presented to the National Academy of Sciences its overview of ISRU architecture and confirmed the following objectives (NASA 2007a,b):

- To identify and characterise resources on the moon;
- To demonstrate ISRU concepts, technologies and hardware;
- To use the moon for operational experience and mission validation for Mars;
- To develop and evolve lunar ISRU capabilities to enable lunar exploration;
- To use ISRU for space commercialisation.

One key aspect of ISRU is the production of oxygen on the moon. NASA proposed an initial target of 5 tonnes of oxygen and 1 tonne of water production per annum by 2022. Oxygen is essential in supporting human life on the moon, and is required to burn fuel during the return journey of a rocket from moon to earth and during the use of a lunar landing module. Figure 7.1 shows NASA's Saturn rocket and Eagle lunar lander from the Apollo 11 mission. Notably, more than 80% of the initial weight of the rocket is fuel and liquid oxygen. If oxygen could be generated through ISRU, then the weight of the rocket could be reduced because oxygen would not need to be carried from the earth to the moon for use there or in subsequent journeys to other destinations such as Mars. Producing oxygen on the moon therefore has the potential to provide huge cost savings.

(a)   (b)

Fig. 7.1 (a) Saturn V rocket from the Apollo 11 mission at the launch on 16 July 1969; (b) Eagle lunar landing module from the same mission. (Images reproduced with permission from NASA).

Another aspect of ISRU is the production of metallic construction materials on the moon. These materials are needed for the development of a lunar infrastructure, and again supply from earth would be exceedingly costly. A particular attraction therefore lies in processes that permit the combined production of oxygen and metals from lunar resources.

This chapter gives an overview of technologies that have been assessed by NASA with respect to their capability to produce oxygen from materials occurring on the moon. The subject is discussed on the basis of economic considerations, lunar geology, as well as scientific and engineering aspects. Particular emphasis is given to the Ilmenox process. This is a variant of the FFC-Cambridge process (Chen et al. 2000), which was invented for the removal of oxygen from metals and oxides through electrochemical processing in molten salts.

7.2 Economic Considerations of In Situ Resource Utilisation

Space exploration cannot be viewed in isolation from the turn of economic and political events. Since President Kennedy's speech in 1961 committing the US to join the Space Race, NASA's budget has risen and fallen in response to the economic climate. In 2004 President Bush gave the vision for US space exploration as both exciting and affordable, which led to the inception of the Constellation Program.

"Returning to the moon is an important step for our space program. Establishing an extended human presence on the moon could vastly reduce the costs of further space exploration, making possible ever more ambitious missions. Lifting heavy spacecraft and fuel out of the earth's gravity is expensive. Spacecraft assembled and provisioned on the moon could escape its far lower gravity using far less energy, and thus, far less cost. Also, the moon is home to abundant resources. Its soil contains raw materials that might be harvested and processed into rocket fuel or breathable air. We can use our time on the moon to develop and test new approaches and technologies and systems that will allow us to function in other, more challenging environments. The moon is a logical step toward further progress and achievement." President George W. Bush, 14 January 2004 (The White House 2004).

Unfortunately, by mid-2007 the Credit Crunch had gripped the economies of both America and Europe, stock markets did not bottom out till March 2009, and economic priorities were strictly short term. In 2010, in the wake of these events, President Obama cancelled the Constellation Program for a return of man to the moon (NASA 2010a). Although NASA's budget provision is being maintained at the level of $19 billion through to the 2016 financial year (NASA 2010b), this appears to be insufficient to pay for the Constellation Program's Orion capsule, a more advanced and spacious version of the Apollo lunar landing module, as well as the proposed Ares 1 and Ares V launchers (Augustine Commission 2009).

However, new factors have emerged to sustain hope for the future:

- The recognition that the financial burden of space exploration is better affordable if shared with international consortia (GES 2007);
- The spur to competition given by advances in space from Russia, India, China and Europe, and the fear of the US being seen to be left behind;
- The outstanding success of NASA's LCROSS mission that provided evidence of water ice near the lunar south pole (NASA 2009; Colaprete et al. 2010).

The economic benefits of independence to a spacefaring nation have been illustrated by NASA's agreement to pay Russia $753 million to deliver twelve astronauts to the International Space Station from 2014 to 2016 (NASA 2011). With NASA's Space Shuttle fleet retired, Russia as the sole supplier can charge what the market can bear. The same argument might apply to the nation pioneering lunar ISRU.

The British National Space Centre, now UK Space Agency, commissioned London Economics, their specialist economics and policy consultant, to prepare an economic analysis to support a study on the options for UK involvement in space exploration. In this it is estimated that the current indicative cost for transporting 1 tonne of oxygen to the moon is between \$25 million and \$100 million. If the oxygen were to be generated on the moon, then for the cargo version of the Altair lunar lander which requires 9.4 tonnes of oxygen to return to earth, the oxygen cost for one return trip would be between \$235 million and \$940 million. At the time of the analysis it was assumed that there might be between four and six return trips to the moon from 2020, with ISRU requirements for rocket propulsion between 37.6 and 56.4 tonnes of oxygen. In four trips, between \$0.9 billion and \$3.8 billion might be saved, less of course the cost of lunar generation (UK Space Agency 2009).

7.3 Lunar Geology

The lunar surface is divided into two main geological units: the ancient, light-coloured lunar highlands, and the darker, generally circular, lunar maria ('seas') which fill the large impact basins, predominantly on the nearside. The compositions of these materials are now reasonably well characterised, as a result of samples returned by the Luna and Apollo missions, and subsequent remote sensing measurements (Heiken et al. 1991; Jolliff et al. 2006). This information is summarised in Fig. 7.2.

The lunar highlands represent the original crust of the moon, and are composed predominantly of anorthositic rocks, i.e., rocks containing more than 90% plagioclase. Lunar plagioclase is invariably calcium based, so to a first approximation the dominant mineralogy of the highlands is anorthite ($CaAl_2Si_2O_8$), with iron and magnesium-bearing minerals, principally pyroxene ($(Mg,Ca,Fe)SiO_3$) and olivine ($(Mg,Fe)_2SiO_4$), typically contributing only a few percent by volume. Thus, the lunar highlands are rich in calcium and aluminium oxides, but poor in magnesium and iron oxides.

The lunar maria, in contrast, are composed of basaltic lava flows. Their mineralogy is dominated by a combination of four minerals, plagioclase, pyroxene, olivine, and ilmenite ($FeTiO_3$), with the proportions of these minerals differing in different lava flows. The principal classification of lunar basalts is based on their titanium content. They are classed as 'high-Ti' if their titanium dioxide (TiO_2) abundances are above 6% by weight, and 'low-Ti' if they are below 6% (Neal and Taylor 1992).

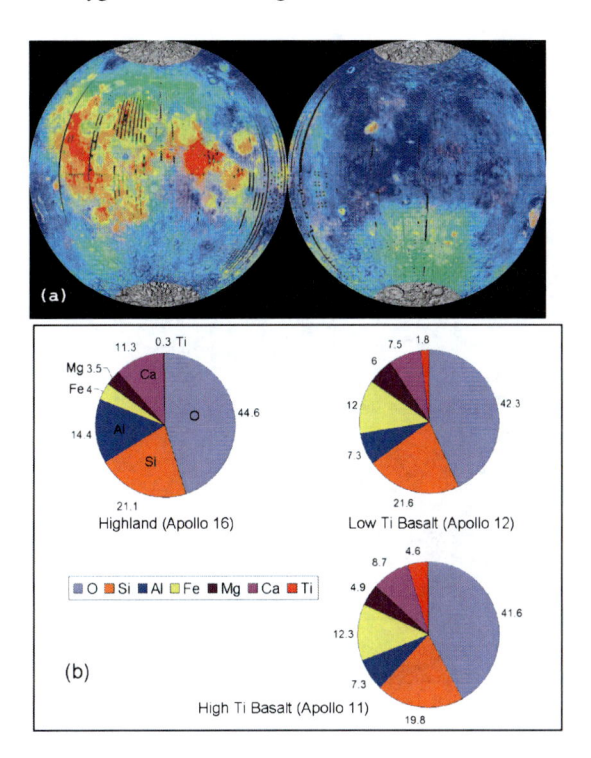

Fig. 7.2 (a) Maps showing the distribution of major lithologies on the moon (Spudis et al. 2002), with the nearside on the left and the farside on the right. The maps are colour coded as follows: anorthositic highlands (blue), low-Ti basalts (yellow), high-Ti basalts (red). (Image courtesy of Spudis); (b) pie diagrams showing representative major element compositions of lunar highlands (Apollo 16), low-Ti basalts (Apollo 12), and high-Ti basalts (Apollo 11). (Diagrams based on data from Stoeser et al. 2010).

The main source of titanium dioxide in lunar rocks is the mineral ilmenite which, as discussed below, is of particular importance for some proposed oxygen-extracting ISRU processes. It is therefore crucial to understand that the ilmenite-bearing high-Ti basalts are in no sense a globally distributed resource, but are confined to geographically restricted mare basins at low latitudes on the nearside (Fig. 7.2a). As a consequence, an ISRU process able to extract oxygen from anorthositic source material, as opposed to ilmenite, would have a significant advantage in the context of global lunar exploration.

In a return to the moon, a main objective will be extended periods of residence. An ideal location will be the south polar highlands, in the vicinity of the Shackleton crater, so that advantage may be taken of the near-permanent daylight throughout the year for illumination and generation of solar electricity. The latter is an essential part of lunar architecture and can readily be met with existing technology.

Figure 7.3 shows an image of the south polar region. As the south polar highlands are predominantly anorthositic in nature (e.g., Spudis et al. 2008), an ISRU process for oxygen extraction from anorthite-rich soils would be required here.

Fig. 7.3 The south pole of the moon as imaged by the Japanese Kaguya spacecraft. The 20 km diameter Shackleton crater is located to the lower right of centre. Favourable illumination conditions exist on the rim of the Shackleton crater and the surrounding hills. (Image courtesy of JAXA/NHK).

A number of geochemical simulants have been produced for lunar materials, which may be used for ISRU studies. The chemical composition of the first simulant produced for experimental purposes, termed JSC-1 (NASA 2005), approximates that of lunar mare samples collected during the Apollo landings. The chemical composition of a new simulant, NU LHT (NASA 2008; Stoeser et al. 2010), approximates that of the anorthositic highlands. The compositions of these simulants are compiled in Table 7.1. It should be noted that the difficulty in preparing simulant materials is not only to establish a representative chemical composition, but also to mimic the particle size distribution and the extent of particle agglomeration as well as to provide a defined proportion of glassy phase.

The oxygen content in the lunar materials is at approximately 45% (see also Fig. 7.2b), and the challenge for ISRU is to release the oxygen completely and leave behind a metal alloy.

Table 7.1 Chemical compositions of lunar simulants JSC-1 (NASA 2005) and NU LHT (NASA 2008). For JSC-1 a concentration range is given because there are marginally different types and specifications.

Oxide	JSC-1 simulant, wt. %	NU LHT simulant, wt. %
SiO_2	46 – 49	46.7
Al_2O_3	14.5 – 15.5	24.4
CaO	10 – 11	13.6
MgO	8.5 – 9.5	7.9
Na_2O	2.5 – 3	1.26
K_2O	0.75 – 0.85	0.08
TiO_2	1 – 2	0.41
MnO	0.15 – 0.20	0.07
FeO	3 – 4	–
Fe_2O_3	7 – 7.5	4.16
Cr_2O_3	0.02 – 0.06	–
P_2O_5	0.6 – 0.7	0.15

7.4 Oxygen Production Processes from Lunar Materials

The different experimental approaches of liberating oxygen from lunar materials may conveniently be placed in the categories of chemical reduction, pyrolysis, aqueous solvent processing, and electrochemical reduction. More specifically, the following approaches were evaluated positively by NASA and investigated experimentally in the past.

7.4.1 Hydrogen Reduction

Hydrogen reduction is applicable to lunar feedstocks rich in iron oxide such as ilmenite. It is a relatively simple process that works at a moderate operating temperature of around 900°C (Gibson and Knudsen 1985; Knudsen and Gibson 1992):

$$FeTiO_3 + H_2 = Fe + TiO_2 + H_2O \qquad (7.1)$$

The water vapour formed is condensed and subjected to electrolysis to generate oxygen and hydrogen. The hydrogen is then re-used in the reduction. The disadvantage of the process is that the oxygen yield depends on the iron oxide content of the feedstock, which is variable on the lunar surface and only amounts to a few percent in the highland soils of the lunar poles. A problem consists in the possible presence of sulphides in the feedstock, as these give rise to the formation of toxic hydrogen sulphide. Under such conditions an additional purification step prior to reduction would be necessary.

7.4.2 Carbothermal Reduction with Methane

Carbothermal reduction by means of methane is applicable to a bulk regolith feedstock and can achieve higher reduction levels than using hydrogen because iron oxide and silicon oxide may be reduced. The reduction step is performed at a temperature above 1600°C (Rosenberg et al. 1992, 1996):

$$FeTiO_3 + CH_4 = Fe + TiO_2 + CO + 2H_2$$
$$MgSiO_3 + 2CH_4 = Si + MgO + 2CO + 4H_2 \qquad (7.2a,b,c)$$
$$CaSiO_3 + 2CH_4 = Si + CaO + 2CO + 4H_2$$

The theoretical oxygen yield is at around 50%, although this may not be reached experimentally. In the next step the carbon monoxide and hydrogen formed and a quantity of additional hydrogen are reacted over a nickel catalyst to produce water:

$$CO + 3H_2 = CH_4 + H_2O \qquad (7.3)$$

The methane is then re-used and the water electrolysed. Drawbacks of the process are those of multiple processing steps with reactant recycling and the very high operating temperature. It is noted that ESA, much later, recommended carbothermal reduction as their preferred oxygen extraction method (ESA-ESRIN 2009a,b).

7.4.3 Vapour Phase Pyrolysis

Vapour phase pyrolysis also uses a bulk regolith feedstock. The process is conceptually simple but requires temperatures in excess of 2000°C to vaporise and decompose some of the metal oxides (Steurer 1992; Senior 1993):

$$FeTiO_3 = Fe + TiO_2 + 0.5O_2$$
$$MgSiO_3 = SiO + MgO + 0.5O_2 \qquad (7.4a,b,c)$$
$$CaSiO_3 = SiO + CaO + 0.5O_2$$

Depending on experimental conditions such as temperature and duration, the oxides of magnesium, aluminium and calcium may also be decomposed to some degree so that oxygen yields of around 50% become achievable. In order to recover the oxygen, the gas formed must be cooled and condensed very quickly to avoid recombination of the metals and suboxides with the oxygen. The clear disadvantage of the process is the exceedingly high operating temperature.

7.4.4 Sulphuric Acid Reduction

Sulphuric acid reduction uses an ilmenite-rich feedstock and produces iron as a side product (Sullivan 1992). The first step is the reaction of the feed material with hot sulphuric acid to form iron sulphates and titanium sulphates as well as water:

$$FeTiO_3 + 2 H_2SO_4 = FeSO_4 + TiOSO_4 + 2 H_2O \qquad (7.5)$$

The reaction product is mixed with water and then cooled, whereupon the iron sulphate is removed by crystallisation and filtration, while the titanium sulphate is hydrolysed to form titanium dioxide and regenerate part of the sulphuric acid:

$$TiOSO_4 + H_2O = H_2SO_4 + TiO_2 \qquad (7.6)$$

The iron sulphate is then electrolysed in aqueous solution. The overall electrolysis reaction is to form iron and oxygen and regenerate the remainder of the sulphuric acid:

$$FeSO_4 + H_2O = Fe + 0.5 O_2 + H_2SO_4 \qquad (7.7)$$

The disadvantages of the method are that it involves a complex sequence of processing steps including digestion, dissolution, filtration, electrolysis and acid recycling, and only achieves low yields of oxygen for materials with low contents of ilmenite.

7.4.5 The Magma Process

The most straightforward approach of winning oxygen and metal on the moon is through the electrolysis of molten lunar regolith (Colson and Haskin 1992, 1993). The electrolytic cell is operated at temperatures of up to 1600°C with an applied potential that enables metal deposition at the cathode and oxygen evolution at the anode:

$$Fe^{2+} + 2e^- = Fe \quad \text{and} \quad Si^{4+} + 4e^- = Si \qquad (7.8a,b,c)$$
$$O^{2-} = 0.5 O_2 + 2e^-$$

The advantages of the Magma process are that no chemical reagents supplied from earth are required and that unbeneficiated lunar feedstock may be employed. Since the feed is multi-component, increasing the applied potential in steps offers the possibility of selective winning of metals in a sequence corresponding to the stabilities of their oxides, i.e., iron, silicon, titanium, magnesium, aluminium, and calcium (Standish 2010). The products of particular interest are iron and silicon, the former being a versatile construction material, the latter being the raw material for the manufacture of solar cells. The remaining oxides of magnesium and aluminium form a spinel of composition $MgAl_2O_4$, which is not electrolysed any further; and the same is true for the calcium oxide. This limits the degree of extractable oxygen to around 50%. In fact, care has to be taken to preclude the formation of free magnesium, as this would alloy with the silicon and degrade its quality and/or evaporate from the melt and react with the oxygen in the gas phase.

An obvious disadvantage of the Magma process is the extremely high operating temperature. This poses various problems with respect to the useable cell and electrode materials. The selection of the anode material is a particular challenge, and thus far only the use of exceedingly expensive anodes made from platinum group metals (Curreri et al. 2006), and especially iridium (Shchetkovskiy et al. 2010), has been met with some success. The presence of iron in the melt is detrimental as it can set up parasitic currents, but current efficiencies close to 100% may be reached once the iron has been removed during the early stage of electrolysis (Sibille et al. 2010). Other disadvantages are the corrosive nature of the silicate melt and the difficulty of removing the highly viscous unreacted spinel from the electrolytic cell (Standish 2010).

7.5 The Ilmenox Process

7.5.1 Fundamentals of Electro-deoxidation

The Ilmenox process is the most recent approach that NASA considered worthwhile testing for liberating oxygen from lunar materials. The Ilmenox process is derived from the FFC-Cambridge process (Chen et al. 2000), which was developed at the University of Cambridge for the electro-deoxidation of metals and metal oxides. The FFC process was invented in a search for ways of using molten salt electrochemistry to remove the oxide scale from titanium artefacts that forms when titanium is heated in air. It warrants a brief excursion through the metallurgy of titanium making to fully appreciate the relevance and scope of the FFC process and hence realise the potential of the Ilmenox process.

The existing process for titanium extraction was invented and developed by Kroll in the 1930s to 1940s and later named after him. The process comprises two steps, the carbo-chlorination of titanium dioxide to titanium tetrachloride, and the magnesium reduction of the titanium tetrachloride to titanium metal (e.g., Habashi 1997). The process is laborious and time-consuming and, by modern standards, grossly environmentally unfriendly. Kroll himself was reported to have stated that within fifteen years an electrolytic route would replace his process.

Conventional thinking was that in order to win titanium via an electrolytic process, it would be necessary to dissolve a titanium salt in a molten salt, such as sodium chloride, and electrolyse it. At the cathode, titanium ions are reduced to titanium, which deposits; and at the anode, chloride ions are oxidised to chlorine, which exits as a gas. However, it has proven to be exceptionally intricate to win solid titanium from a molten salt, because the metal product always occurs in the form of a very fine powder which is both difficult to harvest and prone to oxidation. Moreover, the efficiency of this type of electrolytic process is poor, owing to titanium existing in several oxidation states in the molten salt and being able to set up parasitic shuttle reactions. Many tens of millions of dollars have been spent in vain on attempts to develop an electrolytic process for titanium winning.

The concept of the FFC process is favourable in that the starting material is not a titanium salt but titanium dioxide that is abundantly and cheaply available. The FFC process uses an electrolytic cell at around 900°C, in which a porous sintered titanium dioxide body is made the cathode in an electrolyte of molten calcium chloride. A potential is applied, which is sufficiently high to decompose the cathode such that oxide ions are expelled into the electrolyte, but also sufficiently low not to decompose the electrolyte. This means the reduction of the oxide in the cathode proceeds without the deposition of metal from the electrolyte onto the cathode (Fray et al. 1999; Chen et al. 2000). The cathodic reaction in the FFC process may hence be construed as the direct ionisation of oxygen, and the cathodic product is the de-oxidised metal that is left behind:

$$TiO_2 + 4\,e^- = Ti + 2\,O^{2-} \tag{7.9}$$

The anodic reaction in the FFC process is the discharge of oxide ions from the electrolyte. When a carbon material is used as the anode, the anodic reaction product is carbon oxides that are released from the cell as a gas:

$$C + O^{2-} = CO + 2\,e^- \quad \text{and/or} \quad C + 2\,O^{2-} = CO_2 + 4\,e^- \tag{7.10a,b}$$

The key features of the FFC process are that the titanium is always in the solid state and forms within the cathode, while the molten salt electrolyte is a transport medium for oxide ions and is not consumed. A schematic of an FFC cell is shown in Fig. 7.4. Usefully, as the titanium product is not deposited from a salt, it is relatively compact and inert and can readily be retrieved from the cell. Overall, the FFC process is not an incremental improvement of pre-existing technologies but a novel approach in the field of electrowinning. Furthermore, it is a generic method with manifold applications for the winning of other metals and the synthesis of alloys and intermetallics from suitable oxide blends (Fray 2001, 2002).

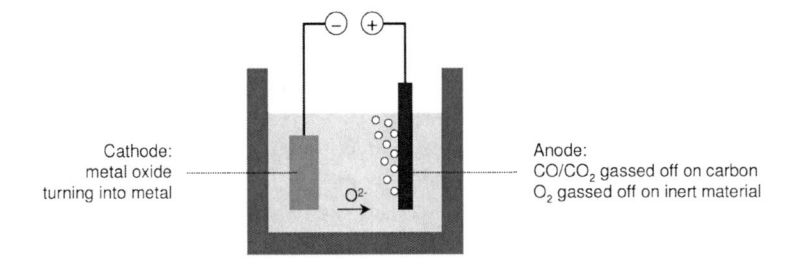

Fig. 7.4 Schematic of an FFC cell, showing conversion of oxide to metal at the cathode, oxide ion transport through the molten salt electrolyte, and gas evolution at the anode.

In order to scale-up the FFC process toward the production of industrial quantities of titanium, it was necessary to identify and optimise all the relevant process parameters. To that end, a series of fundamental studies was conducted at the University of Cambridge that extended over a period of several years. This rigorous research program led to a full understanding of the individual reaction steps in the cathode, the transference properties of the electrolyte, and the reactions at the anode. These studies are now published (Schwandt and Fray 2005; Fray et al. 2006; Alexander et al. 2006, 2011; Schwandt et al. 2009). The scientific results, along with the practical experience, gained in these studies have made possible the successful production of titanium in pilot plants (Schwandt et al. 2010).

7.5.2 Scope of the Ilmenox Project

In 2004 NASA's Exploration Systems Mission Directorate held an open competition for contracts in the field of ISRU. Of the 3,700 submissions received, 70 were accepted. Only two grants were awarded to companies outside the US. Among these was the submission by the team from British Titanium plc and the University of Cambridge, partnered by Florida Institute of Technology and NASA's Kennedy Space Centre. This contract, given the name 'Ilmenox', was initially awarded $14.3 million in staged payments. Its aim was to demonstrate the feasibility of stripping oxygen from ilmenite and regolith via the FFC process. As discussed in Section 7.3, ilmenite is a mixed oxide of iron and titanium, and lunar regolith is a complex mixture of rock and mineral fragments, consisting of various combinations of the oxides of calcium, aluminium, silicon and other metals. By 2005, when the Ilmenox project commenced, the FFC process was considered a method of choice, since its ability to reduce titanium dioxide and several mixed oxides had been proven, and since significant progress toward the scale-up to pilot plant level had already been achieved. It was anticipated that this knowledge could be passed to the adaptation of the technology for extracting oxygen from lunar materials.

7.5.3 Oxygen-Evolving Anodes

The initial objective of the FFC process was to electrowin metals via the cathodic reduction of oxide to metal. The only purpose of the anodic reaction was to remove the cathodically liberated oxide ions from the electrolyte. The simplest way of achieving this in a high-temperature molten salt electrolytic cell was by means of a graphite anode and the formation of carbon oxides. The Ilmenox project now required the production of oxygen at the anode. Consequently, the oxygen-consuming carbon anode had to be replaced for an oxygen-evolving, i.e., inert, non-carbon anode in order to enable the desired anodic reaction:

$$2\,O^{2-} = O_2 + 4\,e^- \tag{7.11}$$

A first guess was to test doped tin oxide (SnO_2) as the anode, considering that this material is widely used as an electrode in silicate melts in the glass industry. It was found that tin oxide is indeed able to generate oxygen when anodically polarised in molten calcium chloride containing a quantity of dissolved calcium oxide (Kilby et al. 2006, 2010; Barnett et al. 2009). However, it was also observed that tin oxide had a limited lifetime as it gradually erodes over several hours of polarisation. Moreover, tin oxide combines with calcium oxide in the melt and slowly forms an insulating layer of calcium stannate which may lead to passivation of the electrode. Overall, the use of tin oxide anodes allowed for proof of concept of the Ilmenox process but continuous processing was impossible.

Significant progress was achieved when testing solid solutions of calcium titanate and calcium ruthenate ($CaTi_xRu_{1-x}O_3$) as the anode material (Fray and Doughty 2009; Jiao et al. 2009; Jiao and Fray 2010). Depending on the content of calcium ruthenate in the mixture, these materials are metallic conductors or semi-conductors. Several laboratory experiments indicated that these mixtures could be used for oxygen production for more than 100 h without showing any sign of erosion or passivation. The solid solutions of calcium titanate and calcium ruthenate therefore became a key material in the subsequent work on the FFC process.

It should be emphasised that, while the use of oxygen-evolving anodes was compulsory for the Ilmenox process, this type of electrode is also imperative for the industrialisation of the FFC process on earth. The reason is twofold. Firstly, the generation of oxygen from inert anodes brings green credentials to the process, in contrast to the emission of greenhouse gases in the steel and aluminium industries. Secondly, the absence of carbon at the anode improves the efficiency of the process and the chemical purity of the cathodic product. This is because carbon dioxide formed at a carbon anode may dissolve in a calcium chloride melt containing calcium oxide and form carbonate ions:

$$C + 3\,O^{2-} = CO_3^{2-} + 4\,e^-$$ (7.12)

These carbonate ions can diffuse to the cathode, where they decompose back to carbon and oxide ions (Schwandt and Fray 2007; Schwandt et al. 2009). The net result of this parasitic reaction is that energy is wasted and carbon from the anode reaches the cathode. This may be a terminal event for metals like high purity titanium, where titanium and carbon can react to titanium carbide that acts as a source of cracks in structural applications, or for tantalum, where the market is mainly for carbon-free materials in the electronics industry.

7.5.4 Electro-deoxidation of Lunar Materials

The experimental rigs used for the Ilmenox project are displayed in Figs. 7.5 and 7.6. Figure 7.5 shows a typical laboratory reactor employed for the processing of gram quantities of oxide feed material, and Fig. 7.6 shows a pilot plant designed for the processing of kilogram quantities.

Fig. 7.5 Laboratory rig used in the Ilmenox project, showing electric furnace with tubular metal retort and electrochemical equipment.

Fig. 7.6 Pilot plant rig used in the Ilmenox project, showing electric furnace with rectangular metal retort supporting the anode chamber (left) and cathode chamber (right).

The first feed material used for electro-deoxidation experiments within the Ilmenox project was ilmenite. This was selected for two reasons. Firstly, the earlier lunar survey by NASA's remote sensing Clementine satellite had discovered several ilmenite-rich areas on the moon, specifically the high-Ti mare basalts on the nearside (Fig. 7.2a). Secondly, extensive research and development expertise was at hand on the processing of pure titanium dioxide, which could be directly used for the work on ilmenite. The second feed material was lunar simulant JSC-1 supplied by NASA.

Figure 7.7 presents original data that were recorded in a typical laboratory experiment, in which gram quantities of oxide feed material were used as the cathode in combination with an oxygen-evolving anode and a molten calcium chloride electrolyte. The cell was polarised at around 3 V for several hours at a temperature of around 900°C under a flow of argon gas. In the course of the experiment both the current through the cell and the oxygen content in the argon gas stream at the reactor outlet were recorded. Oxygen monitoring was done with a commercial lambda-probe. It is seen that the current is high at the beginning of the run, indicating that the oxide-to-metal reduction proceeds at a high rate, and then falls gradually during the remainder of the run, indicating that the reduction approaches completion. In the experiment, oxygen is removed from the cathode and transported through the electrolyte to the anode, released from there into the gas stream and then measured with the oxygen gas sensor. Consequently, there is a good correlation between the time-dependencies of the current through the cell and the oxygen pressure in the outlet gas. Overall, good reproducibility was achieved between the various experiments, indicating the proper functioning of the Ilmenox process.

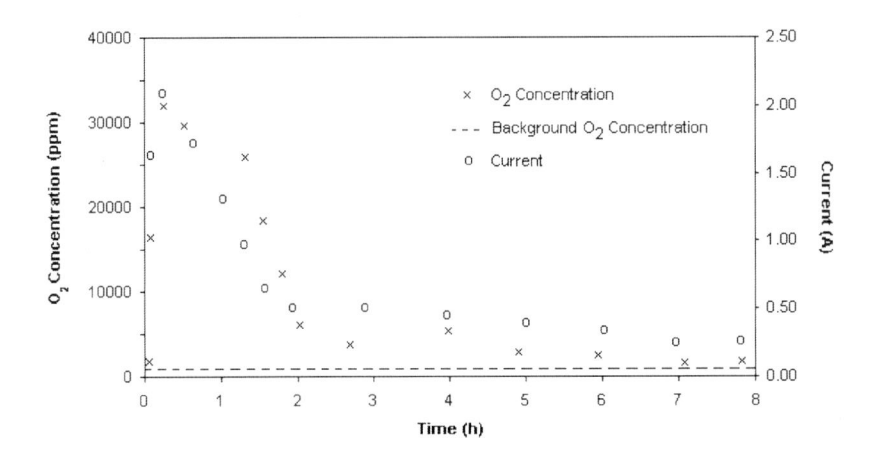

Fig. 7.7 Gas oxygen content versus time and cell current versus time for an FFC reduction using an oxide cathode and an oxygen-evolving anode in molten calcium chloride electrolyte at a temperature of 900°C under a flow of argon gas.

Figures 7.8 and 7.9 present some of the materials from the Ilmenox project. Figure 7.8 shows a photograph of the ilmenite powder, the sintered ilmenite tiles used in the electrolyses, the reduced ilmenite tiles, and a button-melted alloy sample made from the reduced material. Figure 7.9 shows scanning electron micrographs of the original JSC-1 feed material and of a polished section of the reduced JSC-1 material.

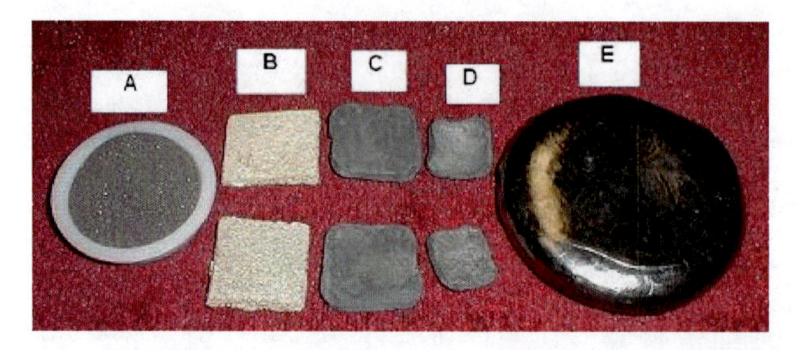

Fig. 7.8 Materials from the Ilmenox project: (A) original ilmenite powder, (B) pressed and sintered ilmenite tiles, (C) 95% reduced ilmenite tiles, (D) >99% reduced ilmenite tiles, (E) a button-melted solid alloy body made from reduced tiles.

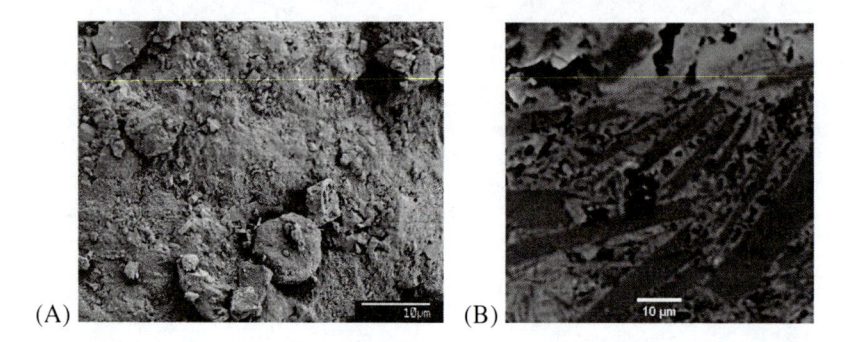

Fig. 7.9 Materials from the Ilmenox project: (A) electron micrograph of the as-supplied lunar simulant JSC-1; the sample has a coarse microstructure with particles of irregular size and shape; (B) electron micrograph of reduced and polished JSC-1; the sample is of a metallic nature with some porosity and elemental analysis confirmed the presence of the expected metallic constituents Ca, Mg, Al, Si and Fe; the darker lath-shaped areas are particularly rich in Al and Si, the lighter areas are rich in Mg and Fe, and the black areas are pores.

7.5.5 Envisaged Future Development

The Ilmenox process as demonstrated thus far is a batch process. This has two fundamental disadvantages. Firstly, it requires the preparation of shaped and sintered precursors for use as the cathode, which would not be a simple undertaking in the lunar environment. Secondly, it involves the inevitable removal of some of the molten salt electrolyte from the electrolytic cell during cathode retrieval, which would necessitate a constant replenishment from earth.

The next step forward is the design and operation of a continuous reactor, which is able to accept a powder oxide feed material and yield a liquid metal product. In this type of reactor, regolith is simply fed into a hopper without the need for any selective mining, beneficiation or other processing; oxygen is gassed off for storage; and the metal alloy of mainly aluminium and silicon is removed in the liquid state via a spout arrangement. The alloy product is poured into moulds and stacked or, more advantageously, subjected to electrolytic refining to separate its metal constituents. The dimensions of such a furnace can be tailored to fit into a rocket, making it an attractive project for future space architecture. This novel concept is currently being patented by Green Metals Ltd.

7.6 Water on the Moon

In October 2009 NASA's LCROSS mission detected water ice in significant quantities in the permanently shadowed Cabeus crater near the lunar south pole (NASA 2009; Colaprete et al. 2010; for a brief review see Anand 2010).

In this mission, the empty Centaur upper stage of an Atlas V rocket was steered into the 100 km wide crater at 9,000 km h^{-1} and punched a hole of about 25 m in diameter. In the rising plume, sensors on board the Lunar Crater Observation and

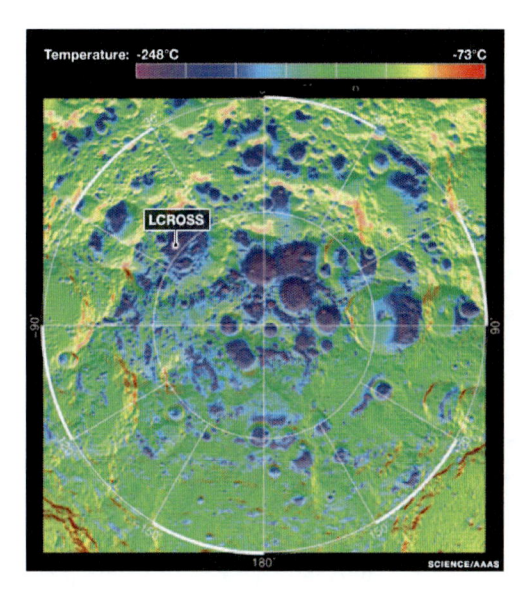

Fig. 7.10 Location map of the impact of the Centaur upper stage of an Atlas V rocket near the lunar south pole. (Image reproduced with permission from NASA).

Fig. 7.11 Impact plume arising from the impact of the Centaur upper stage of an Atlas V rocket near the lunar south pole. (Image reproduced with permission from NASA).

Sensing Satellite (LCROSS) and the Lunar Reconnaissance Orbiter (LRO) made observations of the debris and vapour cloud formed. A map of the location of the impact near the lunar south pole and an image of the resulting impact plume are shown in Figs. 7.10 and 7.11, respectively.

Based on the LCROSS results it is estimated that 5.6±2.9% of the total mass of the targeted lunar crater's soil consists of water ice (Colaprete et al. 2010), although this value is model-dependent and remains to be confirmed. At these levels, astronauts could extract around 100 litres of water from every cubic metre of lunar soil. Water from craters would provide essential life support for a manned moonbase and could be electrolysed into oxygen and hydrogen.

The presence of water on the moon may have implications on the further development of the various processes for oxygen extraction from lunar materials including the Ilmenox process. However, it should be borne in mind that surface operations for ice-mining will be technologically challenging, energy-consuming and very expensive, perhaps even unfeasible, within the permanently shadowed craters where the averaged near-surface temperature is of the order of -230°C or lower (Paige et al. 2010). Moreover, the transport of water for exploration elsewhere on the moon will be problematic in the mountainous terrain of the polar highlands, as is evident from the topography, slope and roughness data in Fig. 7.12. In contrast, oxygen extraction from regolith has the advantage of promising viability wherever electricity is available and, importantly, also yields a metal co-product.

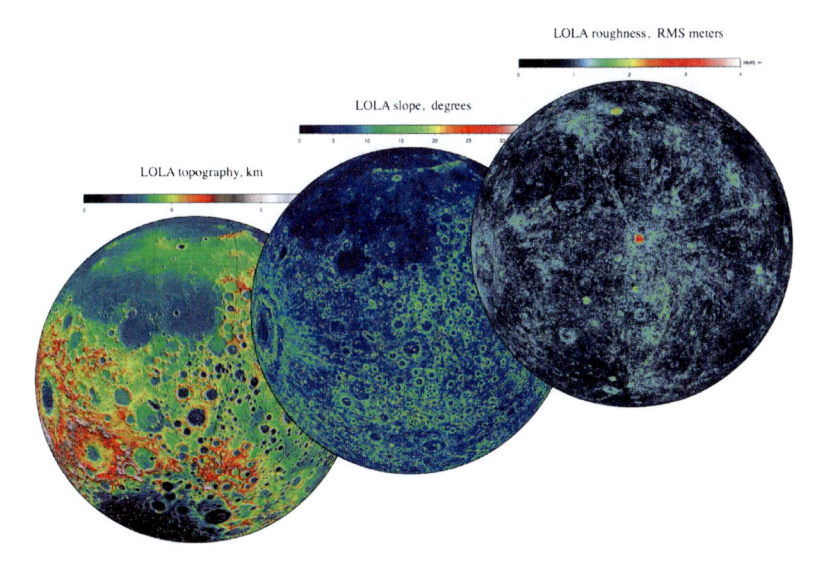

Fig. 7.12 Topography, slope and roughness data of the nearside of the moon as taken by the Lunar Reconnaissance Orbiter (LRO) through Lunar Orbiter Laser Altimeter (LOLA) investigation. (Image reproduced with permission from NASA).

7.7 Outlook - The Case for ISRU

"This cause of exploration and discovery is not an option we choose; it is a desire written in the human heart." President George W. Bush, 4 February 2003 (NASA 2004).

The vision for space exploration has clearly not been met over the past seven years, although the challenge to 'live off the land' has achieved great strides. Notable progress has been made with the discovery of water ice on the moon and the demonstration of the separation of regolith simulant into its constituents of oxygen and metal. This may open up new possibilities for space exploration in the future.

It will need to be evaluated what the arguments are for the further development of ISRU techniques such as the Ilmenox process. The particular attractions of the Ilmenox process clearly lie in its ability to support human operations well beyond the water ice-holding craters near the poles, and to produce metals for a variety of purposes in addition to oxygen. Few other lunar ISRU proposals combine both these benefits to the same extent as the Ilmenox process. These advantages will also apply to any other airless bodies in the solar system which are covered by oxide-rich regoliths. In the particular case of martian exploration, where local resources of water may be widespread, the key argument for the Ilmenox process will not lie in being able to compete with these local resources but in being able to make metals in situ. The economic benefits of this will rise as human activity in the solar system becomes more widespread.

As a final thought, it is worthwhile noting that some of the research and development work triggered through space-related projects is bound to give new momentum to the further development of terrestrial technologies. In particular the utilisation of inert anodes will contribute to making metallurgical extraction processes, including the FFC process, more environmentally friendly. Also the winning of a liquid metal as the primary product from a powder oxide feedstock in the FFC process is attractive, because it would allow dispensing with several processing steps that are currently required for compact oxide feed materials and solid metal products. While all these concepts are very attractive and rewarding, they also require substantial future funding.

Acknowledgements. The authors acknowledge the great contributions of Dr Greg Doughty, Dr John Haygarth, Dr Stuart Mucklejohn, Mr Shaan Oosthuizen, Dr Kamal Tripuraneni Kilby, Dr Lilia Centeno-Sánchez, and Dr Shuqiang Jiao to the Ilmenox project. Mr Richard Stevens is thanked for critically reading the manuscript.

References

Alexander, D.T.L., Schwandt, C., Fray, D.J.: Microstructural kinetics of phase transformations during electrochemical reduction of titanium dioxide in molten calcium chloride. Acta Materialia 54, 2933–2944 (2006)

Alexander, D.T.L., Schwandt, C., Fray, D.J.: The electro-deoxidation of dense titanium dioxide precursors in molten calcium chloride giving a new reaction pathway. Electrochimica Acta 56, 3286–3295 (2011)

Anand, M.: Lunar water: a brief review. Earth, Moon, and Planets 107, 65–73 (2010)

Augustine Commission, Review of US human spaceflight plans committee: seeking a human spaceflight program worthy of a great nation (2009),
http://www.nasa.gov/pdf/
397898main_HSF_Cmte_FinalReport_High.pdf

Barnett, R., Kilby, K.T., Fray, D.J.: Reduction of tantalum pentoxide using graphite and tin-oxide-based anodes via the FFC-Cambridge process. Metallurgical and Materials Transactions B 40, 150–157 (2009)

Chen, G.Z., Fray, D.J., Farthing, T.W.: Direct electrochemical reduction of titanium dioxide to titanium in molten calcium chloride. Nature 407, 361–364 (2000)

Colaprete, A., Schultz, P., Heldmann, J., Wooden, D., Shirley, M., Ennico, K., Hermalyn, B., Marshall, W., Ricco, A., Elphic, R.C., Goldstein, D., Summy, D., Bart, G.D., Asphaug, E., Korycansky, D., Landis, D., Sollitt, L.: Detection of water in the LCROSS ejecta plume. Science 330, 463–468 (2010)

Colson, R.O., Haskin, L.A.: Oxygen from the lunar soil by molten silicate electrolysis. In: McKay, F.A., McKay, D.S., Duke, M.B. (eds.) Space Resources, vol. 3, Materials, NASA, Lyndon B Johnson Space Center, Houston, TX, pp. 195–209 (1992)

Colson, R.O., Haskin, L.A.: Producing oxygen by silicate melt electrolysis. In: Lewis, J., Matthews, M.S., Guerrieri, M.L. (eds.) Resources of Near-Earth Space, University of Arizona Press, Tucson, AZ, pp. 109–127 (1993)

Crawford, I.A.: The scientific case for renewed human activities on the moon. Space Policy 20, 91–97 (2004)

Curreri, P.A., Ethridge, E.C., Hudson, S.B., Miller, T.Y., Grugel, R.N., Sen, S., Sadoway, D.R.: Process demonstration for lunar in situ resource utilization – molten oxide electrolysis. MSFC Independent Research and Development, Project No 5-81, NASA Center for AeroSpace Information, Hanover, MD (2006)

ESA-ESRIN, Sustained moon surface operations, automated surface infrastructure elements (2009a), http://esamultimedia.esa.int/ docs/exploration/ReferenceArchiteture/Final%20ReviewJan09/ 07_Final_Presentation_CGS_Automated%20SurfaceInfrastructure Elementsv3.pdf

ESA-ESRIN, Initially identified areas of interest for Europe, utilisation of in-situ resources (2009b), http://esamultimedia.esa.int/docs/exploration/ ReferenceArchitecture/Final%20ReviewJan09/ 12_Final_Presentation_CGS_ISRUv2.pdf

Fray, D.J.: Emerging molten salt technologies for metals production. JOM 53(10), 26–31 (2001)

Fray, D.J.: Anodic and cathodic reactions in molten calcium chloride. Canadian Metallurgical Quarterly 41, 433–439 (2002)

Fray, D.J., Doughty, G.R.: Improvements in electrode materials. Patent WO2009010737 (2009)

Fray, D.J., Farthing, T.W., Chen, Z.: Removal of oxygen from metal oxides and solid solutions by electrolysis in a fused salt. Patent WO9964638 (1999)

Fray, D.J., Schwandt, C., Doughty, G.R.: Improved electrolytic method, apparatus and product. Patent WO2006027612 (2006)

GES, The global exploration strategy: framework for coordination (2007), http://esamultimedia.esa.int/docs/GES_Framework_final.pdf

Gibson, M.A., Knudsen, C.W.: Lunar oxygen production from ilmenite. In: Mendell, W.W. (ed.) Lunar Bases and Space Activities of the 21st Century. Lunar and Planetary Institute, Houston, TX, pp. 543–550 (1985)

Habashi, F.: Handbook of Extractive Metallurgy. Titanium, vol. 2, ch. 22. Wiley-VCH, Heidelberg (1997)

Heiken, G.H., Vaniman, D.T., French, B.M. (eds.): The Lunar Sourcebook. Cambridge University Press, Cambridge (1991)

Jiao, S.Q., Fray, D.J.: Development of an inert anode for electrowinning in calcium chloride-calcium oxide melts. Metallurgical and Materials Transactions B 41, 74–79 (2010)

Jiao, S.Q., Kumar, K.N.P., Kilby, K.T., Fray, D.J.: Preparation and electrical properties of x CaRuO3 / (1-x) CaTiO3 perovskite composites. Materials Research Bulletin 44, 1738–1742 (2009)

Jolliff, B.L., Wieczorek, M.A., Shearer, C.K., Neal, C.R. (eds.): New Views of the Moon. Reviews in Mineralogy & Geochemistry, vol. 60. Mineralogical Society of America and the Geochemical Society, Chantilly, VA (2006)

Kilby, K.C.T., Centeno, L., Doughty, G., Mucklejohn, S., Fray, D.J.: The electrochemical production of oxygen and metal via the FFC-Cambridge process. Space Resources Roundtable VIII, 64–65 (2006)

Kilby, K.T., Jiao, S.Q., Fray, D.J.: Current efficiency studies for graphite and SnO2-based anodes for the electro-deoxidation of metal oxides. Electrochimica Acta 55, 7126–7133 (2010)

Knudsen, C.W., Gibson, M.A.: Processing lunar soils for oxygen and other materials. In: McKay, F.A., McKay, D.S., Duke, M.B. (eds.) Space Resources, vol. 3, Materials, NASA, Lyndon B Johnson Space Center, Houston, TX, pp. 186–194 (1992)

NASA, The vision for space exploration (2004), http://www.nasa.gov/pdf/ 55583main_vision_space_exploration2.pdf

NASA, Material safety data sheet JSC-1A (2005),
    `http://isru.msfc.nasa.gov/lib/Documents/`
    `JSC-1a_MATERIAL_SAFETY_DATA_SHEET.pdf`
NASA, Overview of in-situ resource utilization (ISRU) architecture (2007a),
    `http://www7.nationalacademies.org/ssb/`
    `1LunarWorkshop_Sanders.pdf`
NASA, In-situ resource utilization (ISRU), development & incorporation plans (2007b),
    `http://www.nasa.gov/pdf/`
    `203084main_ISRU%20TEC%2011-07%20V3.pdf`
NASA, Material safety data sheet NU-LHT-2M (2008),
    `http://isru.msfc.nasa.gov/lib/Documents/`
    `PDF%20Files/LHT-2M_MSDS_7-31-08.pdf`
NASA, LCROSS impact data indicates water on moon (2009),
    `http://www.nasa.gov/mission_pages/LCROSS/`
    `main/prelim_water_results.html`
NASA, Remarks by the President on space exploration in the 21st century (2010a),
    `http://www.nasa.gov/news/media/trans/obama_ksc_trans.html`
NASA, Fiscal year 2012 budget estimates (2010b),
    `http://www.nasa.gov/pdf/`
    `516674main_NASAFY12_Budget_Estimates-Overview-508.pdf`
NASA, NASA extends crew flight contract with Russian Space Agency (2011),
    `http://www.nasa.gov/home/hqnews/2011/mar/`
    `HQ_C11013_Soyuz_Contract.html`
Neal, C.R., Taylor, L.A.: Petrogenesis of mare basalts: a record of lunar volcanism. Geo-
    chimica et Cosmochimica Acta 56, 2177–2211 (1992)
Paige, D.A., Siegler, M.A., Zhang, J.A., Hayne, P.O., Foote, E.J., Bennett, K.A., Vasavada,
    A.R., Greenhagen, B.T., Schofield, J.T., McCleese, D.J., Foote, M.C., DeJong, E., Bills,
    B.G., Hartford, W., Murray, B.C., Allen, C.C., Snook, K., Soderblom, L.A., Calcutt, S.,
    Taylor, F.W., Bowles, N.E., Bandfield, J.L., Elphic, R., Ghent, R., Glotch, T.D., Wyatt,
    M.B., Lucey, P.G.: Diviner lunar radiometer observations of cold traps in the moon's
    south polar region. Science 330, 479–482 (2010)
Rosenberg, S.D., Beegle Jr., R.L., Guter, G.A., Miller, F.E., Rothenberg, M.: The onsite
    manufacture of propellant oxygen from lunar resources. In: McKay, F.A., McKay, D.S.,
    Duke, M.B. (eds.) Space Resources, vol. 3, Materials, NASA, Lyndon B Johnson Space
    Center, Houston, TX, pp. 162–185 (1992)
Rosenberg, S.D., Musbah, O., Rice, E.E.: Carbothermal reduction of lunar materials for
    oxygen production on the moon: reduction of lunar simulants with methane. Lunar and
    Planetary Science 27, 1105–1106 (1996)
Schwandt, C., Fray, D.J.: Determination of the kinetic pathway in the electrochemical re-
    duction of titanium dioxide in molten calcium chloride. Electrochimica Acta 51, 66–76
    (2005)
Schwandt, C., Fray, D.J.: The electrochemical reduction of chromium sesquioxide in mol-
    ten calcium chloride under cathodic potential control. Zeitschrift für Naturforschung
    A 62, 655–670 (2007)
Schwandt, C., Alexander, D.T.L., Fray, D.J.: The electro-deoxidation of porous titanium
    dioxide precursors in molten calcium chloride under cathodic potential control. Electro-
    chimica Acta 54, 3819–3829 (2009)
Schwandt, C., Doughty, G.R., Fray, D.J.: The FFC-Cambridge process for titanium metal
    winning. Key Engineering Materials 436, 13–25 (2010)

Senior, C.L.: Lunar oxygen production by pyrolysis. In: Lewis, J., Matthews, M.S., Guerrieri, M.L. (eds.) Resources of Near-Earth Space, University of Arizona Press, Tucson,
AZ, pp. 179–197 (1993)

Shchetkovskiy, A., McKechnie, T., Sadoway, D.R., Paramore, J., Melendez, O., Curreri,
P.A.: Development and testing of high surface area iridium anodes for molten oxide
electrolysis. In: Song, G.B., Malla, R.B. (eds.) Earth and Space 2010: Engineering,
Science, Construction, and Operations in Challenging Environments. ASCE, New York
(2010), doi:10.1061/41096(366)96

Sibille, L., Sadoway, D., Tripathy, P., Standish, E., Sirk, A., Melendez, O., Stefanescu, D.:
Performance testing of molten regolith electrolysis with transfer of molten material for
the production of oxygen and metals on the moon. NASA Technical Reports Server,
NASA, Kennedy Space Center, Report No KSC-2009-310 (2010)

Spudis, P.D.: The moon and the new Presidential space vision. Earth, Moon, and Planets 94, 213–219 (2005)

Spudis, P.D., Zellner, N., Delano, J., Whittet, D.C.B., Fessler, B.: Petrologic mapping of
the moon: a new approach. In: 33rd Annual Lunar and Planetary Science Conference,
Houston, TX, Abstract 1104 (2002)

Spudis, P.D., Bussey, B., Plescia, J., Josset, J.-L., Beauvivre, S.: Geology of Shackleton
crater and the south pole of the moon. Geophysical Research Letters 35, L14201, 1–5
(2008)

Standish, E.: Design of a molten materials handling device for support of molten regolith
electrolysis. Masters Thesis, Ohio State University (2010)

Steurer, W.: Vapor phase pyrolysis. In: McKay, F.A., McKay, D.S., Duke, M.B. (eds.)
Space Resources, vol. 3, Materials, NASA, Lyndon B Johnson Space Center, Houston,
TX, pp. 210–213 (1992)

Stoeser, D., Wilson, S., Rickman, D.: Design and specifications for the highland regolith
prototype simulants NU-LHT-1M and -2M. NASA Technical Reports Server, NASA,
Marshall Space Flight Center, Report No NASA/TM-2010-216438 (2010)

Sullivan, T.A.: A modified sulfate process to lunar oxygen. In: Sadeh, W.Z., Sture, S., Miller, R.J. (eds.) Engineering, Construction, and Operations in Space III, pp. 641–650.
ASCE, New York (1992)

The White House, Remarks by the President on U.S. space policy (2004),
    http://history.nasa.gov/Bush%20SEP.htm

UK Space Agency, Economic analysis to support a study on the options for the UK involvement in space exploration. Final Report by London Economics (2009)

8 In-Situ Water Production by Reducing Ilmenite

Yang Li[1,2], Xiongyao Li[1], Shijie Wang[1], Hong Tang[1],
Hong Gan[1,2], Shijie Li[1], Guangfei Wei[1,2], Yongchun Zheng[3],
Kang T. Tsang[4], and Ziyuan Ouyang[1]

[1] Institute of Geochemistry, Guiyang, China
[2] University of the Chinese Academy of Sciences, Beijing, China
[3] The National Astronomical Observatories, Beijing, China
[4] Hong Kong University of Science and Technology, China

8.1 Introduction

Water is considered to be a fundamental condition of human's colonization of the moon. The supply of oxygen is essential to the utilization of most lunar resources. According to the result of rough calculation, the cost for delivering oxygen from earth to moon for a ten-people lunar base will be nearly 5 to 9 billion dollars a year (Schrunk 1999; Taylor and Carrier 1993; Duke 2003). The cost is so high that water cannot be entirely supported by transportation from the Earth. Nevertheless, there are considerable mineral reserves in lunar soil. Compared to transporting water from earth directly, it is more economical to extract it from the lunar soil. The effects of meteorites, solar wind and cosmic ray make most of the lunar surface covered with a layer of lunar soil. The thickness of lunar soil is approximately 4~5 meters at the mare and >10 meters on the highland. The only practical source of water in the lunar soil is igneous minerals which contain typically 40 to 50% oxygen as oxides. The major minerals are ilmenite, anorthite, and olivine. All these oxides can provide oxygen and water to the lunar base even though some of them cannot be easily reduced. Compared with oxides of silicon, aluminum, titanium, calcium or magnesium, it is much more easily to extract oxygen from iron oxide such as ilmenite.

In this chapter, experiment simulating in-situ water production was designed and operated. The principal aim is to describe the reduction of ilmenite, selection of reducing agent, reaction condition and influencing factors. Finally, various conclusions are given in Sect. 8.4.

8.2 In-Situ Water and Oxygen Production Processes

The processes for oxygen production on the Moon were listed in Table 8.1 (Taylor and Carrier 1993). Before choosing the proper experimental process, three preconditions should be reached. The first is the raw material must be supplied by the surface materials on the Moon. Secondly, the reaction must be highly efficient. That is the least amount of raw material and the lowest energy consumption make the maximum oxygen output. Thirdly, the reduction reaction must be simple and direct in order to reduce the cost and risk.

Table 8.1 Processes for oxygen production on the Moon

Solid/Gas Interaction	References
Ilmenite reduction with hydrogen	Gibson and Knudsen (1988a)
Ilmenite reduction with C/CO	Chang (1959); Zhao and Shadman(1990)
Ilmenite reduction with methane	Friedlander (1985)
Glass reduction with hydrogen	McKay et al. (1991)
Reduction with hydrogen sulfide	Dalton and Hohman (1972)
Extraction with fluorine	Burt (1988); Seboldt et al. (1991)
Carbochlorination	Lynch (1989)
Chlorine plasma reduction	Lynch (1989)
Silicate/Oxide Melt	
Molten silicate electrolysis	Haskin (1985);Colson and Haskin(1990)
Fluxed molten silicate electrolysis	Dalton and Hohman (1972)
Carbothermal reduction	Rosenberg et al. (1966);Cutler and Krag (1985)
Magma partial oxidation	Waldron (1989)
Li or Na reduction of ilmenite	Semkow and Sammells (1987)
Pyrolysis	
Vapor phase reduction	Steurer and Nerad (1983)
Ion(Plasma) separation	Steurer and Nerad (1983)
Plasma reduction of ilmenite	Allen et al. (1988)
Aqueous Solutions	
HF acid dissolution	Waldron (1985)
H_2SO_4 acid dissolution	Sullivan (1990)
Co-Product Recovery	
Hydrogen/Helium water production	Christiansen et al. (1988)

Energy efficiency is the main shortcoming of Silicate/Oxide Melt and Pyrolysis processes. High energy consumption makes it difficult to last for a long time. There is latent risk of the container or other accessories for the erosion of aqueous solutions. In addition, acids needed in the reaction are difficult to produce on the Moon. Feeding the chemical reaction with supplies from space vehicle will increase the cost of production inevitably. Compared with other oxygen production processes, H_2 reducing processes, with their much reduced reaction temperature and lower energy consumption, and is simpler and director. What's more, H_2 can be fed by collecting solar wind proton particles in the lunar regolith or the propellant of lunar spacecraft. So H_2 is chosen as the reducing agent in our process.

Choosing the appropriate oxide is as important as the reducing agent. The oxide must be utilized in-situ on the Moon and can be easily reduced by H_2. Pyroxene, olivine, feldspar and ilmenite are the most common oxides on the surface of the Moon. The free energy of cations in the lunar oxides can be gradated as $Ca^{2+} > Mg^{2+} > Al^{3+} > Ti^{2+} > Ti^{4+} > Si^{4+} > Na^+ > Fe^{2+} > Fe^{3+}$ (Burt 1988). So we conclude that ilmenite can be reduced more easily than silicates.

Heating methods can be divided into two kinds: traditional heating (blaze, hot blast, electric, steam heating) and microwave heating. The band of microwave frequency is from 300 MHz to 300 GHz. Microwave heating is preferable for the production of oxygen on the Moon due to the following reasons: firstly, the heating rate of microwave for typical lunar regolith parameters is roughly $50°C/s$ at the regolith temperature of $1000°C$ in Taylor's simulation experiments (Taylor and Meek 2003), assuming a microwave electric field of 300V/cm, relative dielectric constant about 5 and at a frequency of 2.45 GHz. Thus, microwave can greatly shorten the heating time. Secondly, microwave can broaden the extent of reaction with its penetrating power to inside of the sample rather than conducting from the surface to the core, as in traditional heating. Thirdly, the energy efficiency of microwave heating is much higher than other heating methods. Microwave heating can save 60% of energy compared with traditional heating methods.

In conclusion, according to the current research results of oxygen in-situ production on the Moon and properties of essential lunar minerals, we finally focus our approach to ilmenite reduced by hydrogen with microwave heating.

8.3 The General Planning of H_2O In-suit Produces by Reducing Ilmenite with H_2

The equation of H_2 reducing ilmenite is as follows:

$$FeO \cdot TiO_2(solid)+H_2(gas) \rightarrow Fe(solid)+TiO_2(solid)+H_2O(gas) \qquad (8.1)$$

Ilmenite can be separated and concentrated from the lunar regolith if only its content exceed 5 wt% (Gibson and Knudsen 1985). Electrostatic and magnetic schemes for beneficiation have been proposed by Williams et al (1979). The natural content of ilmenite in lunar regolith is between 2~20wt% (Taylor 1990; Taylor et al. 1993; Taylor and Higgins 1994). The content of ilmenite is as high as 10~24 wt. % in Apollo 11 high Ti lunar soil sample (Cameron 1992; Taylor 2004; Taylor and McKay 1992). It has been evaluated that the total reserves of ilmenite in the lunar mare basalt may be $1.3 \times 10^{15} \sim 1.9 \times 10^{15}$ tons with the content of 8 wt.%.

The main sources of H_2 on the moon are the solar wind protons which are contained on the surface layer of lunar soil grains. They are originated from solar corona and composed of H^+, He^+ and other particles. The average speed of solar wind ions is nearly 420 km/s, so it can penetrate ~100nm into the surface layer of lunar soil grains. H^+ is one of the chief constituents of solar wind. The proton flux in solar wind is ~3.8×10^8 protons $cm^{-2}s^{-1}$ at 1AU and with accumulation up to 4 billion years the total content of proton on the lunar surface is ~80g cm^{-2}.

While it is advantageous to heat the reactants by microwave for its high efficiency, it is difficult to measure the temperature under the condition of microwave heating, because the thermocouple will be damaged in the microwave field. So muffle furnace was temporarily used to heat the reactants before this problem was resolved in our simulation experiment. Nevertheless, microwave should be used in the future in-situ water and oxygen production in lunar base for its high energy efficiency and high heating rate.

In conclusion, theoretically ilmenite can be reduced by hydrogen and there is an abundant of ilmenite and protons on the surface layer of the Moon. In the following sections, experimental simulation was designed to verify this idea and key parameters of the experiment as well as its energy efficiency are optimized. Results of our experiment are presented.

8.4 The Simulation Experiment of Oxygen and Water Production

8.4.1 Experimental Procedure

The flow chart of the simulation experiment is shown in Fig. 8.1. For the convenience to collect the reduced ilmenite powder and get rid of the potential contaminant, ilmenite powder was contained in glass tube and inserted into the

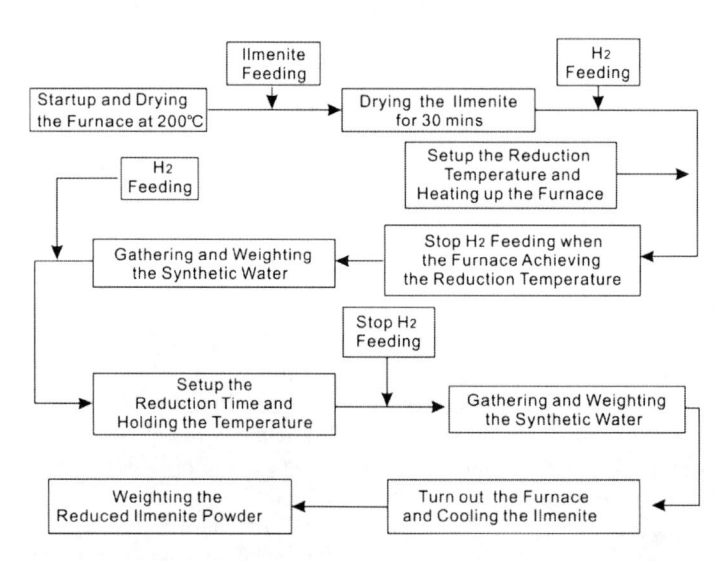

Fig. 8.1 Technological processes of the in-situ water and oxygen production simulation experiment

cavity of muffle furnace to be heated and reduced. Due to the possible high temperature involved, quartz tube that can endure up to 1300°C must be used in the reduce reaction. Water was gathered in a U tube which was refrigerated by liquid nitrogen to obviate the contamination of residual H_2. Samples and produced water were weighted by Sartouris BSA 224S, the resolution is 0.0001 g.

The experimental steps in sequential order are:

(1) Preheat the stove (200°C) and open the valve of hydrogen to remove the air and vapor in the stove - keep it for half an hour;

(2) Place the ilmenite powder in the quartz tube and weight it. Heat up the stove to the preset temperature, holding the hydrogen flux with 0.03 L/h;

(3) When the preset temperature is reached, stop the hydrogen supply and weight the water gathered in the U tube;

(4) If other higher preset temperatures is needed, repeat step 3;

(5) Pull the quartz tube to the cooling zone, weight the quartz tube and calculate the loss of ilmenite.

8.4.2 Experimental Result

In this section, we describe the preliminary results of our water and oxygen production simulation experiment. In order to get more information about the specimen, the reduced ilmenite powder was submitted to the analysis of SEM and XRD. Reduced ilmenite powder was consolidated by resin and thin sections were prepared by diamond cutting machine. Using the SEM, we can see that some of the ilmenite grains are surrounded by irregular rims. The outer part of the rim is shinier than the inner part. Those rims are analyzed by EDX, the main chemical composition of the outer rim is Fe and the inner rim is TiO_2, as shown in Fig.8.2.

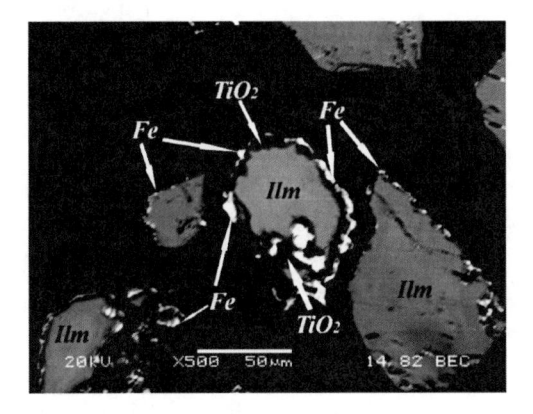

Fig. 8.2. BSE image of specimen 41501. The ilmenite grain is surrounded by the rime of Fe and TiO_2.

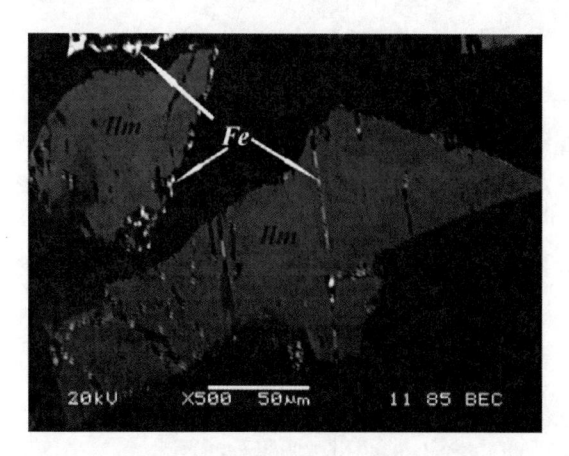

Fig. 8.3 BSE image of specimen 41501. Fe and TiO_2 were produced along apertures

In addition, linear traces of Fe appear in the apertures of an ilmenite grain (Fig.8.3). Likewise, for the fine grain of abraded ilmenite powder, Fe and TiO_2 are also detectable in the spectrum of XRD. Obviously, ilmenite has been reduced by hydrogen and water was produced in the process of reduction. We explain the distribution regularity of Fe and TiO_2 on the reduced ilmenite grain as follows. For the ilmenite grain with no aperture, hydrogen just reduced the surface layer of ilmenite grain. This is because the coating of Fe and TiO_2 can prevent hydrogen to penetrate into the inner part of ilmenite and protect the residuary ilmenite to be reduced. Nevertheless, apertures provide channels for hydrogen to penetrate into the inner part of the ilmenite grain, producing Fe and TiO_2 along the apertures.

8.4.3 Influencing Factors

We conclude that temperature, reaction time, ilmenite supply and grain size are the four main parameters that affect the output of the simulation experiment. Their separated influences are deduced and confirmed by the designed experiment.

8.4.3.1 Temperature

Temperature is one of the main parameters in the reducing reaction. It is believed that raising the temperature can improve the efficiency of reduce reaction distinctly. In order to confirm this hypothesis, separated experiments were designed for 10g ilmenite with terminal reaction temperatures from $400°C$ to $1000°C$ with steps of $100°C$. Each time, the terminal temperature was maintained for 15 mins. Amounts of water produced were obtained by repeating the experiments 3 times and taking average. The result is displayed in Fig.8.4, which shows the roughly linear rise of water production with temperature.

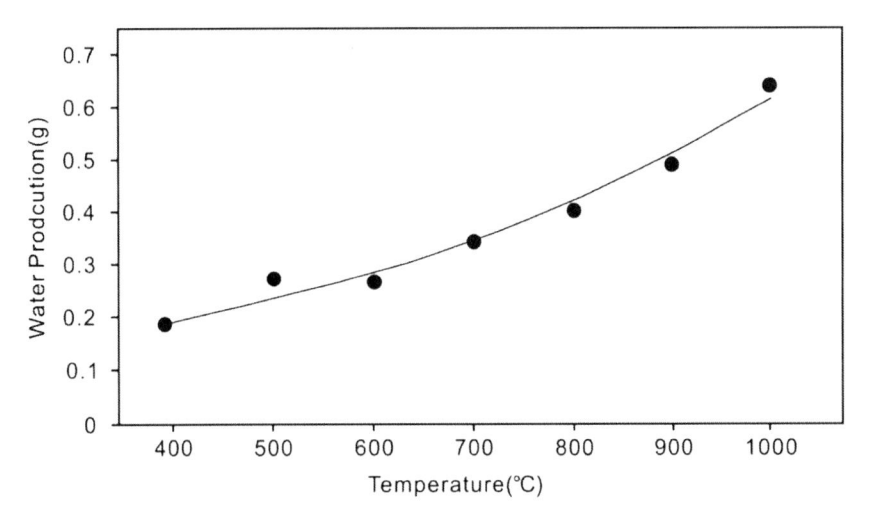

Fig. 8.4 Water production with 10 g of ilmenite powder from $400^{\circ}C$ to $1000^{\circ}C$

The increase in production at high temperature is more pronounced than at lower temperature. The result of this set of simulation confirms the hypothesis that higher reaction temperature improves the efficiency of water production.

8.4.3.2 Reaction Time

Extending time will increase water production. Simulation experiments were designed to increase reaction time from 30 mins to 180 mins with steps of 15 mins with a sample of 10 g of ilmenite. The reaction temperature is fixed at $1000^{\circ}C$. The result of this set of simulation experiments were showed in Fig.8.5. We know that water production increases linearly with the reaction time, the potential maximum

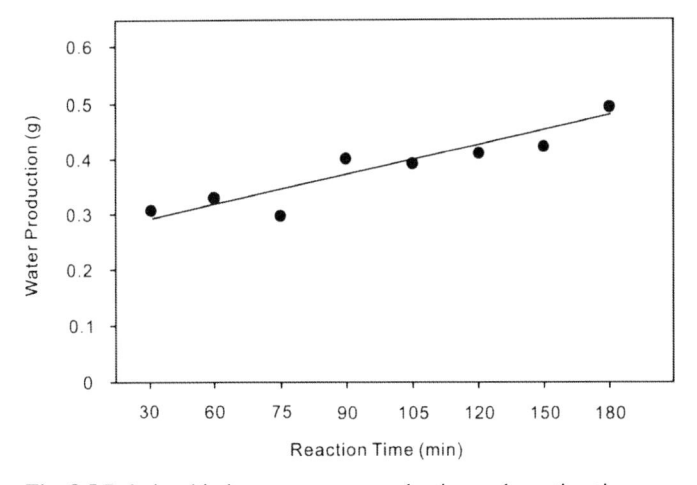

Fig. 8.5 Relationship between water production and reaction time

water production do not appear in the range of reaction time from 30 mins to 180 mins. In our simulantion experiments, the relationship between water production and reaction time has similar behavior with different amount of ilmenite supply.

8.4.3.3 Ilmenite Supply

Another set of simulation experiments was designed with ilmenite supply increasing from 5 g to 40 g, in steps of 2.5 g. The reaction temperature was constant at $1000\,^{\circ}C$ for half an hour. Nine data points of water production were collected and displayed in Fig. 8.6, which shows the maximum water production was obtained when ilmenite's supply exceeded 20 g and remains roughly constant with the increasing of ilmenite supply. The potential explanation is hydrogen cannot penetrate into the inner cores of the ilmenite grains. So the reducing reaction just takes place on the outer layer of the ilmenite grains.

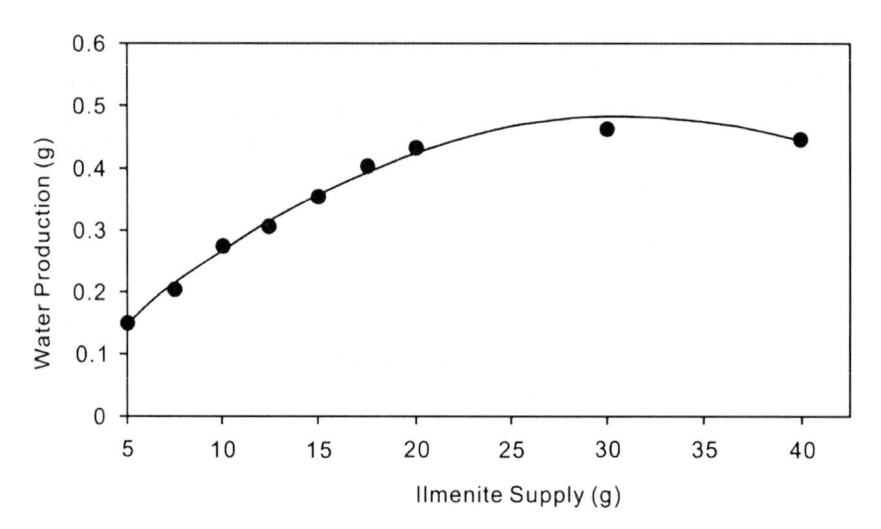

Fig. 8.6 Relationship between water production and ilmenite supply

8.4.3.4 Grain Size

The size of ilmenite grain remains constant in our simulation experiment. However, as we know the grain surface area change inversely with its size. That is the smaller grains have higher surface to volume ratios. So for ilmenite powders with the same quality and weight, those with smaller grain sizes provide more contact surface in reaction. The potential influence of grain size should be taking into account. The result of our experiment and ZHAO Y's experiment were compared with grain sizes of 75 μm and 10 μm, respectively. It is clear that the smaller grain size sample used in ZHAO Y's experiment has been reduced more notably than the larger grain size sample in our experiment (Zhao and Shadman 1993).

8.4.4 Energy Efficiency

The muffle furnace power source used in our experiment is 2 kW. The water production per kWh (WPW) was calculated and analyzed for the relationship with ilmenite supply as well as temperature. When the temperature of ilmenite powder of 1000°C, we observed that WPW increases with the ilmenite supply and reaches the peak when ilmenite supply is 20 g. Subsequently, WPW decrease with the ilmenite supply. So, 20 g ilmenite powder can provide the highest value of WPW at 1000°C. Those results were showed in Fig.8.7 and Fig.8.8.

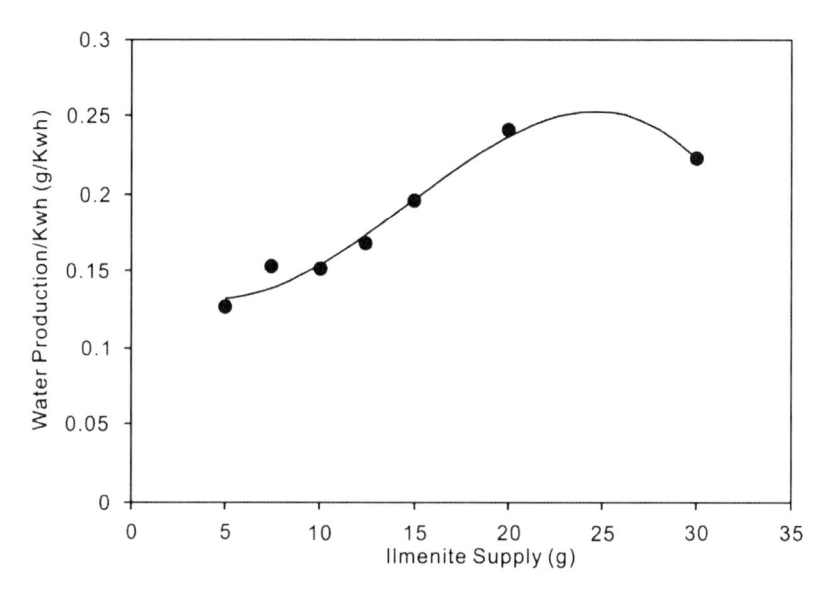

Fig. 8.7 Relationship between water production/kWh and ilmenite supply at a temperature of 1000°C

In another set of experiments of 10 g ilmenite we obtained the variation trend of WPW with temperature. Our results are summarized in Fig.8.8 which shows that 800°C is the most inefficient temperature for the reducing reaction and 600°C is apparently the highest point. However, there may be some interfering factors that have not been considered thoroughly. For instance, high temperature can drive the absorbed water out from the ilmenite sample. Meanwhile, heating up the ilmenite powder is a very slow process in a muffle furnace. So the absorbed water which was driven out of the ilmenite powder gradually and was included in the water production of 600°C. This problem will be much less serious with the increasing of temperature. So comparing the WPW of different temperature, it can be concluded that 1000°C produce the highest WPW.

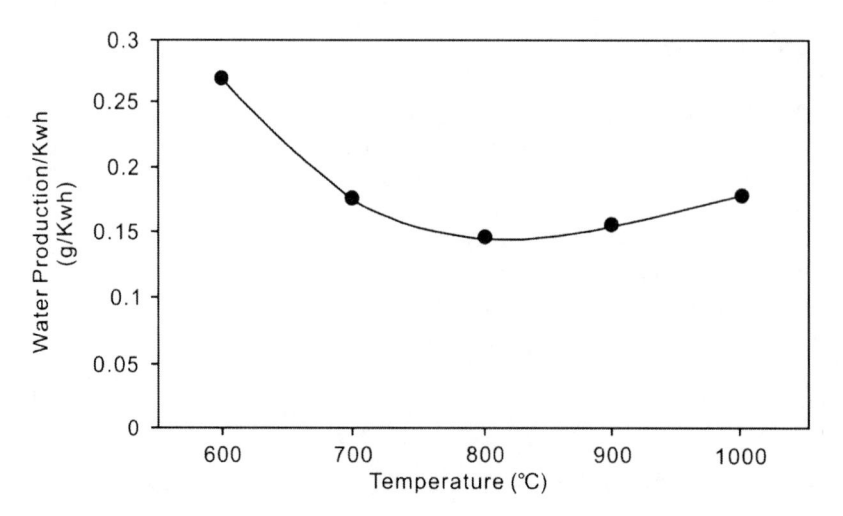

Fig. 8.8 Relationship between water production/kWh and temperature of 10 g ilmenite

In our simulation experiments, the water production increases continually with the reaction time until the ilmenite powder cannot be reduced by H_2 anymore. However the optimized reaction time should be clarified in order to select the economical experimental programs. Nine simulation experiments were carried out with different reaction time and varied data was obtained. The relationship between water production/kWh and reaction time was presented in Fig. 8.9. It shows that WPW decreases with the increasing reaction time. Decrease of WPW

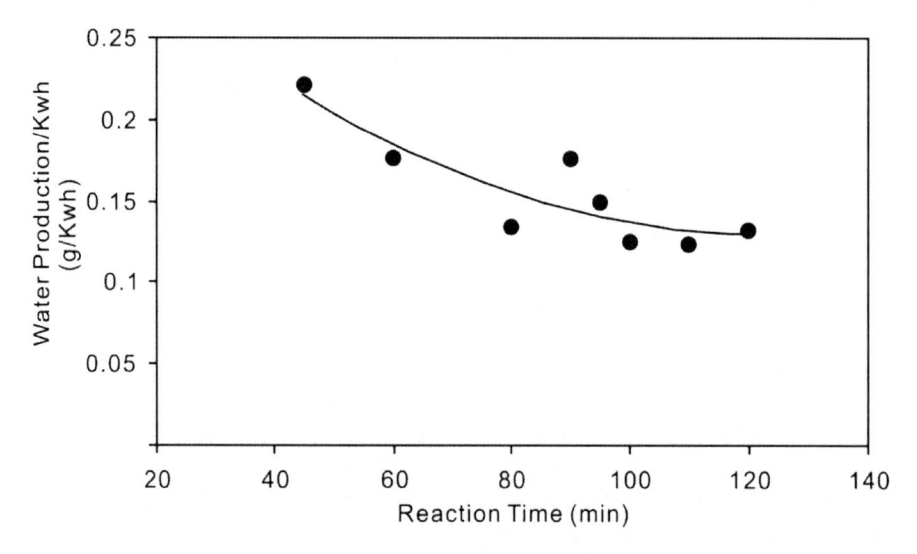

Fig. 8.9 Relationship between water production/kWh and reaction time of 10 g ilmenite at a temperature of $900\,^\circ$C

becomes weaker at the long end of the reaction time. The optimized reaction time was not obtained in our simulation experiments. From the consideration of the energy consumption, what we know is that we should set the reaction time as short as possible if only the water production can meet our need.

8.4.5 Optimizing the Parameters of the Simulation Experiment

Summarizing the influencing factors and energy efficiency consideration, we obtain the optimized set of parameters for the simulation experiment: a sample of 20 g ilmenite which reduced by hydrogen at a temperature of $1000^\circ C$ can provide the maximum water production per unit energy consuming. However, an optimized reaction time was not obtained. We know that the longer the reaction time, the smaller the WPW. So the reaction time should be set as short as possible provided that the water production can meet our need. In addition, reducing the grain size can greatly improve the level of water production.

8.5 Conclusion

Water and oxygen production is one of the key technologies for establishing a lunar base. With the microwave source available in our simulation experiments, reducing 20 g ilmenite powder can maximally produce 0.25 g water and consuming energy of 1 kWh. The ratio of water produced to ilmenite consumed is 0.0125. So it can be calculated that 120 kg of ilmenite can produce 1500 g of water which can meet one astronaut's need for a day. However, the energy consumed for producing 1500 g water is 6000 kWh and may be too high to be supported by the lunar base. Lower pressure, smaller grain size and heating by microwave can distinctly improve the efficiency of the reaction and reduce the energy consumed. With a surface atmospheric pressure of 1×10^{-9} Pa, the low pressure lunar environment makes the reduction taken place under lower temperature possible and improves the energy efficiency. Smaller the grain size sample provides larger reaction area, and so reducing the grain size can greatly enhance the reduction rate of ilmenite. In addition, the application of microwave in the calefaction of ilmenite should also be considered seriously. Therefore, the influence of H_2 pressure, grain size and the microwave calefaction are the main topics for future research work.

References

Burt, D.M.: Lunar mining of oxygen using fluorine. In: Mendell, W.W. (ed.) The Second Conference on Lunar Bases and Space Activities of the 21st Century, vol. 2, pp. 423–428. NASA Conference Pubulication 3166, Houston (1988)

Cameron, E.N.: Apollo 11 ilmenite revisited. In: Stein, S., Willy, Z.S., Russell, J.M. (eds.) Engineering, Construction and Operations in Space III, pp. 1–13. American Society of Civil Engineers, Denver (1992)

Duke, M., Hoffman, S., Snook, K.: The lunar surface reference mission: A description of human and robotic surface activities. NASA Technical Report (TP-2003-210793). Johnson Space Center, Houston, pp. 1–136 (2003)

Gibson, M.K., Knudsen, C.W.: Lunar oxygen production from ilmenite. In: Mendell, W.W. (ed.) Lunar Bases and Space Activities of the 21st Century, pp. 543–550. Lunar and Planetary Institute, Houston (1985)

Schrunk, D., Sharpe, B., Cooper, B., Thangavelu, M.: The Moon: Resources Future Development and Colonization. Wiley Praxis Series In Space Science and Technology, Hoboken, New Jersey, pp. 239–278 (1999)

Williams, R.J., McKay, D.S., Giles, D., Bunch, T.E.: Mining and beneficiation of lunar ores. In: Billingham, J., Gilbreath, W., O'Leary, B., Gosset, B. (eds.) Space Resource and Space Settlements. NASA SP-428, NASA, Washington DC, pp. 275–288 (1979)

Taylor, L.A., Oder, R.R.: Magnetic beneficiation of hi-Ti mare soils: concentrations of ilmenite and other components. In: Ryder, G., Sharpton, V.L. (eds.) Abstracts of the 21st Lunar and Planetary Science Conference, Houston, pp. 1245–1246 (1990)

Taylor, L.A., McKay, D.S.: An ilmenite feedstock on the moon: beneficiation of rocks verus soil. In: Lyndon, B. (ed.) Proceedings of the 23th Lunar and Planetary Science Conference, Lunar and Planetary Institute, Houston, pp. 1411–1412 (1992)

Taylor, L.A., Chambers, J.G., Patchen, A.: Evaluation of lunar rocks and soils for resource utilization: detailed image analysis of raw materials and beneficiated products. In: Proceedings of 24th Lunar and Planetary Science Conference, Lunar and Planetary Institute, Houston, pp. 1409–1410 (1993)

Taylor, L.A., Carrier, W.D.: Oxygen production on the moon: an overview and evaluation. In: John, S.L., Mildred, S.M., Mary, L.G. (eds.) Resources of Near-Earth Space, pp. 60–108. University of Arizona Press, Tucson (1993)

Taylor, L.A., Higgins, S.J.: Lunar mineral feedstocks from rocks and soils: X-Rays digital imaging in resource evaluation. In: Proceedings of 25th Lunar and Planetary Science Conference, Part 1: A-G, pp. 235–236. Lunar and Planetary Institute, Houston (1994)

Taylor, L.A., Meek, T.T.: Microwave processing of lunar soil. In: Steve, M.D., Charles, T.B., Christopher, G.T. (eds.) Proceedings of the International Lunar Conference, pp. 109–124. American Astronautical Society and Space Age Publishing Co., Hawaii (2003)

Taylor, G.J.: Cosmochemistry and human exploration. Planetary Science Research Discoveries, Colorado, pp. 1–12 (2004)

Zhao, Y., Shadman, F.: Production oxygen from lunar ilmenite. In: John, S.L., Mildred, S.M., Mary, L.G. (eds.) Resources of Near-Earth Space, pp. 149–178. University of Arizona Press, Tucson (1993)

9 Potential ISRU of Lunar Regolith for Planetary Habitation Applications

Eric J. Faierson and Kathryn V. Logan

Virginia Polytechnic Institute and State University, USA

9.1 Introduction

When humans return to the Moon, In-Situ Resource Utilization (ISRU) of lunar regolith will allow a more efficient, less costly, and thus, a more sustainable human presence on the Moon to be achieved. Maintaining a human presence on the Moon will require methods to mitigate lunar dust, provide protection from micrometeoroid impact, and reduce astronaut exposure to radiation. It would also be desirable to grow plants at the outpost, both for food and other life support purposes. Furthermore, extraction of resources such as Helium-3, metals, and oxygen from lunar regolith would also be of value.

This chapter describes using a solid state, self-propagating chemical reaction and in-situ lunar regolith to form structural materials for potential planetary habitation applications. The foundation of the chemical reaction is the combination of in-situ lunar regolith with powdered aluminum, and the addition of a relatively small amount of thermal energy to initiate an oxidation-reduction reaction that is self-propagating without any further addition of thermal energy after the reaction initiates. The resulting product is a near-net-shape, refractory and structurally stable shape that will take the form of the original unreacted mixture. Reactions were initiated by the authors in both a standard Earth atmosphere and a vacuum environment, utilizing regolith simulant instead of actual lunar regolith in order to demonstrate the reaction feasibility. The influence of the mass ratio of regolith simulant and aluminum on reaction propagation, reaction product composition, and reaction product strength are several subjects described in this chapter. Minimizing the quantity of aluminum required for the geothermite reaction would contribute to optimal use of resources available on the Moon. The term "geothermite" is used to refer to chemical reactions that exhibit thermite-type behavior, and incorporate unrefined minerals and glass reacting with a reducing agent. Thermite reactions are described in the subsequent section. One possible source for aluminum on the lunar surface would be from expended vehicular parts such as fuel pods.

The reaction is of interest due to the potential for use in lunar dust mitigation, micrometeoroid protection, structural applications, and radiation shielding on the lunar surface. Lunar dust mitigation could be realized by using the reaction product for landing surfaces, roads, and/or blast berms. Micrometeoroid protection and radiation shielding could be achieved by utilizing the reaction product as an outer barrier in lunar structures. In addition, the thermal energy released by the reaction could be harnessed for other purposes, such as extraction of volatiles from lunar regolith.

9.2 Background

This section will give a brief background on issues pertinent to the research discussed in this chapter. An overview of thermite and Self-propagating High-temperature Synthesis (SHS) reactions will be presented. The composition and characteristics of both the JSC-1A series of lunar regolith simulant and actual lunar regolith will be discussed. In addition, prior studies utilizing regolith simulant, and problems encountered during Apollo missions as a result of lunar dust will be summarized briefly.

9.2.1 Thermite and Self-propagating High-temperature Synthesis (SHS) Reactions

Thermite reactions are exothermic chemical reactions between a metal and metal oxide. The reaction causes the formation of a more stable metal oxide from the starting metal and a reduced metal that was initially bound in the starting metal oxide. Thermite reactions are a type of Self-propagating High-temperature Synthesis (SHS) reaction (Mei et al. 1999).

After reaction initiation, SHS reactions produce sufficient heat during the reaction to allow propagation without the addition of further external energy. One advantage of SHS reactions is that a lower input energy is needed for material processing when compared with conventional furnaces. There are a wide range of applications for materials produced using SHS reactions; some of which include abrasives, high-temperature intermetallics, precursor material for subsequent ceramic processing, and composite materials (Mossino 2004).

SHS reactions can be achieved in chemical mixtures that are able to sustain a combustion wave during exothermic reactions. Metastable phases can be formed in the reaction product due to large thermal gradients and fast cooling rates created by the SHS reaction. Several variables influence the SHS process: stoichiometry of the reactants, green density, particle size, atmospheric pressure, reactant volume, and method of reaction initiation (Mossino 2004).

Prior research has shown that an SHS reaction can be initiated using lunar regolith simulant and powdered aluminum. Whiskers containing aluminum, nitrogen, and oxygen were observed in reaction products formed in a standard atmosphere (Faierson et al. 2008, 2010; Faierson and Logan 2010).

A study conducted by Guojian et al. (2000a,b) has shown that whiskers, specifically aluminum nitride (AlN) whiskers, can be produced using an SHS reaction. Several different whisker morphologies were observed, including hexagonal, branched, dendritic, layered, and star-like. Some whiskers had a stacked structure of AlN single crystals. The whiskers had diameters between 0.01-20 μm. X-ray diffraction (XRD) analysis identified the presence of hexagonal AlN, corresponding to powder diffraction file (PDF) # 25-1133 (Guojian et al. 2000a,b). Oxygen has high solubility within the AlN structure and can substitute for nitrogen. Magnesium and silicon also have high solubility within the AlN structure and can substitute for aluminum (Kumashiro 2000).

Several mechanisms can be used to explain AlN whisker growth. Whisker growth by the VS (Vapor-Solid) mechanism involves super-saturation of vapor phase atoms, followed by deposition on a substrate. Growth by the VLS (Vapor feed gases, Liquid catalyst, and Solid whisker) mechanism involves a liquid catalyst and is substantially quicker than the VS growth process. Vapor is deposited on the catalyst since it is a preferred deposition site. Growth occurs by precipitation from the super-saturated liquid. Impurities play an important role in growth by the VLS mechanism. Upon cooling, a spherical shape forms to terminate whiskers formed by the VLS mechanism. High temperatures can result in evaporation of a droplet terminating an AlN whisker, resulting in a change in growth mechanism from VLS to VS (Guojian et al. 2000a,b).

A study conducted by Sen et al. (2005) investigated carbothermal reduction of lunar regolith simulant. A mixture of regolith simulant and 20 wt.% graphite powder was placed in an alumina crucible and heated for one hour at 1500 °C in an argon atmosphere. A metallic alloy was formed from the reduction of the components of the stimulant (Sen et al. 2005).

Research performed by Martirosyan and Luss (2008) investigated an SHS reaction between nanoparticles of titanium and boron, and JSC-1 regolith simulant in an argon atmosphere. Mixtures contained up to 60 wt.% titanium and boron, having particle sizes ranging from ~60-100 nm. Reactions were also conducted with titanium and boron particles having sizes on the order of microns. The regolith simulant particle size was on the order of ~1 μm. Mixtures were pressed into cylindrical samples having a porosity of around 35%. Temperature measurements collected from a thermocouple indicated that mixtures containing nano-particles achieved maximum temperatures of around 1600° C; whereas, mixtures using micro-particles had maximum temperatures around 1200° C (Martirosyan and Luss 2008).

9.2.2 Lunar Regolith Simulant

Lunar regolith simulants are used to simulate the characteristics of actual lunar regolith, such as chemical composition and geotechnical properties. One common type of regolith simulant used in recent studies to simulate the chemical composition of lunar regolith from mare regions on the Moon is the JSC-1A series of simulant, which was commissioned by NASA to support future lunar surface exploration. The simulant series includes the JSC-1AC, JSC-1A, and JSC-1AF lunar regolith simulants, consisting of very coarse, coarse, and fine particles, respectively. The term regolith refers to the layer of loose, unconsolidated rock of all sizes overlying bedrock. The average particle size of the JSC-1AF regolith simulant was 24.89 μm. The median particle size was 23.72 μm. Around 80% of the JSC-1AF particles were between 5 and 46 μm (NASA-MSFC 2006). The average particle sizes for three samples of JSC-1A simulant were 196.8 μm, 184.7 μm, and 181.6 μm. The median particle sizes were 105 μm, 103.5 μm, and 99.85 μm, respectively. Around 80% of the JSC-1A particles were between 19 μm and 550 μm (NASA-MSFC 2007).

The JSC-1A series of simulants are composed of minerals and glass, and include many of the silicate minerals present on the Moon. The simulant is mined from a volcanic ash deposit in Arizona and primary constituents include the plagioclase solid solution series and basaltic glass. The simulants also contain components of the pyroxene and olivine solid solution series, along with various trace minerals. The average plagioclase content of the simulant is 70% anorthite, 29% albite, and 1% orthoclase ($KAlSi_3O_8$). The calcium-pyroxene mineral content averages 45% wollastonite ($CaSiO3$), 38% enstatite, and 22% ferrosilite. Olivine minerals are present in an average concentration of 73% forsterite and 27% fayalite (NASA-MSFC 2006).

Data from bulk composition analysis of JSC-1AF regolith simulant compared with actual lunar regolith samples obtained in the Apollo and Luna missions are shown in Table 9.1.

Table 9.1 Bulk Composition of Lunar Regolith and JSC-1AF Simulant

	Maria		Highlands		Simulant
	Apollo 14	Luna 16	Luna 20	Apollo 16	JSC-1AF
SiO_2	47.93	41.70	45.40	44.94	47.1
Al_2O_3	17.6	15.33	23.44	26.71	17.1
CaO	11.19	12.50	13.38	15.57	10.3
FeO	10.37	16.64	7.37	5.49	7.57
MgO	9.24	8.78	9.19	5.96	6.9
Fe_2O_3	-	-	-	-	3.41
TiO_2	1.74	3.38	0.47	0.58	1.87

(Compositions in weight percent) (Sen et al. 2005; NASA-MSFC 2006)

Allen et al. (1994) sintered regolith simulant contained in graphite molds using both radiant and microwave heating. The sintered product showed potential for use in lunar construction applications (Allen et al. 1994). Another method of ISRU that was investigated by Curreri et al. (2006) utilized Molten Oxide Electrolysis (MOE). When regolith is used in MOE, the process is referred to as Molten Regolith Electrolysis (MRE). A bulk heating method was used with an electric potential to extract materials such as iron, silicon, and oxygen from regolith (Curreri et al. 2006). Most lunar regolith compositions have a liquidus around 1200° C and a low viscosity melt can be formed at temperatures around 1400° C (Fabes et al. 1992).

Lunar rocks consist of both igneous rocks and breccias. Igneous rocks form through the crystallization of silicate melts, and breccias form through meteoroid impacts. Breccias are a mixture of rocks and regolith that are cemented together by a process of shock-metamorphism induced by meteorite impacts. Most minerals within lunar rocks are silicates. Three classes of silicate minerals found on the Moon include pyroxene, plagioclase feldspar and olivine solid solution series. End-members of the pyroxene series include diopside ($CaMgSi_2O_6$), hedenbergite ($CaFeSi_2O_6$), ferrosilite ($Fe_2Si_2O_6$), and enstatite ($Mg_2Si_2O_6$). The minerals anorthite ($CaAl_2Si_2O_8$), and albite ($NaAlSi_3O_8$) are end-members of the plagioclase feldspar series. The olivine series end-members are forsterite (Mg_2SiO_4) and fayalite (Fe_2SiO_4) (Taylor 1992). An end-member is a pure compound at the end of a solid solution series. Typical lunar regolith less than 1 cm in size consists of over 95 wt.% particles smaller than 1 mm, around 50 wt.% particles smaller than 50 μm, and 10-20 wt.% particles under 20 μm (Taylor and Meek 2005).

Lunar dust is abrasive and tends to adhere to surfaces of objects. Lunar dust was responsible for creating a number of hazards during the Apollo missions including: obscured vision, loss of surface traction, degraded radiators, coated surfaces, clogged equipment, inhaled dust, errant instrument measurements, abrasion, and failed seals (Gaier and Sechkar 2007). One example of the adverse effects of lunar dust was demonstrated by the Surveyor 3 spacecraft, which had been 183 m away from the Apollo 12 landing site for 2.5 years before Apollo 12 landed. Upon examination by the Apollo 12 crew, the spacecraft appeared to have been sandblasted with lunar dust propelled by exhaust gases from the descent stage of the lander. The optical mirror on the Surveyor 3 was damaged from lunar dust accumulation and pitting (Heiken et al. 1991). Problems with lunar dust that were encountered during the Apollo missions indicate that a method to mitigate adverse effects of lunar dust will be necessary for future, longer duration missions.

9.2.3 Discussion

The first part of this section discusses sample preparation of the geothermite reactant mixture and reaction initiation and propagation in different reaction environments. The next part of the section describes X-ray diffraction (XRD) and scanning electron microscopy (SEM)/energy dispersive spectroscopy (EDS) analysis of the regolith simulant and reaction products, as well as thermodynamically favorable chemical reactions potentially occurring within the geothermite reaction. The last portion of this section discusses compressive strength tests performed on reaction products, fabrication of near-net-shape Voissoir dome elements, and potential applications for the geothermite reaction.

9.2.4 Sample Preparation and Reaction Environments

The authors investigated chemical reactions between lunar regolith simulant and aluminum using both a standard atmosphere and a vacuum environment. The experimental matrix used for the investigation is shown in Table 9.2. At least three samples were fabricated for each set of reaction parameters in a standard atmosphere. At least one sample was fabricated for each set of reaction parameters under vacuum conditions.

Table 9.2 Experimental Matrix of Reaction Parameters

Environment	Reactants			
	(Wt.% Aluminum and Simulant Type)			
	19.44%	24.45%	28.85%	33.33%
Standard Atmosphere	JSC-1AF	JSC-1AF	JSC-1AF	JSC-1AF
	JSC-1A	JSC-1A	JSC-1A	*
Vacuum	JSC-1AF	JSC-1AF	JSC-1AF	JSC-1AF
(0.600 Torr)	JSC-1A	JSC-1A	JSC-1A	*

*JSC-1A did not support the desired reaction propagation repeatability

Two particle size distributions of lunar regolith simulant were utilized in experiments described in this chapter: The JSC-1AF simulant had fine particles, and the JSC-1A had coarse particles. Four different reaction mass ratios, chosen both theoretically and experimentally, were utilized. The objective was to find the smallest quantity of aluminum that would produce a reaction product with acceptable properties. Two of the four reactant mass ratios were determined experimentally in order to set an upper and lower boundary of aluminum required to achieve repeatable SHS reactions that propagate completely through a sample, using JSC-1AF simulant in a standard atmosphere. It is emphasized that the upper and lower aluminum mass ratio are not discrete limits on reaction propagation, but boundaries needed to achieve adequate reaction repeatability. Reactions have propagated to completion outside the highest and lowest quantities of aluminum described in this chapter.

Two mass ratios were calculated using the bulk composition of oxides found within the JSC-1AF regolith simulant. The 28.85 wt.% aluminum mass ratio represented the proportion of aluminum required to reduce all oxides found within the regolith simulant, regardless of thermodynamic favorability. The 24.45 wt.% aluminum mass ratio represented the proportion of aluminum required to reduce the selected thermodynamically favorable oxides, consisting of SiO_2, TiO_2, Fe_2O_3, and FeO.

Sample preparation was the same for both standard atmosphere and vacuum environments. The aluminum used in the reactions had a -325 mesh particle size. The four primary reactant mass ratios consisted of 80.56% simulant and 19.44% aluminum; 75.55% simulant and 24.45% aluminum; 71.15% regolith and 28.85% aluminum; and 66.67% regolith and 33.33% aluminum by weight. All reactant mass ratios were measured in wt.%. Each of the mass ratios was used for both JSC-1AF and JSC-1A simulants, with the exception of the 33.33% aluminum mass ratio. The 33.33% aluminum mass ratio did not support adequate reaction propagation repeatability using the JSC-1A simulant. The powders were manually mixed in total quantities of 80 g until a uniform dispersion of reactant particles was achieved. The mass of each constituent was determined using an Ohaus Adventurer Pro AU313 Scale. Aluminum foil crucibles were fabricated by rolling a sheet of aluminum foil over a cylinder mandrel. Foil extending off the edge of the cylinder was folded to form the crucible bottom.

9.2.5 Standard Atmosphere SHS Reaction Initiation

A length of u-shaped nickel-chromium (NiCr) wire was pushed into the sample mixture stopping just above the crucible bottom.

Standard atmosphere reactions were performed in a plexiglass chamber lined with refractory bricks and insulation. The aluminum foil crucible was placed on a bed of sand. The chamber was continuously purged of airborne particulate by means of an air filter and fan assembly. A Variac power supply was used to induce joule heating. In order to minimize the creation of voids in the mixture, the crucible was tamped down twice after connection to the Variac terminals. The power supplied by the Variac was controlled manually, and was increased to ~22.5 Amperes over a period of ~4.5 minutes.

In a standard atmosphere, reaction propagation typically began at the uppermost surface of the cylinder. A horizontal planar combustion wave propagated from the top of the cylinder to the bottom. Images of the standard atmosphere reaction process using JSC-1AF simulant and 28.85 wt.% aluminum are shown in Figs. 9.1-9.4.

It was observed that reactions conducted with mixtures having a larger quantity of aluminum in the reactants tended to evolve more gas than those with lower quantities of aluminum. Reactions utilizing JSC-1AF simulant evolved more gas than those using JSC-1A simulant. It is possible that formation of larger quantities of Al_2O gas, a chemical species that will be discussed later, were induced by the larger aluminum quantities.

Fig. 9.1 (left) Heat is applied to a mixture of JSC-1AF regolith simulant and aluminum powder; (right) Reaction propagation begins, no further external energy is applied.

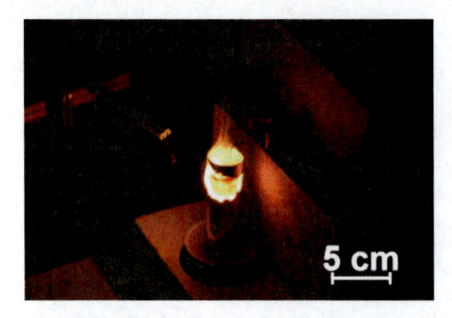 

Fig. 9.2 (left) ~40 seconds after initiation; (right) ~1 minute 20 seconds after initiation.

Fig. 9.3 (left) ~2 minutes after initiation; (right) ~2 minutes 30 seconds after initiation.

Fig. 9.4 (left) Reaction propagation has completed (~3 minutes 10 seconds after initiation); (right) Reaction product after cooling (~ 5 minutes 40 seconds after initiation)

As the combustion wave passed through a portion of the reactant mixture, slight expansion of that portion of the cylinder was observed. After the wave passed, the region returned to approximately the initial size. It was frequently observed in reactions using either simulant that excess aluminum from the crucible would melt down the side of the sample and accumulate. The accumulation of aluminum was observed to slow reaction propagation as the combustion wave reached the bottom of the sample. All but one reactant mass ratio produced enough heat to complete the reaction at the bottom of the sample. Reactions conducted with JSC-1A simulant and 19.44% aluminum typically had an unreacted layer 0.25-0.5 cm thick at the bottom of the specimen. An example of reaction completion is shown in Fig. 9.4 (left). Some of the residual aluminum from crucible melting can be seen at the bottom of the reaction product in Fig. 9.4 (right).

9.2.6 Preliminary Reactions Initiated in Vacuum

A vacuum chamber was designed in order to perform SHS reactions in a vacuum environment. The chamber was made from Type 304 Stainless Steel, and was a horizontally oriented cylinder. A glass viewport was installed on the chamber door, and a quartz viewport was installed at the top center of the chamber. A two conductor, 30 amp electrical feedthrough, was centered on the left side of the chamber. A y-adaptor was attached to the exhaust valve on the chamber with additional valves attached to each side of the adaptor. One valve allowed air to enter the chamber for re-pressurization and the other side was connected to an exhaust fan and air filter. The exhaust fan and air filter were used to remove any airborne contaminants within the chamber after re-pressurization.

A manual valve was attached to the vacuum port on the chamber to allow isolation of the vacuum chamber from the vacuum pump. Three vacuum traps were attached in series between the valve and the vacuum pump in order to prevent fine particulate from the SHS reactions from entering the vacuum pump. The trap nearest the chamber used a stainless steel gauze filter, which was followed by a sodasorb filter and a five micron polypro filter. A Leybold D8B dual-stage vacuum pump was attached to the output of the five micron polypro filter. A digital vacuum gauge, with a range of 0.001-760 Torr, was connected to a thermocouple-type sensor installed at the top of the chamber.

Initial experiments utilized a single NiCr wire for reaction initiation in vacuum, with the same configuration that was used in a standard atmosphere. It was observed that failure of the NiCr wire occurred more quickly in vacuum than in a standard atmosphere. In addition, reaction initiation was not observed when failure occurred, which resulted in an open condition within the circuit. Examination of the wire after attempting a reaction revealed significant pitting. It was hypothesized that the reducing conditions found in the vacuum allowed a chemical reaction to occur between a component in the NiCr wire and component(s) in the reactant mixture. In particular, it was hypothesized that elemental silicon produced from aluminum oxidation during heating was interacting with a component in the NiCr wire and causing the degradation of the wire. In order to slow the degradation process, the NiCr wire was heated in a standard atmosphere for a given amount of time prior to use in the vacuum environment, allowing an oxide layer to form on the exterior of the NiCr wire, which slowed the pitting process. Sufficient repeatability was not achieved using a single strand of NiCr wire. Repeatable reaction propagation was achieved by utilizing two pieces of NiCr wire that were braided together. Use of the double-braided NiCr wire decreased the resistance in the Variac circuit, allowing a larger number of amps to flow at lower Variac settings than in a standard atmosphere.

9.2.7 SHS Reactions in Vacuum

After sample preparation was accomplished, as described for the reactions conducted in a standard atmosphere, the u-shaped length of heat-treated, double-braided 18 awg NiCr wire was pushed into the sample mixture. The tip of the NiCr wire was pushed to a depth just above the crucible bottom. The ends of the NiCr wire were connected to terminals on the electrical feedthrough within the vacuum chamber. In order to minimize creation of voids in the mixture, the crucible was tamped down twice after connecting the NiCr wire to the Variac terminals. The vacuum chamber door was closed, and all valves leading to the chamber were closed. The vacuum pump was turned on, and the valve located between the vacuum traps and the vacuum chamber was opened to allow evacuation of the chamber to begin. When the pressure in the chamber reached 0.600 Torr, the Variac power supply was turned on and the Joule heating process began. Current was increased to ~23 Amperes over a period of ~0.5 minutes.

A pressure of 0.600 Torr is roughly 0.08% of a standard Earth atmosphere (760 Torr). The pressure was selected to avoid substantially longer vacuum chamber pump down times, while eliminating substantial quantities of atmospheric gases, and reaching a significantly lower pressure than that which exists in a standard atmosphere. The vacuum pump continuously evacuated the chamber throughout the entire reaction process.

Visual observations during the reaction process were far more difficult to make for vacuum reactions due to the limited size and location of the viewport on the vacuum chamber. From the vantage point of the viewport, reactions were observed to begin near the midpoint between the upper and lower surface of the cylinder. Combustion wave propagation did not occur in a horizontal, planar manner as it did in a standard atmosphere. The combustion wave propagation could be described as radiative spreading from the center of the sample. Images of the reaction process in a vacuum environment are shown in Figs. 9.5-9.8.

Reaction products produced in vacuum generally had regions of loose powder at the upper and lower surfaces of the cylinder. The lower surface was most likely unreacted simulant and aluminum. The uppermost surface was generally a white to brown-grey color, distinctly different from the lower surface. It is thought that the upper surface may be composed of elements that are easily volatized in vacuum, such as magnesium and sodium. The region just below the uppermost surface was likely to be unreacted simulant and aluminum.

Fig. 9.5 (left) Heat is applied to reactant mixture; (right) Energetic emission of particles, propagation begins.

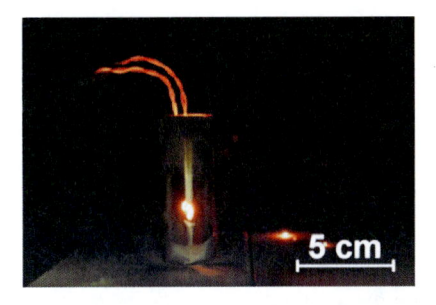 

Fig. 9.6 (left) ~20 seconds after initiation; (right) ~1 minute after initiation.

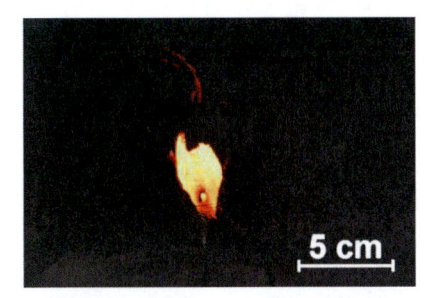 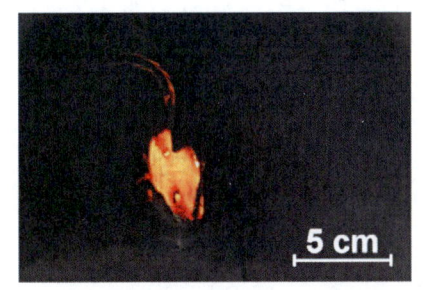

Fig. 9.7 (left) ~ 2 minutes after initiation; (right) ~4 minutes after initiation.

 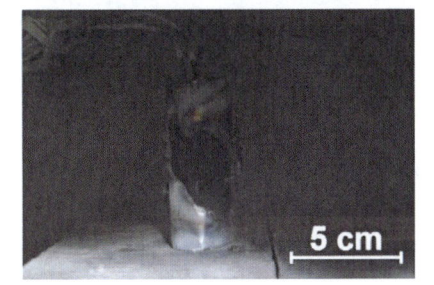

Fig. 9.8 (left) ~6 minutes after initiation; (right) The vacuum reaction product after cooling (~7 minutes 30 seconds after initiation)

Pressure within the vacuum chamber was generally observed to increase over two intervals during the synthesis process, even with continuous chamber evacuation using a vacuum pump. Without continuous evacuation, pressure within the chamber increased substantially due to outgassing of volatiles from the sample mixture. The first increase in chamber pressure occurred shortly after heating commenced. The second increase occurred when reaction propagation began, the time of which was dependent on the reaction parameters. The upper limit of measured chamber pressures was ~2.5 Torr. The 19.44% and 28.85% aluminum mass ratios using JSC-1A, and the 24.45% aluminum mass ratio using JSC-1AF exhibited significantly longer reaction times.

9.2.8 XRD Analysis

A portion of each sample, located near the center lengthwise, was used for X-ray diffraction (XRD) analysis. Each sample was ground with a mortar and pestle, and sieved to -200 mesh in order to obtain particle sizes suitable for XRD analysis. XRD analysis was also performed on unreacted JSC-1AF and JSC-1A regolith simulant. XRD analysis for standard atmosphere reaction products was conducted using a Panalytical X-Pert Pro diffractometer. A continuous goniometer scan was conducted from 10° to 120°. A current of 40 mA was used with an accelerating voltage of 45 kV. The sample tray was rotated during the scan in order to decrease effects of preferred orientation of particles. XRD scans for vacuum synthesized products were conducted on a Siemens D5000 X-ray diffractometer. Analysis of the diffraction patterns generated by each of the diffractometers using the same sample indicated suitable correlation between diffraction peaks generated by the two diffractometers. A 2-theta / theta locked scan was conducted on the Siemens D5000 from 10° to 120°, with a step of 0.01° and a step interval of 2.5 seconds.

XRD patterns for unreacted JSC-1AF simulant and reaction products formed using JSC-1AF simulant in a standard atmosphere are shown in Fig. 9.9. The diffraction pattern at the bottom of Fig. 9.9 is that of unreacted JSC-1AF simulant. The diffraction pattern above the one of the simulant is of a reaction product formed using 19.44% aluminum. Moving further up the figure are diffraction patterns of reaction products formed using 24.45% aluminum, 28.85% aluminum, and 33.33% aluminum, respectively. Major diffraction peaks are identified within the patterns, and the legend for the diffraction peaks labeled in Fig. 9.9, as well as Figs. 9.10-9.12 is given in Table 9.3.

Chemical species that appeared to be present in the diffraction patterns for reaction products formed in a standard atmosphere using JSC-1AF simulant, shown in Fig. 9.9, included Si (A), $MgAl_2O_4$ (B), $CaAl_4O_7$ (C), Al_2O_3 (D), AlN (E, F), $FeSi_2$ (G), $Ca_2Al_2SiO_7$ (H), and plagioclase minerals (I).

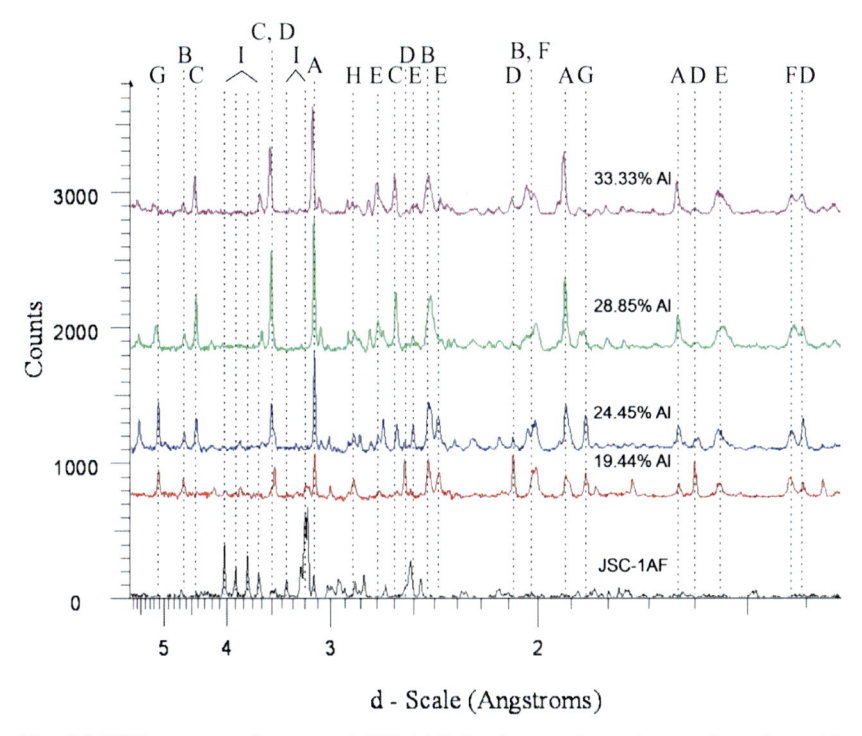

Fig. 9.9 XRD patterns of unreacted JSC-1AF simulant, and reaction products formed in a standard atmosphere utilizing 19.44%, 24.45%, 28.85%, and 33.33% aluminum, from bottom to top, respectively

Table 9.3 Legend of Chemical Species Identified in X-ray Diffraction Patterns

Label	Chemical Name	Chemical Formula	Powder Diffraction File Number(s)
A	Silicon	Si	27-1402
B	Spinel	$MgAl_2O_4$	21-1152
C	Grossite	$CaAl_4O_7$	46-1475; 23-1037
D	Corundum	Al_2O_3	10-0173; 46-1212
E	Aluminum Nitride	AlN	25-1133
F	Aluminum Nitride	AlN	46-1200
G	Ferdisilicite	$FeSi_2$	35-0822
H	Gehlenite	$Ca_2Al_2SiO_7$	35-0755
I	Plagioclase Solid Solution Series	$CaAl_2Si_2O_8$- $NaAlSi_3O_8$	41-1481; 41-1486 10-0393; 18-1202; 41-1480
J	Aluminum	Al	04-0787

Diffraction peaks for unreacted minerals in the plagioclase solid solution series
(I) can be observed in the diffraction pattern for the reaction product that utilized
19.44% aluminum. As the aluminum content in the reactant mixture increased,
diffraction peaks for unreacted minerals in the plagioclase solid solution series
were no longer observed.

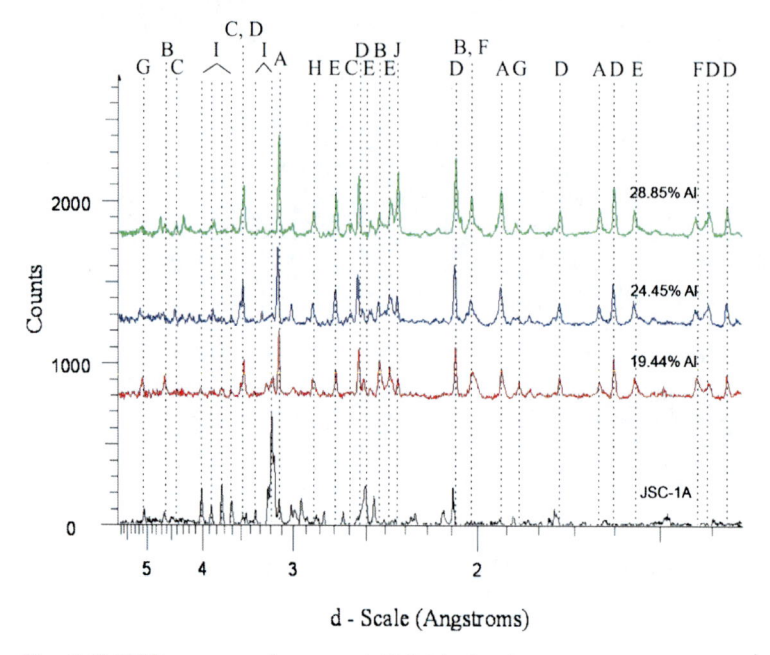

Fig. 9.10 XRD patterns of unreacted JSC-1A simulant, and reaction products formed in a
standard atmosphere utilizing 19.44%, 24.45%, and 28.85% aluminum, from bottom to top,
respectively.

XRD patterns for unreacted JSC-1A simulant and reaction products formed using JSC-1A simulant in a standard atmosphere are shown in Fig. 9.10. The diffraction pattern at the bottom of Fig. 9.10 is that of unreacted JSC-1A simulant. The
diffraction pattern above the one of the simulant is of a reaction product formed
using 19.44% aluminum.

Moving further up the figure are diffraction patterns of reaction products
formed using 24.45% aluminum, and 28.85% aluminum, respectively. Major diffraction peaks are identified within the patterns, and the legend for the diffraction
peaks labeled in Fig. 9.10 is given in Table 9.3.

Chemical species that appeared to be present in the diffraction patterns for reaction products formed in a standard atmosphere using JSC-1A simulant, shown in
Fig. 9.10, included Si (A), $MgAl_2O_4$ (B), $CaAl_4O_7$ (C), Al_2O_3 (D), AlN (E, F), Fe-
Si_2 (G), $Ca_2Al_2SiO_7$ (H), plagioclase minerals (I), and Al (J). Diffraction peaks are
observed for unreacted minerals in the plagioclase solid solution series in the diffraction pattern for the 19.44% aluminum composition. Diffraction peaks for unreacted aluminum are present within the diffraction patterns for all mass ratios of
reaction products using JSC-1A simulant.

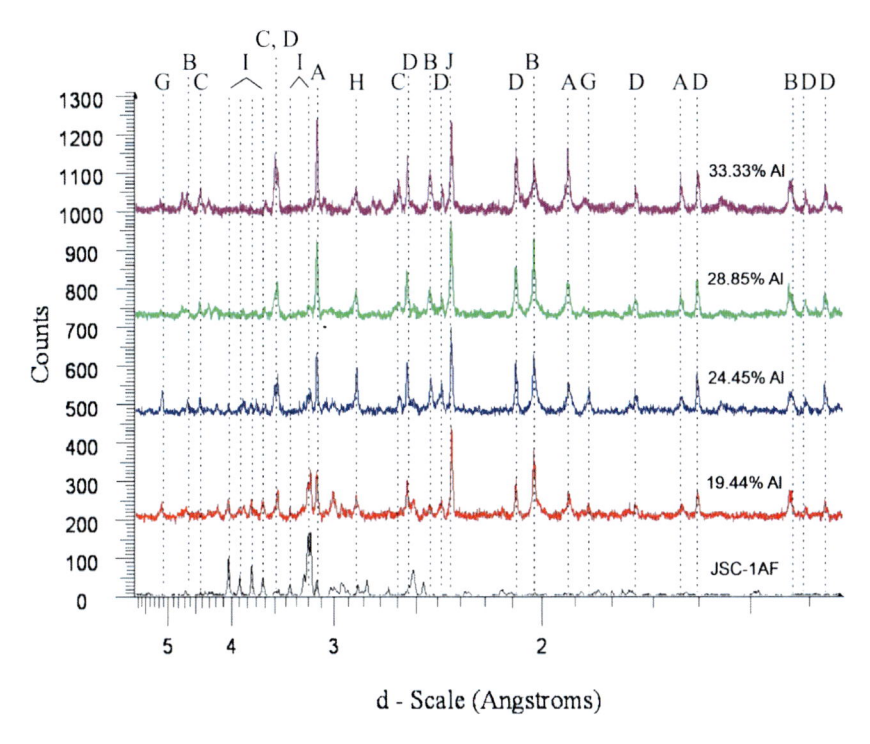

Fig. 9.11 XRD patterns of unreacted JSC-1AF simulant, and reaction products formed in a vacuum environment utilizing 19.44%, 24.45%, 28.85%, and 33.33% aluminum, from bottom to top, respectively.

XRD patterns for unreacted JSC-1AF simulant and reaction products formed using JSC-1AF simulant in a vacuum environment are shown in Fig. 9.11. The diffraction pattern at the bottom of Fig. 9.11 is that of unreacted JSC-1AF simulant. The diffraction pattern above the one of the simulant is of a reaction product formed using 19.44% aluminum. Moving further up the figure are diffraction patterns of reaction products formed using 24.45% aluminum, 28.85% aluminum, and 33.33% aluminum, respectively. Major diffraction peaks are identified within the patterns, and the legend for the diffraction peaks labeled in Fig. 9.11 is given in Table 9.3.

Chemical species that appeared to be present in the diffraction patterns for reaction products formed in a vacuum environment using JSC-1AF simulant, shown in Fig. 9.11, included Si (*A*), $MgAl_2O_4$ (*B*), $CaAl_4O_7$ (*C*), Al_2O_3 (*D*), $FeSi_2$ (*G*), $Ca_2Al_2SiO_7$ (*H*), plagioclase minerals (*I*), and Al (*J*). Significant diffraction peaks for unreacted minerals in the plagioclase solid solution series (*I*) are observed in the diffraction patterns for the reaction products containing 19.44% and 24.45% aluminum. Diffraction peaks for aluminum are present in the diffraction patterns for all mass ratios.

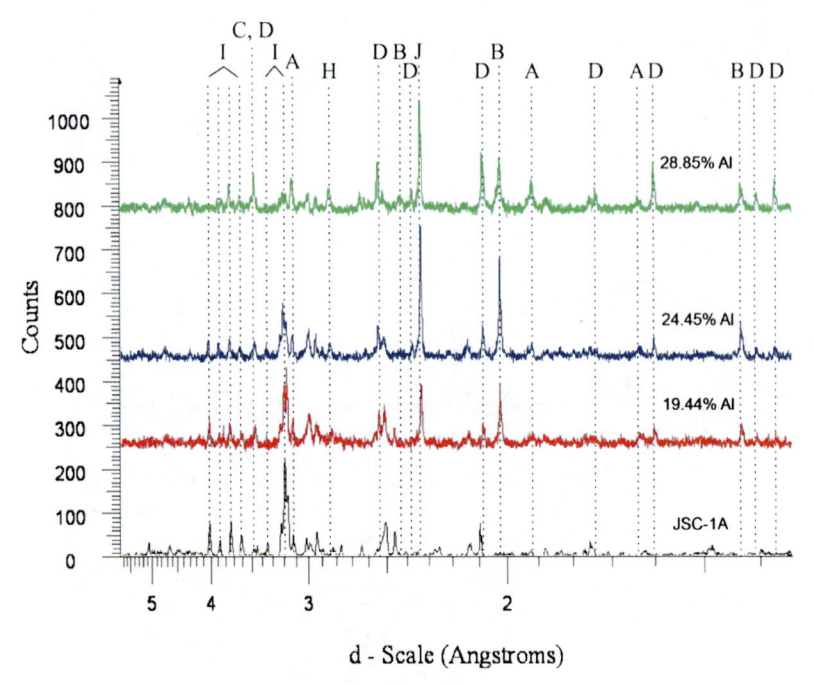

Fig. 9.12 XRD patterns of unreacted JSC-1A simulant, and reaction products formed in a vacuum environment utilizing 19.44%, 24.45%, and 28.85% aluminum, from bottom to top, respectively

XRD patterns for unreacted JSC-1A simulant and reaction products formed using JSC-1A simulant in a vacuum environment are shown in Fig. 9.12. The diffraction pattern at the bottom of Fig. 9.12 is that of unreacted JSC-1A simulant. The diffraction pattern above the one of the simulant is of a reaction product formed using 19.44% aluminum. Moving further up the figure are diffraction patterns of reaction products formed using 24.45% aluminum and 28.85% aluminum, respectively. Major diffraction peaks are identified within the patterns, and the legend for the diffraction peaks labeled in Fig. 9.12 is given in Table 9.3.

Chemical species that appeared to be present in the diffraction patterns for reaction products formed in a vacuum environment using JSC-1A simulant, shown in Fig. 9.12, included silicon (*A*), $MgAl_2O_4$ (*B*), $CaAl_4O_7$ (*C*), Al_2O_3 (*D*), $FeSi_2$ (*G*), $Ca_2Al_2SiO_7$ (*H*), plagioclase minerals (*I*), and aluminum (*J*). Diffraction peaks are present for unreacted plagioclase solid solution series minerals and aluminum in the diffraction patterns for all mass ratios of reaction products. The diffraction counts for the unreacted plagioclase minerals decrease as aluminum increases within the reactant mixture. Diffraction counts for Al_2O_3 tend to increase with a higher mass ratio of aluminum.

9.2.9 SEM/EDS Analysis

The microstructure and element composition of fracture surfaces of each reaction product were examined using Scanning Electron Microscopy (SEM) and Energy

Dispersive Spectroscopy (EDS). Analyses were also performed on unreacted JSC-1AF and JSC-1A simulant. A Hitachi 3700SN SEM and a LEO 1550 SEM were used to perform the analyses. An Oxford EDS system was used to obtain chemical compositions of selected microstructures. Samples were coated with a 15 nm 60/40 gold/palladium coating using a Cressington 208HR sputter coater.

9.2.10 Unreacted JSC-1AF Simulant

Micrographs of a sample of unreacted JSC-1AF simulant are shown in Fig. 9.13-9.15 at magnifications of 500x, 1kx, and 2.5kx, respectively. Many sub-angular and angular particles are observed in Figs. 9.13 and 9.14 having sizes ranging from ~10 μm to 50 μm. Particles ranging in size from ~30 μm to less than 1 μm are present in Fig. 9.15.

Fig. 9.13 SEM micrograph of unreacted JSC-1AF simulant at 500x.

Fig. 9.14 SEM micrograph of unreacted JSC-1AF simulant at 1kx.

Fig. 9.15 SEM micrograph of unreacted JSC-1AF simulant at 2.5kx.

EDS analyses were performed on the JSC-1AF and JSC-1A simulants in order to establish, in a semi-quantitative manner, that the simulants had reasonable uniformity of elemental composition. Although JSC-1AF is just a fine particle size distribution of JSC-1A, it is possible that the particle size refining process could have preferentially eliminated components, such as chemical species with a greater hardness. Table 9.4 compares the mean atomic percentage and deviation ($\pm$ 1 σ) for each element in the JSC-1AF and JSC-1A simulants. Four regions of JSC-1AF were analyzed and five regions of JSC-1A were analyzed. The data in Table 9.4 does not indicate a substantial difference in elemental composition between simulants.

Table 9.4 Mean EDS Analysis of JSC-1AF and JSC-1A Regions

	JSC-1AF	JSC-1A
O	64.82 ± 0.30	67.73 ± 0.26
Na	2.12 ± 0.07	2.23 ± 0.10
Mg	2.16 ± 0.05	2.35 ± 0.13
Al	6.80 ± 0.11	6.19 ± 0.12
Si	14.06 ± 0.08	13.64 ± 0.21
P	0.18 ± 0.03	0.19 ± 0.02
K	0.37 ± 0.02	0.34 ± 0.03
Ca	4.54 ± 0.09	3.53 ± 0.09
Ti	0.56 ± 0.03	0.46 ± 0.02
Mn	0.04 ± 0.04	0.01 ± 0.03
Fe	4.35 ± 0.24	3.32 ± 0.13
Mo	0.02 ± 0.04	0.00

(Values in Atomic %)

9.2.11 Microstructural Analysis of Reaction Products Formed in a Standard Atmosphere

The microstructure of a reaction product synthesized using JSC-1AF simulant and 19.44% Al is shown in Fig. 9.16. Table 9.5 shows the chemical data obtained from analysis of Spectrum 1. Spectrum 1 is located on a whisker cluster that likely formed through a chemical reaction involving a vapor phase, due to the small whisker diameters and the termination of many of the whiskers in a spherical

7µm   Electron Image 1

Fig. 9.16 SEM micrograph of the microstructure of a reaction product synthesized in a standard atmosphere using JSC-1AF simulant and 19.44% aluminum with the regions of EDS analysis identified.

Table 9.5 EDS Analysis of Spectrum 1 shown in Fig. 9.16

Element	Wt.%	At.%
O	49.80	63.05
Na	0.99	0.87
Mg	4.51	3.75
Al	35.98	27.01
Si	5.19	3.74
Ca	2.03	1.03
Fe	1.51	0.55

shape. Chemical analysis indicated high concentrations of aluminum and oxygen within the region. Silicon and magnesium were also present in smaller quantities. No nitrogen was detected by EDS. The data from EDS indicate that the whiskers were likely composed of aluminum oxide.

The microstructure of a reaction product synthesized from JSC-1AF simulant and 19.44% Al is shown in Fig. 9.17. Table 9.6 shows the chemical data obtained from analysis of spectrum 1. Spectrum 1 is located on a cluster of whiskers that

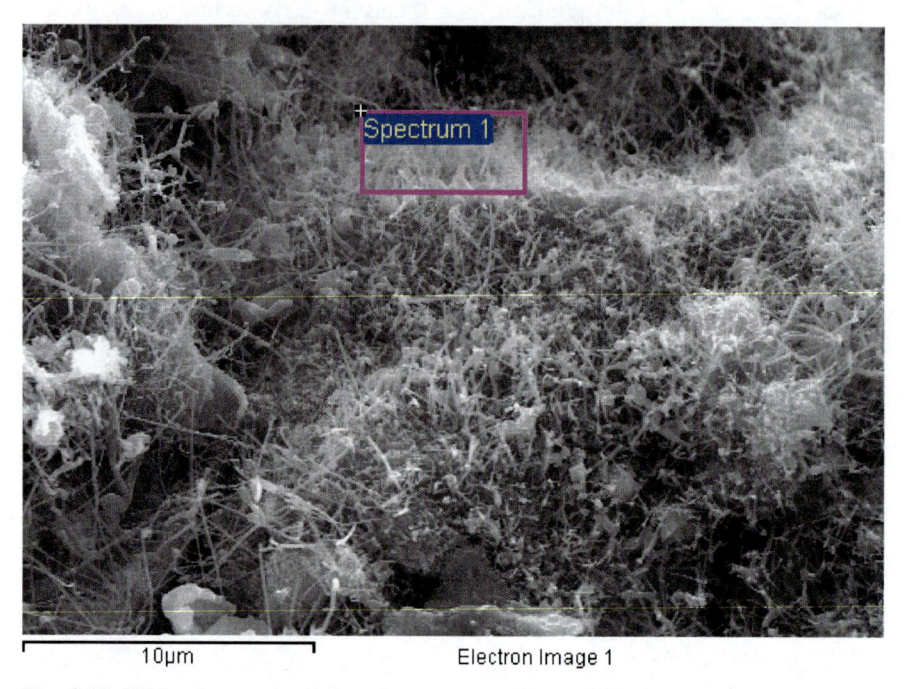

10µm   Electron Image 1

Fig. 9.17 SEM micrograph of the microstructure of a reaction product synthesized in a standard atmosphere using JSC-1AF simulant and 19.44% aluminum with the regions of EDS analysis identified.

Table 9.6 EDS Analysis of Spectrum 1 shown in Fig. 9.17

Element	Wt.%	At.%
N	8.91	12.33
O	46.82	56.74
Na	0.99	0.84
Mg	1.97	1.57
Al	32.43	23.30
Si	5.66	3.91
Ca	1.37	0.67
Fe	1.85	0.64

likely formed through a chemical reaction involving a vapor phase due to the small whisker diameters and that many of the whiskers terminate in spherical shapes. The chemical analysis data likely indicate that the whiskers are composed of phases of aluminum nitride, aluminum oxynitride, and/or aluminum oxide.

Individual whiskers from a reaction product synthesized using JSC-1AF simulant and 33.33% Al are shown in Fig. 9.18. Most of the whiskers shown have diameters less than 100 nm.

The microstructure of a reaction product synthesized using JSC-1A simulant and 19.44% Al is shown in Fig. 9.19, as are EDS Spectra 1 and 2. Many spherical shapes are observed around the interface between the particle and the whisker networks.

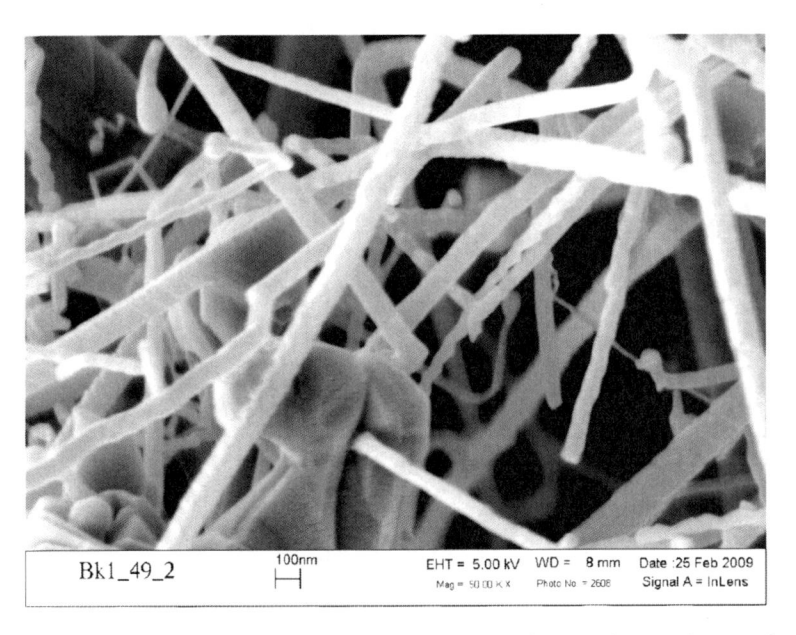

Fig. 9.18 SEM micrograph of the microstructure of a reaction product synthesized in a standard atmosphere using JSC-1AF simulant and 33.33% aluminum.

The spherical shapes at the particle surface may be evidence of an initial stage of whisker growth. Table 9.7 shows the chemical data obtained from analysis of Spectrum 1 and 2. Spectrum 1 contains a smooth surface, likely indicative of melting, containing no whiskers. The chemical composition is primarily aluminum and oxygen, the region also has significant calcium and silicon. Spectrum 1 also has some magnesium. EDS data likely indicate that Spectrum 1 is composed of spinel, grossite, corundum, and reduced silicon. Spectrum 2 is located in a network of whiskers, most of which terminate in spherical shapes. Elemental analysis indicates that Spectrum 2 has over 25 atomic percent nitrogen, and contains large amounts of aluminum and oxygen. Spectrum 2 also contains significant silicon content. It is likely that the region is composed of aluminum nitrides, aluminum oxynitrides, and/or aluminum oxides.

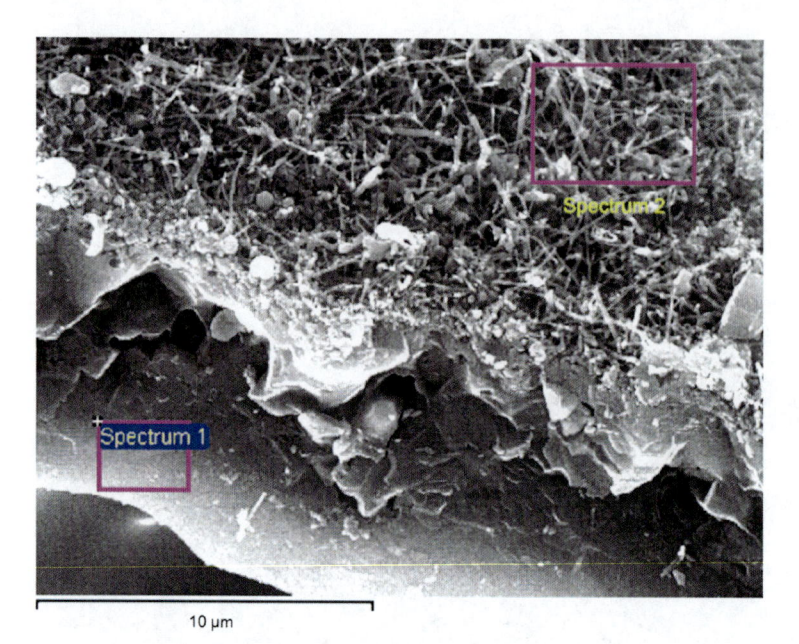

Fig. 9.19 SEM micrograph of the microstructure of a reaction product synthesized in a standard atmosphere using JSC-1A simulant and 19.44% aluminum with the regions of EDS analysis identified.

Table 9.7 EDS Analysis of Spectra 1 and 2 shown in Fig. 9.19

	Spectrum 1		Spectrum 2	
Element	Wt.%	At.%	Wt.%	At.%
N	-	-	18.30	26.56
O	33.96	48.98	26.37	33.49
Na	1.46	1.46	0.58	0.52
Mg	4.87	4.62	0.87	0.72
Al	29.85	25.53	38.40	28.92
Si	10.79	8.87	11.16	8.08
Ca	16.39	9.43	0.75	0.38
Ti	-	-	0.48	0.20
Fe	2.69	1.11	3.09	1.13

9.2.11.1 Microstructural Analysis of Reaction Products Formed in a Vacuum Environment

The microstructure of a reaction product formed from JSC-1AF simulant and 19.44% Al in a vacuum environment is shown in Fig. 9.20. The spherical shapes are potentially iron silicides or aluminum.

The microstructure of a reaction product formed from JSC-1AF simulant and 33.33% Al in a vacuum environment is shown in Fig. 9.21. Very tiny whiskers can be seen on some of the surfaces of the particles, but they are negligible compared to the whisker growth observed in the standard atmosphere reaction products.

The microstructure of the reaction product using JSC-1AF simulant and 19.44% Al mass ratio is shown in Fig. 9.22, as are EDS Spectra 1, 2, 3, and 4. Table 9.8 shows the chemical data obtained from analysis of Spectra 1, 2, 3, and 4. Spectra 1 and 4 are located on spherical particles.

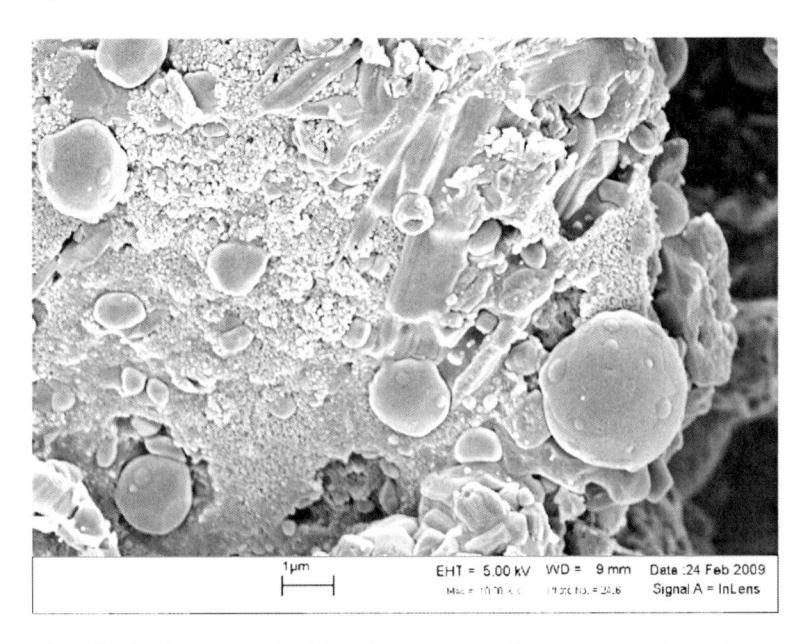

Fig. 9.20 SEM micrograph of the microstructure of a vacuum-synthesized reaction product using JSC-1AF simulant and 19.44% aluminum.

Elemental analysis indicated that the particles have significant silicon and iron. Data from XRD indicate that the particles are likely to be composed of intermetallic iron-silicides. The spherical particles of iron-silicides may have been formed by phase separation from the larger particle; the spherical shape suggests a low degree of surface wetting with regards to the larger particle surface. Spectrum 2 is located on a flat surface of a particle with cubic geometry. Chemical analysis of the region indicates large concentrations of aluminum, magnesium, and oxygen. EDS data from Spectrum 2 coupled with XRD data indicate that the region is likely composed of spinel. Spectrum 3 is located on another spherical particle. EDS indicated large concentrations of silicon, iron, aluminum, and oxygen. The data from chemical analysis and XRD likely indicate the particle in Spectrum 3 is composed of iron-silicides and aluminum oxides.

Fig. 9.21 SEM micrograph of the microstructure of a vacuum-synthesized reaction product using JSC-1AF simulant and 33.33% aluminum.

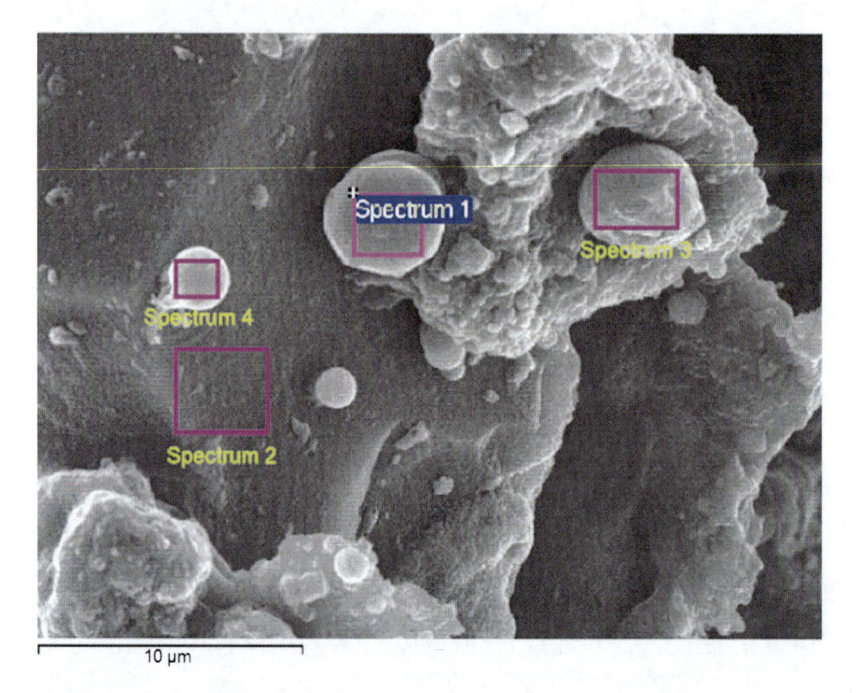

Fig. 9.22 SEM micrograph of the microstructure of a vacuum-synthesized reaction product using JSC-1AF simulant and 19.44% aluminum with the regions of EDS analysis identified.

The microstructure of a reaction product created using JSC-1AF simulant and 28.85% Al is shown in Fig. 9.23, as are EDS Spectra 1, 2, and 3. Table 9.9 shows the chemical data obtained from analysis of Spectra 1, 2, and 3.

Table 9.8 EDS Analysis of Spectra 1, 2, 3, and 4 shown in Fig. 9.22

	Spectrum 1		Spectrum 2		Spectrum 3		Spectrum 4	
	Wt.%	At.%	Wt.%	At.%	Wt.%	At.%	Wt.%	At.%
O	8.45	16.99	31.87	43.97	12.95	25.98	14.02	26.28
Mg	-	-	15.42	14.00	1.28	1.69	1.35	1.68
Al	2.89	3.43	45.03	36.84	8.44	10.05	6.31	7.01
Si	49.80	57.09	4.11	3.23	29.89	34.18	43.23	46.14
Ca	0.40	0.32	3.56	1.96	1.03	0.82	0.36	0.26
Ti	-	-	-	-	6.21	4.16	-	-
Mn	-	-	-	-	-	-	0.74	0.40
Fe	38.46	22.17	-	-	40.22	23.13	33.99	18.25

A pitted particle is observed within the micrograph shown in Fig. 9.23. Elemental analysis indicated that there is little difference in elemental composition of the different regions of the particle. All had significant aluminum and oxygen, along with some calcium. Data from elemental analysis and XRD indicate that the region is likely composed of grossite and aluminum oxide. It is possible that the pitting of the particle could have occurred due to volatization of sodium, potassium, or magnesium from the region.

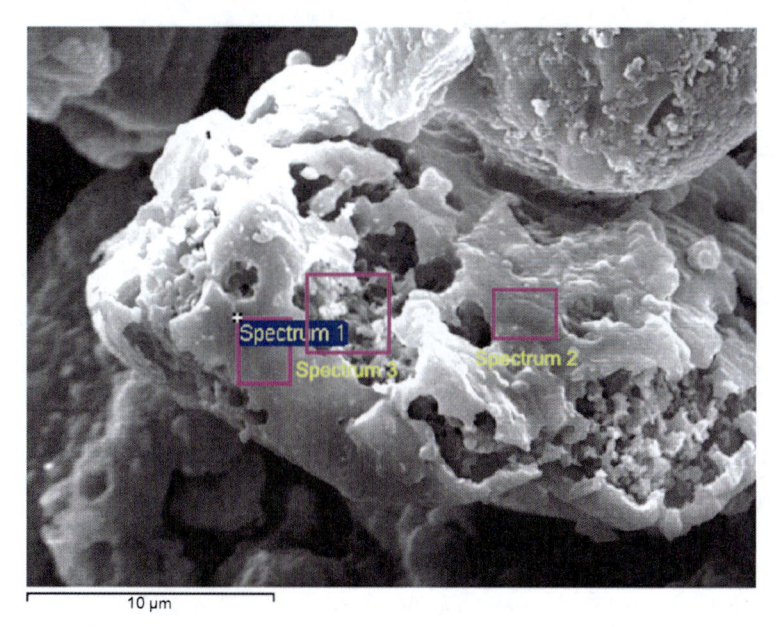

Fig. 9.23 SEM micrograph of the microstructure of a vacuum-synthesized reaction product using JSC-1AF simulant and 28.85% aluminum with the regions of EDS analysis identified.

Table 9.9 EDS Analysis of Spectra 1, 2, and 3 shown in Fig. 9.23

	Spectrum 1		Spectrum 2		Spectrum 3	
Element	Wt.%	At.%	Wt.%	At.%	Wt.%	At.%
O	43.19	57.88	50.52	64.86	45.84	60.53
Mg	1.84	1.62	1.47	1.24	2.26	1.97
Al	39.60	31.46	33.54	25.53	35.32	27.66
Si	4.81	3.66	5.51	4.02	6.90	5.19
Ca	8.65	4.63	7.22	3.71	6.53	3.44
Ti	-	-	-	-	0.38	0.17
Fe	1.91	0.73	1.75	0.65	2.76	1.04

The microstructure of a reaction product created using JSC-1A simulant and 24.45% Al is shown in Fig. 9.24, as are EDS Spectra 1, 2, and 3. Table 9.10 shows the chemical data obtained from analysis of Spectrum 1, 2, and 3. Spectrum 1 is located on a flat and smooth particle surface. The chemical composition indicates significant quantities of aluminum, magnesium, and oxygen, along with some silicon and iron.

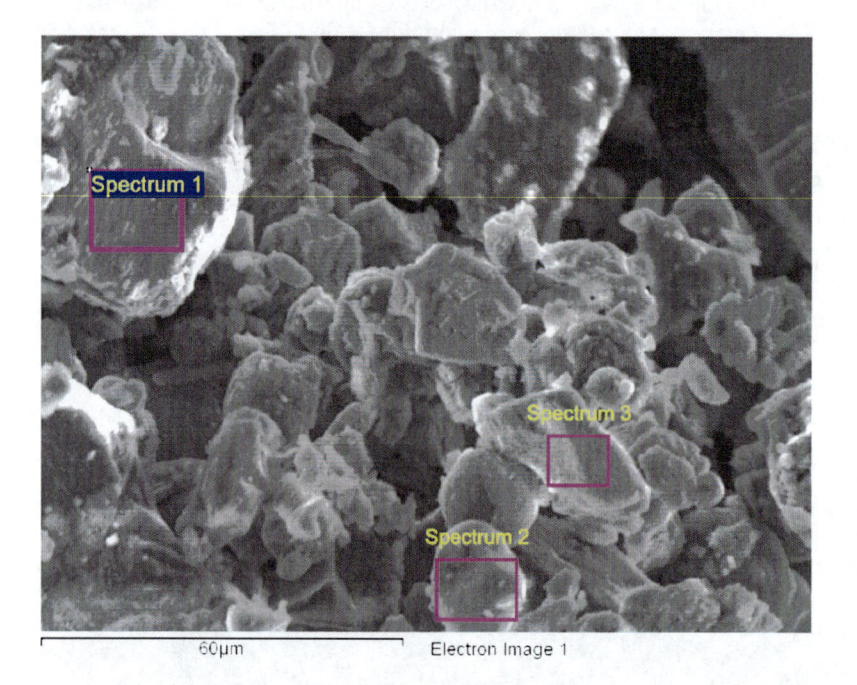

Fig. 9.24 SEM micrograph of the microstructure of a vacuum-synthesized reaction product using JSC-1A simulant and 24.45% aluminum with the regions of EDS analysis identified

Table 9.10 EDS Analysis of Spectra 1, 2, and 3 shown in Fig. 9.24

	Spectrum 1		Spectrum 2		Spectrum 3	
Element	Wt.%	At.%	Wt.%	At.%	Wt.%	At.%
O	40.56	55.76	10.23	16.27	44.59	58.55
Mg	16.26	14.71	0.53	0.55	0.25	0.22
Al	19.33	15.75	75.49	71.22	47.43	36.92
Si	11.13	8.71	12.52	11.35	3.48	2.61
Ca	0.43	0.24	0.23	0.15	0.65	0.34
Ti	-	-	-	-	0.13	0.06
Fe	12.29	4.84	1.00	0.46	3.48	1.31

EDS data from Spectrum 1, coupled with data from XRD, likely indicate the region is composed primarily of spinel and iron silicides. Spectrum 2 is located on a sub-spherical particle. EDS analysis indicates that large quantities of aluminum and some silicon are present. Data from chemical analysis and XRD indicate that the particle is likely composed of elemental aluminum and silicon. Spectrum 3 is located on an angular particle with a flat and smooth surface. Elemental analysis indicates large concentrations of aluminum and oxygen. The chemical analysis of Spectrum 3 and data from XRD likely indicate the presence of corundum.

9.2.11.2 Comparison of Reaction Products Formed in Standard Atmosphere Conditions and Vacuum Conditions

Networks of whiskers were characteristic of standard atmosphere reaction products, and were almost entirely absent from vacuum reaction products. Very small whiskers were observed on a few particles in vacuum, but did not compare to the quantity observed in a standard atmosphere. Most of the whiskers formed in a standard atmosphere reaction product terminated in a spherical shape, possibly indicating a vaporization process occurred. The nanoscale diameters of the whiskers also support the theory that whisker formation involved a vapor phase. The presence of whiskers was likely to greatly increase the interparticle bonding and overall strength within the standard atmosphere reaction product.

Two types of whiskers were found using EDS analysis, nitrogen-bearing whiskers and whiskers containing no nitrogen. Small quantities of silicon appeared to be present in both type of whiskers. EDS analysis indicated that the nitrogen-bearing whiskers were most prevalent. Typically nitrogen-bearing whiskers had over 20 at.% nitrogen. The whiskers also had large quantities of aluminum and oxygen present. It was hypothesized that the whiskers were primarily composed of the chemical species of aluminum nitride. The presence of oxygen likely indicates incorporation of aluminum oxide or aluminum oxynitride into the whiskers. Prior research by Guojian et al. (2000a,b) indicated that ambient nitrogen can combine with Al_2O gas and cause AlN whisker growth. It is suggested that a similar process is occurring within the reaction investigated in this chapter. The termination of whiskers in a spherical shape indicates that growth occurred by the VLS mechanism. Growth by the VLS mechanism would indicate that a liquid phase was present at the whisker tip, creating a preferential deposition surface for gas molecules. Precipitation from the super-saturated liquid phase would have caused whisker growth at the solid-liquid interface. Impurities such as silicon, magnesium, and oxygen could have been incorporated within the AlN whiskers and assisted in the growth process. The other type of whiskers observed have no bound nitrogen; EDS analysis indicated that aluminum and oxygen were major constituents. It was hypothesized that these whiskers are composed of aluminum oxide, also found in the XRD patterns. The presence of silicon in most of the whiskers could indicate that silicon acts as the liquid catalyst in the VLS whisker growth process. Evidence of partial melting was observed in some micrographs, indicating a liquid phase sintering process occurred during the reaction. However, the prevalence of whiskers in the microstructures largely concealed evidence of liquid phase sintering.

9.2.11.3 Reaction Thermodynamics in a Standard Atmosphere

It was determined that the following reactions were thermodynamically favorable over various temperature regimes ranging from 298 K to 1998 K in a standard atmosphere reaction:

$$20Al + 5CaAl_2Si_2O_8 \rightarrow 5CaAl_4O_7 + 10Si + 5Al_2O(g)$$
$$6Al_2O(g) + 6N_2(g) \rightarrow 12AlN(s) + 3O_2(g)$$
$$Al_2O(g) + O_2(g) \rightarrow Al_2O_3$$
$$4Al + 3O_2(g) \rightarrow 2Al_2O_3$$
$$8Al + 2NaAlSi_3O_8 \rightarrow Na_2O + 6Si + 5Al_2O_3$$
$$8Al + 3CaMgSi_2O_6 \rightarrow Ca_3Mg(SiO_4)_2 + 2MgAl_2O_4 + 4Si + 2Al_2O(g) \quad (9.1)$$
$$8Al + 3CaMgSi_2O_6 \rightarrow Ca_3Al_2O_6 + 3MgAl_2O_4 + 6Si$$
$$10Al + 3CaFeSi_2O_6 \rightarrow Ca_3Al_2O_6 + 3FeSi_2 + 4Al_2O_3$$
$$Fe + Si \rightarrow FeSi$$
$$8Al + 2CaFeSi_2O_6 \rightarrow Ca_2Al_2SiO_7 + 2Al_2O + 2FeSi + Si + Al_2O_3$$
$$Fe + 2Si \rightarrow FeSi_2$$
$$8Al + 3CaAl_2Si_2O_8 \rightarrow 3CaAl_4O_7 + 6Si + Al_2O_3$$
$$12Al + 2CaMgSi_2O_6 + 2CaAl_2Si_2O_8 \rightarrow 4CaAl_4O_7 + 8Si + 2Mg$$

From the chemical equations above, chemical species of $CaAl_4O7$, Si, AlN, Al_2O_3, Na_2O, $MgAl_2O_4$, $Ca_3Al_2O_6$, $FeSi_2$, FeSi, and Mg could be formed during reactions conducted in a standard atmosphere.

9.2.12 Reaction Thermodynamics in a Vacuum

It was determined that the following reactions were thermodynamically favorable over various temperature regimes ranging from 298 K to 1998 K in a reaction conducted in a 0.600 Torr vacuum:

$$20Al + 5CaAl_2Si_2O_8 \rightarrow 5CaAl_4O_7 + 10Si + 5Al_2O(g)$$
$$Fe + 2Si \rightarrow FeSi_2$$
$$Fe + Si \rightarrow FeSi$$
$$8Al + 2CaFeSi_2O_6 \rightarrow Ca_2Al_2SiO_7 + 2Al_2O(g) + 2FeSi + Si + Al_2O_3 \quad (9.2)$$
$$8Al + 2NaAlSi_3O_8 \rightarrow Na_2O + 6Si + 5Al_2O_3$$
$$4Al_2O(g) + Mg \rightarrow MgAl_2O_4 + 6Al$$
$$8Al + 3CaAl_2Si_2O_8 \rightarrow 3CaAl_4O_7 + 6Si + Al_2O_3$$
$$12Al + 2CaMgSi_2O_6 + 2CaAl_2Si_2O_8 \rightarrow 4CaAl_4O_7 + 8Si + 2Mg$$

From the chemical equations above, chemical species of $CaAl_4O7$, Si, $FeSi_2$, FeSi, $Ca_2Al_2SiO_7$, Al_2O_3, Na_2O, $MgAl_2O_4$, and Mg could be formed during reactions conducted in a vacuum environment.

9.2.13 Mechanical Strength Measurements

General handling of the reaction products indicated that the outer diameter of the vacuum reaction products was substantially weaker than the standard atmosphere product. The area inside the vacuum reaction product, in the vicinity of the NiCr wire, was significantly stronger than the exterior areas in many instances.

Compressive strength tests were performed on specimens fabricated in a standard Earth atmosphere. The ends of each reacted sample were leveled prior to testing. A ceramic tile saw utilizing a diamond blade was used to machine the samples, without use of a lubricating medium. A Marathon digital caliper was used to make three measurements of sample diameter and three measurements of sample length after each sample was cut. The mean diameter for each sample was used to calculate the surface area, which was used in the calculation of the compressive strength of the material. ASTM standard C 1424-9 was used as a guide for the compressive strength testing. An Instron 4468 testing machine with a 50kN load cell was used to perform compressive strength testing using a platen displacement rate of 1 mm/min.

Figure 9.25 shows the ultimate mean compressive strength of various reactant mass ratios synthesized in a standard atmosphere using JSC-1AF and JSC-1A simulants. The first number within each bar on the chart refers to the weight percent aluminum used in the reaction, the letters that follow denote whether JSC-1A simulant or JSC-1AF simulant was used, and the last number denotes how many compressive tests were run on the given mass ratio. The error bars represent ± 1 standard deviation from the mean.

Mean Compressive Strength of JSC-1A and JSC-1AF Standard Reaction Products

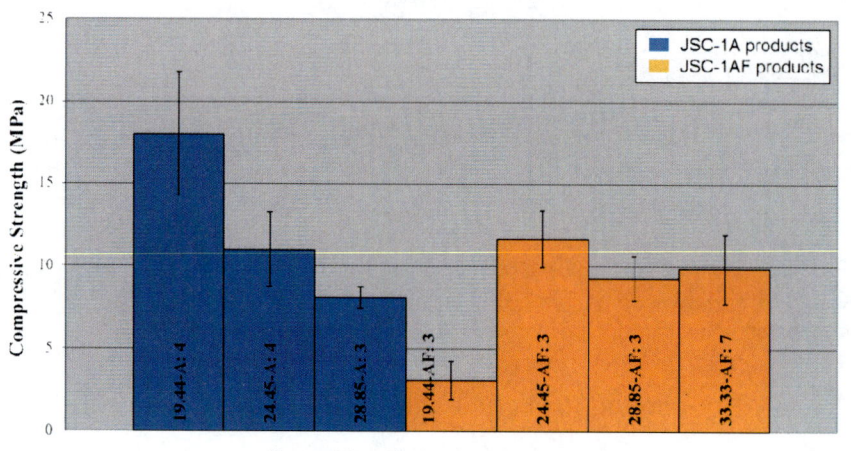

Fig. 9.25 Mean compressive strength of JSC-1A and JSC-1AF reaction products formed in a standard atmosphere.

With the exception of the 19.44% aluminum mass ratios, data from Fig. 9.25 indicate that the compressive strengths of the reactions products for the various compositions and simulants were typically within +/- 1 standard deviation of each other. The average strengths typically ranged from 8-12 MPa. The reaction product utilizing the 19.44% mass ratio of aluminum and JSC-1A simulant yielded significantly higher compressive strengths than all other samples. The average compressive strength was around 18 MPa. However, the reaction product using the same mass ratio of aluminum and JSC-1AF simulant exhibited the lowest compressive strength (~ 3 MPa) of any of the samples. Unsteady combustion was observed in some of the reactions performed using a mass ratio of 19.44% aluminum with JSC-1AF simulant. The unsteady combustion could have produced defects and voids within the sample, resulting in a lower compressive strength. Although compressive strength tests were not performed on vacuum reaction products, specimen handling indicated that reaction products formed in a standard atmosphere had higher strengths. The lack of appreciable whiskers in the vacuum reaction product would indicate that the quantity of whiskers present within reaction products influenced the strength.

9.2.14 Fabrication of Voissoir Dome Elements

One potential concept for utilizing the reaction product is as a lunar structure in the form of a Voissoir dome. A Voissoir dome consists of multiple courses of Voissoir elements stacked on top of one another to form a dome. A Voissoir dome does not require mortar between each Voissoir element. In order to provide proof of concept, a silica slip crucible was fabricated into the negative shape of a dome element. Each dome element within a course of the dome would have the same shape. However, each course within the dome would require a slightly different shape of crucible. The silica crucible was filled with a mixture of regolith simulant and aluminum, and a NiCr wire was immersed within the mixture. The reaction was initiated through Joule heating of a nickel-chromium wire, as discussed in an earlier section. The reaction product took the shape of the crucible, and formed an element that could be utilized in a dome. A more detailed discussion of the Voissoir dome construction can be found in our study (Faierson et al. 2010). Images of the reaction process are shown in Figs. 9.26 – 9.29.

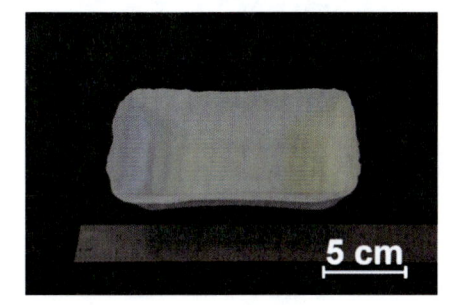 

Fig. 9.26 (left) A silica-slip crucible fabricated in the shape of a Voissoir dome element; (right) Silica-slip crucible filled with mixture of aluminum and regolith simulant, with NiCr wire immersed within the mixture.

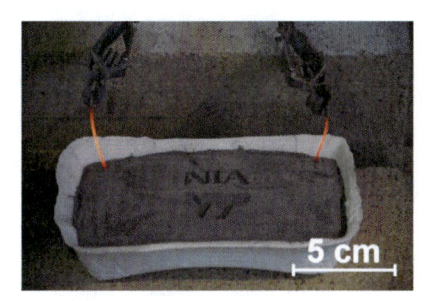 

Fig. 9.27 (left) Heat is applied to the reactant mixture through joule heating; (right) Reaction initiation begins.

 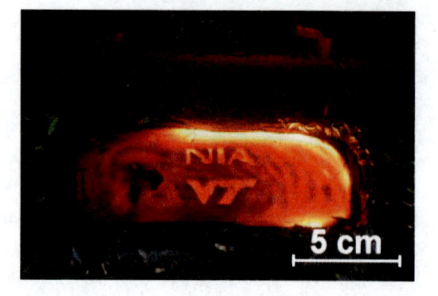

Fig. 9.28 (left) Reaction propagation continues; (right) Reaction completion.

 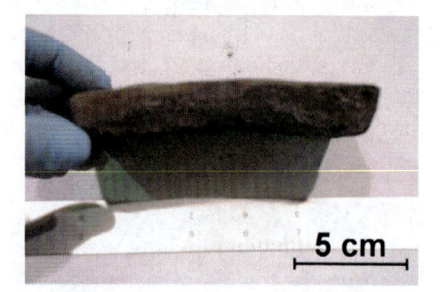

Fig. 9.29 (left) Top-view of near-net-shape reaction product; (right) Side-view of near-net-shape reaction product, illustrating curved surfaces.

9.3 Conclusion

To summarize, Si, Al_2O_3, $MgAl_2O_4$, $CaAl_4O_7$, $FeSi_2$, and $Ca_2Al_2SiO_7$ were common chemical species identified by XRD in both standard atmosphere and vacuum environment reaction products. XRD analyses indicate that aluminum nitrides are present within all reaction products formed in a standard atmosphere, indicating that the reaction is interacting with atmospheric gases. Unreacted aluminum and unreacted plagioclase minerals were commonly identified in XRD patterns of reaction products synthesized in a vacuum environment.

SEM analysis showed networks of whiskers formed in the products of all reaction products formed in a standard atmosphere. Many whiskers terminated in a spherical shape, indicating growth by a VLS mechanism. EDS analysis of the samples indicated that nitrogen was present within all mass ratios of reaction products formed in a standard atmosphere. Whiskers are hypothesized to be composed of aluminum nitride, aluminum oxide, and/or aluminum oxynitride.

With the exception of the 19.44% aluminum mass ratio, average compressive strengths for all specimens synthesized in a standard atmosphere ranged from 8-12 MPa. Reaction products synthesized using a 19.44% aluminum mass ratio and JSC-1A simulant yielded the highest compressive strengths (~ 18 MPa), while samples synthesized using the same mass ratio and JSC-1AF simulant yielded the lowest compressive strengths (~3 MPa). The lower compressive strengths produced using 19.44% aluminum and JSC-1AF simulant are thought to be a result of defects caused by unsteady combustion wave propagation that was observed during synthesis.

A near-net-shape Voissoir dome element was formed within a reusable silica-slip crucible. The reaction product has potential for use in lunar dust mitigation, micrometeoroid protection, structural applications, and radiation shielding on the lunar surface. Lunar dust mitigation could be realized by using the reaction product for landing surfaces, roads, and/or blast berms. Thermal energy released by the reaction could be harnessed for other purposes, including extraction of useful volatiles such as Helium-3, which has potential for use in nuclear fusion reactors.

Another potential use for this reaction is binding nitrogen to lunar regolith. This could be useful for producing a soil capable of sustaining plant growth. Preliminary experiments have indicated that the addition of water to reaction products can produce an ammonia-type smell. One specimen that was exposed to water and sealed in a bag produced sufficient gas to rupture the sealed bag. It is hypothesized that the high surface areas of the aluminum nitride/oxynitride nano-whiskers are facilitating the production of aluminum hydroxide and ammonia from water.

9.3.1 Future Work

Research is underway to initiate a geothermite reaction in an argon environment. Utilizing an argon environment will provide gas pressure, potentially allowing growth of aluminum oxide whiskers, while eliminating nitrogen interaction in the reaction. The inert nature of argon will prevent its incorporation into the reaction product, allowing a more efficient in-situ resource utilization process for use on the Moon.

Acknowledgements. This study was funded by National Institute of Aerospace (NIA) contract number VT-03-01. The authors thank NASA Langley Research Center (LaRC) for the use of laboratory facilities. The authors also thank Steve McCartney at ICTAS and Jim Baughman at NASA LaRC for assistance with SEM and EDS analysis, as well as David Hartman at NASA LaRC for assistance with XRD.

References

Allen, C.C., Graf, J.C., McKay, D.S.: Sintering Bricks on the Moon. In: Engineering, Construction, and Operations in Space IV, pp. 1220–1229 (1994)

Curreri, P.A., Ethridge, E.C., Hudson, S.B., Miller, T.Y., Grugel, R.N., Sen, S., Sadoway, D.R.: Process Demonstration For Lunar In Situ Resource Utilization-Molten Oxide Electrolysis, NASA Marshall Space Flight Center, pp. 1–32 (2006)

Fabes, B., Poisl, W., Beck, A., Raymond, L.: Processing and Properties of Lunar Ceramics. In: AIAA Space Programs and Technology Conferences, Huntsville, Al (1992)

Faierson, E., Logan, K., Hunt, M., Stewart, B.: Lunar Habitat Construction Utilizing In-Situ Resources and an SHS Reaction. In: AIAA Space 2008 Conference and Exposition, San Diego (2008)

Faierson, E.J., Logan, K.V.: Geothermite Reactions for In-Situ Resource Utilization on the Moon and Beyond. In: ASCE Earth and Space Conference 2010, Honolulu, HI (2010)

Faierson, E.J., Logan, K.V., Stewart, B.K., Hunt, M.P.: Demonstration of concept for fabrication of lunar physical assets utilizing lunar regolith simulant and a geothermite reaction. Acta Astronautica 67, 38–45 (2010)

Gaier, J.R., Sechkar, E.A.: Lunar Simulation in the Lunar Dust Adhesion Bell Jar, NASA Glenn Research Center (2007)

Guojian, J., Hanrui, Z., Jiong, Z., Meiling, R., Wenlan, L., Fengying, W., Baolin, Z.: Morphologies and Growth Mechanisms of Aluminum Nitride Whiskers by SHS Method-Part 1. Journal of Materials Science 35, 57–62 (2000a)

Guojian, J., Hanrui, Z., Jiong, Z., Meiling, R., Wenlan, L., Fengying, W., Baolin, Z.: Morphologies and Growth Mechanisms of Aluminum Nitride Whiskers by SHS Method-Part 2. Journal of Materials Science 35, 63–69 (2000b)

Heiken, G.H., Vaniman, D.T., French, B.M., Schmitt, H.H. (eds.): Lunar Sourcebook. Cambridge University Press (1991)

Kumashiro, Y.: Electric Refractory Materials. CRC Press (2000)

Martirosyan, K.S., Luss, D.: Nanoenergetic Fabrication of Dense Ceramics for Lunar Exploration. Lunar and Planetary Science XXXIX (2008)

Mei, J., Halldearn, R.D., Xiao, P.: Mechanisms of the Aluminum-Iron Oxide Thermite Reaction. Scripta Materialia 41(5), 541–548 (1999)

Mossino, P.: Some Aspects in Self-propagating High-temperature Synthesis. Ceramics International 30, 311–332 (2004)

NASA-MSFC, Characterization Summary of JSC-1AF Lunar Mare Regolith Simulant (2006)

NASA-MSFC, Characterization Summary of JSC-1A Bulk Lunar Mare Regolith Simulant (2007)

Sen, S., Ray, C.S., Reddy, R.G.: Processing of Lunar Soil Simulant for Space Exploration Applications. Materials Science and Engineering A 413-414, 592–597 (2005)

Taylor, L.A.: Resources for a Lunar Base: Rocks, Minerals, and Soil of the Moon. In: 2nd Conference on Lunar Bases and Space Activities (1992)

Taylor, L.A., Meek, T.T.: Microwave Sintering of Lunar Soil: Properties, Theory, and Practice. Journal of Aerospace Engineering 18(3), 188–196 (2005)

10 Lunar Drilling, Excavation and Mining in Support of Science, Exploration, Construction, and In Situ Resource Utilization (ISRU)

Kris Zacny

Honeybee Robotics Spacecraft Mechanisms Corporation, New York

10.1 Space and Lunar Exploration: Historical Review

Exploration of almost any extraterrestrial body follows a path from low complexity, low science pay-off to high risk, high pay off. The Moon, being the closest extraterrestrial body, was the first body to be examined with a naked eye by ancient astronomers and philosophers from Babylonia, Greece, and Egypt. The invention of a telescope by Hans Lipperhey in 1608 allowed much more detailed observation of the Moon. With the aid of telescopes, Galileo Galilee could not only view details of the lunar surface, but also discovered the four largest moons of the Jupiter: Io, Europa, Ganymede and Callisto (now called Galilean satellites). The next leap in exploration of space was possible thanks to the development of rockets. Initially, rockets were built for military purposes only. During World War II, these included infamous V2 rockets and soon after, during the Cold War; they included Inter Continental Ballistic Missiles (ICBMs) capable of carrying nuclear warheads across the oceans. The ICBMs later formed the foundation for space rockets.

Even today, commercial rocket launching companies, such as Orbital Sciences Corporation are using various stages of the ICBMs to launch satellites into the Earth orbit. For example, the Taurus first stage is based on a Peacekeeper ICBM first stage. Back in the 1950s, the first space rockets were used to launch Soviet Sputnik satellite around the Earth orbit, followed by the US Explorer 1 satellite that discovered the Van Allen radiation belt. Rockets allowed launching of spacecraft into the Earth orbit and point them towards the universe, and see images that were not distorted by our atmosphere. Observing planets from space also allowed viewing them across the entire range of electromagnetic spectrum, and not just the frequency portions of the spectrum that pass through our atmosphere. For example, water molecules in our atmosphere preferentially absorb electromagnetic radiation at wavelengths around 2900, 1950 and 1450 nanometers (Carter and

McCain 1993). Hence, we can not view the Moon or any other object in space in these particular wavelengths from Earth. (The frequency is inversely proportional to wavelength, with c (the speed of light) being proportionality constant as follow: c=frequency x wavelength). Another good example is Ozone, an isotope of Oxygen. Ozone absorbs harmful UV radiation. Hence people living in the certain parts in the Southern Hemisphere, where the atmosphere is mostly free of Ozone (this is termed the Ozone hole), need to take precautions when going outside. However, these Ozone-free areas would be ideal for observing space in this particular UV wavelength that would otherwise be absorbed by the Ozone.

Once we learned and mastered the launch of rockets into Earth orbit, we quickly learned that the easiest way to see the planetary surface up close is using so-called flybys. This is accomplished by spacecraft that zoom past the Moon (or other planets) and take pictures at much closer distances. This step was much easier to accomplish than trying to orbit the Moon and also allowed taking pictures of the far side of the Moon. In 1959, Soviet flyby spacecraft Luna 3 was the third spacecraft to be sent to the Moon, and it returned never-before-seen views of the far side of the Moon. However, a scientific benefit of flybys is minimal because the spacecrafts have only a short time to investigate the surface as it zooms past the Moon. Another type of spacecraft that could fall into this 'short-lived' category is an impactor. Impactors take photos as they are speeding towards the surface. The first lunar impactor took some magnificent photos and returned them back to Earth before being destroyed by the collision with the lunar surface. For example in 1962, the US Ranger 4 spacecraft transmitted pictures of the lunar surface for 10 minutes before crashing onto the Moon. As technology advanced and our knowledge of celestial mechanics improved, we learned how to launch lunar orbiters. The orbiters could orbit the Moon for a very long time and allowed investigation of a much larger surface area and at much closer distances. In 1966, the Soviet Luna 10 spacecraft became the first artificial satellite of the Moon. A few months before the successful Luna 10 mission, the Luna 9 spacecraft, the twelfth attempt at a soft-landing, finally achieved a soft landing on the Moon and transmitted the first photographs from the lunar surface. At the same time, Luna 9 showed that the lunar surface would support a lander, that is the lander would not sink into the lunar dust as many people were afraid of. Landing on the Moon opened up an entire array of research fields related to in-situ exploration. Thanks to lunar landers we could actually touch lunar surface, analyze it, take microscopic pictures and deploy surface instruments, and later on subsurface instruments.

The third spacecraft to successfully land on the Moon, Luna 13 was equipped with spring loaded booms, one of which carried a penetrometer designed to measure the forces required to penetrate the lunar regolith. This was the first geotechnical measurement of the lunar surface. A few months later, the US Surveyor 3 spacecraft used its scoop to dig the lunar regolith and used the excavation data to determine properties of lunar soil.

Of course, there is a limit to what kind of instrument can be sent to the Moon, and hence, the next step was to bring some samples back to Earth and analyze them with a large suite of instruments. In fact, one of the best science investigations was done, not when we returned the regolith during the Apollo or Luna missions, but many years later, when technological advances allowed building of much more sophisticated and sensitive instruments.

The Apollo and Luna programs were the main lunar exploration programs undertaken by the United States and Soviet Union, respectively. During the Apollo program, 12 astronauts walked on the surface of the Moon and returned safely back to Earth. Of the 7 missions (Apollo 11-Apollo 17), only Apollo 13 never reached the surface of the Moon (the mission was dubbed a successful failure). During each of the Apollo three day surface missions, the astronauts' schedule was packed with exploration tasks. These included taking photographs and movies, deploying instruments, drilling, collecting rocks and soil (referred to as regolith). In fact the deepest hole on the Moon at 292 cm was drilled during Apollo 17 mission (Heiken et al. 1991).

Around the same time, the Soviet Luna program deployed robotic rovers (Lunokhod 1 and 2) as well as landers. The two Lunokhods were remote controlled rovers and traversed tens of kilometers, taking pictures, collecting geotechnical data and performing scientific observations. Another great achievement by the Soviets that has not been repeated since then, was a series of sample return missions. The Soviet Luna 16, Luna 20 and Luna 24 landers drilled and successfully returned subsurface regolith. The first two spacecrafts had a short drill for acquiring lunar regolith from shallow depths, while Luna 24 had a 2 meter long drill. Luna 16 returned 100 grams of regolith from up to 35 cm depth, Luna 20 returned 50 grams from up to 27 cm depth, and Luna 24 returned 170 grams from up to 160 cm depth (Heiken et al. 1991). A fraction, 3 grams to be exact, of the Luna returned sample was shared with US scientists. In exchange, the US gave Soviet scientists 3 grams of samples returned by the Apollo astronauts.

The analysis of samples returned by the US and Soviet missions revealed that the regolith contains a lot of elements that can be used to sustain human presence on the Moon. For example, regolith has Oxygen locked in a mineral called Illmenite. Mining and processing of this mineral using the Hydrogen Reduction process, could hence liberate the Oxygen (Gibson and Knudsen 1985).

The most recent discovery of large deposits of hydrogen suggests there are large amounts of water-ice on the Moon. On 9 October 2009, the Lunar Crater Observation and Sensing Satellite (LCROSS) mission, consisting of a mostly empty Centaur rocket, impacted the shadowed area inside Cabeus crater, at the lunar South Pole. The ejected debris, dust, and vapor were then analyzed by a second "shepherding" spacecraft. Based on the analyzed data, the concentration of water ice in the regolith is estimated to be ~5.6% by mass (Calaprete 2010). These water reservoirs could be an enabling resource, not only to support human presence on the Moon, but also to support human exploration of other Solar System bodies. In fact, if we were to mine this water and turn it into hydrogen and oxygen (i.e. rocket fuel and oxidizer, respectively) we could refuel the rocket tanks for the journey back home or to travel to other planetary destinations. If the water-ice extends to 80° latitude, there may be enough ice to launch thousands of space shuttle sized spacecrafts.

The commercial sector is also very interested in these water deposits (Chandler et al. 2007; Blair and Zacharias 2010). Currently, one of the major reasons the satellites around the Earth need to be replaced is because they are running out of fuel. The fuel is used to keep the satellites in correct positions with respect to the Earth (i.e. for attitude control or station keeping). Normally, during the majority of the life of the satellite, the revenue from a particular satellite (being it communication) is used to pay for building, launching, and operating of the satellite. Only in the last few years, when around 10% of fuel is left, the revenue is almost pure profit. Extending the life of the satellite, by refueling it in space, could therefore be very profitable. A big question, however, is whether launching fuel from Earth or mining it, processing and launching from the Moon is more economically enticing. No doubt launching anything from the Moon towards the Earth orbit, because of much lower lunar gravity, is less expensive. However, robust In Situ Resource Utilization (ISRU) technology has to be developed for mining and processing of the water and the regolith on the Moon. How much this ISRU is going to cost is still unknown. It is just a matter of time, though, before the technological advances will enable economical exploitation of the Moon.

Table 10.1 Keeping the Score: number of successful lunar missions (orbiters, flybys, impactors, and landers) to date (S=success; F=Failure)

Years	US		Soviet Union		Other Countries	
	S	F	S	F	S	F
1958		4		3		
1959		3	2	2		
1960		2		2		
1962	1	2				
1963				3		
1964	2			2		
1965	2		1	6		
1966	3	1	5	1		
1967	4	3				
1968	2		2	2		
1969	5		1	5		
1970	1	1	4	1		
1971	4		1	1		
1972	4		1			
1973			2			
1974			1	1		
1976			1			
1990 - 1999	2				1	
2000 - 2009	3				5	
2010 -					1	
Total	33	16	21	29	7	

Looking at the past lunar mission and keeping the score, one thing becomes clear. Our technology has been steadily improving and the success rate of landing on or Orbiting the Moon went from 0% in 1958 to 100% since 1990 (Table 10.1). This is especially encouraging as we prepare ourselves for the next decade of exploration. Past decades of lunar exploration gave us enough knowledge to make the decision that the Moon is worth continuing exploration. The Moon offers a wealth of resources that can sustain human presence, beyond Earth.

The topic of In-situ Resource Utilization is quite broad and covers all the aspects of assuring sustainable human presence on the Moon (and other planets) and includes mining and processing of local resources. The focus of this chapter is on mining and excavation technologies that will enable human exploration of the Moon and possibly lunar tourism. A number of examples of past, present and future lunar excavation technologies are presented.

10.2 How the Moon Is Different from Earth

The three day lunar surface exploration during each of the Apollo 11-17 (except for Apollo 13) missions, was probably the longest that could safely be achieved. We learned that highly abrasive lunar regolith combined with vacuum and low and high temperature adversely impacted equipment and operation on the Moon. For example, the wire in the mesh wheels of the Lunar Roving Vehicle (LRV) was breaking and the lunar dust got into the seals of the spacesuits and caused sealing problems. Every time astronauts fell, fine lunar dust would cover the spacesuits and in turn reduce radiative capabilities of the initially white spacesuit (white surface reflects more than black surface). This caused astronauts to overheat as the cooling system was unable to keep up with a heat generated by constantly moving astronauts.

In order to develop a correct approach to dealing with lunar problems, we first need to thoroughly understand the lunar environment and its implications. Only then, we can develop the technologies and hardware that will enable sustaining human or robotic presence on the Moon, not just for a couple of days, but for months and years.

The following subsections will dig deeper into understanding of the lunar environmental and its implication to human exploration.

10.2.1 The Effect of Temperature

Unlike the thermal environment on the Earth, which is influenced by the Sun, geothermal heating and atmosphere, the temperature of the lunar surface is affected almost entirely by the solar radiation and to a lesser extent by geothermal heating. During the lunar day, which lasts for two weeks, the surface temperature can reach up to +123°C (253°F). However during the lunar night, which also lasts approximately two weeks, the surface temperature drops to -153°C (-243°F). The temperature of the subsurface, because of geothermal heating, low thermal conductivity of lunar regolith and vacuum, does not see these large temperature fluctuations. In fact, the temperature just one meter below the surface is almost constant at -19°C (-3°F).

Recent data from the Diviner Lunar Radiometer Experiment, one of seven instruments aboard NASA's Lunar Reconnaissance Orbiter (LRO), revealed that some parts at the lunar pole reach a temperature as low as -238°C or -397°F (Paige et al. 2010). This temperature is among the lowest that has been recorded anywhere in the solar system, including the surface of Pluto. Figure 10.1 shows the daytime and nighttime temperatures on the surface of the Moon recorded by the Diviner instrument.

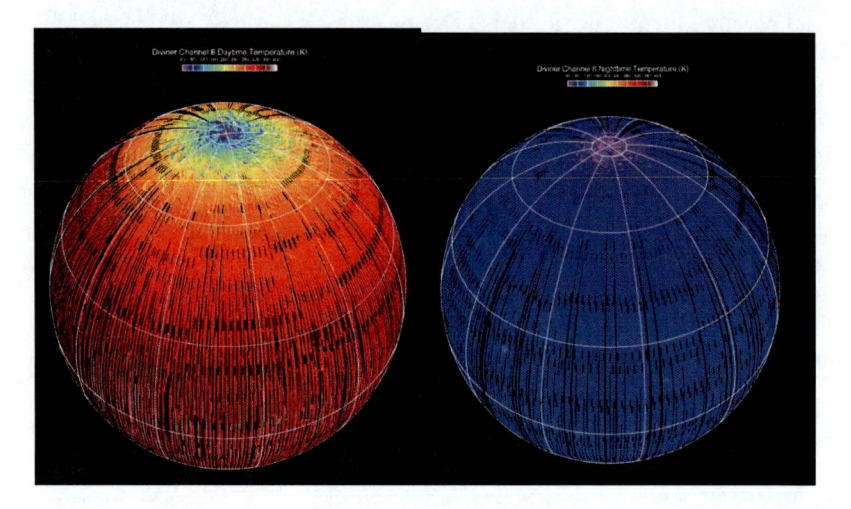

Fig. 10.1 Daytime (Left) and nighttime (right) thermal maps of the Moon acquired by the Diviner instrument aboard NASA's Lunar Reconnaissance Orbiter. (Courtesy NASA/GSFC/UCLA)

The large surface temperature fluctuations between the day and the night necessitate complicated thermal management. During a day, the hardware generating heat (for example electric motors driving rover wheels) need large radiators, while during the night, the hardware (and in particular, the electronics) need to be kept warm, which requires an energy source (batteries or nuclear based power systems). When the temperature is very low, solder joints within the electronics board may break due to uneven thermal expansion between the metal joint itself and the silicon wafer. Silicon based electronics also have other problems at low temperatures.

Thermal management is very costly and hence a number of private, government and academic institutions are developing Silicon Germanium (Si-Ge) based devices and circuits that will enable construction of low temperature electronic components (Olvera-Cervantes et al. 2008).

At present, the way around thermal management is to keep all electronics components within a warm box. The necessary heat could be generated by radioisotope heater units (RHU). An RHU is a small device, about 3.2 cm long by 2.6 cm in diameter and weighs approximately 40 grams. It provides one watt of thermal power (heat), derived from the decay of a few grams of a radioactive element such as Plutonium 238. Another source of heat could be a Radioisotope Thermal Generator (RTG). RTG also uses decay of Plutonium 238 to produce heat and this heat is converted to electricity using an array of thermocouples. The conversion of heat to electricity is very inefficient and hence the waste heat energy can be transferred using heat pipes to keep the fragile components warm.

An alternative approach to surviving the lunar night is the use of the 'thermal wadi,' a concept for storing solar energy as sensible heat in a thermal mass for later use (Wegeng et al. 2007; Sacksteder et al. 2010; Balasubramaniam et al. 2011). The thermal wadi (Fig. 10.2) utilizes engineered thermal mass made of regolith processed to increase its bulk thermal diffusivity (about x100) to that of consolidated rock or better. The increased thermal diffusivity, achieved by sintering or melting, allows surface heat to penetrate far deeper than the few centimeters solar heating penetrates native regolith during the two-week lunar day, and passively stores heat below grade. Solar heating the wadi on successive lunar days can be accomplished by either simple ambient solar illumination or through a suntracking reflector, depending on the lunar latitude. During the lunar night, deploying a reflective umbrella that reduces radiation heat loss to space helps maintain the wadi surface temperature above 0°C (32°F) for the entire two weeks of darkness. Mobile surface assets such as rovers can be parked on the thermal mass inside the reflective umbrella and stay warm passively. In comparison, the surface temperature of native regolith is nominally below 0°C before the sun sets on the moon and falls as low as -150°C (-238°F) during the night. The reflective umbrella would not protect native regolith in the same way because insufficient heat is stored during the day.

The utility of stored thermal energy provided to lunar surface assets could be substantial. Recent studies of thermal management of lunar rovers (Thornton et al. 2010; Jones et al. 2011) found that minute radiative heat leaks accumulate over two weeks of darkness such that it is unlikely that a rover could survive with anything less than near-perfect thermal insulation and substantial stored energy. Additionally, as the size of rovers decreases, the mass associated with sufficient stored energy to stay warm during a two week night overwhelms the payload capacity of the rover. The thermal wadi concept can effectively offload the nighttime thermal survival challenge onto a simple element of lunar surface infrastructure.

Unfortunately, not all components can be maintained within a specific temperature range. Wheels of a rover or a bucket of an excavator will experience these extreme low and high temperatures and must not break. Hence, these components require appropriate choice of materials and proper design. Some of the critical aspects that need to be considered when developing these systems are lubrication, thermal expansion and material embitterment.

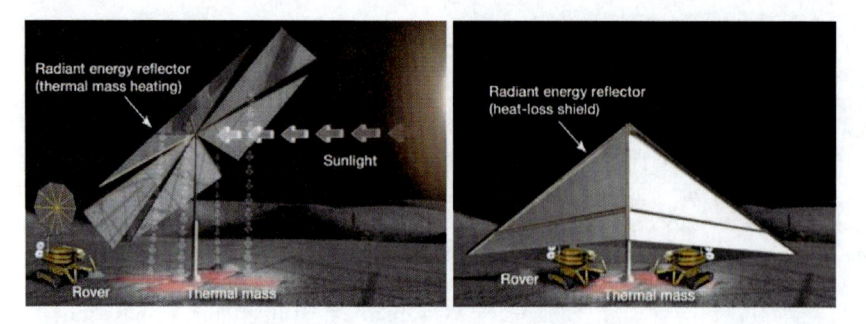

Fig. 10.2 Thermal Wadis use modified (fused) regolith as a thermal mass to store reflected solar energy during the lunar day (Left). Aided by a reflective umbrella, the heat stored in the thermal mass can sustain mobile surface assets parked there for the entire lunar night (Right). Picture courtesy Gayle Dibiasio, NASA Glenn Research Center.

Viscosity (defined as resistance to flow) of lubricants increases as temperature is reduced. At certain temperature, lubricants just freeze solid. For this reason those who live in cold areas such as Alaska or Northern Canada pay particular attention to what oil they put in their cars. On the Moon, all fluid-based lubricants will freeze. Hence the possible solution could be a dry lubricant or low friction material such as PTFE, or anti-galling material such as beryllium copper.

Thermal expansion becomes important when two materials of different coefficients of thermal expansion are in contact with each other. As the temperature changes, the two materials will expand or contract at different rates and hence large stresses may develop at the contact area. These stresses could cause one of the materials to break and the entire component to stop working properly.

Materials with Body Centered Cubic (BCC) lattice structure, such as steel undergo transition from ductile to brittle as the temperature drops. The transition temperature is referred to as the Ductile to Brittle Transition Temperature (DBTT). For example, the hull of the Titanic was made of steel with the DBTT of +32°C, which means, the steel was very brittle during the passage between Southampton, England and New York, US. When the hull hit an iceberg and created a small crack, the crack propagated very fast along the rest of the hull and caused much larger damage than engineers would have anticipated. If the steel had been ductile, it would just deform plastically. Some materials, such as Aluminum, Titanium, and composites do not have the ductile to brittle transition and are therefore better choice for the cold lunar environment.

10.2.2 Vacuum

The hard vacuum on the surface of the Moon creates many problems. For example many materials such as plastics and rubber loose flexibility and become brittle due to outgassing. Cooling of any systems generating heat (motors, computer boards) can be done only via inefficient radiation since gases conduction or convection is non-existent. On Earth, the surface of all materials is coated by a thin layer of oxides and moisture. On the Moon, these layers would quickly be lost (would de-gas) and cause a change in the frictional properties of the surface. In extreme cases, welding of two materials in contact may be even possible, since "clean" surfaces, i.e. surfaces without any oxides have unsatisfied bonds (are charged either positively or negatively).

Photoelectric ionization from solar ultra-violet (UV) radiation and ion/electron bombardment from solar wind charges lunar regolith and anything else exposed to this radiation. Since in vacuum this charge can not be dissipated easily, the net charge accumulated on a component can be on the order of several volts and is positive when in sun-lit areas and negative in shaded areas. This charge can cause electrostatic forces, contributing to adhesion and cohesion of the lunar fine regolith. In addition, charge build up can cause an electric short, which could damage electronic components.

10.2.3 Communication Delay

The distance from Earth to the Moon is relatively short, on the order of 382,500 kilometers. Hence it takes around 3 seconds for a round trip communication. The short communication delays make the teleoperation from the Earth quite feasible. In fact, the Soviet rovers, Lunokhod 1 in 1970/1971 and Lunokhod 2 in 1973 were remotely controlled on the Moon from Earth. Two five-man teams used images from the two low resolution cameras mounted on the front of the rover to sent driving commands to the rover. The cameras sent single frames at intervals that varied from seven to 20 seconds on Lunokhod 1. In 1970 and 1971 during its 322 days operations, Lunokhod 1 traversed over 10 kilometers, returned more than 20,000 images and high-resolution panoramas. In addition, it performed 25 lunar soil analyses with an x-ray fluorescence spectrometer and 500 soil penetrometer measurements. In 1973, Lunokhod 2 covered over 37 km in 4 months and sent back 80,000 pictures. Lunokhod 2 essentially had driven three times as far in half the lifetime as Lunokhod 1. This was attributed to the experience and confidence, the controllers gained during Lunokhod 1 operation, and also due to better cameras and faster refresh rate (every three seconds) on Lunokhod 2.

Note that the Lavochkin Association sold Lunokhod 2 and its lander, the Luna 21, for $68,500 to Richard Garriott, video game developer and entrepreneur (Chang 2010). Both items are still on the Moon.

10.2.4 Abrasive Lunar Regolith

John Young, Apollo 16 mission commander commented that "Dust is the number one concern in returning to the Moon." Other Apollo astronauts made similar statements, observing that lunar dust presented significant challenges to all lunar activities requiring operation of a mechanical device (Wagner 2006). Gene Cernan, Apollo 17 mission commander noted, "By the middle or end of the third Extra Vehicular Activity (EVA), simple things like bag locks and the lock which held the pallet on the Rover began not only to malfunction but to not function at all...You have to live with it but you're continually fighting the dust problem both outside and inside the spacecraft."

A detailed analysis of the Apollo data helped to categorize adverse affects of lunar dust into the following nine categories: vision obscuration, false instrument readings, dust coating and contamination, loss of traction, clogging of mechanisms, abrasion, thermal control problems, seal failures, and inhalation and irritation (Gaier 2005). Of these nine, all except for inhalation and irritation, have to be solved in order for excavation system to survive the required life time. Even a simple device, such as a locking collar of the shovel was difficult to operate after just a single EVA. In addition, it seems the end of the Soviet Lunokhod 2 mission was caused by lunar dust. On May 9, 1973, the rover's open lid during a traverse in a narrow trench, touched its wall and became covered with lunar dust. This dust fell on to the radiators while the lid was closed for the lunar night. The next lunar day, when the rover woke up and started its daily operations, the electronics components started to warm up. Since lunar dust is a great insulator, the electronics temperature increased until the rover eventually failed (Chaikin 2004).

Astronauts also noted that some latching mechanisms, such as Velcro, did not work after they were coated with dust.

Lunar regolith is unlike anything we have here on Earth. It is highly abrasive (particles can scratch hard materials), adhesive (particles stick and coat materials), and cohesive (particles stick to each other) due to the way it was formed and the lunar environment (vacuum, radiation). On Earth, soil particles tend to be rounded off due to the weathering actions of the wind and water. Since the Moon doesn't have any of these, lunar particles are extremely jagged. Lunar regolith also consists of particles we call agglutinates (Fig. 10.3). These were formed during meteorite impacts and consist of fused glass and rock fragments. In some matured soils, agglutinates may make up as much as 60% of the soil by volume (Heiken et al. 1991). Agglutinates, because of their highly jagged nature and small size are extremely abrasive and can penetrate and coat mechanisms and seals causing them to malfunction.

Hence, in order to survive on the surface of the Moon, the hardware has to be dust tolerant. There are groups within US that are looking at ways to accurately reproduce lunar regolith and to develop dust tolerant technologies. NASA Marshall Space Flight Center (MSFC), together with United States Geological Survey (USGS) has been developing highlands regolith simulant (NU-LHR-XX), while Orbital Technologies Corporation (Orbitec) has been developing Mare regolith simulant (JSC-1a). Other countries involved in lunar exploration are also developing their own simulant. For example, in Canada, NORCAT developed OB-1 and CHENOBI, while in Japan, JAXA with Shimizu developed FJS-1. Testing lunar-focused hardware in these simulants and also in vacuum chambers is of paramount importance. For example, vacuum testing of dust tolerant electrical and mechanical connectors developed by Honeybee Robotics for Lunar Surface Systems showed significant increase in friction and abrasion compared to the ambient testing at 760 torr (or 1 atmosphere) (Herman et al. 2011).

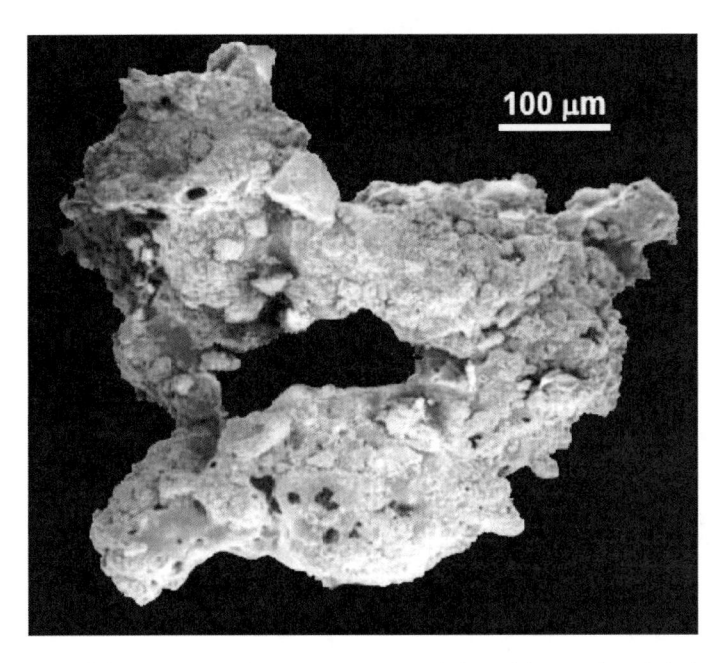

Fig. 10.3 Lunar agglutinates are glass-welded particles produced during impacts of meteoroids. Image credit: David McKay, NASA/JSC.

10.2.5 One Sixth the Gravity of Earth

Gravity on the Moon is one sixth of that on Earth and that means that everything weighs six times less. For example, a 60 kg person, weighing 600 N on Earth, would weigh only 100 N on the Moon. Alternatively, to apply the same vertical force, the mass of an object on the Moon has to be six times greater. This large reduction in gravitational acceleration has far reaching consequences on the choice of excavation methods.

Terrestrial excavators are massive and use brute force to rip through the ground. On the Moon, to generate the same excavation forces, the excavator would have to be up to six times heavier. The size of the excavation blade actually makes a huge difference to the gravity scaling factor, hence, the forces scaling is not simple 1:6 (Zacny et al. 2010b). Bringing a very heavy excavator to the Moon, not only will be very expensive, but may also not be feasible. This is because we no longer have a rocket capable of launching large payloads. The launch capacity of the largest available rocket, Delta IV Heavy, is four times lower than that of the Apollo Saturn V during the 1960s and 1970s.

10.3 Space Heritage: Examples of Excavation Systems Deployed on the Moon

The space heritage of excavation systems that were deployed on the Moon is quite limited and dates back to the late 1960s and early 1970s. The excavation tools included scoops, drills, rakes, soil samplers. Scoops are very useful for collecting loose soil but have an inherent drawback in that they cannot dig into hard rocks, and struggle in highly compacted soils. Drills, on the other hand, can penetrate even the hardest rocks known to man; however, they may require additional sample collection and transfer mechanisms and are not suitable to collect large quantities of material.

10.3.1 Lunar Scoops

The first scoops deployed on the Moon were onboard Lunar Surveyor landers. Surveyor landers were launched to the Moon between 1966 and 1968 as precursors to the Apollo missions and to demonstrate the feasibility of soft landings on the Moon (Fig. 10.4). Of the seven Surveyor crafts, only two failed: Surveyor 2 crashed and contact with Surveyor 4 was lost.

Fig. 10.4 Apollo 12 landed near the Surveyor 3 landing site. The scoop is in front of the astronaut, while the Apollo 12 Lunar Module is in the background. Credit: NASA, Kennedy Space Center.

All seven Surveyor spacecrafts are still on the Moon, though some parts (including a camera) of Surveyor 3 were returned to Earth by the Apollo 12 astronauts.

The purpose of a Surveyor scoop shown in Fig. 10.5 was to perform geotechnical experiments on the surface of the Moon with the goal of measuring bearing strength of the top soil and also to determine the related excavation forces. The scoop was approximately 12 cm long and 5 cm wide and consisted of a container and a sharpened blade. The scoop could hold a maximum of 100 cm³ of granular material. It was mounted on a pantograph arm that could extend 1.5 m or retracted close to the spacecraft motor drive. The arm could move from an azimuth of +40° to -72° and be elevated 13 cm using an electric motor. The scoop also had a flat footpad for bearings tests. The scoop on the Surveyor 3 performed seven bearing tests, four trench tests, and 13 impact tests. Only Surveyor 3 and 7 directly measured actual digging and trenching forces. By pulling the scoop toward the spacecraft while pushing down to dig trenches, excavation forces were measured.

Fig. 10.5 Surveyor 3 scoop, photographed by the Apollo 12 astronauts. Courtesy NASA.

Recently a number of bearing tests were conducted on JSC-1a lunar soil simulant with a replica of the returned Surveyor 3 excavation scoop (Zeng et al. 2010). In particular, tests conducted in the loosest JSC-1a beds and the data from the Surveyor 7 Bearing Test #2 seem to be in agreement (Bucek et al. 2008). This suggests that the Surveyor 7 bearing tests were probably conducted in low density or loose regolith. Examination of the lunar surface by Apollo astronauts showed that the top layer of the lunar regolith is quite loose. From the excavation stand point, it therefore makes sense to excavate the top loose layer rather than try to dig deeper into hard compacted regolith.

10.3.2 Lunar Drills

Just two countries deployed drills on the Moon; these were the United States and the Soviet Union.

The US Apollo program used an astronaut operated rotary percussive drill, called the Apollo Lunar Surface Drill (ALSD), to acquire a continuous regolith core from up to 3 meters and also to deploy two heat flow probes to a depth of 2.4

meters (Fig. 10.6). The drill was manufactured by Martin Marietta (now Lockheed Martin), and was one of the first battery powered drills. The motor driving the drill operated on 23 Volt and 430 Watt. It rotated the drill at 280 revolutions per minute. The hammer systems impacted at 2270 blows per minute, (i.e. 8.1 blows per revolution) and delivered an impact at energy per blow of 4.4 Joules. Only the last three missions, Apollo 15, 16 and 17, used the ALSD.

Fig. 10.6 Apollo astronaut practices drilling using the Apollo Lunar Surface Drill. Image courtesy: NASA

The entire ALSD drill package weighed 13.4 kg, not including the drill string and caps. Drilling a hole was actually relatively easy and took on average between 5 to 15 minutes per hole. The individual drill stems came in short (and manageable) sections and were screwed to each other to drill deeper.

Each of the Apollo 15-17 missions drilled three holes: two for the two heat flow probes and one for a regolith core. The heat flow probe hollow stems (or rather casings) were made of a boron fiberglass material to reduce thermal conductivity and used a full faced bit. Once the target depth was reached, the heat flow probe was manually lowered into the top ended, hollow composite stem. The stem for acquisition of a core had a core bit at the end and allowed acquisition of a continuous regolith core. Once a target depth was reached, the drill string would be pulled out of the hole, individual stems unscrewed and capped at both ends, and stored. In this way the regolith core was preserved in each stem.

During the first drilling mission, i.e. Apollo 15, problems were encountered both with a heat flow probe drilling and a core drilling. In particular, the composite casings lacked auger flutes at the stem joints, which prevent clearing of dense regolith. Both casings essentially ceased to penetrate at just over 1 meter depth. The stem for Apollo 16 and 17 were redesigned to include a continuous auger and allow clearing the dense soil from the hole.

Drilling a 3 meter core was relatively easy but the drill stem was hard to remove from the hole. The core drill was actually left in a hole while the other tasks were completed. At the end of the second EVA both astronauts had to work at the limit of their strength to pull the drill stem out of the hole. Scott during this time sprained his shoulder. The solution on Apollo 16 and 17 was to use a jack for removal of drill stems. The Apollo 16 and 17 crews had little difficulty in drilling to ~ 3 m depth and extracting the deep core. In addition, a rack was supplied which held the drill stems for the heat flow probe off the ground and made them easy to reach by a suited astronaut (Fig. 10.7). The core stems for the deep core were stored on the hand tool carrier on the back of the Lunar Roving Vehicle.

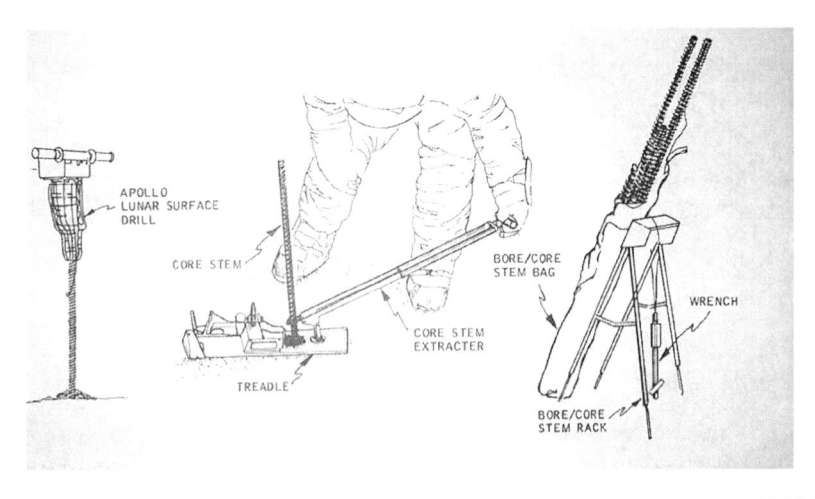

Fig. 10.7 Apollo Lunar Surface Boring and Coring Hardware. Image courtesy NASA.

The three Soviet missions, Luna 16, 20 and 24 performed a feat that has not been repeated so far, either. All three missions bore into subsurface, acquired regolith cores and returned them to Earth. And all this was done in the early 1970s (Johnson 1979).

The Luna 16 drill penetrated to a depth of 35 cm before encountering probably a hard rock or large fragments of rock. The column of regolith in the drill tube was then transferred to the soil sample container, and hermetically sealed. The 101 grams of collected material was returned to Earth on 24 September 1970. Luna 20 returned 30 grams of similarly collected samples in 1972.

In 1976 Luna 24 was the last of the Luna series of spacecraft and the third mission to retrieve lunar soil samples. However, instead of a short, arm deployable drill deployed on the Luna 16 and 20, it had a 2 meter drill. The mission success-fully collected 170.1 grams of lunar samples from a depth of up to 1.6m and deposited them into a collection capsule.

10.4 Examples of Various Lunar Excavation Technologies

Over the years there have been dozens of technology development projects related to planetary drilling and excavation (Mueller and van Susante 2011; van Susante and Dreyer 2010; Zacny and Bar-Cohen 2010c; Bar-Cohen and Zacny 2009; Zacny et al. 2008a; Bernold 1993). Many of the systems utilize approaches employed in the terrestrial mining and drilling applications but have been scaled down and redesigned for tackling harsh lunar conditions. Figure 10.8, for example, shows some of the excavators developed over the years at the Colorado School of Mines (van Susante and Dreyer 2010). The sections that follow include examples of not only more conventional approaches but also some radical designs that for various reasons are quite well suited for lunar environment.

Bucket Wheel   **Backhoe**   **Bucket Ladder**

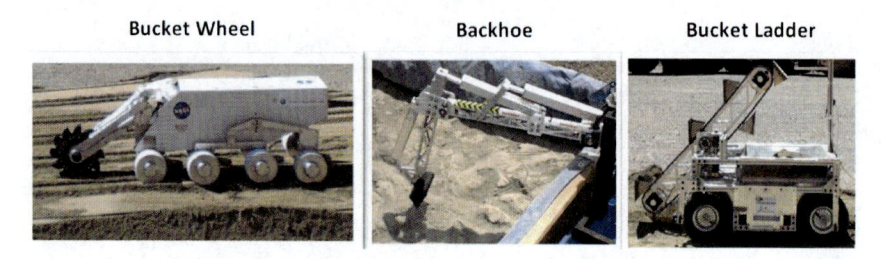

Fig. 10.8 Example of various types of excavators developed at the Colorado School of Mines. Photo courtesy Paul van Susante.

10.4.1 Lunar Drill: The MoonBreaker

Recent discovery of large deposits of water-ice makes lunar exploration even more enticing. This water could be mined and split into Oxygen and Hydrogen and used as rocket propellant for journey back home, halving the costs of lunar expeditions. Lunar propellant could also refuel the International Space Station, Earth orbiting satellites and planetary spacecraft. This refueling approach, because of the very shallow lunar gravity well, is more cost effective than sending the fuel from Earth.

However, before sending larger In Situ Resource Utilization plants to the Moon for processing of water ice and lunar regolith, we first need to send smaller and less expensive reconnaissance missions to prospect on the ground and confirm orbital findings. Such missions would use drills to penetrate at least one meter below the surface and acquire samples for analysis. There are at least two lunar development projects with a goal of building a drill for groundtruthing on the Moon. The first one includes a core drill developed by Northern Centre for Advanced Technology Inc. (NORCAT) and is funded by the Canadian Space Agency. The second one, called the MoonBreaker, is being developed by Honeybee Robotics and is funded be NASA.

The MoonBreaker, pictured in Fig. 10.9, is a rotary-percussive system, similar to a very successful Apollo Lunar Surface Drill. The main difference is that the MoonBreaker requires less power and does not need the drill head to be contained within a pressure vessel. The ALSD rotary-percussive drill head was enclosed within a pressurized container, filled with Nitrogen gas. The gas helped with heat dissipation and lubrication of moving parts. The MoonBreaker drill solved these problems by selecting different materials (not available in the 1960s when the Apollo drill was built) and the use of dry lubricants.

This state of the art drill does not acquire a core. Instead, the cuttings are being moved up the auger flutes all the way to a sampling system where they fall into a cup or an inlet to an instrument (Fig. 10.10). Although a core is more scientifically interesting than cuttings, a system that acquires cuttings is robotically much simpler and hence more reliable.

The Apollo drill acquired long regolith cores but these cores were not analyzed on the Moon. Individual drill stems were capped at both ends to prevent regolith loss and brought back to Earth where they were manually opened and carefully analyzed.

Of note is also the Mars Science Laboratory drill to be flown to Mars in November of 2011. The MSL was initially base-lined with a core drill. However, engineers soon realized the complexity of handling a core on Mars and the added complexity of crushing a core in-situ and delivering of samples of correct particles sizes to the onboard instruments. Hence, the core drill was descoped and substituted by a drill that acquires drilled cuttings for analysis. The drill is now called the Powder Acquisition Drill System (Okon 2010).

Fig. 10.9 The Honeybee Robotics MoonBreaker Rotary-Percussive drill undergoing testing in the Antarctic lunar analog site.

Fig. 10.10 Sampling system allows bringing up of regolith samples directly to a cup or analysis instrument.

The MoonBreaker drill has been tested in a large vacuum chamber in various lunar analog simulants, such as JSC-1a (Bucek et al. 2008), as well as in the Antarctic lunar analog (Paulsen et al. 2011).

There are advantages and disadvantages to testing in both locations. Testing in a vacuum chamber, for the most part is very convenient because it is done in a laboratory and hence if something goes wrong a quick fix can be made. In addition, conditions within a chamber can be continually regulated and monitored. During the chamber tests, a sample of JSC-1a was saturated with water and then frozen to $-80°C$ by running liquid nitrogen in the cooling coils. The strength of such a frozen soil is around 50 MPa, i.e. same as a strength of sandstone or limestone rock (Zacny et al. 2006). Drilling data revealed that the MoonBreaker managed to penetrate such a formation down to 1 meter depth in approximately 1 hour, with around 100 Watts of power and at 100 Newton Weight on Bit (WOB). Hence, the drill for a first time demonstrated penetration at 1-1-100-100 level (1m in 1hr with 100W and 100N). Since energy is proportional to power and time, drilling to 1 meter depth required 100 Whr – that is as much as burning a 100 Watt bulb for just one hour. The main achievement, however, was drilling at very low WOB. This term indicates the force that is required for the drill to actually penetrate a rock. Naturally, if a rock is harder; one has to push much harder than if a rock is softer. On Earth this is actually taken for granted since the gravity is relatively high. However, on the Moon, the same drill platform would be able to apply a force that is six times lower. For example if a drilling platform on Earth weighs 60kg, we can apply (in theory) 600 Newton WOB. On the Moon, the same drilling platform would be able to apply only 100 Newton – six times less. This is quite significant. The fix to this problem is to use a percussive drill (as opposed to a purely rotary drill). A percussive drill not only rotates but also hammers and it is this hammering action that chips a rock or breaks up soil. The main disadvantage of a percussive drill is that it requires more power than a purely rotary drill, because of the additional actuator required to drive the percussive mechanism. However, because percussive drill penetrates formation much faster than rotary drill, it is a much more efficient.

To illustrate this point, let's consider two drills: one is purely rotary and requires 20 Watts to drill 1 meter in 10 hours, and another one, a rotary-percussive that needs 100 Watts to drill 1 meter in 1 hour. Since energy is a product of power and time, for the rotary drill, the energy to drill to 1 meter will be 200 Whr (20 Watt x 10 hours) while for the rotary percussive the corresponding energy would be 100 Whr (100 Watts x 1 hr).

The Antarctic lunar analog has many disadvantages mainly related to logistics. Firstly, one needs a permit to get there, it is expensive to get there, and if something goes wrong, unless it is a simple fix, a spare part has to be shipped in and this takes a long time and most probably will never arrive in time. However, what the Antarctic environment offers is cold environment (both frozen ground and cold atmosphere) as well as geological uncertainty. The cold means that not only the drilled formation is frozen but also the drill itself is cold soaked and hence it can be tested for survivability in cold environments. The geological uncertainly means that no one really knows what the drill will be penetrating. Unlike in a laboratory, where a sample can be carefully prepared to either avoid any rocks or to add known number of rocks, ice etc., in the Antarctic the drill has to be able to penetrate what the Mother Nature created. This in a way is one of the best environments to test drilling algorithms.

The MoonBreaker drill was deployed on the slopes of Mount Erebus, still an active volcano. The formation was made of re-worked glacial deposits and tephra (cinder and ash) and hence analogous to the lunar regolith (Fig. 10.9). The ice cemented ground was found at the depth of around 15-20 cm. The MoonBreaker performed flawlessly and acquired samples from 1 meter depth in one hour (Paulsen et al. 2011). The power was just over 100 Watt (at 130 Watt) but the WOB was much less than 100 Newton (i.e. 80 Newton). Hence, the MoonBreaker again demonstrated drilling at the 1-1-100-100 level.

A danger of drilling in icy formations is that ice could locally thaw around a warmer drill bit and soon re-freeze around the bit trapping it in a hole. If this occurs, the drill is lost forever since trying to pull it out would most probably break the drill rather than free it from the frozen ground. For that very reason, the MoonBreaker has a temperature sensor embedded inside a bit. This sensor monitors the temperature during the drilling process and outputs the data to the drilling algorithm. If the bit temperature increases too much, the drill algorithm slows down the drill or in the extreme case pulls the drill entirely out of the hole and lets the drill and the ground cool off.

Probably the most exciting experience occurred when the drill was teleoperated from California, in the same way it could one day be teleoperated on the Moon, from Earth. The only thing missing was a two to three second communication delay. The MoonBreaker team assembled the drill in the Antarctic and connected it to an internet. On the other side of the globe, a class of 5th graders assembled at the Valley View School, in Pleasanton, CA. The class was initially briefed about the drilling procedures and then they set off to run the drill entirely by themselves. About ten minutes later, the drill penetrated first 10 centimeters and acquires a first sample. The drilling progress continued this way until a target depth of 30 cm was reached (Zacny 2011). This demonstration was not only very exciting for both the team in the Antarctica and also in the school in California, but also showed that teleoperating a drill (even by 5th graders) is feasible. Who knows, maybe one these kids will one day operate a real lunar drill on the Moon?

10.4.2 Lunar Pneumatic Excavator

The Soviet Lunokhod 2 mission ended because of lunar dust (Chaikin 2004). During one of the traverses, the Lunokhod 2 descended into a narrow crater, some 15 feet wide. During this traverse, the lid touched the crater wall and scraped some of the lunar dust onto it. When the lid was closed for the lunar night, this dust fell onto the radiator, which was designed to get rid of heat during the day. The lunar dust, combined with space vacuum forms a great insulator. When the day arrived and the rover began its operation, the temperature aboard Lunokhod 2 started to increase until the electronics overheated and the rover ceased to operate.

The lunar dust also caused a lot of problems to the Apollo astronauts and the hardware deployed during the Apollo missions. It was not a question if but when something would break. Lunar soil consists of very abrasive particles and a large fraction of lunar soil is so small that it can be considered dust. These tiny abrasive particles can penetrate any seal and mechanism and cover and stick to all surfaces because of electrostatic. Hence, mechanisms operating on the surface of the Moon have to be dust tolerant or somehow sealed from dust for the duration of the mission. Designing and building such a system is not only difficult to do but also expensive and does not guarantee success (Herman et al. 2011).

On Earth, vacuum cleaners work in dusty environments and they do just fine. They are actually designed to collect dirt and dust. The reason for their successful operation lies in the fact they use air and not a mechanical system, to move particles across large distances (from the nozzle to a collection bin). There are no moving parts, no seals, no bushings involved – just a tube with an open inlet on one side and a bag on the other side. Vacuum cleaners of course work on the principle of pressure difference. Suction at the bag side creates low pressure. Hence the air from the outside is 'sucked' through the nozzle and into the hose, and eventually exits the exhaust port on the other side of the bag. This air picks up dust and dirt, carries it through the tube and dumps it into a collection bag. Of course a vacuum cleaner as is will not work on the Moon, since there is no atmosphere on the Moon to 'suck in' but rather an extremely high vacuum of 10^{-12} Torr. However, instead of trying to create suction at the nozzle inlet, why not create a high pressure? If a nozzle is sufficiently well sealed, injected gas will accelerate the soil particles along the tube and move them to a collection bin (Zacny et al. 2004). A cyclone separator like the ones commonly used on Earth would easily separate particles from carrier gas (Zacny et al. 2008b). An excavator that operates on such a principle was built and tested in GRC-1 lunar soil simulant (Oravec 2009) and inside a vacuum chamber. The pneumatic excavation hardware was build on top of the NASA Ames Research Center K10 mini rover.

Initially, though, during the development stage, a number of nozzles were developed to identify the most significant nozzle characteristics. The nozzle actually is the most critical component. The purpose of the nozzle was to acquire and retain soil and assist with moving a part of it up the tube and into the cyclone via injected gas. The idea was for the remaining soil at the mouth of the nozzle to act as a plug and prevent the gas from escaping into the outside. The final nozzle configuration worked very well and little or no gas was lost to the outside (Fig. 10.11). The cyclone separator mounted above the soil bin was used to separate the gas from soil particles. Soil would drop to the bin underneath while gas would be vented into the chamber, above. This gas, however, could be captured and recycled back into the system making this approach even more efficient. The main advantage of this method of sampling is that there are virtually no moving parts. A deployment arm just needs to lower the nozzle into the soil, and the remaining action of collecting the soil is done with gas, which is inherently dust tolerant.

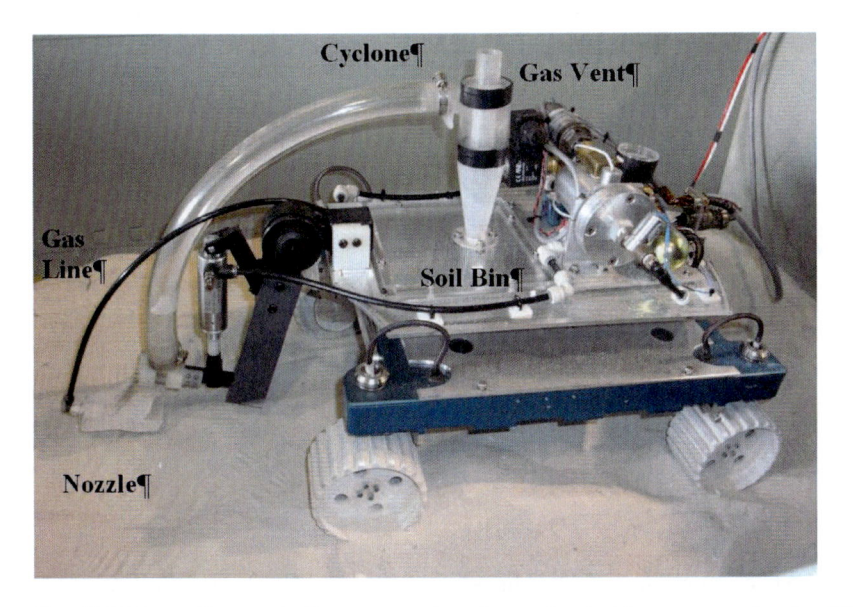

Fig. 10.11 Lunar Pneumatic Excavator undergoing testing inside a vacuum chamber (Zacny et al. 2008b)

The efficiency of the pneumatic system was also tested during microgravity flights. The experiment was designed to measure the mass of soil particles lofted as a function of gas pressure and mass of gas. A large vacuum chamber was mounted on the inside of the airplane to enable testing at both, vacuum and lunar gravity. The test data revealed that just one gram of gas at 7 psia (i.e. 50% of the atmospheric pressure) could loft 6000 grams of JSC-1a lunar soil simulant at speeds of around 5 m/s (Zacny et al. 2010a). Thus, with 1 kg of gas at 7 psia, around 6 tons of soil can be mined on the Moon. These high efficiencies are mainly attributed to the presence of vacuum rather than lower gravity. In vacuum, gas exit velocity is proportional to a pressure ratio between the pressure in a cylinder and pressure on the outside. If the pressure on the outside is virtually zero (i.e. vacuum), the exit velocity reaches a choking velocity. Since gas momentum is proportional to velocity, the high momentum the gas particles achieve while exiting a pressure tank into vacuum is exchanged with soil particles, which are literally being propelled away.

The benefit of pressurized gas in vacuum has been already applied to rocket thrusters. A number of cold gas thrusters are replacing conventional hydrazine thrusters for attitude control, for example. A cold gas thruster is a rocket engine/thruster that uses a (typically inert) gas as the reaction mass. Cold gas thrusters are used mainly due to their simplicity and reliability; they consist of a pressurized gas tank, a valve, and a nozzle, and plumbing connecting these components. Hence there is not much that can fail.

In 1993, Sullivan evaluated the feasibility of pneumatic transfer for the movement of regolith (lunar soil) at a lunar base at lunar gravity conditions on NASA's KC-135 aircraft. He found that the choking velocity (in the vertical transfer) and the saltation velocity (in the horizontal transfer) at lunar gravity were reduced to only 1/2-1/3 of the velocity required at 1 g (Sullivan et al. 1993). Choke and saltation velocities are minimum gas velocities required to keep particles aloft. Hence this experiment conducted at atmospheric pressure showed that gravity does not have as much of an effect as vacuum does.

Although gas can be viewed as consumable, there are a number of potentially large sources of gas on the Moon making this approach quite sustainable. For example, landers use high pressure pressurant such as Helium to pressurize propellant tanks. Helium is kept at pressures in excess of 1000 psi. This gas is vented after touch down to prevent propellant from leaking out, but instead it could be used for pneumatic excavation. In addition, propellant itself is a good source of gas. Propellant can be burned in a small rocket thruster and exhaust gases could be used for pneumatic mining system. In addition, there are a number of by-products of In Situ Resource Utilization reactions that would be vented, and these could be captured and used for mining instead. Hence, this particular approach would be not only effective but also sustainable.

10.4.3 Lunar Percussive Excavator

Terrestrial excavators, such as backhoes, use brute force to dig up soil with larger excavators used for harder soils and even rocks, while smaller excavators used for softer soils (Bernold 1993). However, this approach would not be very efficient on the Moon having gravity six times lower than on Earth. Bringing large excavators will be prohibitively expensive. An alternative to the use of brute force is to use percussive systems. Percussive diggers can excavate soils by applying up to 95% less force (Zacny et al. 2008c). This translates into large reduction in the mass of the excavator. Others have also found that vibrating of bulldozer blades helped to reduce draft force up to 71%-93% (Shabo et al. 1998).

The principle behind force reduction is in fact that percussive systems tend to vibrate the soil that is in contact with the vibrating blade. This essentially reduces friction between the blade and the soil particles. In addition, vibrating soil particles have lower density and in turn can rearrange themselves easier than compacted particles. The density effect is quite pronounced.

Lin et al. (1994), for example conducted a number of digging tests in soils of various densities. He used small explosives to loosen up soil. He found that specific excavation energy in soils at relative density of 80% loosened by explosions was 40% lower than in soils at 100% relative density (Lin et al. 1994). Relative density is a term to describe the extent of soil density. At a relative density of 0%, soil is at its loosest state, while at the relative density of 100%, soil is fully compacted.

As with any approach, there are advantages and disadvantages. The advantage of the percussive system is clear – lower excavation forces, lower mass of an excavator, lower mass at launch, smaller rocket, less expensive rocket, less expensive mission. The main disadvantage of a percussive or vibratory excavator is that it requires additional power to run the percussive motor and that it needs another mechanism (i.e. percussive mechanism) which is just one more thing that can break. However, given the fact that solar energy can be harvested using solar panels for very low mass, in a tradeoff between lower mass or high power, lower mass will always win. To deal with a potential risk of losing a percussive actuator, it might be worthwhile to have a redundant system (i.e. fly two actuators with the second one taking over once the primary fails).

10.5 Lunar Excavation Competitions

Competition drives innovation. In most cases the competition is driven by geopolitical environment, for example wars. Some major technological developments occurred during Cold War and World War II. During WWII we saw rapid developments of rocketry and invention of radars. If it wasn't for the Cold War, we may have never set foot on the Moon. Of course a war, in itself is detrimental to everyone and causes great losses to everyone involved. But is it possible to bring just the competitive aspect and leave everything else out? Well, the answer is yes.

One of the early competitions with prize money was called the Longitude Prize and was offered by British government in 1714. The prize money of 10,000 British Pounds was to be awarded to a first person who could devise a method of determining ship's longitude within 60 nautical miles (111 km). This prize helped in development of chronometers, i.e. a very precise clock. Knowing the time at a known fixed location, such as Greenwich Mean Time (GMT) in England, and at a local noon (for example in the middle of the Atlantic Ocean) allowed calculating of the time difference between the two locations. Since the Earth rotational period is 24 hours, the time difference could then be used to calculate the longitude of the ship's position in relation to the Greenwich Meridian ($0°$ longitude) using spherical trigonometry. Precise chronometers allowed navigation at sea and in turn helped with the discoveries of new lands.

In 1919, New York hotel owner Raymond Orteig established the Orteig Prize of $25,000 to the first aviator to fly non-stop from New York City to Paris or from Paris to New York. His business incentive was to increase the number of tourists. Eight years later, Charles Lindbergh became a national hero for winning the prize in his aircraft Spirit of St. Louis. Public interest in air travel skyrocketed and the rest is history.

In 1996 the "X Prize" was established. The name was changed to "Ansari X Prize" in 2004 after a multi-million dollar donation from entrepreneurs Anousheh Ansari and Amir Ansari. The prize of $10,000,000 was offered to the first non-government organization to launch a reusable manned spacecraft into space (defined at above 100,000 km altitude) twice within a two week period. The prize was eventually won by the Burt Rutan and Paul Allen team in October of 2004 though the first successful launch was achieved on June 21, 2004. The spaceplane was called SpaceShipOne and it was carried to a high altitude by its mother ship White Knight before being released and firing its own rocket engine. Although the team won $10 million, the total investment in this venture is estimated to be an order of magnitude more. As a result of this success, Richard Branson created Virgin Galactic with a goal to provide suborbital spaceflights to the public, for various space science missions, and to provide orbital launches of small satellites. Current cost per person is $200,000.

10.5.1 NASA Regolith Excavation Challenge

In 2003, NASA followed this path of inventions development through competitions by establishing the prize contests called Centennial Challenges (named in honor of the 100 years since the Wright brothers' first flight). One of the first challenges included the Regolith Excavation Challenge. The goal of that challenge was to encourage development of new lunar regolith excavation technologies, required for sustainable human presence on the Moon. Mining of the lunar regolith is a first step in the multi-stage process of In Situ Resource Utilization (ISRU). The goal of ISRU is to process lunar regolith and water-ice into elements that can be used to sustain human presence on the Moon and make planetary exploration more affordable.

A number of teams consisting of professional engineers with tens of years of aerospace and robotics experience, university students, and even high school students entered the competition. To win the prize, the robotic diggers had to navigate around rocks, collect at least 150 kg of lunar regolith simulant (JSC-1a), and deliver it to a collection bin within a 30-minute period. To make things even more challenging, the operation of the robots was made more difficult by imposing a two second communication delay. The operators thus could have the first hand experience of how the operation of the mining robots would take place on a far away lunar surface.

After two initial attempts whereby no team managed to win the prize, finally in 2009, three teams succeeded in winning the first ($500,000), second ($150,000), and third ($100,000) prizes. Paul's Robotics of the Worcester Polytechnic Institute took home the top prize by beating 22 other teams (Fig. 10.12). The excavator, called the Moonraker, managed to collect and dump 439 kg of regolith. The team used a number of cameras placed in strategic locations to navigate around the soil bin and to dump the regolith into a collection bin. A few days before the competition, the team performed a dress rehearsal in sand and noticed that a process of mining creates a lot of dust that obscures view from cameras. Hence to aid in their navigation, the team installed four blue LEDs in each of the four corners of the miner. These four LEDs proved indispensable during the actual competition where

lunar-like regolith rather than beach sand was used (Fig. 10.13). Since lunar regolith contains a large fraction of small, micron and sub-micron size particles, it
creates a lot of dust.

Fig. 10.12 Paul's Robotics lunar excavator competes at the NASA Regolith Excavation Challenge on October 18th, 2009. The team won the first prize of $500,000. Photo courtesy K. Zacny.

Fig. 10.13 Paul's Robotics lunar excavator ready for its first dump. During the 30 minute
period, teleoperated excavator, called the Moonraker, collected and dumped 439 of regolith.
The four blue LEDs were installed a few days before the competition to aid in navigation.
Photo courtesy K. Zacny.

It should be noted that some of the mining approaches employed during the competition (e.g. bucket ladder system) are not suitable for dusty lunar environment. It would be just a matter of time before some of the mechanisms would essentially clog with dust and cease to operate.

From the business perspective, the competition generated an immense pay-off to the field of planetary robotics. Assuming that each robotic excavator would cost $500,000 to build by independent contractors, the total financial investment over the course of three years could be valued at over $10 million dollars. That's at least ten times less than the total cash prize.

10.5.2 Lunabotics Mining Competition at NASA Kennedy Space Center

Inspired by the success of the NASA Centennial Challenges Regolith Excavation Competition, Robert Mueller of NASA Kennedy Space Center masterminded the Lunabotics Mining Competition at NASA Kennedy Space Center (Lunabotics 2011). The Lunabotics Mining Competition is tailored for university teams across the world with an overall aim to spur student's interest in science, technology, engineering and mathematics (STEM). The first competition held in 2010 attracted 30 university American teams and provided wealth of innovative lunar excavation ideas and concepts; some of them may one day be applied to actual lunar excavators (Fig. 10.14).

Fig. 10.14 In 2010 NASA Kennedy Space Center Lunabotics competition attracted over 30 university teams. Shown is an excavator built by Georgia Tech team. Photo courtesy K. Zacny.

The 2010 competition required students to design and build a mining system that would collect a minimum of 10 kg of lunar simulant, Black Point -1 (BP-1), and dump it into a collection bin. This had to be achieved by remote operation or full autonomy in a 15 minute period. The competition was so successful, that the 2nd Lunabotics competition had to be moved to a larger venue to accommodate over 60 US and international teams. Although the competition prizes are much smaller than what the NASA Regolith Excavation Centennial Challenge offered, the Lunabotics Mining Competition is not about the money. It is about the innovation, and teamwork in the friendly and yet competitive environment. In fact the teams, even though they competed against each other, helped one another with advice, spare parts and labor. This is an example of a win-win competition for all. Students get hands-on experience by applying their innovative ideas to solving practical problems and NASA receives dozens of tested concepts it can choose from once the lunar landing gets a go ahead.

10.5.3 *Google Lunar X Prize*

Currently, a number of teams are competing fearlessly to win the $30 million Google Lunar X Prize. A prize will be given to the first privately funded team to land a robot on the surface of the Moon, traverse 500 meters and send images and data back to the Earth (XPRIZE 2011). One of the teams competing for the prize is Astrobotic, a spin-off from Carnegie Mellon University in Pittsburg, PA.

Fig. 10.15 Astrobotic lunar rover with Honeybee Robotics MoonBreaker drill

The Astrobotic rover is actually large enough to be able to carry and deploy a number of payloads, of which the most enticing is the drill. Hence, the Honeybee Robotics lunar drill, the MoonBreaker has been redesigned to fit the rover (Fig. 10.15). The drill is scheduled to fly to the Moon in 2014 and this will be the first lunar drilling missions since the Soviet Luna 24 sample return mission in 1976.

10.6 Final Thoughts

Subsurface exploration of the Moon is a natural step in the exploration process. Apart from the science driven questions, there is a large impetus for utilizing lunar resource in support of future human and robotic exploration not only of the Moon, but also other Solar System bodies. Recent discovery of large water deposits makes the utilization of the Moon even more enticing. Water could be mined, processed and split into oxygen and hydrogen. These two elements could be used as LOX/LH2 fuel for rocket propulsion or in fuel cells to provide electricity. The fuel could also be sent to the Earth orbit to refuel Earth orbiting satellites and International Space Station. Commercialization of In Situ resource Utilization technologies on the Moon would also enable establishment of lunar hotels for adventurous tourists and settlement.

The technology development enabling ISRU will, no doubt be expensive, and the lunar ISRU in itself is very risky. Hence, a partnership between a government space agencies and commercial sector is required to achieve the final goal, whether it be permanent human presence on the Moon, or lunar mining company, exploring local resources for supporting lunar habitats, solar system exploration or Earth orbiting satellites and space stations.

Acknowledgements. Some of the research reported in this Chapter was conducted by Honeybee Robotics Spacecraft Mechanisms Corporation under various contracts with National Aeronautics and Space Administration (NASA) and the Department of Defense (DoD). Information about Thermal Wadis was provided by Kurt Sacksteder of NASA Glenn Research Center and Robert Wegeng of Pacific Northwest National Laboratory. Robert Mueller of NASA Kennedy Space Center provided information about the NASA KSC Lunabotics Competition, while Paul van Susante of Colorado School of Mines provided information about the CSM excavator systems. The author would like to thank the reviewers of this chapter: Professor Leonhard Bernold of the University of News South Wales in Sydney, and Dr. Paul van Susante of Colorado School of Mines. Thanks must also go to my wife Izabella Zacny, for her help with editing of this chapter.

References

Astrobotic Inc. (2011), http://astrobotic.net/ (accessed March 17, 2011)

Balasubramaniam, R., Gokoglu, S., Sacksteder, K., Wegeng, R., Suzuki, N.: Analysis of Solar-Heated Thermal Wadis to Support Extended-Duration Lunar Exploration. J. Thermophysics and Heat Transfer 25(1), 130–139 (2011)

Bar-Cohen, Y., Zacny, K. (eds.): Drilling in Extreme Environments Penetration and Sampling on Earth and Other Planets. Wiley, New York (2009)

Bernold, L.: Motion and Path Control for Robotic Excavation. Journal of Aerospace Engineering 6(1), 1–18 (1993)

Bernold, L.: Compaction of Lunar-Type Soil. Journal of Aerospace Engineering 7(2), 175–187 (1994)

Blair, B., Zacharias, M.: The Value of Fuel Transfer to a Space Network. Presented at the NASA/USAF Advanced Space Propulsion Workshop (ASPW 2010), Colorado Springs, CO, November 15-17 (2010)

Bucek, M., Agui, J., Zeng, X., Wilkinson, R.A.: Experimental Measurements of Excavation Forces in Lunar Soil Test Beds. In: ASCE Conf. Proc. doi:10.1061/40988(323)5, Proc. of the 11th Int. Conf. on Engineering, Science, Construction, and Operations in Challenging Environments (2008)

Chandler, F., Bienhoff, D., Cronick, J., Grayson, G.: Propellant Depots for Earth Orbit and Lunar Exploration. In: AIAA SPACE 2007 Conference & Exposition, Long Beach, California, AIAA 2007-6081, September 18-20 (2007), http://pdf.aiaa.org/preview/CDReadyMSPACE07_1808/PV2007_6081.pdf

Colaprete, A., Schultz, P., Heldmann, J., Wooden, D., Shirley, M., Ennico, K., Hermalyn, B., Marshall, W., Ricco, A., Elphic, R.C., Goldstein, D., Summy, D., Bart, G.D., Asphaug, E., Korycansky, D., Landis, D., Sollitt, L.: Detection of Water in the LCROSS Ejecta Plume. Science 330(6003), 463–468 (2010)

Carter, G.A., McCain, D.C.: Relationship of leaf spectral reflectance to chloroplast water content determined using NMR microscopy. Remote Sensing of Environment 46(3), 305–310 (1993)

Chaikin, A.: The Other Moon Landings. Air & Space Magazine, 30–37 (February/March 2004)

Chang, K.: After 17 Years, a Glimpse of a Lunar Purchase. New York Times (2010), http://www.nytimes.com/2010/03/31/science/space/31moon.html?ref=science (retrieved March 1, 2011)

Gaier, J.: The Effects of Lunar Dust on EVA Systems During Apollo Missions. NASA TM-2005-213610/REV1 (2005)

Gibson, M., Knudsen, C.: Lunar Oxygen Production from Ilmenite. In: Mendell, W.W. (ed.) Lunar Bases and Space Activities of the 21st Century, Lunar and Planetary Institute, Houston, p. 543 (1985)

Heiken, G., Vaniman, D., French, B. (eds.): Lunar Sourcebook: A User's Guide to the Moon. Cambridge University Press (1991)

Herman, J., Sadick, S., Maksymuk, M., Chu, P., Carlson, L.: Dust-Tolerant Mechanism Design for Lunar & NEO Surface Systems. Paper #1401, IEEE Aerospace Conference, Big Sky, Montana, March 5-12 (2011)

Johnson, N.L.: Handbook of soviet lunar and planetary exploration. Science and technology series, vol. 47. Amer. Astronaut Soc. Publishing (1979)

Jones, H.L., Thornton, J.P., Balasubramaniam, R., Gokoglu, S., Sacksteder, K., Whittaker, W.L.: Enabling Long-Duration Lunar Equatorial Operations with Thermal Wadi Infrastructure. In: 49th AIAA Aerospace Sciences Meeting (January 2011)

Lin, C., Goodings, D., Bernold, L., Dick, R., Fourney, W.: Model Studies of Effects on Lunar Soil of Chemical Explosions. Journal of Geotechnical Engineering 120(10), 1684–1703 (1994)

Lunabotics Mining Competition, NASA Kennedy Space Center (2011), http://www.nasa.gov/offices/education/centers/kennedy/technology/lunabotics.html

Mueller, R.P., van Susante, P.A.: Review of Lunar Regolith Excavation Robotic Device Prototypes. In: AIAA Space 2011, Long Beach, CA, September 26-29 (2011)

Okon, A. (2010) Mars Science Laboratory Drill. In: Proc. 40th Aerospace Mechanism Symposium, NASA KSC, May 12-14 (2010)

Olvera-Cervantes, J., Cressler, J., Medina-Monroy, J.-L., Thrivikraman, T., Banerjee, B., Laskar, J.: A New Analytical Method for Robust Extraction of the Small-Signal Equivalent Circuit for SiGe HBTs Operating at Cryogenic Temperatures. IEEE Trans. Microwave Theory and Techniques 56, 568–574 (2008)

Oravec, H.: Understanding Mechanical Behavior of Lunar Soils for the Study of Vehicle Mobility. PhD Thesis, Case Western Reserve University, Civil Eng. (2009), http://etd.ohiolink.edu/view.cgi?acc_num=case1233521118

Paige, D.A., Siegler, M.A., Zhang, J.A., Hayne, P.O., Foote, E.J., Bennett, K.A., Vasavada, A.R., Greenhagen, B.T., Schofield, J.T., McCleese, D.J., Foote, M.C., De Jong, E.M., Bills, B.G., Hartford, W., Murray, B.C., Allen, C.C., Snook, K.J., Soderblom, L.A., Calcutt, S., Taylor, F.W., Bowles, N.E., Bandfield, J.L., Elphic, R.C., Ghent, R.R., Glotch, T.D., Wyatt, M.B., Lucey, P.G.: Diviner Lunar Radiometer Observations of Cold Traps in the Moon's South Polar Region. Science 330, 479–482 (2010)

Paulsen, G., Zacny, K., McKay, C., Glass, B., Szczesiak, M., Craft, J., Santoro, C., Shasho, J., Davila, A., Marinova, M., Pollard, W., Jackson, A.: Field Testing of the IceBreaker Mars Drill in the Antarctic. In: LPSC 2011, Abstract #190 (2011)

Sacksteder, K.R., Wegeng, R.S., Suzuki, N.H.: Lunar Prospecting Using Thermal Wadis and Compact Rovers Part A: Infrastructure for Surviving the Lunar Night. In: AIAA Space 2010 Conference (August 2010)

Shabo, B., Barnes, F., Sture, S., Ko, H.: Effectiveness of vibrating bulldozer and plow blades on draft force reduction. Trans. of the ASAE 41(2), 283–290 (1998)

Sullivan, T., Koenig, E., Knudsen, C., Gibson, M.: Pneumatic conveying of materials at partial gravity. J. of Aerospace Engineering 7(2), 199 (1994)

Thornton, J., Whittaker, W., Jones, H., Mackin, M., Barsa, R., Gump, D.: Thermal Strategies for Long Duration Mobile Lunar Surface Missions. In: 48th AIAA Aerospace Sciences Meeting (January 2010)

van Susante, P., Dreyer, C.: Lunar and Planetary Excavation Prototype Development and Testing at the Colorado School of Mines. In: ASCE Earth and Space 2010, Honolulu, HI, March 5-8 (2010)

Wagner, S.: The Apollo Experience Lessons Learned for Constellation Lunar Dust Management. NASA TP-2006-213726 (2006)

Wegeng, R.S., Mankins, J.C., Taylor, L.A., Sanders, G.B.: Thermal Energy Reservoirs from Processed Lunar Regolith. In: 5th Int. Energy Conv. Eng. Conf. (July 2007)

Zeng, X., He, C., Oravec, H., Wilkinson, A., Agui, J., Asnani, V.: Geotechnical Properties of JSC-1A Lunar Soil Simulant. J. Aerosp. Engrg. 23, 111 (2010)

Zacny, K., Huang, K., McGehee, M., Neugebauer, A., Park, S., Quayle, M., Sichel, R., Cooper, G.: Lunar Soil Extraction Using Flow of Gas. In: Proc. of Revolutionary Aerospace Systems Concepts - Academic Linkage (RASC-AL) Conference, Cocoa Beach, Florida, April 28-May 1 (2004)

Zacny, K., Glaser, D., Bartlett, P., Davis, K., Wilson, J.: Test Results of Core Drilling in Simulated Ice-Bound Lunar Regolith for the Subsurface Access System of the Construction & Resource Utilization eXplorer (CRUX) Project. In: 10th Int Conf on Engineering, Construction, and Operations in Challenging Environments, Earth & Space 2006 Conference, League City, TX, March 5-8 (2006)

Zacny, K., Bar-Cohen, Y., Brennan, M., Briggs, G., Cooper, G., Davis, K., Dolgin, B., Glaser, D., Glass, B., Gorevan, S., Guerrero, J., McKay, C., Paulsen, G., Stanley, S., Stoker, C.: Drilling Systems for Extraterrestrial Subsurface Exploration. Astrobiology Journal 8(3), 665–706 (2008a)

Zacny, K., Mungas, G., Mungas, C., Fisher, D., Hedlund, M.: Pneumatic Excavator and Regolith Transport System for Lunar ISRU and Construction. Paper No: AIAA-2008-7824 and Presentation, AIAA SPACE 2008 Conference & Exposition, San Diego, California, September 9-11 (2008)

Zacny, K., Craft, J., Wilson, J., Chu, P., Davis, K.: Percussive Digging Tool for Lunar Excavation and Mining Applications. Abstract 4046, LEAG-ICEUM-SRR, Cape Canaveral, FL, October 28-31 (2008c)

Zacny, K., Craft, J., Hedlund, M., Chu, P., Galloway, G., Mueller, R.: Investigating the Efficiency of Pneumatic Transfer of JSC-1a Lunar Regolith Simulant in Vacuum and Lunar Gravity During Parabolic Flights. In: AIAA Space 2010, AIAA-2010-8702, Anaheim, CA, August 31-September 2 (2010a)

Zacny, K., Mueller, R.P., Craft, J., Wilson, J., Hedlund, M., Cohen, J.: Five-Step Parametric Prediction and Optimization Tool for Lunar Surface Systems Excavation Tasks. In: ASCE Earth and Space, Honolulu HI, March 15-17 (2010b)

Zacny, K., Bar-Cohen, Y.: Drilling and excavation for construction and in situ resource utilization. In: Badescu, V. (ed.) Mars: Prospective Energy and Material Resources, ch. 14. Springer, Heidelberg (2010c)

Zacny, K.: IceBite Blog: Remote Control (2011),
http://www.astrobio.net/index.php?option=com_expedition&task=detail&id=3692 (accessed March 21, 2011)

XPRIZE, The Google Lunar X PRIZE (2011),
http://www.googlelunarxprize.org/ (accessed March 21, 2011)

11 Challenges in Transporting, Handling and Processing Regolith in the Lunar Environment

Otis Walton

Grainflow Dynamics, Livermore, CA, USA

11.1 Introduction

It is well known that powders become more 'cohesive' as their mean particulate size decreases. This phenomenon is evidenced by such characteristics as poor flowability, clumping, avalanching, difficulty in fluidizing, and formation of quasi-stable, low-density configurations that are easily compacted. Gravity is often the primary driving force for powder movement in common powder processing and transfer operations. Because of this, gravity plays a role in how the flow behavior of powders is typically characterized. As a result, the 'cohesiveness' of a powder varies with gravity-level, with a powder appearing more 'cohesive' as the effective gravity level is decreased. A factor of four reduction in gravity level is enough to dramatically change the apparent character of a powder from 'free-flowing' to 'cohesive' in the same size apparatus. The well-known Geldart classification of the fluidization characteristics of powders also changes at reduced gravity-levels, causing the effective boundary between Geldart-C (cohesive) and Geldart-A (aeratable) powders to shift to larger particle sizes under low gravity conditions. In addition, electrostatic effects influence material handling operations with fine particulates terrestrially, especially under dry conditions when the surrounding air is least conductive so that triboelectrically acquired charges decay very slowly. Both the reduced gravity and vacuum conditions on the moon raise electrostatic concerns to a much higher level. Triboelectric charging can occur just as on earth (dependent on the relative electron work functions of the contacting materials); however, in vacuum there is no conductive atmosphere to allow charge to dissipate. In addition, any surface exposed to the exterior lunar environment will acquire surface charge either directly from space (via photoelectric solar incident rays on the sunlit side, or from the solar plasma on the dark side) or indirectly from the ultra-low density plasma within a Debye length of the lunar surface (e.g., of photoelectrically released electrons on the sunlit side). The lack of an atmosphere also means that particle surfaces have never been exposed to an oxidizing gas and are likely to exhibit an effective surface energy which is significantly higher than the same minerals would exhibit terrestrially. Thus, it is nearly impossible to consider regolith handling operations without taking into account the potential for charged fine dust particles adhering to exposed surfaces. [A comment on notation, in this chapter g_o is meant to be the acceleration of gravity on the earth's surface, e.g. $g_o \sim 9.8$ m/s^2; while g-level, generally indicates a multiple of g_o.]

Most of the lunar surface is covered with regolith, a mixture of fine dust and rocky debris produced by meteor impacts and varying in thickness from about 5 m on mare surfaces to about 10 m on highland surfaces. The bulk of the regolith has been described as a fine gray soil with a bulk density of about 1.5 g/cm^3, but it also contains breccia and rock fragments from the local bed rock (Carrier et al. 1991, Taylor et al. 2005). The large number of very fine particles increases the surface area per unit mass, and thus the surface energy per unit mass available for cohesive forces to act in the bulk material. Also, the absence of air and water has allowed the fines to remain in the regolith as a greater percentage of the mass than would be typical of terrestrial geologic deposits. A typical granular material, like dry sand, has a void volume of around 40% (this is also referred to as having a solid-fraction of around 60%). Although the solids fraction of in situ Lunar regolith is relatively high (often 50% solids or greater) it is expected that, if this material were disturbed (as in ISRU operations) it could form low-density, stable configurations ('fluff') typical of any fine cohesive powder under 'sifted' conditions. Very fine (micron-scale) cohesive powders often exhibit bulk solids fractions that are as low as 15% and the pharmaceutical industry often uses the high void fraction of sifted fine powders as an indication of how cohesive a powder is. At the base of some Lunar crater-wall avalanches, where relatively recent flows have occurred (on a geologic time scale), very-low-density regolith might be expected in situ; however, such a low-density condition is generally only expected to be a likely state for lunar regolith after it has been excavated (and there has been speculation that some permanently shadowed craters might, also, contain such highly unconsolidated material).

The size distribution of a granular material is an important factor in determining its bulk physical behavior. Lunar regolith contains a significant portion of submicron particles (several percent of the total mass) and the distribution of fine particles extends all the way down to as small as tens of nanometers. Terrestrially, sand is a common granular material with which almost everyone of familiar. Typical sand deposits, however, have a relatively narrow size range from the smallest to the largest particles because the sand was transported or deposited, in the desert or on the beach, through the action of a fluid medium (i.e., wind and/or water). For example, desert sand is distinguished from dust at a particular location by the prevailing wind conditions – dust stays suspended in air long enough to be carried away with the wind. The sand is left behind (and is transported around, close to the desert surface, by a process known as saltation (Bagnold 1941). The phenomena which create sand deposits on earth do not exist on the moon. Likewise there is no liquid water, thus no living organisms are present to cause organic decomposition or to create soil on the moon. The mineral composition of lunar regolith is similar to some terrestrial volcanic ash deposits, but there are chemical differences (described elsewhere in this book) and the physical characteristics, size distribution, and mechanical properties of lunar regolith differ significantly from terrestrial fine silty soils which, at first glance, might appear to be most like regolith. Compared to any natural deposits of granular material on earth, lunar regolith has an extraordinarily wide range of sizes – from millimeter to nanometer scale. This very wide size distribution, and especially the presence of a significant amount of extremely fine particles, contributes to the unusual, and perhaps problematic, nature of handling lunar regolith.

11.2 Dust

Fine dust is an active constituent of the Lunar surface environment and any ongoing lunar operation involving moving parts or mechanical machinery will find that fine dust and dust mitigation will be major challenges to overcome. As scientist-engineer Harrison Schmidt commented rather recently "Dust is the Number-One environmental problem on the Moon" (Schmidt 2005). A microscopically thin top layer of the entire Moon's surface becomes positively charged by photoelectric ionization from sunlight during the lunar daylight, and negatively charged from the electrons of the solar plasma at night (Colwell et al. 2007). Anytime exposed lunar regolith is disturbed, some of the charged fine surface dust particles will be attracted to and adhere to any nearby surface (Walton 2007). It is a widely held truism that fine regolith particles are jagged and irregular in shape; however this notion is based on observations of larger particles. In contrast, SEM images of the finest size fractions of lunar regolith demonstrate that this is not the case (Greenberg et al. 2007). Nonetheless, the fine regolith particles are abrasive, and in addition to partially coating surfaces they contact, they can also lodge in small crevasses, seals, or gaps between moving parts and severely affect functional operation of mechanical equipment. Partial coating of surfaces can reduce performance of solar cells, increase heat absorption and reduce radiation, causing overheating of instruments or equipment. During Apollo astronaut expeditions, it was found that easily-disturbed dust stuck to any surface, damaged seals, caused slip joints to bind, threads to become unusable, clogged holes, decreased flexibility in space suit gloves and arm joints, and reduced visibility. The astronauts found that the finest dust was not easily removed from their space suits or equipment and it was carried into crew quarters, where it also became a potential health hazard.

Various approaches have been proposed to deal with potential dust problems (Taylor et al. 2005); the most viable-appearing approaches involve having multilayered dust defenses, or dust prevention, measures in place, each designed to handle a potential 'overload' if the layer above experiences a failure. Special surface coatings, electrostatic brushes/wands, CO_2-snow nozzles and a whole host of other technologies have been proposed to prevent dust from depositing, or to remove fine lunar dust from surfaces. Research into potential dust prevention and dust mitigation methods is continuing; however, at this time no single approach has been demonstrated to be superior at either preventing dust from accumulating/depositing on surfaces, or for removing it from them.

Analyses and laboratory experiments have demonstrated that it is physically possible for small quantities of charged very fine dust particles to become levitated above the charged lunar surface, and travel horizontally in response to the changing surface charge as the sun rises or sets (Colwell 2007). There is also some evidence that surfaces exposed to the lunar environment degrade over time, possibly from accumulating a dust coating; however, such degradation is a slow process taking decades. During Apollo 11, 14 and 15 Laser Ranging Retroreflector (LRRR) arrays with surfaces generally facing up were deployed (see Fig. 11.1). Those units still function, but have had a gradual reduction in the reflectance of coherent light from their surface. Recent Apache Point Observatory data has been compared to historical data collected by earlier observatories involved in lunar laser ranging. For the period 1973 to 1976, no significant reduction in reflected laser light intensity (at full moon) was apparent in data records, but it began to emerge clearly in a 1979 to 1984 data set. The efficiency of the three Apollo reflector arrays has diminished by a factor of ten at all lunar phases and by an additional factor of ten when the lunar phase is near full moon. While heating effects may play a part in the performance degradation of the LRRRs, lunar dust is suggested to be the more likely candidate, as this would be consistent with the very gradual performance degradation – and where the most substantial performance loss occurs right on Full Moon (Murphy et al. 2010).

Recent evaluation of Apollo era data on Dust Detector Experiments (DDE's) (O'Brien 2011) suggest that dust splashed onto vertical surfaces (and initially adhered to those surfaces) subsequently exhibited reductions in dust density on those surfaces with successive lunations. Such changes in level of adhered dust on vertical surfaces exposed to sunlight may be the result of the changing charge levels in that dust (from subsequent exposure to sunlight and photo-ionization) either reducing adhesion to the surface, or creating repelling electrostatic forces within the dust, and/or with the surface, which could cause some of the dust to be shed from the surface during subsequent recharging (Johnson 2009). O'Brien's (2011) evaluation of the Apollo Lunar Ejecta and Meteoroids (LEAM) experimental data has also raised some questions about whether that experiment ever directly detected any low-velocity, horizontally traveling dust particles (Colwell et al. 2007; Stubbs et al. 2007), leaving only the LRRR as the most viable direct evidence of gradual dust deposition on exposed surfaces.

Fig. 11.1 Corner Cube Prisms and the Apollo 15 LRRR comprised of 300 (3.8 cm) fused quartz corner-cube prisms in a hexagonal array1 Corner Cube Prisms and the Apollo 15 LRRR comprised of 300 (3.8 cm) fused quartz corner-cube prisms in a hexagonal array

11.3 Vacuum Effects on Electrostatics and Adhesion

In addition to the natural charging of the topmost thin layer of particles on the lunar surface exposed to sunlight (or the night sky), regolith which is excavated, transported or handled may acquire charges triboelectrically. Regolith is non-conducting, and the lack of a humid atmosphere means that any acquired charge is likely to remain on the regolith particles for extended periods of time. Any charge on fine particles, whether acquired through UV radiation, or from handling, will contribute to long-range electrostatic forces that can cause disturbed dust particles to be attracted to and to adhere to surfaces, charged or not, and whether the surfaces are conductors or not. It is often overlooked that image-charge forces attract charged non-conducting particles to conducting surfaces (as can be readily demonstrated by rubbing a balloon against one's hair, to triboelectrically charge it, and then placing the balloon in contact a metal surface, where it will adhere). Once fine particles are deposited on a surface, they adhere not only because of the existence of electric charges, but also due to inherent surface-energy-related (van der Waals) adhesion forces, which all materials exhibit. The surface-energy adhesive/cohesive forces are very short range, so they have little effect in movement or long-distance attraction of fine particles, but once contact is established, the surface-energy forces can be much greater magnitude than electrostatic forces (depending on the particle size, charge, and surface energy of the materials involved).

One of the more challenging uncertainties in attempting to 'predict' cohesive/adhesive forces on fine particles in a lunar environment is how to account for the potentially dramatic effect the lunar vacuum can have on various important phenomena. A number of effects are immediately obvious, such as the insignificance of aerodynamic drag forces, lack of permeability concerns, or out-gassing, etc. (except, perhaps, during lunar avalanches when bound solar-wind gas molecules might be released, causing local fluidization). The lack of water vapor or oxygen, can also affect the surface chemistry of many materials. As already described, the hard-vacuum above the lunar surface provides no shielding from UV radiation or bombardment by solar wind particles. Also, the surface energy of materials can be dramatically affected by being in a vacuum. When a new surface is created (as by fracturing) in the presence of a foreign vapor, like laboratory air, some adsorption of vapor molecules (e.g., water, or hydrocarbons) can take place on the newly created surface and lower the surface energy from its value in a vacuum (Israelachvili 1991; Adamson 1976). For example, when mica is cleaved in high vacuum the surface energy is $\gamma_s \approx 4500$ mJ/m^2, but when cleaved in humid laboratory air it falls to $\gamma_s \approx 300$ mJ/m^2 (Bailey et al. 1970). As noted by Israelachvili (1991) "as a general rule of thumb we may say that the van der Waals interaction is dominated by properties of the bulk or substrate materials at large separations and by the properties of the adsorbed layers at separations less than the thickness of the layers. In particular, this means that the adhesion energies (i.e., at contact) are largely determined by the properties of any adsorbed films, even when these are only a monolayer thick." The lack of an atmosphere does not change the fundamental surface energy of the materials; however, the lack of adsorbed molecules like oxygen or water may mean that materials in a lunar environment will have much less surface contamination, and thus, exhibit a much higher effective surface energy than the contaminated surfaces we are used to dealing with terrestrially.

Changes in particle surface energies with vacuum can result in soil mechanics tests of cohesive-strength being different under vacuum than under usual laboratory conditions. Experimentally the cohesion of Lunar soil sample #10084-93 from Apollo 11 (sealed under N_2) (Grossman et al. 1970, cited in Lee 1995) was measured under vacuum of 5×10^{-9} Torr, and then exposed to O_2, O_2 + 3.5% H_2O at 2 Torr, 500 Torr and 760 Torr, at 27°C and 200°C. Reduction of cohesion after exposure to the vapor was observed in all cases. Similar changes in the cohesive behavior of fine silicate powder were also observed under high vacuum by Salisbury et al. (1964). Loss of cohesion for silicates was observed in the presence of air but the cohesion was restored when the system was evacuated. Electrostatic effects were demonstrated not to be the cause of the enhanced cohesion under vacuum. These observation should be a warning that cohesion tests and low-stress consolidation tests on lunar soil samples should not be taken as representative of in situ behavior, unless they are performed under high vacuum conditions.

11.4   Properties of Granular Solids Relevant to Regolith

During the Apollo era, and in many subsequent evaluations of the physical properties of lunar regolith, the emphasis was on typical 'geotechnical' concerns related to construction of facilities, such as its in situ load-bearing capacity and the slope-stability of berms, trenches or piles of excavated regolith. These properties were summarized in Figure 9.39 (among others) of the Lunar Sourcebook (Carrier et al. 1991), reproduced as Fig. 11.2 here along with its original descriptive caption. Several researchers have commented on the high cohesion of lunar regolith evidenced by the stability of steep slopes and trench walls of lunar regolith (as described in the caption of Fig. 11.2) (e.g. Lee 1995, Taylor 2005) and have indicated a need to understand the reason for that high cohesion. Given the reduced driving force of lunar gravity and with the degree of consolidation found in-situ, these predictions are not unusual for a material with the size-distribution of lunar regolith. It is primarily the pre-existing state of consolidation that is surprising, and which results in the observed cohesive strength. Carrier (2005) has relatively recently produced a summary of Chapter 9 from the Lunar Sourcebook emphasizing the importance of the Relative-Density of in situ regolith.

The angle-of-repose is an often misunderstood parameter, especially if a material exhibits cohesive properties, yet it is often considered to be a reasonable characterization of a material's flow behavior, and there is confusion as to how this quantity might scale with reduction in gravity-level. For non-cohesive materials the angle-of-repose is nearly the same as the geotechnical internal friction angle of the material (except that there are multiple angles of repose, 'drained', 'poured', 'dynamic', etc (Brown and Richards 1970). A diverging geometry, like a conical volcano, will have a shallower value than a converging

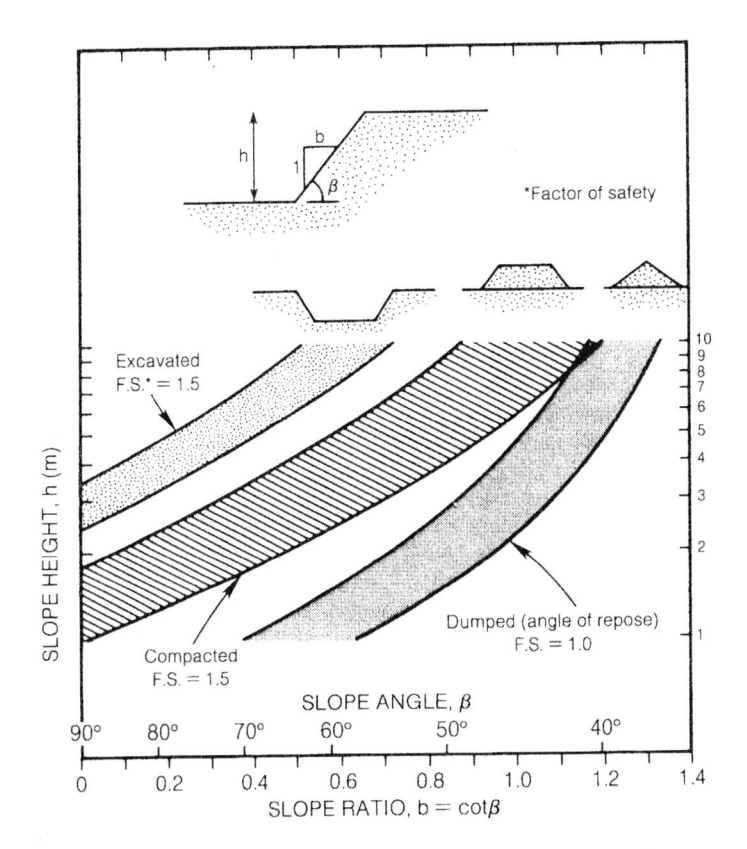

Fig. 11.2 [From Carrier et al. 1991] "Calculated stability of artificial slopes constructed in lunar surface material. Data are presented in terms of slope height (vertical axis) as a function of slope angle (or slope ratio) (horizontal axis). Inset (upper left) diagrammatically shows a slope and the quantities involved. Data are presented for three situations: (1) an excavation in lunar soil (stippled area); (2) a compacted pile of excavated lunar soil (ruled area); and (3) a dumped pile of lunar soil (shaded area). The data show that a vertical cut can safely be made in lunar soil to a depth of about 3m, while an excavated slope of 60° can be maintained to a depth of about 10m."

geometry, like a hole in the bottom of a flat container. Half-way in between those is the planar drained angle-of-repose which results if a vertical supporting wall is removed from one side of a container filled with the material. Since the angle-of-repose is essentially a 'frictional' quantity (like the ratio of shear to normal stress) it will not vary significantly with gravity (which would just increase the average normal stress) unless cohesive forces are present. As soon as cohesion is present (even just an unconfined cohesive strength due to overconsolidation of a material, with no significant interparticle cohesive forces) the angle-of-repose no longer has any meaning. The angle that a cohesive material will exhibit, depends on the system size (as well as on gravity). Only if the system gets large enough for gravity to overcome the cohesion or unconfined strength does the angle-of-repose begin to have meaning. For 'as poured' lunar regolith (i.e., unconsolidated) any system larger than maybe a meter, or so, would probably be large enough to overcome the small interparticle cohesion that exists; however, for consolidated lunar regolith (as found in situ) the unconfined strength is very high (because it has to dilate in order to shear). It is the consolidation of lunar soil that gives it the appearance of being highly cohesive (it does have true interparticle cohesion, but not nearly as much as damp sand). In-situ Lunar regolith would be expected to be able to support a vertical excavated 'wall' that is 3 or 4 meters tall without collapsing, under lunar gravity. If the vertical height of an excavation is deeper than that, then the wall could collapse to the 'angle of repose' (probably around 35 to 40 degrees or so). That 'collapsed' angle would be similar to terrestrial angles-of-repose, and could be over 40 degrees, but probably not much. Figure 11.2 indicates that dumped regolith would have a slope angle of around 37-42 degrees if it is 10 meters high. For lower heights the dumped material would exhibit steeper angles – say 45 to 55 degrees if it is only 2 m high (these are for lunar gravity and vacuum conditions).

A limited number of particle-scale measurements have been done on the physical properties of lunar regolith particles that contribute to inter-particle cohesion, like surface energy, hardness, surface morphology. On the other hand, a significant number of bulk soil mechanics tests have been performed on Lunar soil, and simulants, including penetrometer, direct shear, one-dimensional oedometer, and triaxial tests which produced the data to prepare a summary like that of Fig. 11.2. Most of these tests were interpreted in terms of the classic Mohr-Coulomb expression, relating shear-stress, τ, to normal-stress, σ

$$\tau = c + \sigma \tan \varphi_f, \tag{11.1}$$

where the parameter, φ_f, is known as the friction angle (or internal-angle-of-friction), and the zero-normal-stress intercept, c, is known as the cohesion (or cohesive strength) of the soil. Materials with a finite cohesive strength, c, can support a vertical free-surface, or wall, until the height is great enough for stresses to exceed the strength of the material. While the Mohr-Coulomb relation is a reasonable representation of the response of a particular sample starting at a particular consolidation, there is not just a single unique Mohr-coulomb curve for a typical fine cohesive powder. Such materials exhibit a family of Mohr-Coulomb curves, each one unique to a particular degree of pre-consolidation. Thus, the values of the fitting parameters, c, and φ_f depend on the previous consolidation of the soil sample. As outlined by Carrier (1991) (at least for one basaltic stimulant of lunar soil) the values of c range from 0.03 to 3 and φ_f ranges from from 28° to 55°, with the variation in these values resulting from differences in the relative-density of the material (see Fig. 11.3), where relative-density is the percent of compaction between the minimum bulk density (the lowest bulk density "at which the soil can be placed") and the maximum bulk density "at which the soil can be placed" (a standard ASTM designation). The 'best-estimate' value for cohesion of Surveyor soils was 0.35 – 0.70 kPa, while cohesion values for Apollo 12 samples ranged from 0.1 to 3.1 kPa depending on the initial degree of consolidation. The Apollo Model 'best estimate' for lunar soil cohesion was 0.1 – 1.0 kPa, with the friction angle ranging from 30° to 50°.

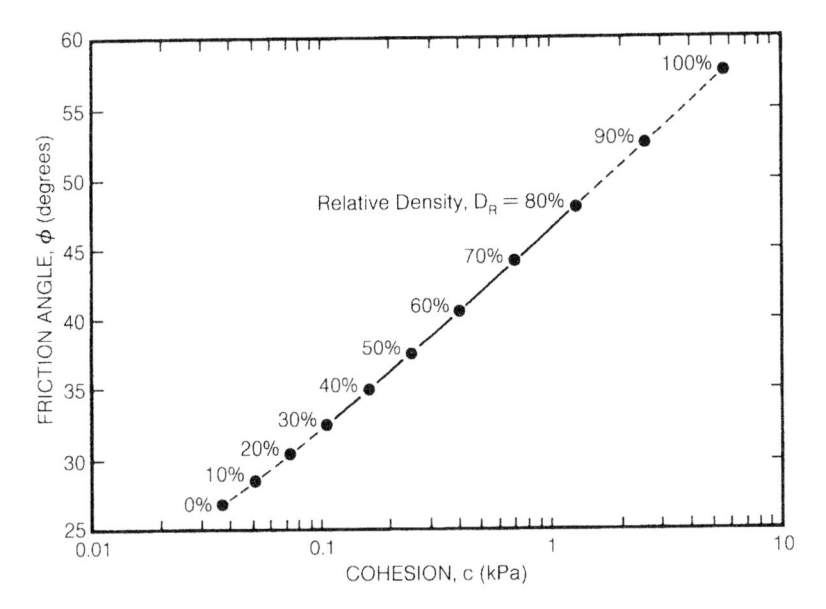

Fig. 11.3 Measured shear strengths of a basaltic lunar regolith simulant showing the internal friction angle (ratio of shear-stress to normal stress at failure) and the cohesive strength for different relative densities (Carrier et al. 1991; Mitchell et al. 1972, 1974)

A few words about terminology may be beneficial to the reader, especially since researchers in different specialties use the same words to mean different things relating to material properties. Interparticle friction contributes to the shear resistance of a granular material, however, the soil-mechanics quantity friction-angle, φ_f, is not a direct measure of interparticle friction. It is simply the arctangent of the slope of the shear-stress normal-stress line for a material obeying a Mohr-Coulomb relation, Eq. (11.1). Similarly, the unconfined shear strength, or the c intercept of the Mohr-Coulomb curve, is non-zero for cohesive materials; however, c is not a direct measure of the interparticle cohesive force or energy. Nonetheless, the fact that lunar regolith exhibits a measurable soil-mechanics cohesion, is an indication that interparticle cohesive forces may be significant for that material.

The cohesive behavior of a dry powder or its cohesiveness or cohesivity is less well defined than the cohesion or cohesive-strength (i.e., c in Eq. 11.1). The qualitative terms of cohesiveness, cohesive behavior, or coheisvity, are really descriptions of a system behavior (not a material property) where the system being described is a combination of the powder, the equipment or container holding or processing the powder, and the driving force – usually gravity. Generally a powder is said to exhibit cohesive behavior when it fails to flow freely and smoothly from a container (like dry sand), or if it agglomerates, or clumps, or exhibits shear-strength properties somewhat like a solid. As one example, the Aero-Flow avalancher (Kaye 1997) determines cohesiveness by analyzing the random nature of the time sequence of clumping avalanches in a slowly rotating horizontal cylinder, partially filled with powder. A variety of other measures also exist for flowability (Prescott 2000; Carr 1965; ASTM 1978) or cohesiveness (Schwedes 1996; 2003). Many of these are examples where 'measurement' of cohesive behavior is an index-test dependent on the characteristics of a measurement system (involving a container of a specific size, a specific driving force, often gravity, and a specific type of flow 'expected'). If the size of the container changes, the driving force changes, the type of flow examined, or the material properties change, then the characterization of the powder's cohesiveness or flowability may be different.

It is generally recognized that the flow behavior of a powder is related to its shear-strength, a property which is both a state and history-dependent variable (as is a powder's cohesive-strength). In soil mechanics the cohesive-strength of a bulk material is well defined and it can be measured with standard geotechnical laboratory methods such as a triaxial test. There are a variety of other methods to measure the shear strength of a powder, utilizing direct shear, such as the Jenike (1964), Schulze (2003), or Peschl (Svarovsky 1987) shear cells which can measure the yield loci or failure envelope of a powder and, in some cases, how it varies with precompaction, or previous stress-strain history. The cohesive-strength of a weakly cohesive powder can also be measured directly (e.g. the Sevilla powder tester (Castellanos 2004; Arslan et al. 2010)).

The compressibility of a powder is often associated with its cohesiveness and has also formed the basis for a variety of additional index tests for ranking the cohesiveness or cohesivity powders. Cohesive powders existing in low-density 'sifted' states, are quite compressible, and can easily be compacted to higher densities by external loads, or by handling or jarring their containers. The pharmaceutical industry routinely deals with fine cohesive powders and often characterizes how cohesive a powder is by its Hausner ratio, that is, the ratio of the density after being 'tapped' repeatedly (up to thousands of times in a controlled tapped-density test) to initial sifted density (Hausner 1967; Abdullah 1999). Such tapped-density tests serve as index tests to classify the cohesiveness of powders. For a more quantitative measure of cohesive powder compaction, one can examine the stress vs. density behavior under controlled uniaxial or isotropic compression. Low-stress-level uniaxial compaction of cohesive powders has been found by several researchers to be a reproducible and reliable indicator of the cohesive nature of a powder. The size-dependence of compaction behavior of typical cohesive powders is illustrated in Fig. 11.4 which shows the solids volume fraction as a function of applied axial load in a series of low stress uni-axial compression tests of spray dried raffinose powders in the size range from 0.8 μm to 4 μm (Walton et al. 2003; Miller et al. 2002) and similar compaction (oedometer) data for Russian lunar regolith samples Luna 16 and 20 (Leonovich et al. 1975, in Carrier et al. 1991) and the compaction of 1 μm silica spheres (Güttler et al. 2009). Raffinose is a common soluble carbohydrate found in vegetables, its behavior is cited here simply because data was readily available on the compaction behavior of several different sized fine raffinose powders with sizes relevant for comparison with the fine-fraction of lunar regolith. The Luna samples are comprised of regolith with perhaps a somewhat higher fraction of agglutinates than most Apollo samples. They were not as fine nor as cohesive as the raffinose, however, they did exhibit an unusually low initial solids fraction compared to typical dry sand or terrestrial silt – a feature often associated with very cohesive powders (although there is some speculation that the low solids fraction of the Luna samples is due to their unusually high (irregularly shaped) agglutinate content (Nakagawa 2006)). The uniaxial compaction curve for 1μm mono-disperse silica spheres (Güttler et al. 2009) has a shape similar to that of the lunar regolith, but with a lower solids fraction for a given applied load – a result of the larger effect of interparticle cohesion in the finer powder. The slope of the raffinose powder compaction curves is somewhat shallower than that of the comparably-sized silica, indicating that the raffinose offers greater resistance to the deformation occurring during compaction (an indirect indication that the raffinose is probably more cohesive than the silica).

Van der Waals surface forces affect all materials. The pull-off force required to separate two spherical particles in contact is roughly $F_c \sim 3\pi\gamma R$, where γ is the surface energy per unit area, and R is the particle radius. The relation stated is from the JKR theory for elastically deformed spheres (Johnson 1971); direct integration

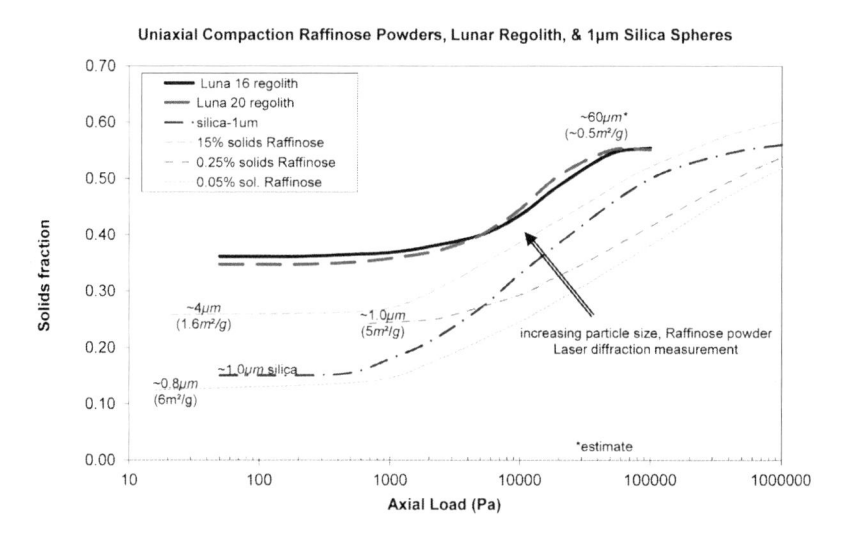

Fig. 11.4 Uniaxial compaction of cohesive powders. Horizontal portion at left end of each curve is reestablishing stress conditions under which the sample was prepared (any unloading-reloading would occur along nearly horizontal lines branching to the left off of each curve). The knee to less compressible behavior at high pressure probably indicates a change in the mode of deformation (i.e., possibly changing from collapse of an initially 'open' structure by particle rearrangement, to a more-usual compact configuration where further compaction occurs by much smaller-scale particle rearrangements, or by particle deformation or damage). (Leonovich et al. 1975; Miller et al. 2002; Walton et al. 2003, Güttler et al. 2009.)

of the van der Waals forces for undeformed spheres, on the other hand, changes the numeric value in the relation from 3 to 4 (Israelachvili 1991). This force has a very short range, and reduces in magnitude by more than an order of magnitude at a separation between the particle surfaces of only 10 nm. The effective cohesion between particles is reduced if the surfaces are held apart by even a very small distance. Surface roughness has the effect of keeping part of the apparent contact region from actually touching. Thus, most naturally occurring particles have lower 'pull-off' forces than would be predicted by the JKR theory (reduced by as much as 2-orders of magnitude). Countering this reduction in effective cohesion due to surface roughness is the previously mentioned increase in effective surface energy for lunar regolith under hard vacuum. The actual magnitude of interparticle cohesive forces acting between regolith particles under lunar conditions is unknown, but it is expected that the particles would exhibit somewhat higher cohesion than similar sized particles would be under terrestrial conditions (even without accounting for lunar electrostatic charge effects, discussed earlier).

As the particles in powders get finer, the surface area per unit mass increases, contributing to an increased effect of interparticle cohesion on the bulk behavior of the powder. As previously mentioned, cohesionless dry sand or glass beads poured into a container typically form a 'bed' of material with a solid fraction occupying around 60% of the volume (the remainder is the interstitial air space between the particles). Fine cohesive powders, on the other hand resist rotation at points of contact and can form a light fluffy bed with the particles occupying as little as 15% of the available airspace. Such fluffy powder beds are often easily deformed in their as-poured (or sifted) state. The fluffy beds are very compressible, and once compacted, they can become much more difficult to deform. This is mentioned, because the degree of consolidation of a powder depends on how it is handled or even how it is poured into a container. Under lunar gravity the velocities of material movement in gravity flow will be much slower and the body forces compacting powders in beds will also be lower. Thus, regolith in bins and hoppers will be less compacted under lunar conditions than it would be under terrestrial conditions – this can affect on how easily it flows out a hopper.

The importance of the compaction behavior occurs because of the strong correlation of the tensile and shear strengths with the compaction state. Figure 11.5 illustrates this correlation; it is a schematic of a constitutive relation used to fit experimentally measured compaction and tensile bulk behavior of a powder comprised of micron sized-spherical silica particles (Güttler et al. 2009). Just as the resistance to further compaction (i.e., the positive pressure compressive-regime curve) increases as the density of the powder increases, so does its tensile strength, shown schematically as the lower tensile-regime curve. It is this increase in tensile strength with degree of compaction that makes cohesive powders sensitive to handling, and difficult to deform or disperse once compacted.

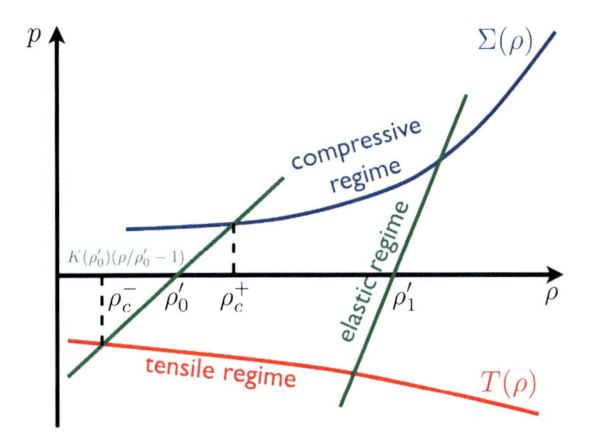

Fig. 11.5 Schematic illustration of constitutive model used to fit the measured compaction, unloading, and tensile strength behavior of micron-scale silica spheres (Güttler et al. 2009)

11.5 Regolith Handling and Processing

The following sections describe the effects of cohesion and gravity on systems that are likely encountered in regolith handling and processing. Bins and hoppers are used for almost all bulk material storage and transfer, and a major concern is often the reliability of initiating flow when a valve or gate is opened. Rotating drums are often used in processing, whether to achieve uniform heat sharing, in granulation or mixing. Fluidized beds are a very standard method used for a wide variety of granular material processes requiring exchange of gasses with solid particles. The effect of gravity and cohesion on these configurations is discussed.

11.5.1 Hopper Flow

Jenike's cohesive arch theory for hopper design has been utilized to design mass flow hoppers with opening sizes that produce reliable flow, ever since it was published (Jenike 1964). As outlined below, the gravity scaling in that theory would imply that the size of hopper openings needed for reliable flow will scale inversely with g-level (i.e., increase as 1/g). Because of the nature of cohesive powders; however, such 1/g scaling, while robust, might be overly conservative, and testing under reduced-g conditions may well be needed to determine how regolith properties (especially consolidation, of as-poured powder) change with gravity-level. Figure 11.6 shows a schematic of an arch formed near the opening of a hopper. According to Jenike's cohesive arch theory, the maximum stress in the arch (due to the weight of the material in the hopper) is proportional to its span, B, and to the material's bulk density times gravity; the location of this maximum stress is at the abutment of the arch at point L and can be approximated as (Tardos 1999; Jenike 1964):

$$\sigma_a = \rho_b g B = 2r\rho_b g \sin \alpha \tag{11.2}$$

Equation (11.2) can be generalized (Shamlou 1990) for a plane and a 3D conical arch,

$$\sigma_a = (2r\rho_b g \sin \alpha)/m \tag{11.3}$$

where m = 1 for a 2D slot-hopper and m = 2 for a conical hopper, B = 2r sin α, and g is the acceleration of gravity. Rearranging this we obtain a relation for the gravity dependence of the arch span, B, corresponding to fixed compressive stress level in the arch, σ_a,

$$B = m\sigma_a /(\rho_b g). \tag{11.4}$$

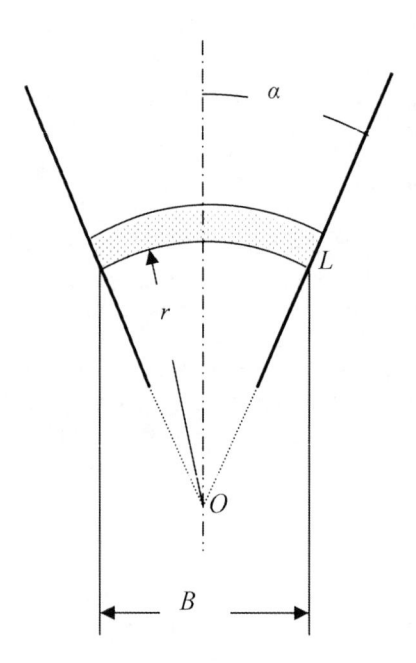

Fig. 11.6 Arch in a hopper (2D – plane slot, or 3D – conical)

If the arch with stress, σ_a, is located at the bottom surface of the material in the hopper (i.e., at the hopper opening), then the maximum compressive stress which the powder can support (next to this free surface) is the unconfined yield strength of the powder, σ_c. If $\sigma_a > \sigma_c$ then the arch will fail due to the stress from the weight of the powder above it, and flow will occur. The full Jenike theory is slightly more complex than this (including the effect of wall friction, etc), but the general trends w.r.t. changes in gravity level are captured in these relations.

Based on the Jenike relations above, one would expect that if gravity were reduced by a factor of six, then, since the stresses are reduced by that same factor, the maximum stable arch size would increase by a factor of six. That is probably generally true; however, there are some confounding factors which might make this an overly conservative assumption. For cohesionless materials σ_c is zero, so there is no stable arch, but for most cohesive materials the unconfined yield strength can be approximately fit by a linear relation, $\sigma_c \approx K\sigma_1 + L$ (see Sect. 6 of Tardos (1999)) where K and L are fit to measured material behavior. The actual shape of the unconfined yield curve, σ_c, varies somewhat non-linearly with σ_1, and since the unconfined strength of the powder depends on the stress (e.g., the '$K\sigma_1$' term) it is not a constant value, and will have a lower value in a hopper under reduced gravity than it has under terrestrial gravity conditions. As stress increases, a greater portion of the total strength arises from the stress-dependent part of the strength curve (i.e., the '$K\sigma_1$' term), and thus, will not contribute to a need for larger arch diameters to ensure breakage (and flow). It is at low stresses where the fixed 'L' term contributes the greatest, and thus, where the factor of '6' scaling in opening size might occur. Another important powder property not to be overlooked is the fact that, the more cohesive a powder is (or the lower gravity is, so that the powder 'acts' cohesive) the more likely the powder is to form a relatively low density, loose, bed when it is initially placed in a hopper (and to be at a lower consolidation, and thus have a lower 'L' intercept value, but also have a lower bulk density, ρ_b – these two changes have opposite effects on estimates of the minimum opening size for reliable flow).

Typical powder characterization tests for hopper flow attempt to measure the change in yield strength with pre-consolidation. They typically characterize this as a sensitivity of the powder to overconsolidation effects. Unfortunately, these tests do not usually deal with extremely low consolidation stresses as might be found under lunar conditions. Thus, until the properties of regolith are measured at very low stresses, to see how sensitive its unconfined strength is to low consolidation stresses, it remains difficult to predict the minimum size opening to ensure robust/reliable gravity driven flow under lunar conditions. Tests at reduced gravity may be necessary to quantify the size of openings that can support arches in hoppers, (i.e., to determine what size hopper openings are needed for unassisted reliable/robust gravity flow of regolith under lunar gravity conditions).

Laboratory tests of JSC-1A simulant to determine minimum outlet diameter required to avoid a cohesive arch in a conical mass-flow bin determined that minimum diameter to be ~5 cm under normal terrestrial conditions (Carson 2010) (and did not show a tendency to increase due to additional preconsolidation). Scaling that by a factor of six would imply a need for conical hopper openings ~30 cm in diameter for reliable flow of material with those properties under lunar gravity. Similar terrestrial tests on lunar simulant NU-LHT-2M indicated that minimum outlet diameter required to avoid a cohesive arch in a conical mass flow bin was ~6 cm (Khambekar 2009) (in addition, this material exhibited a consolidation effect, increasing the minimum opening size to ~10 cm if the material was overconsolidated above the hopper opening). If these values are scaled by a factor of six, then the minimum opening sizes become ~36 cm, or 60 cm, under lunar gravity conditions. A caveat regarding these measurements is that they were performed at load-levels comparable to those found in terrestrial industrial-scale material handling equipment. Under lunar conditions, and in smaller-scale equipment, the load-levels would be lower, and the degree of consolidation would be correspondingly lower, presumably with a correspondingly reduced cohesive strength for the regolith (and thus, possibly smaller opening sizes required to avoid cohesive arches).

11.5.2 Flow in Rotating Drums

Rotating drums are a common configuration for many mineral or powder processing operations. They also form the basis for various powder flow characterization instruments. Here we use flow in a rotating drum as a convenient method of demonstrating the changes in powder flow behavior as the effective gravity-level changes.

Flow of cohesionless granular materials in partially filled rotating drums exhibit a range of behaviors depending on the rotation rate of the drum (Brown and Richards 1970), including:

- Periodic avalanches at low rotation rates
- A nearly-linear top surface in the 'dynamic angle of repose' range
- Bi-linear or S-shaped top surface at higher rotation rates

Cohesive powders, on the other hand, exhibit large build-up and collapse of steep structures in the upper half of the top surface layer, with a chaotic time dependence – these will be discussed in more detail later in this chapter.

The nearly-linear dynamic-angle-of-repose regime for cohesionless powders occurs when the duration of an individual avalanche is close to the period between avalanches, so that the motion becomes nearly continuous. Thus, the rotation rates at which this flow regime exists depend not only on the gravity driving force, but also, on the physical size of the drum and the characteristics of the specific granular material. Experiments and Discrete Element Method (DEM) simulations have demonstrated that for a variety of different flow-modes in rotating drums, the same flow mode is obtained at various g-levels if the rotation rate is scaled with the square-root of the g-level (Walton and Baun 1993; Walton et al. 2007). Although it was recognized that the modes of flow are also dependent on the physical size of the drum, the size-scaling relation was not determined in those studies.

11.5.3 Micro-Avalancher Observations of Cohesive Powder Flow

When a non-flowing, cohesive powder is placed in a slowly rotating drum at elevated g-levels (i.e., at the end of a centrifuge arm) the additional driving force of the elevated g-level can cause the powder to flow. If the g-level is elevated high enough, then the cohesive forces in the powder become completely overwhelmed by the elevated body forces, and the powder can flow like a typical cohesionless powder at 1-g_0. Walton et al. (2007) tested a series of cohesive powders in a micro-avalancher consisting of a small rotating drum (either 3/8 inch diameter or ¼ inch diameter) and 1/8 inch thick (along the axis of rotation) located at the end of a centrifuge arm. The rotation rate of the small drum was maintained at a fixed fraction of the rotation rate of the centrifuge arm in order to scale the drum rotation properly to produce self-similar flow conditions at all g-levels.

Figure 11.7 shows representative changes in a cohesive powder's flow behavior in a drum that is slowly rotated while at the end of a centrifuge arm. Each column of pictures in Fig. 11.7 has snapshots of the flowing powder at the same g-level, varying from $12.5g_0$ on the left to $400g_0$ on the right hand side. The powder shown in Fig. 11.7 consisted of 20 to 70 micron lactose 'carrier' particles blended with 3.8wt% of a fine 1-micron size active pharmaceutical ingredient (API) powder which was quite cohesive. The resulting powder was cohesive enough that at 1-g_0, the 42 mg of powder in the cell did not move inside of the small drum (~ ¼" dia) when the drum rotated slowly. Once the g-level exceeded about $5g_0$ the powder began to move, and by $10g_0$ the behavior was quite consistent. The first column of snapshots on the left of Fig. 11.7 is at a g-level of $12.5g_0$. As can be seen in Fig. 11.7, at $12.5g_0$ the powder behaved like a typical cohesive powder in a larger drum at 1-g_0; that is, it exhibited periodic avalanches when large chunks fell off at random time intervals. As the g-level increased to higher levels the powder behavior gradually changed to that of a free-flowing, or non-cohesive powder, exhibiting a smooth top surface with a single 'dynamic angle of repose.'

In addition to the change in flow behavior, the bulk density of the powder in the centrifuging drum tests also changed as the gravity level increased. The change in bulk density of the powder can be seen in Fig. 11.7 by observing that the fraction of the area of the circle covered by powder in the snapshots is larger in the columns on the left hand side than in the columns on the right hand side. Because of this compaction at high g-levels, the flow characteristics of a powder can be different after being processed at each g-level. Powder removed after being processed at only $12.5g_0$ would be expected to have a lower yield strength and to flow more easily (at 1-g_0) than a powder that had been exposed to $400g_0$ body forces, and densification – collapsing some of the open structure existing because

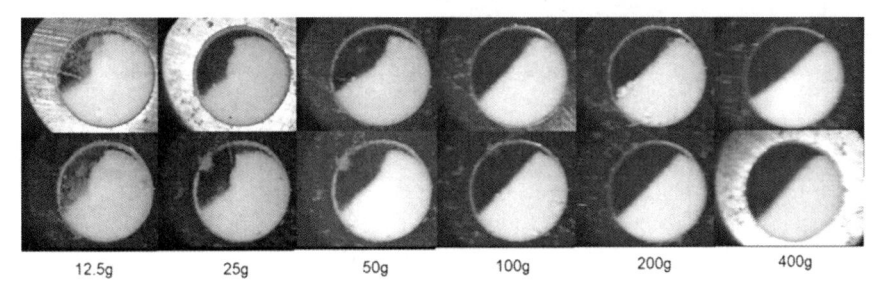

12.5g   25g   50g   100g   200g   400g

Fig. 11.7 Snapshots of a 42mg powder sample, comprised of (20-70 μm) lactose particles blended with fine particles (~1 μm) of a cohesive active pharmaceutical ingredient (API) in a small planetary-mill configuration, with the centrifugal acceleration on the periphery of the cell, a_c, maintained at a fixed fraction of the main centrifuging acceleration, A_c, (a_c/A_c = 0.0001). The second row of images at each g-level is another frame from the video of the material motion in the same test as the upper image (showing a little of the variety of shapes during flow).

cohesive interparticle forces keep the powder from becoming a close-packed structure. Thus, a powder might flow better while at $400g_o$ than at $12.5g_o$, but, if it has been processed at $400g_o$ and then is brought back to terrestrial, or lunar, conditions, it may be compacted, and more difficult to remove from the processing container and/or to transfer to subsequent processing steps. This is one example of how the recent deformation and stress history of a powder contributes to its flow behavior.

Even very sticky, cohesive powders will flow if the g-level is raised high enough. Figure 11.8 shows the behavior of a powder comprised of only the micron-sized API powder which had been blended with the larger lactose particles for the powder samples of the previous figure. This powder was so cohesive that it stuck to the cell walls at effective gravity-levels below $100\text{-}g_o$, and it was still exhibiting periodic avalanches at g-levels of 200 and 400, but behaved nearly like a free-flowing powder at $1200\ g_o$.

Similar changes in angle of repose and density were observed for other powders tested, including BCR Limestone, and cornstarch. All cohesive powders tested exhibited a transition from large avalanches to a smooth single-angle-of-repose linear top surface as the gravity level increased, with the transition occurring over a g-level change of approximately a factor of four. Since regolith simulants typically exhibit small vertical 'cliffs' on the order of a centimeter or so in height under terrestrial conditions, one would expect that in process equipment smaller than something like 10 cm of so in size, regolith would behave like a cohesive powder under lunar conditions. If the system size is increased significantly, say to a scale of 1 m, or so, then regolith would probably behave like terrestrial free-flowing powders. Exact sizes for such transitions in behavior are difficult to predict because of the lack of knowledge of the low-compaction strength properties of regolith.

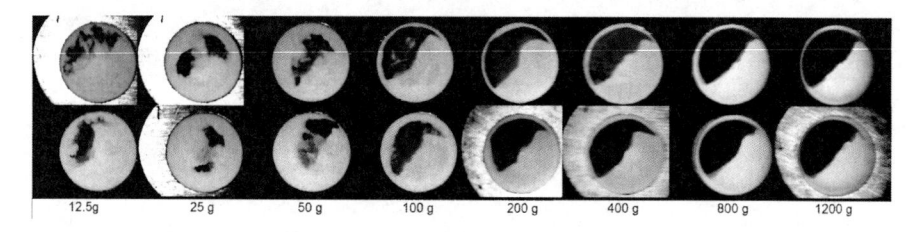

| 12.5 g | 25 g | 50 g | 100 g | 200 g | 400 g | 800 g | 1200 g |

Fig. 11.8 Snapshots of a very fine (~1 μm) and very cohesive active pharmaceutical ingredient (API) powder in a slowly rotating micro-avalancher at effective gravity levels from 12.5 to 1200. As in Fig. 11.7, the lower images are each an additional frame from the video of the same test conditions as the upper image.

11.5.4 Fluid Bed Processing

A common mode of processing granular solids is in fluidized beds, and they are also under consideration for a various resource recovery processes on the moon. Powders are typically categorized for fluidization according to the Geldart classification (Geldart 1973) as: C – cohesive (very fine), A – aeratable (non-bubbling, uniform fluidization w/air under ambient conditions), B – bubbling fluidization, D – spouting bed (large particles) see Fig. 11.9.

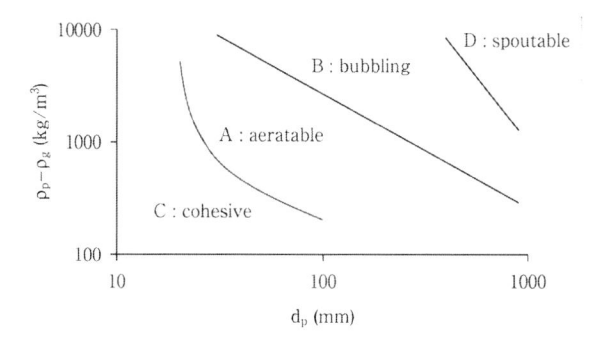

Fig. 11.9 Geldart (1973) classification of powder fluidization properties, by size and density

Qian et al. (2004) examined the effects of the body-forces acting on particles in fluidized beds through the use of a rotating cylindrical chamber with a porous frit on the outer wall, so that fluidizing gas traveled from the outer radius to the center region of the rapidly rotating (centrifuging) cylindrical fluidized bed. They found that increasing the effective g-level (i.e., the centrifugal force acting in the shallow bed near the cylinder wall) had the effect of shifting a powder's Geldart classification boundary locations toward smaller sizes as the g-level increased. In other words as the g-level increased the boundaries shown schematically in Fig. 11.9 would move to the left so that a fine cohesive powder (Group-C) could behave like a Group-A (aeratable), and group A powders could be shifted to behave like group B powders, etc. Williams et al. (2006) examined an extrapolation of Qian et al's theoretical relationship for the shift in fluidization behavior of powders with g-level, in the opposite direction, by looking at fluidization of glass and alumina powders under reduced gravity conditions (during parabolic flights). Their results showed 'fair agreement' with Qian et al's model; generally corroborating that reducing gravity can shift the boundaries between Geldart classifications in the opposite direction so that larger particles act more cohesive as gravity is reduced (i.e., the boundaries in Fig. 11.9 move to the right under reduced gravity). The experiments aboard the reduced gravity aircraft, however, had limited precision, and additional tests with an alternative approach were proposed. This gravity-scaling, or shifting of Geldart classification boundaries with g-level means that a given powder will behave more cohesively in a fluidized bed at reduced gravity than it does under terrestrial fluidization.

A physical explanation of the causes for this behavior is straightforward. Under reduced gravity the terminal velocity for a particle settling in a gas is reduced. Also, the superficial gas velocity corresponding to minimum fluidization (i.e., the gas velocity when the pressure drop of the gas passing up through the bed just balances the weight per unit area of the particle bed) is reduced if gravity is lower. The net effect is that fluidization under reduced gravity is less energetic, and results in lower particle accelerations, lower collision velocities and lower shear rates in the gas field – all of these effects mean that cohesive interparticle forces are less likely to be overcome by the slow moving gas flow, and the powder will 'behave like a more-cohesive powder' than if it were fluidized under terrestrial gravity.

11.6   Regolith Transport

In addition to excavation concerns related to the highly consolidated in situ state of lunar regolith (e.g., see Carrier 2005), decisions on the best method to use for transporting regolith may involve a different set of constraints than typical terrestrial mineral recovery operations. For example, if the interest is in recovery of volatiles, like water, which may be in subsurface deposits, or might escape if exposed to sunlight and vacuum, the transport of the regolith may need to be done in closed systems. Such closed systems could be sealed transport containers, or tubes or pipes running from the excavation location to the processing unit's entry point. The following discussion of transport methods is presented as an example of how the constraints on operations might be significantly different under lunar conditions – and thus unconventional transport means might need to be utilized.

11.6.1   Pneumatic Conveying

Dilute-phase pneumatic conveying is a viable means of moving fine granular solids through small tubes. Under reduced-gravity both the settling-velocity and the saltation-velocity of particles in a gas are significantly reduced (for the same gas pressure). The minimum gas velocities to achieve vertical or horizontal pneumatic conveying are directly related to the settling and saltation velocities, respectively. Thus, under lunar gravity, pneumatic conveying can function at much lower gas velocities and be much more efficient than it is terrestrially. Mechanical separation of the solids from the gas usually requires either a cyclone separator or a baghouse filter and, under lunar conditions, it may be important to recover the gas for reuse. Under lunar gravity significantly lower gas densities (and thus pressures) can be utilized; however, at low-enough pressures the mean free path of the gas starts to become so large that it becomes less efficient at transferring momentum to particles (e.g., Cunningham slip corrections apply to drag coupling terms at low gas densities). When evaluating pneumatic conveying features against the features of mechanical conveying system, the compressor combined with the gas-solid separation system could be compared to the equipment required for producing and delivering mechanical conveying forces (e.g., power) for mechanical screw, or tubular-drag, conveyors. The system size dependences of the pneumatic and/or mechanical conveying under extra-terrestrial conditions have not been firmly established, so it is difficult to estimate which approach would be the most efficient at specific target transfer rates. Pneumatic systems offer many advantages related to flexibility of use. A potential drawback of pneumatic conveying is the potential mixing of desirable volatiles with the conveying gas – which may then require additional separation processes to recover the desired volatile gas. Another disadvantage of pneumatic conveying is the need to carry a gas source for the transport or obtain the gas from a waste stream of some other process (even if carrying a gas source might eliminate the need for a compressor, it could still put an absolute limit on the total regolith conveying/transport capacity for the mission).

11.6.2 Mechanical Conveying

The most versatile (or mission-flexible) conveying method is probably a nearly autonomous robotic rover which can go to a desired location, collect a sample and return with it. This is also likely one of the most energy intensive and operationally complex approaches. For fixed-location operations terrestrial experience has shown that the most energy-efficient mechanical systems are those which minimize friction (through the use of bearings on axles, or by floating on water). For dry terrestrial material transport rail cars or conveyor belts are the most energy efficient – they also, however, involve significant investment in capital (i.e., mass and infastructure). Recognizing that frictional losses related to the weight of a material are greatly reduced under lunar gravity, can significantly reorder the efficiency evaluation of various modes of motion, and make a number of transport means viable that would seldom be considered terrestrially. For example, ballistic transfer (i.e., tossing the material to its destination, like a precision snow-blower, or catapult) can become relatively energy efficient, but may require significant technology development to become robust (and minimize dust and/or other hazards). Since, under lunar gravity, frictional resistance to dragging something over a surface is greatly reduced, a sled pulled back and forth along a tether line (or on a cable loop, with a pulley at each end) might be nearly as energy-efficient as any terrestrial conveyor belt or rail car, but could require a very minimal infrastructure to put in place. Such a sled-on-tether approach could enable a very small excavating rover to obtain regolith from a variety of locations (sending material back on a tether-traveling sled) and the only limit to range would be the length of the tether cable, and the processing capacity of the base or lander facility. The combination of surface obstacles and a tether cable, might put restrictions on such an excavating rover's path between sample collection sites, and other minor complications could arise in providing mechanical means to fill and empty regolith containers (since gravity-flow may not be a reliable material transfer means). Dust, of course, may be an issue with a sled based system.

If mission flexibility requirements are modest – needing only to excavate or mine from a fairly limited number of locations within 10's of meters of a regolith-processing facility, then an additional variety of mechanical means may also become viable energy-efficient options, allowing selection decisions to be based on other important factors (like retention of volatiles, ease of implementation, robustness and flexibility of system, etc). For example, tubular drag conveyors and screw conveyors are enclosed, dust-free, granular material conveying systems which drag material along inside of pipes or tubes with weight-dependent friction being a major contributor to the total conveying power required. That weight-dependent component of power-loss is significantly reduced under lunar gravity, so other factors or features of the conveying become more important in determining the viability of such conveying methods in meeting mission requirements.

11.6.3 Tubular-Drag Conveyors

Terrestrially a variety of commercial suppliers provide tubular-drag conveyors for granular solids (e.g., Hapman's Disc-in-Tube systems on chains or cables; Spiroflow's CABLEflow cable-drag conveyors; MPE's Chain-vey tubular drag-chain conveyors; and Cablevey's cable-conveyors, etc.). As the name suggests the disc-like pushing plates for these systems are linked together along a central chain or cable with the discs spaced within about 2 tube-diameters of each other along the tensile-force delivering cable or chain. Special turn-around pulleys, located before the fill-point and after the emptying location, accomplish relatively tight-radius turns, outside of the tube enclosure, at each end of the conveying line. Filling is generally accomplished via gravity-flow through the bottom of a feed hopper, and emptying usually involves having an opening in the bottom half of the tube, again allowing gravity-flow to occur. Achieving gravity-flow filling and emptying may be problematic for small systems under lunar-gravity, so that if tubular-drag conveyors were utilized, they might need to incorporate some flow-assistance devices at fill and empty locations.

For tubular-drag conveyors it is perhaps important to note a significant caveat regarding potential reduction in frictional conveying losses with reduced gravity. Although the material weight-related frictional drag (a major source of the conveying power requirement of such conveyors) is reduced at low-g, the friction of the discs rubbing on the tube walls includes an additional factor which depends on the direction changes, or bends, in the conveying line. The conveying tension in the central cable or chain undergoes changes in direction by applying a perpendicular rubbing force on the walls (applied through the discs) – each bend adds to the overall power requirements of the conveying line. Most of the bend-related friction is caused by the friction of the conveying discs against the wall, and thus it exists whether the line is empty of full (this component increases somewhat with material loading in the tube, but will do so less under reduced gravity than terrestrially). An additional infrastructure factor also needs to be considered in evaluating overall efficiency of tubular conveyors – they have an inherent mass disadvantage because there is a need to provide both conveying and return paths, typically making the tube and cable (or chain) lengths twice the overall conveying distance.

The linear speed in tubular drag conveyors typically is kept below around 10 cm/s (and the same linear speeds are used for systems ranging from 7.6 cm to 30.5 cm in diameter). If those same linear speeds are appropriate for drag tube conveyors that are scaled down in size, then a 2 cm diameter tubular-drag conveyor could convey excavated regolith at a rate of ~50 kg/hr. The most compelling argument against tubular drag conveyors is the challenge to keep the mechanisms functioning for extended periods without jamming (and the effort required to restore the system to full functioning status if a jam occurs). While tubular-drag conveyors can be very robust for certain materials they have not been proven to be as reliable for some materials with very broad size ranges. Because of this, and the continuous frictional rubbing nature of their transport, they might be most suitable for consideration for short duration applications.

11.6.4 Screw Conveying

The designs for typical mechanical conveying of solids via a rotating central screw represent only minor modifications from the 2000 year old Archimedes screw designed for lifting water. As a central screw turns slowly, granular material rises partly up the screw but continuously slides down the screw face (due to the gravitational force), and thus slides axially along the pipe wall. The frictional sliding-losses from this type of screw conveying are inherently higher than those of tubular-drag conveyors, because the granular material slides not only along the pipe wall, but it also continuously slides down the tilted screw face (with a reduced normal force and lower friction, because of the tilt angle of the screw). For vertical conveying, or for steep inclines, this mode of conveying does not function at all, however. For such cases the rotation rate of the inner screw has to be high enough to cause the granular material to be slung against the pipe wall (all the way around the circumference) and have high enough friction to prevent sliding of the material back down the tube. In this centrifuging-swirl, or vortex-flow, mode the path the granular material takes is roughly three times the conveying length (because of its helical path along the pipe wall) and thus, most vertical screw conveyors seldom achieve conveying efficiency greater than about 1/3 of that of horizontal screw conveyors (Roberts 1999, 2005). Because screw conveyors have this inherent inefficiency for vertical transport, one might conclude that they would make poor candidates for solids conveying on extraterrestrial missions, where efficiency is a premium. Under lunar gravity, however, the energy efficiency of conveying is quite different than it is under terrestrial conditions. For example, the slow-flow mode of horizontal screw conveying may only exist under lunar gravity for fairly large-diameter screw conveyors, but the magnitude of the normal-force from centrifuging material on the duct-wall that is needed to make the screw function in a vortex flow mode, is much smaller than in terrestrial conveying. The duct-wall forces facilitating rapid-rotation transport of material under lunar gravity would be independent of conveying orientation or tilt angle. Another important feature of screw conveying is that the conveyed material can be 'pushed' a short distance (about a pipe diameter or so) out of a side opening, or Y-stub tube, at the end of a screw conveyor, without depending on the force of gravity to remove material from the conveying duct. The robust nature of screw feeders operating in a vortex-flow mode, allows this method of conveying to be considered in trade-off studies of conveying methods for cohesive materials under lunar gravity. A 5 cm diameter screw conveyor rotating at 200 rpm could probably transport regolith at a rate of ~50 kg/hr under lunar conditions.

There is ample anecdotal evidence that screw conveyors also suffer from occasional jamming, with sometimes considerable difficulty to restore full functionality. It is not entirely clear, however, whether such problems could be greatly minimized by appropriate selection of the ratio of screw-to-casing gap size compared to the maximum feedstock particle size, and/or through selection of other operational parameter settings. For example, while some researchers state that their experience demonstrates that screw conveyors simply don't function for vertical conveying, commercial systems, like Cargotec's Siwertell ship unloaders, regularly move up to

2400 ton/hr using vertical screw conveyors (albeit of materials within well controlled size ranges). Any such purely mechanical, friction-based conveying method would, of course, need to be thoroughly vetted with a wide range of operational condition testing before being considered appropriate for space or lunar applications.

11.7 Discussion and Conclusion

Experiments with micron-scale pharmaceutical powders in a centrifuging, rotating-drum micro-avalancher, covering g-levels from 12.5 to 1200 (a factor of 100 variation in g-level) clearly demonstrate the changes from clumping (with no flow), to avalanching flow, to free-flowing behavior as the effective g-level is increased. A mere factor of four change in effective g-level (from $25g_o$ to $100g_o$) was sufficient to show a significant change from avalanching behavior to free flowing behavior for more than one powder tested. Extrapolation of this same behavior to gravity levels below our terrestrial level (such as to the 1/6 $-g_o$ conditions on the moon) would indicate that Lunar regolith will exhibit more 'cohesive' behavior in processing, transfer and handling equipment than the same powder would exhibit terrestrially. Thus, Lunar in-situ resource utilization (ISRU) processes may need to use larger size openings, steeper slopes or non-gravity driving forces in processing, transfer and handling equipment than would be used for comparable powders and processes on earth.

Most of the lunar surface is charged to a significant potential (of at least several volts) due to the UV photoelectric emissions on the sunlit side, and the solar wind bombardment on the dark side (Colwell 2007) (resulting in a net positive/negative surface charge on the lit/dark side, respectively). The net charge on the regolith particles, and the lack of a conductive ground to neutralize charges, will lead to substantial static-electric-charge effects. Both the increase in cohesion due to the higher surface energies, and the high static-charges will increase interparticle forces over those exhibited by the powders tested in this research. Thus, in addition to the effects associated strictly with changes in the g-level, increased cohesion, and static charges will compound problems associated with powder flow in ISRU operations. Care needs to be taken to design ISRU equipment which can compensate for these incompletely understood effects.

Under lunar gravity the cohesive nature of powders will start to have an influence on the bulk deformation and flow properties of granular materials at much larger particle sizes than is typical of terrestrial powder flow operations. There are a variety of reasons to expect powders made from lunar regolith to be cohesive. First, decreasing particle size increases cohesion relative to other forces acting on particles in a powder, and lunar regolith contains a very significant fraction of very fine particulates. Second, the lack of humidity and an atmosphere has left the surfaces more chemically reactive than typical terrestrial materials, thus the surface energy is likely higher than is typical in mineral processing operations terrestrially. Third, the lack of an atmosphere eliminates aerodynamic drag forces which are a major factor acting on particles in powders terrestrially. Elimination of aerodynamic forces may allow the effects of interparticle cohesion to be more apparent than would be the case if aerodynamic forces overwhelmed interparticle cohesion. Fourth, the lunar surface is likely charged to several volts, and thus, some of the surface regolith particles will have a net charge, and exhibit static-electric interactions with each other and with other surfaces. Finally, decreasing gravity decreases the major driving force acting on materials in many processing operations, thus causing assemblies of particles (i.e., powders) to appear to be more cohesive in their bulk behavior than they would on earth.

Some points to keep in mind when considering designs for regolith processing or handling equipment:

- Dust will be a factor for any mechanical equipment (all small crevasses will collect dust).

- Electric charges will be present on fine dust particles (compounding dust issues).

- Gravity-flow of bulk material will be slower (and probably less reliable) than terrestrial systems.

- Materials will appear to be more cohesive than they would terrestrially.

- If equipment utilizes gravity to function, it will probably need to be larger in size than comparable terrestrial designs.

- Proof-testing of processing and handling equipment (or components) in an appropriate environment (reduced gravity and/or vacuum) is necessary.

Acknowledgments. Portions of this work were supported under various NASA contracts with Glenn Research Center and Kennedy Space Center. That support is gratefully acknowledged. The careful reading and suggestions of reviewers are also greatly appreciated.

References

Abdullah, E.C., Geldart, D.: The use of bulk density measurements as flowability indicators. Powder Technol. 102, 151–165 (1999)

Adamson, A.W.: Physical Chemistry of Surfaces, 3rd edn. Wiley, New York (1976); 5th edn. Wiley, London (1990)

Arslan, H., Batiste, S., Sture, S.: Engineering Properties of Lunar Soil Simulant JSC-1A. ASCE J. Aerospace Engng 23(1), 70–83 (2010)

ASTM, The time that a given mass of powder takes to discharge through a hopper. Book of ASTM Standards, Part 9, American Society for Testing and Materials, Philadelphia, p. 45 (1978)

Bagnold, R.: The Physics of Blown Sand and Desert Dunes, p. 265. Methuen, London (1941)

Bailey, A.I., Price, A.G., Kay, S.M.: Interfacial energies of clean mica and of monomolecular films of fatty acids deposited on mica, in aqueous and non-aqueous media. Spec. Discuss. Faraday Soc. 1, 118–127 (1970)

Brown, R.L., Richards, J.C.: Principles of Powder Mechanics. Pergamon Press, Oxford (1970)

Carr, R.L.: Evaluating flow properties of solids. Chem. Engng 18, 163–168 (1965)

Carrier III, W.D.: The four things you need to know about the geotechnical properties of lunar regolith. Lunar Geotechincal Institute, Lakeland, FL, USA (2005)

Carrier III, W.D., Olhoeft, G.R., Mendell, W.: Physical Properties of the Lunar Surface. In: Heiken, G.H., Vaniman, D.T., French, B.M. (eds.) Lunar Source Book, pp. 475–594. Cambridge University Press (1991)

Carson, J.: Flow Properties Test Report Lunar Simulant JSC-1A. Jenike and Johanson, Rept. 10583-1 (private communication) (2010)

Castellanos, A., Valverde, J.M., Quintanilla, M.A.S.: The Sevilla Powder Tester: A tool for Characterizing the Physical Properties of Fine Cohesive Powders at Very Small Consolidations. Kona 22, 66–80 (2004)

Colwell, J.E., Batiste, S., Horanyi, M., Robertson, S., Sture, S.: Lunar regolith: dust dynamics and regolith mechanics. Reviews of Geophysics 45, RG2006 (2007), doi:10.1029/2005RG000184

Greenberg, P.S., Chen, D., Smith, S.: Aerosol Measurements of the Fine and Ultrafine Content of Lunar Regolith. NASA TM 2007-214956 (2007)

Geldart, D.: Types of gas fluidization. Powder Technology 7, 285–292 (1973)

Grossman, J.J., Ryan, J.A., Mukerjee, N.R., Wegner, M.W.: In: Levinson, A.A. (ed.) Proc. of the Apollo 11 Lunar Science Conference, vol. 3, p. 2171. Pergamon Press, New York (1970)

Güttler, C., Krause, M., Geretshauser, R.J., Seith, R., Blum, J.: The Physics of Protoplanetesimal Dust Agglomerates, Towards a Dynamical Collision Model. Astrophysical J. 701, 130–141 (2009)

Israelachvili, J.N.: Intermolecular and Surface Forces. Academic Press, London (1991)

Hausner, H.H.: Friction Conditions in a Mass of Metal Powder. Int. J. Powder Metallurgy 3(4), 7–13 (1967)

Jenike, A.W.: Storage and flow of solids. Bulletin 123, Engineering Experiment Station, University of Utah (1964)

Johnson, K.L., Kendall, K., Roberts, A.D.: Surface energy and the contact of elastic solids. Proc. R Soc. London, Ser. A 324, 301, 131 (1971)

Johnson, S.M., Walton, O.R.: A Model for Surficially Variant Electrostatic Charges on Dielectric Powders for Particle Method Applications. In: Powders & Grains 2009: Proc. 6th Int. Conf. on Micromechanics of Granular Media, Golden CO, AIP (2009)

Kaye, B.H.: Characterizing the flow of metal and ceramic powders using the concepts of fractal geometry and chaos theory to interpret the avalanching behavior of a powder. In: Battle, T.P., Henein, H. (eds.) Processing and Handling of Powders and Dusts, Minerals Metals & Materials Society, Warrendale, PA, pp. 277–282 (1997)

Khambekar, J.: Flow Properties Test Report Lunar Simulant NU-LHT-2M. Jenike & Johanson, Rept 10643-1 (private communication) (2009)

Leonovich, A.K., Gromov, V.V., Semyonov, P.S., Penetrigov, V.N., Shvartov, V.V.: Luna 16 and 20 investigations of the physical and mechanical properties of lunar soil. In: COSPAR Space Research XV, pp. 607–616. Akademie-Verlag, Berlin (1975) (Luna 16 & 20)

Lee, L.-H.: Adhesion and cohesion mechanisms of lunar dust on the moon's surface. In: Rimai, D.S., DeMejo, L.P., Mittal, K.L. (eds.) Fundamentals of Adhesion and Interfaces, VSP Utrecht, Netherlands, pp. 73–94 (1995)

Miller, D.P., Lechuga-Ballesteros, D., Williams, L., Tan, T., Kanda, J., Foss, W., Walton, O., Mandel, A., Cai, X.: Dispersibility of Spray-dried Raffinose: Effects of Particle Size and Relative Humidity. In: AAPS Annual Meeting, Toronto, Canada, November 10-14 (2002)

Mitchell, J.K., Houston, W.N., Scott, R.F., Costes, N.C., Carrier III, W.D., Bromwell, L.: Mechanical properties of lunar soil: Density, porosity, cohesion, and angle of friction. In: Proc. of 3rd Lunar Sci. Conf., pp. 3235–3253. MIT, Cambridge (1972)

Mitchell, J.K., Houston, W.N., Carrier III, W.D., Costes N.C.: Apollo Soil Mechanics Experiment S-200. Final report, NASA Contract NAS 9-11266, Space Sciences Laboratory Series 15, Issue 7, Univ. of California, Berkeley (1974)

Murphy, T.W., Adelberger, E.G., Battat, J.B.R., Hoyle, C.D., McMillan, R.J., Michelsen, E.L., Samad, R.L., Stubbs, C.W., Swanson, H.E.: Long-term degradation of optical devices on the moon (2010), arXiv:1003.0713v1 (astro-ph.EP)

Nakagawa, M.: Personal communication, Colorado School of Mines, Golden, CO (2006)

O'Brien, B.J.: Review of measurements of dust movements on the Moon during Apollo. Planet Space Sci. (2011), doi:10.1016/j.pss, 04.016

Prescott, J.K., Barnum, R.A. (2000) On Powder Flowability. Pharmaceutical Technology, pp. 60–84 (October 2000),
http://www.jenike.com/Articulos/on-powder-flowability.pdf

Qian, G.-H., Bágyi, I., Burdick, I.W., Pfeffer, R., Shaw, H., Stevens, J.G.: Gas–solid fluidization in a centrifugal field. AIChE Journal 47(5), 1022–1034 (2001)

Roberts, A.W.: The Influence of Granular Vortex Motion on the Volumetric Performance of Enclosed Screw Conveyors. Powder Technology 104, 56–67 (1999)

Roberts, A.W.: Design Considerations And Performance Evaluation Of Screw Conveyors. Centre for Bulk Solids and Particulate Technologies, U. Newcastle, Australia (2005)

Salisbury, J.W., Glaser, P.E., Stein, B.A., Vonnegut, B.: Adhesive Behavior of Silicate Powders in Ultrahigh Vacuum. J. Geophys. Res. 69(2), 235–242 (1964)

Shamlou, P.A.: Handling of bulk solids - Theory and practice. Butterworths, London (1990)

Schmidt, H.: The Apollo Experience: Problems Encountered with Lunar Dust. In: Biological Effects of Lunar Dust, Workshop, Sunnyvale, CA, USA, March 2005, pp. 29–31 (2005)

Schulze, D., Wittmaier, A.: Flow Properties of Highly Dispersed Powders at Very Small Consolidation Stresses. Chem. Eng. Technol. 26(2), 133–137 (2003)

Schwedes, J.: Review on testers for measuring flow properties of bulk solids. Granular Matter 5, 1–43 (2003); also, Schwedes, J.: Measurement of flow properties of bulk solids. Powder Technology 88 (3), 285–290 (1996)

Stubbs, T.J., Vondrak, R.R., Farrell, W.M.: Impact of dust on lunar exploration. In: Proc. of Int. Conf. Dust in Planetary Systems, Kauai, HI, ESA SP-643 (2007)

Svarovsky, L.: Powder Testing Guide: Methods of Measuring the Physical Properties of Bulk Powders. Elsevier Applied Science, London (1987)

Tardos, G.: Arching in Hoppers: Jenike's Method of Hopper Design. A tutorial at Univ. Florida, Engineering Ressearch Center for Particle Technology (1999),
http://www.erpt.org/992Q/tard-00.htm

Taylor, L.A., Schmidt, H.H., Carrier, W.D., Nakagawa, M.: The Lunar Dust Problem: From Liability to Asset. In: AIAA 1st Space Exploration Conference: Continuing the Voyage of Discovery, Orlando, Florida, paper 2510, January 30- February 1 (2005)

Walton, O.R., Braun, R.L.: Simulation of Rotary-Drum and Repose Tests for Frictional Spheres and Rigid Sphere Clusters. In: Joint DOE/NSF Workshop on Flow of Particulates and Fluids, Ithaca, NY, USA, September 29-October 1 (1993), paper,
http://www.grainflow.com/index_files/
Rotary_Drum_Simulation_DOE-NSF-1993f

Walton, O.R., De Moor, C.P., Miller, D.P.: Simulation of Low-Stress Compaction of Cohesive Micron-Scale Powders. In: AIChE 2003 Annual Meeting (Session T4-35a), San Francisco, CA, USA, November 16-21 (2003)

Walton, O.R.: Adhesion of Lunar Dust. NASA/CR—2007-214685 (2007),
http://gltrs.grc.nasa.gov/reports/2007/CR-2007-214685.pdf

Walton, O.R., De Moor, C.P., Gill, K.S.: Effects of gravity on cohesive behavior of fine powders: implications for processing Lunar regolith. Granular Matter 9(5), 353–363 (2007)

Williams, R., Shao, R., Overfelt, R.A.: The flowability of fine powders in reduced gravity conditions. Granular Matter 10(2), 139–144 (2006)

12 Power System Options for Lunar Surface Exploration: Past, Present and Future

12.1 Introduction

Robotic and human surface exploration of the Moon has been, and still is, a primary goal in the short- as well as long-term exploration program of many space exploration and exploitation agencies.

The ambient conditions present on the lunar surface, however, are very challenging or even prohibitive for many of the proven and innovative power system elements considered for energy storage and/or power generation applications in future space exploration missions. Spacecraft power system design for lunar surface applications therefore requires a very careful evaluation of the possibilities and the limitations of each individual power system element in order to find a single element - or a combination of multiple power system elements - satisfying the energy storage and power generation requirements of a specific lunar surface exploration mission element in the best possible way; and this not only with respect to gravimetric and volumetric figures, but also with respect to operational safety, performance degradation, and compatibility with operating conditions, for instance.

The slow revolution period of the Moon of approximately 27.3 terrestrial days results in long and cold lunar nights as well as long and hot lunar days. Temperature-sensitive power system elements such as batteries can therefore hardly be used without an insulating compartment or an active thermal control system. The same is also true for low-temperature hydrogen/oxygen fuel cells where liquid water is normally produced as the electrochemical reaction product. Sub-zero degree Celsius operation of low-temperature hydrogen/oxygen fuel cells is therefore possible within certain limits, but is certainly a challenge with respect to water ice formation during start-up and deactivated storage.

Due to the fact that the Moon does only have a very thin atmosphere – the surface pressure at night is in the order of $3 \cdot 10^{-15}$ bar and the total mass of the atmosphere is only approximately 25,000 kg (NASA Moon Fact Sheet 2010) - the solar irradiance available on the surface of the Moon is equal to the air-mass-zero irradiance available to satellites operated in Earth orbit, but with the limitation of a long day and night cycle. Solar power is therefore a good option and predictable in output due to the lack of cloud formation and dust storm conditions, as faced with the Earth and Mars, for instance. Intermediate energy storage options, however, have to be considered for nighttime operation and thus add a considerable mass penalty to an all-solar power system.

Intermediate energy storage in a day/night cycle could be considerably reduced or even eliminated if the base or surface-bound mission element was operated in a region receiving as much sun as possible, and thus minimizing or even eliminating the need for intermediate nighttime energy storage. Such regions could be located near the poles of the Moon. The Moon's spin axis is nearly perpendicular to the ecliptic plane. Polar regions having a low elevation, such as the floors of impact craters, may therefore never be directly illuminated, whereas regions of high elevation may be permanently illuminated (Bussey and Spudis 2004).

Solar illumination therefore has to be investigated not only as a function of seasonal and diurnal illumination conditions, but also a function of topographical factors (Zuber and Smith 1997; Lia et al. 2008). Lia et al. introduced a theoretical model estimating solar illumination conditions as a function of solar altitude and topographical factors. The model predicts an illumination of the lunar surface if the solar altitude is non-zero and all the elevations within 210 km in solar irradiance direction are smaller than the critical elevations; otherwise, the surface is shadowed (Lia et al. 2008). This theoretical analysis can be combined with experimental data to provide a comprehensive understanding of diurnal and seasonal illumination conditions.

The Clementine mission recently imaged both lunar poles during summer for the northern hemisphere. One of the main conclusions drawn from these investigations is that no areas near the south pole are constantly illuminated, although several regions exist that receive more than 50% sunlight in winter. Certain regions on rims of craters and ridges were collectively illuminated for approximately 98% of the time. At the north pole, several areas on the rim of Peary crater were constantly illuminated for an entire summer day (Bussey 1999; Bussey 2005). Diurnal energy storage would certainly be a different issue with mission elements operating in these locations than with systems operated in equatorial region, for instance. It will certainly have to be investigated whether or not these (almost) constantly illuminated regions represent good candidate sites for future landing and surface operations, though.

Power systems for lunar surface applications generally have to provide a very high degree of robustness, should be extremely reliable and on-site installation should be very easy. In addition, the power systems should be very flexible to enable high-output power applications even during nighttime or in case of an emergency, a high predictability of the diurnal and seasonal output power availability is also desirable to enable time-, application- and energy-optimized mission planning.

Thus, there are many different requirements and challenges that power systems for lunar surface exploration have to meet, and there are a number of power system technologies either readily developed and at a level of technical maturity that enables near-future applications, or currently still under development, that can actually satisfy these requirements either alone or in a combined operation with other power system elements.

This chapter first presents how the power system requirements of spacecraft and subsystems have been satisfied in past soft-landing lunar missions. Based on an assessment of the possibilities and limitations of these proven approaches in power system design, as well as recent advances in innovative power system technologies, different power system options are discussed for specific applications in future lunar surface exploration scenarios.

12.2   The Timeline of Lunar Exploration

Within this chapter, the power systems developed and applied in lunar surface exploration applications are discussed in five different periods:

- Early flyby and impact missions (1959-1965)
- Soft landing and orbiter missions (1966–1969)
- Crewed landing, rover and sample return missions (1969-1975)
- Recent missions (1990-2010)
- Future lunar exploration (2010+)

At first, an overview of the lunar exploration missions conducted within the different periods is given. Each of the lunar exploration missions launched within the period is briefly presented with launch date, mission name and mission outcome. This overview includes flyby probes, orbiters, landers, return probes and sample return missions. The term *lander* is exclusively used for soft-landing mission elements designed to operate on the surface of the Moon over a certain period of time upon landing. Mission elements intentionally - or unintentionally - reaching the lunar surface without performing a soft-landing operation are referred to as mission elements *impacting* the lunar surface.

The focus of this chapter is put on power system options for lunar surface exploration applications. Only the power systems of selected landers as well as autonomously operating surface exploration mission elements such as automated rovers or Earth-return capsules are subsequently further discussed in detail.

Power systems for flyby and return probes as well as orbiters and mission elements intentionally impacting the lunar surface are subject to completely different requirements and operating conditions than spacecraft designed for lunar surface operation. The power systems of these mission elements are therefore neither considered nor discussed in detail within this chapter.

The timeline presented in the following is modified from the National Space Science Data Center (NSSDC) lunar exploration chronology (NSSDC Chronology 2010).

12.3 Early Flyby and Impact Missions (1959-1965)

The first man-made object to be successfully launched into the direction, and actually reaching the vicinity, of the Moon was the Soviet Luna 1 spacecraft. Luna 1 was soon followed by other U.S. and Soviet spacecraft designed and operated to conduct first scientific investigations as well as to test concepts and technologies required with subsequent exploration missions.

The first man-made object to reach the lunar surface was the Soviet Luna 2 probe, which impacted on the Moon in September 1959. The far side of the Moon was first photographed in October 1959 by the Soviet Luna 3 probe, launched less than a month after Luna 2.

A number of impact missions were subsequently conducted in the first half of the 1960s. These missions were designed to achieve a lunar impact trajectory, and to transmit high-resolution photographs of the lunar surface during the final minutes of flight up to the impact onto the lunar surface, and thus the inevitable destruction of the spacecraft. This approach was extensively applied with the U.S. series of Ranger missions launched until 1965, whereas the Soviet space program rather focused on developing and testing technologies and spacecraft intended for soft landing procedures.

The Soviet Luna 5 through 8 spacecraft were all launched in 1965 and were intended for tests of soft landing procedures. Different problems with altitude control systems, airbags and retrorocket systems, however, still prevented the first successful soft landing of a spacecraft on the lunar surface in 1965. This was only to succeed in 1966 with the Luna 9 spacecraft.

Zond 3 was another Soviet spacecraft launched in 1965. Following the Zond 2 spacecraft launched in 1964 attempting a flyby of Mars, Zond 3 performed a successful flyby and took a series of pictures of the Moon.

The chronology of lunar exploration missions launched between 1959, beginning with the first lunar flyby, and 1965, prior to the first soft landing, modified from the NSSDC Lunar Exploration Timeline (NSSDC Chronology 2010), is presented in Table 12.1.

Table 12.1 Mission timeline from 1959-1965: early flyby and impact missions

Launch Date	Mission Name	Outcome
January 2, 1959	Luna 1	Flyby
March 3, 1959	Pioneer 4	Flyby
September 12, 1959	Luna 2	Impact
October 4, 1959	Luna 3	Probe
August 23, 1961	Ranger 1	Attempted test flight
November 18, 1961	Ranger 2	Attempted test flight
January 26, 1962	Ranger 3	Attempted impact
April 23, 1962	Ranger 4	Impact
October 18, 1962	Ranger 5	Attempted impact
April 2, 1963	Luna 4	Flyby
January 30, 1964	Ranger 6	Impact
July 28, 1964	Ranger 7	Impact
February 17, 1965	Ranger 8	Impact
March 21, 1965	Ranger 9	Impact
May 9, 1965	Luna 5	Impact
June 8, 1965	Luna 6	Attempted lander
July 18, 1965	Zond 3	Flyby
October 4, 1965	Luna 7	Impact
December 3, 1965	Luna 8	Impact

12.3.1 Luna 2

Luna 2 was the second in a series of Soviet spacecraft successfully launched into the direction of the Moon on September 2, 1959. On September 13, the spacecraft released sodium gas on translunar course. This bright orange cloud of sodium gas was used for tracking as well as for studying the behavior of gas in space (NSSDC Luna 2, 2010).

Luna 2 was the first man-made object to impact on the lunar surface on September 14, 1959, when the communication with the spacecraft suddenly ceased, indicating that the hard-landing on the surface of the Moon has taken place. Luna 2 did not carry a propulsion system; a soft landing was therefore neither intended nor possible. Luna 2 was based on a spherical spacecraft design with exterior antennae and instrument parts. An image of the Luna 2 spacecraft is shown in Fig. 12.1.

Due to the fact that Lunar 2 neither carried a propulsion system for a controlled descent nor was designed for surface operations, the power system is not further discussed as the power system design issues were different from landers designed for surface operation upon performing a successful soft-landing.

Fig. 12.1 Luna 2 (Image credit: NASA/NSSDC ID: 1959-014A)

12.4 Soft-Landing and Orbiter Missions (1966-1969)

The second period in the timeline of lunar exploration missions applied within this chapter started early in 1966 when Luna 9, a Soviet automatic lunar station that successfully performed the first soft landing of a man-made object on the Moon, was launched.

The chronology of lunar exploration missions launched between 1965, beginning with the first soft landing, and 1969, prior to the first crewed landing, modified from the NSSDC Lunar Exploration Timeline (NSSDC Chronology 2010), is presented in Table 12.2.

This first soft landing was followed by a series of U.S. and Soviet orbiters and landers launched in the years 1966-1969. The Soviet Luna series sent three orbiters (Luna 10-12), a lander (Luna 13) and finally another orbiter (Luna 14) and a lander (Luna 15). Luna 15 was a member of the advanced Soviet Ye-8-5 series that was designed to return samples drilled on the surface of the Moon back to the Earth. Luna 15 was launched just three days prior to the U.S. Apollo 11 mission, to be performing the first crewed landing on the Moon, and impacted on the Moon just prior to the launch of Apollo 11.

Table 12.2 Mission timeline 1966–1969: soft landing and orbiter missions

Launch Date	Mission Name	Outcome
January 31, 1966	Luna 9	Lander
March 31, 1966	Luna 10	Orbiter
May 30, 1966	Surveyor 1	Lander
August 10, 1966	Lunar Orbiter 1	Orbiter
August 24, 1966	Luna 11	Orbiter
September 20, 1966	Surveyor 2	Attempted lander
October 22, 1966	Luna 12	Orbiter
November 6, 1966	Lunar Orbiter 2	Orbiter
December 21, 1966	Luna 13	Lander
February 4, 1967	Lunar Orbiter 3	Orbiter
April 17, 1967	Surveyor 3	Lander
May 8, 1967	Lunar Orbiter 4	Orbiter
July 14, 1967	Surveyor 4	Attempted lander
August 1, 1967	Lunar Orbiter 5	Orbiter
September 8, 1967	Surveyor 5	Lander
November 7, 1967	Surveyor 6	Lander
January 7, 1968	Surveyor 7	Lander
April 7, 1968	Luna 14	Orbiter
September 15, 1968	Zond 5	Return probe
November 10, 1968	Zond 6	Return probe
December 21, 1968	Apollo 8	Crewed orbiter
May 18, 1969	Apollo 10	Orbiter
July 13, 1969	Luna 15	Orbiter

The Soviet lunar exploration program also included further Zond missions. Zond 5 was an unmanned version of the Soyuz spacecraft and successfully performed a flyby and return to Earth in 1968 in preparation of the first Soviet crewed flight to the Moon. The capsule safely reentered the Earth's atmosphere and was successfully recovered. Although the first Soviet circumlunar Earth-return mission was a success, the program was ultimately cancelled because the N1 rocket required to take it to the Moon was never successful.

The U.S. lunar exploration program consisted of the Surveyor and Lunar Orbiter spacecraft series. The Surveyor program sent a total of seven landing craft to the Moon. The program was a success, although Surveyor 2 and 4 were a failure due to engine problems in case of Surveyor 2 and a loss of contact during the descent phase with Surveyor 4.

The Surveyor program successfully demonstrated the technology for soft landing and surface operation, and provided a comprehensive set of scientific data.

The U.S. Lunar Orbiter program consisted of a total of five orbiters launched between 1965 and 1966. The orbiters were used to take photographs of the near and far side of the Moon with resolutions down to a single meter. All of the orbiters were intentionally crashed onto the lunar surface at the end of their lifetime.

In 1968, the first Apollo flights were made to the Moon. Apollo 8 was the first crewed mission to leave Earth orbit and was successfully placed into a lunar orbit. Whereas Apollo 9 remained in Earth orbit, Apollo 10 was again sent into a lunar orbit and included extensive testing procedures of the lunar module.

12.4.1 Luna 9

Luna 9 was the first spacecraft to achieve a soft landing on the Moon on February 3, 1966, after being launched three day earlier from Baikonur Cosmodrome in the Soviet Union. Luna 9 successfully transmitted photographic data taken from the lunar surface (NSSDC Luna 9 2010).

Luna 9 was an automatic lunar station having a spherical body with a diameter of just below 60 centimeters and a mass of roughly 100 kg. The central element was a hermetically sealed container pressurized to 1.2 atmospheres. Electronics, batteries, thermal control system and scientific equipment were installed into this pressurized container. The thermal control system was designed to maintain the interior near room temperature, approximately between 19 and 30°C.

After touching down with the help of retrorockets and an airbag system, four petals opened out and four antennas were automatically deployed on the outside of the pressurized container.

The power system of the Luna 9 automatic station consisted of solar panels installed on the inside of the four petals and a battery system. Operations were powered by the battery system. The solar panels were only used to recharge the battery module (NSSDC Luna 9 2010). A picture of the Luna 9 spacecraft is shown in Fig. 12.2.

Fig. 12.2 Luna 9 (Image credit: NASA/NSSDC ID: 1966-006A)

12.4.2 Surveyor Program

Surveyor 1 was the first in a series of U.S. spacecraft launched on May 30, 1966, and achieved the first U.S. soft landing on the Moon on June 2, 1966 (NSSDC Surveyor 1 2010).

The Surveyor program was designed to demonstrate the technology necessary to achieve landing and surface operations on the Moon. In a total of seven individual missions, the technology for soft landing and operating on the surface of the Moon was successfully demonstrated. In addition, important scientific knowledge about the Moon - and the conditions encountered on the surface of the Moon - could be gained. This includes TV pictures transmitted after landing as well as information about topography, radar reflectivity and surface as well as spacecraft temperatures (NASA SP-184 1969).

Surveyor was designed on a tripod of aluminium tubing. Power, communications, propulsion, flight control, and payload systems were mounted onto this tripod. A central mast was installed in the middle of the structure, hinged landing legs were attached to the corners of the tripod. The spacecraft was about 3 meters tall, 4.5 meters in diameter and weighed just below 1,000 kg at launch and just below 300 kg in landed configuration.

Electrical power was provided by a planar solar panel installed on top of the central mast and a secondary silver-zinc main battery. The main battery was used to provide power to the spacecraft during transit, touchdown, and during nighttime. In addition, an auxiliary battery was installed with the first four spacecraft (Surveyor 1-4) to provide redundant energy storage capacity for the transit and landing phases. The auxiliary battery was a primary silver-zinc battery. A picture of the Surveyor 1 spacecraft is shown in Fig. 12.3.

Fig. 12.3 Surveyor 1 (Image credit: NASA/NSSDC ID: 1966-045A)

The silver-zinc couple, or, more appropriately, the silver oxide-zinc couple, was selected as battery chemistry of choice at the time as it provided the highest energy density (80 Wh/lb = 176 Wh/kg) and best met the mission requirements.

Secondary silver-zinc batteries, playing an important role not only within the Surveyor but, for instance, also with the Apollo program, can be described by the following electrochemical half-cell reaction equations:

$$2AgO + Zn + H_2O \underset{\text{charge}}{\overset{\text{discharge}}{\rightleftharpoons}} Ag_2O + Zn(OH)_2 \quad [1.86 \text{ V}] \qquad (12.1)$$

$$Ag_2O + Zn + H_2O \underset{\text{charge}}{\overset{\text{discharge}}{\rightleftharpoons}} 2Ag + Zn(OH)_2 \qquad (12.2)$$

with zinc hydroxide $Zn(OH)_2$ decomposing into zinc oxide and water.

The potentials cited in the equations represent open circuit values for the upper and lower plateaus, respectively. Under load, potentials of 1.7 and 1.5 V are typical levels for this cell chemistry (Moses et al. 1970).

Nominal energy requirements of the Surveyor missions were between 7,400 Wh (Surveyor 1 & 3) and 6,180 Wh (Surveyor 5). These energy requirements had to be satisfied by the combined operation of main and auxiliary battery, if included, and the central solar panel. The net energy provided by the solar panel was between 4,400 Wh (min. with Surveyor 1-3) and 5,460 Wh (max. with Surveyor VII). Net battery energy supplied by the battery was between 1,344 Wh (min. with Surveyor 5) and 3,000 Wh (max. with Surveyor 1&2).

Main battery energy for all seven Surveyor-missions was 3,450 Wh. Nominal energy of the primary auxiliary battery installed in Surveyor 1-5 was 1,060 Wh (Moses et al. 1970).

12.5 Crewed Landing, Rover and Sample Return Missions (1969-1975)

The era of crewed landings started on July 16, 1969, with the launch of Apollo 11, and the successful soft landing of the lunar module Eagle on July 20, 1969. The U.S. continued their crewed Apollo program until Apollo 17, with only Apollo 13 being a partial failure due to an oxygen tank rupture on translunar coarse. The U.S. lunar exploration program ceased with Apollo 17, whereas the Soviet program was continued until 1976. The Soviet program successfully soft-landed and operated two surface rovers, Lunokhod 1 & 2, on the Moon. Three sample return missions, safely returning up to more than 100 g of material collected in the vicinity of the descent stage, were also sent to the Moon before the lunar exploration program ceased in 1976.

The chronology of lunar exploration missions launched between 1969, beginning with the first crewed landing, and 1975, when the last lunar mission of the 1970s was launched, modified from the NSSDC Lunar Exploration Timeline (NSSDC Chronology 2010), is presented in Table 12.3.

12.5.1 Apollo Program Lunar Module

Within the U.S. Apollo program, the Lunar Module (LM) and the Lunar Roving Vehicle (LRV) are of particular interest with respect to lunar surface power system design and operation.

The Lunar Module was the landing element developed and successfully operated within the Apollo program. It was designed to soft-land a crew of two from lunar orbit, and to safely return them back into lunar orbit after the surface exploration stay.

The Lunar Module consisted of an ascent and a separate descent stage and was docked to the Command and Service Module during translunar coarse. After completing the surface visit and transferring the crew back to the command/Service Module, the Lunar Module was discarded.

Table 12.3 Mission timeline 1969-1975: crewed landings, surface rovers and sample return missions

Launch Date	Mission Name	Outcome
July 16, 1969	Apollo 11	Crewed landing
August 7, 1969	Zond 7	Return Probe
November 14, 1969	Apollo 12	Crewed landing
April 11, 1970	Apollo 13	Aborted landing
September 12, 1970	Luna 16	Sample return
October 20, 1970	Zond 8	Return probe
November 10, 1970	Luna 17	Rover
January 31, 1971	Apollo 14	Crewed landing
July 26, 1971	Apollo 15	Crewed landing
September 2, 1971	Luna 18	Impact
September 28, 1971	Luna 19	Orbiter
February 14, 1972	Luna 20	Sample return
April 16, 1972	Apollo 16	Crewed landing
December 7, 1972	Apollo 17	Crewed landing
January 8, 1973	Luna 21	Rover
June 2, 1974	Luna 22	Orbiter
October 28, 1974	Luna 23	Lander
August 14, 1976	Luna 24	Sample return

The original plan for the Lunar Module power system was to include fuel cell power plants in the descent stage to provide power during descent and surface operations, while the ascent module was initially to be powered by battery modules providing power during the short ascent and rendezvous phase with the Command and Service Module. Due to various problems with the fuel cell power plant, the plan was altered and battery modules were also used with the descent module power system. The relatively short surface operation (less than 65 hours until Apollo 14, up to 72 hours with Apollo 15-17) and the low average power consumption of approximately 1,100 Watts made an all-battery-powered solution possible (Polsgrove et al. 2009).

An overview of the Lunar Module power system is given in the following. The information is primarily taken from the Apollo mission summary report (NASA JSC-09423 1975) published three years after the final Apollo mission number 17 had ended.

The Lunar Modules were powered solely by silver-zinc (AgZn) batteries. Until Apollo 14, the power system consisted of two ascent batteries rated at 296 Ah each and four descent batteries rated at 400 Ah each.

The increased lunar stay of up to 72 hours demanded with the Apollo 15-17 missions required a re-design of the descent module batteries to deliver 415 Ah instead of 400 Ah each and a total of five instead of four batteries.

Each ascent battery weighed 56.25 kg and measured 127 x 203.2 x 914.4 mm. The ascent batteries were used during lift-off, rendezvous and docking with the command and service module. In addition, the ascent batteries could also be used to provide online-support of the descent batteries in the event of an abort during lunar descent. In case of one of the two ascent battery modules failing, the other module could provide sufficient power to accomplish safe rendezvous and docking.

Each descent battery weighed 60.33 kg and measured approximately 228.6 x 254 x 431.8 mm. The descent batteries provided small heater loads early in the mission, lunar descent power and lunar surface stay power. In terms of total energy requirements for both, the four- and the five-battery configuration missions, there was energy margin of approximately one battery. However, in terms of available output power, one battery could, under emergency conditions, meet the power demands of the entire lunar module. Two batteries could nominally supply power to the limits of their specified capacity at total spacecraft loads.

The Lunar Module batteries performed above the specified requirements when emergency power was needed during Apollo 13 after the loss of the Command and Service Module fuel cell power plant. However, postflight analysis revealed that an unexpected current spike occurred during transearth coast. The spike was associated with the occurrence of a "thump and snowflakes" reported by the crew. The postulated cause was the venting of potassium hydroxide by one of the descent batteries creating a short circuit, igniting the mixture of hydrogen and oxygen normally produced by a silver-zinc battery. The resulting explosion blew the battery cover off and vented the electrolyte into space, thus causing the aforementioned "thump and snowflakes".

Another flight problem occurred during the translunar coast period of the Apollo 14 mission. A lunar module ascent battery indicated a lower-than-expected open circuit voltage (0.3 Volt decay). Real-time battery testing (both on the ground and in the lunar module) supplied the necessary confidence that the battery would perform the required flight functions and it was decided to proceed as planned. No differences between the test battery and the flight battery were observed throughout the mission. An illustration of the Apollo Lunar Module is shown in Fig. 12.4.

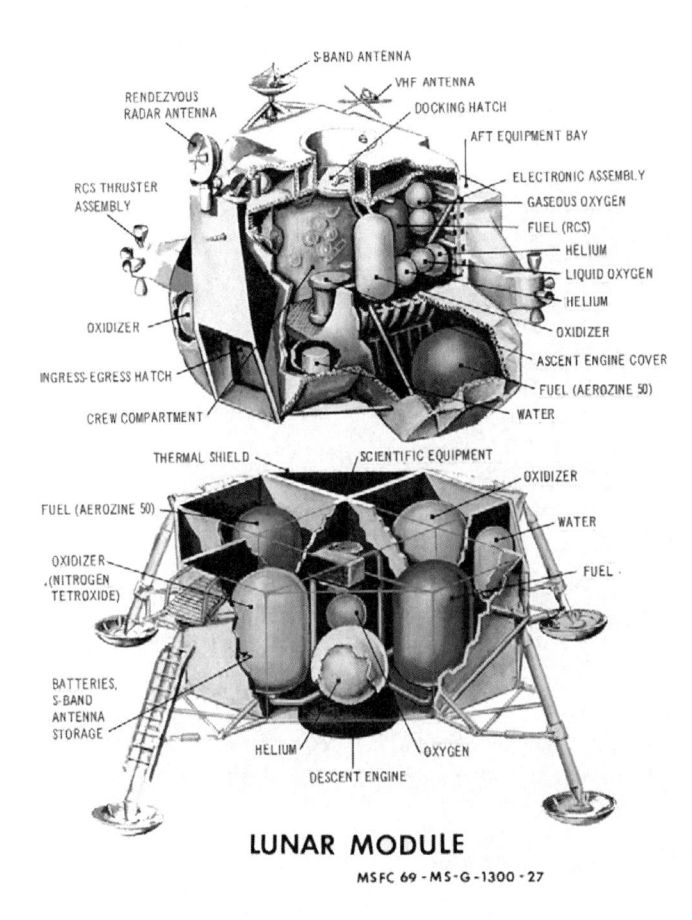

LUNAR MODULE

MSFC 69 - MS - G -1300 - 27

Fig. 12.4 Lunar Module illustration (Image credit: NASA)

12.5.2 Apollo Program Lunar Roving Vehicle

The Lunar Roving Vehicle (LRV) was a battery-powered four-wheeled rover developed within the Apollo program. The Lunar Roving Vehicle was used on the Moon during the last three missions of the Apollo Program (Apollo 15, 16 and 17) flown in 1971 and 1972. The LRVs were used to provide the astronauts greater surface mobility during their extravehicular activities. In the previous Apollo missions, the astronauts were restricted to short walking distances around the landing site; this operating range was significantly increased with the LRVs, although the range was still restricted to remaining within walking distance of the Lunar Module in case of a failure of the LRV.

The LRVs weighed 209 kg and had a length of 310 cm. The operational lifetime of the LRVs on the moon was 78 hours. They could make any number of sorties up to a cumulative distance of 92 kilometers (Boeing 1972). A picture of the Apollo Lunar Roving Vehicle is shown in Fig. 12.5.

Fig. 12.5 Lunar Roving Vehicle (Image credit: NASA)

An overview of the LRV power system is given in the following. The information is primarily taken from the Apollo 17 LRV Technical Information (Boeing 1972).

The LRVs' only power source were two 36 Volt non-rechargeable silver-zinc batteries. Each battery had a nameplate capacity of 120 Ah and contained individual 23 cells. Each battery weighed 26.8 kg.

During lunar surface exploration, both batteries were normally used simultaneously on an approximate equal load basis. The batteries were located on the forward chassis and were enclosed by a thermal blanket and dust covers. If one battery failed, the entire electrical load could be switched to the remaining battery.

The drive motors were direct-current series which operated from a nominal input voltage of 36 VDC and provided 0.25 horse powers each (190 Watt).

Drive times of the LRVs during Apollo 15/16/17 were 3:02, 3:26 and 4:29. The surface distances traveled with Apollo 15/16/17 were 27.9, 26.7 and 33.8 km (NASA JSC-09423 1975).

The ampere-hours consumed during Apollo 15/16/17 were 52.0, 88.7 and 73.4. From the total of 242 Ah available, this represents some 21.5, 36.7 and 30.3% of the total amperage.

12.5.3 Luna 16/20/24: Sample Return Missions

Luna 16, 20 and 24 were three successful Soviet sample return missions. Luna 16 was launched on September 12, 1970, and soft-landed on September 20, 1970. After 26 hours and 25 minutes on the lunar surface, the ascent stage successfully lifted off from the Moon carrying 101 grams of collected material in a hermetically sealed sample container. The Luna 16 re-entry capsule returned directly to the Earth and landed on September 24, 1970 (NSSDC Luna 16). Luna 16 thus managed the third retrieval of lunar soil samples and was only preceded by the U.S. Apollo 11 and 12 missions.

Luna 20 was launched approximately one and a half years after Luna 16 on February 14, 1972, and soft-landed on February 21, 1972. The ascent stage of Luna 20 was launched from the lunar surface on February 22, 1972, carrying 30 grams of collected lunar samples in the sealed Earth-return capsule. It landed in the Soviet Union on February 25, 1972 (NSSDC Luna 20).

Luna 24 was launched more than four years after Luna 20 on August 14, 1976 and soft landed on August 18, 1976. The ascent stage of Luna 24 was launched from the lunar surface on August 19, 1976, carrying 170.1 grams of collected lunar samples in the sealed Earth-return capsule. It landed in the Soviet Union on August 22, 1976 (NSSDC Luna 24). An image of the Luna 16 landing craft is shown in Fig. 12.6.

Fig. 12.6 Luna 16 landing craft (Image credit: NASA/NSSDC ID: 1970-072A)

The Luna 16/20/24 spacecraft featured a two-stage design consisting of a small ascent stage mounted on top of a descent stage. The descent stage also acted as a launch pad for the ascent stage. The ascent stage was a smaller cylinder with a rounded top. It carried a cylindrical and hermetically sealed soil sample container inside a small re-entry capsule. The spacecraft descent stage was equipped with a television camera, radiation and temperature monitors, telecommunications equipment, and an extendable arm with a drilling rig for the collection of the lunar soil sample.

The complete spacecraft had a launch mass of 5,750 kg and a height of 3.96 m. The ascent stage had a weight of 520 kg (of which 245 kg were propellant) and a height of approximately 2 m. The return capsule had a weight of 39 kg and a diameter of 50 cm (Harvey 2005).

Due to the short mission durations of less than two weeks, the spacecraft could be equipped with a rather simple power system. Both, descent and ascent stage were powered by batteries (Engelhardt 2005). Battery power also enabled landing and operating the spacecraft in lunar night conditions. Ascent stage and return capsule were also both battery-powered.

12.5.4 Luna 17/21: Lunokhod 1 & 2 Rovers

Luna 17 and 21 were two successful Soviet missions landing surface rovers on the Moon. Luna 17 was launched on November 10, 1970, and soft-landed on November 15, 1970 (NSSDC Luna 17). Luna 21 was launched roughly two years later on January 8, 1973 and soft-landed on January 15, 1973 (NSSDC Luna 21).

The landing modules of Luna 17 and 21 had ramps by which the payload, the Lunokhod 1 and 2 rovers, descended onto the lunar surface. The Lunokhod surface rovers were based on a central tub-like compartment standing on eight independently powered wheels. The tub-like compartment could be closed by a lid to prevent excess cooling during night. An array of different antennae, television cameras, devices for analyzing soil samples and scientific equipment, such as an x-ray spectrometer, were installed.

The rovers were approximately 135 cm high and had a mass of 840 kg. They were about 170 cm long and 160 cm wide and had 8 wheels, each with an independent suspension, motor and brake. The rovers had two speeds: approximately 1 and 2 km/h. Images taken by multiple TV camera systems were used by a team of controllers on Earth to drive the rover in real time (NSSDC Luna 21).

Power was supplied by a solar panel installed on the inside of a hinged lid covering the tub-like instrument bay. Solar power received by the solar panel when the lid was opened was used to recharge the batteries. The main container was kept warm by the decay of a polonium-210 heat source (Harvey 2005). A heat exchanger was used to cool the rover during daytime; the lid was closed during nighttime to prevent the interior from cooling down too low. The interior of the compartment could thus be kept at near room temperature within a temperature range of 15-20°C (Engelhardt 2005).

Lunokhod 1 was intended to operate through three lunar days, but actually operated for eleven lunar days. The operations of Lunokhod 1 officially ceased on October 4, 1971, almost eleven Months after the landing.

The science payload of the Lunokhod rovers also included a French-built laser reflector; the reflector of Lunokhod 1 recently generated a renewed interest into the almost 40-year-old rover.

The laser reflector was used in a couple of range measurements made within the first three months of the landing of Lunokhod 1, but these results are both unpublished and unavailable. Apart from a couple of glimpses of the reflector seen in the 1970s, Lunokhod 1 was effectively lost until March of 2010, when images from the Lunar Reconnaissance Orbiter determined its coordinates with uncertainties of about 100 m. This allowed the Apache Point Observatory Lunar Laser-ranging Operation to quickly acquire a laser signal of the reflector. The reflector appears to be in excellent condition, delivering a signal roughly four times stronger than its twin reflector on the Lunokhod 2 rover. Based on the measurements, the position of Lunokhod 1 could be determined with an uncertainty of a few centimeters. The reflector will thus significantly advance the precision of lunar laser-ranging and the resulting gravitational and lunar science (Murphy et al. 2010).

Lunokhod 2 was operated for about 4 months; the mission was officially completed on June 4, 1973. The rover covered a total of 37 km of terrain. A picture of the Lunokhod 1 Rover landed with the Luna 17 mission is shown in Fig. 12.7.

Fig. 12.7 The Lunokhod 1 Rover / Luna 17 (Image credit: NASA/NSSDC ID: 1970-095A)

12.5.5 LK Lunar Lander

The LK (Lunniy Korabl, Russion for "lunar craft") was a Soviet lunar lander, and thus the counterpart of the U.S. Lunar Module. The LK was significantly smaller than the Lunar Module and only designed to land a single cosmonaut instead of two astronauts on the surface of the Moon. It had only a third of the weight of the Lunar Module and had no docking tunnel.

The cosmonaut would thus have to space-walk from the LK to the lunar orbiter. Unlike the Lunar Module, the LK did not use separate descent and ascent stages when travelling from lunar orbit to landing on the surface and vice versa. The LK was rather to use the same engines for descent and ascent. The LK had a launch weight of approximately 5,500 kg, a height of 5.2 m and an estimated span across the deployed landing gear of 4.5 m (Portree 1995).

Although developed and even tested in simplified form, the LK never reached the Moon because the N1 rocket required to take it to the Moon was never successful and finally cancelled. The electrical power system of the LK lander was to be an all-battery system providing an average output power of 0.5 kW. This would have been less than half of the average output power of the Lunar Module (1.1 kW average for the LM versus 0.5 kW average for the LK). The total energy available would have been 30.0 kWh (Wade 2010). An illustration of the LK lunar lander is shown in Fig. 12.8.

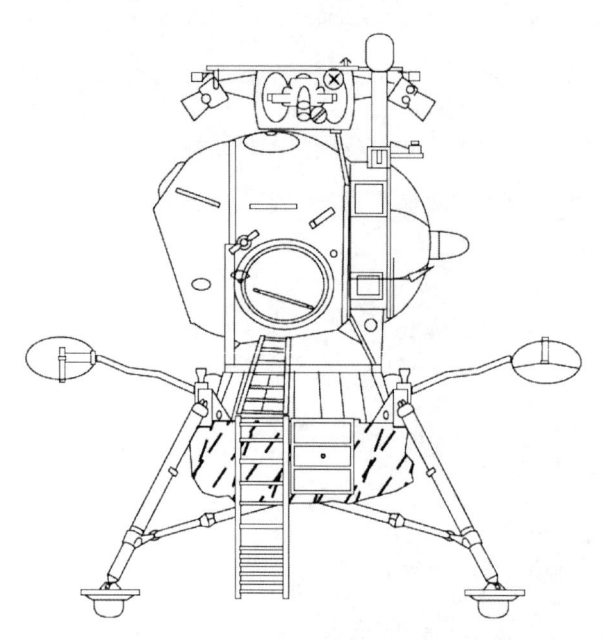

Fig. 12.8 LK lunar lander (Image credit: NASA/Portree 1995)

12.6 The Recent Missions (1990-2010)

The Luna 24 mission of 1976 put an end to the intense lunar exploration program of the United States and the Soviet Union for almost two decades. The first "modern" mission flown to the Moon was the Hiten mission of the Institute of Space and Aeronautical Science, University of Tokyo, Japan. The objective of this mission was to test and verify technologies for future lunar and planetary missions.

The first U.S. mission flown to the Moon was the orbiter Clementine launched in 1994. This mission had the objective to map most of the lunar surface carrying five different imaging systems operating at a number of resolutions and wavelengths from UV to IR. The mission lasted merely half a year and ended when the spacecraft left lunar orbit on May 3 (NSSDC Clementine 2010).

A number of missions have been flow by different space agencies since the late 1990s. AsiaSat 3/HGS-1/PAS 22 was a communication satellite that did not achieve the proper orbit and made two flybys of the Moon on the way into a new geosynchronous orbit.

The Lunar Prospector was a U.S. spacecraft used for polar orbit investigations of the Moon.

The chronology of recent lunar exploration missions, modified from the NSSDC Lunar Exploration Timeline (NSSDC Chronology 2010), is presented in Table 12.4.

Table 12.4 Mission timeline 1990-2010: the recent missions

Launch Date	Mission Name	Outcome
January 24, 1990	Hiten	Flyby and orbiter
January 25, 1994	Clementine	Orbiter
December 24, 1997	AsiaSat 3/HGS-1/ PAS 22	Lunar flyby
January 7, 1998	Lunar Prospector	Orbiter
September 27, 2003	SMART 1	Lunar orbiter
September 14, 2007	Kaguya (SELENE)	Lunar orbiter
October 24, 2007	Chang'e 1	Lunar orbiter
October 22, 2008	Chandrayaan 1	Lunar orbiter
June 17, 2009	Lunar Reconnaissance Orbiter and LCROSS	Lunar orbiter and impactor
October 1, 2010	Chang'e 2	Lunar orbiter
2011 (tbd)	Gravity Recovery and Interior Laboratory (GRAIL)	Lunar orbiter
2013 (tbd)	Lunar Atmosphere and Dust Environment Explorer (LADEE)	Lunar orbiter

SMART-1 (Small Missions for Advanced Research in Technology 1) was a lunar orbiter of the European Space Agency (ESA) launched in 2003 designed to test spacecraft technologies, primarily the solar-powered ion drive, for future missions.

Kaguya (formerly SELENE, for SELenological and ENgineering Explorer), was a Japanese Space Agency (JAXA) lunar orbiter mission launched in 2007.

The Chang'e 1 orbiter launched in 2007 was the first Chinese mission to the Moon. Chandrayaan-1, launched in 2008, was the first Indian Space Research Organization (ISRO) mission to the Moon. Chang'e 1 as well as Chandrayaan-1 were both designed to develop, test and demonstrate the respective space agencies' capabilities for future lunar and planetary exploration.

The Lunar Reconnaissance Orbiter (LRO), the first mission of NASA's Robotic Lunar Exploration Program was launched in 2009, is designed to map the surface of the Moon and characterize future landing sites in terms of terrain roughness, usable resources, and radiation environment with the ultimate goal of facilitating the return of humans to the Moon.

At the time of writing this chapter, two missions are planned in the near future: the 2011 Gravity Recovery and Interior Laboratory (GRAIL) and the 2012 Lunar Atmosphere and Dust Environment Explorer (LADEE). Neither of these two missions will include soft-landing mission elements.

Due to the fact that none of the missions launched since 1990 was designed for soft landing and surface operation, the missions are not directly relevant with respect to the issue of power system design for lunar surface exploration. The data produced with these missions, however, is a very relevant source of information for future surface exploration mission planning and power system design.

12.7 Future Lunar Exploration (2010+)

Future lunar exploration can be sub-divided into short- and medium/long-term exploration. Short-term exploration includes orbiting missions such as NASA's Gravity Recovery and Interior Laboratory (GRAIL) and the Lunar Atmosphere and Dust Environment Explorer (LADEE) mission. Both missions are scheduled for launch in 2011/2012.

Missions to the Moon are currently also planned by other space agencies. The China National Space Administration (CNSA), for instance, plans a series of missions to the moon following the 2007 orbiter Chang'e 1. Chang'e 2 will be a second orbiter and is to be followed by a soft lander and surface rover (Chang'e 3) and a sample return mission (Chang'e 4). SNSA's ambitious lunar exploration plans even include a manned mission in the 2025-2030 time range.

NASA has also developed plans for robotic and crewed surface exploration not only, but also, under the NASA Authorization Act of 2005. These plans will be briefly discussed in the following.

12.7.1 NASA Authorization Act of 2005

The medium-term exploration program of NASA is currently governed by the Authorization Act of 2005, an act of the United States Congress (NASA Authorization Act of 2005). It was signed by the then President George W. Bush and became Public Law 109-155, 109th Congress on December 30, 2005.

The act requires NASA to carry out a balanced set of programs in human spaceflight, in aeronautics research and development, and in scientific research. The act, among other objectives, also directs NASA to send robotic spacecraft to study the Moon.

The main objectives of the program are formulated in the following sentence directly taken from the NASA Authorization Act of 2005:

> The Administrator shall establish a program to develop a sustained human presence on the Moon, including a robust precursor program, to promote exploration, science, commerce, and United States preeminence in space, and as a stepping-stone to future exploration of Mars and other destinations. The Administrator is further authorized to develop and conduct appropriate international collaborations in pursuit of these goals.

The act makes into law, and establishes milestones for, the United States Vision for Space Exploration. Specifically it directs the NASA Administrator to develop a sustained human presence on the Moon with a lunar precursor program, and authorizes international collaborations in pursuit of these goals.

The NASA Administrator is directed to "strive to achieve" the following milestones:

- Return Americans to the Moon no later than 2020.
- Launch the Crew Exploration Vehicle as close to 2010 as possible.
- Use the International Space Station to study the impacts of long duration stays in space on the human body.
- Enable humans to land on and return from Mars and other destinations on a timetable that is technically and fiscally possible.

The Constellation Program was developed in response to the goals defined by the Vision for Space Exploration and the NASA Authorization Act of 2005, to send astronauts back to the Moon and possibly to Mars, as well. Constellation includes the development of spacecraft and booster vehicles to replace the Space Shuttle. A series of booster is being developed under the name Ares. Orion is being designed as the crew compartment for the Constellation program and Earth orbit missions. Altair will be the main transport vehicle for lunar-bound astronauts.

Vision and objectives of the NASA Authorization Act of 2005 have been widely discussed. The vision was often dubbed as "Apollo 2.0" or "Apollo on steroids", and many aspects of the program were suggested to be overworked or replaced.

On February 1, 2010, President Barack Obama announced a proposal to cancel the program, effective with the U.S. 2011 fiscal year budget, but later announced changes to the proposal in a major space policy speech at Kennedy Space Center on April 15, 2010. He committed to increasing NASA funding over five years and completing the design of a new heavy-lift launch vehicle by 2015 and to begin construction thereafter. He also predicted a U.S. crewed orbital Mars mission by the mid-2030s, preceded by an asteroid mission by 2025 (Chang 2010).

As of today, the future of the Constellation Program is uncertain and major revisions are to be expected. The current status of the spacecraft and their power system design is nevertheless discussed in the following. Even with major revisions in objectives and timeline, the state-of-the-art in spacecraft power system design is nevertheless reflected very well in the concepts developed for the Constellation spacecraft.

12.7.2 Altair Spacecraft

The Altair spacecraft, previously known as the Lunar Surface Access Module (LSAM), is the element of the Constellation Program designed for landing on the surface of the Moon. The basic concept of Altair resembles the Apollo Lunar Module and is again based on a separate descent and an ascent stage. Unlike the Apollo Lunar Module, Altair is designed to carry a crew of four astronauts to the surface, leaving the Orion crew module temporarily unoccupied in lunar orbit.

The Altair lunar landers are designed to enable astronauts to live and work on the Moon for extended periods of time, providing the experience needed to expand human exploration farther into the solar system.

A trade study was performed during the first Altair design analysis cycle in 2007 to determine the optimal power source, or a combination of different power sources, respectively, for the new vehicle (Polsgrove et al. 2009). Based on estimated average power requirements, various power system options such as photovoltaic, battery, and fuel cell systems were evaluated. It was quickly realized that an all-battery or solar array/battery power source would not be possible. After analyzing the remaining power system options, the most optimal solution was found to be a fuel cell on the decent module and a battery on the ascent module, which is identical to the baseline Apollo design (which was later changed to an all-battery system also with the descent module with the Lunar Module). Artist's renderings showing the Altair spacecraft landed on the Moon and docked to the Orion spacecraft are shown in Figs. 12.9. and 12.10.

The fuel cell system considered with Altair is a hydrogen/oxygen system offering the possibility of operating on the same hydrogen and oxygen propellants also used with the descent module engine. Thus, Altair will be landed with significant quantities of propellants remaining in the large propulsion tanks. Utilizing these unused propellants upon landing as resources for surface power generation

Fig. 12.9 Artist's rendering showing crew members working in the vicinity of their Altair spacecraft (Image credit: NASA)

Fig. 12.10 Artist's rendering showing Orion (right) docked with Altair (Image credit: NASA)

suggests itself. Thus, the propellant safety margin not used in the descent and additional residuals which cannot be used by the engine (e.g. because they cannot be transported to the engine in proper conditions) would not be wasted.

The residual propellant is currently estimated to be 3% of the maximum propellant load, which would be approximately 130 kg of hydrogen and 706 kg of oxygen. Using a conservative conversion factor of 2.2 kWh of energy from each kg of water produced by the fuel cell, this remaining propellant represents 1,746 kWh of electrical energy storage and 794 kg of reaction product water (Polsgrove et al. 2009).

Whereas Apollo completed a 65 hour mission using only 70 kWh of energy, current estimates for Altair show almost 1 MWh of energy requirements for the 225 hour surface stay mission. The average power consumption of Altair is expected to be much greater than that of Apollo Lunar Module. Current estimates for Altair are in the order of 4 kW average power consumption (Polsgrove et al. 2009).

12.8 Power System Options for Lunar Surface Exploration

Power system options for (medium/long-term) surface exploration of the Moon can essentially be broken down into three main options: surface solar energy utilization, landed resources, and In-Situ Resource Utilisation (ISRU) options. The first two options have been used in previous missions and are also exclusively considered with the current Altair spacecraft. The third option has been extensively discussed with Mars surface missions, but may also be a very interesting option with lunar surface exploration missions.

Due to the near absence of a lunar atmosphere, the only resources available are surface regolith and sub-surface materials, other than with Mars, where the presence of a comparably thick and carbon-dioxide-rich atmosphere provides additional possibilities in ISRU technology.

ISRU focuses on technologies necessary to extract consumables (e.g. oxygen, water, nitrogen, helium) and using them in various applications. Possible applications include the use as propellants (In-Situ Propellant Production – ISPP), the use as human life-support system replenishment (Environmental Control and Life Support System - ECLSS, Extra-vehicular activity - EVA, etc.), source materials for In-Situ Fabrication and Repair (ISFR) technologies, and source materials for radiation shielding and shelters.

The beauty of all of these ISXX concepts is the fact that the reduction of each kilogram of Earth launch mass also reduces the launch costs. By avoiding the launch of a full cache of oxygen for the Earth-return flight, for instance, the Earth launch mass could thus be significantly reduced at the cost of including the equipment necessary for the ISRU-based production of oxygen.

A typical example for this kind of approach is the 1993 NASA/JSC LUNOX proposal where the cost reduction provided by producing liquid oxygen propellant for the Earth-return flight from lunar regolith was investigated. The study revealed that this approach would permit drastic downsizing of the manned space vehicles and would result in cost reductions of up to 50%. Central element of this concept was a lunar oxygen production and storage plant powered by a nuclear power reactor. An artist's image of the ISRU plant processing regolith for oxygen production is shown in Fig. 12.11.

Fig. 12.11 ISRU plant processing regolith for oxygen production (Image credit NASA/JSC)

Although oxygen extraction from lunar regolith currently remains the primary option, even the extraction of water might be possible in the future. Preliminary data from the Lunar CRater Observation and Sensing Satellite (LCROSS) indicates that the mission successfully uncovered water during the October 9, 2009 impacts into the permanently shadowed region of Cabeus cater near the Moon's south pole (Dino 2009). Extraction of water would offer completely new possibilities, as it could not only be used in life support applications, but also be easily split into hydrogen and oxygen, thus providing fuel and oxidizer to be used as propellants for Earth-return vehicles as well as for surface power generation.

Leaving the question of the actually extracted resources aside (oxygen alone or in combination with hydrogen; other minerals or materials), a general overview of energy technologies and pathways for lunar surface power generation applications is presented in Fig 12.12.

This schematic highlights the three main options for providing and/or generating storable energy resources on the Moon. The three main options are surface solar energy utilization, landed resources, and ISRU-derived resources. Landed and ISRU-derived options are often, but not necessarily, linked to fuel-to-electric energy conversion technologies. A description of the synonyms used in Fig. 12.12, is given in Table 12.5.

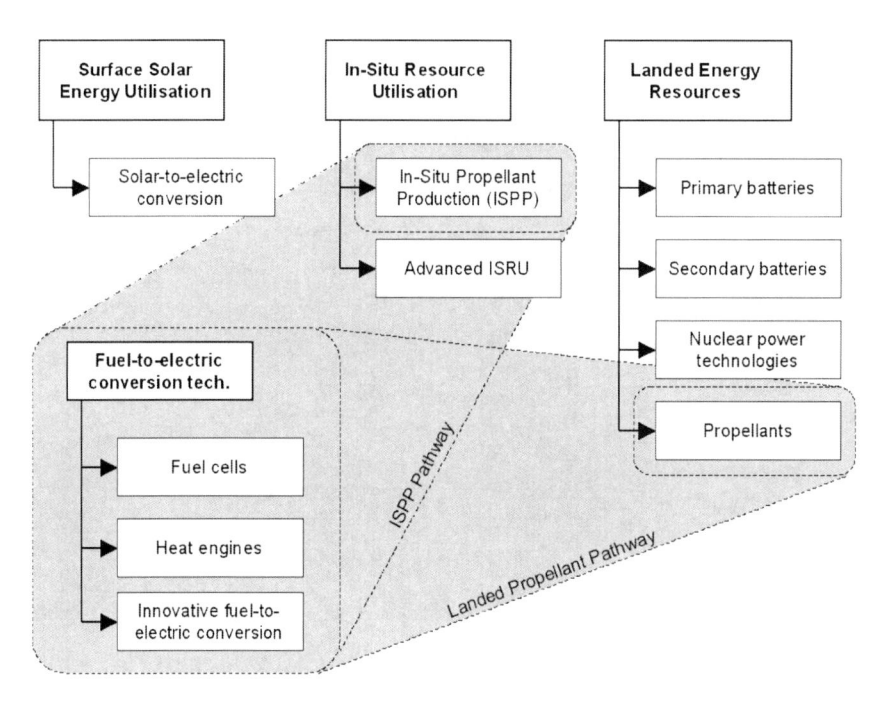

Fig. 12.12 Overview of energy conversion and power generation technologies as well as pathways for lunar surface power generation applications

A shift from battery-based to fuelled energy storage and power generation systems is to be expected with surface exploration mission elements which cannot be operated by batteries and/or solar panels primarily due to power requirements and space limitations. Many of these compact and high-power applications could also be powered by nuclear power systems, primarily static and dynamic isotope power systems. A nuclear option is, however, limited not only by technical issues, but also considered problematic with respect to public opinion for reasons not to be discussed within the limited scope of this chapter.

The utilization of solar panels is possible with lunar surface exploration applications far easier than with Mars surface applications. The lack of a thick – or from an engineering point of view even noteworthy - atmosphere prevents dust storms, and thus also problems associated with reduced surface solar energy availability and dust deposition faced on Mars. The available solar energy is generally higher in magnitude that with Mars due to the more beneficial orbital parameters

Table 12.5 Description of synonyms used in Fig. 12.12

Abbreviation used in Fig. 12.12.	Description
Surface solar energy utilisation	Technologies converting solar into electric energy
Solar-to-electric conversion	Photovoltaics, solar dynamic power systems, etc
In-Situ Resource Utilisation	Technologies for utilising resources available on the lunar surface
In-Situ Propellant Production (ISPP)	Technologies for utilising resources available on the lunar surface for propellant production (e.g. LUNOX – lunar-derived oxygen)
Advanced ISRU	e.g. production of solar cells from regolith materials, etc.
Landed energy resources	Directly landing energy resources on the lunar surface
Primary batteries	Non-rechargeable batteries
Secondary batteries	Rechargeable batteries
Nuclear power technologies	Nuclear power plants, radioisotope generators, dynamic isotope power systems, …
Propellants	Fuels (e.g. hydrogen, methane) and oxidizers (primarily oxygen)
Fuel-to-electric conversion tech.	Technologies for converting fuel and oxidizer into electric energy
Fuel cells	Electrochemical conversion of fuel and oxidizer into electric energy
Heat engines	Conversion of chemical into mechanical energy via combustion
Innovative fuel-to-electric conversion tech.	Thermoelectric and thermophotovoltaic converters, etc.
ISPP pathway	Energy storage and/or power generation from propellants (partly) produced from in-situ resources
Landed propellant pathway	Energy storage and/or power generation from propellants landed onto the Moon

of the Earth, and thus also the Moon orbiting around the Earth. Solar panels are nevertheless limited in terms of nighttime power generation and integration into lightweight, portable or mobile systems. In this case, a fuelled energy storage and power generation system would perform very favorably, as energy can thus not only be stored with a very high specific energy and energy density, but also installed into a compact package in terms surface area requirements.

This would also be true with nuclear-generated power options, where radioisotope units are now combined with more efficient energy conversion systems to provide higher output power levels. The utilization of nuclear power system elements is, however, not only a technical or engineering issue, but also a political issue, as briefly mentioned above.

Fuelled power system elements can essentially be supplied from landed as well as from ISRU-derived resources. A hydrogen/oxygen fuel cell, for instance, can be equally operated with residues from the descent engine hydrogen and oxygen tanks, a landed cache of hydrogen and ISRU-produced oxygen extracted from lunar regolith, or ISRU-produced hydrogen and oxygen extracted from water-containing minerals in cold and permanently-shaded craters.

The flexibility of having the ISPP as well as the landed propellant pathway thus strongly favors the utilization of fuel-to-electric conversion technology in future lunar surface exploration applications. Whether this fuel-to-electric conversion technology will be a conventional internal combustion engine or a turbine, a fuel cell, or an innovative technology such as an alkali metal thermoelectric converter will have to be determined on the basis of a specific mission profile with defined energy storage and output power generation requirements on the basis of an overall mission risk analysis.

12.9 Conclusions

Primary and secondary batteries as well as photovoltaic modules have been the power system elements of choice with mission elements designed for lunar surface operation in the past.

Fuel cells, although already considered with the Apollo Lunar Module, are an interesting option if power consumption and surface stay are increased, and the comparably low specific energy of batteries becomes a strong limitation. By including a fuel cell system operating on hydrogen and oxygen, the same propellants also used with the descent module engine, the propellant safety margin not used in the descent and additional residuals which cannot be used by the engine (e.g. because they cannot be transported to the engine in proper conditions) would not have to be wasted but could be used for surface power generation. In addition, fuel cells also provide the possibility of including In-Situ Resource Utilization (ISRU) technology in the power system design. In-Situ Resource Utilization and In-Situ Propellant Production are innovative concepts aiming at a strong reduction of the overall mission Earth launch mass by partly or fully replacing launched with in-situ-produced materials or propellants. ISRU can not only be considered with missions to the surface of Mars, but is also a very interesting option with various different applications in lunar surface exploration.

In the future, power system engineers will be able to select and combine an increasing number of power system technologies in order to satisfy the power and energy demand of mission elements. Depending on average and peak power requirements, mission profile and mission duration, a prudent combination of different power generation and energy storage technologies will provide power systems that are lightweight and compact, and at the same time safe and robust. Thus, future spacecraft power systems for lunar surface operation will not be *limiting*, but rather *enabling* technologies.

References

Boeing, Apollo 17 LRV Technical Information (1972),
    http://history.nasa.gov/alsj/a17/A17_LunarRover2.html
Bussey, D.B.J., Spudis, P.D., Robinson, M.S.: Illumination conditions at the lunar South
    Pole. Geophys. Res. Lett. 26, 1187–1190 (1999)
Bussey, D.B.J., Spudis, P.D.: The Lunar Polar Illumination Environment: What We Know
    & What We Don't Space Resources Roundtable VI (2004),
    http://www.lpi.usra.edu/meetings/roundtable2004/pdf/6022.pdf
Bussey, D.B.J., Fristad, K.E., Schenk, P.M., Robinson, M.S., Spudis, P.D.: Constant illu-
    mination at the lunar north pole. Nature 434, 842–842 (2005)
Chank, K.: Obama Vows Renewed Space Program. The New York Times April 16, 2010,
    on page A18 of the New York edition (2010),
    http://www.nytimes.com/2010/04/16/science/space/
    16nasa.html?_r=1
Dino, J.: LCROSS Impact Data Indicates Water on Moon (2009),
    http://www.nasa.gov/mission_pages/LCROSS/main/
    prelim_water_results.html
Engelhardt, W.: Galileo, Cassini, Giotto. Raumsonden erforschen unser Planetensystem.
    Wissenschaftlicher Verlag Harri Deutsch, Frankfurt am Main (2005) (in German)
Harvey, B.: Soviet and Russian Lunar Exploration: Comparisons of the Soviet and Ameri-
    can Lunar Quest. Springer, Heidelberg (2005)
Lia, X., Wang, X., Zheng, Y., Cheng, A.: Estimation of solar illumination on the Moon: A
    theoretical model. Planet. Space Sci. 56, 947–950 (2008)
Moses, A.J., Hetherington, W.M., Weinberger, D., Uchiyama, A.A., Bogner, R.D., Long,
    W.L.: Surveyor batteries Final engineering report. Technical Memorandum, pp. 33–432
    (1970)
Murphy, T.W., Adelberger, E.G., Battat, J.B.R., Hoyle, C.D., Johnson, N.H., McMillan,
    R.J., Michelsen, E.L., Stubbs, C.W., Swanson, H.E.: Laser Ranging to the Lost Lunok-
    hod 1 Reflector (2010), submitted to Icarus arXiv:1009.5720v2
NASA Authorization Act of 2005, PUBLIC LAW 109–155, 109th Congress (2005)
NASA JSC-09423, Apollo program summary report (1975)
NASA Moon Fact Sheet (2010),
    http://nssdc.gsfc.nasa.gov/planetary/factsheet/moonfact.html
NASA SP-184, Surveyor program results. Scientific and Technical Information Division,
    National Aeronautics and Space Administration (1969),
    http://ia700108.us.archive.org/3/items/
    surveyorprogramr00unit/surveyorprogramr00unit.pdf

NSSDC Clementine (2010),
   http://nssdc.gsfc.nasa.gov/planetary/lunar/clementine1.html
NSSDC Luna 2, NSSDC ID: 1959-014A (2010)
NSSDC Luna 9, NSSDC ID: 1966-006A (2010)
NSSDC Luna 16, NSSDC ID: 1970-072A (2010)
NSSDC Luna 20, NSSDC ID: 1972-007A (2010)
NSSDC Luna 14, NSSDC ID: 1976-081A (2010)
NSSDC Luna 17, NSSDC ID: 1970-095A (2010)
NSSDC Luna 21, NSSDC ID: 1973-001A (2010)
NSSDC Surveyor 1, NSSDC ID: 1966-045A (2010)
NSSDC Chronology, Lunar Exploration Chronology (2010),
   http://nssdc.gsfc.nasa.gov/planetary/chrono.html
Polsgrove, T., Button, R., Linne, D.: Altair Lunar Lander Consumables Management. Document ID: 20100002861, Report Number: DRC-05-01 (2009)
Portree, D.S.F.: Mir Hardware Heritage. NASA RP 1357 (1995)
Wade, M. (2010), http://www.astronautix.com/craft/lk.htm
Zuber, M.T., Smith, D.E.: Topography of the lunar south polar region: implications for the size and location of permanently shade areas. Geophys. Res. Lett. 24, 2183–2186 (1997)

13 The Use of Lunar Resources for Energy Generation on the Moon

Alex Ignatiev and Alexandre Freundlich

University of Houston, Texas, USA

13.1 Introduction

Energy is fundamental to nearly everything that humans would like to do in space, whether it is science, commercial development, or human exploration. If indigenous energy sources can be developed, a wide range of possibilities emerges for subsequent development. Some of these will lower the cost of future exploration, others will expand the range of activities that can be carried out, and some will reduce the risks of further exploration and development. This picture is particularly true for the Moon where significant electrical energy will be required for a number of lunar development scenarios; including science stations, lunar resource processing, and tourism. As example, the presence of water in the permanently shadowed craters at the poles of the Moon (Colparete et al. 2010; Heldman et al. 2011) will allow for propellant production: oxygen and hydrogen through electrolysis which will require significant amounts of electrical energy. Further, proposed large radio telescopes on the back side of the Moon would require over 1 MW of power for echo microwave astronomy. The availability of solar cell raw materials in the surface regolith of the Moon, e.g., silicon, and the fact that the surface of the Moon is an ultra-high vacuum (1×10^{-10} Torr) environment allows for the direct fabrication of thin film solar cells directly on the surface of the Moon.

Processing of the lunar regolith can result in the extraction of silicon, aluminum, magnesium, and other elements of importance for solar cell structures, and the ultra-high vacuum surface environment of the Moon negates the need for vacuum chambers within which to undertake thin film deposition processing. Armed with these facts, and the fact that the costly transport and installation of an immense number of terrestrial solar cells to support lunar energy needs will not be required, one can now develop an off-Earth thin film solar cell fabrication scenario that utilizes the Moon's resources to manufacture extensive solar arrays on the surface of the Moon in the development of a lunar power system. This lunar energy advancement can also benefit cis lunar space through the possibilities of power beaming lunar-generated electrical energy to either lunar satellites or earth satellites. Further, the possibility of fabrication of solar cells on the Moon can allow for low gravity well launch of these cells into space for use on space satellites, asteroids, and solar power satellites servicing the Moon or the Earth.

13.2   Thin Film Solar Cell Fabrication

In order to initiate solar cell fabrication on the Moon, the necessary 'tools' need to be emplaced on the surface of the Moon. Two specific types of 'tools'are required: i.) a tool to extract solar cell raw materials from the lunar surface regolith (the fine surface dust prevalent on the Moon to depths of from 1 –20 meters); and ii.) a tool to deposit thin film solar cells on the surface of the Moon.

Regolith processing on the Moon to extract metallic and semiconducting elements needed for solar cell fabrications can be accomplished by a number of processes (Rosenberg et al. 1964; Eagle Engineering 1988; Rao 1979; Knudsen et al. 1990; Filsinger et al. 1990; Rosenberg et al. 1996). Of the prevalent elements at the surface of the Moon, the most abundant semiconductor is silicon (Table 13.1). Hence, it is most direct to focus on the fabrication of silicon solar cells using silicon extracted from the regolith.

Table 13.1 Abundances of Solar Cell Materials in the Lunar Regolith

Si	~155,000 ppm
Al	~53,000 ppm
Ga	~ 1 ppm
As	~ 1 ppm
Dr	~ 3 ppm
S	~ 1,000 ppm
Te	~ 200 ppm

For silicon extraction, carbothermal reduction (Knudsen et al. 1990; Filsinger et al. 1990; Rosenberg et al. 1996) has been proposed for several abundant silicates, including anorthite ($CaAl_2Si_2O_8$) and pyroxene ($(Ca,Mg,Fe)SiO_3$). Anorthite is abundant in both maria and highlands rocks, pyroxene is most abundant in the maria. Additional processing of lunar regolith for extraction of oxygen, a process which can also result in the extraction of silicon) has included hydrogen reduction (Lin et al. 2003; Briggs et al. 1990) and fluorine reduction (Burt 1989). All of these processes require reagents to be brought to the Moon, as well requiring a closed cycle process on the Moon to reduce the resupply of reagents from Earth.

Electrochemical processing (Keller 1988) is also a possible approach to extraction of elements from regolith. Electrochemical processing does not required sealed volumes, but does require highly stable anodes (so as to reduce/eliminate replacements brought from Earth), and a significant amount of energy to melt the regolith to a magma, and to electrolyze it. Electrochemical processing of lunar regolith simulant has been done for the extraction of silicon (Keller 1988). This has resulted in nominal quality silicon which was used in preliminary studies of the development of silicon solar cells from the extracted silicon (Ignatiev et al. 1998). The studies showed that such silicon can be used to fabricate thin film silicon solar cells through vacuum deposition (Ignatiev and Freundlich 1998). Hence to avoid the cost and logistics of transporting reagents from Earth, a regolith processing 'tool', i.e., a Regolith Processor, incorporating magma electrolysis will need to be deployed on the surface of the Moon to supply the raw materials.

On Earth, silicon solar cells are not typically vacuum deposited on glass substrates but are principally fabricated from single crystal wafers not unlike those used for semiconductor device fabrication. However, as noted, the Moon possesses an ultra-high vacuum surface environment, hence vacuum deposition of silicon can be well applied on the Moon without the need for vacuum chambers. Thin film silicon solar cells, when terrestrially vacuum deposited, are typically deposited in crystalline form on single crystal substrates. Vacuum deposition of silicon on glass is problematic however, due to atomic disorder in the grown films, and when cells are achieved they typically have low efficiencies (a few %). This, though, may be acceptable for the lunar environment since large areas of low efficiency solar cells can be fabricated on the Moon to give the required *total* power capacity needed for use, i.e., quality can be traded off for quantity.

For a thin film solar cell, the substrate represents the most massive part of the structure. A common terrestrial substrate for thin film cells is glass. Nominal SiO_2 glass is not readily available on the surface of the Moon; however we have shown that lunar regolith can be *melted* to form a glass that is quite suitable as a solar cell substrate. Melted and re-solidified lunar regolith simulant (JSC-1) exhibits an electrical resistance of greater than 10^{11} Ω, and shows a smooth surface morphology consistent with good solar cell substrate material (Fig. 13.1).

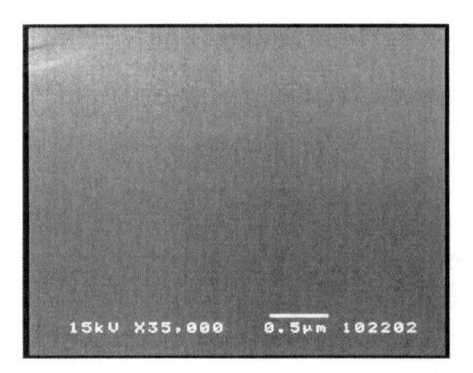

Fig. 13.1 SEM micrograph of a thermally melted JSC-1 regolith simulant

Thus, there is a requirement for melting lunar regolith. Lunar regolith does not have uniform composition at all places on the Moon; hence it does not have a singular melting temperature. Lunar regolith simulant nominally softens at approximately 1300°C and melts at about 1500°C. Such temperatures can be obtained on the surface of the Moon by solar-thermal heating. This has been demonstrated by Nakamura et al. (2009) utilizing solar concentrators and light pipes to heat regolith simulant as illustrated in Fig. 13.2.

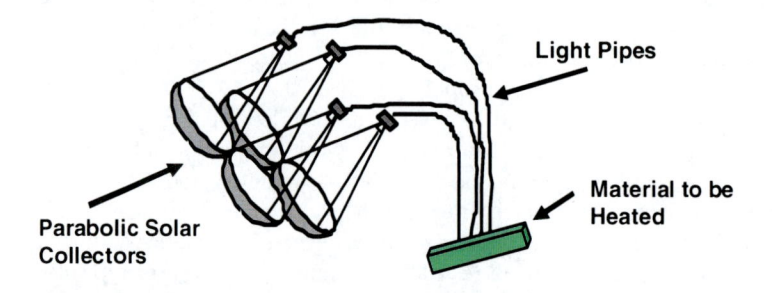

Fig. 13.2 Schematic of multiple parabolic collectors connected to optic fibers channeling solar energy to heat regolith for melting

Regolith melting under actual lunar environment conditions may result in melted /solidified regolith with an uneven, non-flat surface as has been anecdotally noted for melting tests of regolith simulant in terrestrial environments. For thin film solar cells, flatness, however, is not a requirement for the substrate glass surface as long as surface roughness remains $\leq$ 100 nm which is well below the values obtained in laboratory melting of regolith simulant under vacuum conditions (Ignatiev et al. 1998). Further, contraction of the molten regolith upon solidification should not impact substrate integrity as the regolith glass substrate is situated on powdered regolith and is thus not 'clamped' to the lunar surface negating any differential thermal expansion effects.

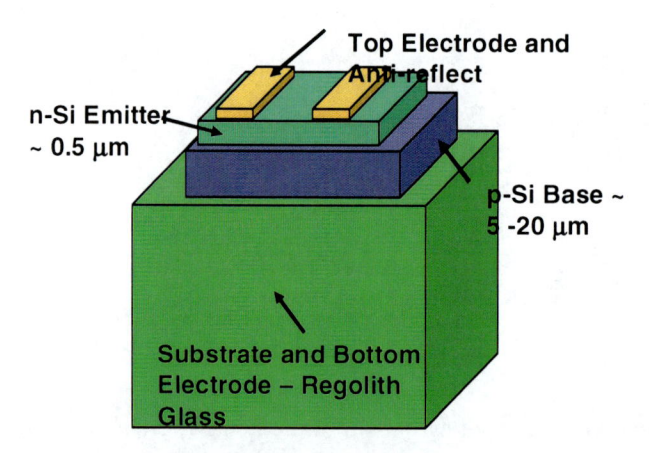

Fig. 13.3 Proposed structure of a thin film silicon solar cell with melted regolith glass substrate

Once the regolith glass substrate is fabricated, a metal bottom contact layer can be sequentially solar-thermally evaporated onto the lunar glass followed by evaporation of p- and n-doped silicon to form the solar cell. Such a thin film deposition of silicon on glass results in microcrystalline silicon solar cell the structure of which is depicted in Fig. 13.3, and the individual silicon solar cell layers and their raw material origin on the Moon is given in Table 13.2.

The efficiency of microcrystalline thin film silicon solar cells is not expected to very high due to a high number of defects; however, recent calculations (SBIR Report 2006) have shown that as the grain size of the microcrystalline silicon thin film is increased the efficiency increases, and the required thickness of silicon solar cell decreases (Fig. 13.4).

Table 13.2 Lunar Si Cell Structures, Source Materials, Indigenous Minerals, and Fabrication Techniques

Layer	Thickness range	Type	Source materials	Indigenous minerals	Fabrication technique
Top electrode	0.1-1 μm	Metallic Ca, FeSi, or Al	Lunar Ca, Al, FeSi	From anorthite, imenite	Vacuum solar thermal evaporation
Antireflection coating	0.1-0.2 μm	TiO or SiO or AlO	regolith	From $(FeTiO_2)$ ilmenite, $(CaAl_2Si_2O_8)$ anorthite	Vacuum solar thermal evaporation
n-type Si	0.1-0.3 μm	Si doped with As, P or S (doping about 100-200 ppm)	Lunar Si n-dopant	From (anorthite) PO_x and S are present in low quantities in minerals	Co-evaporation of Si and dopant
p-type Si	1-10 μm	Si doped with Al (20-50 ppm)	Lunar Si p-dopant	From anorthite, pyroxene	Vacuum solar thermal evaporation
Bottom electrode	1-2 μm	Al, Ca, FeSi	Lunar Al, Ca, FeSi	From anorthite, pyroxene	Vacuum solar thermal evaporation
Substrate	2-5 μm	glass	regolith	regolith	Solar thermal melting

This indicates that research on increasing grain size during the growth of the microcrystalline silicon to at least 1 micron will not only increase efficiency to as high as possibly 10-11%, but will require less silicon material to grow the thin film cell, and hence faster times for growth, i.e., faster fabrication and higher production rates. As noted above, all of the materials required to fabricate a silicon solar cell are present in the lunar regolith.

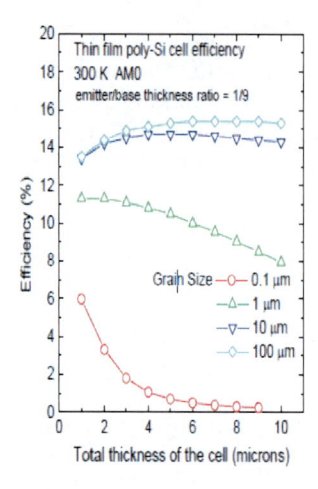

Fig. 13.4 Dependence of microcrystalline thin film silicon solar cell efficiency on grain size and cell thickness

With the microstructure of the thin film silicon cell set, we now define the 'tool' that will be needed to fabricate the solar cells on the surface of the Moon – the Cell Paver (Fig. 13.5). The Cell Paver is a self-contained, self-reliant mobile fabricator that clears or circumvents rocks and boulders and smoothes the terrain directly in front of it, thus preparing a bed for the fabrication of solar cells.

Fig. 13.5 Artist's drawing of Cell Paver fabricating thin film solar cells on the surface of the moon

The Cell Paver is projected to be of ~200 kg mass, and would traverse the lunar surface depositing solar cells on melted regolith glass as it moves. The Cell Paver would operate in lunar daylight, hibernate at lunar night, and would be periodically replenished from the Regolith Processor with the small amounts of raw material that it expended in solar cell fabrication process. i.e., FeSi or Al for the bottom and top electrodes and the metal wire interconnects, and n-doped and p-doped silicon (doped in the Regolith Processor with As (n) and Al (p)).

The Cell Paver would utilize solar thermal energy for the evaporation tasks in the fabrication of the solar cells, and would therefore house a large number of parabolic solar collectors with integrated light pipes to support the evaporation task. As noted previously, the solar collectors and light pipes would also be used to locally melt the top few millimeters of the regolith to form a lunar glass substrate for the solar cell deposition directly onto the lunar surface. The Cell Paver would also house solar cells for direct electrical energy generation for Paver command, control and motive power. As the Paver moves across the lunar surface it would utilize a small plow at the front to move larger rocks out of its path, and to smooth the regolith for solar cell fabrication.

As part of the solar cell fabrication process, individual cells would be connected in series by deposited metal film wires as part of the cell growth process resulting in a multi-cell array. These arrays operating at from 100 V to over 500 V would then be interconnected by deposited thick film metal wires along with p-n blocking diodes to form a power system. The blocking diodes will effectively disconnect an array form the system if damage (e.g., from micrometeorites) occurs to the array. Based on the projected thermal energy output of the Cell Paver solar thermal system (~1,500 W maximum thermal energy available for melting/evaporation from the 1.6 m^2 solar thermal collectors), the evaporation and deposition/fabrication rates would yield approximately 1 m^2/hr of deposited silicon solar cells, i.e., the Cell Paver would move at a rate of about 1 m/hr – this low transit rate is important as it would minimize any dust kick-up that would occur from the motion of the Cell Paver thus reducing the defect formation effect of surface dust in the fabricated solar cells. The operating Cell Paver would yield about 65 W/hr of fabricated solar cells, and after one year of operation would result in a solar cell array capacity of about 200 kW at a nominal 5% thin film solar cell efficiency.

As noted above, the Cell Paver will be fed raw materials extracted by the Regolith Processor which will require a significant electrical power level (several hundred watts to several kilowatts) to extract the silicon and metals needed for the solar cell fabrication on the Moon. This can be accomplished by invoking the concept of "bootstrapping", i.e., the Cell Paver would be deployed on the Moon with a full load of raw materials. It would then operate on the lunar surface fabricating solar cells for some nominal time and generate a nominal capacity (~40 kW) of thin film silicon solar cells on the surface of the Moon. The Regolith Processor could then be connected to this initial array of fabricated solar cells so that it could use the generated electrical energy to start producing the additional raw materials required by the Cell Paver for it to continue to fabricate silicon solar cells on the Moon - now from indigenous raw materials.

The projected lifetime for the Cell Paver (based on past performance of deployable planetary robotic vehicles, e.g., Spirit and Opportunity Mars rovers) is ~5 years. Hence, the deployment of the Cell Paver/Regolith Processor system for a five-year period will result in a total fabricated electrical power capacity from this one system of ~1 MW. If improvements of microcrystalline silicon solar cell efficiency can be made as projected in Fig. 13.4, then with a possible 10% efficiency the lunar fabricated solar array capacity will reach 1 MW in 2½ years and 2 MW in 5 years. This brings a highly energy-rich environment to the Moon, and promises significant advantages in the utilization of the Moon. Of course the ability to fully utilize this electrical energy rich environment on the Moon would require additional infrastructure in the form of electrical energy transmission systems. These could include transmission wires fabricated in thick film form via a modified Cell Paver (depositing only metal), or by microwave power beaming from point to point or from a lunar power satellite to the lunar surface.

The Cell Paver and the Regolith Processor make up a nominal payload to the surface of the Moon of ~ 350 kg. If this payload was relegated to only the transport of terrestrially fabricated solar cell arrays and their support frames and deployment mechanisms, it would account for approximately 150kW of solar cell capacity on the Moon. This is over 10 times less energy generation capacity than the optimized fabricated solar cells will deliver for the same payload mass. Further, adding the cost of the terrestrial solar cell procurement, the cost of the support frames, and the cost of the mechanized deployment mechanism, it is projected that even for the 5% efficient lunar fabricated solar cells, the cost of attaining solar electric energy on the Moon by fabricating the solar cells there as compared to bringing them there is less than 1/10.

An additional benefit of the fabrication of solar cells on the Moon for lunar power is the possibility of mitigating solar cell damage events such as micrometeorite strikes or space radiation damage by simply making more solar cells. The deployment of terrestrial solar cells would require launch of additional solar cells (mass) to the Moon to compensate for the damaged cells, whereas fabrication of the solar cells on the Moon would allow for additional production and direct replacement of damaged cells without the requirement of any additional mass launched to the Moon. Further, the use of lunar materials instead of terrestrial materials and the manufacturing of the solar cell systems off-planet, adds to the reduction of contamination of the terrestrial environment.

Finally, the promised energy-rich environment for the Moon through the fabrication of lunar solar cells should not be underestimated. Uses of the Moon for science (observatories), for technology development (living of the land), for commerce (resource exploitation), and for colonization will require many MW of power. This will only be possible through the availability of extensive electrical energy on the Moon. Fabricating solar cells on the Moon will yield that electrical energy resource.

13.3 Conclusion

The ability to fabricate solar cells on the Moon for use on the Moon as well as in cis-lunar space can result in an extremely energy–rich environment for the Moon. The use of the Moon's vacuum environment and its elemental resources can yield thin film silicon solar cells that are deposited directly on the surface of the Moon. This can be accomplished by the deployment of a moderately-sized Cell Paver/Regolith Processor system on the surface of the Moon with the capabilities of fabricating thin film silicon solar cells. The power system will support extraction of needed raw materials from the lunar regolith, preparation of the regolith for use as a substrate, evaporation of the silicon semiconductor material for the solar cell structure directly on the regolith substrate, and deposition of metallic contacts and interconnects to finish off a complete solar cell array. This on-Moon fabrication process would also result in an electric power system that was repairable/replaceable through the simple fabrication of more solar cells, and allow for the expansive use of the Moon.

Acknowledgements. The authors acknowledge partial support from the State of Texas through the Center for Advanced Materials, and the R.A. Welch Foundation.

References

Briggs, R.A., Sacco Jr., A.: Hydrogen reduction mechanism of ilmenite between 823 and 1353 K. J. Mater. Res. 6(3), 574–584 (1990)

Burt, D.M.: Lunar Mining of Oxygen Using Fluorine. Amer. Sci. 77, 574–579 (1989)

Colaprete, A., Schultz, P., Heldmann, J., Wooden, D., Shirley, M., Ennico, K., Hermalyn, B., Marshall, W., Ricco, A., Elphic, R.C., Goldstein, D., Summy, D., Bart, G.D., Asphaug, E., Korycansky, D., Landis, D., Sollitt, L.: Detection of Water in the LCROSS Ejecta Plume. Science 22, 463–468 (2010)

Eagle Engineering, Conceptual Design of a Luna Oxygen Pilot Plant Lunar Base Study, NASA-JSC NAS9-17878, Task 4.2(1988)

Filsinger, D.H., Bourrie, D.B.: Silica to Silicon: Key Carbothermic Reactions and Kinetics. J. Am. Ceram. Soc. 73(6), 1726–1732 (1990)

Heldman, J., et al.: LCROSS (Lunar Crater Observation and Sensing Satellite) Observation Campaign: Strategies, Implementation, and Lessons Learned, Space Science Reviews (2011), doi:10.1007/s11214-011-9759-y

Ignatiev, A., Kubricht, T., Freundlich, A.: Solar Cell Development on the Surface of the Moon. In: Proc. 49th International Astronautical Congress, IAA-98-IAA. 13.2.03 (1998)

Ignatiev, A., Freundlich, A.: Lunar Regolith Thin Films: Vacuum Evaporation and Properties. In: AIP Proceedings, vol. 420, p. 660 (1998)

Keller, R.: Lunar Production of Aluminum, Silicon and Oxygen. In: Sohn, H.Y., Geskin, E.S. (eds.) Metallurgical Process for the Year 2000 and Beyond, pp. 551–562. Mineral, Metal and Materials Soc., Warrendale, PA (1988)

Knudsen, C., Gibson, M.: Development of the Carbotek Process™ for Lunar Oxygen Production. Engineering, Construction and Operations in Space II, American Society of Civil Engineers, p. 357 (1990)

Lin, H.Y., Chen, Y.W., Li, C.: The mechanism of reduction of iron oxide by hydrogen. Thermochimica Acta 400, 61–67 (2003)

Nakamura, T., Van Pelt, A.D., Smith, B.K.: Solar Thermal System for Oxygen Production from Lunar Regolith. In: 47th AIAA Aerospace Sciences Meeting, AIAA-2009-0660, Orlando, January 5-8 (2009)

Rao, D.B.: Extraction Processes for the Production of Aluminum, Titanium, Iron, Magnesium and Oxygen from Non-terrestrial Sources. Space Resources and Space Settlements, NASA SP-428 (1979)

Rosenberg, S.D., Guter, G.A., Miller, F.R.: The On-Site manufacture of Propellant Oxygen Utilizing Lunar Resources. Chem. Engr. Prog. 62, 228 (1964)

Rosenberg, S.D., Hermes, P., Rice, E.E.: Carbothermal Reduction of Lunar Materials for Oxygen Production on the Moon. Final Report, In Space Propulsion, Ltd., Contract NAS 9-19080 (1996)

SBIR Final Report, Lunar In-Situ Fabrication: The Manufacturing of Thin Film Solar Cells on the Surface of the Moon. NASA #NNM04AA65C (2006)

14 Perpetual Sunshine, Moderate Temperatures and Perpetual Cold as Lunar Polar Resources

James D. Burke

The Planetary Society,
Pasadena, USA

14.1 Introduction

Because the Moon's spin axis is nearly perpendicular to the plane of the ecliptic, sunlight is always horizontal at the poles. This results in a thermal environment that is almost constant and is benign relative to any other parts of the lunar surface. Also in the polar regions, topography provides places, such as crater bottoms, where sunlight never strikes the surface. In the perpetual darkness there, equilibrium temperatures are extremely low. At mountain tops and on crater rims sunlight is nearly continuous, though not truly perpetual: The Moon does have seasons because of the 1.5 degree tilt of its polar axis from the ecliptic normal, and also sunlight is greatly reduced (though not totally extinguished, because of refraction through Earth's atmosphere) for a few hours during a total lunar eclipse. Observations and topographic analysis show that the maximum percentage of surface illumination at any one place over the year is about 70 per cent; if two sites are considered there can be grazing sunlight more than 90 per cent of the time.

These solar illumination conditions, together with the resulting surface and near-subsurface thermal environment, constitute a natural resource independent of any consequences such as the retention of frozen volatiles in cold traps. *In-situ* resource uses (ISRU) hold the key to advancing lunar settlement without having to bring everything up from Earth. By having solar energy collected at a tall tower, with heat rejection at very low temperatures nearby, one could design efficient thermodynamic systems to provide steady energy-conversion functions in a lunar base, including regulated periscope-distributed sunshine for plant growth, continuous heating and cooling for environmental control, and high-temperature heat for materials processing, all the way from food preparation to solar-furnace smelting.

This chapter treats first, an eclectic history of efforts to imagine and measure the thermal properties of the Moon; second, the modern assembly of techniques and results leading to our present understanding of the polar natural environmental resource; third, some possible ways to exploit that resource given the assumption of human near-polar habitats; fourth, a few of the desires and options for learning more; and finally, the need for an agreed regime for the protection and equitable use of the unique natural situation existing in the polar regions of the Moon.

14.2 Speculation about, and Investigation of, Lunar Thermal Conditions

Plutarch noted that moonlight delivers no perceptible warmth. Johannes Kepler, in the *Somnium,* posthumously published in 1634 and only in part a fantasy, visualized both extreme heat and extreme cold as inevitable aspects of the Moon. Actual attempts to measure lunar surface temperatures began with the work of Macedonio Melloni. In 1846 Melloni, using a newly-invented thermopile and a galvanometer on Mount Vesuvius, showed a tiny amount of warming due to moonlight. In Ireland, Lord Rosse pursued the subject and concluded that, despite his best efforts, lunar temperatures remained undetermined. Other 19th-Century investigators included Sir John Herschel and Alexander von Humboldt. In 1883-1887 S.P. Langley took up the effort at the Allegheny Observatory and, even though his experimental and instrumental designs were ingenious, concluded erroneously that the illuminated lunar surface must be very cold. All of these early efforts were impacted by the difficulty of separating the lunar infrared signal from that of Earth's atmosphere, but they did confirm the primary concept that moonlight consists partly of reflected sunlight and partly of emitted lunar heat. In 1924 Pettit and Nicholson, using a bolometer at Mount Wilson, at last obtained meaningful disk-integrated thermal estimates for the Moon's near side. In the same year Bernard Lyot, based on polarimetric observations from the Pic du Midi Observatory in the Pyrenees, concluded that the very top surface of lunar material must be extremely rough on a microscopic scale, an observation borne out by the backscatter peak in the Moon's photometric function, the full Moon being eleven times as bright as the half Moon. Gradually it was becoming evident that what later came to be called the regolith must have very strange physical properties, including thermal ones.

In 1964 Saari and Shorthill, using an infrared detector at the Kottamia Observatory in Helwan, Egypt, scanned the Moon during a total lunar eclipse, showing very rapid and slower cooling as likely characteristics of loose soil and denser materials respectively.

By 1960 the idea that volatiles could be cold-trapped in permanently shaded crater bottoms near the lunar poles began to lead to analyses and proposals for investigating these environments. Watson, Murray and Brown published a prediction (Watson et al. 1961) that temperatures as low as 40 K might prevail in these locations. Arnold took up the quest, making additional calculations about the balances among lunar heat flow, scattered sunlight and cosmic radiations and particles driving polar cold-trap temperatures and also proposing various ways, including gamma-ray spectrometry, for detecting the ices if they existed near the surface. A representative sample of this work is (Arnold 1979).

The first definitive Earth-based thermal observations of the Moon through a lunation were those of Gary, Stacey and Drake in 1965, using a three-millimeter-wavelength radiometer and a fifteen-foot parabolic antenna at the Aerospace Corporation's observing site in El Segundo, California (Gary 1967). These observations are interpreted as showing actual temperatures in about the top centimeter of the regolith.

The landings of Luna 9 and Surveyor 1 in 1966, with close-up imaging, confirmed the fluffy nature of the top regolith and its increasing compaction with depth. Some hours after landing Luna 9 shifted its tilt, giving an unexpected bonus in the form of an additional imaging viewpoint, and the Surveyor footprints, plus a pressure-pad experiment, gave an indication of soil mechanics in the top few centimeters of depth. Though these were not thermal measurements, they did contribute to the growing understanding that in general the regolith must have a near-surface layer that is underdense and highly insulating. Whether or not this is true for a cryogenic polar soil saturated with frozen volatiles is a question for the future.

During the 1960's five Lunar Orbiters were launched to provide surface mapping for Apollo. The first three were so successful in imaging candidate low-latitude landing sites that the last two were dedicated to a broader scientific goal and placed in polar orbits, giving a first look at high-latitude topography and shadowing, the governing properties for illumination and surface temperatures.

Infrared remote-sensing measurements resumed with an IR radiometric investigation from the Apollo 17 orbiting Command and Service Module (CSM), mapping thermal properties of the top of the regolith over the band of low latitudes covered by the CSM's orbit (Low and Mendell 1973). Meanwhile, down on the surface, a variety of Apollo investigations had begun to elucidate the true physical, and hence thermal, properties of the materials that cover the Moon.

The scientific goals of Apollo 15 and 17 included determination of the Moon's intrinsic heat flow (Langseth and Keihm 1972-73). For this it was essential to make precise temperature measurements at different depths in the regolith. These *in situ* observations constitute a vital checkpoint for the model-derived and remotely-sensed temperature-versus-depth, temperature-versus-time of day, and temperature-versus-latitude findings of other methods.

With the ensemble of Earth-based and space-based data in hand it was possible to provide a first understanding of the Moon's surface and near-surface thermal conditions (Heiken et al. 1991). While the diurnal (i.e., monthly) temperature of the equatorial top surface swings between 400 K in daytime and 90 K at night, at a depth of about a meter the temperature is constant at about 220 K - an invitation from nature for the burial of long-lived archival data capsules (Phoenix ISU 2007).

14.3   Modern Missions over the Poles and New Environmental Knowledge

Following the Apollo missions and the contemporary robotic Soviet Luna landers, orbiters, rovers, sample-returners and Zond circumlunar human-flight precursors, mission sequences ending in 1972 and 1976 respectively, there was a long hiatus in direct exploration of the Moon. Nevertheless, interest continued to build in the polar regions as people began to consider the effect of natural lunar environments on the design and operation of human-occupied, permanent lunar habitats (Burke 1978, 1985). Spurred on not only by the ideas and scientific findings covered in this chapter but also by the tantalizing prospect of a cold-trapped materials bonanza, planning returned to consideration of the next logical exploration step; namely, remote sensing of lunar characteristics from instruments on robotic polar-orbiting spacecraft.

Eventually this has resulted in the flights of Clementine (1994), Lunar Prospector (1998), SMART-1 (2003), Kaguya (2007), Chang-E 1 (2007), Chandrayaan-1 (2008), Lunar Reconnaissance Orbiter (LRO) (2009) and Chang-E 2 (2010). Each of these polar orbiters has contributed in one way or another to our understanding of the natural resources in the polar regions. With specific relevance to the illumination and thermal topics of this chapter, the most significant investigations are the imaging of Clementine, Kaguya, LRO and Chang-E 1 and the infrared radiometry of the Diviner instrument on LRO. Chang-E 1 also made a global microwave survey whose data will in the near future enhance our knowledge of near-subsurface temperatures over the entire Moon.

Another interesting way to investigate the near subsurface is to measure the flux and energy of neutrons emitted from the Moon, due either to radioactive decay of lunar materials or to subsurface interactions with cosmic rays. The Lunar Prospector thermal/epithermal neutron data gave the first persuasive evidence of excess hydrogen, thus probably water, at both poles. Another ingenious use of the neutron data is, with the help of modeling, to derive global near-subsurface temperature indications (Little et al. 2003).

Let us turn now to examining the application of these results. Spacecraft laser altimetry and precise orbit determination give detailed knowledge of near-polar topography. The aggregate classical knowledge of the Moon's motions, first elucidated by J.D. Cassini in 1667 and continuing with modern astrometry and selenodesy, including laser ranging from Earth, allows the Moon's cyclic monthly and annual and secular millennial movements to be precisely known. This enables us to predict illumination at locations to be considered for the installation of polar habitats and energy-conversion systems.

14.4   Lunar Polar Base Illumination and Thermal Environmental Considerations

Given all of the observations and analyses described above it became clear that, even ignoring any possible material resources such as cold-trapped volatiles, the constant illumination and benign thermal conditions near the poles are a natural bonus for locating an inhabited lunar base there. At lower latitudes, base designers would have to contend with the scorching noonday and frigid two-week night. Indeed many earlier studies called for the installation of nuclear power plants and other advanced systems to cope with these challenging environments, but by the time when serious lunar human-flight planning resumed in the 1990's attention had shifted to the poles. Based on a combination of topography, illumination and expected access to material resources, the South polar region came to be preferred.

As mentioned above, in specific reference to illumination and thermal conditions near the poles, the most relevant orbital investigations are laser altimetry, visible imaging and visible, infrared and microwave radiometry. Figure 14.1 is a representative chart showing illumination percentages superimposed on topography for a South polar area (Bussey 2011).

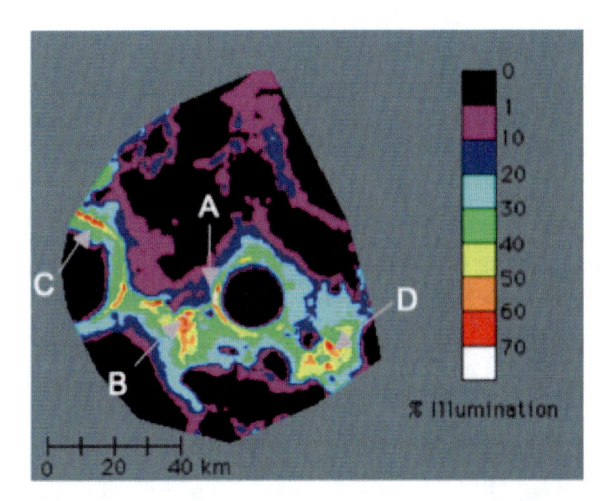

Fig. 14.1 Illumination and topography chart in false color for a South polar area (Bussey 20911)

Polar base siting analyses have been based on a collection of such information, taking into consideration changes occurring over time as well as needs for maintaining operations through seasonal darkness and eclipses. For the thermal aspect of base planning, the observations of the LRO Diviner investigation and, in the near future, of the Chang-E 1 microwave survey are the most important.

Diviner, whose scientific objectives, design and operation are thoroughly described in (Paige et al. 2010) is a nine-channel visible and infrared filter radiometer instrument whose wavelength channels are selected to determine thermal and mineral characteristics, using both reflected solar and emitted lunar radiation. Figure 14.2 is a typical Diviner data product showing derived South polar region temperatures superimposed on topography.

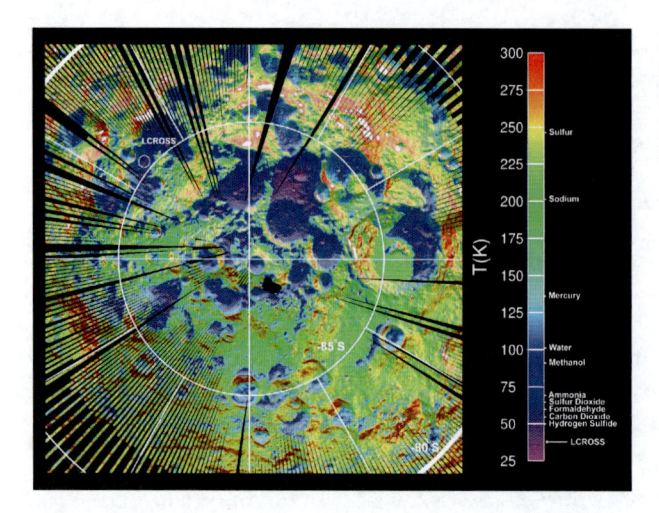

Fig. 14.2 Diviner-observed South polar temperatures superimposed on LRO-measured topography

The nearly-uninterrupted sunlight at some elevated locations encourages the conceptual design of energy-conversion systems intended to supply continuous base power, controllable piped-in sunlight for photosynthesis and habitat environmental control.

Fig. 14.3 A painting by Maralyn Vicary Flynn, made for The Planetary Society at the request of this chapter author, intended to portray the power tower concept. The tower, rotating continuously at about one-half degree per hour to face the Sun, has three main elements: a photoelectric array, a heat absorber to provide medium-temperature energy for base environmental control, and the primary mirror admitting direct sunlight for various purposes including periscope-delivered interior illumination, photosynthesis and a solar furnace. Also shown are the antennas for a cryogenic infrared and millimeter-wave astrophysical observatory.

From a near-polar crater bottom it is possible to view only less than half of the sky, but from the South lunar pole this view includes the galactic center and many other interesting celestial regions. Not shown in the painting are radiators for heat rejection at the low-temperature ends of thermodynamic cycles. These could be on the shaded side of the tower, where their equilibrium temperature would be somewhat raised by reflected sunlight from nearby illuminated lunar surfaces, or they could be in the bottoms of craters in permanent shadow, rejecting heat to dark space and also possibly providing some warmth to aid in the recovery of frozen volatiles.

14.5 Desired Next Polar Exploration Steps

Once the orbital observations have been analyzed and fully exploited for scientific, mission-precursor engineering and resource-evaluation purposes, the obvious next need is for detailed robotic exploration on the surface. For missions in nearly continuous sunlight the design tradeoffs and constraints for landers, rovers and subsurface exploration with ground-penetrating radar or drilling are familiar, as they have been examined in many previous studies for lower-latitude lunar surface exploration. The Soviet Lunokhods and the American Mars rovers exhibited practical solutions for robotic surface travel. Also, drilling and drive-tube core collection were demonstrated during the Apollo and Soviet sample-returning investigations.

For exploring the shadowed cold traps this body of experience may or may not be entirely valid. According to some analyses, including that resulting from the neutron observations, at least the top few centimeters of the cold-trap regolith may have been dried out by sublimation during micrometeorite bombardment - but this is not proven; the only way to be sure is to drill.

Operating a robotic rover in a dark crater bottom poses new challenges. For extended-time sorties the needed power sources are a radioactive thermoelectric generator (RTG) or a Stirling engine driven by radioactive heat, referred to as an Advanced Stirling Radioisotope Generator (ASRG). Command and data links will have to use some form of relay, either via orbiter or via an emplaced surface transponder with line-of-sight to the rover. Imaging will need a rover-borne light source and/or a searchlight following the rover. Ground simulations of such traverses, as commonly done in preparing for Mars roving missions, will require special dark facilities and even then there will be uncertainties about real lunar obstacles and surface navigation.

Examining these possibilities and retiring perceived risks will need a substantial technology-development program ahead of any commitment to a lunar polar cold-trap robotic roving mission. The launch of such a program is unlikely until further advances in lunar planning. Nevertheless, some mission proposals are in work. One example, called Luna-Resurs, would include an Indian orbiter carrying a Russian lander with a micro-rover (Roscosmos/ISRO 2011).

14.6   Polar Policy and Legal Matters

The structure of international policy and law relating to lunar resources is incomplete. The Outer Space Treaty of 1967 does assert some broad general principles as discussed below, and there have been decades of discussion of managing lunar resources in the UN's Committee on the Peaceful Uses of Outer Space (COPUOS) and the International Institute of Space Law (IISL), with as yet no actual international action.

The absence of binding multinational commitments does not mean a complete lack of potential for effective regulation. In addition to treaties, other mechanisms exist for governing equitable access to, and exploitation of, off-Earth resources. One example is the set of agreements among members of the International Telecommunications Union (ITU) on allocation of spectrum and orbital slots. Though unenforceable, these regulations are mostly accepted because they operate for the mutual benefit of all users.

In addition to treaties and multilateral agreements, a variety of other measures can be applied among governments, non-governmental organizations and businesses using natural resources. A discussion of those relating to the Moon is given in (Pelton and Bukley 2009).

Investigations already completed convincingly show that the lunar polar natural illumination and thermal resource is limited to two small portions of the Moon's surface. The areal and in-depth distribution of the cold-trapped material resource is not known in detail, and what quantity of useful volatiles can be economically extracted is as yet completely unknown. Nevertheless it is clear that all of the polar resources are confined to two small regions. Therefore, when exploitation is considered, it is necessary to examine existing and proposed means for governing the activity.

A logical policy goal is to assure that these natural resources be made available on an equitable basis, with due consideration of their limited locality and quantity. This is not a new problem in either terrestrial or celestial law, but its solution with regard to the Moon has proved to be vexing.

The Outer Space Treaty of 1967 has some relevant provisions, including ruling out "appropriation" of territory, but it is not specific with regard to lunar mining. It does have some language relating to the collection and removal of samples for scientific purposes, and it also recognizes the possibility of in situ resource use (ISRU) at least to the extent of applications within and supporting space missions. The prospect of ISRU for other uses, such as delivery into orbit or to Earth, is left unexamined. One of many thoughtful discussions of lunar ISRU is given in (Rapp 2009).

The Moon Agreement of 1979, intended to reduce some of the 1967 treaty's uncertainties, was ratified in a few countries, but it has not come broadly into force because no major space-faring government would support it. Nevertheless, its provisions are a reasonable guide to a possible future regulatory regime.

In the Policy and Law curriculum of the annual Space Studies Program (SSP) of the International Space University, the governing of lunar resource exploitation is always a subject of much debate. Analogies are drawn with the Antarctic

Treaty, the law of the deep sea, the unenforceable but generally agreed-upon regulations of the ITU and other such terrestrial examples. One ISU alumnus, Virgiliu Pop, writes on the subject of ownership of lunar resources; see (Pop 2008) and Chap. 23 of this book. A useful summary providing background for these discussions is given in Chap. 9 of (Pelton and Bukley 2010). In that chapter a quotation from the Moon Agreement reads in part as follows: "... [W]hen exploitation of the natural resources of the Moon is about to become feasible, states, parties to the Agreement, must establish an international regime, including appropriate procedures, to govern such exploitation." Among the functions of such a regime would be equitable sharing of the product, with recognition of the effort and contribution of developers as well as of the needs of beneficiaries.

With regard to the particular polar resources that are the subject of this chapter, namely continuous sunlight and useful thermal conditions, the regulatory needs are (a) protection of sunlit real estate and (b) equitable use of cold crater-bottom regions. For the peaks of nearly-perpetual light, as shown in Fig. 14.1, an example is illustrative: Suppose that users want to erect not just one, but two power towers such as the one shown in Fig. 14.3. Each will periodically shadow the other. To keep this from being a major problem they must be far enough apart so that, as seen from one, the other subtends only a small portion of the Sun's one-half-degree disk. This requirement, as expressed in a governing agreement, would be analogous to the terrestrial doctrine of Ancient Lights, a long-established concept intended to reduce the shadowing of one building by another in high latitudes on Earth.

With regard to permanently-shadowed crater bottoms, the thermal policy need is modest because radiators are likely to be small relative to the available land area. Therefore this part of the legal regime will be mainly concerned with cold-trapped material resources.

Another aspect of lunar policy and legal regulation is sustainable use with environmental protection so that future generations of humans may enjoy the same benefits as those now potentially available at the Moon. When the total of the Moon's minable surface materials is considered, even very large industrial extraction does not seem to be a dominant problem. For example, G.K. O'Neill, in imagining huge in-space settlements built using lunar raw materials, made calculations showing a relatively minor impact on the Moon (O'Neill 2000). For the material resources unique to the polar regions this is likely not to be the case, so sustainable use and recycling will probably be main concerns. For the resources considered in this chapter, sunlight and temperature, it seems likely that human uses, at least in the early stages of lunar settlement, will not pose important sustainability or environmental-impact questions.

14.7 Conclusion

The nearly-uninterrupted solar energy, the benign surface thermal environment and the cold traps in the lunar polar regions are natural resources conducive to placing human-occupied outposts there. To make this possibility a reality and also to make that reality an achievement reflecting the best character aspects of humanity in executing a large-scale technological program, a variety of precursor activities will be needed.

Here we have examined some of those needs. First, it is necessary to exploit fully the findings of the many lunar polar orbiting missions now in progress or planned, in the context of prior Earth-based and space-based knowledge of the Moon. Next, polar robotic exploration on the surface is essential. As a part of this program, precursor demonstrations of solar energy collection at favorable sites could be made. Also the thermal environment at both illuminated and permanently-dark sites should be measured at the surface and in the near subsurface. Surface electrical charging in sunlit and dark regions could also be investigated. The landing and roving missions to achieve these goals could at the same time begin to answer today's outstanding questions about the quantity and accessibility of polar material resources including cold-trapped volatiles.

While these exploratory missions are proceeding, a parallel development of applicable policy and law should be underway, so that when the time comes for both robotic and human *in situ* resource uses (ISRU) appropriate and equitable regulatory regimes will be in place.

Given these developments, humanity could be resuming the movement, long explored in science fiction and briefly begun by Apollo and its failed Soviet competitor, toward permanent residence off Earth. With sustainable lunar living and a commitment to preserving the lunar environment and its natural resources, humans would be setting out markers on a road toward becoming a two-world society with its knowledge, traditions and wisdom established and saved for the future on both Earth and Moon.

References

Arnold, J.R.: Ice in the Lunar Polar Regions. J. Geopyhs. Res. 84, 5659–5668 (1979)

Burke, J.D.: Energy Conversion at a Lunar Polar Site. In: Billman, K., Summerfield, M. (eds.) Radiation Energy Conversion. In Space, American Institute of Aeronautics and Astronautics, New York (1978)

Burke, J.D.: Merits of a Lunar Polar Base Location. In: Mendell, W. (ed.) Lunar Bases and Space Activities of the 21st Century, Lunar and Planetary Institute, Houston (1985)

Bussey, D.B.J., et al.: Determining Lunar Polar Illumination Conditions Using Kaguya and LRO Topography. In: Geophysical Research Abstracts, vol. 13. European geosciences Union, Vienna, Paper No. 2011-12692 (2011)

Gary, B.: Results of a Radiometric Moon Mapping Investigation at Three Millimeters Wavelength. Ap. J. 147, 245–254 (1967)

Heiken, G., et al. (eds.): Lunar Sourcebook: A User's Guide to the Moon. Cambridge University Press (1991)

Langseth, M.G., et al.: Heat Flow Experiment. In: Apollo 17 Preliminary Science Report, US Government Printing Office, NASA, Washington (1973)

Little, R., et al.: Latitude Variation of the Lunar Subsurface Temperature: Lunar Prospector Thermal Neutrons. J. Geophys. Res. 108, 5046 (2003)

Low, F., Mendell, W.: Infrared Scanning Radiometer. In: Apollo 17 Preliminary Science Report, US Government Printing Office NASA, Washington (1973)

O'Neill, G.K.: The High Frontier. Space Studies Institute, Princeton (2000)

Paige, D.A., et al.: Diviner Radiometer Experiment (2010),
    http://diviner.ucla.edu

Pelton, J., Bukley, A. (eds.): The Farthest Shore: A 21st Century Guide to Space. Apogee
    Books (2010)

Phoenix Project Team. A Lunar Archive. International Space University, Phoenix (2007),
    http://www.isunet.edu

Pop, V.: Who Owns the Moon? Springer, Heidelberg (2008)

Rapp, D.: Lunar In-Situ Resource Utilization. In: Badescu, V. (ed.) Mars: Prospective
    Energy and Material Resources. Springer, Heidelberg (2009)

Roscosmos/ISRO, Luna-Resurs (2009),
    http://russianspaceweb.com/luna_resurs.html

Watson, K., et al.: The Behavior of Volatiles on the Lunar Surface. J. Geophys. Res. 66,
    3033 (1961)

15 Condition of Solar Radiation on the Moon

Xiongyao Li[1], Wen Yu[1,2], Shijie Wang[1], Shijie Li[1], Hong Tang[1],
Yang Li[1,2], Yongchun Zheng[3], Kang T. Tsang[4], and Ziyuan Ouyang[1]

[1] Institute of Geochemistry, Guiyang, China
[2] University of the Chinese Academy of Sciences, Beijing, China
[3] The National Astronomical Observatories, Beijing, China
[4] Hong Kong University of Science and Technology, China

15.1 Introduction

Solar radiation is an exterior heat source of the Moon and represents a key resource with respect to returning to the Moon. It controls the variation of lunar-surface temperature during the lunation, and changes the thermal radiation properties of the lunar surface. In lunar Earth-based exploration, orbital exploration, and manned and unmanned lunar surface activities, solar radiation is an important factor which should be considered.

Multispectral imaging, thermal radiation imaging, and microwave radiation measurement are the main techniques used to explore the Moon. IIM and MRM on the CE-1, MI and SP on the KAGUYA, HySI and M^3 on the CHANDRAY-AAN-1, and DLRE and LORC on the Lunar Reconnaissance Orbiter (LRO) had been launch to explore the Moon on the lunar orbit in recent years. The signal received by these instruments is affected by reflection of the solar radiation and emission of the lunar surface. In manned and unmanned lunar surface activities, the solar radiation is also considered in the design of heat control.

Lighting condition is another aspect in study on solar radiation of the Moon. Because the Moon's spin axis is nearly perpendicular to the ecliptic plane, it results in different lighting conditions at the lunar poles. Areas which have low elevation, such as the floors of impact craters, may never be directly illuminated by the Sun, i.e. they are permanently shadowed, whilst regions of high elevation, relative to the local terrain, may be permanently illuminated (Bussey and Spudis 2004). The presence of hydrogen has been detected near the lunar poles (Feldman et al. 1998,2000). The areas with enhanced hydrogen content are generally coincident with areas permanently in shadow (Margot et al. 1999; Feldman et al. 2001). The temperature in the permanently shaded regions is low enough to trap water ice and other volatiles over the age of the solar system (Vasavada et al. 1999), and the lunar rotation axis has been stable for at least 2 billion years. If the hydrogen is in the form of water ice, it would confirm the idea postulated by Watson et al. (1961) and debated by many scientists since that ice might exist in lunar cold traps (Arnold 1979; Butler 1997; Starukhina 2001; Hodges 2002). Beside water ice, other volatiles might also be enriched in permanent shadow at the lunar poles. Reservoirs of volatiles in permanent shadows are scientifically valuable and maybe a potential in situ resource for future exploration.

The solar illumination condition is also an important parameter for site selection, when considering a landing on the lunar surface and building a lunar base. It impacts heavily the landing of the lunar rover and the human activities on the Moon. It should be considered in mission timing, duration and return, as well as design of subsystem (Kruijff and Ockels 1995).

15.2   Theoretic Model

Previous studies showed the solar constant (solar irradiance at 1 AU) varies between 1361.8 and 1369.2 $W \cdot m^{-2}$ (Willson 1997, 2003; Lean et al. 1995; Fröhilich and Lean 1998; Fröhilich 2002).

In consideration of the limiting conditions of perihelion, aphelion, perigee and apogee in the Sun-Earth-Moon system, the lunar-surface solar irradiance ranges from 1308.9 $W \cdot m^{-2}$ to 1425.7 $W \cdot m^{-2}$ with a range exceeds 100 $W \cdot m^{-2}$ at the equator.

In order to describe accurately the variations of lunar-surface solar irradiance, Sun-Moon distance, and solar radiation incidence angle, we calculated the basic parameters for the Sun-Earth-Moon system at time t, such as geocentric ecliptical longitude of the Sun, Sun-Earth distance, geocentric ecliptical latitude and longitude of the Moon, and Earth-Moon distance, according to the Variations Séculaires des Orbites Planétaires 87(VSOP87) planetary orbital theory and the Éphémérides Lunaires Parisiennes 2000-82(ELP2000-82) lunar orbital semi-analytic theory (Bretagnon and Francon 1988; Chapront-Touzé and Chapront 1983). From the location relationship between the Sun and the Moon in the geocentric ecliptical coordination system, we could establish the relationship among Sun-Moon distance, solar radiation incidence angle, selenographic latitude and longitude of the sub-solar point, and time. We have further constructed the lunar-surface effective solar irradiance real-time model.

15.2.1   Model Description

The intensity of solar radiation is usually expressed in solar irradiance. And the effect of solar radiation on the Moon is mainly caused by solar irradiance normal to the lunar surface, which named effective solar irradiance. The geometric relationship among lunar-surface solar irradiance (LSI), lunar-surface effective solar irradiance (LESI) and solar radiation incidence angle (i) is shown in Fig 15.1. The LESI could be expressed by the following equation:

$$LESI = LSI \cdot \cos(i) \tag{15.1}$$

With the assumption that the solar energy has any loss in space transmission, the LSI could be expressed as follows (Maxwell 1998; Owczarek 1977):

$$LSI = S_0 R_{sm}^{-2} \tag{15.2}$$

where, S_0 is the solar constant, R_{sm} is the dimensionless Sun-Moon distance relative to 1 AU.

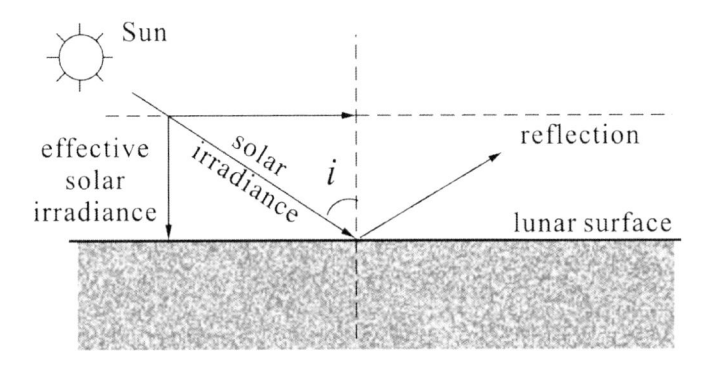

Fig. 15.1 Schematic diagram of LSI

According to the relationship among the location of the Sun, the Earth and the Moon (Fig. 15.2), the Sun-Moon distance could be expressed as:

$$R_{sm} = R_{em} \sin\varphi_{em} / \sin\varphi_{sm} \tag{15.3}$$

where, R_{em} is the dimensionless Earth-Moon distance relative to 1 AU, φ_{em} is the geocentric ecliptical latitude. They could be obtained from the lunar orbital semi-analytic theory ELP2000-82 (Bretagnon and Francon 1988).

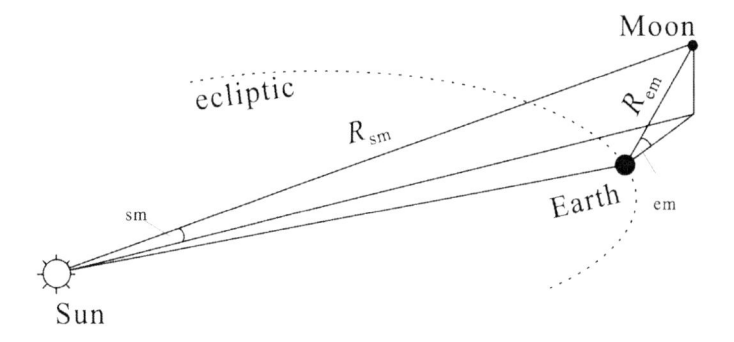

Fig. 15.2 Geometric relationship in the Sun-Moon system

φ_{sm} is the heliocentric ecliptical latitude. It was given by Meeus (1991) as follows:

$$\varphi_{sm} = \varphi_{em}\, R_{em}/R_{se} \tag{15.4}$$

where, R_{se} is the dimensionless Sun-Earth distance relative to 1AU. It could be deduced from the planetary orbital theory VSOP87 (Chapront-Touzé and Chapront 1983). And, the LSI could finally be expressed as:

$$LSI = \frac{S_0 \sin^2\left(\varphi_{em}\, R_{em}/R_{se}\right)}{R_{em}^2 \sin^2 \varphi_{em}} \tag{15.5}$$

Furthermore, it is assumed that the Moon is an ideal spheroid, and the topographic effect on the solar radiation incidence angle is neglected. According to the geometric relationship among the location of the Sun, the Moon, the sub-solar point and the measured point in the geocentric ecliptical coordinates system, we could establish the relationship between the solar radiation incidence angle and the radius of the Moon, the Sun-Moon distance, the selenographic longitude and latitude of the sub-solar point, and the selenographic longitude and latitude of the measured point (Fig. 15.3).

In order to correspond to the exploring data of Apollo, Clementine and Lunar Prospector, we adopted the definition of selenographic coordinates before 1974. It is defined that the north is the top and the south is the bottom of the Moon, and the prime meridian refers to the longitude across the line between the center of the Earth and the Moon. The selenographic longitude is expressed in terms of 0~+180 degree toward the east and 0~-180 degree toward the west. The selenographic latitude is expressed in terms of 0~+90 degree toward the north and 0~-90 degree toward the south (Davies and Colvin 2000; Heiken et al.1991). In Fig. 15.3., it could seen,

$$i = \alpha + \beta \tag{15.6}$$

where,

$$\alpha = \arccos\left[\pm\sin\varphi_n \sin\varphi_d + \cos\varphi_n \cos\varphi_d \cos(\psi_n - \psi_d)\right] \tag{15.7}$$

$$\beta = \arcsin\left[R_{moon}\sin\alpha \Big/ \left(\frac{R_{em}^2 \sin^2 \varphi_{em}}{\sin^2\left(\varphi_{em}\, R_{em}/R_{se}\right)} + R_{moon}^2 - \frac{2R_{em}R_{moon}\sin\varphi_{em}\cos\alpha}{\sin\left(\varphi_{em}\, R_{em}/R_{se}\right)} \right)^{1/} \right] \tag{15.8}$$

Here R_{moon} is the radius of the Moon (1.161363636×10^{-5} AU; Duke 1999; Heiken et al. 1999). ψ_n, φ_n, ψ_d and φ_d are the selenographic longitude and selenographic latitude of the measured point and sub-solar point, respectively.

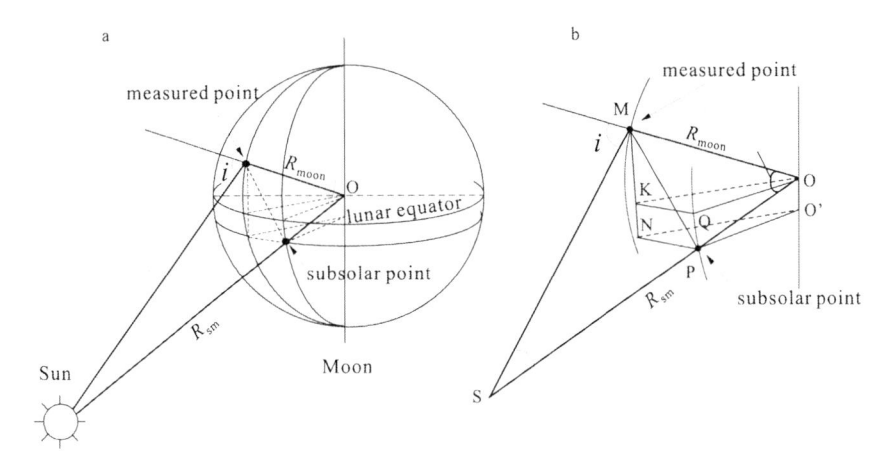

Fig. 15.3 Schematic diagram of solar radiation incidence angle on the lunar surface. O is the center of the Moon, plane KOQ is located on the equatorial plane of the Moon, plane N O'P is parallel to the equatorial plane of the Moon, K and N are the projections of M on each plane, Q is the projection of P on the equatorial plane of the Moon.

According to the analysis above, the LESI could be finally expressed as:

$$LESI = \frac{S_0 \sin^2\left(\varphi_{em}\, R_{em}/R_{se}\right)\cos(i)}{R_{em}^2 \sin^2 \varphi_{em}} \tag{15.9}$$

where, i could be worked out from Eqs. (15.6-15.8). Parameters in Eqs. (15.7) to (15.9) could be got from astronomical algorithms. From Eq. (15.9), it could be seen that the solar constant, the geocentric ecliptical latitude of the Moon, the Earth-Moon distance, the Sun-Earth distance and the solar radiation incidence angle are the factors affecting the lunar-surface effective solar irradiance. Because the geocentric ecliptical latitude of the Moon is very small ($0\leq\varphi_{em}\leq 5.15°$) (Duke 1999; Heiken et al. 1991), the LESI could be reduced as follows:

$$LESI \approx \lim_{\varphi_{em}\to 0} \frac{S_0 \sin^2\left(\varphi_{em}\, R_{em}/R_{se}\right)\cos(i)}{R_{em}^2 \sin^2 \varphi_{em}} = \frac{S_0\cos(i)}{R_{se}^2} \tag{15.11}$$

Previous studies showed that the S_0 ranges from 1361.8 W·m^{-2} to 1369.2 W·m^{-2}, the variation range is about 7.4 W·m^{-2} (Willson 1997, 2003; Lean et al. 1995; Fröhilich and Lean 1998; Fröhilich 2002).

The Sun-Earth distance ranges from 0.98AU to 1.02 AU (Guo 1988). The variation of LESI caused by the variation of Sun-Earth distance is larger than 100 W·m^{-2}. The solar radiation incidence angle ranges from 0° to 90°. As a result, the LESI ranges from 0 to 1425.7 W·m^{-2}. Therefore, the Sun-Earth distance and solar radiation incidence angle are the main factors affecting the variation of LESI.

15.2.2   Error Analysis

According to the theory of error propagation, the system error equals to the sum of absolute values of partial derivatives multiplying the relative error of each variable. From Eqs. (15.1) and (15.2), we could obtain:

$$LESI = S_0 R_{sm}^{-2} \cos(i) \tag{15.11}$$

From Eq. (15.11), it could be shown that the error of LESI is caused by the error propagation of S_0, R_{sm} and i. It could be expressed as follows:

$$\sigma_{LESI} = \left| \frac{\partial I}{\partial S_0} \sigma_{s_0} \right| + \left| \frac{\partial I}{\partial R_{sm}} \sigma_{R_{sm}} \right| + \left| \frac{\partial I}{\partial i} \sigma_i \right| \tag{15.12}$$

where, σ_{s0} is the relative error of solar constant. σ_{Rsm} is the relative error of Sun-Moon distance. σ_i is the relative error of solar radiation incidence angle. By substituting Eq. (15.11) into Eq. (15.12), it could be changed as:

$$\sigma_{LESI} = \frac{\cos(i)}{R_{sm}^2} \sigma_{s_0} + \frac{2 S_0 \cos(i)}{R_{sm}^3} \sigma_{R_{sm}} + \frac{S_0 \sin(i)}{R_{sm}^2} \sigma_i \tag{15.13}$$

The variation of solar constant is closely related to solar activity (Willson 1997,2003; Fröhlich 2002; Solanki and Fligge 1998). Previous studies showed that the solar constant has a tendency of variation in an 11-year cycle with strong/weak alternation. It ranges from 1361.8W.m^{-2} to 1369.2W.m^{-2}(Willson 1997, 2003; Lean et al. 1995; Fröhlich and Lean 1998; Fröhilich 2002). Therefore, the S_0 constant could be taking the average of both maximum and minimum values, i.e., $S_0=1365.5\pm3.7$ W·m^{-2}. The relative error of S_0 is,

$$\sigma_{S_0} \le 3.7 \qquad [\text{W·m}^{-2}] \tag{15.14}$$

According to the theory of VSOP87, the relative error of R_{sm} is,

$$\sigma_{R_{sm}} \le 9.13 R_{sm} \times 10^{-6} \quad [\text{cy}^{-1}] \tag{15.15}$$

Because φ_d is very small in Eq. (15.7), the errors of α and ψ_d should be on the identical order of magnitude, i.e., 10^{-9} cy^{-1} (Meeus 1991). β is also very small because the Sun-Moon distance is much larger than the radius of the Moon. So, the error of solar radiation incidence angle results mainly from α, approximately on the order of magnitude of about 10^{-9} cy^{-1}, too:

$$\sigma_i \le 10^{-9} \qquad [\text{cy}^{-1}] \tag{15.16}$$

By substituting Eqs. (15.14-15.16) into Eq. (15.13), we obtain:

$$\sigma_{LESI} = \frac{3.7 \cos(i)}{R_{sm}^2} + \left[\frac{18.26 S_0 \cos(i)}{R_{sm}^2} \times 10^{-6} + \frac{S_0 \sin(i)}{R_{sm}^2} \times 10^{-9} \right] \times N \tag{15.17}$$

where, N is the number of year relative to 12 o'clock on January 1, 2000. According to the geometric relationship in the Sun-Earth-Moon system, when the Earth is located at the perihelion and the Moon is on the line between the Sun and the Earth, R_{sm} would reach the minimum, and be equal to 0.980466 AU (Guo 1988; Duke 1999; Heiken et al. 1991). i ranges from 0° to 90°, the error of LESI in 100 years from 1950 to 2050 could be expressed as follows:

$$0 \leq \sigma_I \leq 3.89 \qquad [\text{W·m}^{-2}] \tag{15.18}$$

During this period, the percentage error of LESI is smaller than 0.28%.

15.3 LESI on Flat Lunar Surface

According to the model mentioned above, we have developed a computational program on the Linux operation system in C++, and simulated LESI of flat lunar surface. The simulation focused on the change of LESI at different selenographic latitude and longitude in a month and an 18.6-year lunar cycle.

15.3.1 LESI Change with Latitude

Due to the Moon's axial tilt, the sub-solar latitude changes from about 1.5°S to 1.5°N in a year (Fig. 15.4). For the same selenographic longitude, the solar incidence angle increases from the sub-solar latitude to the two sides. So, LESI shows a decrease trend from the sub-solar latitude to the two sides. With the simulation of the first month in 2008 (JDE from 2454466.5 to 2454496.5), LESI on the prime meridian achieves the maximum value near the lunar equator in the latter half month (Fig. 15.5). In the month, the sub-solar latitude varies in southern hemisphere from 1.18°S to 0.48°S. It leads to a distinct difference of LESI between lunar South Pole and North Pole. There is a small LESI changed from 29.15 W·m^{-2} to

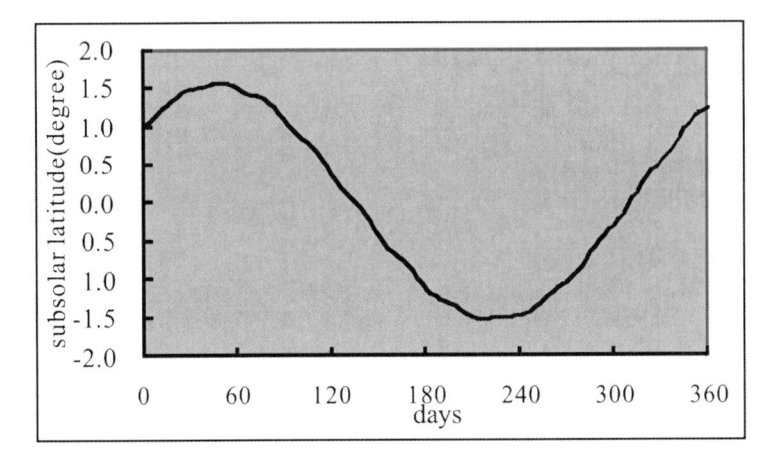

Fig. 15.4 Sub-solar latitude changes in a year. The simulation represents the change in 1994 (JDE from 2449353.5 to 2449718.5).

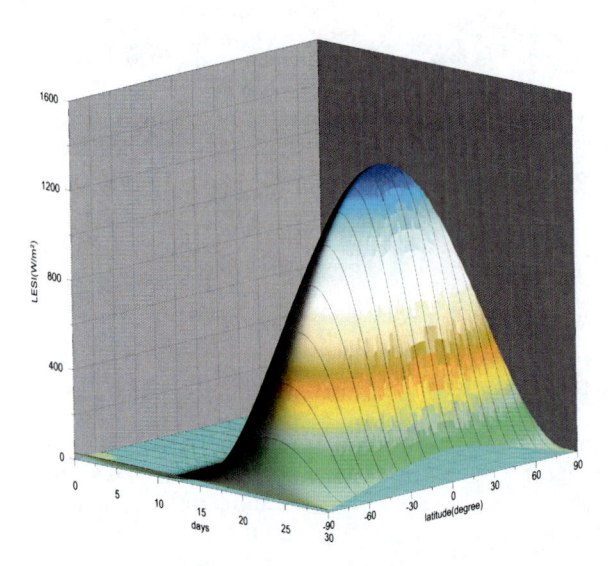

Fig. 15.5 LESI on the prime meridian changes with time and latitude in a month. The simulation represents the change in first month in 2008 (JDE from 2454466.5 to 2454496.5). The sub-solar latitude changes from 1.18°S to 0.48°S in this month, and LESI decreases from the lunar equator towards the two sides. The South Pole has a small LESI in the whole month and LESI is 0 at the North Pole.

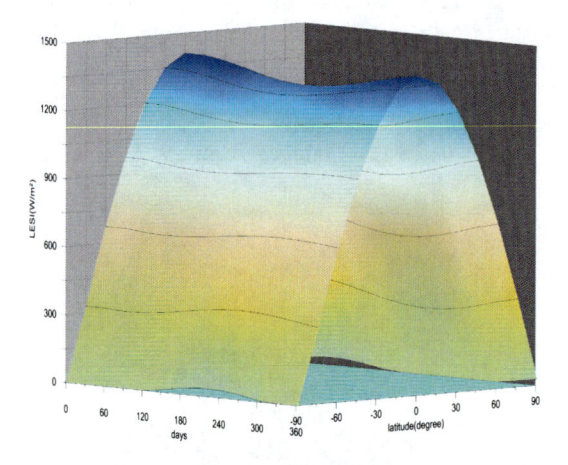

Fig. 15.6 The maximum monthly LESI on the prime meridian changes with time and latitude in a year. The simulate ion represents the change in 1994 (JDE from 2449353.5 to 2449718.5). The maximum monthly LESI got a larger value near the lunar equator, and decreased from the equator to the poles.

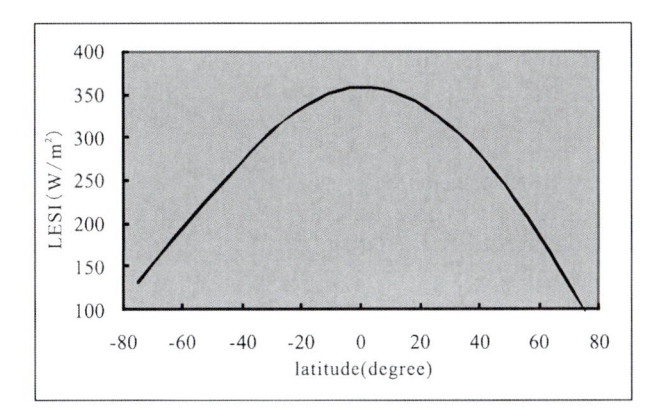

Fig. 15.7 Annual variation of the maximum monthly LESI at different latitude. It is about 360 W·m^{-2} at the lunar equator and only 100 W·m^{-2} at 80°N and 80°S on the prime meridian in 1994 (JDE from 2449353.5 to 2449718.5).

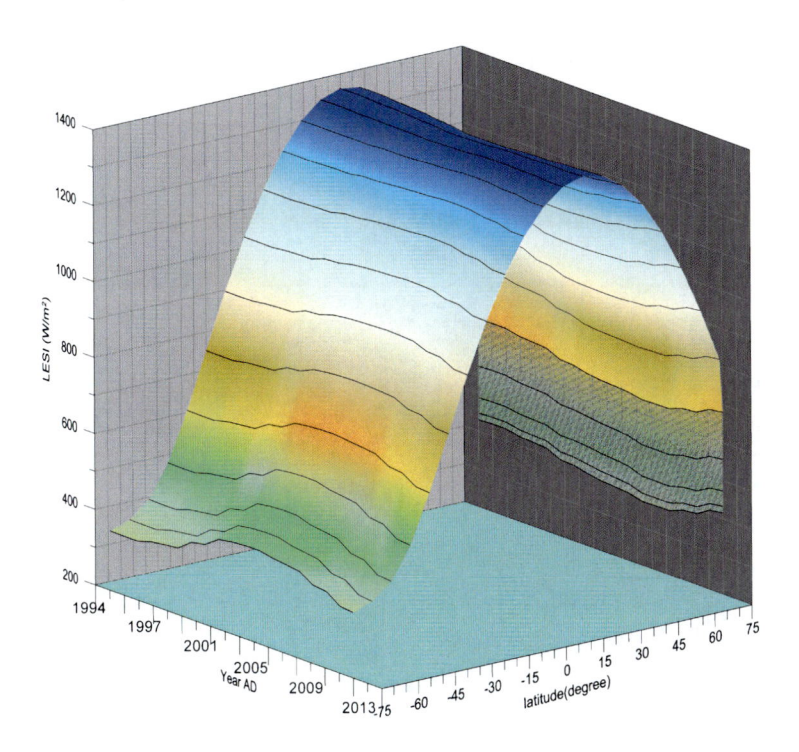

Fig. 15.8 The maximum annual LESI on the prime meridian changes with time and latitude in 18.6-year lunar cycle. The simulation represents the change during the period from 1994 to 2013 (JDE from 2449353.5 to 2456146.5).

11.69 $W \cdot m^{-2}$ in the whole month at lunar South Pole. In contrast, the lunar North Pole is in the dark night with a zero LESI. Like the variation of length of day and night on the earth, the southern hemisphere has longer lunar day and shorter lunar night than the northern hemisphere during the month.

With the movement of the Moon, the maximum monthly LESI and the maximum annual LESI of different latitude change in the 18.6-year cycle. To find the variation of the maximum monthly LESI at different latitude, LESI on the prime meridian in 1994 have been simulated (Fig. 15.6). The maximum monthly LESI got a larger value near the lunar equator, and decreased from the equator to the poles. Comparing the change of LESI in the southern hemisphere to the northern hemisphere, an opposite tendency is apparent. The annual variation of the maximum monthly LESI at low latitude is larger than those at high latitude (Fig. 15.7). From the simulation, the range is about 360 $W \cdot m^{-2}$ at the lunar equator and only 100 $W \cdot m^{-2}$ at 80°N and 80°S in 1994 (JDE from 2449353.5 to 2449718.5).

In the simulation of LESI during the 18.6-year lunar cycle (Fig. 15.8), the maximum annual LESI shows a similar change to the maximum monthly LESI. It got a larger value near the lunar equator, and decreased from the equator to the poles. An opposite change tendency is also shown in the southern hemisphere and the northern hemisphere. But the variation of the maximum annual LESI shows an opposite trend in different latitude compared to the maximum monthly LESI. It is much smaller at low latitude than at high latitude (Fig. 15.9). During the period from 1994 to 2013 (JDE from 2449353.5 to 2456146.5), the variation is only about 4 $W \cdot m^{-2}$ at the lunar equator and about 80 $W \cdot m^{-2}$ at 80°N and 80°S.

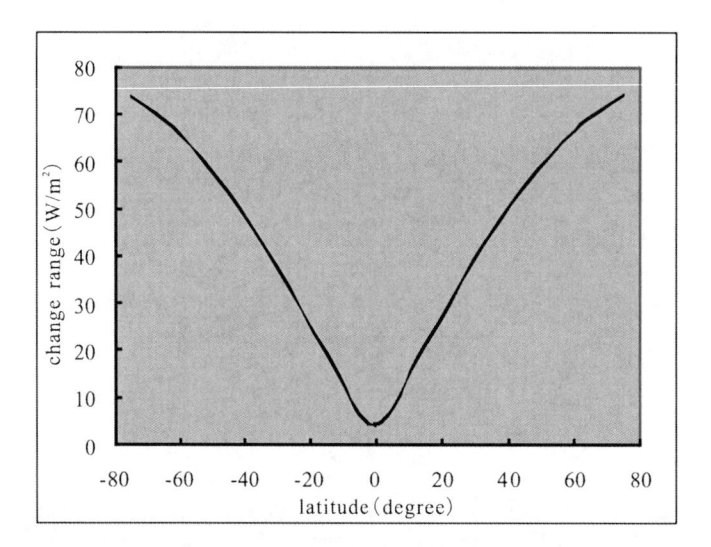

Fig. 15.9 Variation of the maximum annual LESI at different latitude. It is only about 4 $W \cdot m^{-2}$ at the lunar equator and about 80 $W \cdot m^{-2}$ at 80°N and 80°S on the prime meridian during the period from 1994 to 2013 (JDE from 2449353.5 to 2456146.5).

15.3.2   LESI Change with Longitude

At the same latitude, LESI get a same maximum value and has a same tendency at different selenographic longitude. But a time delay is shown at different seleno-graphic longitude (Fig. 15.10). It is mainly caused by the lunar rotation. It might indicate that the effect of selenographic longitude must be considered only in the investigation of LESI at a particular time or in a short time scale.

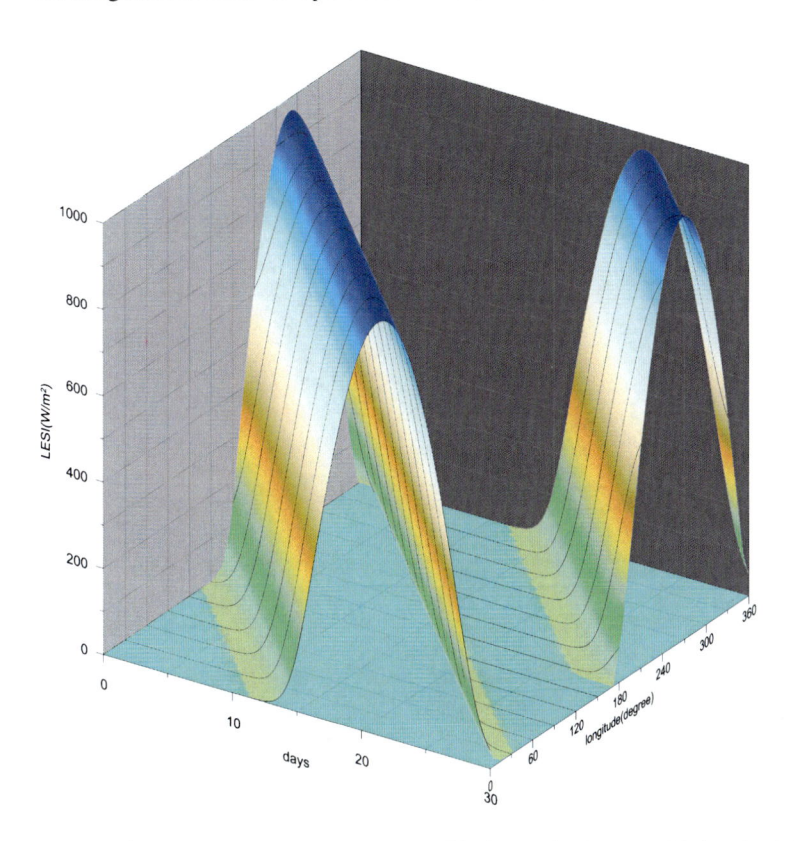

Fig. 15.10 LESI at lunar equator changes with time and selenographic longitude in a month. The simulation represents the change in first month in 2008 (JDE from 2454466.5 to 2454496.5).

15.3.3   Global Distribution of LESI

To investigating the distribution of LESI, simulation has been done with a spatial resolution of 1°×1°. Figure 15.11 shows the global distribution of LESI at Jan 1st, 2008 (JDE 2454466.5). A ring-like distribution of LESI is shown, and it decreases gradually from the centre outwards with a sub-solar point at 1.18°S and 93.08°W. Due to the tilt of axis of the Moon, polar day and night appears at the lunar poles and it is verified by the simulation. A non-zero LESI appears alternately in a year.

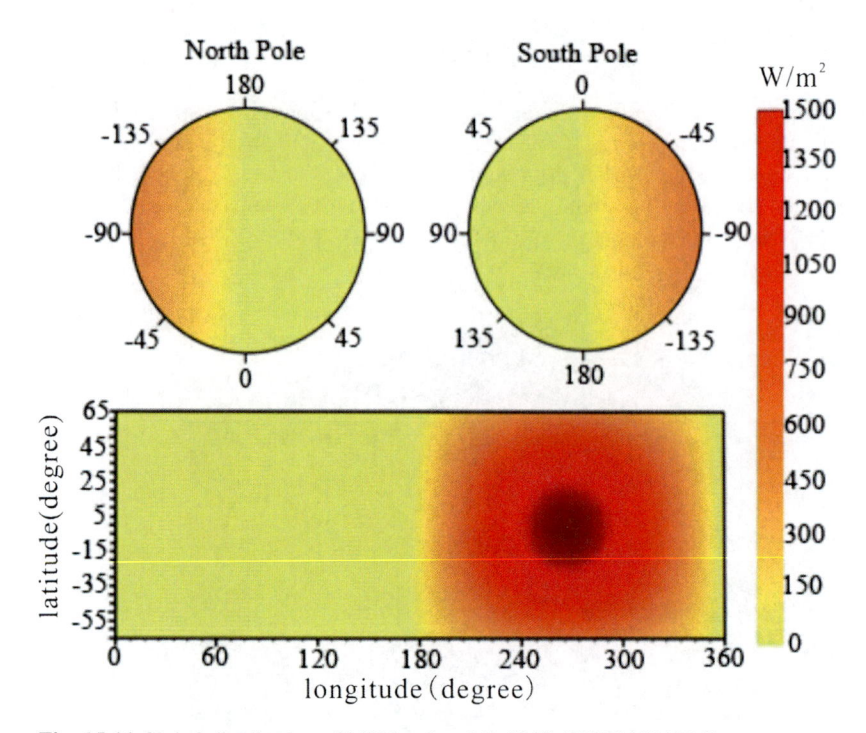

Fig. 15.11 Global distribution of LESI at Jun 1th, 2008 (JDE 2454466.5).

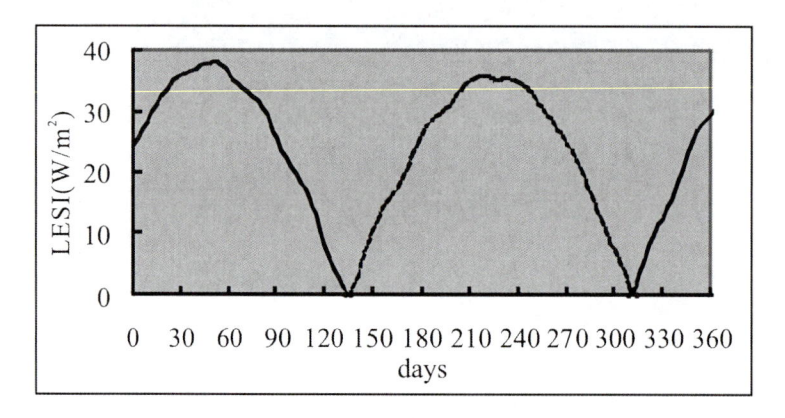

Fig. 15.12 LESI at the lunar poles changes in 1994 (JDE from 2449353.5 to 2449718.5).
Solid line represents the lunar North Pole, dot line for the lunar South Pole.

In the first 135 days and the last 50 days of the year, the lunar North Pole has a non-zero LESI, and the lunar South Pole has a zero LESI. That means the North Pole is in the polar day and the South Pole is in the polar night. And it is the opposite in the middle 180 days (Fig. 15.12). The maximum of LESI at lunar North Pole and lunar South Pole are 38.1 $W \cdot m^{-2}$ and 35.9 $W \cdot m^{-2}$, respectively.

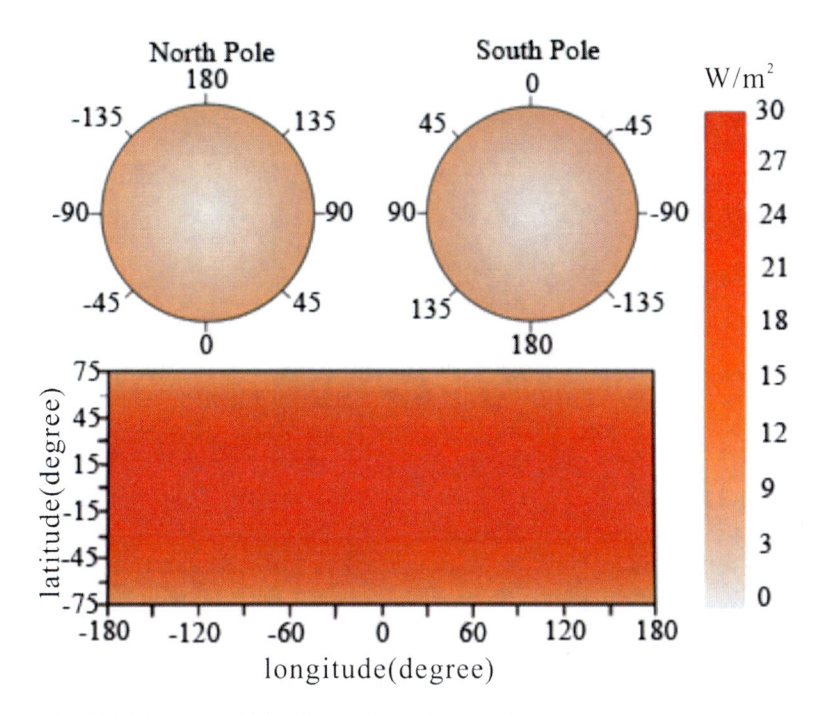

Fig. 15.13 Average of LESI in 18.6-year lunar cycle.

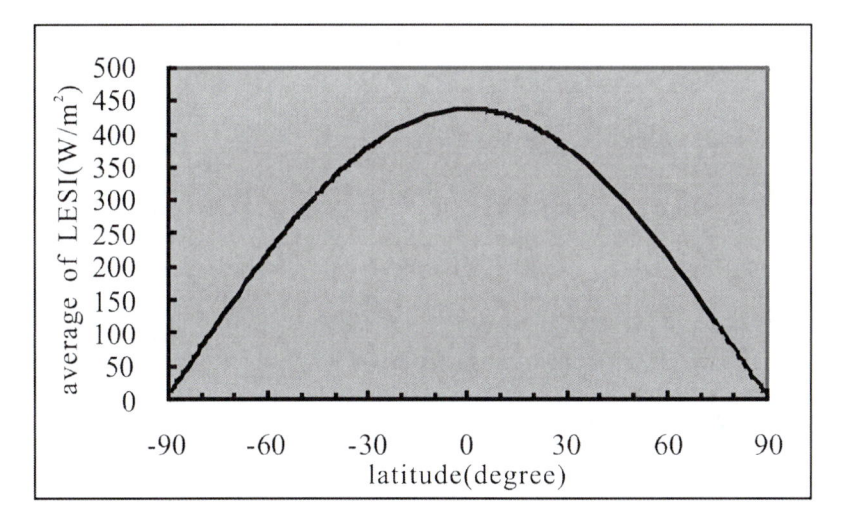

Fig. 15.14 An 18.6-year average of LESI changes with latitude.

Compared to the instantaneous distribution of LESI, the average of LESI in 18.6-year lunar cycle shows a stripe distribution (Fig. 15.13). It decreases gradually from the lunar equator to the lunar poles (Fig. 15.14).

15.4 Effect of Lunar Topography

Lunar topography makes an important effect on the LESI by shadowing and changing the incidence angle. To discussing the effect of lunar topography, a topography shadow model and a method calculated the incidence angle on an inclined surface by an equivalent point is introduced.

15.4.1 Topographic Shadow Model

Solar illumination on the Moon is related to the solar altitude and the lunar topography. The solar altitude is required to determine the solar illumination condition at a lunar site. It could be expressed as a function of solar incidence angle, as follows:

$$\delta = \frac{\pi}{2} - i \qquad (15.19)$$

where, δ is the solar altitude; i is the solar incidence angle on a flat lunar surface.

The solar illumination condition of a flat lunar surface could be estimated from the equations above. A flat lunar surface is illuminated when the solar altitude is non-zero and otherwise shadowed when the solar altitude is zero.

Once the solar altitude has been obtained, the solar illumination conditions of lunar surface could be estimated from lunar topography. In the extreme case with a zero solar altitude, the high elevations of lunar surface in solar incidence direction would shadow the lunar sites with low elevation behind them (Fig. 15.15).

As shown in Fig. 15.15, the relationship between the radii of site A and B could be written as:

$$\cos\theta = \frac{R_A}{R_B} \qquad (15.20)$$

where, R_A and R_B are the lunar radii at site A and B, respectively; θ is the angular distance between two lunar sites in solar incidence direction.

Archinal et al. (2007) have developed a unified lunar control network (ULCN2005) and lunar topographic model based on Clementine images and the previous ULCN network, which had been derived from Earth-based, Apollo photographs, Mariner 10, and Galileo images of the Moon. Their model infers that the maximum difference of radius is about 17.53 kilometers, with the minimum lunar radius of 1724.64 kilometers at the lowest lunar site (48.90°N, 3.83°E) and 1747.49 kilometers for the maximum lunar radius at the highest lunar site (13.91°N, 97.68°E) relative to geometric center of the Moon. So, the maximum θ is about 8°, when R_1 and R_2 are 1724.64 and 1747.49 kilometers respectively. The distance on the lunar surface corresponding to 8° of the θ is about 240 kilometers, as estimated by the relationship in Fig. 15.15. Thus the maximum distance between two lunar sites is about 240 kilometers in order that one site may shadow the other.

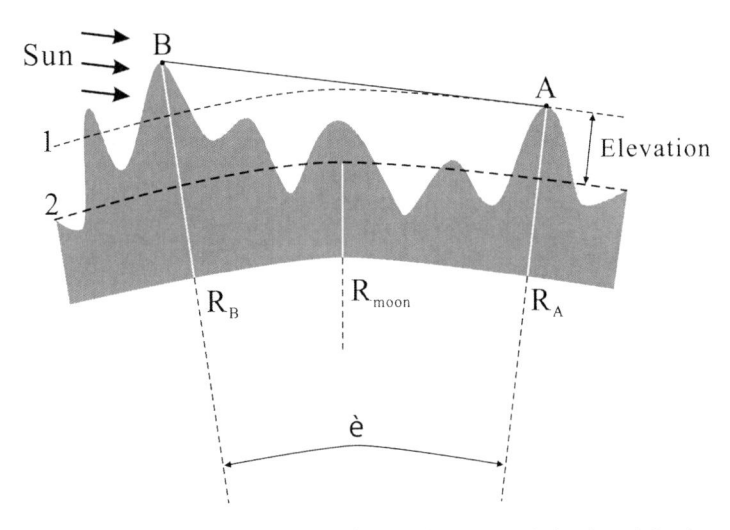

Fig. 15.15 Arc 1 refers to a part of great circle passed site A and B, the sun light transfers from site B to site A; arc 2 is a part of great circle with a mean radius, line BA is a tangent to arc 1 at site A; θ refers to the angular distance between the lunar sites A and B.

In practical application, it would be unnecessary to calculate all the topographical influence within 240 kilometers radius. Statistical analysis of Clementine topography data shows that the maximum difference of elevation within 240 kilometers radius is about 13323.6 m, with the low elevation of -4613.4m at 4.625°S, 93.625°E and the high elevation of 4807.0 m at 6.375°S, 97.875°E. Considering the angular diameter of the Sun, such a difference of elevations inplies that about 6° of θ is enough for estimating lunar topographical shadow. Inside the area with θ < 6°, the sites might be shadowed by the front-sites in solar irradiance direction, if elevations of the front-sites are high enough. Corresponding to 6° of θ, the distance on lunar surface is about 180 kilometers. Therefore, it is enough to take the elevations within 180 kilometers in solar irradiance direction into account when estimating the lunar topographical shadow.

In the common case with a non-zero solar altitude and 0.27° of the apparent radius of the Sun, the critical elevation at the site with an angular distance of θ away could be expressed as follows:

$$H = (R_{moon} + H_A)\frac{\sin(180.27^{\circ} - i)}{\sin(i - \theta - 0.27^{\circ})} - R_{moon} \tag{15.21}$$

where, H_A is the elevation at lunar site A; H is the critical elevation at the site B which the angular distance is θ, i is the solar incidence angle which could be worked out from Eqs. (15.6-15.8), and θ could be determined by the selenographic longitude and latitude of lunar sites A and B, as follows:

$$\cos\theta = \pm\sin\varphi_A \sin\varphi_B + \cos\varphi_A \cos\varphi_B \cos(\lambda_A - \lambda_B) \tag{15.22}$$

Here, λ_A, φ_A, λ_B and φ_B are the selenographic longitude and latitude of lunar site A and B, respectively.

Equation (15.22) implies that H is a function of H_A, λ_A, φ_A, λ_B and φ_B. By comparing the elevation at site B to the critical elevation, it could be estimated whether the lunar site A is shadowed by the frontal site B or not. The site A would be shadowed when the elevation at site B is higher than H, otherwise it is illuminated.

According to the above analysis, the solar illumination condition on the lunar surface is controlled by the solar altitude and the topographical shadow. To describe the solar illumination condition mathematically, the solar altitude factor (fs) and the topographical factor (ft) are defined as follows:

$$fs(\lambda_A, \varphi_A, t) = \begin{cases} 0 \cdots\cdots\cdots\cdots \delta = 0 \\ 1 \cdots\cdots\cdots\cdots \delta > 0 \end{cases} \tag{15.23}$$

$$ft(\lambda_A, \varphi_A, H_A, \lambda_B, \varphi_B, H_B) = \begin{cases} 0 \cdots\cdots\cdots\cdots H_B \geq H \\ 1 \cdots\cdots\cdots\cdots H_B < H \end{cases} \tag{15.24}$$

where, δ and H are determined by Eqs. (15.19) and (15.21), respectively; H_B is the elevation at lunar site B. fs is a function of the time, the selenographic longitude and the selenographic latitude at the lunar site A. And ft is a function of the selenographic longitude, the selenographic latitude, and the elevation at the lunar site A and B.

Finally, the solar illumination model could be simply expressed as follows:

$$F = fs(\lambda_A, \varphi_A, t) \cdot ft(\lambda_A, \varphi_A, H_A, \lambda_B, \varphi_B, H_B) \tag{15.25}$$

The solar illumination condition could be inferred from Eq. (15.25). It is a function of the time, the selenographic longitude, the selenographic latitude, and the elevation at the lunar site A and B. The lunar surface is illuminated by the Sun if F is 1, and shadowed when F is zero.

15.4.2 LESI on Inclined Lunar Surface

To conveniently estimating the LESI on inclined lunar surface, the inclined angles in direction of longitude and latitude are use to described the inclination of lunar surface. The inclination of the surface mainly leads to the change of the solar incidence angle. For an inclined surface with an inclined angle in direction of longitude of ε and an inclined angle in direction of latitude of μ, the solar incidence angle could be determined by an equivalent point. The relationship among the inclined angles and the longitude and latitude of the inclined surface and equivalent point could be expressed as following:

$$\lambda_e = \lambda_i + \varepsilon \tag{15.26}$$

$$\varphi_e = \varphi_i + \mu \tag{15.27}$$

where, λ_e, λ_i, φ_e, and φ_i are the selenographic longitude and latitude of equivalent point and inclined lunar surface. ε is the inclined angle in direction of longitude with a positive sign for eastern inclination and negative sign for western inclination. μ is the inclined angle in direction of latitude with a positive sign for northern inclination and negative sign for southern inclination. With these longitude and latitude of equivalent point, the incidence angle on the inclined lunar surface could be worked out by Eqs. (15.6-15.8). Then, LESI could be estimated with the computed solar incidence after the estimation of topography shadow.

15.5   Solar Radiation on the Moon

The solar illumination conditions represent a key resource with respect to returning to the Moon, and are also an important parameter for site selection, when considering a landing on the lunar surface and building a lunar base. For these reasons, analysis of the lunar polar lighting conditions using Clementine image data had been mainly made by Bussey and co-authors (Bussey et al. 1999, 2001, 2003, 2004, 2005; Fristad et al. 2004). In order to definitively understand the lunar illumination environment, more data is required. Wide area imaging with coverage over an entire year is necessary to identify all regions of illumination extremes. Current data is not enough for the global analysis. Additionally lighting simulations using high resolution topography could produce quantitative illumination maps.

Zuber and Smith (1997) had proposed a theoretical model based on the geometric characters of crater. They term the horizontal elevation required for permanent shadow at the lunar South Pole in the elevation angle, the co-latitude, and the angular distance to the crater rim. To simplifying the model, we use only the selenographic coordinates, elevations, and time to discuss the LESI and illumination condition at the lunar South Pole.

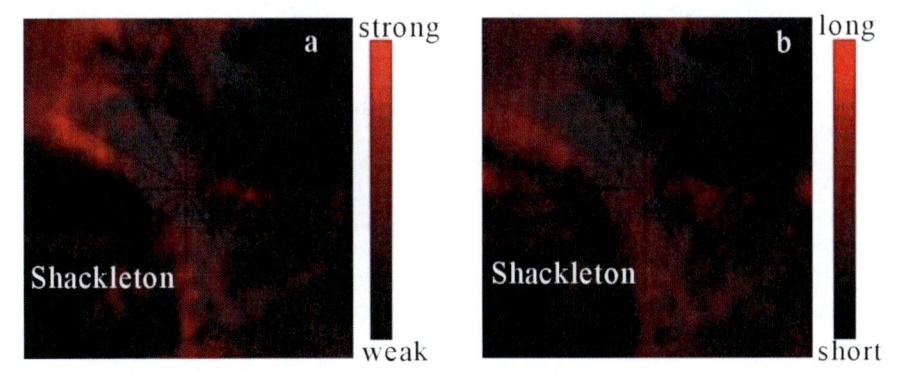

Fig. 15.16 Variation of the average of LESI (a) and illumination time (b) at the lunar South Pole region

With the Laser Altimeter data of Chang'E-1 and topography data deduced from Clementine, the solar radiation at a small region around the lunar South Pole have been simulated based on the above discussion. Figure 15.16 shows the change of average of LESI and illumination time at lunar South Pole region in 1994 (JDE from 2449353.5 to 2449718.5). It is easily identified that stronger LESI and longer illumination time appear at the Shackleton Crater rim. This region is reported to be permanent illuminated in previous studies (De Weerd et al. 1998; Kruijff 2000).

References

Archinal, B.A., Rosiek, M.R., Kirk, R.L., Hare, T.L., Redding, B.L.: Final completion of the unified lunar control network 2005 and lunar topographic model. In: Proceedings of Lunar and Planetary Science XXXVIII, League City, Texas, p. 1904 (2007)

Arnold, J.R.: Ice in the lunar polar regions. J. Geophys. Res. 84, 5659–5668 (1979)

Bretagnon, P., Francon, G.: Planetary theories in rectangular and spherical variables VSOP 87 solutions. A&A 202, 309–315 (1988)

Bussey, D.B.J., Spudis, P.D., Robinson, M.S.: Illumination conditions at the lunar south pole. Geophys. Res. Lett. 26, 1187–1190 (1999)

Bussey, D.B.J., Robinson, M.S., Edwards, K., Cook, A.C., Watters, T.: Simulation of illumination conditions at the lunar south pole. In: Proceedings of Lunar and Planetary Science XXXII, League City, Texas, p. 1907 (2001)

Bussey, D.B.J., Lucey, P.G., Steutel, D., Robinson, M.S., Spudis, P.D., Edwards, K.D.: Permanent shadow in simple craters near the lunar poles. Geophys. Res. Lett. 30, 1278 (2003)

Bussey, D.B.J., Robinson, M.S., Fristad, K., Spudis, P.D.: Permanent sunlight at the lunar north pole. In: Proceedings of Lunar and Planetary Science Conference XXXV, League City, Texas, p. 1387 (2004)

Bussey, D.B.J., Spudis, P.D.: The Lunar Polar Illumination Environment: What We Know & What We Don't. In: Proceedings of Space Resources Roundtable VI, Colorado, p. 6022 (2004)

Bussey, D.B.J., Fristad, K.E., Schenk, P.M., Robinson, M.S., Spudis, P.D.: Constant illumination at the lunar north pole. Nature 434, 842 (2005)

Butler, B.J.: The migration of volatiles on the surfaces of Mercury and the Moon. J. Geophys. Res. 102, 19,283-19,291 (1997)

Chapront-Touzé, M., Chapront, J.: The Lunar ephemeris ELP 2000. A&A 124, 50–62 (1983)

Davies, M.E., Colvin, T.R.: Lunar coordinates in the regions of the Apollo landers. J. Geophys. Res. 105, 20277–20280 (2000)

De Weerd, J.F., Kruijff, M., Ockels, W.J.: Search for Eternally Sunlit Areas at the Lunar South Pole From Recent Data: New Indications Found. In: Proceedings of 49th International Astronomical Congress, IAF-98-Q.4.07 (1998)

Duke, M.B.: The lunar environment. In: Eckart, P. (ed.) The Lunar Base Handbook, pp. 105–151. McGraw-Hill Primis Custom Publishing, New York (1999)

Feldman, W.C., Maurice, S., Binder, A.B.: Fluxes of fast and epithermal neutrons from Lunar Prospector: Evidence for water ice at the lunar poles. Science 281, 1496–1500 (1998)

Feldman, W.C., Lawrence, D.J., Elphic, R.C.: Polar hydrogen deposits on the Moon. J. Geophys. Res. 105, 4175–4195 (2000)

Feldman, W.C., Maurice, S., Lawrence, D.J.: Evidence for water ice near the lunar poles. J. Geophys. Res. 106, 23,231–23,251 (2001)

Fristad, K., Bussey, D.B.J., Robinson, M.S., Spudis, P.D.: Ideal landing sites near the lunar poles. In: Proceedings of Lunar and Planetary Science Conference XXXV, League City, Texas, p. 1582 (2004)

Fröhlich, C.: Total solar irradiance variations since 1978. Adv. Space Res. 10, 1409–1416 (2002)

Fröhlich, C., Lean, J.: The Sun's total irradiance: Cycles, trends and related climate change uncertainties since 1976. Geophys. Res. Lett. 25, 4377–4380 (1998)

Ruitao, G.: Introduction to the Earth. Normal Untversity Press Beijing, Beijing (1988) (in Chinese)

Heiken, G., Vaniman, D., French, B.M.: Lunar Sourcebook: A User's Guide to the Moon. Cambridge University Press, Cambridge (1991)

Hodges, R.R.: Ice in the lunar polar regions revisited. J. Geophys. Res. 107 (2002)

Kruijff, M., Ockels, W.J.: Lunar visibilities and lighting conditions. In: Proceedings of the Lunar Science Conference XXVI, pp. 807–808 (1995)

Kruijff, M.: Peaks of Eternal Light on the Lunar South Pole: How They Were Found and What They Look Like. In: Proceedings of Fourth International Conference on Exploration and Utilization of the Moon. ESTEC, ESA SP-462, p. 333 (2000)

Lean, J., Beer, J., Bradley, R.: Reconstruction of solar irradiance since 1610: Implications for climate change. Geophys. Res. Lett. 22, 3195–3198 (1995)

Margot, J.L., Campbell, D.B., Jurgens, R.F., Slade, M.A.: Topography of the lunar poles from radar interferometry: A survey of cold trap locations. Science 284, 1658–1660 (1999)

Maxwell, E.L.: Metstat-the Solar Radiation Model Used in the Production of the National Solar Radiation Data Base(NSRDB). Solar Energy 26, 263–279 (1998)

Meeus, J.: Astronomical algorithms. Willmann-Bell Inc., Virginia (1991)

Owczarek, S.: Vector model for calculation of solar radiation intensity and sums Incident on tilted surfaces: Indentification for the three sky condition in Warsaw. Renewable Energy 11, 77–96 (1977)

Solanki, S.K., Fligge, M.: Solar irradiance since 1874 Revisited. Geophys. Res. Lett. 3, 341–344 (1998)

Starukhina, L.: Water detection on atmosphereless celestial bodies: Alternative explanations of the observations. J. Geophys. Res. 106, 14,701–14,710 (2001)

Vasavada, A.R., Paige, D.A., Wood, S.E.: Near-surface temperatures on Mercury and the Moon and the stability of polar ice deposits. Icarus 141, 179–193 (1999)

Watson, K., Murray, B.C., Brown, H.: The behavior of volatiles on the lunar surface. J. Geophys. Res. 66, 3033–3045 (1961)

Willson, R.C.: Total solar irradiance trend during solar cycles 21 and 22. Science 26, 1963–1965 (1997)

Willson, R.C.: Secular total solar irradiance trend during solar cycles. Geophys. Res. Lett. 5, 3(1)–3(4) (2003)

Zuber, M.T., Smith, D.E.: Topography of the lunar south polar region: Implications for the size and location of permanently shade areas. Geophys. Res. Lett. 24, 2183–2186 (1997)

16 Photovoltaic Power Generation on the Moon: Problems and Prospects

T.E. Girish and S. Aranya

University College, Trivandrum, India

16.1 Introduction

Photovoltaic power is important for the current and future Lunar space missions. Alternating fortnights of bright sunshine offers a clean and unlimited energy resource on the Moon. Apollo (Bates and Fang 2001) and Lunokhod (Torchynska and Polupan 2002) missions conducted earliest solar cell experiments on the lunar surface during the 1970's. Space solar cell technology has significantly evolved during these forty years (Markvart and Castner 2003; Partain and Fraas 2010). Keeping in mind the recent renewed interest in lunar studies (Sridharan et al. 2010) the advantages and disadvantages of photovoltaic power generation on the Moon will be discussed in this chapter which is a modified version of our earlier paper (Girish and Aranya 2010).

After a brief review of the vital features of lunar physical environment we have addressed some aspects of solar cell performance expected during future lunar missions (see http://www.lpi.usra.edu). The maximum degradation in efficiency for different types of solar cells under extreme proton irradiation conditions and high lunar daytime temperature is calculated. Long term degradation in solar cell efficiency in the lunar radiation background is also estimated from our current knowledge of long term space weather variability near 1 AU and radiation resistance data of space solar cells. The prospects of PV power generation in lunar bases are discussed with some suggestions for improving the efficiency of the silicon solar cell operation on the Moon. This includes solar cell cooling, sunlight concentration and solar proton radiation shielding possibilities. We have also addressed the prospects of PV power generation during lunar night and features of solar cell operation in the lunar polar regions using very large LILT (Low Intensity Low Temperature) solar cell arrays.

16.2   Lunar Physical Environment and Its Variability

Understanding and prediction of the lunar radiation environment is essential for efficient photovoltaic power generation on the Moon. Sun is an important modulator of lunar physical environment. Hence prediction of solar activity and associated space weather changes are very much relevant for efficient PV power generation on Moon. The important characteristics and the observed solar cycle variability (Adams et al. 2007; De Angelis et al. 2007; see http://www.villaolmo.mib.infn.it/ICATPP) of the Moon's radiation environment is summarized in Table 16.1.

Table 16.1 Lunar radiation environment and its solar cycle variability

Type	relative contribution	energy	Solar cycle Change
Protons	80%	1-100 MeV	X 15
Neutrons	18%	MeV-GeV	X 3
Other particles	2%	MeV-GeV	X 3
EUV/X-rays	0.1%	KeV-MeV	X 10

Energy (E) and fluence (F) of solar proton events (SPE's) are available from direct (using satellites) and indirect (e.g. ionospheric effects) observations for the past five sunspot cycles (Dodson et al. 1975; Shea et al. 2006; SPE data 1976–2007). We could obtain a statistically significant correlation between the annual occurrence of SPE's ($E > 10$ MeV) and yearly mean sunspot number during the years 1955-2006 (Girish and Aranya 2010). This result implies that from the knowledge of the probable time of occurrence of future sunspot maxima (Kim et al. 2006) we will know periods of most frequent occurrence of energetic SPE's. The fluence of solar proton events however show irregular changes and cannot be easily predicted.

The period of occurrence of large fluence solar proton events ($E > 10$ MeV, $F > 2 \cdot 10^9$ particles/cm^2), has been inferred for the past 450 years (1561–1950 AD) using nitrate deposition in polar ice cores (McCracken et al. 2001a). Girish and Gopkumar (2010) found a relation between secular changes in occurrence of such large fluence solar proton events near 1 AU and solar magnetic flux amplification factor changes during the sunspot cycles 9–23. During the long interval from 1561 to 2010 AD the most severe solar proton event occurred during September 1859 (McCracken et al. 2001b; Shea et al. 2006) in association with the historic Carrington solar flare event with a fluence of nearly $2 \cdot 10^{10}$ particles/cm^2. This can be considered as an extreme limit of the solar proton fluence which may occur during future lunar mission periods.

The temperature of the lunar surface near equator varies from -150 °C during mid night to 100 °C during noon. Temperature generally decreases on Moon from equator to poles following a cosine law. Both these changes can be approximated by the relation (see http://www.nss.org):

$$T = 373.9\cos^{0.25}\Theta\sin^{0.167}\phi \quad [\text{K}] \qquad (16.1)$$

where Θ is solar angle above lunar horizon and ϕ is lunar latitude, both in degrees. During a lunar day of about 14.5 earth days duration near equator, about 70% of the time surface temperature exceeds 70°C. Other factors in the lunar environment which is likely to affect solar cell operation are the lunar dust and micrometeorites.

16.3 Problems of Photovoltaic Power Generation on the Moon

Photovoltaic power generation on the Moon has three important problems to overcome: (i) radiation related degradation of solar cells including sudden decrease in efficiency related to specific events such as large fluence solar proton events and long term degradation in the lunar radiation environment, (ii) higher operational temperatures during the lunar day can result in significant lowering of efficiency of the solar cells and (iii) non availability of direct sunlight for photovoltaic power generation on the Moon during the long lunar night (14.5 days near equator).

16.3.1 On the Nature of Radiation and Thermal Degradation of Solar Cells on Moon

Both energetic solar protons and high lunar temperature will affect the solar cell operation on Moon. Extensive data is available on the radiation and temperature associated performance loss of solar cells under laboratory and space born conditions (Henry et al. 1987; Sumita et al. 2004; Bates and Fang 1992; Landis et al. 2005; Fatemi et al. 2000; see http://www. pvlab.ioffe.ru; http://www. emcore.com/assets/photovoltaics). Under the extreme limits of solar proton irradiation ($E > 30$ MeV, $F > 2 \cdot 10^9$ particles/cm^2) and maximum lunar day time temperature (100 °C) we have calculated the probable reduction in efficiency and relative power output per unit area of different solar cells on the Moon. The results are shown for Si, GaAs, Group IIIi-V multijunction and CIGS solar cells in Table 16.2. Radiation hardness data of these solar cells are obtained from published studies (Torchynska and Polupan 2002; Henry et al. 1987; Sumita et al. 2004).

Crystalline Si solar cells has a reported temperature coefficient in power as 0.45 % /K, GaAs has 0.21 % /K, and multijunction Group III-V cells made of materials like GaInP has 0.18 %/K (Landis et al. 2005). Recent studies suggests for CIGS solar cells a value comparable to that of c-Si solar cells which we can adopt here (CIGS Solar Cell Module Data, see http://www.globalsolar.com; Preliminary Specification Sheet – CIGS Thin Film Solar Modules, see http://www.xsunx.com). Using the above temperature coefficient data for different solar cells we have estimated the probable decrease in efficiency due to maximum lunar day time temperature and these results are also given in Table 16.2. For these calculations a temperature change (ΔT) of 75° C on the Moon is assumed for all solar cells. BOL (Beginning of Life) efficiencies of different solar cells given in Table 16.2

Table 16.2 Solar cell degradation under extreme solar proton irradiation and high lunar temperature. Relative power output of different solar cells on Moon is also calculated.

Type of Solar cell	Degradation due to extreme solar proton irradiation (>30 MeV)	Degradation due to maximum lunar temperature	Net Degradation	BOL Efficiency	EOL Efficiency and Relative Power
Si	20-25%	34%	55%	17	7.65 (1)
Ga As	5-10%	16%	25%	25	19 (2.48)
Gr III-V multijunction	5-10%	13.5%	20%	30	24 (3.14)
CIGS	0-5%	34%	39%	17	10.4 (1.4)

are typical values selected from relevant publications (Partain and Fraas 2010; Fatemi et al. 2000; Chopra et al. 2004). The radiation and temperature associated performance loss of solar cells is considered to be addictive and the net decrease in efficiency for all solar cells on the Moon is then calculated. Using the above data EOL efficiencies of different solar cells are estimated and also given in Table 16.2.

Relative power per unit area of different solar cells (which is proportional to the EOL efficiency) under extreme limits of temperature and solar proton irradiation on the Moon is given in brackets in the last column of Table 16.2. Here we have assumed that relative power output for Si solar cells on the Moon is unity. We can find from Table 16.2 that relative power that can be derived out of Si solar cells on Moon is minimum and that from Group III-V Cells is maximum. The Ga As solar panels used in Lunokhod space missions during early seventies had an average efficiency of 11% on Moon and the power output from them was found to be two times that of Si solar cells tested under identical conditions on Moon (Torchynska and Polupan 2002). The Lunokhod results are in agreement with our results in Table 16.2.

16.3.2 Long Term Degradation of Solar Cells in the Lunar Radiation Environment

Long term solar cell degradation on the Moon is controlled by several factors such as solar proton event occurrences, galactic cosmic rays, UV/X-ray radiations from the sun, thermal recycling effects (lunar day and night temperature extremes). The contribution of SPE's is a dominant factor affecting solar cell performance on Moon. Large solar proton events like that occurred during 1972 August, 1989 March etc can cause a solar cell degradation in efficiency which is equivalent to degradation during 0.5 to 1 year exposure to lunar radiation background (Aburaya et al. 2001).

Thus the occurrence number of major SPE'S with $F > 2 \cdot 10^9$ particles/cm^2 during a space mission period becomes important to calculate the EOL (End of Life) efficiency of solar cells. The maximum frequency of occurrence of such large fluence SPE's per sunspot cycle can be up to six to eight (McCracken et al. 2001b). For a lunar mission lasting 5–6 years we can expect up to a maximum of four major SPE events similar to that occurred during the years 2000–2005 (Shea et al. 2006).

Table 16.3 Average solar cell degradation in the radiation environment per year estimated for different space missions. Type of solar cells used is also indicated

Period	Space Mission	Number of major SPE's	Average solar cell degradation per year
1971-76	Apollo 14/15 (Lunar)	1	2% (Si)
1987-97	ETS-V	1	2.5% (Si), 3% (Ga As)
1986-2000	MIR station	2	2% (AlGa As/Ga As)
2001-05	CLUSTER	3	3%

In Table 16.3 we have shown our calculations of the average solar output decrease observed per annum for four different space missions including the Apollo 14/15 solar cell experiments on Moon (Bates and Fang 2001; Bates and Fang 1992; Aburaya et al. 2001; Wolfgang 2007). The number of large fluence solar proton events ($F > 2 \cdot 10^9$ particles/cm^2) observed during these mission periods are also shown in this table. We can expect a solar cell efficiency degradation of 2–3% per year during future long term space missions to Moon.

The fluence of galactic cosmic ray (GCR) particles is at least two orders of magnitude less than that of solar energetic protons (see http://www.lip.pt/events/2006/ecrs). Hence GCR contribution to solar cell efficiency degradation on Moon is likely to be small. Laboratory studies of solar UV radiation exposures of Si solar cells suggested 1.5–2% degradation of solar cell output per year (Matcham et al. 1998). Apollo solar cell experiments on Moon did not report any significant effects of lunar dust on solar cell performance (Bates and Fang 1992, 2001).

16.4 Prospects of Photovoltaic Power Generation in the Lunar Bases

Photovoltaic systems can partially meet the power needs of a long term lunar base (Petri et al. 1990). Manufacturing solar cells on Moon, making use of lunar materials is a viable option for this purpose (see http://www.asi.org/adb). Silicon is abundant on Moon and Si solar cell technology (Ignatiev et al. 2004) is easy to implement. The low radiation resistance and relatively large temperature related performance loss of Si solar cells is a matter of concern. However, it may be possible to overcome these defects of Si solar cells and ensure efficient photovoltaic (PV) power generation on Moon by adopting one or more of the following techniques: (i) PV concentrated systems (ii) solar cell cooling and (iii) magnetic shielding of solar proton radiation.

16.4.1 Methods to Enhance the Efficiency of Silicon Solar Cells in the Lunar Environment

By adopting active or passive cooling techniques (Fortea 1981; Royne et al. 2005) the high solar cell temperature at lunar noon time (about 100°C) can be brought down to ambient conditions in earth (25°C) which can improve the efficiency of Si solar cells on Moon by 34% as seen from Table 16.2. Alternatively one can use concentrated photovoltaic systems for power generation during the lunar day. When Si solar cells is subjected to severe proton irradiation under laboratory conditions (37 MeV, 10^{10} particles/cm^2) and the cell temperature is also increased from 25°C to 80°C, the output power decreased only by 10% (Henry et al. 1987) when illuminated with concentrated light by a factor of hundred ($\times 100$). However, extra cooling arrangements are preferable for photovoltaic power generation during lunar noon time with concentrated systems.

Cooling of Si solar cells below 90 K during the lunar day using cryogenic techniques has several advantages. LILT Solar cell operation with enhanced efficiency may become feasible (Girish 2006). Since helium isotopes are abundant in the lunar regolith and nitrogen is present in lunar soil (Becker 2006) production of cryofluids such as liquid nitrogen and liquid helium may be feasible on Moon. Magnetic shielding of harmful lunar radiations becomes essential for manned lunar bases. This can protect not only life but also solar cells from severe solar proton irradiation. The magnetic shielding of protons (Cocks and Watkins 1993) within a spherical region of radius R_p on Moon is governed by the equation

$$R_p(\text{metres}) = \sqrt{\left(\frac{kM}{p}\right)} \qquad (16.2)$$

where $k=(\mu q)/(4\pi)$, q is proton charge, μ is permeability of free space, M is the moment of the electromagnet used for radiation shielding and p is the momentum of the solar proton. It is estimated that for protons of 200 MeV energy we require a magnetic moment of 10^{10} A/m^2 for radiation shielding. Superconducting magnets with T_c lying in the liquid nitrogen temperature range with critical current density of 10^6 A/cm^2 can be used for this purpose (Lee et al. 1999) Si solar cells placed in a circular tank of liquid nitrogen over the roof of the lunar base can be protected from severe solar proton radiations by employing such high T_c superconducting wire magnets. But the effects of strong magnetic fields on solar cell operation needs to be considered here (Serafettin 2002, 2008; Deng et al. 2011). Notable changes in electrical parameters of Si solar cells including performance losses are reported in these studies.

16.4.2 Photovoltaic Power Generation during the Lunar Night

Reflected or albedo radiation from Earth (earth shine) with a power of 0.1 W/m^2 can be used for photovoltaic power generation during the long lunar night (Girish 2006; Girish et al. 2007). Since lunar night time temperature falls below $-100°C$ highly efficient LILT solar cells can be used for this purpose. From a 1000 m^2 LILT solar cell array with an average efficiency of 15% near lunar equator we can expect a maximum power of only 15 W. If light concentration by hundred times ($\times 100$) is applicable for LILT conditions we can enhance this albedo photovoltaic power up to 1 kW which is sufficient to run a small lunar base near lunar mid night.

16.4.3 Solar Cell Operation on the Lunar Polar Regions

There are specific locations on the lunar polar regions which receive nearly continuous solar illumination during an year. One such region is near the rim of Shackleton crater on the Moon's south pole (88.75° S) which receives solar illumination for nearly 313 days an year (Bussey et al. 2010) The average temperature around this crater is found to be 90 K (Shackleton crater, see http://en.wikipedia.org). Assuming a cosine latitudinal variation of solar irradiance from lunar equator to pole (see http://www.lunarpedia.org) we can estimate a solar energy of 30 W/m^2 around this region. Using a 1 km^2 LILT solar cell array we can generate at least a power of 2–3 kW on the polar regions of the Moon.

16.4.4 The Need for Establishing a Solar Cell Test Facility on the Moon

Before planning a long term lunar base or establishing a very large solar cell arrays for transport of lunar photovoltaic power to earth it will be worthwhile to consider the establishment of a solar cell test facilities at preferred locations on Moon. This will have also the following merits:

- Scientific data related to long term operation of solar cells on the Moon is available only for Si and Ga As solar cells collected forty years ago. We have to study performance of new generation solar cells (Partain and Fraas 2010) on Moon made of other materials such as CIGS, Group III-V multijunction, CdTe etc. in the evolving lunar environment.
- Night time lunar environment is mainly dominated by galactic cosmic ray effects especially during sunspot minima. Studies of the performance of solar cells on the Moon during lunar night when illuminated with natural/artificial lights will be a problem of scientific interest because solar particle/radiation effects on the solar cells is nearly absent during these periods.
- Chemical composition of lunar materials favour only manufacturing of Si solar cells on Moon (Ignatiev et al. 2000). Further for the use in long term lunar bases with moderate power demands (below 10 MW) transport of earth manufactured solar cells to Moon is likely to be less expensive than lunar

manufacturing of solar cells using the existing technology (Lunar solar cell manufacturing, see http://www.cam.ub. edu).

- More detailed scientific information related to the effects of lunar dust (Taylor et al. 2005) and micrometeorites on solar cell operation (Letin et al. 2005) near lunar surface needs to be collected.

16.5 Concluding Remarks

1. Since temperature related performance loss and radiation related degradation in Gr III-V multijunction solar cells is relatively low they can be used for photovoltaic power generation in short term lunar missions. Annual degradation in efficiency of 2–3% can be expected for these solar cells while operated in the lunar environment.
2. Since Si solar cells can be easily manufactured on Moon using local resources they are preferable for power generation in long term lunar bases. To enhance the efficiency of Si solar cell performance on Moon methods such as sunlight concentration, solar cell cooling and radiation shielding from proton irradiation (employing ultra-strong magnetic fields) can be adopted.
3. To meet power requirements during lunar night and in small lunar bases in the polar regions (near Shackleton crater) very large LILT solar cell arrays can be used.
4. Performance of different types of new generation space solar cells needs to be studied in the hazardous and variable lunar physical environment by establishing solar cell test facilities on the Moon. Without this knowledge massive power generation projects on Moon such as the Luna ring (see http://www.lunarscience.arc.nasa.gov) cannot be successfully implemented.

References

Aburaya, T., Hisamatsu, T., Matsuda, S.: Analysis of 10 years flight data of cell monitor on ETS-V. Solar Energy Mat. Solar Cells 68, 15–22 (2001)

Adams, J.H., Bhattacharya, M., Lin, Z.W., Pendleton, G., Walts, J.W.: The ionizing radiation environment on the Moon. Adv. Space Res. 40, 338–341 (2007)

Bates, J.R., Fang, P.H.: Results of solar cell performance in lunar base derived from Appolo missions. Solar Energy Mat. Solar Cells 26, 79–84 (1992)

Bates, J.R., Fang, P.H.: Some astronomical effects observed by solar cells from Apollo missions on lunar surface. Sol. Energy Mat. Sol. Cells 68, 23–29 (2001)

Becker, R.H.: Nitrogen on the moon. Science 290, 1110–1111 (2006)

Bussey, D.B.J., Mc Govern, J.A., Spudis, P.D., Neish, C.O., Noda, H., Yishihora Sorensen, S.A.: Illumination conditions of the south pole of the moon derived using Kaguya topography. Icarus 208, 558–564 (2010)

Chopra, R.L., Paulson, P.D., Dutta, V.: Thin-film Solar Cells-An Overview. Prog. Photovolt. Res. Appl. 12, 69–92 (2004)

Cocks, F.H., Watkins, S.: Magnetic shielding of interplanetary spacecraft against solar flare radiation. NASA CR-195539. NASA, Washington (1993)

De Angelis, G., Badavi, F.F., Clem, J.M., Blattnig, S.R., Clowdsley, M.S., Nealy, J.E., Tripathi, R.K., Wilson, J.W.: Modelling of the lunar radiation environment. Nuclear Physics B (Proc. Suppl.) 166, 69–183 (2007)

Deng, A., Thiam, N., Thiam, A., Maiga, A.S., Sissoko, G.: Magnetic field effect on the electrical parameters of a polycrystalline silicon solar. Res. J. Appl. Sci. Engg. Techn. 3(7), 602–611 (2011)

Dodson, H.W., Simon, P., Svestka, Z.: Catalog of solar particle events 1955–1969. D. Reidel Publishing Co., Dordrecht (1975)

Fatemi, N.S., Polland, H.E., Hon, H.D., Sharps, P.R.: Solar array trades between very high efficiency multijunction and space solar cells. In: Proceedings of 28th IEEE PVSC Symposium, Alaska (2000), http://www.encore.com

Fortea, J.P.: A Study of different techniques for cooling solar cells in centralized concentrator photovoltaic power plants. PhD Thesis. University of Toulouse (1981)

Girish, T.E.: Nighttime operation of photovoltaic systems in planetary bodies. Solar Energy Mat. Solar Cells 90, 825–831 (2006)

Girish, T.E., Aranya, S.: Moon's radiation environment and expected performance solar cells in the future Lunar missions (2010), arXiv Preprint: 1012.0717

Girish, T.E., Aranya, S., Nisha, N.G.: Photovoltaic power generation using albedo and thermal radiations in the satellite orbits around planetary bodies. Sol. Energy Mat. Sol. Cells 91, 1503–1504 (2007)

Girish, T.E., Gopkumar, G.: Secular Changes in Solar Magnetic Flux Amplification Factor and Prediction of Space Weather (2010), arXiv preprint: 1011.4639

Henry, B., Crutis, Swar, C.K.: Performance of GaAs and Si concentrator cells under 37 Mev proton irradiation. NASA Technical memorandum 100144. NASA, Washington (1987)

Ignatiev, A., Freundlich, A., Duke, M., Rosenberg, S.: New architecture for Space solar power systems: Fabrication of silicon solar cells using in-situ resources (2000), http://www.niac.usra.edu

Ignatiev, A., Freundlich, A., Horton, C.: Solar cell development on the surface of moon from in-situ lunar resources. In: Proc. IEEE Aerospace Conference, vol. 1, p. 318 (2004), doi:10.1109/AERO.2004.1367615

Kim, M.Y., Wilson, J.W., Cucinotta, F.A.: A solar cycle statistical model for the projection of space radiation environment. Adv. Space Res. 37, 1741–1748 (2006)

Landis, G.A., Merrit, D., Raffale, R.P., Scheiman, P.: High Temperature Solar Cell Development, pp. 241–247. NASA CP-2005-213431, NASA, Cleveland (2005)

Lee, L.F., Paranthaman, M., Mathis, J.M., Goyal, A., Kroeger, D.M., Specht, E.D., Williams, R.K., List, F.A., Martin, P.M., Park, C., Norton, D.P., Christen, D.K.: Alternative Buffer Architectures for High Critical Current Density YBCO Superconducting Deposits on Rolling Assisted Biaxially-Textured Substrates. Jpn. J. App. Phys. 38, L178–L180 (1999)

Letin, V.A., Nadiradze, A.B., Novikov, L.S.: Analysis of solid microparticle influene on spacecraft solar arrays. In: Proc. 31st Photovoltaic Specialists Conf., IEEE/PVSC 2005-1488269, pp. 862–865 (2005), doi:10.1109/PVSC.2005.1488269

McCracken, K.G., Dreschoff, G.A.M., Zeller, E.J., Smart, D.F., Shea, M.A.: Solar cosmic ray events for the period 1561-1994: 1 Identification in polar ice 1561-1950. J. Geophys. Res. 106, 21585–21598 (2001a)

McCracken, K.G., Dreschoff, G.A.M., Zeller, E.J., Smart, D.F., Shea, M.A.: Solar cosmic ray events for the period 1561-1994: 2 The Gleissberg periodicity. J. Geophys. Res. 106, 21599–21609 (2001b)

Markvart, T., Castner, L.: Practical handbook of Photovoltaics. Fundamentals and applications, pp. 418–420. Elsevier, Amsterdam (2003)

Matcham, J., Eesbeck, M.V., Gerlach, L.: Effects of simulated solar UV radiation in the solar cells efficiency and transparent cell components. In: Proceedings of 5th European Space Power Conference, ESA SP-416, pp. 643–650 (1998)

Partain, L.D., Fraas, L.M.: Solar Cells and their Applications, 2nd edn., pp. 397–424. Wiley, Singapore (2010)

Petri, D.A., Cataldo, R.L., Bozek, J.M.: Power system requirements and definition for Lunar and Mars outposts. In: Proceedings of the 25th IECEC, American Institute of Chemical Engineers, New York, vol. 5, pp. 18–27 (1990)

Royne, A., Dey, C., Mills, D.: Cooling of photovoltaic cells under concentrated illumination; a critical review. Solar Energy Mat. Solar Cells 86, 451–483 (2005)

Serafettin, E.: The effect of electric and magnetic fields on the operation of a photovoltaic cell. Sol. Energy Mat. Sol. Cells 71, 273–280 (2002)

Serafettin, E.: Comparing the behaviours of some typical solar cells under external effects. Teknoloji 111(3), 233–237 (2008)

Shea, M.A., Smart, D.F., McCracken, K.G., Dreschoff, G.A.M., Prince, H.E.: Solar proton events for 450 years: The Carrington event in perspective. Adv. Space Res. 38, 232–328 (2006)

SPE data (1976-2007), http://www.swpc.noaa.gov/ftpdir/indices/SPE.txt

Sridharan, R., Ahmed, S.M., Das, T.P., Sreelatha, P., Pradeepkumar, P., Neha, N., Supriya, G.: Direct evidence for water (H_2O) in the sunlit lunar ambience from CHACE on MCP of Chandrayan I. Planet Space Sci. 58, 947–950 (2010)

Sumita, T., Imaizumi, M., Kawakitta, S., Matsuda, S., Kuwajima, S., Obshima, T., Kamiya, T.: Terrestrial Solar Cells in Space. In: 2004 IEEE Radiation Effects Data Workshop (IEEE Cat. No.04TH8774), Georgia, July 22-22 (2004), doi:10.1109/REDW.2004.1352887

Taylor, L.A., Schmitt, H.H., Carrier, W.D., Nakagawa, M.: The lunar dust problem: from liability to asset. In: Proc. 1st Space Exploration Conference, pp. 1–8. AIAA publication, Florida (2005)

Torchynska, T.V., Polupan, G.P.: III-V material solar cells for space application. Semicond. Phys. Quant. Electr. Optoelect. 5, 63–70 (2002)

Wolfgang, K.: Radiation effects on space craft and counter measures selected cases. In: Liensten, J. (ed.) Space Weather, pp. 231–240. Springer, Berlin (2007)

17 Fuel Cell Power System Options for Lunar Surface Exploration Applications

Simon D. Fraser

Graz University of Technology,
Austria

17.1 Introduction

The first fuel cell has been demonstrated by Sir William Grove in 1839. More than 170 years later, fuel cells are still considered an innovative and emerging technology, although they have already proven their possibilities and advantages in a considerable number of terrestrial demonstration and early commercial applications as well as a number of successful space missions flown since the 1960s.

One of the reasons for this is that fuel cells are primarily being considered as a substitute for well-established and cost-optimised technologies in commercial terrestrial mass-market applications. The comparison of advantages and possibilities of fuel cell technology on the one hand, and still present financial drawbacks as well as limitations with respect to technical maturity and applicability in commercial mass market applications on the other hand is obviously still not in favour of a wider market penetration of terrestrial fuel cell technology.

Another reason is that fuel cell technology, although inherently simple in design, nevertheless still requires extensive and profound research and development efforts in order to understand design and material challenges still faced in state-of-the-art fuel cell stack and system design.

The financial aspect of fuel cell technology is primarily relevant with terrestrial mass market applications, where the substitution of an established technology by an innovative technology can only be achieved if the advantages clearly outweigh the disadvantages as long as no regulatory body forces a change in technology.

The financial dimension in terms of Euros or Dollars per kilowatt of fuel cell power is not such a one-dimensional aspect with spaceflight applications where the overall system costs include, among others, a significant contribution of the launch and maintenance costs, for instance. Fuel cell systems have therefore already been an interesting option for spaceflight applications in the past. In fact, space industry has been one of the few branches where fuel cell systems have already been economically competitive in niche applications in the past (Fraser 2001).

Fuel cells have therefore played a key role in power system design for manned space exploration applications since the 1960s. Alkaline fuel cells have powered the Apollo spacecraft, safely taking men to the Moon and bringing them back home to the Earth again, and they still provide electrical energy to the Space Shuttle Orbiters today (Kordesch and Simader 1996; Larminie and Dicks 2000).

Recent advances in fuel cell research have provided promising new cell materials and fuel cell system engineering technologies. It will only be a question of further research and development funding as well as time until these scientific advances can be successfully transferred into durable and reliable fuel cell systems designed for even the most challenging applications in space exploration.

This chapter provides the reader with a short overview of principles, types and configurations of fuel cells, and discusses their possibilities in future lunar surface exploration applications.

17.2   An Overview of Fuel Cell Technology

Fuel cells are electrochemical conversion devices that are capable of directly and continuously converting fuels and oxidants into electrical energy. This energy conversion process does not proceed with the intermediate step into thermal and mechanical energy, as it is has to be done with heat engines, but directly and in a very efficient way. A comparison of the energy conversion process of a heat engine and a fuel cell is shown in Fig. 17.1.

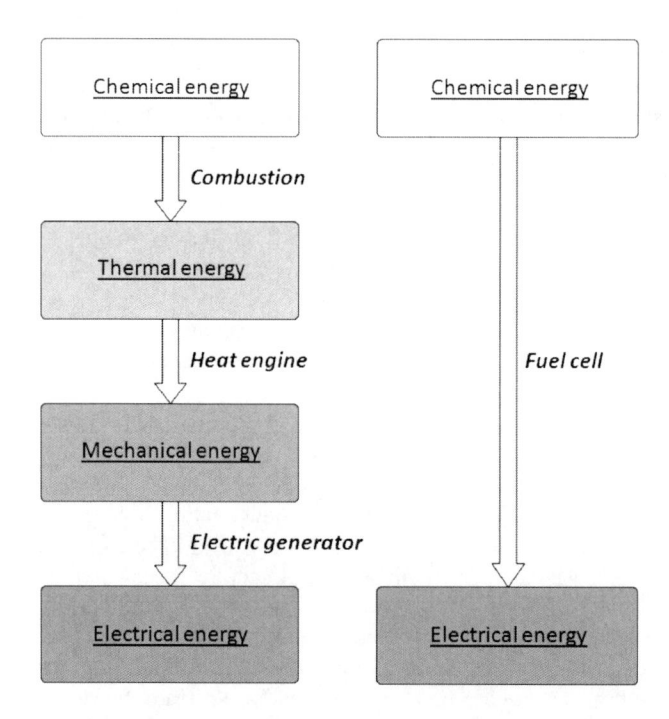

Fig. 17.1 Energy conversion process for a heat engine (left) and a fuel cell (right)

Co-generation, also referred to as combined heat and power (CHP), is the use of waste heat to further increase the efficiency of the energy conversion process. With low-temperature fuel cell systems, the waste heat will normally be available at temperature below or just above 100°C. This waste heat can be primarily used in low-temperature heating and thermally-driven cooling processes. Waste heat of intermediate and high-temperature fuel cells is available at considerably higher temperature levels of up to 1000°C (typical operating temperature ranges are given in Table 17.2).

The idea of operating a high-temperature SOFC with gas and steam turbines in a combined cycle with a fuel-to-electric conversion efficiency in excess of the fuel cell system efficiency suggests itself and has been investigated and successfully demonstrated in terrestrial applications (Palsson et al. 2000; Kuchonthara et al. 2003).

Many different technologies for fuel-to-electrical conversion have been developed and are now available with different degrees of technical maturity. This includes dynamic conversion technologies such as internal combustion engines, microturbines or Stirling engines as well as static conversion technologies such as thermoelectric and thermophotovoltaic generators or alkali metal thermal-to-electrical converters (Fraser 2001; Baker 2002).

One of the most promising options for fuel-to-electric conversion are fuel cells, as they combine the high specific energy and stored energy density of a fuel/oxidant combination with an inherently simple energy conversion technology capable of operating at very high conversion efficiency levels. Fuel cells therefore seem perfectly suited for some of the most challenging applications not only, but in particular, in future lunar surface exploration.

The returned efficiency of fuel cell system can be up to 70% without co-generation, and even higher in combined fuel cell / heat engine cycles. This level of fuel-to-electric conversion efficiency is very high compared to combustion engine technology. The roundtrip efficiency of a fuel cell system in terms of storing and releasing electric energy, however, is a limiting factor, as even an efficient energy storage and release process providing 70% efficiency in storing (i.e. conversion of electricity into chemical energy of fuel/oxidant within an electrolyser, for instance) and releasing energy (i.e. fuel-to-electric conversion in the fuel cell), for instance, results in a roundtrip efficiency of only 49%, whereas lithium-ion battery technology is capable of providing roundtrip efficiencies in excess of 90%.

In a very simplified view, fuel cells are similar to batteries where electrochemical oxidation and reduction reactions also provide a difference in electric potential and thus a stream of electrons if the electrodes are electrically connected. In comparison to batteries, however, do fuel cells not have to be replaced or recharged as the electrodes are not consumed or intentionally modified in operation, but only provide the catalytically active sites required to engage the continuous flow of reactants in the electrochemical half-cell reactions. As long as an input of fresh reactants is therefore available, fuel cells can therefore constantly produce electrical energy without being subject to intermittent charge and discharge cycles.

A fuel cell can therefore be operated more or less just like an internal combustion engine driven generator, but with a totally different and more efficient energy conversion process. As long as fresh fuel and oxygen are available, electricity can be continuously produced. If the fuel and/or oxygen tanks are depleted, they can be easily replaced, and the fuel cell system can be immediately re-started, again providing an electrical energy output as long as fresh fuel and oxygen are available.

17.2.1 Single Cell Design

A single cell essentially consists of an electrolyte sandwiched between the anode and the cathode electrode. During operation, a stream of electrons flows from the anode electrode to the cathode electrode. Liquid or gaseous fuel is supplied to the anode utilising a system of gas channels and a porous gas diffusion electrode design, whereas oxygen or ambient terrestrial air are fed to the cathode electrode utilising a similar gas distribution system.

A single cell is shown in Fig. 17.2 with the anode and cathode half-cell reactions for a fuel cell with acidic electrolyte being operated on hydrogen and oxygen.

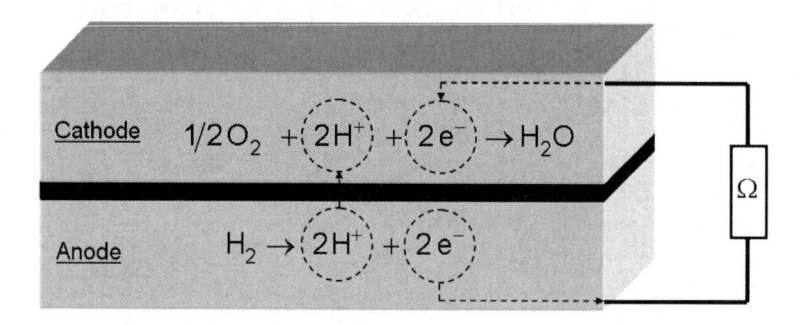

Fig. 17.2 Half-cell reactions and mobile ions of the five main types of fuel cells

Catalyst materials are applied at the interface regions between the electronically conducting electrodes and the electronically isolating electrolyte to increase the electrochemical reaction kinetics, and thus to increase the current and power output of a cell.

A single fuel cell run on hydrogen/oxygen will normally produce operational output voltages of less than one Volt. In order to provide a useful output voltage, actual fuel cell systems normally consist of a number of individual cells electrically connected in series. Such a module consisting of a number of individual cells is referred to as a fuel cell stack. When referring to a fuel cell, a fuel cell stack is normally considered and not a single cell as even very small fuel cell applications in the sub-Watt output range are normally being operated with a number of single cells, thus a stack and not a single cell.

17.2.2 Fuel Cell Types

A number of different types of fuel cells have been developed and are available in different stages of technical maturity. These different types are normally distinguished by their electrolyte and their operating temperature domain. Five main types of fuel cells are normally distinguished (Kordesch and Simader 1996; Larminie and Dicks 2000):

- Alkaline Fuel Cells (AFCs)
- Polymer Electrolyte Membrane Fuel Cells (PEMFCs)
- Phosphoric Acid Fuel Cells (PAFCs)
- Molten Carbonate Fuel Cells (MCFCs)
- Solid Oxide Fuel Cells (SOFCs)

These main types of fuel cells can be further sub-divided into the range of modern fuel cells types presented in Table 17.1. Each type of fuel cell offers specific possibilities and specific limitations, which makes the choice of the fuel cell type a strongly application-oriented task.

Table 17.1 Main fuel cell types with abbreviation and electrolyte; fuel cell types being of particular relevance with respect to future spaceflight applications are marked with white lettering on black background

Type	Abbreviation	Electrolyte
Alkaline Fuel Cell	AFC	Potassium hydroxide solution (KOH), liquid or in matrix
Alkaline Anion Exchange Membrane Fuel Cell	AAEM-AFC	Alkaline anion exchange membrane
Phosphoric Acid Fuel Cell	PAFC	Phosphoric acid (H_3PO_4) in solid matrix
(Low-Temperature) Polymer Electrolyte Membrane Fuel Cell[a,b]	(LT-)PEMFC[a,b]	Perfluorosulfonic acid membrane[c]
Direct-Methanol Fuel Cell	DMFC	Perfluorosulfonic acid membrane[c]
High-Temperature Polymer Electrolyte Membrane Fuel Cell	HT-PEMFC	HT membrane[d]
Molten Carbonate Fuel Cell	MCFC	Molten mixture of alkali metal carbonates
Low-Temperature Solid Oxide Fuel Cell	LT-SOFC	Ion-conducting ceramic material
Intermediate-Temperature Solid Oxide Fuel Cell	IT-SOFC	Ion-conducting ceramic material
(High-Temperature) Solid Oxide Fuel Cell[e]	(HT-)SOFC[e]	Ion-conducting ceramic material

[a]normally referred to as PEMFC without the prefix LT
[b]PEMFCs are also known as PEFCs and SPFCs (Solid Polymer Fuel Cells)
[c]e.g. NAFION membranes (registered trademark of Dupont)
[d]e.g. membranes based on Polybenzimidazole (PBI)
[e]normally referred to as SOFC without the prefix HT

Two of the most important fuel cell characteristics are operating temperature and possible fuels as well as poisonous species. Fuel cell types that are of particular relevance with respect to spaceflight applications are marked in Table 17.1 with white lettering on black background.

Space-qualified alkaline fuel cells (AFCs) are available and currently operated aboard the Space Shuttle Orbiter fleet. Polymer electrolyte membrane fuel cells (PEMFCs) are currently dominating the market of small and mobile terrestrial applications, and are considered as primary candidates for future applications in space missions. Direct-Methanol Fuel Cells (DMFCs), derived from direct-hydrogen PEMFC technology, will most likely only be relevant in very small applications, where small liquid fuel cartridges may be preferred over pressurized hydrogen cylinders. Solid oxide fuel cells (SOFCs) have reached a degree of maturity in stationary terrestrial application where a wider market introduction is within reach. This makes SOFCs also a relevant technology for future applications in space missions. This is particularly true with Mars surface missions, where SOFC technology can be mission-enabling, as carbon monoxide, directly produced from the carbon dioxide rich Martian atmosphere (95.32% carbon dioxide in Martian atmosphere; NASA 2010), can be used as fuel. Carbon dioxide decomposition has been successfully demonstrated with high-temperature solid oxide electrolyser (Förstner 1998) or photocatalytic carbon dioxide decomposition processes (Pipoli et al. 2000, 2002); SOFC operation with carbon monoxide has also been successfully demonstrated, and is currently being investigated with respect to SOFC performance and degradation (Inui et al. 2006; Homel et al. 2010).

The operating temperature generally has a very strong impact on system design and application, considering that start-up and thermal material compatibility will obviously rather be an issue with a fuel cell stack operating at 800°C than at room temperature.

Compatibility of fuel cells and fuels is also strongly dependent on the operating temperature regime. Low-temperature fuel cells such as PEMFCs require noble metal catalyst materials to provide sufficient reaction kinetics and thus high current and power densities given in A/cm² W/cm². Platinum group metal (PGM) catalysts such as Platinum, Palladium or Ruthenium therefore have to be used with many low-temperature fuel cell systems in order to provide sufficiently high values of current and power density. Some of these catalysts can be poisoned by carbon monoxide, a common by-product of a hydrocarbon reforming process. A careful selection of catalyst materials as a function of operating temperature, cell chemistry and fuel is therefore prudent not only with respect to the durability of the fuel cell system, but also with respect to the lifetime costs of the system.

High-temperature fuel cells such as SOFCs, on the other hand, do not require these noble metal catalysts as sufficient reaction kinetics can also be achieved with non-noble metal catalysts due to the elevated operating temperatures. Carbon monoxide is therefore not poisonous for high-temperature fuel cells. A carefully selection of fuel cell and operating temperature regime as a function of fuel and possible impurities and contaminants present within the fuel is therefore necessary in order to provide a sufficient lifetime of the system.

An overview of operational characteristics and technological status of the different types of fuel cells is provided in Table 17.2. Fuel cell types that are of particular relevance for future spaceflight applications are again marked with white lettering on black background.

Table 17.2 Overview of operating parameters and technological status; fuel cell types that are of particular relevance for future spaceflight applications are marked with white lettering on black background

Fuel Cell type	Operating temperature	Fuel options (poisonous species)	Technological status
AFC	60 - 90°C	H_2 (poison: CO_2)	available (space)
AAEM-AFC	60 - 90°C	H_2 (poison: CO_2)	current R&D
PAFC	160 - 220°C	H_2 (poison: CO/mildly)	outdated
(LT-)PEMFC	25 - 95°C	H_2 (poison: CO/strongly)	available
DMFC	25 - 95°C	CH_3OH[1]	available
HT-PEMFC	120 - 200°C	H_2 (poison: CO/mildly)	current R&D
MCFC	620 - 660°C	H_2, CO, reformate	available
LT-SOFC	< 650°C	H_2, CO, reformate	current R&D
IT-SOFC	650 - 800°C	H_2, CO, reformate	current R&D
(HT-)SOFC	800 - 1000°C	H_2, CO, reformate	available

[1]Methanol (CH_3OH), often provided in liquid state

Half-cell reactions and mobile ions of the five major types of fuel cells are presented in Fig. 17.3.

Hydrogen is the only species directly engaged in the electrochemical anode half-cell reactions of low- and medium-temperature fuel cells (AFCs, PEMFCs, and PAFCs). High-temperature fuel cells (MCFCs and SOFCs), on the other hand, can directly utilise hydrogen and carbon monoxide as fuel, and are thus particularly well-suited for operation with hydrocarbon fuel sources. Hydrocarbons can normally not be directly utilised in fuel cells, with the only presently relevant exception being methanol in DMFCs, but have to be converted into a hydrogen and carbon monoxide rich synthesis gas. This can be done with the well-known steam reforming reaction, for instance.

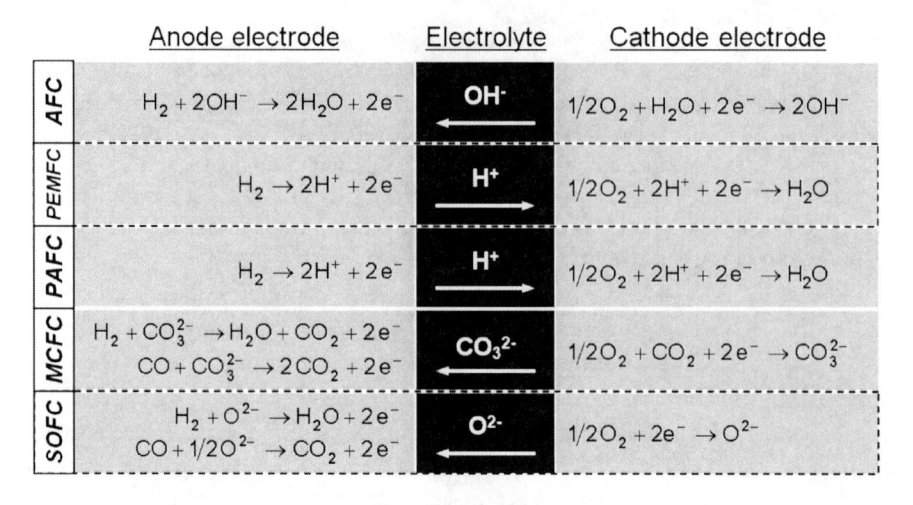

Fig. 17.3 Half-cell reactions and mobile ions of the five main types of fuel cells

17.2.3 Fuel Cell Operation

Fuel cells and electrolysers are essentially based on very similar or even the same electrochemical cell technology. The only difference between fuel cells and electrolysers is that the overall and half-cell reactions - as well as mass transport and consumption of species and charges - are reversed when switching from fuel cell to electrolyser operation and vice versa.

Two figures determine whether an electrochemical cell is operated as a fuel cell or as an electrolyser: the operational cell voltage and the reversible cell voltage.

The *operational cell voltage* is the voltage available at the electrical ports of the electrochemical cell during operation.

The *reversible cell voltage*, on the other hand, is the theoretical voltage available if the conversion process proceeded without any losses. The reversible cell voltage can be derived from the molar change in Gibbs free energy of formation.

Thus, three electrochemical cell voltage regimes can be distinguished:

$E_{cell} > E_{reversible}$:   Cell is operated as electrolyser

$E_{cell} = E_{reversible}$:   Theoretical Open Circuit Voltage (OCV); actual OCV is normally smaller than $E_{reversible}$ due to fuel crossover, internal currents and other overpotentials (Fisher 1996; Hamann and Vielstich 1998; Fraser 2004).

$E_{cell} < E_{reversible}$: Cell is operated as a fuel cell

Although an electrochemical cell can theoretically be operated bi-directionally as fuel cell and as electrolyser, most practical cells are nevertheless only operated either as fuel cell or as electrolyser, as electrode structure, materials and catalysts can thus be specifically optimised for one specific application instead of applying a "best of both worlds" solution.

A coupled electrolyser and fuel cell operation can then also be realised with two separate units, a dedicated fuel cell and a dedicated electrolyser. There are, however, very specific electrochemical cells that are designed for bi-directional operation as fuel cell and as electrolyser in one single unit. These cells are called unitized regenerative or reversible fuel cells.

17.2.4 Fuel Cell System Configurations

As discussed above, an electrochemical cell can be operated as fuel cell converting the chemical energy of a fuel into electrical energy, or as an electrolyser, converting electrical energy into chemical energy of fuel and oxygen. Three basic fuel cell system configurations can be distinguished:

- Primary fuel cell systems
- Regenerative (secondary) fuel cell systems
- Unitised regenerative or reversible (secondary) fuel cell systems

The primary fuel cell system offers a very simple design. These systems have to be refuelled from an external source, as suggested by the word *primary* borrowed from primary (i.e. non-rechargable) batteries as opposed to secondary (i.e. rechargable) systems.

Primary fuel cell systems can either be directly operated on fuels such as hydrogen, or pre-processed and pre-conditioned fuels. A typical example of the former is a direct hydrogen fuel cell system, a typical example of the latter is a fuel cell operated on reformed hydrocarbon fuel, where the stored (liquid) hydrocarbons have to be converted into a hydrogen-rich synthesis gas prior to feeding them to the fuel cell. A typical application for such a system is if a liquid hydrocarbon fuel is to be utilised in a low-temperature fuel cell. The only exception where a direct utilisation of a liquid hydrocarbon fuel in a low-temperature fuel cell is possible are direct-methanol fuel cells (DMFCs). These fuel cells have special Platinum/Ruthenium catalysts that enable a direct electrochemical utilization of (liquid) methanol without prior pre-reforming.

Secondary regenerative fuel cell systems can be recharged by operating the integrated electrolyser module, thus providing similar electrical/electrical energy storage characteristics as secondary battery systems. Two basic designs are possible: regenerative fuel cell systems where fuel cell and electrolyser are separate units, and unitised regenerative fuel cell (URFC) systems where fuel cell and electrolyser are identical and sequentially operated as fuel cell and electrolyser.

The former split-system approach offers the advantage of separately optimising fuel cell and electrolyser to their specific applications, whereas the latter offers the full functionality of fuel cell and electrolyser in one potentially compact and lightweight unit by avoiding the installation of two separate electrochemical cell modules, but normally at the cost of a reduced efficiency in fuel cell and/or in electrolyser operation.

Simplified fuel cell system layouts of the three basic fuel cell system configurations are shown in Fig. 17.4. The flow direction of the electrons is indicated as a function of fuel cell and/or electrolyser mode of operation.

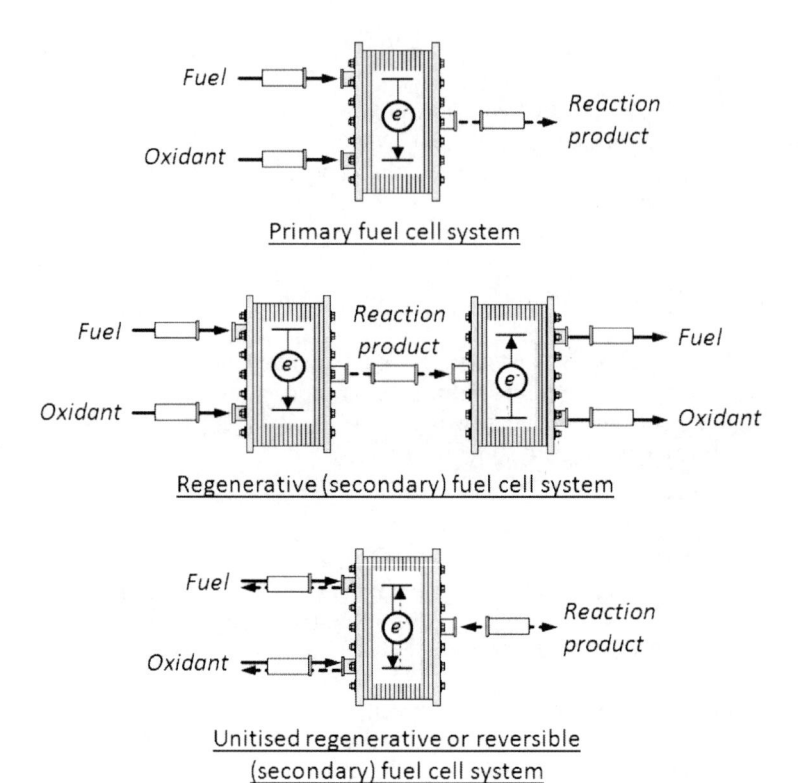

Fig. 17.4 Overview of fuel cell system configurations

17.3   Fuel Cells and Lunar In-Situ Resource Utilisation

One of the primary objectives in space mission design is to keep the overall system as well as each subsystem as light as possible in order to reduce the mission launch costs. In case of power systems for lunar and Martian surface applications, a promising approach of how to provide propellants to ascent and Earth-return spacecraft, as well as to other surface-bound mission elements, has been proposed. This so-called In-Situ Resource Utilisation (ISRU) aims at producing propellants and/or consumables directly on location, thus on the surface of the Moon or Mars, for instance, rather than launching a full supply from Earth (Zubrin and Wagner 1996; Hoffman and Kaplan 1997).

Different ISRU options have been considered in the past. Taking a feedstock of hydrogen, water or hydrocarbons from Earth and producing methane and oxygen on the planetary surface is one of the most promising options with future Mars missions. Directly splitting carbon dioxide taken from the Martin atmosphere into carbon monoxide and oxygen is another very interesting option discussed with Mars.

In case of the Moon, the primary option in ISRU technology is the extraction of oxygen from regolith. Elements known to be present in lunar regolith include, among others, oxygen (44%), silicon (22%), aluminium (9%), iron (8%), calcium (8%), magnesium (5%), manganese (5%), titanium (1%) and others (3%), as derived from elemental analysis of lunar material from Apollo and Luna landings (Tripuraneni-Kilby et al. 2006).

Four different approaches in ISRU technology are briefly discussed in the following; three of them aim at oxygen extraction from lunar regolith, the fourth approach is aiming at a direct utilisation of water resources.

17.3.1 Hydrogen Reduction of Lunar Regolith

The hydrogen reduction process essentially aims at extracting oxygen from lunar regolith by exposing the granular regolith material to a hydrogen atmosphere in a reactor chamber at elevated temperatures (Taylor and Carrier 1992).

This is done in four steps: in step one, granular regolith material is fed into the reduction reactor. In step two, hydrogen is pumped into the reactor and the reduction reaction produces a mixture of hydrogen and steam. This steam can then be condensated in step three, separating the gaseous hydrogen and the liquid water. The liquid water is finally pumped into an electrolyser unit in step four, splitting the water into hydrogen and oxygen. The hydrogen is recycled and again used for a new reduction process, whereas the oxygen can be stored for later usage.

A simplified schematic of the hydrogen reduction ISRU system is shown in Fig. 17.5. The heat exchanger used for cooling the reduction reactor output gas and condensating the water vapour is not shown for the sake of simplicity.

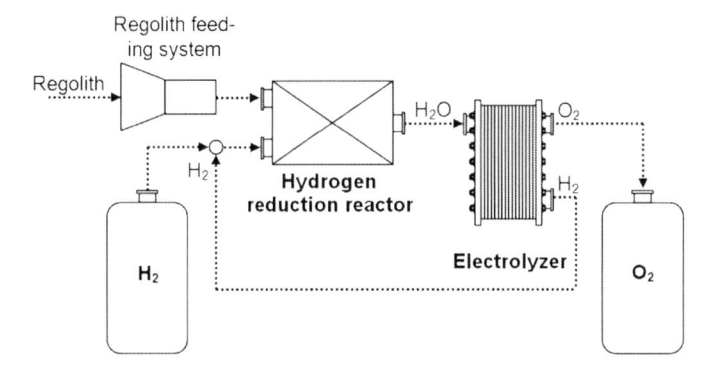

Fig. 17.5 Basic design of a hydrogen reduction ISRU system

17.3.2 Carbothermal Reduction of Lunar Regolith

A second option for lunar oxygen extraction is the carbothermal reduction process. The carbothermal reduction process extracts oxygen from lunar regolith by intensely heating a portion of regolith via laser or with highly concentrated solar light, thus melting the regolith. This molten regolith is then exposed to a stream of methane. The methane reduces silicon, iron and titanium oxides, among others, via a complex reduction reaction possibly via pyrolysis of the methane into hydrogen and solid carbon. Carbon deposition on the surface of the melt and a subsequent reduction of the metal oxides available in the molten regolith with this deposited carbon produces gaseous carbon monoxide (Balasubramaniam et al. 2009).

The reduction reactor output flux of hydrogen and carbon monoxide can then be further processed e.g. by means of the Methanation reaction (Clark et al. 2005). The Methanation reaction is shown in Eq. (17.1) and produces one mole of methane and water out of one mole of carbon monoxide and three moles of hydrogen.

$$CO + 3\,H_2 \rightarrow CH_4 + H_2O \tag{17.1}$$

Oxygen can then again be generated by condensating the steam and a subsequent electrolysis, as previously described with the hydrogen reduction process.

A simplified schematic of the carbothermal reduction ISRU system is shown in Fig. 17.6. The heat exchanger used for cooling the methanation reactor output gas and condensating the water vapour is not shown for the sake of simplicity.

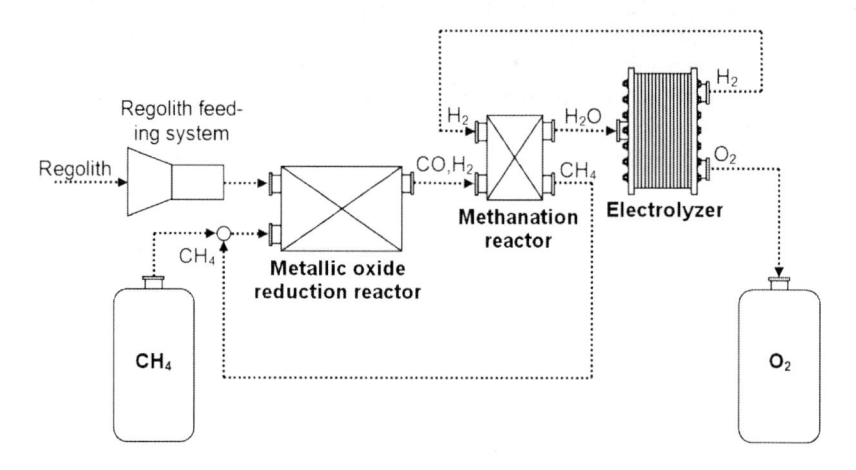

Fig. 17.6 Basic design of a carbothermal reduction ISRU system

17.3.3 Molten Salt Reduction of Lunar Regolith

This oxygen extraction approach is based on the Fray-Farthing-Chen (FFC) Cambridge process discovered in 1997 (Chen et al. 2006). The FFC Cambridge process is an electrochemical method operated at elevated temperatures in which metal oxides are cathodically reduced to the respective metals or alloys in molten salts.

The FFC Cambridge process is currently being considered as an option for future terrestrial metal production, and has also been recently considered in the context of ISRU as an option for lunar oxygen extraction (Tripuraneni-Kilby et al. 2006). LiCl-KCl, NaCl-KCl and $CaCl_2$ have been discussed as possible salts with the FFC-Cambridge process in the context of ISRU applications (Clark et al. 2005).

Bench-scale testing of the FFC-Cambridge process with JSC-1 lunar regolith stimulant material proved that the process was suitable for oxygen extraction (Tripuraneni-Kilby et al. 2006). Further experimental and theoretical analysis will show whether or not the FFC Cambridge process might be a both interesting and practical option for lunar oxygen extraction.

17.3.4 Direct Utilisation of Water Resources

The recent discovery of water in near-surface regions provides at least the theoretical possibility of water extraction. NASA's Moon Mineralogy Mapper, aboard the Indian Space Research Organization's Chandrayaan-1 spacecraft, as well as NASA's Cassini spacecraft and NASA's EPOXI spacecraft have recently confirmed the discovery of water molecules in the polar regions of the Moon (NASA 2009). Cassini passed the Moon several times to increase its cruising speed by using the Moon's gravitational force for acceleration, EPOXI passed on the way to the comet Hartley 2. The data measured during the flybys of Cassini and EPOXI confirmed the data previously generated with the Moon Mineralogy Mapper, an imaging spectrometer contributed to the Indian Chandrayaan-1 mission launched in 2008 by NASA.

A direct ISRU of water would avoid the landing of a feedstock of hydrogen or hydrocarbons and thus enable a direct ISRU production of both, fuel and oxygen.

17.3.5 The Relevance of Fuel Cells in ISRU Applications

Each of the ISRU approaches described above, as well as additional concepts and approaches considered in engineering studies and not to be described in detail within this overview, provides certain possibilities and limitations. The choice of the ISRU-related fuel/oxidant combination is generally no straight-forward decision, but rather depends on various mission parameters and has to be determined on the basis of a comprehensive evaluation of mission profile and local availability of resources under careful consideration of an overall mission risk analysis.

Assuming that an ISRU system is available for propellant production, the fuels and oxidants produced could not only be used as propellants for ascent and Earth-return vehicles, but also for power generation applications with surface-bound mission elements. Mobile pressurised rovers designed to act as mobile habitat and laboratory for scientists, for instance, require substantial quantities of energy when operated on a continuous surface exploration trip for hundreds of hours. This energy demand cannot be easily satisfied by photovoltaic arrays and/or batteries, as the external surface of a rover is normally too small to efficiently mount the large solar panels required, and mass estimations presented within this chapter will show that the estimated battery mass would also be prohibitively high. The installation of a light-weight fuel cell system would thus increase range and duration of the surface exploration trips.

A fuelled system, however, could easily provide the energy storage capacity if an efficient energy conversion technology is utilised. And fuel cells may be the technology of choice for this application.

17.4 Possibilities and Limitations of Fuel Cells Operated in Lunar Surface Applications

Any spacecraft or other electrically-powered space mission element requires a sufficient supply with electrical energy to fulfil the mission objectives, and at the same time needs to be able to provide the average and peak electrical output power required during operation.

The energy can essentially be provided by landing sufficient resources with the spacecraft, on-site production or generation utilising resources such as the atmosphere, the sun or regolith material directly available at the landing site, or by landing a separate power plant. The first option has e.g. been applied with the Apollo Lunar Module, the second option is e.g. applied with solar energy utilisation, the third option is considered with large manned lunar surface base missions, where a small nuclear reactor with some tens or hundreds of kilowatts of electric output power is considered as stationary base supply.

Fuel cells provide interesting options in each of these three scenarios. In a landing operation with sufficient energy resources, fuel cells could be the energy conversion option of choice if their conversion efficiency combined with their gravimetric and/or volumetric energy densities are better than those provided by other power system options such as batteries, for instance. Synergies in terms of a utilisation of fuel and/or oxygen resources available from the descent engine propellant system suggest themselves.

In an ISRU or solar power generation scenario, fuel cells could again be the option of choice if their gravimetric and/or volumetric energy density is better than those provided by other energy storage options such as secondary batteries, for instance. In addition, the possibility of using ISRU-derived fuels and/or oxygen provides interesting options in terms of launch mass reduction and mission integration. A high energy conversion efficiency is again mandatory in order to be able to store energy as efficiently as possible.

In a scenario where a stationary (nuclear or non-nuclear) power plant is used for power generation, fuel cells could again be the option of choice for intermediate energy storage applications if their gravimetric and/or volumetric energy density is better than the performance provided with other energy storage options. A fuel cell powered surface rover being recharged by the stationary power plant could thus provide an increased operating range and mission duration to the rover, or be designed with a reduced mass and/or a higher performance.

The key to a critical assessment of the possibilities of fuel cells in lunar surface exploration applications is therefore the specific energy and/or the energy density in comparison to other power generation and storage technologies. Considering the available net output energy and power of the system, the conversion efficiency is implicitly included in this comparison of performance data.

17.4.1 Specific Energy and Energy Density of Hydrogen/Oxygen Fuel Cells versus Secondary Batteries

The following discussion on possibilities and limitations of fuel cell operation in lunar surface exploration applications is based on the two key values of *specific energy* and *energy density*. The definition of these two values is:

- Specific energy: energy content per unit of mass in kWh/kg
- Energy density : energy content per unit of volume in kWh/liters

By comparing estimates of these two values computed for a hydrogen/oxygen reference fuel cell system and a secondary battery system, the possibilities and limitations of fuel cell technology in lunar surface exploration applications is discussed in the following. A hydrogen/oxygen fuel cell system is chosen as a reference in this evaluation; other fuel/oxygen-couples are possible, but a difference in single- or two-digit percentages of either specific energy and/or energy density does not affect the general conclusions drawn from this assessment (Fraser 2009).

Secondary batteries are the energy storage option of choice with current and near-future space missions. A comparative analysis of fuel cells and secondary batteries therefore suggests itself. The capability of modern battery technology is reflected in the goals for electric vehicle batteries defined by the U.S. Advanced Battery Consortium (USABC) very well. The USABC is an organization whose members are Chrysler Group LLC, Ford Motor Company and General Motors Company. The objective of the USABC is to jointly develop advanced battery systems for hybrid and electric vehicle applications.

An overview of the USABC goals for electric vehicle batteries is presented in Table 17.3.

Table 17.3 Goals for advanced batteries for electric vehicles defined by the U.S. Advanced Battery Consortium (USABC 2010)

Property	Minimum goals for commercialisation	Long term goal
Specific energy (C/3 discharge rate)	150 Wh/kg	200 Wh/kg
Energy density (C/3 discharge)	230 Wh/l	300 Wh/liter

The specific energy and energy density estimates of the hydrogen/oxygen fuel cell system presented in the following are derived from properties of hydrogen, oxygen and water presented below. An overview of properties utilised in deriving these gravimetric and volumetric energy storage and fuel cell system performance estimates is presented in Table 17.4.

Table 17.4 Properties of hydrogen, oxygen and water utilised with the specific energy and energy density estimates of the hydrogen/oxygen fuel cell system

Species	Property	Value
Hydrogen	Molar mass	2.016 kg/kmol
Hydrogen	Density	0.071 kg/liter
Hydrogen	Lower heating value	33.33 kWh/kg
Oxygen	Molar mass	31.999 kg/kmol
Oxygen	Density	1.130 kg/liter
Water	Molar mass	18.015 kg/kmol

The lower heating value (LHV) is defined as the amount of heat released by combusting a unit of mass of fuel, hydrogen in this case, with oxygen. LHV values assume that the water component of a combustion process is in vapour state at the end of combustion, as opposed to the higher heating value (HHV), which assumes that all of the water produced in a combustion process is in a liquid state.

Specific energy and energy density of hydrogen excluding oxygen and storage system versus the long term battery goals for the USABC electric vehicle program are shown in Fig 17.7. The figures for the secondary battery module are directly taken from the USABC electric vehicle battery development goals (0.2 kWh/kg and 0.3 kWh/liter), the specific energy of hydrogen is the LHV (33.33 kWh/kg), and the energy density is computed by multiplying the LHV given per unit of mass of hydrogen fuel with the density of cryogenic hydrogen of 0.071 kg/liter. This provides an energy density of cryogenic hydrogen in the order of 2.4 kWh/liter.

Assuming a compressed hydrogen storage vessel, the energy density would obviously be different. Cryogenic storage of reactants, however, is exclusively considered within this discussion.

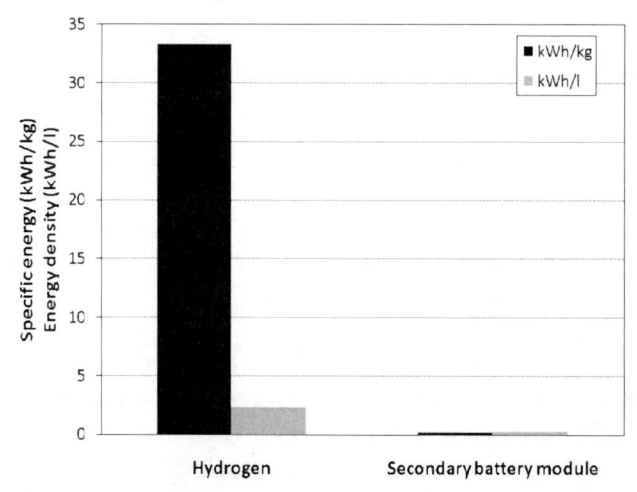

Fig. 17.7 Comparison of specific energy and energy density of liquid hydrogen (left) versus the USABC long term goal for electric vehicle batteries (right)

Hydrogen alone can obviously not be combusted without oxygen. The stoichiometric combustion reaction of hydrogen with oxygen is:

$$H_2 + \tfrac{1}{2}O_2 \rightarrow H_2O \tag{17.2}$$

Half a mole of oxygen is thus consumed in the combustion of each mole of hydrogen. The mass balance of the reaction is shown in Eq. (17.3) per unit of mass of hydrogen:

$$1 \text{ kg } (H_2) + 7.936 \text{ kg } (O_2) \rightarrow 8.936 \text{ kg } (H_2O) \tag{17.3}$$

The values are computed from the molar mass values of the three species given in Table 17.4.

In order to utilise the energy stored in one kilogram of hydrogen, a total of 8.936 kg of species have therefore got to be transported and stored. This reduces the specific energy from 33.33 kWh/kg for pure hydrogen down to 3.73 kWh/kg for stoichiometric quantities of hydrogen and oxygen. The energy density is also reduced to a value of 1.579 kWh/liter, as the volume of cryogenic oxygen also has to be considered. The density of liquid oxygen is high compared to the density of liquid hydrogen, so the reduction in energy density is not as significant as the reduction in specific energy.

Specific energy and energy density of hydrogen and oxygen excluding storage system versus the long term battery goals for the USABC electric vehicle program are shown in Fig 17.8.

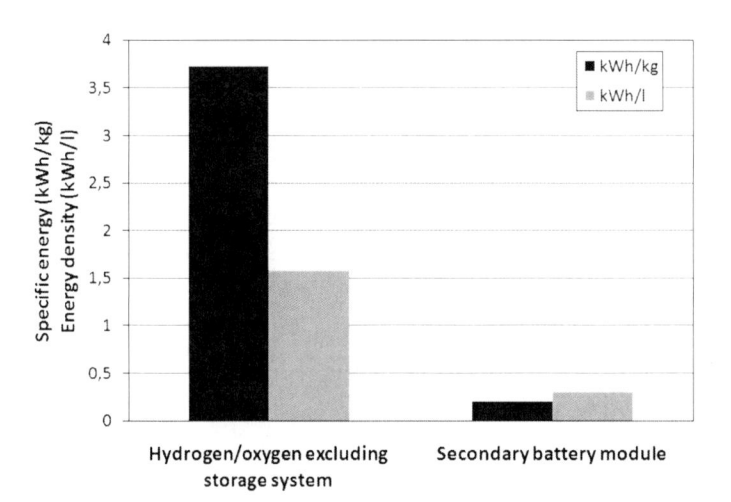

Fig. 17.8 Comparison of specific energy and energy density of liquid hydrogen including the amount of oxygen required for a stoichiometric combustion reaction and excluding storage system (left) versus the USABC long term goal for electric vehicle batteries (right)

The specific energy and energy density estimates presented above were computed for cryogenic hydrogen and oxygen excluding storage vessels. The storage vessels required for cryogenic hydrogen and oxygen storage, however, add a significant mass and volume penalty to the overall system. Mass and volume of the storage system are thus also considered in the following.

A gravimetric storage efficiency of 15 wt% (weight-percent) is assumed with the hydrogen storage system. This means that a vessel capable of storing one kilogram of hydrogen has a mass of 6.67 kilograms. Identical vessels are assumed for oxygen storage. This seems reasonable for an initial mass and volume estimation, as both vessels are considered for cryogenic storage. The gravimetric storage efficiency of the oxygen storage system is therefore 240 wt%, as the density of liquid oxygen is approximately 16 times the density of liquid hydrogen (1.13 kg/liter for liquid oxygen versus 0.071 kg/liter for liquid hydrogen).

A volumetric storage efficiency of 50% is assumed. This implies that the storage system itself (i.e. storage vessel plus isolation and tubing) has the same volume as the reactants stored inside of the vessel. A storage system capable of storing one cubic meter of liquid hydrogen would therefore add another cubic meter to the overall system volume. A total of two cubic meters would therefore be required to store one cubic meter of hydrogen. The same volumetric storage efficiency is also assumed with the (considerably smaller) oxygen storage vessel.

Specific energy and energy density estimates presented in Fig 17.9 were computed by dividing the LHV of hydrogen through the mass and volume estimates for cryogenic hydrogen and oxygen including storage system presented in Table 17.5.

Table 17.5 Mass and volume estimates for cryogenic hydrogen and oxygen including storage systems, computed per kilogram of hydrogen

Property	Mass estimate	Volume estimate
Mass of cryogenic hydrogen	1.00 kg	14.08 liters
Mass of cryogenic hydrogen storage vessel (15 wt%, 50% volumetric storage efficiency)	6.67 kg	14.08 liters
Mass of cryogenic oxygen	7.95 kg	7.02 liters
Mass of cryogenic oxygen storage vessel (240 wt%, 50% volumetric storage efficiency)	3.31 kg	7.02 liters
Total	**18.91 kg**	**42.21 liters**

Specific energy and energy density of hydrogen and oxygen including storage system versus the long term battery goals for the USABC electric vehicle program are shown in Fig 17.9.

In addition, a fuel-to-electric conversion efficiency has to be considered with the fuel cell, and discharge efficiency as well as a depth of discharge limit the actual amount of energy drawn from the secondary battery system have to be included with figures for the battery module.

The fuel-to-electric conversion efficiency of a fuel cell defines how much of the LHV of the fuel is actually available as electric energy at the electrical output ports of the fuel cell. A fuel-to-electric conversion efficiency of 67% is considered within the specific energy and energy density estimates in the following. This implies that a third of the LHV of the hydrogen fuel is lost in the fuel-to-electric conversion process. Two-thirds of the LHV of the fuel are available as electric energy; this is an ambitious figure for terrestrial fuel cell systems, but nevertheless assumed for an optimised high-performance fuel cell system designed for spaceflight applications.

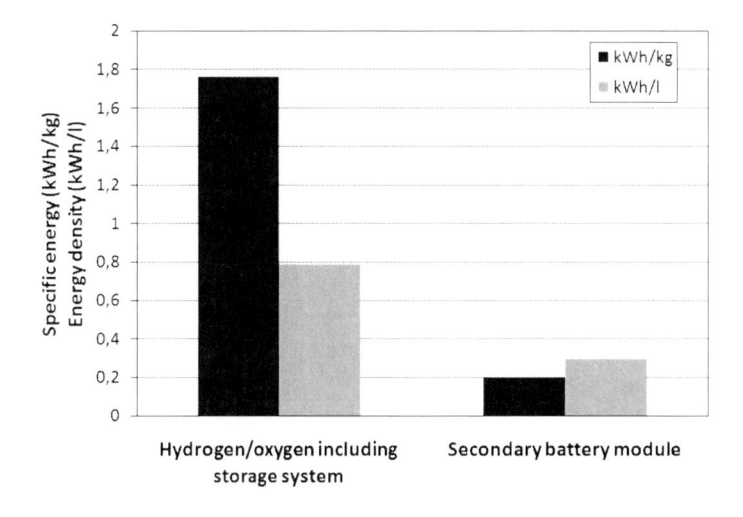

Fig. 17.9 Comparison of specific energy and energy density of liquid hydrogen including the amount of oxygen required for a stoichiometric reaction and including storage system (left) versus the USABC long term goal for electric vehicle batteries (right) (USABC 2010)

By introducing this fuel-to-electric conversion efficiency, the hydrogen and oxygen mass required to provide a certain energy output thus has to be increased correspondingly.

The values for depth-of-discharge and discharge efficiency of the secondary battery system considered within this study are presented in Table 17.6 along with the fuel cell related energy conversion parameters.

Table 17.6 Performance parameters applied with the fuel cell and secondary battery system

Property	Value
Fuel-to-electric conversion efficiency of the hydrogen/oxygen fuel cell system	67 %
Specific power of the hydrogen/oxygen fuel cell	1000 W/kg
Power density of the hydrogen/oxygen fuel cell	1000 W/liter
Depth-of-discharge limit assumed with the secondary battery system	80 %
Energetic discharge efficiency of the secondary battery system	90 %

A comparison of the specific energy and energy density of liquid hydrogen including the amount of oxygen required for a stoichiometric reaction and including storage system and conversion efficiencies are shown in Fig. 17.10 in comparison to the USABC long term goal for electric vehicle batteries.

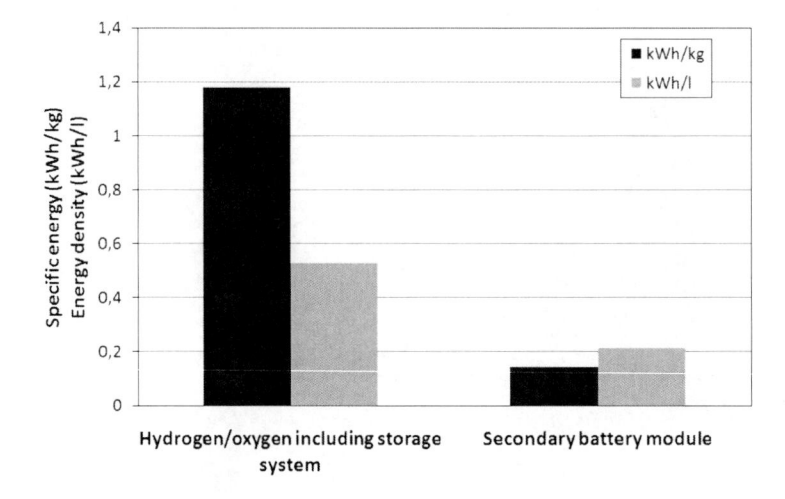

Fig. 17.10 Comparison of specific energy and energy density of liquid hydrogen including the amount of oxygen required for a stoichiometric reaction and including storage system and conversion efficiencies (left) versus the USABC long term goal for electric vehicle batteries (right)

Specific energy and energy density of the hydrogen/oxygen fuel cell system (excluding fuel cell stack) and secondary battery system are given in Table 17.7. These values are going to be used for the comparative system analysis presented in the following.

Table 17.7 Summary of the energy storage parameters computed with the fuel cell and the secondary battery system

Property	Hydrogen/oxygen including cryogenic storage system and fuel cell conversion efficiency excluding mass and volume of the fuel cell stack	Secondary battery system including depth-of-discharge limit and energetic discharge efficiency
Specific energy	1.18 kWh/kg	0.144 kWh/kg
Energy density	0.53 kWh/liter	0.216 kWh/liter

A value of 2.2 kWh of energy per kilogram of water produced was presented within a document about consumables management with the upcoming Altair lunar lander of NASA's constellation program (Polsgrove et al. 2009). This energy content can be derived if the mass of the storage system is neglected, and only the electrochemical conversion efficiency is considered, as dividing the LHV of hydrogen with the mass of water produced (33.33 kWh divided by 8.936 kg) already provides a figure of 3.73 kWh/kg. Assuming a fuel-to-electric conversion efficiency of 67%, the useable electrical energy output would be 2.46 kWh/kg. Assuming that the propellant storage system, inevitably required with the decent engine, can be used for hydrogen/oxygen storage, these figures are correct. If a dedicated surface storage system for hydrogen/oxygen has to be included, as it would be the case with a surface rover, for instance, these optimistic figures of energy content cannot be achieved.

The specific energy of the hydrogen/oxygen system is therefore estimated as being 8.2 times the specific energy of the secondary battery system. The energy density of the hydrogen/oxygen system is computed as 2.5 times the energy density of the secondary battery system.

Mass, volume and efficiency of the power management and distribution system are neglected in this technology assessment as they would be similar and thus rather independent from the specific power source and more a function of the required average and peak output power.

Mass and volume of ancillary systems such as the thermal control system, for instance, would have to be computed in detail as a function of the specific application and cell chemistry and are therefore not further considered in the following. In general, the thermal subsystem of a fuel cell will normally be significantly larger than the one of a battery power system designed for the same peak output power. Utilizing the efficiency values introduced with the mass and volume estimations above (fuel-to-electric conversion efficiency of the fuel cell system: 67%; energetic discharge efficiency of the secondary battery system: 90%), some 50% of the net electrical output power of the fuel cell system have to be dissipated as waste heat, whereas only 11% of the net output power have to be dissipated as waste heat with the battery system.

Depending on the type of fuel cell and the waste heat generation temperature, thermal subsystem weight – and most importantly radiator weight – will be the mass driver of the whole system. This is particularly true for low-temperature fuel cell systems, where the waste heat generation temperature is in the order of the lunar daytime surface temperature. An increase in fuel cell temperature therefore suggests itself in order to be able to apply a reasonably-sized radiator operating with a correspondingly larger temperature difference to ambient.

17.4.2 Fuel Cell System Utilisation in Lunar Surface Exploration Applications

The main advantage of a hydrogen/oxygen system, in addition to the strong link with ISRU technologies, is that energy can be stored more efficiently in terms of the gravimetric storage efficiency in units of energy per unit of mass, whereas the volumetric storage efficiency in units of energy per unit of volume is still better but similar in magnitude to the data estimated for the secondary battery system. The specific energy and energy density estimations computed in the previous section clearly show the possibilities of hydrogen/oxygen fuel cell technology compared to secondary battery systems.

What has not yet been considered in the analysis presented above were mass and volume of the fuel cell stack itself, as only mass and volume of the stored hydrogen and oxygen were considered. The conversion efficiency of the fuel cell system was included as it directly affects the net energy output available per unit of mass of hydrogen, but mass and volume of the stack have been neglected up to now. Only a fuel cell stack is considered with the hydrogen/oxygen fuel cell system, and not an electrolyser unit. This suggests that the fuel cell system is a primary system and has to be refuelled from an external source e.g. located at the stationary base power plant. If an additional electrolyser module was included, the gravimetric and volumetric requirements of the fuel cell system would be increased unless a unitized regenerative fuel cell system was included; in this case, the specific power and power density values utilized in the following would be too conservative, though.

Specific power and power density of a fuel cell stack are estimated at 1000 W/kg and 1000 W/liter in Table 17.6. The mass and volume penalty imposed by including the fuel cell stack in the overall specific energy and energy density estimations depends on one key parameter, being the ratio of stored energy versus output power. We will refer to this parameter as *energy-to-power ratio*.

A large energy-to-power ratio strongly favors a hydrogen/oxygen system, as the mass and volume penalty of the fuel cell stack is small compared to the significant advantages of the hydrogen/oxygen storage system. In case of a small energy-to-power ratio, however, mass and volume of the fuel cell stack will reduce specific energy and energy density of the hydrogen/oxygen system significantly, as a large and heavy stack has to be included in order to convert the comparably small quantities of fuel and oxygen into electricity.

In order for the hydrogen/oxygen fuel cell system considered within this study to be the power system element of choice in terms of the subsystem mass, Eq. (17.4) has to be true:

$$m_{H2/O2} + m_{stack} < m_{battery} \qquad (17.4)$$

where $m_{H2/O2}$ is the mass of hydrogen/oxygen including storage system e cluding the fuel cell stack (kg), m_{stack} is the mass of the fuel cell stack (kg) while $m_{battery}$ is the mass of the secondary battery system (kg). In this, the mass of hydrogen/oxygen including storage system and of the secondary battery system can be computed according to Eq. (17.5-17.7):

$$m_{H2/O2} = Q/\mu_{H2/O2} \qquad (17.5)$$

$$m_{stack} = P/\mu_{stack} \qquad (17.6)$$

$$m_{battery} = Q/\mu_{battery} \qquad (17.7)$$

where $\mu_{H2/O2}$ is the specific energy of hydrogen/oxygen including storage system and excluding the fuel cell stack (1.18 kWh/kg according to the mass estimation presented above), μ_{stack} is the specific power of the fuel cell stack (1 kW/kg),

$\mu_{battery}$ is the specific power of the battery system (0.144 kWh/kg according to the mass estimation presented above), P is the average output power (kW) while Q is the output energy (kWh). Equation (17.4) can also be written as shown in Eq. (17.8) and rearranged to Eq. (17.9).

$$Q/\mu_{H2/O2} + P/\mu_{stack} < Q/\mu_{battery} \tag{17.8}$$

$$Q \cdot (1/\mu_{H2/O2} - 1/\mu_{battery}) + P/\mu_{stack} < 0 \tag{17.9}$$

The energy-to-power ratio is derived as the ratio of the electrical energy output and average electrical output power according to Eq. (17.10):

$$\eta_{energy-to-power} = Q/P \tag{17.10}$$

where $\eta_{energy-to-power}$ is the energy-to-power ratio (dimensionless).

Due to the fact that the term $(1/\mu_{H2/O2} - 1/\mu_{battery})$ is a negative rational number considering the parameters used in the current assessment of technologies, the inequation is reversed and one gets Eq. (17.11):

$$\eta_{energy-to-power} = Q/P > (-1/\mu_{stack}) / (1/\mu_{H2/O2} - 1/\mu_{battery}) \tag{17.11}$$

Utilising the parameters computed above ($\mu_{H2/O2} = 1.18$ kWh/kg, $\mu_{stack} = 1$ kW/kg, $\mu_{battery} = 0.144$ kWh/kg), the following limiting energy-to-power ratio is derived:

$$\eta_{energy-to-power} > 0.164 \tag{17.12}$$

Equation (17.12) states that the hydrogen/oxygen fuel cell system considered within this study provides a higher specific energy than the secondary battery system if the energy output in units of energy (kWh) is at least 16.4% of the average power output in units of power (kW). If this is not the case, and a high output power and low energy output application is considered, the battery system would be lighter and thus the preferred option with respect to a specific energy optimisation.

A similar estimation can also be made with respect to the power system volume. The overall power system volume can be computed according to Eq. (17.13):

$$V_{H2/O2} + V_{stack} < V_{battery} \tag{17.13}$$

where $V_{H2/O2}$ is the volume of the hydrogen/oxygen system excluding the fuel cell stack (liters), V_{stack} is the volume of the fuel cell stack (liters), $V_{battery}$ is the volume of the secondary battery system (liters). In this, the three subsystem volumes are computed according to Eq. (17.14-17.16):

$$V_{H2/O2} = Q/\rho_{H2/O2} \tag{17.14}$$

$$V_{stack} = P/\rho_{stack} \tag{17.15}$$

$$V_{battery} = Q/\rho_{battery} \tag{17.16}$$

where $\rho_{H2/O2}$ is the energy density of hydrogen/oxygen including storage system and excluding fuel cell stack (0.53 kWh/liter according to the mass estimation presented above), ρ_{stack} is the power density of fuel cell stack (1 kW/liter) while $\rho_{battery}$ is the energy density of a battery system (0.216 kWh/liter according to the mass estimation presented above). Utilising the parameters derived above, Eq. (17.17) is derived:

$$\eta_{\text{energy-to-power}} > 0.365 \qquad (17.17)$$

Equation (17.17) states that the hydrogen/oxygen fuel cell system considered within this study provides a higher energy density than the battery system if the energy output in units of energy (kWh) is at least 36.5% of the average power output in units of power (kW). If this is not the case, and a high output power and low energy output application is considered, the battery system would be more compact and thus the preferred option with respect to a energy density optimisation.

17.4.3 Power System Evaluation

Limiting energy-to-power ratios were computed in the previous section in terms of the specific energy and the energy density of a hydrogen/oxygen fuel cell system versus a battery system. These limits are further discussed with a set of power system examples included in a power and energy map in the following.

A: Power tool (example)
Average power: 2.5 kW
Energy content: 0.21 kWh (sufficient for five minutes operation)
Energy-to-power ratio: 0.083

B: Robotic exploration rover (example)
Average power: 0.1 kW
Energy content: 500 kWh (sufficient for 5000 hours = 208 terrestrial
                          days of operation)
Energy-to-power ratio: 5000

C: Apollo Lunar Module
Average power: 1.1 kW (Polsgrove, Button, Linne 2009)
Energy content: 70 kWh (Polsgrove, Button, Linne 2009)
Energy-to-power ratio: 63.6

*D: Altair (equivalent to Lunar Module developed within the Constellation
Program of NASA)*
Average power: 4 kW (Polsgrove, Button, Linne 2009)
Energy content: 1000 kWh (Polsgrove, Button, Linne 2009)
Energy-to-power ratio: 250

E: Auxiliary power system for habitat (example)
Average power: 10 kW
Energy content: 10 kWh (sufficient for one hour of operation)
Energy-to-power ratio: 1

F: Mobile pressurised laboratory (example)
Average power: 7.5 kW
Energy content: 900 kWh (sufficient for 120 hours of operation)
Energy-to-power ratio: 120

The power system example applications presented above are plotted in a power versus energy map in Fig. 17.11. The low energy part of Fig. 17.11 is shown in Fig. 17.12. The power versus energy region where secondary batteries provide a higher specific energy (dark grey) and energy density (light grey) than the hydrogen/oxygen fuel cell system are highlighted; fuel cell and battery system specification are taken from Table 17.7.

The power versus energy plots clearly show that a fuel cell system performs better than a secondary battery system in a wide range of different applications. This is particularly true in applications with a high energy-to-power ratio. A straightforward example for this kind of application are manned surface rovers. Being operated for a period of 10+ terrestrial days at average output power levels in the order of 7.5 or even more kilowatts of electrical power, in the order of 1 MWh of energy have to be stored. Similar figures are also estimated for the next generation Lunar Module, the Altair lander. Secondary batteries would be just too heavy to be considered for this application, and that is exactly where fuel cells become an interesting option.

Only when high output powers are required over short operational periods do batteries actually perform better than the hydrogen/oxygen fuel cell systems considered in this technology assessment. This would be the case with the high

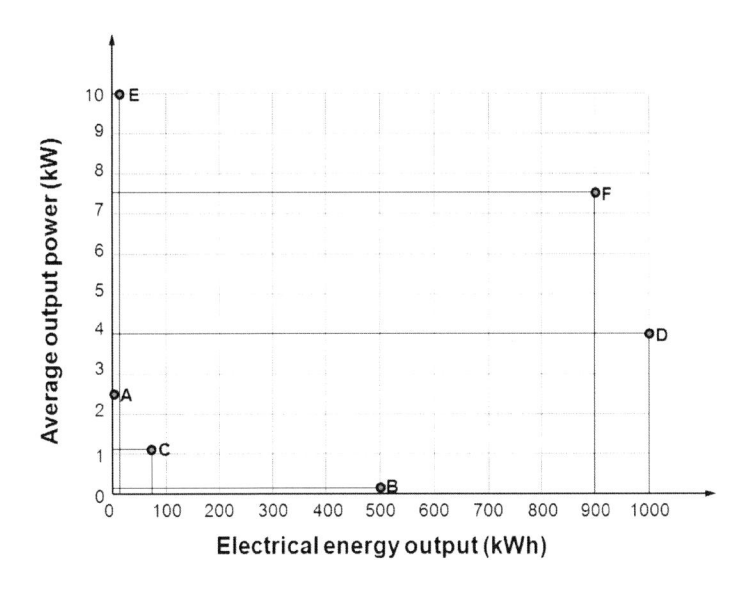

Fig. 17.11 Output power versus energy output for selected applications

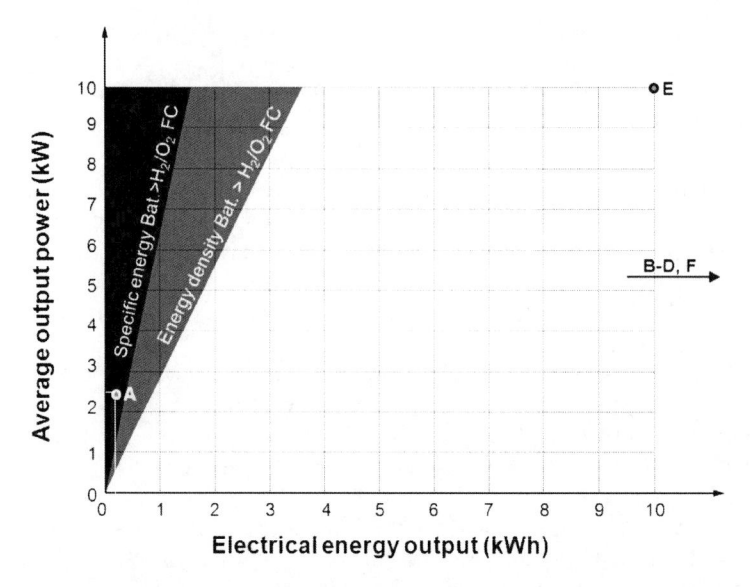

Fig. 17.12 Output power versus energy output domain where secondary batteries provide a higher specific energy (dark grey) and energy density (light grey) than the hydrogen/oxygen fuel cell system considered within this study

performance power tool indicated with "A" in Fig. 17.12, and there will be many more applications where a compact secondary battery is just the better option.

The comparison of fuel cell versus battery technology presented above is based on strong simplifications. The results derived from these power system evaluations are neither generally applicable nor do they reflect the true potential of either technology in full detail. The purpose of this discussion was primarily to present the possibilities of fuel cell. The true potential of fuel cell technology has to be carefully evaluated not only against batteries, but also against other power system technologies such as solar energy utilisation or nuclear systems.

A typical example of an application where a fuel cell performs better than a secondary battery but nevertheless is not the best power system technology is the robotic exploration rover indicated as "B" in Fig 17.11.

A photovoltaic system in combination with a small secondary battery system would most likely be the power system technology of choice for this specific application unless the operation with a full energy supply is favoured over a constant or intermittent recharging during exploration.

17.5 Summary

Fuel cells have been successfully used in space missions in the past. Being electrochemical power sources, they can directly and reasonably efficiently convert the chemical energy of a wide range of different fuels into electrical energy.

Relevant fuels and oxygen are readily available in many different mission scenarios if the descent stage is powered by a hydrogen/oxygen engine, for instance. Upon landing, significant quantities of the hydrogen and oxygen propellants will normally be remaining in the propulsion tanks. This propellant safety margin not used in the descent as well as additional residuals which cannot be used by the engine because they cannot be transported to the engine in proper conditions, for instance, could be used as fuel and oxygen source for a hydrogen/oxygen fuel cell system. The propellants would thus not be wasted, but rather used for surface power generation.

A second and very relevant aspect is the possible link to In-Situ Resource Utilisation Techniques. Being able to utilize propellants partly or fully produced from local resources in surface power generation applications provides very interesting synergies and possibilities in terms of launch mass reduction and on-site flexibility.

Fuel cell technology itself is very compatible with applications in space exploration. Fuel cells are made from simple elements with few moving parts. They can provide a very high degree of reliability and if operated on high-grade fuel show little degradation and almost no mechanical wear in the major subsystem components. Specific energy and energy density of fuel cell systems can be significantly higher than the design targets defined for future battery systems. A simplified power system technology evaluation presented within this chapter has clearly shown the advantages of fuel cell technology, particularly if a high energy-to-power ratio is required.

Although fuel cell technology has – as of yet – not managed to achieve a significant market penetration in terrestrial consumer applications, it has already become a very important asset in spacecraft power system engineering, and will become even more important in future planetary and lunar exploration scenarios.

Acknowledgement. I would like to thank Mr. Martin Lang and Dr. Max Schautz (ESA/ESTEC) for previous collaborations.

References

Balasubramaniam, R., Hedge, U., Gokoglu, S.: Carbothermal processing of lunar Regolith using methane. NASA/TM-2009-215622, Space Technology and Applications International Forum (STAIF–2008). Albuquerque, New Mexico, February 10–14 (2009)

Baker, A.M.: Future power systems for space exploration: executive summary. ESA Contract #14565/00/NL/WL (2002)

Chalk, S.G., Miller, J.F.: Key challenges and recent progress in batteries, fuel cells, and hydrogen storage for clean energy systems. J. Power Sources 159, 73–80 (2006)

Chen, G.Z., Fray, D.J., Farthing, T.W.: Direct Electrochemical Reduction of Titanium Dioxide to Titanium in Molten Calcium Chloride. Nature 407, 361–364 (2006)

Clark, L., Payne, K., Mishra, B., Gustafson, B.: Integrated ISRU for human exploration – propellant production for the Moon and beyond (2005),
http://sci2.esa.int/Conferences/ILC2005/
Presentations/ClarkL-01-PPT.pdf

Fisher, A.C.: Electrode dynamics. Oxford University Press, Oxford (1996)

Förstner, R.: Characterisation of a regenerative solid oxide fuel cell for Mars applications. Diploma thesis. Universität Stuttgart (1998)

Fraser, S.D.: Non-nuclear power system options for a mission to Mars and derived terrestrial applications. Diploma thesis. Graz University of Technology (2001)

Fraser, S.D.: Mathematical modelling of polymer electrolyte membrane fuel cells for stack design and power system analysis. Ph.D. thesis. Graz University of Technology (2004)

Fraser, S.D.: Fuel cell power system options for Mars surface mission elements. In: Badescu, V. (ed.) Mars: Prospective Energy and Material Resources, Springer, Heidelberg (2009)

Hamann, C.H., Vielstich, W.: Elektrochemie, 3rd edn. Wiley-VCH, Weinheim (1998)

Hoffman, S.J., Kaplan, D.I. (eds.): The reference mission of the NASA Mars exploration study team, 6107th edn. NASA Special Publication 6107 (1997)

Homel, M., Gür, T.M., Koh, J.H., Virkar, A.V.: Carbon monoxide-fueled solid oxide fuel cell. J. Power Sources 195, 6367–6372 (2010)

Inui, Y., Urata, A., Ito, N., Nakajima, T., Tanaka, T.: Performance simulation of planar SOFC using mixed hydrogen and carbon monoxide gases as fuel. Energy Convers Manag. 47, 1738–1747 (2006)

Kordesch, K., Simader, G.: Fuel cells and their applications. Wiley-VCH, Weinheim (1996)

Kuchonthara, P., Bhattacharya, S., Tsutsumi, A.: Energy recuperation in solid oxide fuel cell (SOFC) and gas turbine (GT) combined system. J. Power Sources 117, 7–13 (2003)

Larminie, J., Dicks, A.: Fuel Cell Systems Explained, 2nd edn. John Wiley & Sons, Chichester (2000)

NASA, Water Molecules Found on the Moon. NASA Headline News on September 24, 2009 (2009),
http://science.nasa.gov/science-news/
science-at-nasa/\2009/24sep_moonwater/

NASA, Mars Fact Sheet (2010), http://nssdc.gsfc.nasa.gov/planetary/
factsheet/marsfact.html

Palsson, J., Selimovic, A., Sjunnesson, L.: Combined solid oxide fuel cell and gas turbine-next term systems for efficient power and heat generation. J. Power Sources 86, 442–448 (2000)

Pipoli, T., Besenhard, J.O., Schautz, M.: Feasibility of a CO/O2 fuel cell to be used on Mars. In: Proc. 2nd Conf. Academic and Industrial Cooperation in Space Research, Graz, Austria, November 15-17 (2000)

Pipoli, T., Besenhard, J.O., Schautz, M.: In situ production of fuel and oxidant for a small solid oxide fuel cell on Mars. In: Proc. 6th European Space Power Conference (ESPC), Porto, Portugal, May 6-10, pp. 699–704 (2002)

Polsgrove, T., Button, R., Linne, D.: Altair Lunar Lander Consumables Management. Document ID: 20100002861, Report Number: DRC-05-01 (2009)

Taylor, L.A., Carrier, W.D.: production of oxygen on the Moon: which processes are best and why. AIAA Journal 30, 2858–2863 (1992)

Tripuraneni-Kilby, K.C., Centeno, L., Doughty, G., Mucklejohn, S., Fray, D.J.: The electrochemical production of oxygen and metal via the FFG-Cambridge process (2006),
http://www.lpi.usra.edu/meetings/roundtable2006/
pdf/tripuraneni.pdf

USABC, Goals for advanced batteries for EVs. Downloaded from the USABC energy storage systems goals (2008), http://www.uscar.org

Zubrin, R., Wagner, R.: The case for Mars: the plan to settle the red planet and why we must. Touchstone, New York (1996)

18 Theory and Applications of Cooling Systems in Lunar Surface Exploration

Simon D. Fraser

Graz University of Technology,
Austria

18.1 Introduction

The thermal energy balance of any space mission element, whether it is a spacecraft, a permanent and crewed lunar base, or a space suit, is derived as a function of heat released by personnel and/or components installed within the outer shell and heat received from the surrounding environment.

Heat can be gained internally, thus inside of the often thermally insulated exterior shell, either intentionally - by electric or radioisotope heating units, for instance – or unintentionally as waste heat by the various and often electrically-operated subsystems. A human being releases an average of 150 Watts of heat, which is one or even two orders of magnitude smaller than the electrical energy consumption of a crewed spacecraft or lunar surface base per crew member, considering that most of the electrical energy consumed will ultimately have to be removed as waste heat to prevent overheating.

In space mission elements, a wide range of different subsystems produces waste heat at different temperature levels. The electrical energy required for the operation of a computer system, for instance, has to be removed from power electronics and processors to prevent overheating of the system. The same is also true with many other subsystems and components required for spacecraft operation. It is one of the primary objectives of a thermal control system to remove excess waste heat in order to assure a safe and reliable operation of the technical and scientific spacecraft components.

The thermal energy balance with the environment is the second key factor besides internal (waste) heat gain. In terrestrial applications, the temperature of objects at near-ambient temperature is often primarily determined by convective heat transfer with the ambient air. The waste heat of a computer system is often removed by a small fan forcing ambient air over components or designated heat exchanger surfaces integrated into the computer system, for instance.

Radiative heat transfer, the transfer of heat between objects by electromagnetic radiation, also has to be considered, as the temperature of an object exposed to the sun is also determined by the thermo-optical properties of the object. Radiative heat transfer not only has to be considered with the absorption of solar radiation, but also with the emission of thermal radiation.

Conductive heat transfer is the third heat transfer mode to be considered in a thermal energy balance with the environment. Conductive heat transfer occurs through the direct physical contact between different objects, or within an object, along a temperature gradient according to Fourier's law, the law of heat conduction.

In terrestrial applications, the overall thermal energy balance of an object is thus derived as the sum of convective, radiative and conductive heat transfer between the object and the environment.

Due to the fact that the Moon only has a very thin atmosphere - the surface pressure at night is in the order of 3×10^{-15} bar and the total mass of the atmosphere is only approximately 25,000 kg (NASA Moon Fact Sheet 2010) - convection is not a relevant heat transfer mode with the environment. The thermal conduction between the space mission element and other surfaces in direct physical contact as well as the (mostly more important) radiative heat transfer therefore become the main heat transfer modes. A spacecraft operating in free space is obviously not in direct contact with any other objects or an atmosphere, respectively, and conductive heat transfer with other objects is thus not possible. In an object operating on the Moon, conductive heat transfer occurs at the contact surface between the object and the lunar surface. Heat conduction thus occurs between an astronaut standing on the lunar surface and the surface, for instance. Conductive heat transfer will, however, be very small as the astronaut's boot will be thermally insulated to reduce the conductive heat transfer as far as possible. Conductive heat transfer will also not play a major role in the thermal energy transfer between a rover and the lunar surface via the rover's tires, for instance. Crossing out conduction and neglecting conductive heat transfer, radiative heat transfer remains the only relevant heat transfer mode.

On the Moon, radiative heat transfer primarily occurs between the surface of an object and the lunar surface, the sun, the albedo of the Earth or any other object, including the IR shine of the Earth and other objects. The net sum of this radiative heat transfer determines whether heat is accumulated or rejected by the object. The waste heat released within the spacecraft and the net radiative heat flux therefore have to be carefully balanced in order to keep the spacecraft's interior and temperature-sensitive components within their safe operating temperature ranges. This can be achieved by applying a wide range of different active and passive systems. The combination of these systems is referred to as the thermal control system (TCS).

A simplified schematic of the purpose of a thermal control system is shown in Fig 18.1. A layer of thermal insulating material applied to an exterior shell separates the interior from the ambient environment. Heat is generated within the interior, shown as heat source "Q", and a radiative energy balance between the object and the environment is considered.

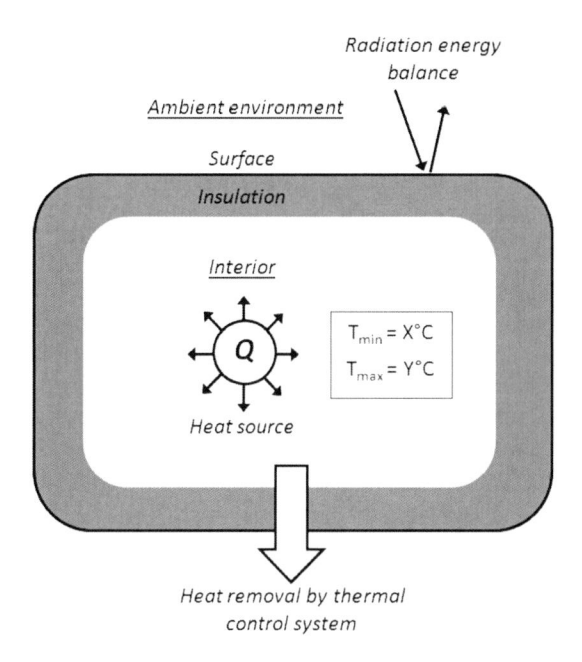

Fig. 18.1 Simplified schematic of a thermal control system

The objective of a thermal control system is therefore to keep temperature-sensitive elements of a spacecraft, lunar surface base or any other space mission element within pre-defined temperature ranges during operation and/or in inactive periods. The subject of TCS engineering, however, is not temperatures directly, but rather heat fluxes. By applying a prudent heat flux optimisation, the operating temperature ranges can be maintained and the spacecraft can thus be safely operated. The primary objective of TCS engineering is therefore to find a single, or normally a combination of a number of different, passive and active systems providing the heat rejection rate required to maintain the temperature in the specific operating range. Particularly in small and low-power applications, heating of the space mission element's interior during lunar nights is also an issue and has to be considered. Historically, this has been done by including radioisotope heating units or electrical heaters.

Heat pumps have been identified as relevant option in spacecraft thermal control systems in the past, and will most certainly be applied in future space and lunar surface mission applications. The objective of this chapter is therefore to discuss the possibilities and limitations of heat pumps in space mission thermal control system applications in general, and with respect to future lunar surface exploitation and exploitation applications in particular.

18.1.1 Goals of Spacecraft Thermal Control System Design

A good overview of the principal goals in spacecraft thermal control system design is given by the European Space Agency Thermal Control System website (ESA 2007a). According to this excellent overview, the principal goals of spacecraft thermal control system design can be defined as:

- to maintain equipment temperature in specified ranges (usually room temperature) during all mission life;

- to guarantee optimum performances when equipment is operating;

- to avoid damage when equipment is not operating;

- to keep the specified temperature stability for delicate electronics, or stable optical components;

- to minimise temperature gradients as specified between units, or along structural elements;

- to maintain boundary temperatures at interfaces between subsystems, to ease interface management;

- to guarantee the correct operation of thermal control subsystem by means of design, analysis and test;

- to determine the most influencing factors, and manage them within the satellite resources and space environment constraints;

Keeping equipment temperatures within specified temperature ranges throughout the relevant mission lifetime is a requirement that has to be evaluated for each and every component installed into the spacecraft or space mission element. Although it normally refers to providing room or near-room temperature conditions of around 293 K, a wide range of temperatures actually has to be considered, particularly if advanced and complex technologies are applied. Examples of different equipment temperature ranges are shown in Fig. 18.2.

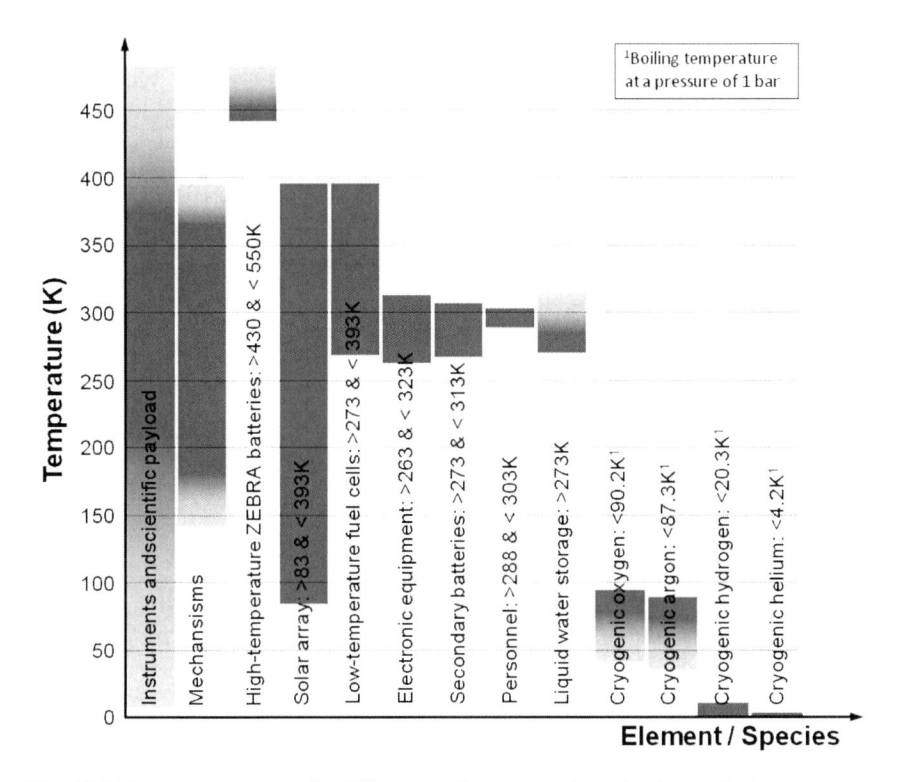

Fig. 18.2 Temperature ranges for different equipments and species (examples)

18.1.2 Ambient Conditions on the Moon

A short overview of bulk lunar parameters is presented in the following. The figures are taken from the NASA Moon Fact Sheet (NASA 2010).

Orbiting around the Earth with a semimajor axis of 0.3844×10^6 km, the Moon has a sidereal revolution period of 27.32 terrestrial days. Lunar days and nights are thus very long, which contributes to the large temperature differences faced not only between different geographical locations on the lunar surface, but also between day- and nighttime. The diurnal temperature ranges normally between a minimum surface temperature of the order of above 100 K and a maximum surface temperature of the order of up to 400 K. Recent investigations have, however, revealed that deep craters located near the polar regions can be even colder than the 100 K figure given above.

The large variations in diurnal temperatures are also enabled by the very thin lunar atmosphere. The lunar atmosphere actually has a pressure of approximately 3×10^{-15} bar at nighttime conditions and can thus hardly provide the degree of diurnal temperature leveling provided by the Earth's atmosphere (NASA 2010), as convective heat transfer is negligible.

Solar surface irradiance is equal to the Earth's at air-mass-zero conditions, being approximately 1367.6 W/m². The combination of a long revolution period and high surface irradiance poses a challenge to thermal control systems, particularly if a solar energy system is considered and diurnal variations in illumination - and thus power output - have to be compensated with intermediate energy storage. Intermediate energy storage in a day and night cycle could be considerably reduced or even eliminated if the base or surface-bound mission element was operated in a region being either constantly illuminated or at least receiving as much sun as possible.

Due to the fact that the Moon's spin axis is nearly perpendicular to the ecliptic plane, polar-adjacent regions having a low elevation, such as the floors of impact craters, may never be directly illuminated, whereas regions of high elevation may be permanently illuminated (Bussey and Spudis 2004).

Solar illumination therefore has to be investigated not only as a function of seasonal and diurnal illumination conditions, but also as a function of topographical factors (Zuber and Smith 1997; Lia et al. 2008). Lia et al. introduced a theoretical model estimating solar illumination conditions as a function of solar altitude and topographical factors. The model predicts an illumination of the lunar surface if the solar altitude is non-zero and all the elevations within 210 km in solar irradiance direction are smaller than the critical elevations; otherwise, the surface is shadowed (Lia et al. 2008).

This theoretical analysis can be complemented with experimental data to provide a comprehensive understanding of diurnal and seasonal illumination conditions. Recent data generated with the Clementine mission imaged both lunar poles during northern hemisphere summer. One of the main conclusions drawn from these investigations is that no areas near the south pole are constantly illuminated although several regions exist that receive >50% sunlight in winter. Certain regions on rims of craters and ridges were collectively illuminated for approximately 98% of the time. At the north pole, several areas on the rim of the Peary crater were constantly illuminated for an entire summer day.

This provides very interesting perspectives with respect to solar energy utilization and may thus be considered with respect to the selection of a future lunar base location. The installation of a surface base in such a region would have certainly also be considered with respect to thermal control system requirements, as heat rejection in daytime conditions is a challenge due to its elevated temperature.

18.1.3 Thermal Control System Overview

Thermal control systems can essentially be broken down into *active* and *passive* systems.

Passive systems are defined by the fact that they are not actively controllable. This means that the magnitude of the heating or cooling load cannot be actively controlled according to specific requirements or following a load profile defined by a user or an automated control system, but is automatically derived as a function of the operating conditions. The thermal energy removed by a specific radiator, for instance, is a function of the radiative energy balance of the radiator. An increase in radiator temperature and constant environmental conditions will automatically result in an increased energy dissipation rate, and a decrease in radiator temperature and still constant environmental conditions will automatically result in a decreased energy dissipation rate. A direct control of the energy dissipation rate as a function of waste heat production of a certain element is not possible, but can only be indirectly established if an increase in waste heat generation also leads to an increase in radiator temperature.

Passive thermal control systems do not require a power input and do not have any (electro)mechanical components or mechanisms. Thus, they are very reliable, normally do not require any maintenance, and impose low mass, volume and cost requirements. The operational lifetime of passive systems are often limited by degradation of the thermo-optical properties. Typical examples of passive thermal control systems are associated with the surface and/or the geometry of the spacecraft, e.g. by applying a special surface coating, thermal insulation, sun shields or radiation fins, for instance. Heat pipes can be used to achieve an efficient heat transfer between the internal waste heat source and the radiator, for instance.

Active thermal control systems, on the other hand, can be actively controlled. This means, that intentional variations in the heating or cooling load are possible. The heat released by an electric heater can, for instance, be actively controlled by applying a higher driving voltage or by changing the frequency of intermittent operating periods.

Active thermal control systems normally require a power input and often apply (electro)mechanical components or mechanisms. Reliability and maintenance issues therefore have to be considered. Mass, volume, and cost requirements often have to be evaluated more carefully than with passive systems.

Typical examples for active thermal control systems are heaters, coolers, shutters/louvers, or fluid loops. The topic of this chapter, heat pumps, in all the different variations and configurations, are also active thermal control systems.

Active thermal control systems can be further broken down into systems with and without temperature lift. A system without temperature lift can only transfer heat available at a certain source temperature along a gradient in temperature to a heat dissipating element, such as a radiator, working at a lower heat sink temperature.

This is indicated with the heat sink temperatures 1 and 2 (T_{sink_1} and T_{sink_2}) shown in Fig. 18.3. Both heat sink temperatures are lower than the heat source temperature (T_{source}). The transfer of heat from the heat source to the heat sinks 1 & 2 can be achieved without a temperature lift. By optimizing the heat transfer from source to sink, e.g. by applying an optimized heat pipe, the temperature difference between source and sink can be reduced. This is considered with the shift from sink 1 to sink 2 and labelled "heat transfer optimisation I" in Fig 18.3.

Finding a good balance between an optimized – and potentially heavier and/or larger – heat transfer system and a reduction in the heat dissipating element size is a classical engineering problem.

Systems applying a temperature lift, on the other hand, can actively increase the temperature at which the waste heat is available for dissipation or for transport to the heat dissipating element. This is indicated with the heat sink temperatures 3 and 4 (T_{sink_3} and T_{sink_4}), shown in Fig. 18.3, being higher than the heat source temperature (T_{source}). By optimizing the overall heat transfer process from source to sink, the sink temperature can again be increased from sink temperature 3 to 4; this is labelled "heat transfer optimisation II" in Fig 18.3. Alternatively, the temperature of the heat sink can also be increased by applying a larger temperature lift with the heat pump.

Systems providing a temperature lift are generally more energy-consuming and complex than systems without temperature lift. This is obvious, considering a no-power (capillary-pumped) or low-power (mechanically-pumped) fluid loop system without temperature lift in comparison to a vapour compression heat pump, requiring a compressor driven by an electric motor, for instance. The big advantage of systems with temperature lift, however, is that the waste heat is available at a higher temperature level, and that the heat dissipating element can thus be correspondingly reduced in size compared to a system without temperature-lift.

Up to date, the solution of choice in space exploration applications was to optimise heat transfer systems without temperature lift in order to reduce the temperature difference between heat source and sink and to keep the radiator as small as possible. This is certainly the preferred option if low heat sink temperatures are available.

If a spacecraft with high waste heat production and/or large surface area is to be operated over longer periods of time in solar day conditions on the Moon, however, a temperature lift system becomes a very relevant and interesting option.

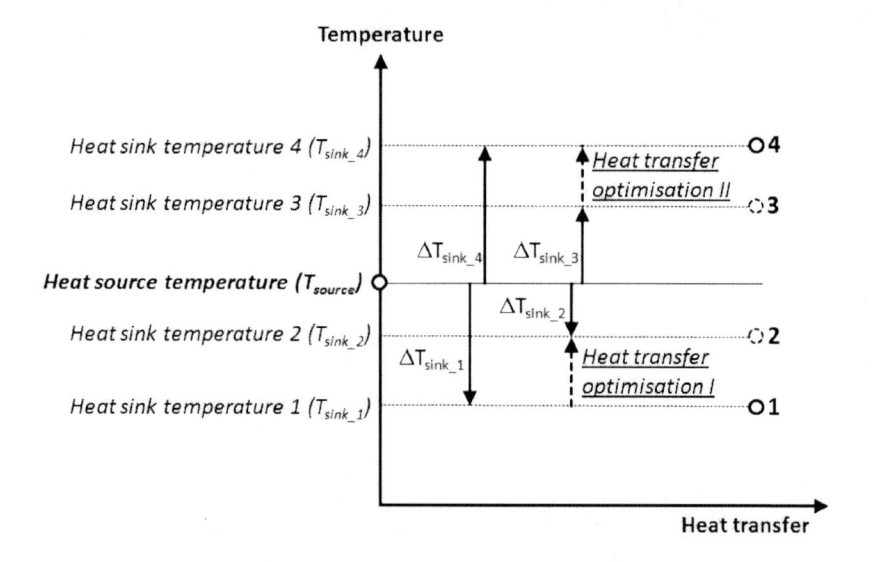

Fig. 18.3 Temperature levels for different heat sinks relative to a common heat source temperature

18.2 State of the Art in Thermal Control System Design

A number of different active and passive systems are currently being applied in spacecraft thermal control system engineering. One of the most common systems is the fluid loop system. A fluid loop system is essentially a closed loop, in which a liquid coolant is circulated. The coolant is first heated by one or more heat sources, and then pumped to the heat sink where it is cooled. Fluid loops are lighter and provide improved heat transfer characteristics than elements designed for conductive heat transfer, and are thus frequently applied in state-of-the-art thermal control system design.

Space-qualified mechanical pumps for fluid loops have been developed at ESA since the late 1970s. Mechanical pump packages suitable for integration into single-phase or two-phase cooling loops compatible with a wide range of fluids, including de-mineralised water, ammonia, and environmentally friendly fluids are now available (ESA 2007b). Fluid loop systems are therefore an established technology.

A simplified schematic of a mechanically-pumped fluid loop system is shown in Fig. 18.4.

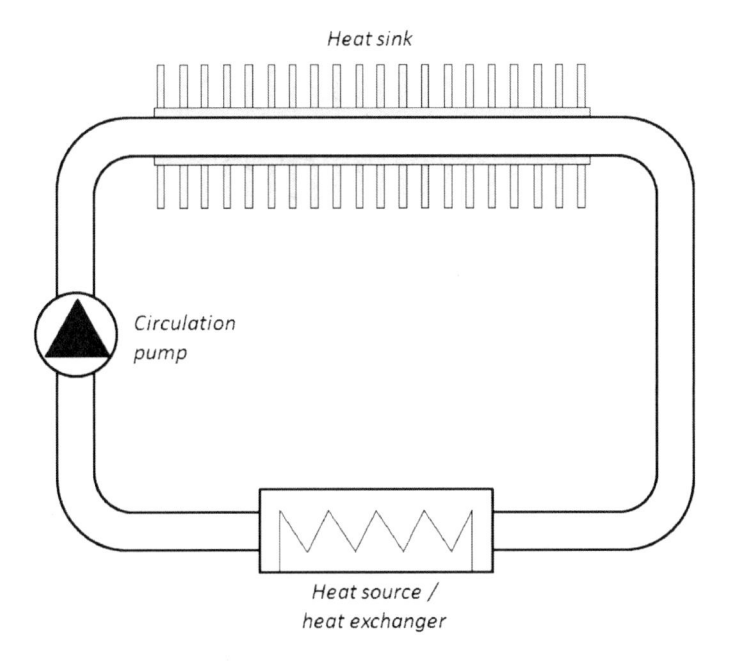

Fig. 18.4 Schematic view of a fluid loop system

A different approach providing very high heat transfer coefficients are heat pipes. A heat pipe is essentially a sealed tube or pipe that is partly filled with a liquid. The two ends of the heat pipe are connected to a heat source and a heat sink, respectively. If the temperature of the heat source end is higher than the heat sink end, liquid available in the heat source reservoir will boil and evaporate, and a constant stream of vapour will flow to the heat sink end, where the vapour condensates and releases the latent heat of condensation (Silverstein 1992).

The return flow of condensate from the heat sink to the heat source end can be achieved by gravitational force, as frequently made in terrestrial applications. A wick can be utilised to facilitate the condensate return flow also in reduced and zero-g-conditions. The intensive gravity draft heat pipe is proposed as a solar power plant on Moon and Mars (Rugescu and Rugescu 2010). A simplified schematic of a heat pipe system is shown in Fig. 18.5.

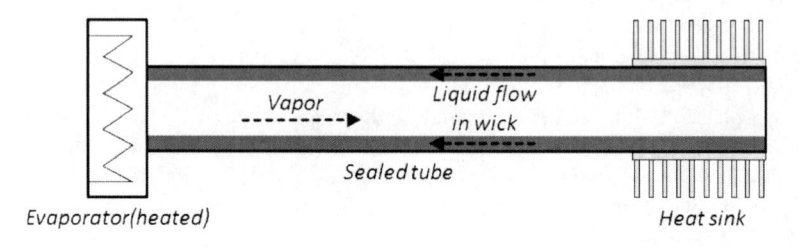

Fig. 18.5 Schematic view of a heat pipe

Heat pipes can also be designed in a closed loop configuration, where the counter-flow of vapour and liquid in classical heat pipes is avoided by transporting the condensate back to the heat source with a dedicated return flow pipe section. A wick can again be utilised to facilitate the transport of the liquid phase.

Hybrid or two-phase systems, where elements of a single-phase fluid loop and a two-phase loop heat pipe are combined, are also possible and currently considered for future applications in space missions.

A simplified schematic of a loop heat pipe system is shown in Fig. 18.6.

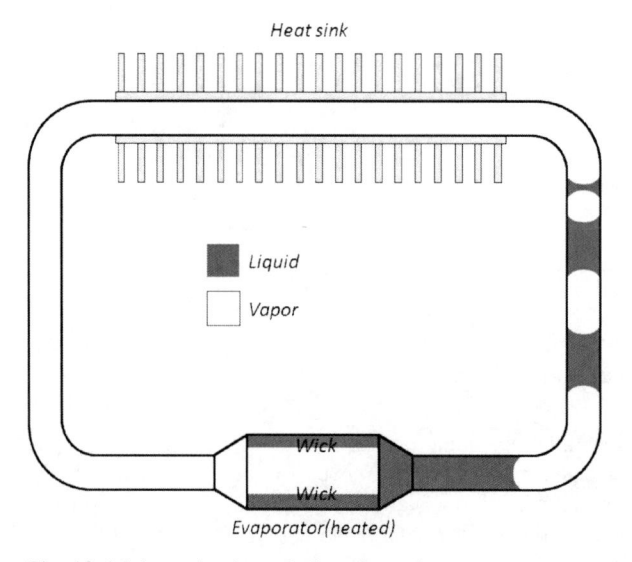

Fig. 18.6 Schematic view of a loop heat pipe

Fluid loop and heat pipe systems - in all different configurations and variations - are designed to transfer heat from a hot end, i.e. the heat source, to a cold end, i.e. the heat sink, as efficiently as possible in terms of the temperature difference between source and sink. None of these systems can be operated if the heat sink temperature exceeds the heat source temperature, as a temperature lift is not possible.

Active thermal control systems providing a temperature lift, and thus enabling heat transport also from a heat source to a higher temperature heat sink, are thermoelectric coolers, referred to as thermoelectric heat pumps in the following. Thermoelectric heat pumps use the Peltier effect to create a temperature difference when an electric voltage is applied (Rowe 1995). Although being an active heat pump, the low efficiency of the system often prevents the actual use as a heat pump. Thermoelectric heat pumps are therefore not further considered in the following.

A simplified schematic of a thermoelectric heat pump is shown in Fig. 18.7.

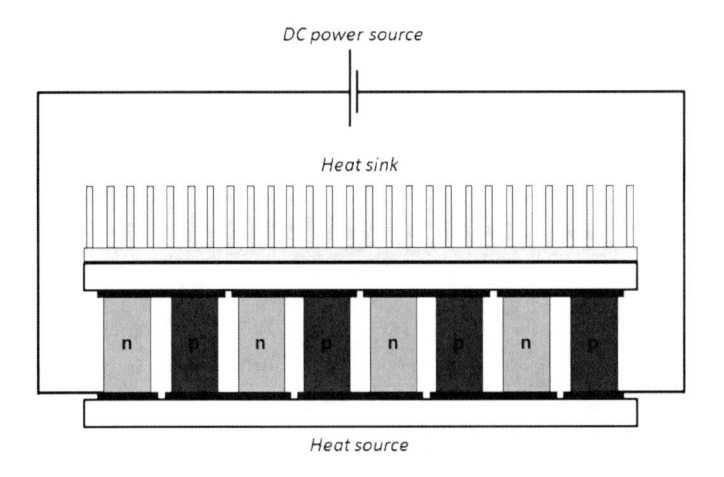

Fig. 18.7 Schematic view of a thermoelectric heat pump

18.2 Overview of Heat Pump Technology

A heat pump is a device that transfers thermal energy from a heat source at a lower temperature to a heat sink at a higher temperature using an external energy input in the form of (shaft) work or a high-temperature heat source, for instance.

Thus, a heat pump can equally be considered as a "heating device" if the objective is to utilize the high-temperature heat released at hot end or as a "cooling or refrigerating device" if the objective is to utilize the low-temperature heat removal for cooling purposes. Regardless of the actual operation as heating or cooling device, the operating principles are identical and briefly described in the following.

The basic principle of a heat pump is shown in Fig. 18.8. There are two temperature levels considered: a low-temperature heat source level (T_{LT}) and a high-temperature heat sink level (T_{HT}). In order to achieve the heat flux from the low-temperature source (Q_{LT}) to the high-temperature sink (Q_{HT}), work (W) is required to satisfy the second law of thermodynamics following the postulation of Clausius, stating that *it is impossible to construct a device that, operating in a cycle, will produce no effect other than the transfer of heat from a cooler to a hotter body* (Abbott and van Ness 1989).

The work required in heat pump operation is normally provided by a compressor driven by an electric motor or a heat engine, as widely applied in terrestrial reverse Rankine cycle heat pumps.

A schematic view of the basic principle of a heat pump, as applied with reverse Rankine cycle heat pumps, is shown in Fig 18.8.

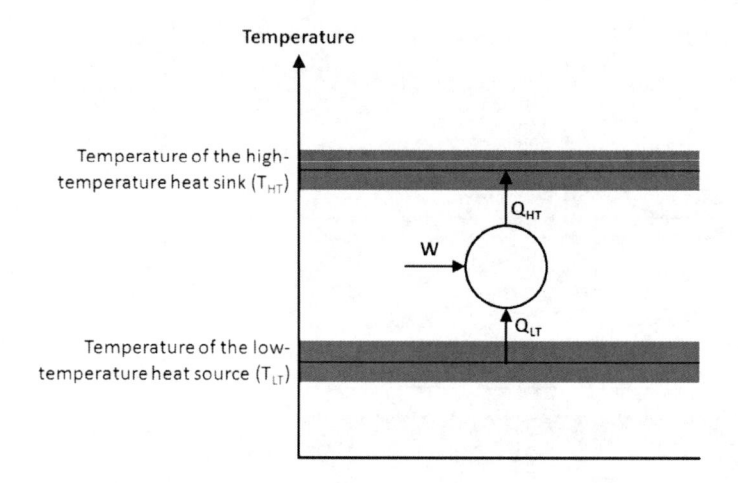

Fig. 18.8 Basic principle of a heat pump

Instead of using mechanical work for driving the heat pump process, a high-temperature heat source can also be utilised. A number of different heat-driven heat pumps have been developed and are available on different levels of technical maturity, with the absorption and adsorption heat pumps being both well-developed and relevant with respect to future applications in space exploration and exploitation missions.

In addition to the low and high temperature levels, an intermediate temperature level (T_{IT}) and (two) intermediate temperature heat fluxes (Q_{IT}) are introduced in order to substitute the mechanical work input of the mechanical-work-driven heat pump system.

Up to date, the basic principles of heat pump technology have been applied in the development of a number of different types of heat pumps. The most common and technologically most mature types of heat pumps are briefly presented and discussed in the following.

18.2.1 Reverse Rankine Cycle Heat Pumps

Reverse Rankine cycle heat pumps, also referred to as vapour-compression heat pumps, are an established technology in terrestrial mass-market applications. Residential and commercial air conditioning is mostly done with electrically-driven vapour compression systems, for instance. Reverse Rankine systems are also applied in automobile air conditioning systems, as shaft work is available as long as the car engine is running. If the car engine is shut down, however, the air conditioning system is thus not available anymore unless a large battery system is installed to operate the compressor electrically, as is normally the case with full hybrid electric or electric vehicles.

Reverse Rankine cycle heat pumps essentially consist of an evaporator, where the refrigerant is heated and finally evaporates, and a condenser, where the refrigerant vapour is condensed, respectively (Kirn and Hadenfeldt 1976; Silberstein 2002). This two-phase technology is the established approach and implemented in many terrestrial air conditioning applications, as opposed to single-phase reverse Brayton cycle heat pumps, which do not play such an important role in residential or automotive air conditioning.

A compressor is required to transfer vapour from the evaporator into the condenser, which is operated at a higher pressure. An expansion valve is therefore normally installed in the condensate backflow line due to the pressure difference and to complete the full cycle.

The evaporator has to be heated by an external heat source in order to accomplish the phase change of the coolant from liquid to vapour; the heat pump therefore removes heat from the external source and can be utilised for cooling purposes. Heat is released in the condenser due to the phase change of the coolant from vapour to liquid; heat can thus be utilised for heating purposes or has to be otherwise dissipated.

A simplified schematic of an electrically-operated reverse Rankine cycle heat pump is shown in Fig. 18.9a. Heat transfer elements receiving thermal energy from an external heat source are labelled "heated"; heat transfer elements that need to be cooled by removing heat to an external heat sink are labelled "cooled".

Based on existing two-phase heat transfer technologies, the vapour compression hybrid two-phase loop technology has been developed. This technology is essentially based on vapour compression heat pumps (thus a two-phase loop) and an additional liquid (thus single-phase) loop (Park and Sunada 2008). The liquid pump essentially circulates liquid between the evaporator and a reservoir. Within the evaporator, the liquid is drawn into the evaporator wick by capillary force and subsequently vaporized. This provides an improved high flux performance, especially during transient heat input conditions (Park and Sunada 2008).

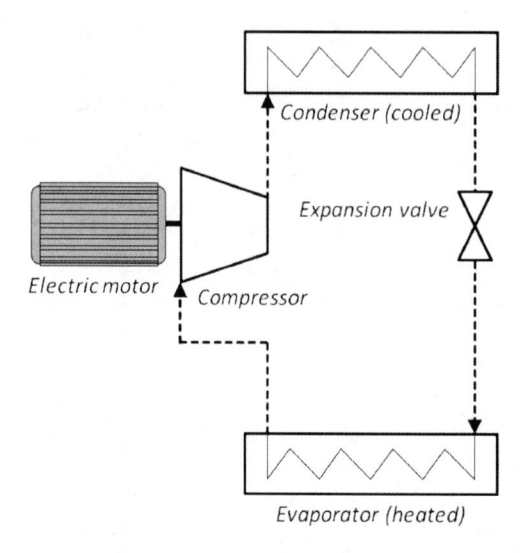

Fig. 18.9a Schematic view of an electric reverse Rankine cycle heat pump

A simplified schematic of a vapour compression hybrid two-phase loop system in basic configuration is shown in Fig. 18.9b.

The temperature lift between evaporator and condenser is a function of the compression rate, and thus the mechanical work provided by the compressor. The higher the work input, the higher the temperature lift, as suggested by the Carnot Cycle. The coefficient of performance (COP) of a mechanically-driven heat pump in cooling mode is the ratio of heat removed (i.e. thermal energy transferred into the evaporator) to input work (i.e. work of the compressor).

Reverse Rankine cycle systems are very flexible in terms of temperature lift and heating/cooling load, either by operating the system discontinuously or with a controlled-speed compressor. In addition, the mechanical work required for driving the compressor can be supplied from an electric motor as well as from any other heat engine or hydraulic and pneumatic motor, for instance. Reverse Rankine cycle heat pump systems can therefore be operated according to application or user requirements as long as a sufficient supply with electrical – respectively mechanical – energy is available.

Reverse Rankine cycle heat pumps are comparably simple in design; an evaporator and a condenser are the only heat transfer elements in the basic heat pump configuration. Other heat pump systems, the ab- and adsorption systems, for instance, require more individual heat transfer processes and therefore more heat transfer area and, of particular relevance with respect to space and lunar applications, an increased heat rejection area and mass. This is a very strong advantage of reverse Rankine cycle heat pumps in space and lunar surface applications, as heat rejection area is often a limiting factor. This is particularly true for lunar daytime applications, as discussed in the heat pump configuration section of this chapter.

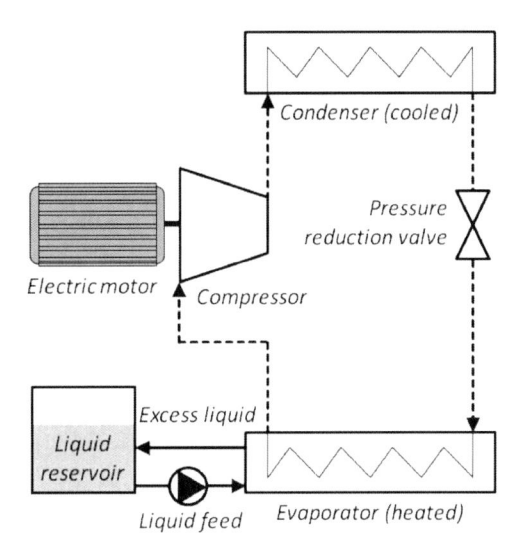

Fig. 18.9b Schematic view of an electric reverse Rankine cycle heat pump

In a study on lunar base thermal control systems using heat pumps, reverse Rankine, absorption and reverse Brayton systems were compared and evaluated (Sridhar and Gottmann 1996). The reverse Rankine system was identified as optimal solution, as the mass was estimated as being considerably smaller than with the other heat pump systems.

The relevance of reverse Rankine cycle heat pumps has, however, not only been confirmed in theoretical studies, but has recently also lead to research and development efforts aiming at the transfer of established terrestrial reverse Rankine cycle heat pump technology into space applications.

Mainstream Engineering Corporation (Mainstream 2010a), a research and development company specialized in advanced thermal control and energy conversion, demonstrated a low-lift heat pump thermal control system for satellites based on an innovative magnetic-bearing centrifugal compressor. The low-lift heat pump experimental results demonstrated that the system has lower total mass than a single-phase pumped loop at the conditions compared in spacecraft thermal control system applications (Mainstream 2010b). In 2006, Mainstream Engineering Corporation had already been awarded a NASA Johnson Space Center contract for the design, fabrication, testing, and flight readiness assessment of a gravity-insensitive heat pump. This vapor-compression cooling system was to be developed for future lunar lander and lunar outpost applications (Mainstream 2006).

In space and lunar surface exploration and exploitation applications, reverse Rankine cycle heat pump technology will certainly provide the benchmark against which other heat pump systems and technologies will have to be evaluated. This is also confirmed in the evaluation of Izutani et al.(1992), which have considered a number of different heat removal technologies and conclude that a fluorocarbon refrigerant heat pump system operated with a temperature lift of 50 K is the optimum choice as it offers the best overall performance with respect to energy consumption, size and weight, and due to the fact that the technology has already been established in terrestrial applications and its reliability is therefore high enough for applications in lunar surface applications.

18.2.2 Absorption Heat Pumps

The basic principle utilised with absorption heat pumps is again to evaporate a fluid in an evaporator, to subsequently transport the vapour to a higher pressure level, and to finally condensate the vapour at this higher pressure and thus temperature level. Other than with reverse Rankine heat pumps, where mechanical work is applied for vapour compression, high-temperature heat is utilised as driving force with absorption heat pumps (Loewer and Bosniakovic 1987; Herold et al. 1996).

This is achieved by first absorbing the refrigerant vapour generated in the evaporator into a carrier fluid. The solution consisting of refrigerant and carrier fluid is then pumped into the generator, a vessel operated at a higher temperature and pressure level than the absorber. Within the generator, the refrigerant is removed via phase change and subsequently condensed in the condenser unit. The condensate is then fed back into the evaporator via a throttling valve, as done with a reverse Rankine cycle heat pump system. A simplified schematic of a basic absorption heat pump is shown in Fig. 18.10.

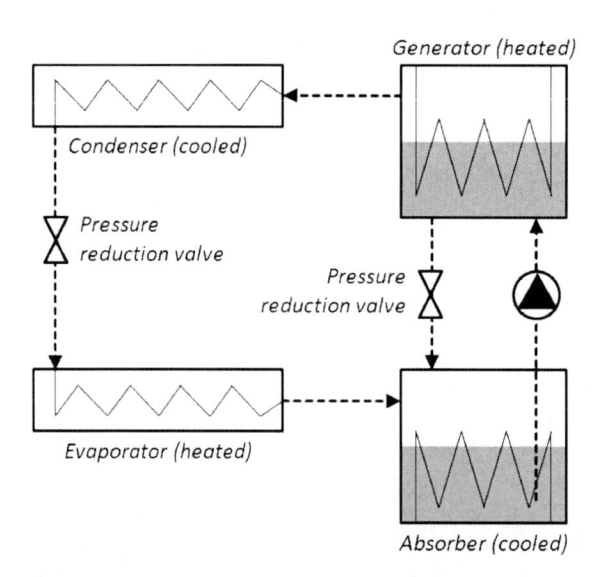

Fig. 18.10 Schematic view of an absorption heat pump

The two most common systems for absorption heat pumps are ammonia/water and water/lithium-bromide. The former system is particularly well-suited for space applications, as ammonia is already utilised as heat transfer medium in spacecraft and often provides a wider useful operating temperature range than water, which is limited to operating temperatures above freezing point. An absorption heat pump where carbon dioxide is absorbed into an alcohol or other organic liquid has also been invented for the Jet Propulsion Laboratory (Jones 2001). This innovative approach combines ongoing research in terrestrial carbon dioxide vapour compression systems with thermally-driven absorption technology.

In a special version of an absorption heat pump, the so-called diffusion-absorption heat pump (DAHP), the mechanical solution pump is replaced by a heat driven bubble pump and auxiliary gas is used to equalize the pressure level throughout the whole system. The efficiency of diffusion-absorption heat pumps, however, is smaller than with other heat-driven heat pump systems, although advances have been achieved with prototype systems in the kilowatt-region. Jakob et al. (2008) reported a maximum coefficient of performance of 0.38 for a 3 kW prototype which is considerably less efficient than with other thermally-driven systems, but may still be enough for certain cooling requirements in remote sensor or deployed system applications. Diffusion-absorption heat pumps are often installed in gas-fired domestic refrigerators due to the fact that they require little or no maintenance and can be operated for long periods without any wear. These features make diffusion-absorption heat pumps also be a relevant option for lunar surface operation, particularly in autonomous applications where the heat output of a solar receiver can be used as the driving force for the cooling process, rather than converting solar energy in electricity and powering an electric-motor-driven reverse Rankine heat pump. In applications where a sufficient supply with electrical energy is available, however, diffusion-absorption heat pumps will most likely not be an option unless an abundant source of high-temperature waste heat is available.

The main advantage of absorption heat pumps is that the system is heat-driven. Only little mechanical or electrical energy is required with the solution pump and the control unit; no electric energy at all is required with diffusion-absorption heat pumps. A temperature lift of waste heat can thus be achieved with very little or no electrical energy input at all, but at the cost of requiring an additional high-temperature heat source providing thermal energy at a temperature well above the intermediate heat rejection temperature present with absorber and condenser.

Disadvantages of absorption heat pumps are that the number of individual heat transfer processes is twice as large as with reverse Rankine cycle heat pumps (four compared to two heat transfer processes), assuming both systems are in the basic configuration.

The coefficient of performance of absorption heat pumps is normally considerably smaller than with reverse Rankine systems. The COP of a heat-driven heat pump in cooling mode is the ratio of heat removed (i.e. thermal energy transferred into the evaporator) and high-temperature heat supplied to the system (i.e. thermal energy transferred into the generator). Absorption heat pumps are therefore more challenging with respect to operating conditions, as a sufficient supply with high-temperature heat as well as roughly twice the heat rejection area has to be installed to dissipate the absorber and condenser heat. This increase in heat rejection area compared to reverse Rankine cycle systems is due to the fact that the number of exothermal processes is twice as large as with a reverse Rankine cycle heat pump, and heat therefore not only has to be removed from the condenser, but also from the absorber.

Heating and cooling of the bulk liquid phase also slows the process down, particularly during start-up and load-change, and consumes high-temperature heat as a significant fraction of the solution pumped into the generator is the carrier fluid, which also has to be thermally cycled. In addition, some of the media used in terrestrial absorption systems can crystallize under certain operating conditions and cause corrosion effects (e.g. lithium bromide - LiBr) whereas others are caustic and hazardous (e.g. ammonia).

Absorption heat pumps in gas-fired configuration are an established terrestrial technology widely applied in large refrigeration applications such as cooling houses or ice skating rinks. Small absorption heat pumps are also often installed in small domestic and hotel refrigerators, as they require little investment and maintenance. More or less maintenance-free diffusion-absorption heat pumps are often used with these small cooling applications; some of them are even driven by an electric heating element, but most system are supplied with a more economical heat source such as a gas burner or long-distance heating from a local combined heat and power plant.

Absorption heat pump technology has also seen a renewed interest in the context of terrestrial cooling applications, where the heat of a solar thermal system is directly used for air conditioning (Florides et al. 2002; Herfurt 2009).

18.2.3 Adsorption Heat Pumps

The basic principles of adsorption heat pumps follow those previously presented with absorption heat pumps. Other than with absorption heat pumps, where the refrigerant vapour is absorbed into a carrier fluid, the refrigerant vapour is rather adsorbed into a solid material with adsorption heat pumps. Due to the fact that solids cannot be handled in a continuous way as can be done with liquids, adsorption heat pumps are rather operated discontinuously. By applying two or more sorption heat exchangers in parallel, as shown in Fig. 18.11, a quasi-continuous operation is possible by always having a least one of the sorption heat exchangers in sorption mode and at least one sorption heat exchanger in desorption mode. In sorption mode, refrigerant vapour is exothermally adsorbed into the solid material; in desorption mode, the adsorbed refrigerant is desorbed endothermally by applying high-temperature heat.

A simplified schematic of an adsorption heat pump is shown in Fig. 18.11. The label "cooled" again indicates elements releasing heat to an external heat sink, whereas the label "heated" indicates elements receiving heat from an external heat source.

The main advantage of adsorption heat pumps is again the primarily heat-driven operation. Very little mechanical or electrical energy is required for operation, as even the solution pump of conventional absorption systems is not required.

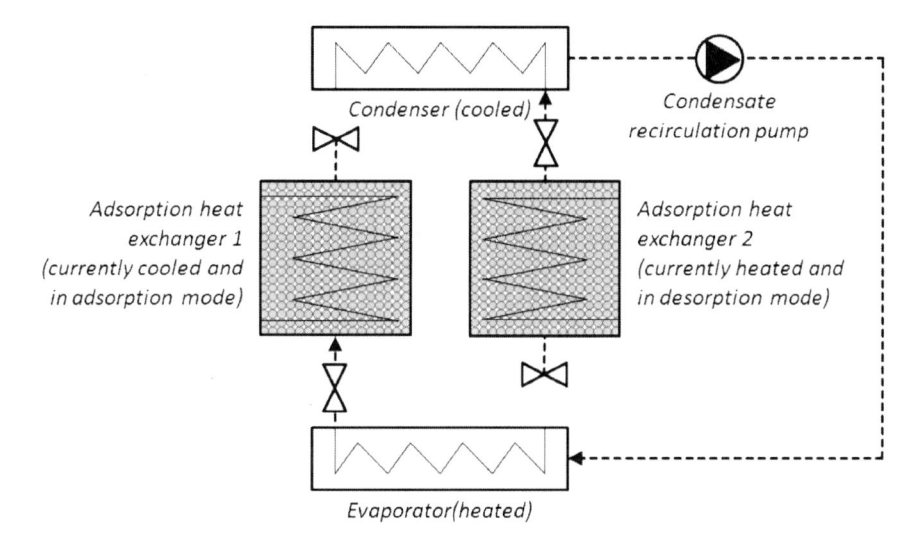

Fig. 18.11 Schematic view of an adsorption heat pump

As with absorption heat pumps, the number of individual heat transfer processes is again twice as large as with reverse Rankine cycle heat pumps (again four compared to two heat transfer processes if both systems are considered in a basic configuration). This again requires the availability of a high-temperature heat source and roughly twice the radiator area required with a reverse Rankine cycle heat pump.

Quasi-continuous operation can be achieved by operating more than one sorption modules in parallel. Periodic thermal cycling of the inert solid sorption material and associated heat transfer equipment (pipes, metal sheets for heat conduction and so on) nevertheless reduces the coefficient of performance which is again computed for cooling operation as presented with absorption heat pumps as ratio of heat removed (i.e. thermal energy transferred into the evaporator) and high-temperature heat supplied to the system (i.e. thermal energy transferred into the sorption module in desorption mode).

Other than with absorption heat pumps, the solid sorption material is normally not associated with crystallisation, segregation or demixing phenomena.

As with absorption heat pumps, adsorption technology has also been considered for a number of potential cooling applications driven by low-grade or waste heat. This includes solar thermal cooling applications as well as automotive cooling applications, for instance (Ziegler 2002). Large systems in the range of 50+ kW cooling load have been commercially available for years. Smaller systems in the range of 10 kW are currently being developed for solar cooling applications.

18.2.4 Reverse Brayton Cycle Heat Pumps

Reverse Brayton cycle heat pumps, also referred to as air refrigeration or Bell Coleman cycle heat pumps, are similar in design to reverse Rankine cycle heat pumps, but are operated without the liquid/vapour phase change, and thus solely in single-phase configuration and gaseous media.

Reverse Brayton cycle heat pumps therefore do not have an evaporator and a condenser, but rather two gas-to-gas or liquid-to-gas heat exchangers. One heat exchanger is used to remove heat from an external heat source; this heat exchanger is operated at low temperature and pressure. The second heat exchanger is used to provide heat to an external heat sink; this heat exchanger is operated at high temperature and pressure. Shaft work provided by compressor is used to transfer the gaseous medium from the low- to the high-pressure side. The energy of expansion can be recovered by a turbine, thus reducing the shaft work required for the compressor. A simplified schematic of a reverse Brayton cycle heat pump is shown in Fig. 18.12.

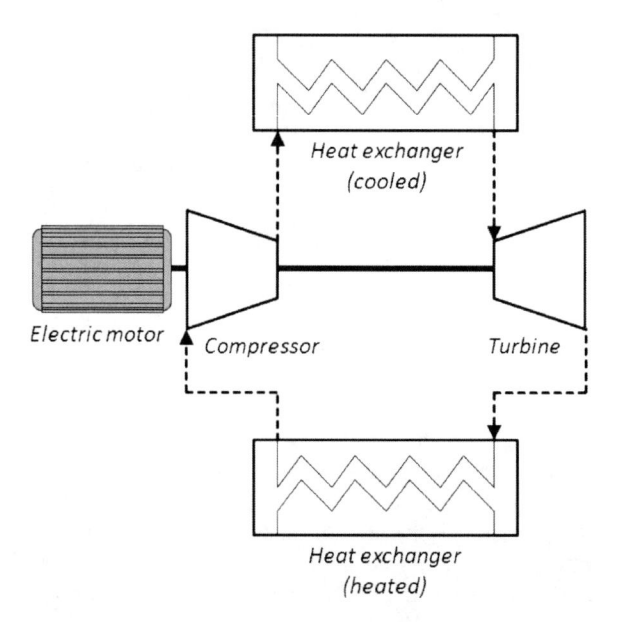

Fig. 18.12 Schematic view of a reverse Brayton cycle heat pump

Brayton cycle heat pumps have to be operated with high mass and volume flows as no phase change is involved. A part of the expansion energy, however, can be recovered with the installation of an expansion turbine.

In addition, the heat transfer processes are not isothermal but isobaric. The deviation of the Brayton process from the Carnot process is therefore larger than with the ideal Rankine process. The efficiency of realised systems is therefore normally smaller than with reverse Rankine heat pumps.

18.2.5 *Jet Ejector Cycle Heat Pumps*

Jet ejector cycle heat pumps are simple and robust devices with a long track record of terrestrial applications. The basic working principle is to generate a constant stream of high-temperature and high-pressure steam, and to feed this steam into a jet ejector. Within the jet ejector, the high-pressure steam is accelerated and the kinetic energy of the molecules increases correspondingly, whereas the pressure decreases. This effect can be used to draw low-pressure steam out of the evaporator into the stream. High-temperature and low-temperature steam are thus mixed within the jet ejector and subsequently condensed in a condenser.

In heat pump applications, the same media are normally used in the high-temperature steam generator and in the low-temperature refrigerant evaporator. Thus, the condensate can be directly pumped back into the generator and evaporator.

The main advantages of jet ejector heat pumps compared to other thermally-driven heat pumps are the simple and robust design. Jet ejector heat pumps can be operated continually and without any considerable degradation or wear. In addition, the cooling load can be easily and quickly controlled by adjusting the temperature – and thus pressure – of the steam generator, or by opening the valve between steam generator and jet ejector only discontinuously.

The main disadvantage of jet ejector cycle heat pumps is the low coefficient of performance, again computed as the ratio of cooling load and high-temperature heat input. This, again, results in more demanding heat source and sink requirements compared to reverse Rankine cycle heat pumps. The number of individual heat transfer processes is three, and thus exactly between reverse Rankine cycle heat pumps with two and ad- as well as absorption heat pumps with normally four heat transfer processes, again assuming all systems in a basic configuration.

The heat to be dissipated from the condenser includes a fraction coming from the evaporator flow (this corresponds to the normal dissipation rate expected with a reverse Rankine cycle heat pump) and the flow coming from the steam generator. Depending on the three temperature levels involved, the heat rejection load can thus easily be in excess of the double amount of a reverse Rankine heat pump.

A simplified schematic of a jet ejector cycle heat pump is shown in Fig. 18.13.

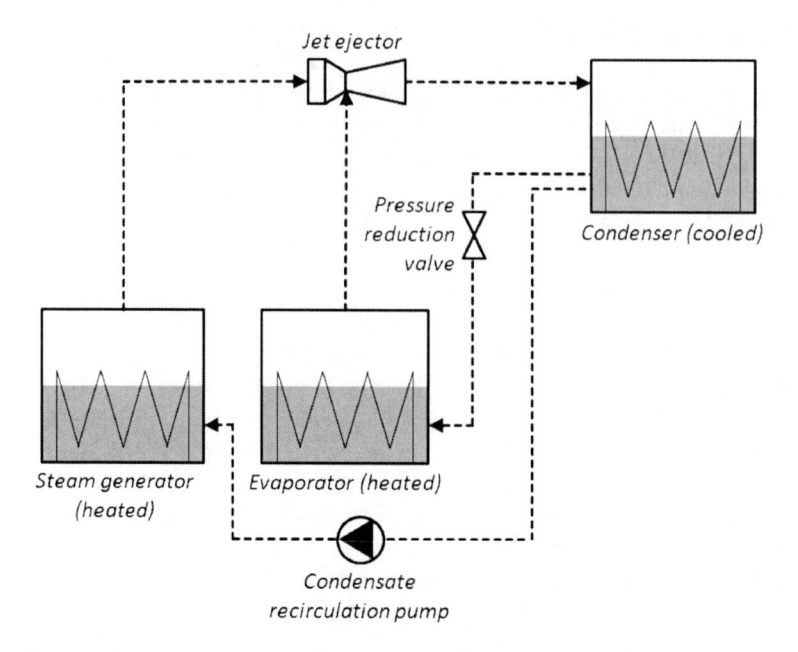

Fig. 18.13 Schematic view of a jet ejector cycle heat pump

Jet ejector heat pumps have been applied in a wide range of different industrial applications for years. The technology is proven and reliable, although not considered efficient enough for many dedicated cooling or refrigeration applications. Recent research and development in the field of solar thermal cooling has proven that jet ejector technology is also a feasible and relevant option, although only few dedicated prototypes have been built and tested with concentrating solar thermal systems (Pollerberg et al. 2009a,b,c).

18.2.6 Stirling Cycle Heat Pumps

A Stirling engine is a heat engine operating by cyclic compression and expansion of a gaseous working fluid. Contrary to internal combustion engines, the gas never leaves the engine. Heat is rather transferred into and out of the engine by heat exchangers. This makes Stirling engine operation independent from the actual heat source or fuel, and thus a principally very flexible energy conversion technology considered, tested and implemented in a wide range of different terrestrial applications.

The ideal Stirling cycle consists of four thermodynamic processes:

1. Isothermal expansion
2. Isochoric (constant-volume) heat-removal
3. Isothermal compression
4. Isochoric heat-addition

These four processes are, however, distinct, because the transitions rather overlap.

Similar to the Rankine and Brayton cycles, the Stirling cycle can not only be used to convert thermal into mechanical energy, but can also be reversed and used as heat pump if mechanical energy is provided. Reverse Stirling cycle heat pumps are thus an alternative to reverse Rankine and Brayton cycle heat pumps. Possibilities and limitations of reverse Stirling cycle heat pumps therefore again have to be considered relative to the benchmark set by reverse Rankine cycle heat pumps.

Stirling engine technology is currently under development for advanced heat-to-electric conversion applications, such as in Stirling radioisotope generators, a combination of a Stirling engine and a radioisotope generator The Department of Energy and NASA are currently developing a radioisotope power system utilizing Stirling power conversion technology for potential future space missions. Triggered by the higher conversion efficiency compared to the thermoelectric generators used with previous Radioisotope Thermoelectric Generators (RTGs), a four-fold reduction in PuO_2 fuel is to be achieved, thereby saving cost and reducing radiation exposure of the servicing personnel (Chan et al. 2007). This development effort is again supported by ongoing research efforts in the field of terrestrial solar thermal energy technology.

Stirling cryocooler technology has also been developed, tested and applied in space missions (Russo and Sugimura 1996). A future development of reverse Stirling cycle heat pumps could therefore be based on a long history of terrestrial and space-oriented technology development. Reverse Rankine systems have been the technology of choice for room temperature level cooling applications.

18.3 Heat Pump System Configurations

The purpose of a heat pump is to transfer heat from an initial temperature level to a higher temperature level. Heat pumps can thus, in principle, be applied in heating as well as in cooling applications, depending on which end of the temperature-lift process is the relevant one. If heat removal is desired, the colder end of the heat pump is used to remove heat from an external (waste) heat source, and to reject this heat at the higher temperature end and a higher temperature level. In heating applications, on the other hand, the higher temperature end of the heat pump is utilised to provide a useable thermal output based on a heat input received at a lower temperature level.

Depending on the specific application, heating as well as cooling can be relevant applications for heat pumps in space mission applications. Possible configurations and applications of electrically and thermally-driven heat pump systems are briefly presented in the following with respect to a future utilisation in lunar surface exploration and exploitation missions.

18.3.1 Limitations in Lunar Daytime Heat Release

In space as well as in lunar surface missions, heat release is primarily achieved by applying a purpose-designed surface finish to in- or decrease the heat release rate between the mission element and the ambient environment and/or by including dedicated heat release elements, so-called radiators, in the system. Radiators require a certain temperature difference with respect to the thermal sink in order to provide a reasonable heat rejection rate in terms of units of energy per unit of time and surface area. A temperature differential of 40 to 50 K between the radiator and the environment is a practical minimum in order to keep the radiator size reasonable (Sridhar and Gottmann 1995).

Transferring a waste heat load to a higher temperature level therefore enables the utilisation of significantly smaller radiators. Radiators size and mass are both an issue in space mission design. Simple tube-fin radiators today measure some 3.5-8.5 kg/m² (Messerschmid and Bertrand 1999). Advanced radiator technology will enable advanced radiators providing a specific mass of less than two, approaching one kg/m² (Juhasz 2002; Messerschmid and Bertrand 1999). These figures are for space radiators and have to be re-evaluated with respect to structure and assembly for lunar gravity conditions, particularly when operated with tens or even hundreds of square meters.

In satellites orbiting the Earth, a low-temperature thermal sink is readily available, at least for a significant part of the time when the radiator is not facing the sun. This is not necessarily the case with operation under lunar daytime conditions, where the surface temperature can reach up to 400 K. In this case, heat release is a challenge and cannot be easily achieved with passive thermal control system elements, and active systems with temperature lift become an interesting and relevant option. Only thus, the desired temperature differential between radiator and environment can be achieved.

Heat release becomes an even greater challenge if a mission element is to be operated in a permanently sunlight region of the Moon. Such a region would be very advantageous with respect to solar power generation, as the intermediate electrical energy storage system required for night operations would not be necessary, but heat release would be a greater challenge than with a location having a normal day and night cycle. Heat pumps operated with a significant temperature-lift could be used to provide a sufficient heat release rate required with reasonably-sized radiators.

18.3.3 Heat Pump Configuration

After specifying the heat source(s) in terms of temperature and desired heat removal rate, the next question is whether a single heat pump should be used, or if multiple heat pumps are to be preferred in a specific application. The former provides the possibility of applying a single heat acquisition loop, or a number of sub-loops, respectively, providing the thermal input to a single heat pump. If multiple heat acquisition loops are applied with a single heat pump, cooling of the individual waste heat sources can be actively controlled by installing valves into the heat acquisition system, thus allowing a differentiation between the individual heat. Due to reasons of redundancy, a central heat pump system may nevertheless be duplicated with a second system operated in parallel, and with each of the heat pumps scaled to provide sufficient heat removal in case of a failure of the other heat pump.

Alternatively, multiple sub-systems with independent heat pumps could be used. Multiple heat pumps will primarily be applied if strongly different cooling temperatures and/or loads are required.

18.3.4 Heat Release Options

The next step after specifying the cooling load(s) and the number of heat pump systems is to determine the heat release technology. Radiators will normally be the option of choice; the question to be answered is if the condenser of the heat pump and the radiator are to be integrated into a single element, or if separate elements are to be applied with a heat release loop applied in between.

A simplified schematic of a reverse Rankine cycle heat pump system with separate condenser and radiator and a liquid heat release cooling loop is shown in Fig. 18.14.

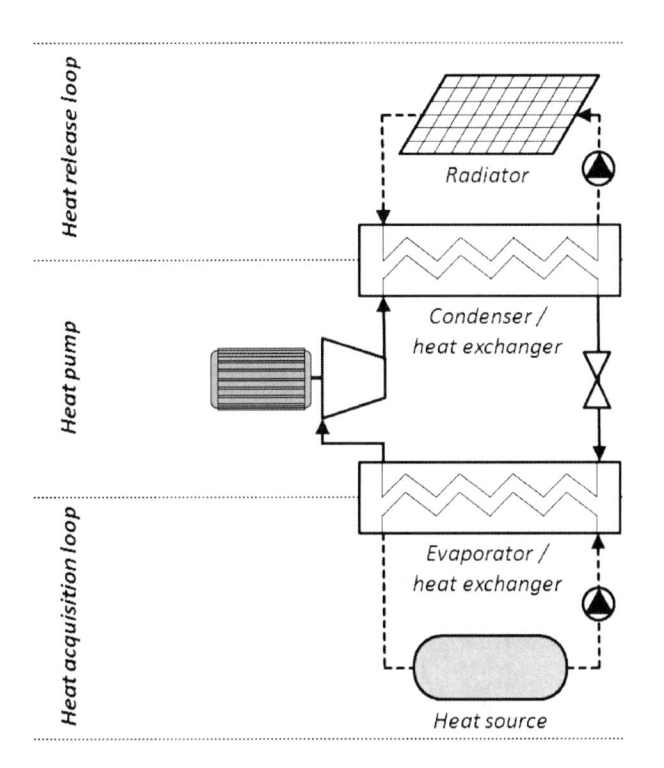

Fig. 18.14 Schematic view of a reverse Rankine heat pump system with external heat acquisition and release loops

Both systems are shown with a heat acquisition loop. As with the heat release loop, also the heat acquisition loop could be eliminated by directly cooling the low-temperature heat source with the evaporator. Due to the fact that there will normally be a number of individual low-temperature heat sources, and that a liquid-based heat acquisition loop will normally be preferred over a multiple evaporator approach, this option is not further considered apart from small or single source applications.

Instead of applying radiators for heat release, thermal energy could also be removed or intermittently stored in a subsurface heat exchanger, sometimes referred to as underground lunar thermal storage system. This option is further addressed in the "Heat Pump System with Thermal Energy Storage" section of this chapter.

18.3.5 Heat-Driven Heat Pump Systems

One of the advantages of reverse Rankine cycle systems is that a single heat transfer element, the evaporator, has to be heated by an external heat source, and a single element, the condenser, has to be cooled by an external heat sink. This is not the case with heat-driven systems, where the number of heat transfer elements is normally doubled and two heat transfer elements have to be cooled, as well as two heat transfer elements have to be heated, assuming both systems in basic configuration.

Heat removal is normally done at a single, intermediate temperature level, whereas the heat input is provided at a low and at a high temperature level. Heating at the low-temperature level corresponds to the desired heat removal in cooling applications, whereas the high-temperature heating is the driving force of the heat pump process and has to be provided by a solar or other high-temperature heat source.

A simplified schematic of an absorption heat pump system with heat release loop for condenser and absorber as well as a heating loop for the generator is shown in Fig. 18.15.

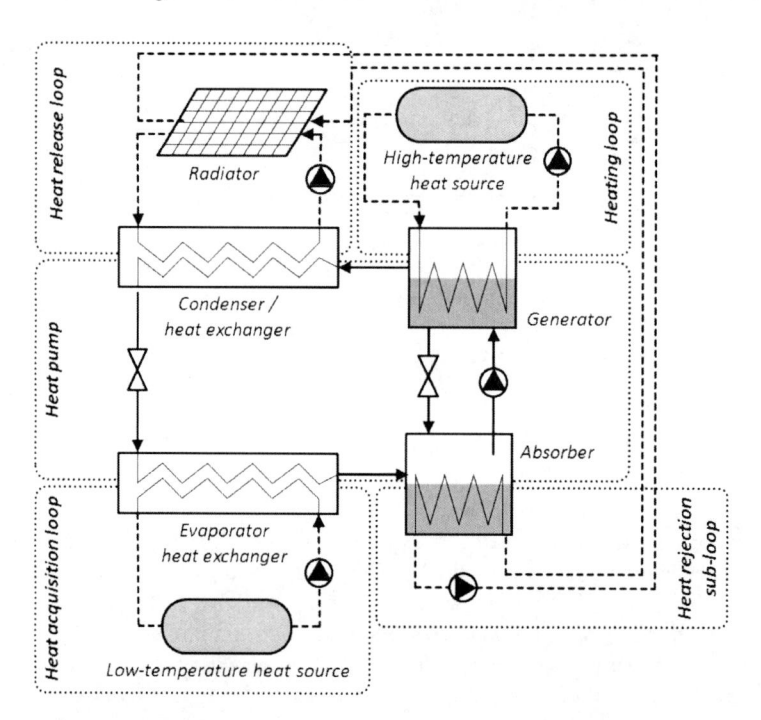

Fig. 18.15 Schematic view of an absorption heat pump system with external heat acquisition and release loops

The additional elements of the heat pump system (four instead of two heat transfer processes) as well as the additional high temperature heat source loop increase the complexity of the overall system. This increase in complexity of the thermal system has to be evaluated against a substantial or even full substitution of the electric energy requirements with heat. The availability of electrical and/or thermal energy (at sufficient temperature levels) will therefore be decisive with respect to the question whether a heat-driven heat pump is a relevant option or not.

In 1996, Hanford and Ewert conducted a study on advanced active thermal control systems for future applications in space exploration, including the International Space Station as well as lunar and Martian outposts (Hanford and Ewert 1996). In their study, they considered electrically-driven reverse Rankine cycle as well as heat-driven heat pumps. According to the baseline missions considered in their study, no useable sources of waste required for a heat-driven cooling system were available. They concluded that in order to generate heat to drive a heat-driven heat pump, at least two other methods were to be considered: one method would be to generate heat from electricity generated by solar photovoltaic power arrays through resistive heating. A second method would use a collector to focus solar radiation on a tube containing a working fluid (solar collector). They conclude that the latter was a more efficient technique and assumed it for their further investigations about heat-driven cycles in this study (Hanford and Ewert 1996).

In general, heat-driven heat pumps are an interesting option if a sufficient supply with high-temperature heat is available. If this is the case, and the increased radiator area is available or can be provided, heat-driven heat pumps do provide interesting options. Heat pump research and development efforts aiming at future applications in space missions, however, are clearly directed towards shaft work driven systems at the moment.

18.3.6 Heat Pump System with Thermal Energy Storage

The introduction of a thermal energy storage system provides a certain degree of independence, e.g. from ambient conditions and application profile, over the long day and night cycle of the Moon. A thermal energy storage system could be used to store excess heat during daytime in order to draw thermal energy from the storage system during night again, or vice versa by cooling the storage medium during a cold lunar night and transferring waste heat into the storage system during daytime. Depending on the actual thermal storage capacity, a reservoir of heat or cold could thus be stored and utilised for emergency cooling or as efficiency boost for the primary thermal management system, for instance.

A simplified schematic of a reverse Rankine cycle heat pump system with external heat acquisition and dissipation loops is shown in Fig 18.16. The thermal energy storage system could be connected to the heat acquisition system, thus providing the possibility of removing heat from the storage system, and/or to the heat release loop, thus providing the possibility of storing heat in the storage system. The system shown in Fig. 18.16 is connected to both, the heat acquisition and the release loop, via valves and thus provides the maximum flexibility with respect to heat removal and storage. The presence of a surface-mounted radiator would theoretically not be required if the subsurface thermal storage system was sized large enough so that heat transferred into a subsurface heat exchanger could be dissipated into the ground by means of thermal conduction within the subsurface layers.

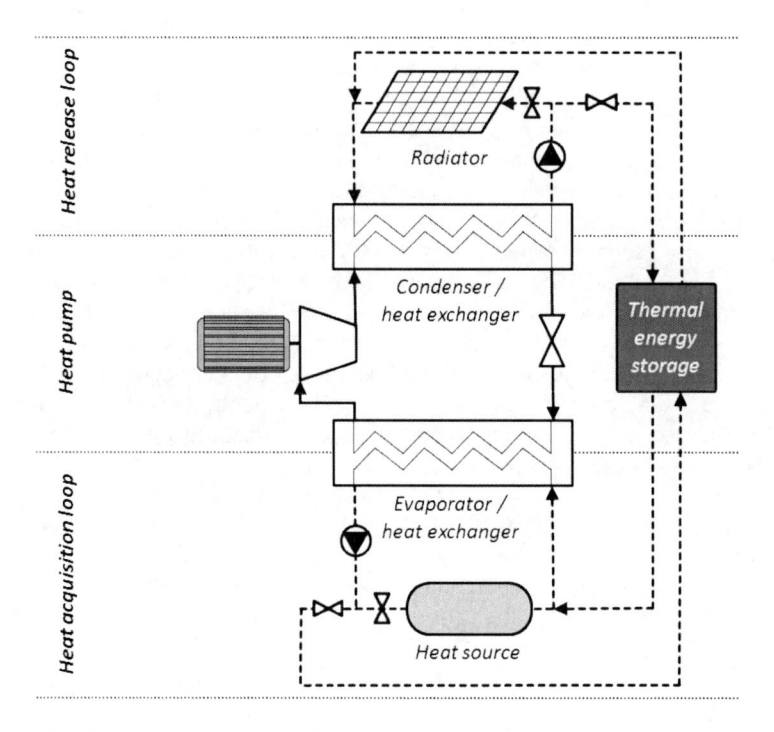

Fig. 18.16 Schematic view of a reverse Rankine heat pump system with external heat acquisition and release loops and a thermal energy storage system

A thermal energy storage system can be considered with a wide range of different systems, such as small and mobile rovers, for instance, up to large static systems. In case of a small rover, a dedicated storage system – or a thermally-insensitive part of the rover that can be used for intermediate thermal energy storage – could be utilised to aid in levelling the interior temperature under the presence of strong ambient temperature conditions, for instance.

In stationary applications, such as a crewed lunar base, lunar soil thermal storage may be considered, as the near-subsurface region already offers constant temperatures independent from the day and night cycles present at the surface. A subsurface collector could therefore be utilised to dissipate heat into the ground, or to draw heat out of the ground, respectively, as it is regularly done with terrestrial heat pump systems. The installation of a subsurface heat exchanger or collector, however, is a technical challenge but might be possible in combination with excavation work made for radiation shielding or in-situ resource utilisation activities. Piling dust and regolith material on top of a ground collector spread onto the surface may also be advantageous to burying a collector in purpose-built trenches, but has to be evaluated in the wider context of base installation and operation. As far as plumbing and working fluids are carried from Earth, the mass of the energy accumulator remains critical in deciding on its implementation.

18.4 Lunar Surface Applications of Heat Pumps

Heat pump technology is very versatile and can be applied in a wide range of different applications. It goes beyond the scope of this chapter to discuss relevant applications in detail. A brief overview of three possible applications - two mobile and a stationary application - is provided in the following. In order to account for the versatility of heat pump technology, one comparably small and one comparably large mobile application and one large stationary application are discussed.

Mobile application (example): robotic surface rover

Two robotic surface rovers have already been applied in lunar surface exploration - Lunokhod 1 landed with Luna 17 on November 15, 1970 (NSSDC Luna 17) and Lunokhod 2 landed with Luna 19 on January 15, 1973 (NSSDC Luna 21) - and more will most likely be applied in future lunar surface exploration and exploitation missions. Possible mission scenarios include mobile surface exploration, by offering a mobile platform to transport scientific payload, collection of soil and rock samples from different locations, preparations of lunar surface base erection, or as robotic work crew and general purpose transport and support vehicle for human extravehicular activities in crewed missions. Depending on the mission scenario, robotic rovers include mini rovers (similar in size to the Mars Exploration Rovers with a mass in the order of 10-150 kg and an electrical energy consumption of up to 1 kW) or larger systems, which would e.g. be utilised as robotic work in base erection or resource utilisation activities.

Thermal management and control with the two previous robotic rovers, Lunokhod 1 & 2, has been made by applying a surface-mounted heat exchanger for cooling and a combination of radioisotope heating elements and a closable lid preventing excess cooling to keep the rover warm at night (Harvey 2005).

The introduction of a heat pump into a robotic surface rover would enable high-power operation with a reasonable radiator area. Surface area is limited with a mobile rover and will often have to be shared between photovoltaic panels, scientific equipment and radiators. A strong reduction in radiator size might therefore be particularly desirable, although high power operation will normally not be possible with a solar power system, and large photovoltaic arrays would therefore not be installed. Elevated output power requirements will most likely be satisfied by the introduction of new power system technologies such as fuel cells or radioisotope systems with advanced fuel- or heat-to-electric conversion technologies. Depending on the type of fuel cell and the desired operational efficiency, operating temperatures in the order of 70-120°C might be required. Considerable quantities of waste heat would thus be generated at this temperature level, as at least a third of the chemical energy content of a fuel is released as waste heat of the electrochemical reaction process. A heat pump could therefore be installed to provide a small temperature-lift with a high coefficient of performance to the waste heat generated by the onboard power systems before transferring it to a reasonably-sized radiator installed on the top of the rover.

Intermediate thermal energy storage could also be considered with robotic rovers. A heat pump could be used to store waste heat in a dedicated thermal storage system, such as a container of phase-change material, or in a thermally-insensitive part of the rover structure. This heat could then be used to warm the interior compartment at night, or to at least reduce the heating load required by resistive heaters, for instance.

A third possible application would also be the transfer of thermal energy within different elements or compartments of the rover. By including a heat pump, waste heat generated at a comparably low temperature level could be used to heat an endothermal process or a piece of equipment requiring elevated operating temperatures. Thus, a dedicated high-temperature heat source would not have to be included, and energy-intensive electric heaters could be avoided.

Mobile application (example): pressurized surface laboratory

Basically, two different approaches in crewed rover design can be distinguished: firstly, an unpressurized rover design similar to the Apollo Lunar Roving Vehicle, and secondly, a pressurized rover design. The latter will be a key element in crewed medium- and long-range lunar surface exploration and will not only serve as means of transportation, but will also be designed as mobile living and working space for the astronaut crew throughout the duration of the surface exploration trip. Pressurized mobile laboratories will enable surface exploration on a wider scale (systems operating at a radius in the order of 500 km are discussed) with a continuous exploration trip lasting some 10-20 terrestrial days, offering full crew accommodation, and being equipped with a basic laboratory infrastructure and scientific payload. In addition, pressurized surface rovers can also be used as safe haven in case of an emergency situation with the stationary lunar base.

Applications of heat pumps in a pressurized surface rover follow those already discussed with the robotic rovers. In addition to waste heat removal from scientific and power system elements, the main objective will also be air conditioning of the crew compartment. As with a stationary lunar base, an adverse temperature gradient between the crew compartment and the surface temperature during daytime operation will make a temperature lift inevitable, again considering the limited surface area for radiator installations provided by the rover structure. Systems without temperature lift, such as lunar subsurface heat exchangers, are not possible with mobile systems. This makes a thermal management system with a temperature lift more or less inevitable for prolonged daytime operations.

Due to the fact that pressurized rovers will have to be equipped with a strong power source in order to provide surface mobility in addition to powering scientific and life support system equipment, an electrically-driven heat pump with a motor power in excess of one kilowatt and a cooling load in the high single- or low two-digit range will most likely be the option of choice.

Stationary application (example): permanent lunar surface base

One of the largest cooling applications will be the permanent surface base. This base will accommodate a crew of astronauts during their stay on the Moon and will provide living area as well as space for scientific investigations and technical equipment. Sridhar and Gottmann assumed an electrical power consumption of 100 kW for a base designed for a crew of four to eight members (Sridhar and Gottmann 1996). They further assumed that the walls of the station would thermally insulate the interior compartment so well, that the heat exchange through the walls was negligible. The total cooling load in this scenario would therefore be just over 100 kW, assuming that all of the electric energy is dissipated as heat. This heat load would have to be removed from the interior and dissipated into the ambient environment, which could be done with a temperature-lift device assuming surface temperature peaks at about 390 K at the near-equatorial region assumed with the investigated scenario (Sridhar and Gottmann 1996).

Independent from the actual figures assumed for the cooling load of a permanent lunar surface base, heat release of such a cooling load is a challenge without temperature lift. Certain scenarios without temperature lift exist - subsurface heat exchangers have been briefly discussed within this study - but a heat pump system based on reverse Rankine cycle technology and providing a temperature lift in the order of 50 K in order to enable the utilization of reasonably-sized surface radiators is currently the baseline scenario and therefore the reference, against which other heat pump and thermal control system concepts have to be evaluated.

18.5 Conclusions

In general, heat pumps are very versatile thermal control system elements and applicable in a wide range of heating and/or cooling applications. Their operating principle is to transfer heat to a higher temperature level by applying (shaft) work or high-temperature heat. By including a heat pump into a spacecraft thermal control system, waste heat generated at a certain temperature does not have to be dissipated at the same temperature or below, but can be transferred to a higher temperature level and then dissipated with a smaller and lighter heat release system.

Elevated heat release temperatures make the thermal control system more robust and thus more independent from environmental conditions and performance-related degradation phenomena. Peak cooling loads can also be removed much easier by intermittently increasing the temperature lift of the heat pump and thus also increasing temperature and energy dissipation rate of the radiator. In addition, degradation of the radiator as a function of time and/or dust deposition can also be easily compensated by a correspondingly higher temperature lift of the heat pump.

Heat pumps also provide the possibility of reducing the radiator - or any other heat release system – in size. This is a very relevant aspect in the ongoing miniaturisation of spacecraft and related subsystems.

Reverse Rankine cycle heat pumps, also referred to as vapour compression heat pumps, are currently being considered as primary heat pump technology in spaceflight applications. They do, however, require a significant shaft work input normally provided by an electrical motor. Alternatively, heat-driven heat pumps, such as the absorption and adsorption systems, are considered with applications in space and lunar surface exploration and exploitation missions. Heat-driven systems are more complex in terms of the heat transfer processes and require an additional high-temperature heat source as well as an increased intermediate-temperature heat sink, but can be operated with a very small - or even without any - electrical energy. Diffusion-absorption heat pumps do not require any electricity or moving parts, for instance. Heat-driven systems are therefore an interesting option if a high-temperature heat source is available.

Heat pump system design generally requires a simultaneous consideration and optimisation of many different aspects, starting from heat acquisition in terms of cooling load and temperature, heat release in terms of sink temperature and radiator size, as well as heat pump operation in terms of temperature lift, thermal storage systems and intermittent or diurnal operating profile.

By utilizing heat pumps in thermal control system engineering, the most important figure of merit in spaceflight applications, the launch mass, can often be reduced. In addition, the thermal control system can be designed in a more robust way by introducing a certain degree of freedom with respect to heat sink operation, and thus with respect to cooling load profile and ambient conditions. As a sufficient heat removal is a prerequisite for safe operation, heat pumps are a very interesting and relevant technology for future thermal control system engineering in general, and for lunar surface exploration and exploitation applications in particular.

References

Abbott, M.M., van Ness, H.G.: Schaum's outline of thermodynamics with chemical applications, 2nd edn. McGraw-Hill (1989)

Bussey, D.B.J., Spudis, P.D.: The Lunar Polar Illumination Environment: What We Know & What We Don't. Space Resources Roundtable VI (2004),
http://www.lpi.usra.edu/meetings/roundtable2004/pdf/6022.pdf

Bussey, D.B.J., Fristad, K.E., Schenk, P.M., Robinson, M.S., Spudis, P.D.: Constant illumination at the lunar north pole. Nature 434, 842 (2005)

Chan, J., Wood, J.G., Schreiber, J.G.: Development of Advanced Stirling Radioisotope Generator for Space Exploration. NASA TM 2007-214806 (2007)

ESA, What does thermal control do? (2007a),
http://www.esa.int/TEC/Thermal_control/SEMZU1T4LZE_0.html

ESA, Mechanically-pumped heat transport technology (2007b),
http://www.esa.int/TEC/Thermal_control/SEM9KLBE8YE_0.html

Florides, G.A., Kalogirou, G.A., Tassou, S.A., Wrobel, L.C.: Modelling, simulation and warming impact assessment of a domestic-size absorption solar cooling system. App. Therm. Eng. 22, 1313–1325 (2002)

Hanford, A.J., Ewert, M.K.: Advanced Active Thermal Control Systems Architecture Study. NASA TM 104822 (1996)

Harvey, B.: Soviet and Russian Lunar Exploration: Comparisons of the Soviet and American Lunar Quest. Springer (2005)

Herfurt, C.: Solarthermische Absorptionskältemaschine (AKM): Auslegung einer solarthermischen Absorptionskältmaschine zur Raumklimatisierung. VDM Verlag Dr. Müller (2009) (in German)

Herold, K.E., Radermacher, E., Klein, S.A.: Absorption chillers and heat pumps. CRC Press (1996)

Izutani, I., Kobayashi, N., Ogura, T., Nomura, I., Kawazoe, M., Yamamoto, H.: Temperature and humidity control system in a lunar base. Adv. Space Res. 12, 41–44 (1992)

Jakob, U., Eicker, U., Schneider, D., Takic, A.H., Cook, M.D.: Simulation and experimental investigation into diffusion absorption next term cooling machines for air-conditioning applications. App. Therm. Eng. 28, 1138–1150 (2008)

Lia, X., Wang, X., Zheng, Y., Cheng, A.: Estimation of solar illumination on the Moon: A theoretical model. Planet Space Sci. 56, 947–950 (2008)

Loewer, H., Bosniakovic, F. (eds.): Absorptionswärmepumpen. C.F. Müller, Karlsruhe (1987) (in German)

Jones, J.A.: Carbon dioxide absorption heat pump. United States Patent 6374630 (2001)

Juhasz, A.J.: Design Considerations for Lightweight Space Radiators Based on Fabrication and Test; Experience With a Carbon-Carbon Composite Prototype Heat Pipe. NASA TP 1998-207427/REV1 (2002)

Kirn, H., Hadenfeldt, A.: Wärmepumpen. C.F. Müller, Karlsruhe (1976) (in German)

Mainstream, Mainstream Awarded NASA Contract to Develop a Gravity-Insensitive Heat Pump for Lunar Applications (2006),
http://www.mainstream-engr.com/company/press/20060530.jsp

Mainstream (2010a), http://www.mainstream-engr.com/ (Mainstream Engineering Corporation Homepage)

Mainstream, Low Lift Heat Pump for Satellites (2010b),
http://www.mainstream-engr.com/research/
turbomachinery/satheatpump.jsp

Messerschmid, E., Bertrand, R.: Space stations: systems and utilization. Springer (1999)

NASA (2010),
http://nssdc.gsfc.nasa.gov/planetary/factsheet/moonfact.html

NSSDC Luna 17, NSSDC ID: 1970-095A (2010)

NSSDC Luna 21, NSSDC ID: 1973-001A (2010)

Park, C., Sunada, E.: Vapor compression hybrid two-phase loop technology for lunar surface applications. Space Technology and Applications International Forum (STAIF). Albuquerque, New Mexico (2008),
http://www.1-act.com/pdf/vchtpllunar.pdf

Pollerberg, C., Ali, A.H.H., Dötsch, C.: Solar driven steam jet ejector chiller. App. Therm. Eng. 29, 1245–1252 (2009a)

Pollerberg, C., Ali, A.H.H., Dötsch, C.: Experimental study on the performance of a solar driven steam jet ejector chiller. Energy Convers Manage. 49, 3318–3325 (2009b)

Pollerberg, C., Heinzel, A., Weidner, E.: Model of a solar driven steam jet ejector chiller and investigation of its dynamic operational behavior. Sol. Energy 83, 732–742 (2009c)

Rowe, D.M.: CRC handbook of thermoelectric. CRC Press (1995)

Rugescu, D.R., Rugescu, D.R.: Solar Energy. In: Rugescu, R.D. (ed.) Potential of the solar energy on Mars, pp. 1–24. INTECH, Vienna (2010)

Russo, S.C., Sugimura, R.S.: NASA IN-STEP Cryo System Experiment flight test. Cryogenics 36, 721–730, (1995 Space Cryogenics Workshop) (1996)

Silberstein, E.: Heat Pumps. Delmar (2002)

Silverstein, C.C.: Design and technology of heat pipes for cooling and heat exchange. Taylor & Francis (1992)

Sridhar, K.R., Gottmann, M.: Lunar base thermal control systems using heat pumps. Acta Astronaut. 39, 381–394 (1996)

Ziegler, F.: State of the art in sorption heat pumping and cooling technologies. Int. J. Refrig. 25, 450–459 (2002)

19 Principles of Efficient Usage of Thermal Resources for Heating on Moon

Viorel Badescu

Polytechnic University of Bucharest, Romania

19.1 Introduction

Understanding lunar surface and subsurface temperatures is critical for future human and robotic exploration (Paige 2010). The fact that the extreme thermal cycling on the Moon takes place over a 28 day cycle, with 14 days of intense sun and 14 days of dark cold, has important implications on the availability of power, operation feasibility and poses a variety of safety and reliability questions (Cohen 2002). Note that the Apollo missions all involved landings which took place in equatorial regions and were conducted during the lunar day. Future exploration is needed to cover a much wider range of latitudes (Paige 2010).

The 28-day cycle may also drive the site selection for a future lunar base. There are many proposals to locate a lunar base at one of the poles where it can receive abundant solar energy (Cohen 2002). Indeed, due to seasonal variations, polar locations can have periods of constant sun light during the summer that can last many months followed by constant night time conditions for the remainder of the year. Some high elevations, such as crater rims, can experience constant daylight during 70 percent of the winter (Christie et al. 2008). There are crater rims near the poles which are nearly "Peaks of Eternal Light" (Christie et al. 2008; Kruijff 2010).

There is another aspect which makes the lunar poles unique environments on Moon. Crater floors may be permanently shielded from the Sun (Bussey at al. 2005; Spudis 2008; Vanoutryve 2010). The temperatures within the shaded regions offer high-vacuum cryogenic environments, which could support cryogenic applications such as high-temperature superconductors (Ryan et al. 2008). Furthermore, the unique cryogenic conditions at the lunar poles provide an environment that could reduce the power, weight, and total mass that would have to be carried from the Earth to the Moon for lunar exploration and research (Ryan et al. 2008).

The extreme thermal environmental conditions on the Moon shape and constrain lunar construction technologies and habitats in far-reaching ways (Cohen

2002). Some types of lunar facilities will have unique thermal control requirements, which restrict the techniques available to the designer (Walker et al. 1995; Cohen 2002). Lunar bases with a long-term human presence (90 days or more) present a very challenging problem. Indeed, typical concepts propose power levels of 12 to 30 kW, which means that a great amount of heat must be provided during the night and rejected during the day (Cohen 2002).

Passive or active heat providing and removal systems may be required depending on the thermal environment of the surrounding (Christie et al. 2008). A significant part of coping with the thermal cycling involves thermal stability in the form of heat sinks or storage (for instance, in cast basalt blocks) for the long lunar night (Sullivan 1990, pp. 6-7; Cohen 2002). In some cases there is a need to find an efficient method for dissipating waste heat (Heiken et al.1991; WOTM 2010). There are other cases, such as for rovers' night operation or exploration and habitation within shaded areas, when efficient heating systems are necessary. Operation of rovers and/or habitation in this cold environment will require additional power for many heaters, or advanced technologies to withstand the cold, such as ultra-low temperature lubrication and materials (Zakrajsek et al. 2005; TeamFrednet-Wiki 2010).

The present chapter shows the principles of efficient usage of thermal resources for heating on Moon. Two cases will be considered, namely heating of habitation spaces and oven operation, respectively.

19.2 Thermal Environments on Moon

The Moon receives the same flux of solar radiation as the Earth. The extremely rarefied atmosphere of the Moon makes the surface distribution of incident solar radiation to be controlled by the moon's shape, the length of the lunar day and the tilt of the lunar spin axis relative to the ecliptic (Paige 2010). This has consequences on the surface temperature, which depends primarily on the site's latitude and longitude and on the Sun's position on the local sky. For equatorial sites, temperatures can range from about 104 K just before sunrise to about 390 K at noon, with rapid (5 K/hr) temperature changes at sunset and sunrise (Seybold 1995; WOTM 2010). At high latitudes, shadowing and other factors become important (WOTM 2010). Note, however, that because of celestial mechanics effects, the measurements or simulated temperatures associated to a given year cannot be exactly applied to any other year. The cause is a complex combination of the Moon's rotation and sidereal periods orbit around the Sun. It is therefore not possible to derive a precise rule to transpose results of one year to another (Vanoutryve 2010).

Temperature measurements made with thermocouples positioned within a few centimeters above the surface are available for the Apollo missions at 20° and 26° N latitude. They ranged from 102 K to 384 K with an average of 254 K. The monthly range was ±140 K (Christie et al. 2008).

The mean temperature measured 35 cm below the surface of the Apollo sites was 40-45 K warmer than the surface. Temperatures measured at depths greater than about 80 cm showed no appreciable day/night variations (Paige 2010). Temperatures measured at 100 cm depth were about 252 K at Apollo 15 site (26° 5' N latitude) and 255 K at Apollo 17 site (20° 10' N) (WOTM 2010). This makes the lunar subsurface a potential place for habitation (Heiken et al. 1991; WOTM 2010; Paige et al. 2010).

Clementine data (Lawson and Jakosky 1999) suggested that the temperature of the polar regions, where the sun rises very little above the horizon, is less than 200 K. Analytical models of Vasavada et al. (1999) have predicted day time surface temperatures at 85° N latitude of 225 K, while the nighttime predictions were 70 K (Christie et al. 2008).

There are many polar crater floors which are in permanent shadow whose temperature is about 80 K (Seybold 1995). Craters within craters, commonly referred to as double-shaded craters, have areas where even colder regions exist with, in many cases, temperatures that should never exceed 50 K (Ryan et al. 2008). Estimates by Carruba and Coradini indicate that the temperatures of single-shaded craters are in the 83-103 K range (Langseth et al. 1976). They also show that double-shaded craters can have temperatures in the 36-71 K range (Ryan et al. 2008). The NASA's Lunar Reconnaissance Orbiter observed the lowest summer temperatures in the darkest craters at the southern pole to be about 35 K. In the north, close to the winter solstice the recorded temperature was about 26 K on the south-western edge of the floor of Hermite Crater (Amos 2010).

19.3 Efficient Heating of Habitation Spaces on Moon

The efficiency of heat conversion into work was intensely studied for over two hundred years. Despite of the fact that a large part of the world's energy resources are used for heating rather than for power generation, the heating efficiency problem, which refers to how much heat at a given temperature can be converted into heat at another temperature, was considered from time to time only (see e.g. Thomson, 1852; Crawford, 1963; Silver, 1981). Historical considerations may be found in Jaynes (2003).

The heating efficiency problem may be shortly described as follows. Consider a hot reservoir (the source) at temperature T_2, a cold reservoir (the ambient) at temperature T_0 and a system at temperature T_1 ($T_0 < T_1 < T_2$). A heat amount Q_2 is extracted from the source. Q_2 can be directly (i.e. irreversibly) used to heat the system at temperature T_1. In this case, the heat received by the system is $Q_1 = Q_2$ and the heat gain factor, defined by $G \equiv Q_1/Q_2$, equals unity. However, a reversible procedure to use Q_2 for the transfer of a heat amount Q_1 to the system at temperature T_1 may be imagined. A treatment based directly on the first and second laws gives in this case (Jaynes 2003):

$$G \equiv \frac{Q_1}{Q_2} = \frac{T_1}{T_2} \cdot \frac{T_2 - T_0}{T_1 - T_0} \tag{19.1}$$

The heating gain factor G in Eq. (19.1) always exceeds unity. In fact, the values of G may be rather large and Jaynes (2003) concluded that (in principle at least) it is possible to heat buildings with an order of magnitude less fuel than one consumes today. This conclusion makes the heating efficiency problem important from a practical point of view.

19.3.1 Two-Temperature Heaters and Jaynes Heaters

A two-temperature heater is a device that transfers heat from a reservoir (body) of temperature T_2 to a body of temperature T_1 (Fig. 19.1a). In the case when $T_2 > T_1$ the device is called a down-heater while when $T_2 < T_1$ one may call it an up-heater.

A Jaynes heater is a three-temperature device that uses a heat engine and a heat pump in contact with the ambient temperature T_0. Figure 19.1 shows the difference between a two-temperature down-heater (Fig. 19.1a) and a Jaynes heater (Fig. 19.1b). Usually, the heat amount Q_2 is allowed to degrade itself directly to the temperature T_1 (Fig. 19.1a). The first law states that the heat amount received by the heated system at temperature T_1 is $Q_1 = Q_2$. The heat transfer is associated to a positive entropy generation $\Delta S = Q_{12}(1/T_1 - 1/T_2) > 0$, which finally means irreversibility and loss (of available work). The Jaynes heater may decrease that entropy generation as described below.

We refer now to the operation of the Jaynes heater shown in Fig. 19.1b (Muschik and Badescu 2008). The heat engine receives the heat flux Q_2 from the hot reservoir and transfers the heat flux Q_1 to the system at temperature T_1 while delivering the work rate W to the heat pump. The heat pump uses the work rate W to extract the heat flux Q_0 from the ambient and provides the heat flux Q'_1 to the heated system.

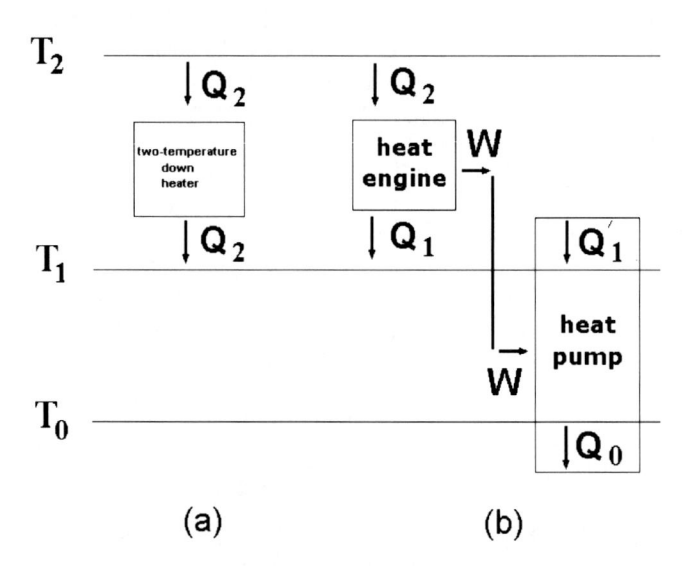

(a) (b)

Fig. 19.1 (a) Two-temperature down heater and (b) Jaynes heater

The first law for the heat engine and the heat pump, respectively, in Fig. 19.1b states:

$$Q_2 + Q_1 + W = 0$$
$$Q'_1 + Q_0 - W = 0 \tag{19.2,3}$$

The following conventions apply: $Q_2 > 0$, $Q_1 < 0$ and $Q'_1 < 0$, $Q_0 > 0$. Also, $W > 0$. Summing up the Eqs. (19.2) and (19.3) yields:

$$Q_1 + Q'_1 + Q_2 + Q_0 = 0 \tag{19.4}$$

The heat supply to the system at temperature T_1 is given by:

$$-Q \equiv Q'_1 + Q_1 \tag{19.5}$$

By convention, $Q > 0$. Use of Eqs. (19.4) and (19.5) yields:

$$Q = Q_2 + Q_0 \tag{19.6}$$

The second law for the heat engine and the heat pump, respectively, states (see Fig. 19.1b):

$$\frac{Q_2}{T_2} + \frac{Q_1}{T_1} \leq 0, \quad \frac{Q'_1}{T_1} + \frac{Q_0}{T_0} \leq 0. \tag{19.7,8}$$

Equations (19.7) and (19.8) may be re-written as:

$$\frac{Q_2}{-Q_1} \leq \frac{T_2}{T_1}, \quad \frac{Q_0}{-Q'_1} \leq \frac{T_0}{T_1}. \tag{19.9,10}$$

We define now two factors of efficiency, α and β, respectively:

$$\alpha \equiv \frac{Q_2}{-Q_1}, \quad \beta \equiv \frac{Q_0}{-Q'_1}. \tag{19.11,12}$$

The first factor of efficiency (α) characterizes heat engine operation and may be easily related to the *reversible* heat engine efficiency η ($\eta = 1 - 1/\alpha$). Also, the second factor of efficiency is a simple function of the coefficient of performance, *COP*, which is a more common performance indicator associated to *reversible* heat pump operation (*COP* = $1/(1-\beta)$). Taking account of Eqs. (19.9-12) one founds the following restrictions for the two efficiency factors:

$$1 \leq \alpha \leq \frac{T_2}{T_1}, \quad 0 \leq \beta \leq \frac{T_0}{T_1} < 1, \quad \alpha > \beta. \tag{19.13-15}$$

Use of Eqs. (19.5),(19.6), (19.11) and (19.12) yields:

$$Q = Q_2 \frac{\alpha - \beta}{\alpha(1 - \beta)} \tag{19.16}$$

Equation (19.16) allows to expressing the heating gain factor as follows:

$$G \equiv \frac{Q}{Q_2} = \frac{\alpha - \beta}{\alpha(1 - \beta)} = 1 + \eta(COP - 1) \geq 1 \qquad (19.17)$$

Various particular situations may be envisaged and two extreme cases are considered now. First, the minimum value of the heat delivered to the body at temperature T_1 is:

$$Q_{min} = Q_2 \qquad (19.18)$$

Inspection of Eq. (19.17) shows that this happens for (i) arbitrary α (i.e. arbitrary η) and $\beta=0$ (i.e. $COP=1$) and (ii) arbitrary β (i.e. arbitrary COP) and $\alpha=1$ (i.e. $\eta=0$). Second, the maximum value of the heat delivered to the body at temperature T_1 is obtained in the reversible limit, for

$$\alpha_{rev} = \frac{T_2}{T_1}, \quad \beta_{rev} = \frac{T_0}{T_1} \qquad (19.19,20)$$

when the two devices in Fig. 19.1b turn into a reversible heat engine and a reversible heat pump, respectively. Then, the heat delivered to the body at temperature T_1 is:

$$Q_{rev} = Q_2 \frac{1 - T_0 / T_2}{1 - T_0 / T_1} > Q_2 \qquad (19.21)$$

and the heating gain factor Eq.(19.17) reduces to the known maximum value Eq. (19.1).

Cases ranging between these two extremes differs each other by the entropy production, which is given by:

$$\Sigma \equiv \frac{-Q_2}{T_1} - \frac{Q_1}{T_1} - \frac{Q'_1}{T_1} - \frac{Q_0}{T_0} \qquad (19.22)$$

One replaces in Eq. (19.22) Q_1, Q'_1 and Q_0 in terms of Q_2 and after some algebra one finds:

$$\Sigma = Q_2 \left[-\frac{1}{T_2} + \frac{\beta}{\alpha T_0} + \frac{\alpha - \beta}{\alpha(1 - \beta)} \left(\frac{1}{T_1} - \frac{\beta}{T_0} \right) \right] \qquad (19.23)$$

The first extreme case described above is associated to maximum irreversibility, i.e.

$$\Sigma_{max} = \frac{1}{T_1} - \frac{1}{T_2} \qquad (19.24)$$

while the second extreme case corresponds to zero entropy generation, i.e.

$$\Sigma_{min} = 0 \qquad (19.25)$$

as expected.

19.3.1.1 Reversible Heating Gain Factor on Moon

Figure 19.2 shows the dependence of the heating gain factor on the common indicators of performance of reversible heat engine and reversible heat pump, respectively. More insight is obtained by looking to some particular cases. We adopt T_1=295 K for the temperature inside the building.

The temperature T_2 of the hot reservoir depends on the technology of providing heat. Heat obtained by high, medium and low solar concentration may be associated to a T_2 value of 3000 K, 700 K and 400 K, respectively. If recovering of some waste heat is considered, the temperature T_2=340 K may be adopted. These values of the temperature T_2 yield different values of the reversible heat engine efficiency η. For instance, η increases from 0.13 in case of using waste heat to 0.26, 0.57 and 0.90 when heat provided by low, medium and high solar concentration, respectively, is considered (Fig. 19.2).

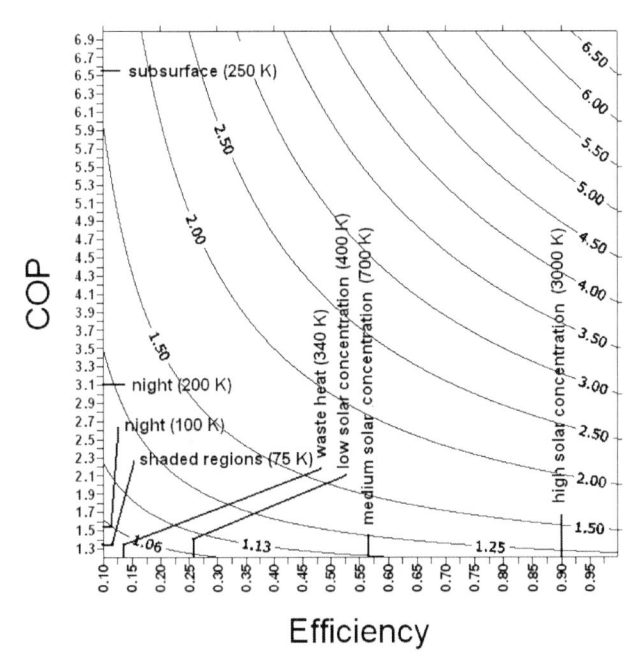

Fig. 19.2 Lines of equal reversible heating gain factor G predicted by Eq. (19.17) as function of the reversible heat engine efficiency η and the coefficient of performance COP of the reversible heat pump. Cases associated with particular values of the hot reservoir temperature T_2 (on the abscissa) and cold reservoir temperature T_0 (on the ordinate axis), are also shown. The temperature inside the building is T_1=295 K.

Different values may be adopted for the environment temperature T_0. For example, for Moon, in case of the shaded or double-shaded surface areas in the polar regions a rough value T_0= 75 K is reasonable. For surface regions at equator or at higher latitudes the ambient temperature T_0 ranges between 100 and 200 K. Finally, an ambient temperature $T_0 = 250$ K may be considered in case of sub-surface habitation. These values of the temperature T_0 yield different values of the reversible heat pump coefficient of performance COP. For instance, COP is valued 1.34 in case of surface shaded areas in the polar regions, ranges between 1.51 and 3.10 when surface heat pump operation during the night is considered and has a value about 6.5 in case of sub-surface habitation (Fig. 19.2).

The sub-surface habitation is characterized by the largest values of the heating gain factor G, which ranges from about 1.75 in case of using waste heat to about 6 in case of using highly concentrated solar radiation. Surface heating systems operating in shaded areas are characterized by small values of the gain factor G, which is less than 1.5, independent of the value of the hot reservoir temperature T_2. Surface heating systems operating during the night are associated to heating gain values ranging between 1.5 and 3, depending on the ambient temperature T_0 and the temperature T_2 of the hot reservoir (Fig. 19.2).

Present-day technology does not allow, however, reaching gain factor values close to those predicted by Eq. (19.1) under the reversible treatment (Silver 1981; Jaynes 2003). The next section shows a more involved treatment.

19.3.2 Endoreversible Model of Heating

A more realistic but still simple approach for the heating efficiency problem is presented now (Badescu 2006). This allows obtaining more accurate upper bounds for the heating gain factor and gives a better perspective for practical applications.

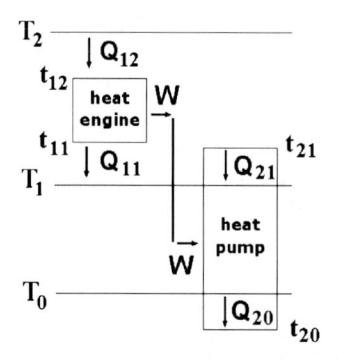

Fig. 19.3 The endoreversible heater. T_2, T_0, T_1 - temperatures of the source, ambient and heated system, respectively; t_{12}, t_{11}, t_{21} and t_{20} - working fluid temperatures; Q_{12}, Q_{11}, Q_{21} and Q_{20} - heat fluxes; W- work rate.

Table 19.1 Equations used to describe the operation of the endoreversible heater of Fig. 19.3. Definitions for the heat fluxes Q_{12}, Q_{11}, Q_{21} and Q_{20} and the entropy fluxes S_1 and S_2 are also given

Endoreversible heat engine		
1	First law	$W = Q_{12} - Q_{11} \equiv k_{12}(T_2 - t_{12}) - k_{11}(t_{11} - T_1)$
2	Reversible internal operation	$S_1 \equiv \dfrac{Q_{12}}{t_{12}} = \dfrac{Q_{11}}{t_{11}} \equiv \dfrac{k_{12}(T_2 - t_{12})}{t_{12}} \equiv \dfrac{k_{11}(t_{11} - T_1)}{t_{11}}$

Endoreversible heat pump		
3	First law	$W = Q_{21} - Q_{20} \equiv k_{21}(t_{21} - T_1) - k_{20}(T_0 - t_{20})$
4	Reversible internal operation	$S_2 \equiv \dfrac{Q_{21}}{t_{21}} = \dfrac{Q_{20}}{t_{20}} \equiv \dfrac{k_{21}(t_{21} - T_1)}{t_{21}} = \dfrac{k_{20}(T_0 - t_{11})}{t_{20}}$

The heating process is treated here by using standard finite-time thermodynamics (for a good introduction to finite-time and endoreversible thermodynamics see Bejan (1996), Hoffmann et al. (1997), Wu et al. (1999)). Figure 19.3 shows an "endoreversible heater". It consists of an endoreversible heat engine coupled to an endoreversible heat pump. Table 19.1 gives relationships associated to endoreversible heater operation. The heat engine receives the heat flux Q_{12} from the hot reservoir and delivers the heat flux Q_{11} to the system at temperature T_1 while delivering the work rate W to the heat pump. The upper and lower fluid temperature in the heat engine are denoted t_{12} and t_{11}, respectively. The heat pump uses the work rate W to extract the heat flux Q_{20} from the ambient and provides the heat flux Q_{21} to the heated system. The upper and lower working fluid temperatures in the heat pump are denoted t_{21} and t_{20}, respectively. Each heat flux Q_{ij} $(i,j=0,1,2)$ depends linearly on a temperature difference, with the appropriate conductance k_{ij} $(i,j=0,1,2)$ as linearity coefficient (Table 19.1). The heat flux received by the heated system, Q_1, and the heating gain factor G, are defined, respectively, by:

$$Q_1 \equiv Q_{11} + Q_{21}, \qquad G \equiv \frac{Q_1}{Q_{12}} \qquad (19.26,27)$$

The aim is to optimize the operation of a given endoreversible heater (i.e. a heater with known conductances). The objective function to be maximized is the heat flux Q_1 delivered to the system at temperature T_1. Two design factors and their ratio are first defined:

$$g_1 \equiv \frac{k_{12}k_{11}}{k_{12}+k_{11}} \qquad g_2 \equiv \frac{k_{21}k_{20}}{k_{21}+k_{20}} \qquad g \equiv \frac{g_2}{g_1} \qquad (19.28\text{-}30)$$

Note that g is a symmetric function of (k_{12}, k_{11}) and of (k_{21}, k_{20}) (i.e. $g(k_{11}, k_{12})= g(k_{12}, k_{11})$ and $g(k_{21}, k_{20})= g(k_{20}, k_{21})$).

19.3.2.1 Optimization Procedure

The optimization procedure is detailed now. The work rate W may be eliminated between Eqs. 1 and 3 in Table 19.1. This yields a relationship between the heat fluxes Q_{12}, Q_{11}, Q_{21} and Q_{20}. Equations 2 and 4 in Table 19.1 allow to writing the fluid temperatures t_{12}, t_{11}, t_{21} and t_{20} in terms of the entropy fluxes S_1 and S_2. Subsequently, they are replaced in the relation just obtained connecting the four heat fluxes. One finds:

$$F \equiv \frac{k_{12}T_2S_1}{k_{12}+S_1} - \frac{k_{11}T_1S_1}{k_{11}-S_1} - \frac{k_{21}T_1S_2}{k_{21}-S_2} + \frac{k_{20}T_0S_2}{k_{20}+S_2} = 0 \qquad (19.31)$$

Also, the objective function Q_1 may be expressed in terms of S_1 and S_2 as follows:

$$Q_1 = \frac{k_{11}T_1S_1}{k_{11}-S_1} + \frac{k_{21}T_1S_2}{k_{21}-S_2} \qquad (19.32)$$

Maximization of Q_1 should take into account the constraint Eq. (19.31). The usual Lagrange method is used. It involves the Lagrange function L defined as:

$$L \equiv Q_1 + \lambda F \qquad (19.33)$$

where λ is a multiplier. Finding the maximum of Q_1 in respect to the independent variables S_1 and S_2 requires solving the system:

$$\frac{\partial L}{\partial S_1} = 0 \qquad \frac{\partial L}{\partial S_2} = 0 \qquad \frac{\partial L}{\partial \lambda} = 0 \qquad (19.34\text{a,b,c})$$

Equations (19.34a) and (19.34b) allow to writing λ as follows:

$$\lambda = \left[1 - \frac{T_2}{T_1}\left(\frac{k_{12}}{k_{11}}\right)^2\left(\frac{k_{11}-S_1}{k_{12}+S_1}\right)^2\right]^{-1} = \left[1 - \frac{T_0}{T_1}\left(\frac{k_{20}}{k_{21}}\right)^2\left(\frac{k_{21}-S_2}{k_{20}+S_2}\right)^2\right]^{-1} \qquad (19.35,36)$$

From Eqs. (19.35) and (19.36) one derives:

$$\left(\frac{k_{12}}{k_{11}}\right)^2\left(\frac{k_{11}-S_1}{k_{12}+S_1}\right)^2 = \frac{T_0}{T_2}\left(\frac{k_{20}}{k_{21}}\right)^2\left(\frac{k_{21}-S_2}{k_{20}+S_2}\right)^2 \equiv C^2 \qquad (19.37,38)$$

The l.h.s and the r.h.s of Eq. (19.37) are functions of different independent variables. Therefore C in Eq. (19.38) must be a constant. Equations (19.37) and (19.38) allow to writing S_1 and S_2 in term of C. However, four sets of solutions (S_1,S_2) are obtained. These solutions are replaced in Eq. (19.31) and a quadratic equation in the unknown C is finally obtained. A rather lengthy analysis is necessary to decide which of the four solutions (S_1,S_2) and which of the two solutions of the associated quadratic equation for C have to be selected. Only those solutions yielding positive heat and entropy fluxes are meaningful. We omit this lengthy analysis and give the results. Two sets of solutions $(S_{1,opt}, S_{2,opt})$ with only one associated solution for C are obtained. They are:

$$S_{1,opt} = \frac{k_{11}k_{12}(1-C)}{k_{11}A + k_{12}} \qquad S_{2,opt} = \frac{k_{20}k_{21}\left(\sqrt{T_0/T_2} - C\right)}{k_{21}A + \sqrt{T_0/T_2}\,k_{12}} \qquad (19.39,40)$$

and

$$S'_{1,opt} = \frac{k_{11}k_{12}(1+C)}{-k_{11}A + k_{12}} \qquad S'_{2,opt} = \frac{k_{20}k_{21}\left(\sqrt{T_0/T_2} + C\right)}{-k_{21}A + \sqrt{T_0/T_2}\,k_{12}} \qquad (19.41,42)$$

Use of Eqs. (19.39),(19.40) and (19.31) yields the following equation in the unknown C:

$$C^2 - \left(\frac{1 + T_1/T_2 + g(T_1/T_2 + T_0/T_2)}{1 + g\sqrt{T_0/T}}\right)C + \frac{T_1}{T_2} = 0 \qquad (19.43)$$

where g is defined in Eq. (19.30). The solution for C in Eq. (19.43) with the sign minus in front of the square root makes sense. The condition $-\infty < C \le (T_0/T_2)^{1/2}$ ensures all heat fluxes are positive in the endoreversible heater. Also, use of Eqs. (19.41), (19.42) and (19.31) yields

$$C^2 + \left(\frac{1 + T_1/T_2 + g(T_1/T_2 + T_0/T_2)}{1 + g\sqrt{T_0/T}}\right)C + \frac{T_1}{T_2} = 0 \qquad (19.44)$$

In this case, the solution for C in Eq. (19.44) with the sign plus in front of the square root makes sense. The condition $-(T_0/T_2)^{1/2} < C < 0$ ensures all heat fluxes are positive.

Use of Eqs. (19.39), (19.40) and (19.43) give the same numerical values for the optimum entropy fluxes and for the optimum temperatures $t_{12,opt}$, $t_{11,opt}$, $t_{21,opt}$ and $t_{20,opt}$ as Eqs. (19.41), (19.42) and (19.44) give. Therefore, only results related to the solution $(S_{1,opt}, S_{2,opt})$ are used here.

One sees that these results depend on the single dimensionless parameter C (defined by the quadratic Eq. (19.44)), incorporating all design and thermal factors. The optimum operation temperatures and heat fluxes are shown in Table 19.2.

Table 19.2 Fluid temperatures and heat fluxes associated to optimum endoreversible heater operation

Optimum temperature		Optimum heat flux	
1	$t_{12,opt} = T_2 \dfrac{Ck_{11} + k_{12}}{k_{11} + k_{12}}$	5	$Q_{12,opt} = T_2 g_1 (1 - C)$
2	$t_{11,opt} = \dfrac{T_1}{C} \dfrac{Ck_{11} + k_{12}}{k_{11} + k_{12}}$	6	$Q_{11,opt} = T_1 g_1 \dfrac{1 - C}{C}$
3	$t_{21,opt} = \dfrac{T_1}{C} \dfrac{Ck_{21} + \sqrt{T_0/T_2}\, k_{20}}{k_{21} + k_{20}}$	7	$Q_{21,opt} = T_1 g_2 \dfrac{\sqrt{T_0/T_2} - C}{C}$
4	$t_{20,opt} = \sqrt{T_0 T_2} \dfrac{Ck_{21} + \sqrt{T_0/T_2}\, k_{20}}{k_{21} + k_{20}}$	8	$Q_{20,opt} = \sqrt{T_0 T_2}\, g_2 \left(\sqrt{T_0/T_2} - C \right)$

The optimum gain factor of the endoreversible heater is given by:

$$G_{opt} = \frac{T_1 / T_2}{C} \left(1 + g \frac{\sqrt{T_0/T_2} - C}{1 - C} \right) \tag{19.45}$$

Another useful form of the optimum gain factor is derived next. First, Eq. (19.44) allows to writing g in terms of C:

$$g = \frac{(C - T_1/T_2)(1 - C)}{\left(C - \sqrt{T_0/T_2} \right)\left(C\sqrt{T_0/T_2} - T_1/T_2 \right)} \tag{19.46}$$

The factor g must be positive (see Eqs. (19.28-30)). A simple analysis shows that the condition $g > 0$ requires $T_1/T_2 < (T_0/T_2)^{1/2}$ (or, in other words, $T_1 < (T_0 T_2)^{1/2}$) and $T_1/T_2 \leq C \leq (T_0/T_2)^{1/2}$. This restriction for C is stronger than, but compatible with, $-\infty < C \leq (T_0/T_2)^{1/2}$ derived above. Use of Eqs. (19.45) and (19.46) yields:

$$G_{opt} = \frac{T_1/T_2 \left(\sqrt{T_0/T_2} - 1 \right)}{C\sqrt{T_0/T_2} - T_1/T_2} \tag{19.47}$$

Equation (19.47) shows that the optimum gain factor depends in a simple manner on C, which in turn depends on the design factor g (see Eq. (19.44)). Taking account Eq. (19.44) and the symmetry properties of g one concludes that $G_{opt}(k_{11}, k_{12}) = G_{opt}(k_{12}, k_{11})$ and $G_{opt}(k_{21}, k_{20}) = G_{opt}(k_{20}, k_{21})$. In other words, G_{opt} is a symmetric function of (k_{12}, k_{11}) and of (k_{21}, k_{20}).

19.3.2.2 Upper Bound on Heating Gain Factor

The maximum heat flux transferred to the system at temperature T_1 (i.e. $Q_{1,max} \equiv Q_{11,opt} + Q_{21,opt}$) depends on the design factors g_1 and g_2 and finally on the conductances

k_{11}, k_{12}, k_{21} and k_{20} (see Eqs. 7 and 8 in Table 19.2 and Eqs. (19.28) and (19.29)). Therefore, $Q_{1,max}$ may be maximized as function of these conductances.

The best heating performance occurs when no constraint exist on conductances. In the limit of infinitely large conductances the operation fluid temperatures tend towards the heat reservoirs and system temperatures (i.e. $t_{12} \to T_2$, $t_{11} \to T_1$, $t_{21} \to T_1$ and $t_{20} \to T_0$) to keep finite the heat fluxes in Table 19.1. As a consequence, the endoreversible heater becomes a reversible heating system and the gain factor riches its maximum value given by Eq. (19.1).

However, a realistic model requires finite conductances. One denotes by k_{tot} the (finite) sum:

$$k_{11} + k_{12} + k_{21} + k_{20} \equiv k_{tot} \tag{19.48}$$

The optimization procedure with Eq. (19.48) as a constraint is presented now. The maximum heat flux $Q_{1,max}$ delivered to the heated system depends implicitly on the four conductances (through the factors g_1 and g_2) according to the relationship:

$$Q_{1,max} = Q_{11,opt} + Q_{21,opt} = T_1 g_1 \frac{1-C}{C} + T_1 g_2 \frac{\sqrt{T_0/T_2} - C}{C} \tag{19.49}$$

Here Eqs. 6 and 7 in Table 19.2 were used. Maximization of $Q_{1,max}$ upon k_{11}, k_{12}, k_{21} and k_{20} should also take account on the constraint Eq. (19.46):

$$g = \frac{g_1}{g_2} = \frac{(C - T_1/T_2)(1-C)}{(C - \sqrt{T_0/T_2})(C\sqrt{T_0/T_2} - T_1/T_2)} \tag{19.50}$$

Equations (19.28) and (19.29) show that g_1 is symmetric in k_{11} and k_{12} and g_2 is symmetric in k_{21} and k_{20}. Therefore, the maximum of $Q_{1,max}$ (say $Q_{1,max,max}$) will be reached for the optimum conductances:

$$k_{11,opt} = k_{21,opt} \qquad k_{21,opt} = k_{20,opt} \tag{19.51,52}$$

Use of Eqs. (19.28),(19.29),(19.51),(19.52) and of constraint Eq. (19.48) allows to write g_1 and g_2 as functions of a single optimum conductance (say $k_{12,opt}$):

$$g_1 = \frac{k_{tot}}{2} \qquad g_2 = \frac{1}{2}\left(\frac{k_{tot}}{2} - k_{21,opt}\right) \tag{19.53,54}$$

Equations (19.53) and (19.54) are replaced in Eqs. (19.49) and (19.50) which become, respectively:

$$Q_{1,max} = T_1 \frac{k_{21,opt}}{2} \frac{1-C}{C} + T_1 \frac{1}{2}\left(\frac{k_{tot}}{2} - k_{21,opt}\right) \frac{\sqrt{T_0/T_2} - C}{C} \tag{19.55,56}$$

$$\frac{k_{21,opt}}{\frac{k_{tot}}{2} - k_{21,opt}} = \frac{(C - T_1/T_2)(1-C)}{(C - \sqrt{T_0/T_2})(C\sqrt{T_0/T_2} - T_1/T_2)}$$

Equation (19.56) is solved for the unknown $k_{12,opt}$, which is subsequently replaced in Eq. (19.55). The result is:

$$Q_{1,\max} = T_1 \frac{k_{tot}}{4} \left(\frac{T_1/T_2 - \sqrt{T_0/T_2}}{C^2 - \left(\sqrt{T_0/T_2} + 1\right)C + T_1/T} - 1 \right) \tag{19.57}$$

Finding the maximum of $Q_{1,\max}$ in respect to the sole variable C requires computing the derivative:

$$\frac{dQ_{1,\max}}{dC} = T_1 \frac{k_{tot}}{4} \frac{\left(\sqrt{T_0/T_2} - T_1/T_2\right)\left[2C - \left(\sqrt{T_0/T_2} + 1\right)\right]}{\left[C^2 - \left(\sqrt{T_0/T_2} + 1\right)C + T_1/T\right]^2} \tag{19.58}$$

Equation (19.58) shows that the derivative $dQ_{1,\max}/dC$ is positive on the definition interval $T_1/T_2 \le C \le (T_0/T_2)^{1/2}$ and vanishes at $C=(1+ (T_0/T_2)^{1/2})/2$, which is outside the definition interval. Therefore, on the definition interval, $Q_{1,\max}$ becomes a maximum for:

$$C_{opt} = \sqrt{T_0/T_2} \tag{19.59}$$

Use of Eqs. (19.56) and (19.58) yields:

$$k_{21,opt} = k_{tot}/2 \tag{19.60}$$

Finally, the other optimum conductances are found by using Eqs. (19.48), (19.51) and (19.52):

$$k_{20,opt} = k_{tot}/2 \qquad k_{11,opt} = k_{12,opt} \to 0 \tag{19.61,62}$$

The resulting optimum conductances are given by Eqs. (19.61) and (19.62):

$$k_{12,opt} = k_{11,opt} \to 0 \qquad k_{21,opt} = k_{20,opt} \to k_{tot}/2 \tag{19.63,64}$$

Equations (19.63) and (19.64) show that the equipartition principle acts for each of the two main components of the endoreversible heater. This is consistent with previous knowledge on endoreversible heat engines and heat pumps (see e.g. Bejan and Tondeur 1998). The equipartition principle does not act at the level of the whole endoreversible heater (in the sense that the four optimized conductances are not equal each other).

Use of Eqs. (19.28),(19.29),(19.63) and (19.64) yield:

$$g_{1,opt} \to 0 \qquad g_{2,opt} \to k_{tot}/4 \qquad g \to \infty \tag{19.65-67}$$

However, the optimum value of the parameter C is finite (see Eq. (19.59)):

$$C_{opt} = \sqrt{T_0 / T_2} \qquad (19.68)$$

Use of Eqs. (19.47) and (19.68) yield the *optimum optimorum* heating gain factor:

$$G_{opt,opt} = \frac{T_1 / T_2 \left(\sqrt{T_0 / T_2} - 1 \right)}{T_0 / T_2 - T_1 / T_2} \qquad (19.69)$$

The factor $(T_0/T_2)^{1/2}$ in Eq. (19.69) is a reminiscent of the similar factor entering the well-known efficiency formula of endoreversible heat engine operating at maximum power (see e.g. Curzon and Ahlborn 1975).

19.3.2.3 Endoreversible Heating Gain Factor on the Moon

Over-unitary values of the heating gain factor $G_{opt,opt}$ defined by Eq. (19.69) are obtained under the condition that $T_1 < (T_0/T_2)^{1/2}$. This means that, for given values of the indoor temperature T_1 and ambient temperature T_0, the hot reservoir must have a temperature $T_2 > T_{2,min}$, where $T_{2,min} = T_1^2/T_0$.

In case of Moon applications, the indoor temperature values around $T_1 = 295$ K while the ambient temperature T_0 ranges between about 75 K (for shaded or double-shaded areas in the surface polar regions) and 250 K (in case of sub-surface habitation). Consequently, the minimum hot reservoir temperature $T_{2,min}$ ranges between less that 400 K for sub-surface habitation to more than 1400 K for surface habitation spaces in shaded polar areas (Fig. 19.4).

The optimum optimorum endoreversible heating gain factor $G_{opt,opt}$ given by Eq. (19.69) is of course smaller than the reversible heating gain factor G given by Eq. (19.1). Figures 19.5a and 19.5b show both quantities.

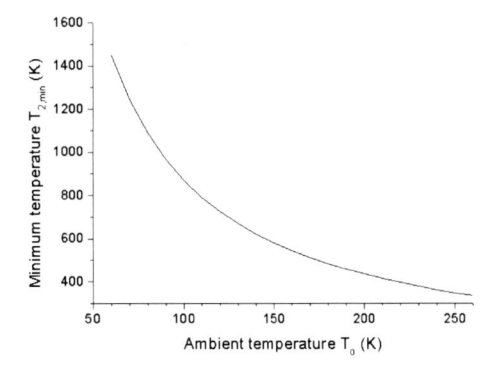

Fig. 19.4 The minimum temperature $T_{2,min}$ associated to $G_{opt,opt}=1$, as function of the ambient temperature T_0. The indoor temperature is $T_1 = 295$ K.

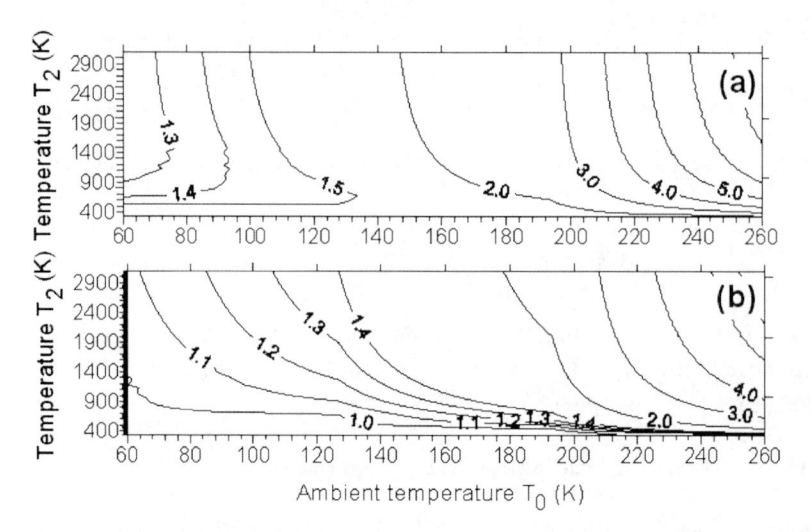

Fig. 19.5 Isolines of heating gain factor as function of the source temperature T_2 and ambient temperature T_0. (a) reversible heating factor G (Eq. (19.1)); (b) optimum optimorum endoreversible heating factor $G_{opt,opt}$ (Eq. (19.69)). The indoor temperature is $T_1 = 295$ K.

The isoline of $G_{opt,opt} = 1$ in Fig. 19.5b corresponds to the curve of minimum temperature $T_{2,min}$ shown in Fig. 19.4. This isoline separates the diagram surface into two areas. For couples of values (T_0, T_2) below that isoline, common two-temperature down heaters (fully irreversible heaters) should be used. In case of pairs of values (T_0, T_2) above that isoline, it makes sense to consider the utilization of Jaynes heaters since they may provide a more efficient heat utilization. Better perspectives for Jaynes heaters utilization characterizes those cases when $G_{opt,opt}$ exceeds 2 or 3. Then, the increased cost of replacing usual heaters with more complex Jaynes heaters may be compensated by increasing the efficiency of using the available thermal resources. From a practical point of view, $G_{opt,opt}$ larger than 2 is associated to Moon sub-surface habitation, where the ambient temperature is about 250 K (Fig. 19.5b).

In practice, any conductance is a product between a heat transfer surface area and a heat transfer coefficient. Therefore, the technology puts lower bounds on heat engine conductance and the limit in Eq. (19.63) should stop to a finite non-null value. The performance is significantly decreased as compared to the case of the optimum optimorum endoreversible heater (Badescu 2006). However, reduction of thermal energy consumption to a half of present day consumption is still possible (Badescu 2006).

19.4 Efficient Oven Operation on Moon

Solving the heating efficiency problem takes account that the second law allows for heat to spontaneously flow from room temperature (say T_1) to a higher temperature T_2, if there is at the same time a compensating heat flow to a system of lower temperature T_0 (the environment). This has important practical applications, among which is the "free oven for Eskimos" proposed in an interesting Note published posthumously by E. T. Jaynes (2003). There, T_2 is the temperature of whatever is heated in the oven. The approach in Jaynes (2003) is based on a traditional reversible thermodynamics treatment. In this section we use a more realistic but still simple finite-time thermodynamics approach. This allows obtaining more realistic upper bounds for the oven heat gain factor and gives a better perspective for practical applications on Moon.

19.4.1 Model of Endoreversible Oven

One or more reversible heat engine - heat pump combinations are necessary to completely avoid the dissipation which is associated to the common heat transfer methods (Jaynes 2003). At first sight this makes the resulting cycles rather complicated for those familiar with the traditional techniques, which are developed on the main idea that the only realistic component which could improve the ovens efficiency is increased thermal insulation. Though the robustness of traditional solutions is out of question, a different approach should be adopted when searching for *upper bounds* on real performances, as we shall do here (Badescu 2009). To this aim, a number of idealizing assumptions will be adopted among which the most important are: (i) steady-state operation and (ii) heat leakage is neglected. These assumptions may be relaxed when a more detailed analysis is performed, as was already done in works involving combinations of heat engines and heat pumps for cooling or power generation purposes (see e.g. Chen et al. 1997a,b; Chen 1999).

Figure 19.6 shows an "endoreversible oven". It consists of two endoreversible heat engines and two endoreversible heat pumps. Table 19.3 gives relationships associated to endoreversible oven operation. The heat engine 1 receives the heat

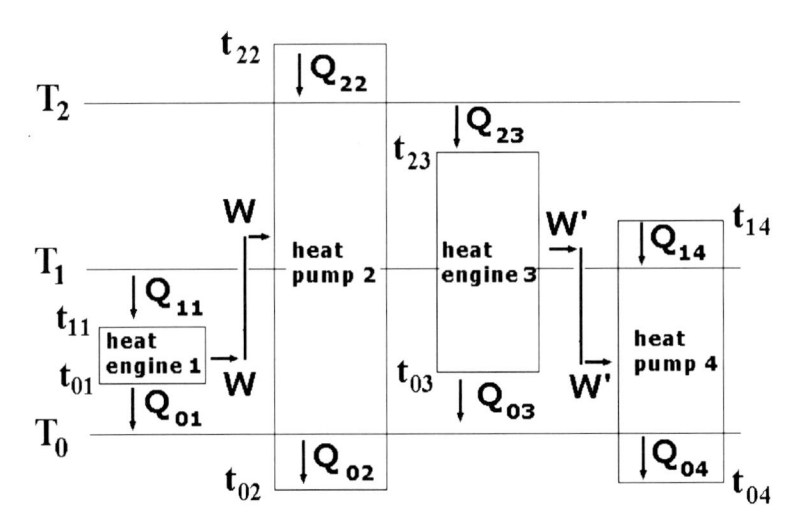

Fig. 19.6 The endoreversible oven. T_2, T_0, T_1 - temperatures of the body inside the oven, environment and room air, respectively; t_{ij} (i,j=0,1,2,3,4) - working fluid temperatures; Q_{ij} (i,j=0,1,2,3,4) - heat fluxes; W, W' - work rates

flux Q_{11} from the room at temperature T_1 and delivers the heat flux Q_{01} to the environment at temperature T_0 and the work rate W to the heat pump 2. The upper and lower fluid temperature in the heat engine 1 are denoted t_{11} and t_{01}, respectively. The heat pump 2 uses the work rate W to extract the heat flux Q_{02} from the environment and provides the heat flux Q_{22} to the food at temperature T_2 in the oven. The upper and lower fluid temperature in the heat pump 2 are denoted t_{22} and t_{02}, respectively. The heat engine 3 extracts the heat flux Q_{23} from the food and transfers the heat flux Q_{03} to the environment while providing the work rate W' to the heat pump 4. The upper and lower fluid temperature in the heat engine 3 are denoted t_{23} and t_{03}, respectively. Also, the upper and lower fluid temperature in the heat pump 4 are denoted t_{14} and t_{04}, respectively. The work rate W' is used in the heat pump 4 to extract the heat flux Q_{04} from the environment and transfer the heat flux Q_{14} to the room. Each heat flux Q_{ij} (i,j=0,1,2,3,4) depends linearly on a temperature difference, with the appropriate conductance U_{ij} (i,j=0,1,2,3,4) as linearity coefficient (Table 19.3).

Table 19.3 Equations used to describe the operation of the endoreversible oven of Fig. 19.6. Definitions for the heat fluxes Q_{11}, Q_{01}, Q_{22}, Q_{02}, Q_{23}, Q_{03}, Q_{14} and Q_{04} and the entropy fluxes S_1, S_2, S_3 and S_4 are also given.

Endoreversible heat engine 1		
1	First law	$W = Q_{11} - Q_{01} \equiv U_{11}(T_1 - t_{11}) - U_{01}(t_{01} - T_0)$
2	Reversible internal operation	$S_1 \equiv \dfrac{Q_{11}}{t_{11}} = \dfrac{Q_{01}}{t_{01}} \equiv \dfrac{U_{11}(T_1 - t_{11})}{t_{11}} \equiv \dfrac{U_{01}(t_{01} - T_0)}{t_{01}}$
Endoreversible heat pump 2		
3	First law	$W = Q_{22} - Q_{02} \equiv U_{22}(t_{22} - T_2) - U_{02}(T_0 - t_{02})$
4	Reversible internal operation	$S_2 \equiv \dfrac{Q_{22}}{t_{22}} = \dfrac{Q_{02}}{t_{02}} \equiv \dfrac{U_{22}(t_{22} - T_2)}{t_{22}} = \dfrac{U_{02}(T_0 - t_{02})}{t_{02}}$
Endoreversible heat engine 3		
5	First law	$W' = Q_{23} - Q_{03} \equiv U_{23}(T_2 - t_{23}) - U_{03}(t_{03} - T_0)$
6	Reversible internal operation	$S_3 \equiv \dfrac{Q_{23}}{t_{23}} = \dfrac{Q_{03}}{t_{03}} \equiv \dfrac{U_{23}(T_2 - t_{23})}{t_{23}} \equiv \dfrac{U_{03}(t_{03} - T_0)}{t_{03}}$
Endoreversible heat pump 4		
7	First law	$W' = Q_{14} - Q_{04} \equiv U_{14}(t_{14} - T_1) - U_{04}(T_0 - t_{04})$
8	Reversible internal operation	$S_4 \equiv \dfrac{Q_{14}}{t_{14}} = \dfrac{Q_{04}}{t_{04}} \equiv \dfrac{U_{14}(t_{14} - T_1)}{t_{14}} \equiv \dfrac{U_{04}(T_0 - t_{04})}{t_{04}}$

The following question may arise quite naturally: Why would one ever construct one heat engine-pump combination to use room temperature heat (T_1) to generate oven temperature heat (T_2) while at the same time has another engine-pump combination uses T_2 to replenish T_1? The answer is that without the second engine-pump combination the heat losses from the oven will be transferred to the room temperature irreversibly. This will be associated of course to available work losses. If the (more complex) cycle proposed here is operating reversibly, there is no dissipation. This double heat pump - heat engine combination is therefore necessary if *upper bounds* on real performances are studied, as already suggested by Jaynes (2003). In other words, a single engine-heat pump combination is enough for oven operation but it is not enough for the most energy efficient oven operation.

To keep a constant temperature T_1 in the room the following relationship must be fulfilled:

$$Q_{11} = Q_{14} \tag{19.70}$$

The energy balance shows that generally a difference exists between the heat flux leaving and entering the oven:

$$Q_2 \equiv Q_{23} - Q_{22} \tag{19.71}$$

To keep a constant temperature T_2 of the body inside the oven, the user has to provide this heat flux difference from other energy sources.

One defines the dimensionless heat gain factor G as the ratio between the heat flux received by the oven from the heat pump 2 and the additional heat flux provided by the user from external sources:

$$G \equiv \frac{Q_{22}}{Q_2} \tag{19.72}$$

Now, the aim is to optimize the operation of a given endoreversible oven (i.e. an oven with known conductances). The objective function to be minimized is the additional heat flux Q_2 delivered to the oven from external sources. Four design factors are first defined:

$$g_1 \equiv \frac{U_{11}U_{01}}{U_{11}+U_{01}}, \quad g_2 \equiv \frac{U_{22}U_{02}}{U_{22}+U_{02}}, \tag{19.73a-d}$$

$$g_3 \equiv \frac{U_{23}U_{03}}{U_{23}+U_{03}}, \quad g_4 \equiv \frac{U_{14}U_{04}}{U_{14}+U_{04}}.$$

These factors have the physical significance of equivalent conductances. For example, the factor g_1 may be thought as the equivalent conductance between the heat sources at temperature T_1 and T_0, respectively. Note that factors g_i (i=1,2,3,4) are symmetrical functions of their arguments (i.e. $g_1(U_{11}, U_{01})= g_1(U_{01}, U_{11})$, etc).

The optimization procedure is described next. The work rate W may be eliminated between Eqs. 1 and 3 in Table 19.3. This yields a relationship between heat fluxes:

$$Q_{11} - Q_{10} - Q_{22} + Q_{02} = 0 \qquad (19.74)$$

Equations 2 and 4 in Table 19.3 allow writing the four fluid temperatures and the four heat fluxes in terms of the entropy fluxes S_1 and S_2:

$$t_{11} = \frac{U_{11}T_1}{U_{11} + S_1}, \quad t_{01} = \frac{U_{01}T_0}{U_{01} - S_1},$$

$$t_{02} = \frac{U_{02}T_0}{U_{02} + S_2}, \quad t_{22} = \frac{U_{22}T_2}{U_{22} - S_1} \qquad (19.75a\text{-}d)$$

$$Q_{11} = \frac{U_{11}T_1S_1}{U_{11} + S_1}, \quad Q_{01} = \frac{U_{01}T_0S_1}{U_{01} - S_1},$$

$$Q_{02} = \frac{U_{02}T_0S_2}{U_{02} + S_2}, \quad Q_{22} = \frac{U_{22}T_2S_2}{U_{22} - S_1} \qquad (19.76a\text{-}d)$$

Use of Eqs. (19.74) and (19.76a-d) yields:

$$F_1 \equiv \frac{U_{11}T_1S_1}{U_{11} + S_1} - \frac{U_{01}T_0S_1}{U_{01} - S_1} - \frac{U_{22}T_2S_2}{U_{22} - S_2} + \frac{U_{02}T_0S_2}{U_{02} + S_2} = 0 \qquad (19.77)$$

Also, the work rate W may be eliminated between Eqs. 5 and 7 in Table 19.3. This yields a relationship between the heat fluxes Q_{23}, Q_{03}, Q_{14} and Q_{04}:

$$Q_{23} - Q_{03} - Q_{14} + Q_{04} = 0 \qquad (19.78)$$

Equations 6 and 8 in Table 19.3 allow writing the fluid temperatures t_{23}, t_{03}, t_{14} and t_{04} and the heat fluxes in terms of the entropy fluxes S_3 and S_4:

$$t_{23} = \frac{U_{23}T_2}{U_{23} + S_3}, \quad t_{03} = \frac{U_{03}T_0}{U_{03} - S_3},$$

$$t_{14} = \frac{U_{14}T_1}{U_{14} - S_4}, \quad t_{04} = \frac{U_{04}T_0}{U_{04} + S_4} \qquad (19.79a\text{-}d)$$

$$Q_{23} = \frac{U_{23}T_2S_3}{U_{23} + S_3}, \quad Q_{03} = \frac{U_{03}T_0S_3}{U_{03} - S_3},$$

$$Q_{14} = \frac{U_{14}T_1S_4}{U_{14} - S_4}, \quad Q_{04} = \frac{U_{04}T_0S_4}{U_{04} + S_4} \qquad (19.80\text{a-d})$$

Use of Eqs. (19.78) and (19.80a-d) yields:

$$F_2 \equiv \frac{U_{23}T_2S_3}{U_{13} + S_3} - \frac{U_{03}T_0S_3}{U_{03} - S_3} - \frac{U_{14}T_1S_4}{U_{14} - S_4} + \frac{U_{04}T_0S_4}{U_{04} + S_4} = 0 \qquad (19.81)$$

Also, the objective function Q_2 given by Eq. (19.71) may be expressed in terms of S_2 and S_3:

$$Q_2 = \frac{U_{23}T_2S_3}{U_{23} - S_3} + \frac{U_{22}T_{21}S_2}{U_{22} - S_2} \qquad (19.82)$$

Equation (19.70) may be re-written in terms of S_1 and S_4 as follows:

$$\frac{U_{11}S_1}{U_{11} + S_1} = \frac{U_{14}S_4}{U_{14} - S_4} = A \qquad (19.83)$$

The l.h.s and the r.h.s of Eq. (19.83) are functions of different independent variables. Therefore, A must be a constant. Use of Eq. (19.83) and Eqs. (19.76a-d) allows writing:

$$Q_{11} = AT_1, \quad Q_{01} = \frac{g_1T_0A}{g_1 - A}, \quad Q_{14} = AT_1, \quad Q_{04} = \frac{g_4T_0A}{g_4 + A} \qquad (19.84\text{a-d})$$

A positive heat flux Q_{01} requires $0 \leq A < g_1$. Usage of Eqs. (19.74), (19.78) and (19.84a-d) allow writing Eq. (19.77) and Eq. (19.81), respectively, in terms of A, S_1 and S_2:

$$F_1 \equiv A\frac{g_1(T_1 - T_0) - AT_1}{g_1 - A} - \frac{U_{22}T_2S_2}{U_{22} - S_2} + \frac{U_{02}T_0S_2}{U_{02} + S_2} = 0 \qquad (19.85)$$

$$F_2 \equiv \frac{U_{23}T_2S_3}{U_{13} + S_3} - \frac{U_{03}T_0S_3}{U_{03} - S_3} - A\frac{g_4(T_0 - T_1) - AT_1}{g_4 + A} = 0 \qquad (19.86)$$

Minimization of Q_2 given by Eq. (19.82), seen as function of the independent variables A, S_2 and S_3, should take into account the constraints Eqs. (19.85) and (19.86). The usual Lagrange method is used. It involves the Lagrange function L defined as:

$$L \equiv Q_2 + \lambda_1 F_1 + \lambda_2 F_2 \qquad (19.87)$$

where λ_1 and λ_2 are multipliers. Finding the minimum of Q_2 requires solving the system:

$$\frac{\partial L}{\partial A}=0, \quad \frac{\partial L}{\partial S_2}=0, \quad \frac{\partial L}{\partial S_3}=0, \quad \frac{\partial L}{\partial \lambda_1}=0, \quad \frac{\partial L}{\partial \lambda_2}=0 \qquad (19.88\text{a-e})$$

Use of Eqs. (19.82), (19.85-87) and (19.88a-c) yields:

$$\lambda_1\left[\frac{T_1}{T_0}-\frac{g_1^2}{(g_1-A)^2}\right]-\lambda_2\left[\frac{T_1}{T_0}-\frac{g_4^2}{(g_4+A)^2}\right] \qquad (19.89)$$

$$\frac{\lambda_1+1}{\lambda_1}=\frac{T_0}{T_2}\frac{U_{02}^2}{U_{22}^2}\left(\frac{U_{22}-S_2}{U_{02}+S_2}\right)^2\equiv C^2 \qquad (19.90)$$

$$\frac{\lambda_2+1}{\lambda_2}=\frac{T_0}{T_2}\frac{U_{03}^2}{U_{23}^2}\left(\frac{U_{23}+S_3}{U_{03}-S_3}\right)^2\equiv D^2 \qquad (19.91)$$

Both the l.h.s and the r.h.s of Eqs. (19.90) and (19.91) are functions of different independent variables. Therefore C and D must be constants. Equations (19.90) and (19.91) allow writing S_2 and S_3 in term of C and D, respectively:

$$S_2=\frac{U_{22}U_{02}\left(\sqrt{T_0/T_2}-C\right)}{\sqrt{T_0/T_2}U_{02}+CU_{22}} \qquad S_3=\frac{U_{03}U_{23}\left(D-\sqrt{T_0/T_2}\right)}{\sqrt{T_0/T_2}+DU_{23}} \qquad (19.92\text{a,b})$$

Positive entropy fluxes require $0\leq C\leq(T_0/T_2)^{1/2}$ and $(T_0/T_2)^{1/2}\leq D<\infty$. Use of Eqs. (19.92a,b), (19.85) and (19.86) yield Eqs. (19.93) and (19.94) below, respectively:

$$g_3T_2\frac{\left(D-\sqrt{T_0/T_2}\right)\left(1-D\sqrt{T_0/T_2}\right)}{D}+A\frac{g_4(T_0-T_1)-AT_1}{g_4+A}=0$$

$$\frac{1}{C^2-1}\left[\frac{T_1}{T_2}-\frac{g_1^2}{(g_1-A)^2}\right]-\frac{1}{D^2-1}\left[\frac{T_1}{T_2}-\frac{g_4^2}{(g_4-A)^2}\right]=0 \qquad (19.93,94)$$

Also, Eqs. (19.90) and (19.91) allow to express λ_1 and λ_2 in terms of C and D. Replacing these expressions in Eq. (19.89) yields the next Eq. (19.95):

$$A\frac{g_1(T_1-T_0)-AT_1}{g_1-A}+g_2T_2\frac{\left(\sqrt{T_0/T_2}-C\right)\left(C\sqrt{T_0/T_2}-1\right)}{C}=0 \qquad (19.95)$$

Once the Eqs. (19.93-95) are solved for A, C and D, the optimum entropy fluxes may be computed by using Eqs. (19.14) and (19.23a,b). Also, the optimum heat fluxes are given by Eqs. (19.76a-d) and (19.80a-d). Finally, the optimum fluid temperatures may be found by using Eqs. (19.75a-d) and (19.79a-d).

19.4.1.1 The Reversible "Free" Oven

One sees that Eqs. (19.93-95) have the solutions $A_{opt}=0$, $C_{opt}=D_{opt}=(T_0/T_2)^{1/2}$. This corresponds to null entropy fluxes and makes the fluid temperatures equal to the constant temperatures T_0, T_1 and T_2 (i.e. $t_{01}= t_{02}= t_{03}= t_{04}=T_0$, $t_{11}= t_{14}=T_1$ and $t_{22}= t_{23}=T_2$). This is a reversible operation regime. One needs, however, infinitely large conductances to keep finite the heat fluxes in Table 19.3. In this ideal case, $Q_2=0$ and the oven gain factor defined by Eq. (19.72) tends to infinity. In other words, there is no need for additional heat for oven operation - except for the energy required to change the state of the matter inside the oven. This is the "free oven" first described in Jaynes (2003), where one concludes that the second law allows us to operate an oven at zero cost, the environment serving as a temporary repository for the entropy flux disposed in heating the oven.

The oven reversible operation regime involves using reversible engines and heat pumps. These highly idealized devices are not accessible to present day technology. More realistic cases are treated in the next section (Badescu 2009).

19.4.2 More Realistic Additional Constraints

In practice, any conductance is a product between a heat transfer surface area and a heat transfer coefficient. A realistic model requires of course using finite conductances. Therefore, the associated equivalent conductances g_i (i=1,2,3,4) are normally finite quantities. The factors g_i (i=1,2,3,4) in Eqs. (19.73a-d) are symmetrical in their arguments. Thus:

$$U_{ii} = U_{01} = 2g_1, \quad U_{22} = U_{02} = 2g_2,$$
$$U_{23} = U_{03} = 2g_3, \quad U_{14} = U_{04} = 2g_4 \qquad (19.96a\text{-}d)$$

Equations (19.96a-d) show that the equipartition principle may be used for any single component of the endoreversible oven.

One denotes by g the (normally finite) sum:

$$g \cong g_1 + g_2 + g_3 + g_4 \qquad (19.97)$$

which may be seen as a "total" oven conductance. In the following, the numerical value of g is of little importance. However, once g is chosen, it remains constant during the mathematical treatment. Thus, g may be thought as a parameter when the final results are obtained. Equation (19.97) is our first additional constraint.

The available technology puts limitations on heat engines and heat pumps performances, apart from the endoreversibility hypothesis we already adopted. As an example, here we shall consider that the coefficient of performance of the endoreversible heat pump 4 is a given under-unitary fraction f of the associated Carnot heat pump, i.e.:

$$COP_4 \equiv \frac{t_{14}}{t_{14} - t_{04}} = f \cdot COP_{4,Carnot} \equiv f \frac{T_1}{T_1 - T_0} \qquad (19.98)$$

Equation (19.98) is our second additional constraint.

In Sect. 19.2 the objective function depends on four independent variables (i.e. it depends explicitly on S_2 and S_3 (see Eq. (19.82)) and implicitly on the variables S_1 and S_4). Three constraints exist there (i.e. two first law relationships - Eqs. (19.77) and (19.81) - and Eq. (19.70)). Therefore, the optimization has a single freedom degree and this allows minimization of the objective function Eq. (19.82).

Adding Eq. (19.98) as a forth constraint removes this single degree of freedom and also removes any possibility of optimization over the variables S_i (i=1,2,3,4). Note that a similar constraint may be defined for any of the other three endoreversible components of the oven but the total number of constraints is limited by the number of freedom degrees.

However, a different sort of optimization is still possible. It refers to the design factors g_i (i=1,2,3,4), which may be seen as independent variables. Equation (19.97) acts as a constraint and three freedom degrees remain. The computation procedure is described next with the parameter f as an entry.

For given value of f, from Eq. (19.98) one computes COP_4 and the temperature ratio:

$$\frac{t_{04}}{t_{14}} = 1 - \frac{1}{COP_4} \tag{19.99}$$

Also, t_{04} and t_{14} are functions of S_4 (see Eqs. (19.79c-d)). Taking into account Eq. (19.96d) one finds:

$$S_4 = \frac{2g_4\left(1 - \dfrac{T_1}{T_0}\dfrac{t_{04}}{t_{14}}\right)}{1 + \dfrac{T_1}{T_0}\dfrac{t_{04}}{t_{14}}} \tag{19.100}$$

From Eq. (19.83) one derives:

$$A = \frac{2g_4 S_4}{2g_4 - S_4}, \quad S_1 = \frac{2g_1 A}{2g_1 - A} \tag{19.101, 102}$$

Use of Eqs. (19.77) and (19.81) allows computation of S_2 and S_3. Any other quantity may be computed as function of the entropy fluxes S_i (i=1,2,3,4) just obtained.

19.4.2.1 Endoreversible Oven Operation on the Moon

The objective function to be minimized is Q_2 given by Eq. (19.71) but this time the independent variables are the factors g_i (i=1,2,3,4). The optimization over g_i (i=1,2,3,4) was performed numerically. A lower bound 0.01 was imposed for the ratios g_i/g (i=1,2,3,4). This is consistent with technological limitations.

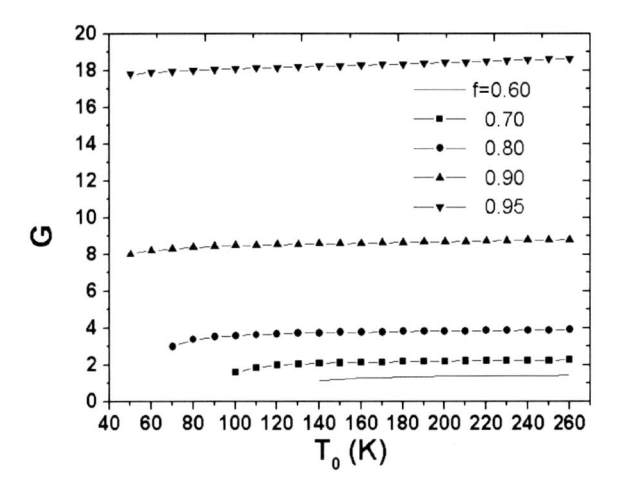

Fig. 19.7 Dependence of the optimum heat gain factor G defined by Eq. (19.72) on the environment temperature T_0 as function of the parameter f entering Eq. (19.98). The standard temperature for oven cooking is $T_2=200°C$ and the room air temperature is $T_1=22°C$.

The heat gain factor G of a constraint optimized endoreversible oven is shown in Fig. 19.7 for several values of the ambient temperature T_0, associated to various environments for habitation on Moon. The gain factor increases by increasing the under-unitary f entering Eq. (19.98), as expected. However, the performance is significantly decreased as compared to the case of a reversible oven, whose heat gain factor tends to infinity. Reduction of thermal energy consumption with an order of magnitude is still possible, but only for rather large values of f.

Note that for given value of f, the constraints ensuring the operation of the endoreversible oven are fulfilled only for values of the ambient temperature T_0 exceeding some minimum threshold value. For example, the oven with $f=0.6$ has a threshold of 140 K. In this case, the oven operation is allowed for sub-surface environments ($T_0=250$ K) or during some "warm" nights with ambient temperature larger than 140 K but is not allowed during "cold" nights ($T_0=100$ K) or for habitation in shaded areas ($T_0=70$ K).

The optimum heat gain shows little dependence on the environment temperature T_0 (whatever the value of f is). This is unexpected at first sight and a brief explanation follows. First, changing the environment temperature T_0 makes the "design" factors g_i (i=1,2,3,4) to change (see Fig. 19.8). Therefore, different values of T_0 in Fig. 19.7 are associated to different design solutions. Second, increasing T_0 (or, in other words, decreasing the temperature difference between room temperature and environment) implies a correspondingly decrease in both the heat flux Q_{11} and the power output W. This finally yields a decrease in the heat flux Q_{22} provided by the heat pump 2. The same decrease in the temperature difference between room temperature and environment implies a decrease in the heat flux Q_{14} (which equals Q_{11} - see Eq. (19.70)) and in the power output W. This finally yields a decrease in the heat flux Q_{23} necessary to drive the heat engine 3.

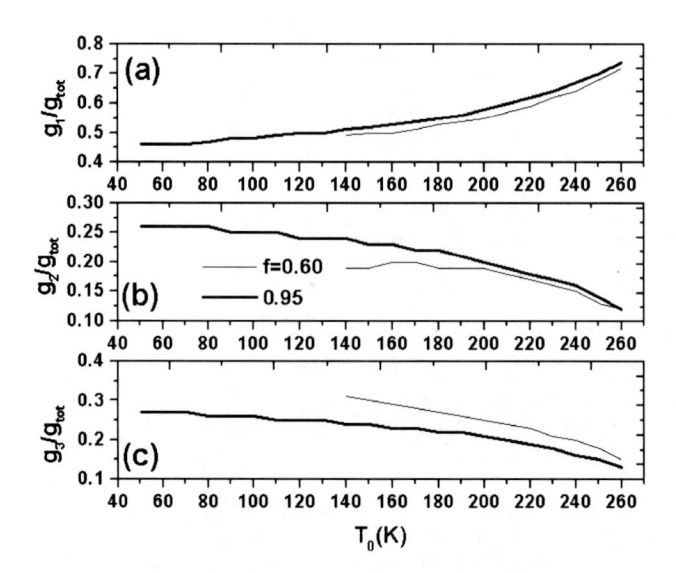

Fig. 19.8 Dependence of the factors g_i (i=1,2,3,4) entering Eq. (19.97) on the environment temperature T_0 for two values of the parameter f entering Eq. (19.98). (a) ratio g_1/g; (b) ratio g_2/g; (c) ratio g_3/g. For other input values see Fig. 19.7.

Therefore, both the nominator and denominator of Eq. (19.72) decrease by increasing the environment temperature T_0 and the gain factor G remains almost constant.

The optimum design factors g_i (i=1,2,3,4) are shown in Fig. 19.8. They depend significantly on the environment temperature T_0 and on the factor f. The equipartition principle does not act at the level of the whole endoreversible oven (i.e. the four factors g_i and the eight optimized conductances, respectively, are not equal each other). The influence of the parameter f on g_i is much smaller than the influence of f on G (compare Fig. 19.8 and Fig. 19.7).

An important aspect refers to the figure of merit for the three remaining components of the optimized endoreversible oven: i.e. the heat engines 1 and 3 and the heat pump 2 in Fig. 19.6. Figure 19.9 shows their indicators of performance in case of f=0.6. Note that the coefficient of performance of heat pump 2 (which is defined by $COP_2 \equiv t_{22}/(t_{22}-t_{02})$) and the heat engine efficiencies (which are defined by $\eta_1 = 1 - t_{01}/t_{11}$ and $\eta_3 = 1 - t_{03}/t_{23}$, respectively) have a weak dependence on f. However, more the parameter f increases, more the performance becomes better.

Both the coefficient of performance and the engine efficiencies in Fig. 19.9 have rather low values. However, these values are only slightly lower than the performance of the associated Carnot heat pump and heat engines (defined by $COP_{2,Carnot} \equiv T_2/(T_2-T_0)$, $\eta_{1,Carnot} \equiv 1 - T_0/T_1$ and $\eta_{3,Carnot} \equiv 1 - T_0/T_3$, respectively). This means that the effectiveness of the heat engines 1 and 3 and of the heat pump 4 (which are defined as ratios of the actual performance indicators and the associated Carnot indicators, respectively) are high. Consequently, important improvements in the present day technology have to be performed in order ovens with high values of the gain factor to be implemented in practice. Sub-surface habitation on Moon is most likely the best place for this implementation.

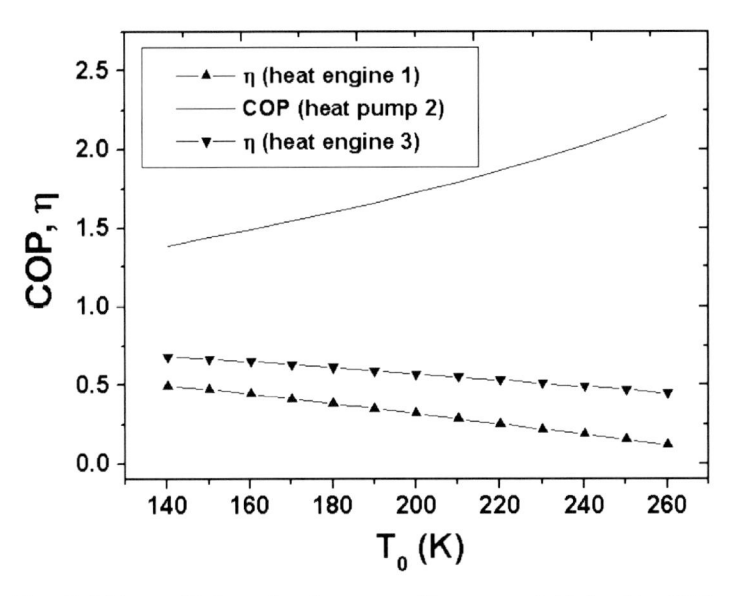

Fig. 19.9 The coefficient of performance of heat pump 2 (defined by $COP_2 \equiv t_{22}/(t_{22}-t_{02})$) and the efficiencies of heat engines 1 and 3 (defined by $\eta_1 = 1-t_{01}/t_{11}$ and $\eta_3 = 1-t_{03}/t_{23}$, respectively) as functions of environment temperature T_0 for $f=0.6$. For other input values see Fig. 19.7.

19.5 Conclusions

This chapter shows the principles of efficient usage of thermal resources for heating on Moon. Two cases were considered, namely heating of habitation spaces and oven operation, respectively.

The endoreversible heater is defined in this chapter as a heat engine coupled to a heat pump, both of them operating endoreversibly. This concept allows obtaining more accurate upper bounds for the heating gain factor than in case the traditional reversible approach is used. The optimum operation temperatures and the heating gain factor obtained under the endoreversible framework depend on a single dimensionless parameter incorporating all design and thermal factors. The quantities associated to reversible operation are recovered in the limiting case of infinitely large conductance. Equation (19.69) gives the upper bound for the heating gain factor under endoreversible operation.

Our results show that important thermal energy savings may be performed on Moon by changing the traditional heating technology. Reducing the energy consumptions of the traditional systems is possible even in the worst case of habitation in the shaded Moon polar areas. From a practical point of view, gain factors larger than 2 are associated to Moon sub-surface habitation, where the ambient temperature is about 250 K. The technology puts lower bounds on heat engine conductance and the performance is significantly decreased as compared to the case of the endoreversible heater. However, reduction of thermal energy consumption to a half of present day level is still possible. Further increasing of the heating performance requires technology improvements for heat pump operation.

The endoreversible oven is defined in this chapter as two sets of heat engines coupled to heat pumps, both sets operating endoreversibly. This concept allows obtaining more realistic upper bounds for the oven heat gain factor than in case the traditional reversible thermodynamics approach is used. The optimum operation temperatures and the oven heat gain factor obtained under the endoreversible framework depend on three parameters incorporating both design and thermal factors. However, this is a matter of choice, not a physical limitation. The quantities associated to reversible operation are recovered in the limiting case of infinitely large conductance. In the more realistic case of finite sizes one shows that the equipartition principle acts for the conductances of both heat engines and heat pumps, but does not act at the level of the whole endoreversible oven.

Our results show that important energy savings may be performed by changing the traditional oven operation technology. Reducing the present days heat consumptions with an order of magnitude is (in principle) possible. The optimum heat gain shows little dependence on the environment Moon temperature T_0. However, the operation of the endoreversible oven is possible only for values of the ambient temperature exceeding some minimum threshold value, depending on the actual performance of heat engines and heat pumps. For example, in the case of present day technology, the oven operation is allowed for Moon sub-surface environments ($T_0=250$ K) or during some "warm" nights with ambient temperature larger than 140 K but is not allowed during "cold" nights ($T_0=100$ K) or for habitation in polar shaded areas ($T_0=70$ K). Consequently, significant technology improvement for the heat engines and heat pumps is necessary to allow practical implementation of ovens with high values of the gain factor. Sub-surface habitation on Moon is most likely the best place for this implementation.

It is important to stress that no physical law stands in the way of realizing these improvements. We agree with Jaynes (2003) that the successful technology will probably avoid the mechanical pumps, depending perhaps on thermoelectric or electrochemical means that avoid all mechanical moving parts.

References

Amos, J.: Coldest place' found on the Moon. BBC News, San Francisco (2010), http://news.bbc.co.uk/2/hi/8416749.stm

Badescu, V.: Endoreversible thermodynamics approach for optimum oven operation. Int. J. Exergy 6, 49–60 (2009)

Badescu, V.: Accurate upper bound for heating gain factor. Europhysics Letters 76(4), 568–574 (2006)

Bejan, A.: Entropy generation minimization: The new thermodynamics of finite-size devices and finite-time processes. J. Appl. Phys. 79, 1191–1218 (1996)

Bejan, A., Tondeur, D.: Equipartition, optimal allocation, and the constructal approach to predicting organization in nature. Rev. Gen. Thermique. 37, 165–180 (1998)

Bussey, D., Ben, J., Fristad, K.E., Schenk, P.M., Robinson, M.S., Spudis, P.D.: The Lunar Polar Environment: An Overview. 2005 Salt Lake City Annual Meeting, Paper No. 154-7, October 16–19 (2005),
http://gsa.confex.com/gsa/2005AM/finalprogram/
abstract_95876.htm

Chen, L., Sun, F., Chen, W.: Optimal Performance Coefficient and Cooling Load Relationship of a Three- Heat-Reservoir Endoreversible Refrigerator. Int. J. Power Energy System 17, 206–208 (1997a)

Chen, J., De Vos, A., Andresen, B.: Analysis of combined systems of two endoreversible engines. Open Syst. & Info. Dyn. 4, 3–13 (1997b)

Chen, J.: The Optimum Performance Characteristics of a Four-Temperature-Level Irreversible Absorption Refrigerator at Maximum Specific Cooling Load. J. Phys. D: Appl. Phys. 32, 3085–3091 (1999)

Christie, R.J., Plachta, D.W., Hasan, M.M.: Transient Thermal Model and Analysis of the Lunar Surface and Regolith for Cryogenic Fluid Storage. NASA/TM—2008-215300. Glenn Research Center, Cleveland (2008),
http://gltrs.grc.nasa.gov/reports/2008/TM-2008-215300.pdf

Cohen, M.M.: Selected Precepts. In: Lunar Architecture, 53rd International Astronautical Congress; IAC-02-Q.4.3.08; The World Space Congress – 2002, Houston, Texas, October 10-19 (2002),
http://adsabs.harvard.edu/abs/2002iaf.confE.716C

Crawford, F.H.: Heat, Thermodynamics and Statistical Physics, pp. 217–219. Harcourt, Brace& World, New York (1963)

Curzon, F.L., Ahlborn, B.: Efficiency of a Carnot engine at maximum power output. Am. J. Phys. 43, 22–24 (1975)

Heiken, G.H., Vaniman, D.T., French, B.M. (eds.): Lunar Sourcebook: A User's Guide to the Moon. Section 3.5. Cambridge University Press (1991)

Hoffmann, K.H., Burzler, J.M., Schubert, S.: Endoreversible thermodynamics. J. Non-Equilibrium Thermodyn. 22, 311–355 (1997)

Jaynes, E.T.: Note on thermal heating efficiency. Am. J. Phys. 71, 180–182 (2003)

Kruijff, M.: The Peaks of Eternal Light on the Lunar South Pole (2010),
http://www.delta-utec.com

Langseth, M.G., Keihm, S.J., Peters, K.: Revised lunar heat flow values. In: Proc. 7th Lunar Sci. Conf., pp. 3143–3156 (1976)

Lawson, S.L., Jakosky, B.M.: Brightness Temperatures of the Lunar Surface: The Clementine Long-Wave Infrared Global Data Set. Lunar and Planetary Science XXX (1999),
http://www.lpi.usra.edu/meetings/LPSC99/pdf/1892.pdf

Muschik, W., Badescu, V.: Irreversible Jaynes Engine for More Efficient Heating. Journal of Non-Equilibrium Thermodynamics 33(4), 297–306 (2008)

Paige, D.: DIVINER Lunar Radiometer Experiment. University of California at Los Angeles (2010), http://diviner.ucla.edu/science.html

Ryan, R.E., Underwood, L.W., McKellip, R., Brannon, D.P., Russell, K.J.: Exploiting Lunar Natural And Augumented Thermal Environments For Exploration And Research. In: 39th Lunar and Planetary Science Conference (Lunar and Planetary Science XXXIX), LPI Construction No. 1391, p. 2346. League City, Texas, March 10-14 (2008),
http://adsabs.harvard.edu/abs/2008LPI.39.2346R

Seybold, C.C.: Characteristics of the Lunar Environment, Texas Space Grant Consortium 3925 West Braker Lane, Austin, Texas 78759 (1995),
http://www.tsgc.utexas.edu/tadp/1995/spects/environment.htm
Silver, R.S.: Reflexions sur la puissance chaleurique du feu. J. Heat Recovery Syst. 1, 205–207 (1981)
Spudis, P.D.: The Poles of the Moon and Their Significance, San Antonio, Texas. AAPG Annual Convention, April 20-23 (2008),
http://www.searchanddiscovery.net/documents/
2008/08147spudis/ndx_spudis.pdf
Sullivan, T.A.: Process Engineering Concerns in the Lunar Environment. In: AIAA Space Programs and Technologies Conference, AIAA- 90-3753, Huntsville AL, September 25-28 (1990)
TeamFrednetWiki, Lunar Environment – TeamFrednetWiki (2010),
http://wiki.xprize.frednet.org/index.php/Lunar_Environment
Tondeur, D., Kvaalen, E.: Equipartition of Entropy Production: An Optimality Criterion for Transfer and Separation Processes. Ind. Eng. Chem. Res. 26, 50–56 (1987)
Vanoutryve, B., De Rosa, D., Fisackerly, R., Houdou, B., Carpenter, J., Philippe, C., Pradier, A., Jojaghaian, S., Espinasse, S., Gardini, B.: An Analysis Of Illumination And Communication Conditions Near Lunar South Pole Based On Kaguya Data. In: 7th International Planetary Probe Workshop, Barcelona, June 15 (2010),
http://www.planetaryprobe.eu/IPPW7/
proceedings/cd/authors.html#v
Vasavada, A., Paige, D., Wood, S.: Near-Surface Temperatures on Mercury and the Moon and the Stability of Polar Ice Deposits. Icarus 141, 179–193 (1999)
Walker, S.T., Alexander, R.A., Tucker, S.P.: Thermal Control on the Lunar Surface. JBIS 48, 27–32 (1995)
WOTM, Working On The Moon - Thermal Environment (2010),
http://www.workingonthemoon.com/WOTM-ThermalEnvrnmnt.html
Wu, C., Chen, L., Chen, J. (eds.): Recent advances in finite-time thermodynamics. Nova Science, New York (1999)
Zakrajsek, J.J., McKissock, D.B., Woytach, J.M., Zakrajsek, J.F., Oswald, F.B., McEntire, K.J., Hill, G.M., Abel, P., Eichenberg, D.J., Goodnight, T.W.: Exploration Rover Concepts and Development Challenges. In: AIAA-2005-2525, NASA/TM-2005-213555, 1st Space Exploration Conference: Continuing the Voyage of Discovery. AIAA, Orlando, January 30-February 1 (2005),
http://ntrs.nasa.gov/archive/nasa/
casi.ntrs.nasa.gov/20050175879_2005173639.pdf

20 Deployable Lunar Habitation Design

S. Haeuplik-Meusburger[1] and K. Ozdemir[2]

[1] Vienna University of Technology, Austria
[2] Yeditepe University, Istanbul, Turkey

20.1 Introduction

Plans to return to the Moon and develop a sustainable human presence there continue to arise and change. In addition to the efforts of space agencies and governments, commercial spaceflight initiatives are taking part in future lunar exploration scenarios (cf. Lunar X-Prize).

The goal of this chapter is to provide an overview to the habitable elements of the human lunar exploration that feature a deployable structural concept. For all human presence scenarios on the Moon, a pressurized habitat is an essential element. Humans need a habitable living and working environment, and simply cannot survive without an additional protective layer. Enclosures have to be designed to withstand large temperature differences, high radiation exposure and micrometeoroid impacts. At the same time, costs have to be minimized and transportation is limited. In these terms, pressurized habitable volume becomes a valuable resource.

Numerous options have been considered for the construction of habitable lunar surface elements. To describe the different approaches to habitat design, a class terminology is used (Smith 1993, Kennedy 2009, Howe et al. 2010). Class I construction is based upon pre-integrated modules, like the Apollo Lunar Module. Class III construction uses in-situ resources to build the structures. Class II construction uses pre-fabricated components that are assembled onsite, such as inflatable or deployable structures. This approach has been consistently reviewed, and numerous projects have been developed as a result.

This chapter presents a selection of historic and recent design studies and projects for deployable lunar habitation. It begins with milestones in the history of deployable structure design (Sect. 20.2) leading to deployable structures that were built and used (Sect. 20.3). The main part outlines design studies, wherein the design objective, the structural and deployment concept are summarized, and habitability aspects are highlighted (Sect. 20.4).

20.2 Milestones in Deployable Structure Design

Deployable and folding structures are widely used in the space environment for space structures such as deployable solar panels for satellites. Inflatable structures, as the most distinctive deployable bodies for use in architecture, have come up in different forms throughout their relatively short history. So far, their use for habitation purposes is still minimal. In the following, the historical progress is summarized.

Fig. 20.1 Early Inflatable Station Concept Model (credit: NASA)

20.2.1 Deployable Structure Design 1950 - 1960

The first principles of pneumatic constructions were developed and patented by the English engineer F. W. Lancaster in 1917 (Kuhn 1991 p. 67). After the development of appropriate materials such as Nylon and Dacron, aeronautical engineer Walter Bird realized the first inflatable structure 'Radome' with a diameter of 15 m in 1947. In 1952, Wernher von Braun introduced a wheel-shaped inflatable space station concept with a diameter of 50 m (NASA [History]). Between 1950 and 1960, Goodyear Aerospace Corporation developed inflatable structures for search radar antennas, radar calibration spheres and inflatable parabolic reflectors (L'Garde 2010). The Echo 1 balloon, a passive communication reflector, was launched by the United States in 1960. Its packaging diameter was 1 m, and when deployed, the balloon had a diameter of 30.5 m (Roberts 1988, NASA Langley Research Center 1959).

20.2.2 Deployable Structure Design 1960 – 1980

In 1960, NASA conducted the Mercury Mark I and II study for a 1-crew and 2-crew space station. It was based on the Mercury capsules (NASA [History];

Vogler 2002). In 1961, NASA and Goodyear built a prototype of an inflatable space station for a crew of 1-2 people with a diameter of 7.3 m (Fig. 20.1) (NASA [GRIN] 1961). In 1965, the Soviets equipped the manned spacecraft 'Voskhod' with an inflatable airlock to successfully perform the worlds' first EVA (World News 2011). During the Apollo program, NASA studied deployment techniques and inflatable structures for an early lunar base. Inflatable or deployable airlocks in conjunction with a rigid lunar base structure were studied but discarded because of the "minimal volume gain compared to the attendant decrease in reliability and increase in operational complexity" (NASA 1967 pp. 4-5). At that time, the only options that remained interesting for an early shelter were inflatable temporary structures in conjunction with the Lunar Module. The 'Stay Time Extension Module – STEM' was developed by Grumman and NASA in 1965. It was a folded module to be carried by the Lunar Module (in place of the LRV) for a crew of two and duration of 8 days. The 'Early Lunar Emergency Shelter' was developed by Goodyear and NASA in 1967. It was an inflatable emergency or temporary shelter and airlock in conjunction with the LM vehicle (NASA 1967 pp. 4-5). Inflatable Re-entry structures and crew escape systems were studied by the US and Russia. During the 1970's, Europe saw the first activities of the inflatable reflectors. Reflector antennas and rigidized truss-type sun shade structures were developed by Contraves and ESA (Freeland et al. 1998).

20.2.3 Deployable Structure Design 1980 – 1990

In 1984, NASA developed an inflatable Personal Rescue Enclosure for Astronauts. It had a diameter of 86 cm and was designed for one person. Prototypes were built, but it was never used. In 1985 to 1987, ILC Dover and the United States Air Force developed a lightweight, collapsible hyperbaric chamber with airlock. In 1989, the Lawrence Livermore National laboratory conducted a feasibility study on inflatable habitats. The Lawrence Livermore Inflatable Space Station had a diameter of 3.5 to 5 m and a length of 17 m. As a result, a concept for a deployable Lunar Base was proposed (ILC Dover 2011).

20.2.4 Deployable Structure Design 1990 – 2000

Between 1997 and 2000, NASA developed the 'TransHab', an inflatable long-duration habitat with central core. It was proposed as additional crew quarters to the ISS. A number of prototypes were built, but the project was discarded. It would expand to 8.2 meters in diameter (NASA 2003). In 1986, the 'IN-STEP Inflatable Antenna Experiment' from L'Garde and NASA was flown on STS-77 (Freeland et al. 1998). In 1997, ILC Dover and NASA developed an inflatable/deployable lunar habitat for a crew of 2 and duration of 6 days. In the same year, deployable airbags were used for the landing of the Mars Pathfinder Rover. The NASA Mars Reference Mission in 1998 included a Mars Surface Inflatable Habitat Concept (NASA 1998).

20.2.5 Deployable Structure Design 2000 – Present

The NASA Mars Exploration Rover Mission in 2004 used deployable airbags for landing. In 2006 Bigelow Aerospace launched the first inflatable space habitat prototype 'Genesis I'. It had a diameter of 2.4 m and a length of 4.3 m. The second prototype, 'Genesis II', was launched in 2007. Future inflatable projects include the inflatable habitat Sundancer and BA 330. Furthermore, numerous inflatable concepts have been developed, including hybrid structures. In 2007, in the frame of the ESA Surface Infrastructure Study, various elements of the surface architecture for Moon and Mars were assessed, including deployable units. Also in 2007, the 'InFlex' Lunar Habitat prototype for a crew of two was built, and in the same year NASA tested an inflatable lunar habitat in the extreme Antarctic environment.

20.3 Building Deployable Structures for Human Exploration

Over the past decade, different deployable space structures with various configurations have been developed. Ranging from inflatable antennas and balloons to spacesuits, inflatable structures have found an extensive array of applications (Fig. 20.2). The development of structural concepts and material developments for deployable structures has advanced. In addition to mechanically deployable structures, inflatable composite and rigidizable structures are being developed.

Fig. 20.2 Test Inflation of a Passive Geodetic Satellite in 1965 (credit: NASA)

Deployable structures have been successfully implemented in space missions. Some distinguished examples are: the Apollo Lunar Roving Vehicle, spacesuits from different eras of human spaceflight, the impact attenuation airbag systems on the Mars Pathfinder & Mars Exploration Rover mission (ILC Dover) and the deployable envelope of Genesis I and II (Bigelow Aerospace). The Spacesuit, the deployable airlock of Voshkod 2 and the deployable lunar roving vehicle of the Apollo missions are introduced below.

20.3.1 The Space Suit

Space suits have been adapted from pressure suits for military aircraft, and are necessary for extra-vehicular activities (EVA) in extra-terrestrial environments. A space suit is the smallest inflatable habitat. The first generation of spacesuits used in the early phases of human spaceflight (Mercury and Gemini missions; Vostok and Voskhod missions) were designed on the basis of pressure suit layouts for high altitude aviation. They were intended only for intra-vehicular activity (IVA). Today, astronauts use three kinds of suits for EVA; the Russian Orlan suit, the American Extravehicular Mobility Unit (EMU) and the Chinese Feitian space suit (Table 20.1). The EMU consists of an upper and a lower torso and is fabricated at ILC Dover in different sizes. It has 14 layers and a mass of about 82 kg (NASA 2010). The Orlan and Feitian suits are semi-rigid one-piece suits with a rear hatch entry.

Table 20.1 Project Summary 'Space Suit'

Design Objective	To enable the astronaut to work outside the space craft. Space suits should further provide protection, life-support and operational mobility
Structural Concept	Early space suits were soft suits; body-like shaped structures with a layered textile shell and joints with rigid components
Deployment Concept	Pressurization with breathable atmosphere at low level after donning
Habitability Aspects	- EMU, Orlan and Feitian suit are operational;
	- Soft-shell spacesuits represent inflatable 'wearable mini habitats';
	- Issues in material, structure, deployment and operation are studied, developed and tested under space mission conditions, providing valuable know-how for inflatable habitat design

Soft-shell Spacesuit; Russian Space Agency; NASA; Chinese Space Agency; 1961 – Present; EMU, Orlan and Feitian suit are operational

In the 1960s, Americans and Soviets developed different types of space suits for lunar missions. The Russian lunar suit 'Krechet' was the first semi-rigid suit. The upper part was made of aluminum alloy and featured an innovative entry concept. Cosmonauts would enter the suit through a rear hatch which also contained the life support equipment. The lunar suit was never used, but the concept was further developed for the 'Russian Orlan suits' (Portree 1997-98). The Apollo spacesuit was individually tailored for each astronaut (Fig. 20.3).

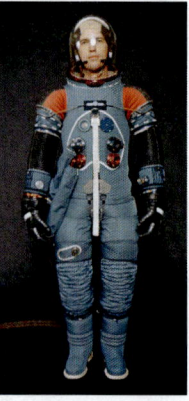

Fig. 20.3 left: Apollo 11 astronaut, Edwin E. Aldrin Jr., wearing the Lunar space suit; right: The A7L pressure garment assembly without the thermal micrometeorite garment (credit: NASA)

The suit consisted of the inner pressurizable envelope – the Torso and Limb Suit Assembly (TLSA) and the outer thermal and protective envelope (NASA 2010). The lunar suits, as their predecessors from Gemini and Voshkod, had to be pressurized, resulting in "suit-stiffening". The design of the joints, as well as the combination of rigid and hybrid space suit elements are among the engineering challenges (cf. Skoog et al. 2003).

20.3.2 Deployable Airlock

The Voskhod 2 mission took place in March of 1965; using a Voskhod 3KD spacecraft with two crew members on board, Pavel Belyayev and Alexei Leonov. The Voshkhod spacecraft was equipped with an inflatable airlock that enabled Alexei Leonov to conduct a 12 minute "spacewalk" in a 'Berkut' space-suit, marking the first EVA in the history of human space-flight (Fig. 20.4).

Fig. 20.4 Inflatable Airlock and Berkut spacesuit, Memorial Museum of Cosmonautics in Moscow (credit: Kucharek)

The inflatable airlock of Voshkod 2 was an external element, designed to be deployed in orbit (Table 20.2). The pressure enclosure with inflatable air beams was connected between two rigid parts; the EVA hatch and the assembly ring which was mounted on the spacecraft. The airlock was deployed by inflating the air beams. Then the airlock was pressurized and Leonov connected his suit to the airlock oxygen system. After depressurization, he opened the airlock and performed the first EVA. Alexei Leonov encountered problems with the photo camera and had difficulties to re-enter the airlock in his inflated space suit. He could only enter through the small hatch after decreasing the pressure in his spacesuit. Following the airlock ingress, he closed the outer EVA hatch and pressurized the airlock. Then he opened the space craft hatch and entered. As the EVA sequence was completed, the airlock was depressurized and detached from the descent module before landing. The whole process took about 1 hour and 40 minutes (Abramov et al. 2003).

Table 20.2 Project Summary 'Deployable Airlock'

Design Objective	To perform an extra-vehicular activity for the Voshkod-2 mission, enabling safe egress and ingress for the first spacewalk in history
Structural Concept	Inflated airlock module that is deployed in orbit. The system contains a textile main body with inflatable air beams and rigid end-caps. After use, the airlock was detached
Deployment Concept	Pressurization of the deployable unit; the spacecraft itself could not be depressurized
Habitability Aspects	- Being the first inflatable structure integrated in a human mission, the Voshkod airlock presented an extended volume, supporting the main operational feature of the mission; - The deployment pattern of the cylindrical structure provided a precursor example for contemporary designs

Inflatable Airlock of Voshkod-2 Mission; Russian Space Agency, Chief Designer G.I. Severin; 1965; Operational in Voshkod-2 Orbital Mission

The deployed unit provided the crew with an extended operational volume. Utilizing inflatable technology, the deployable airlock module asset allowed the mission to be completed without major design changes and kept the mass limits of the Voshkod spacecraft.

20.3.3 Deployable Roving Vehicle

The Lunar Roving Vehicle (LRV) was a deployable, electric four-wheel vehicle (Fig. 20.5 and Table 20.3). It was used during the Apollo missions: 15, 16 and 17. The rover was designed to operate under low-gravity conditions on the Moon and to be capable of traversing the lunar surface. The Rover allowed the Apollo astronauts to extend the range of their surface extra vehicular activities.

The Lunar Rover was folded and stored in one of the triangular-shaped bays with the underside of the chassis facing out. On the lunar surface, it had to be deployed by the astronauts themselves. The astronauts checked the conditions of the cargo bay. Following the general check, the EVA crew walked a few meters away from the Lunar Module while pulling strings to trigger the deployment mechanism. The rear chassis and wheels were the first to deploy and to touch the lunar surface. When lowered to the lunar surface, the astronauts erected the seats and foot restraints. The astronauts checked the configuration, sat into the rover and activated the electronic system (Baker 1971; Eckart 1999; Young 2007).

Fig. 20.5 Installation of the Lunar Roving Vehicle in the Lunar Module at the Kennedy Space Center in 1971 (credit: NASA)

Table 20.3 Project Summary 'Deployable Roving Vehicle'

Design Objective	To develop a surface vehicle for Apollo missions in order to extend the exploration radius of the crew; it had to fit in one of the triangular bays of the Lunar Module
Structural Concept	Lander-integrated deployable light weight structure; featuring aluminum alloy chassis and metal wire-mesh wheels
Deployment Concept	Manual Unfolding/deployment by crew
Habitability Aspects	- The LRV demonstrated a combination of an unpressurized mobility system with spacesuit systems; - As mini-habitats; the folding and packaging concept is elaborated

Apollo Lunar Roving Vehicle; Boeing – NASA, Folding concept by Ferenc Pavlics (Young 2007 p. 29); 1971 – 72; Operational in Apollo 15-16-17

The Lunar Rover is a remarkable folding and packaging concept. The front and the rear chassis portions as well as the suspension system were foldable. The deployed rover on the lunar surface was 3.1 meters long and 1.8 meters wide. When packaged, it was about 1.5 meters by 0.5 meters. The wheels were made of aluminum wire mesh. The foldable seats were made from aluminum and fabric covers (NASA 2005).

20.3.4 Summary: Advantages of Deployable Structures

Aerospace architecture and design is constrained by a number of factors including budget and transportation limits. Very early, the aerospace industry recognized that compared to rigid structures, deployable structures offer a number of advantages that may be essential for the success of long-term missions (Roberts 1988; NASA 1967; SICSA 1988; Cadogan 1999, 2007; Haeuplik-Meusburger et al. 2009). In the following, some of these issues are summarized:

Volume and Weight Efficiency

Total mass and volume for a spacecraft are important determinants. For early missions, lightweight structures that are transported from Earth to low-Earth orbit and planetary surfaces are essential for mission planning. Deployable structures combine low packaging volume with the potential for a greater 'on-planetary-surface volume'. The packaging ratio depends on the selected deployment concept and materials used, whereas "large structures packaged in wide diameter containers (...) offer the greatest deployed-to-packaged volume ratios" (SICSA 1988).

Technical Readiness

Inflatable structures have been built and used in space. They can be "engineered to withstand the rigors of space use" (Cadogan 2007). In 1996, the 'Inflatable Antenna Experiment' was flown on STS-77 to verify that large inflatable space structures can be built at low cost, have high mechanical packaging efficiency, high deployment reliability, and can be manufactured with high surface precision (Freeland et al. 1998, p. 4). In 2004 rovers took images on the mars surface after an airbag-protected landing and in 2006 Bigelow Aerospace launched its first inflatable space habitat prototype.

Enhanced Protection

Compared to traditional rigid structures, multi-layer deployable systems may offer enhanced protection from radiation due to the secondary radiation effect in metallic habitats. According to Bigelow, the shielding of their BA 330 utilizes an innovative micrometeorite and orbital debris shield. It also reduces the impact of secondary radiation (Bigelow 1999-2010). Cadogan pointed out that flexible composite structures tend to be more damage tolerant due to their "Forgiveness" as compared to rigid mechanical systems (Cadogan 1999).

Use of In-Situ Resources

Prefabricated deployable structures could be combined with in-situ lunar materials and resources. Several concepts have been developed using local materials. For example, regolith dust can be used as a concrete ingredient, regolith filled bags serve as outermost protection layer of the habitat, and water as a by-product from He-3 mining can be used as additional radiation protection (SATC 2002; Mohanty et al. 2003).

Habitability and Flexibility

Habitability and human factors are important factors in the design of any inhabited building type, but in the case of an isolated and confined habitat on the lunar surface, habitability becomes a critical issue. Due to the mentioned restraints, habitable volume is limited. Deployable and inflatable structures have the advantages of providing larger volume, but also offer a high degree of potential adaptability. Adaptable structural concepts can further enhance mission success by adapting to changing requirements such as mission objectives, crew condition and technological developments (cf. Häuplik-Meusburger et al. 2009, 2011).

Spin-Offs on Earth

Since the beginning of the human space flight era, a considerable amount of space technology 'spin-offs' were generated. NASA emphasizes that innovative, inflatable technologies designed for space may also generate innovative solutions for Earth use (NASA 2003).

20.4 Deployable Lunar Habitation Design Studies

This chapter introduces a selection of historic and recent studies of inflatable structures for a lunar base. The projects are briefly described and summarized according to their design objective, the proposed structural and architectural concept, and aspects related to habitability. Projects are in chronological order.

20.4.1 The Lunar Stay-Time Extension Module (STEM)

With the goal to extend the surface-stay time of the crews for post Apollo moon missions, the Goodyear Corporation designed an inflatable lunar shelter, called the Stay Time Extension Module (STEM)(Fig. 20.6 and Table 20.4).

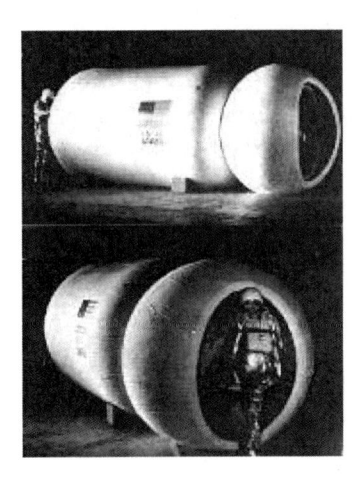

Fig. 20.6 Prototype of the Lunar Stay Time Extension Module (credit: NASA / Goodyear)

Table 20.4 Project Summary 'The Lunar Stay-Time Extension Module'

Design Objective	"To develop, define and evaluate conceptual designs of lunar shelters, for application in the early 1970's, which can, after 6 months of unattended storage, sustain a 2-man crew for at least 14 days"
	(NASA 1967 p. I-I)
Structural Concept	Inflatable cylindrical shelter with integrated airlock
	Dimensions: D: 2.1 m; L: 5.3 m (Shelter: 3,8 m); V: 14.6 m^3 (Shelter: 11.6 m^3); 5 psi; 148 kg
	Materials: 3-layer laminate consisting of nylon outer cover, closed-cell vinyl foam, and inner nylon cloth bonded by polyester adhesive layers
Deployment Concept	Industrial-terrestrial manufacture; packaged in a quadrant of the Lunar Module; assembly and in-situ deployment by the astronauts
Habitability Aspects	- Larger space for an extended surface stay despite vehicle constraints;
	- The Airlock is integrated into the design

Lunar Stay-Time Extension Module; Goodyear Aerospace Corporation for NASA Langley Research Center; 1964-1965; Prototype built

The STEM was designed to support two crew members on the lunar surface from eight to a maximum of 30 days. It was an inflatable cylindrical shelter with an integrated airlock and could be transported with the Apollo Lunar Module in a similar way as the Lunar Roving Vehicle (Project Apollo, 2010; SICSA, 1988). The deployable habitat prototype consisted of a two-piece module, featuring a pressurized habitable module and an airlock unit. In order to deploy the shelter on the lunar surface, astronauts would have to release the shelter from the Lunar Module, unpack, position and pressurize it at the desired location. The composite wall consisted of four layers: the pressure bladder, a structural layer, a foam micrometeoroid barrier, and the outer cover (NASA 1965, 1967). A full-scale mockup named 'Moby Dick' was built in 1968, but the program was eventually terminated (NASA 1991; Cadogan et al. 1999).

20.4.2 The Inflatable Lunar Habitat

The Inflatable Lunar Habitat concept (Table 20.5) was developed during the 'Lunar Base Systems Study' undertaken by the Advanced Programs Office in the Engineering Office at JSC between 1986 and 1989.

Table 20.5 Project Summary 'The Inflatable Lunar Habitat'

Design Objective	Assessment of an inflatable habitat system for the lunar surface in combination with other surface units with the goal of improving knowledge on how to live and work on a planetary surface
Structural Concept	Spherical pneumatic envelope with an interior structural cage to support the floors, walls and equipment
	Dimensions: D: 16 m; Open Volume: 2145 m³
	Materials: Kevlar Dupont; Regolith 'sandbags' for radiation protection
Deployment Concept	Inflation driven deployment and structural consolidation with in-situ material; industrial-terrestrial manufacture, assembly and pre-packaging of elements, in-situ deployment and structural finalizing process
Habitability Aspects	- Very large habitable volume on the lunar surface that is not divided into smaller modules;
	- Radiation protection can be potentially adjusted

The Inflatable Lunar Habitat; NASA JSC; 1986-1989; Concept Study

A multi-storey, sixteen-meter diameter, spherical shaped inflatable habitat was designed in addition to other surface facilities such as pressurized rovers, a solar shield and a construction shack. The habitat was designed to accommodate six to twelve crew members (Roberts 1988). An interior 'structural cage' had to support the floors, walls and equipment. The lower level accommodates stowage and life-support systems. Level two and three are the main levels, which serve for surface missions and base operations. A hydroponic garden connects these two levels. The airlock and dust removing devices for lunar surface work are located on level three. The laboratory is on level four and the viewing and exercising deck is on the top level. The private crew quarters are located on the lowest habitable level, because of radiation protection.

Three-meter thick, regolith-filled bags were intended to be used in shielding the habitable spaces from radiation. The study was conducted by Advanced Programs personnel with contractor support from Eagle Engineering Inc. and Lockheed Engineering and Sciences Co. (NASA 1989).

20.4.3 The Lawrence Livermore Lunar Base

Following the 'Lunar Base Systems Study', ILC Dover developed the Lawrence Livermore Lunar Base - a design study of an inflatable habitat structure for a lunar base (Fig. 20.7 and Table 20.6). This project was part of a greater feasibility study of inflatable structures for low Earth orbit.

Fig. 20.7 Livermore Concept Sketch of Lunar Base (credit: ILC Dover)

The lunar base design concept consists of a series of inflatable modules that form the main structure to accommodate laboratories and habitation areas. The modules are equipped with a secondary pressure containment wall and accommodate different pressurized compartments. Each compartment was about 3.5 m in length with a diameter of 5 m. The design of the outer skin included the functional distinction between several enveloping layers. In addition to a thermal shielding layer, the lunar base was covered with regolith to protect the crew from meteorites and radiation (ILC Dover 2011).

The composition of rigid and flexible components has been assessed as well as material and trade analysis. During the study, unique safety features were developed to close-off damaged chambers of the habitat system, as in submarines (Cadogan 2009).

Table 20.6 Project Summary 'The Lawrence Livermore Lunar Base'

Design Objective	Deployable surface habitat for human lunar exploration (low cost and with a packable configuration to fit in existing launch vehicles)
Structural Concept	Structural concept based on hybrid (rigid and inflatable) system; combination of rigid metal elements and flexible membranes (Tedlar/urethane laminate bladder)
	Dimensions: D: 3.5 m – 5 m (1 m corridor); L: 17 m: 7.5 psi
Deployment Concept	Inflation (Membranes) and Extrusion (Rigid Plates)

Habitability Aspects	- Inflatable modules accommodate different functions
	- Detailed assessments in safety and structural issues were produced during the study, paving the way to advanced modular habitat systems

Lawrence Livermore Inflatable Lunar Habitat Study; ILC Dover under contract to Lawrence Livermore National Laboratory (LLNL); 1989-1990; Concept Study

20.4.4 Expandable Lunar Habitat

In the frame of the 'Human Lunar Return study (HLR)', ILC Dover was commissioned to conduct a design study on an expandable Lunar Habitat. The expandable habitat was intended for a 6-day surface mission with a crew of 2 with EVA capability.

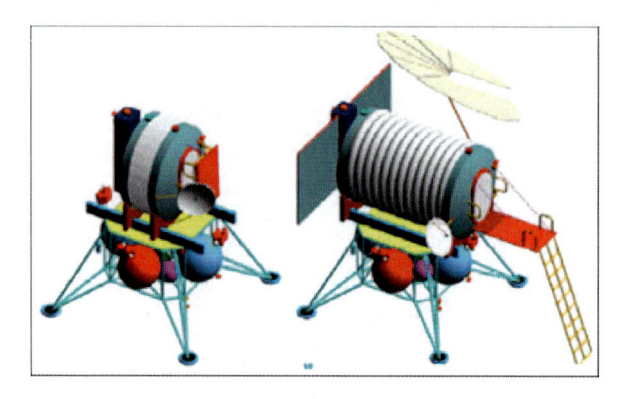

Fig. 20.8 NASA JSC Concept Sketch of the Lunar Lander (credit: NASA / ILC Dover)

The habitat unit would have been landed and deployed automatically prior to the arrival of surface crews (Fig. 20.8 and Table 20.7).

Table 20.7 Project Summary 'NASA-JSC Deployable Habitat'

Design Objective	Configuration and design study of an inflatable habitat unit for a 6 day, 2 crew lunar mission with EVA
	(low cost and packable to fit on existing launch vehicles)
Structural Concept	Structural concept based on hybrid (rigid and inflatable) system:
	flexible composite cylindrical section and rigid composite end-caps; use of Multilayer Insulation
	Dimensions: D: 2.3 m (7.5 ft); L: 3.6 m (12 ft); 8 psi
	Materials: Elastica (fabric, webbing), Vectran (yarn), Coated fabric (multiple bladder), Urethane (coating)
Deployment Concept	Inflatable deployable configuration; "bellow-like expansion"
	Industrial-terrestrial manufacture, assembly and pre-packaging of elements, and automated in-situ deployment
Habitability Aspects	- The automatically landed habitat systems can be used to build up a network of surface assets for an advanced lunar human exploration mission
	- Hybrid structural concept combines different requirements

NASA-JSC Deployable Lunar Habitat (Lunar Return Program Study); ILC Dover under contract to NASA JSC; 1996-1997; Design Study

The deployable structure featured a cylindrical bellow-like system with a flexible mid-body and rigid end-caps housing the technical hardware. The fabric section was built from various layers, including multilayer insulation and a thermal micrometeoroid cover shell (ILC Dover 2011). Out of a number of considered folding concepts for the floor, foldable plates were selected.

The study defined the materials and detailed interfaces between the rigid end-cap, the bladder and the Thermal Micrometeoroid Cover Shell. Mass, volume and stress analysis were performed, and all system components were defined (Cadogan 2007).

20.4.5 Kopernikus Lunar Base

The Kopernikus Lunar Base concept was produced at the Lunar Base Design Workshop, hosted by ESA/ESTEC in 2002, by a group of architecture and engineering students from the Vienna University of Technology, Technical University of Eindhoven and Colorado School of Mines (Fig. 20.9 and Table 20.8).

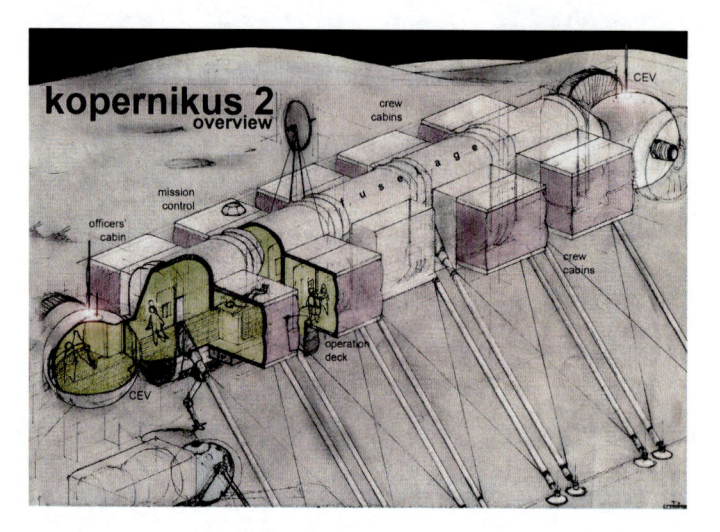

Fig. 20.9 An advanced configuration of the Kopernikus Lunar Base (credit: K. Ozdemir)

The Kopernikus station design project featured a commercial station to provide logistic and other services for other stations in the South Pole region. The station has a linear pressure vessel system, which is elevated from ground by a support structure.

Table 20.8 Project Summary 'Kopernikus Lunar Base'

Design Objective	Design of a Lunar base concept for the settlement phase of the human lunar exploration
Structural Concept	Hybrid, prefabricated, modular structure, to be transported in packaged-configuration and assembled at the mission site
Deployment Concept	Expendable cabin units deploy by pressurization of the interior volume
Habitability Aspects	- Deployable units provide the crew with a good level of habitable space, while housing the essential technical infrastructure (e.g. LSS) - Modular Expansion concept

Kopernikus Mobile Lunar base; K. Ozdemir, R. Waclavicek et al. (HB2 - Vienna University of Technology) with L. Johnson, P. Mayr, G. Pamperl, F. Verschüren; 2002; Concept Study

The main hull of the station is formed by rigid and pressurized habitable units that are attached end-to-end, like the International Space Station. This rigid body forms the backbone of the station, on which the additional units are plugged.

The backbone houses the inner transfer corridor, connecting all the habitable units, as well as life-support and power systems. Crew cabins, command structure and the medical station are attached to the sides of the backbone hull. Crew cabins are partly inflatable modules that are stowed in a folded configuration in transport and expanded after assembly to the backbone (Ozdemir 2009).

The deployable crew cabin units support the expansion of the station within a modular system, while providing mass and volume benefits in transport and assembly phases. The hybrid structure of the units is configured to house the essential hardware components (e.g. technical systems, built-in furniture, etc.) as well as plenty of habitable space for the crew.

20.4.6 Tycho Rolling Lunar Base

The Tycho Lunar base concept was also developed at the Lunar Base Design Workshop and the Vienna University of Technology (Mohanty et al. 2003). The mission scenario for the mobile lunar base Tycho was based on the idea of mining Helium-3. The main task of the crew was scientific research, supervising the mining and scouting of new areas.

Fig. 20.10 Model of the Rolling Lunar base (credit: Haeuplik-Meusburger, Lorenz, Springer - HB2; Foto: F. Schachinger)

The self-imposed goal for the design team was to design a mobile surface habitat that would support the mining process and provide adequate protection from radiation and micrometeoroids. The concept of a rolling habitat was developed (Fig. 20.10 and Table 20.9). The spherical lunar base would roll by mass displacement, a few meters every week. The driving system is based upon three spheres and the inner sphere contains the habitat. The middle sphere is partly filled with water that works as a slider liquid. The outer layer forms the radiation shielding which is provided by lunar-water, a by-product of the mining process. In the parking position all three spheres are connected through the airlock and the system is balanced. To start rolling, air-cushions are inflated on one side, and deflated on the opposite side. Water is pushed from one side to the other and the total centre of gravity is displaced.

Table 20.9 Project Summary 'Tycho Lunar Base'

Design Objective	Design of a mobile lunar surface habitat with integrated radiation shielding using ISRU
Structural Concept	Three spherical layers enclose the habitat; the middle sphere is filled with water; used as radiation shielding and driving system
Deployment Concept	Main structure is deployed and rigidized; 'inflatable windows' are deployed on demand to displace the surrounding water and clear the view to the outside
Habitability Aspects	- A mobile surface habitat that has sufficient radiation protection and makes use of in-situ resources;
	- Development of the Inflatable Window concept
	- Adaptable to a great variety of interior configurations

Tycho Mobile Lunar base; S. Haeuplik-Meusburger, S. Lorenz, M. Springer et al. (HB2 - Vienna University of Technology) with J. Bernabeu, Ch. Nicolas, V. Pop; 2002; Concept Study

The "Inflatable Window" proposes a further layer of adaptability. Inflatable cushions are mounted onto a spherical grid and independently pressurized. Inflating a cushion will displace the surrounding water and clear the view to the outside (Häuplik 2003).

20.4.7 Mobitat

The 'Mobitat' concept was developed by Plug-in Creation Architecture as a self-contained, mobile pressure vessel habitat for use on the Moon and other planetary surfaces (Fig. 20.11 and Table 20.10). It was inspired by the Habot Mobile Lunar Base Concept by John Mankins, NASA (Cohen 2003; Howe 2004).

In contrast to the Habot concept, the Mobitat concept is deployable in order to fit in a variety of launch vehicles. Each lunar base module has its own integrated descent rocket engines and crane. The lunar base system measures 7.2 m in length and 4.2 m in diameter when folded.

The lunar base is composed of a mobile platform and a modular pressure vessel. The mobile platform can be used with or without the pressure vessel. In addition, it supports surface operations such as excavation, drilling and construction implements (Howe 2004). Its height can be adjusted between 1.9 m and 5 m from the surface. The modular pressure vessel can also be assembled into larger bases or used with a rover.

Fig. 20.11 Deployed configuration of the Mobitat on the lunar surface (credit: Concept and design by A. Scott Howe)

Table 20.10 Project Summary 'Mobitat'

Design Objective	Development of a mobile planetary surface base design with an expansion option to larger configuration
Structural Concept	Foldable modular rigid system, Kit-of-parts Theory (Howe 2002)
	Dimensions: W: 4.2 m, L: 7.2 m (pressure vessel W: 4.2 m, H: 3.0 m)
Deployment Concept	The folded package is delivered to lunar orbit and then the structure is deployed; The mobility system is further deployed during the descent to the lunar surface
Habitability Aspects	- The pressure vessel (hexagonal shape) and the mobile platform can be separated; - Both offering various uses (Rover, Modular larger base)

Mobitat; A. Scott Howe (Plug-in Creations); 2004; Design Study

20.4.8 Moonwalker Lunar Base

The Moonwalker lunar base design study is based on the premise of different forms of adaptability and their respective architectural translation (Fig. 20.12 and Table 20.11). Four conceptual levels of adaptability were formulated:

(1) Contextual adaptability: Site changes and mission details must be implemented from the very beginning. Moonwalker possesses legs that function as landing, standing, and locomotion tools, and occasionally as bionic arms. The inflatable rigidized structure offers twelve docking possibilities for plug-ins and expansions thus enabling efficient adaptation to future missions.

Fig. 20.12 Moonwalker – Walking Lunar Base (credit: S. Haeuplik-Meusburger; rendering: Zoom VP)

Table 20.11 Project Summary 'Moonwalker Lunar Base'

Design Objective	Design for a lunar base, presenting maximum flexibility to its users, the environment and the mission objectives
Structural Concept	Inflatable habitat with inflatable and rigid modules, inflatable 'airbags' for additional meteorite protection; the inflated multi-layer membranes are reinforced by a tensile net
	Dimensions: D: 5.5 m, L: 6 m (Packed); D: 11 m, H: 10 m (Deployed)
Deployment Concept	Rigidized inflatable multi-layer structure; the interior membrane can be further adjusted by pneumatic muscles and inflatable cushions
Habitability Aspects	- In addition to adjustments to mission and environmental issues, the lunar base is designed to allow individual adjustment of the user;
	- The interior has built-in-flexibility;
	- The crew quarters consist of two parts, one which relates to the lunar environment and can be expanded to the outside and one inside – that reflects and transports social tendencies

Moonwalker – mobile Lunar base; Sandra Haeuplik-Meusburger (HB2 - Vienna University of Technology); 2005; Design Study

(2) External adaptability: The use of a modular and multi-layered structure permits partial exchange and thus optimal adaptability to specific design challenges e.g. protection from meteorites and solutions for handling lunar dust; a not yet sufficiently investigated radiation environment.

(3)   Internal adaptability: Moonwalker reacts to changing user-preferences. Changes within a society, i.e. crew exchange - imply changes of social and spatial dimensions. The correlation between space and humans is spatially visible and cognitively readable.

(4)   Reflexive adaptability: Cognitive design strategies are implemented. Built-in furniture is designed to react and adapt to different users and uses.

The deployable crew quarter offers adaptability on all four conceptual levels.

20.4.9 *Deployable Structures for a Lunar Base*

The purpose of the study 'Lunar Exploration Architecture - Deployable Structures for a Lunar Base' was to investigate bionic concepts applicable to deployable structures and to interpret the findings for possible implementation concepts (Fig. 20.13 and Table 20.12).

The study aimed at finding innovative solutions for deployment possibilities. Translating folding/unfolding principles from nature, candidate geometries were developed and researched using models, drawings and visualizations. The use of materials, joints between structural elements and construction details were investigated for these conceptual approaches. Reference scenarios were used to identify the technical and environmental conditions, which served as design drivers. Mechanical issues and the investigation of deployment processes narrowed the selection down to six chosen concepts. The applicability of these concerning the time-scale of the mission was evaluated on a conceptual stage (Gruber et al. 2007).

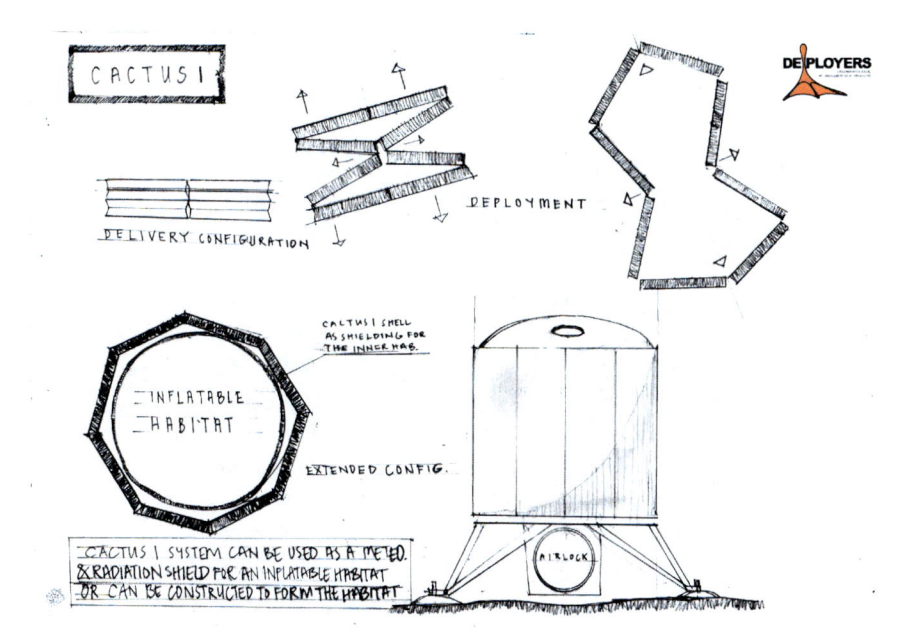

Fig. 20.13 Deployable envelope for a lunar habitat (Image: HB2, TU Vienna; drawing: K. Özdemir)

Table 20.12 Project Summary 'Study Deployable Structures for a Lunar Base'

Design Objective	Deployable structural concepts for the human exploration in the lunar environment; formal transformation based on biological role models
Structural Concept	Different concepts ranging from Rigid Metal Plates to Flexible Membranes
Deployment Concept	Inflation (Membranes) and/or Unfolding (Rigid Plates); Industrial-terrestrial manufacture and pre-packaging of elements; automated in-situ deployment
Habitability Aspects	- A variety of folding / deployment concepts were developed - Deployable structures provide the lunar crews with increased habitable volume and protection from environmental hazards (e.g. radiation, micro-meteorite impacts)

Deployable Structures for a Lunar Base; P. Gruber, B. Imhof, S. Haeuplik –Meusburger, K. Oezdemir, R. Waclavicek, H. Koch, V. Kumer (HB2 - Vienna University of Technology) and G. Jeronimidis (University of Reading) through a contract from ESA and Thales Alenia Space; 2006; Design Study

The deployable structures were intended mainly for structural consolidation of the surface assets of lunar exploration architecture. As an example, 'Cactus 1' adopted the horizontal growth pattern of a cactus plant in order to transform folded plates into a shielding envelope.

20.4.10 Self-deploying Fully Automated Solar Farm on the Moon

Placing the emphasis on the co-existence of 'Robospheres' and 'Biospheres' of the human space flight endeavor, the 'Spaceships That Think! Consortium' presented an advanced concept sketch of a future solar farm on the lunar surface comprising self-deploying robotic units. This farm is designed to be operated and maintained by using robotic systems. Human intervention is thought to be used only if absolutely necessary.

The self deploying architecture of the lunar farm is based on hybrid structure (rigid & inflatable) modules that are delivered in compact form, positioned into the planned location by robotic elements and deployed to their final configuration (Fig. 20.14 and Table 20.13). In order to ease the deployment and assembly, the stowed layout of the elements are mostly based on cylindrical forms. Using a standard configuration for basic units can provide advantages in each step of the exploration architecture. (e.g. design, development, transport, assembly and operation).

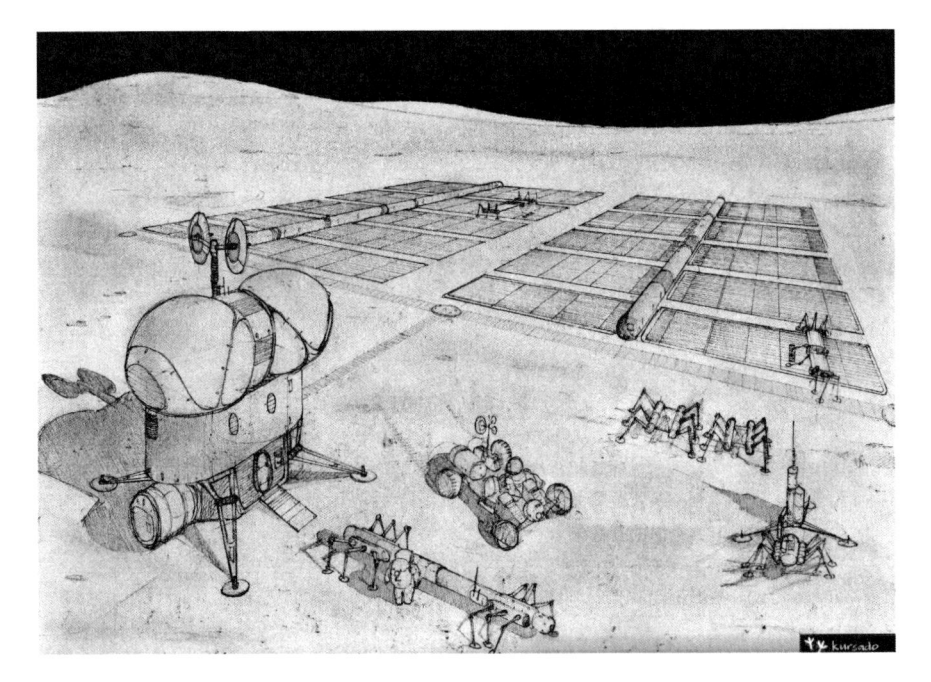

Fig. 20.14 Solar Farm on the Moon (credit: K. Ozdemir)

Table 20.13 Project Summary 'Self-Deploying Fully Automated Solar Farm on the Moon'

Design Objective	Deployable surface elements for a robotic lunar base (pre-human lunar exploration phase)
Structural Concept	Structural concepts based on hybrid (rigid and inflatable) systems
	Rigid Metal Elements, Flexible Membranes
Deployment Concept	Inflation (Membranes) and Unfolding (Rigid Plates)
	Industrial-terrestrial manufacture, assembly and pre-packaging of elements, automated in-situ deployment
Habitability Aspects	- The self-deploying, self-cleaning, self-repairing robotic infrastructure can increase the level of habitability and operational efficiency while reducing the workload and related risks for the crews engaged in surface missions

Design for a robotic Lunar Base; S. Mohanty, K. Ozdemir, S. Songur; 2006; Scenario Design

20.4.11 *InFlex Lunar Habitat*

The 'Intelligent Flexible Lunar Habitat Study (InFlex)' features a demonstrator structure, where various issues in deployable habitat system design, ranging from volumetric layouts to door-sizing factors can be assessed on an optimal test bed.

Fig. 20.15 The planetary surface habitat and airlock unit (credit: NASA / ILC Dover)

Table 20.14 Project Summary 'NASA InFlex Lunar Habitat'

Design Objective	Deployable surface habitat for the human lunar exploration
Structural Concept	Structural concept based on hybrid (rigid and inflatable) system
	Rigid Metal Elements, Flexible Membranes
	(polyester webbings, coated fabric)
Deployment Concept	Inflation (Membranes) and Unfolding (Rigid Plates)
	Industrial-terrestrial manufacture, assembly and pre-packaging of elements, automated in-situ deployment
Habitability Aspects	- The InFlex Habitat Demonstrator provides a test bed for detailed habitability assessments ranging from volumetric layouts to door-sizing factors

InFlex Lunar Habitat; ILC Dover for NASA LaRC; 2007; Prototype

The demonstrator contains two main units, one resembling the habitable module including windows and the other an airlock (Fig. 20.15 and Table 20.14). The two vertical configuration cylinders are arranged to hold a crew of 2 with suits and hardware. The structure is sized to be representative in mass and geometry, but fabricated from low-cost polyester webbings and coated fabric. Research issues include Integrated Health Monitoring Systems, Signal Transfer Systems, Self Healing Bladder Materials, Anti-Microbial Materials, Low Permeation Materials, and Enhanced Radiation Protective Materials. Studies on hatch sizing and outfitting studies were conducted (Cadogan 2007; ILC Dover 2011).

20.4.12 Moon Base Two

'MoonBaseTwo' is an inflatable laboratory located at the Shackleton Crater near the lunar South Pole (Fig. 20.16 and Table 20.15). It is a further development of the inflatable laboratory 'MoonBaseOne'.

Fig. 20.16 Artist's impression of the 'MoonBaseTwo' in operation on the Moon (credit: Architecture and Vision)

Table 20.15 Proejct Summary 'Moon Base Two'

Design Objective	A lunar base for 4-6 astronauts for 6 months
Structural Concept	Inflatable dome-shaped habitat with radiation protection; automatically deployed after landing
	Dimensions: D: 7.5 m, L: 6 m (Packed); D: 20 m, H: 10 m (Deployed)
	Materials: Aluminum, carbon fiber, Kevlar, MLI, Nomex, Regolith
Deployment Concept	Self-deployment and 5-day set-up; 6-month regolith fill by robotic system; hard floor and inflated volume filled by internal pressure corresponding to Earth sea level atmospheric pressure
Habitability Aspects	- Instant deployable habitat;
	- Large habitable volume compared to transport volume ;
	- Adjustable radiation protection

Moon Base Two; Architecture and Vision (Arturo Vittori, Andreas Vogler); 2007; Design Study

Conceived as a long-term base for conducting in-situ research and to explore the surrounding environment; it will also help in the study of permanent human settlements far from the Earth. Designed to be transported to orbit and launched by the Ares V rocket, the station automatically deploys after landing on the lunar surface. After inflation, outside bags will be filled with Moon Regolith to protect the crew from radiation. The Lunar base is designed to accommodate a crew of up to 4 people for 6 months (AV 2011).

20.4.13 IHAB - Lunar Inflatable Habitation System

The inflatable habitation system proposed by Thales Alenia Space is designed for a nominal crew of 4 that will stay for 6 to 12 months. The crew delivery and return is performed by the US Crew Excursion Vehicle CEV/LSAM system (Musso et al. 2006, 2008).

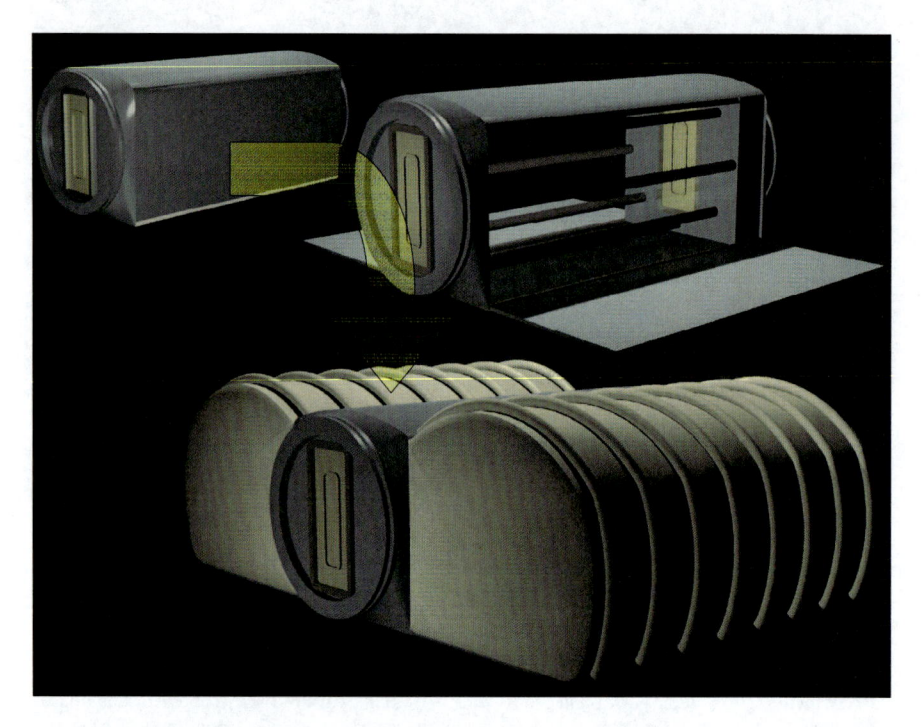

Fig. 20.17 Preliminary Concept (credit: Thales Alenia Space)

Within the study, a variety of configuration concepts were traded for different mission scenarios. The concept shown below consists of a rigid core in combination with inflatable elements (Fig. 20.17 and Table 20.16). After automatic deployment, the habitat will be covered with lunar regolith sacks as radiation protection. Robotic devices excavate regolith, transport and fill the sacks to cover the habitat.

Table 20.16 Project Summary 'IHAB – Lunar Inflatable Habitation System'

Design Objective	A habitat for a crew of 4 with a surface stay for 6 to 12 months
Structural Concept	Rigid Core: L: 5.9 m, D: 5.5 m, V: 126 m³ Inflatable Module: L: 5.9 m, D: 6.0 m, V: 164 m³
Deployment Concept	Structural restraint ribbon net holds pressure load, multilayer bladder layer; different deployment concepts evaluated
Habitability Aspects	- Detailed assessment/development of composite structural system for habitats - Development of hybrid folding / deployment concepts

IHAB-Lunar Inflatable Habitation System; Thales Alenia Space, Italy; 2008; Design Study

20.4.14 Evolutionary Growth Lunar Base

The scope of the Minimum Functionality Habitation Element Systems Concept Study program was wisely conceived to address not only requirements to meet barest living and work essentials, but to consider evolutionary pathways forward as well. This entails a strategic perspective of progressive growth sequences ranging

Table 20.17 Project Summary 'Evolutionary Growth from Minimum Functionality Habitation Element to Deployable Lunar Base'

Design Objective	A multifaceted strategy to link NASA Minimal Functionality Habitable Element (MFHE) requirements to a compatible growth plan; leading forward to evolutionary, deployable habitats including outpost development stages
Structural Concept	Conventional modules and hybrid (hard shell with inflatables attached) structures
Deployment Concept	Arrival-ready operational status of hard shell modules with a fully automated deployment, all major equipment systems in place and functioning, and soft shell structures inflated upon arrival
Habitability Aspects	- Larger capacity modules can accommodate greater internal layout versatility for shared, dedicated and evolutionary uses; - and provide expanded stowage capacities for consumables, maintenance tools, parts, and research/housekeeping supplies

Minimal Functionality Habitation Element study (NASA LSSCS BAA); SICSA (Larry Bell (Director), Olga Bannova (Research Professor), graduate students: Harmon Everett, Jessica Corbett, Frank Eichstadt, Luke Schmick); 2008-2009; Design Study from early expeditionary missions – to operational outposts – to more self-sufficient settlements that process in-situ resources – and potentially forward to envision lunar commercialization industries and launch vehicle propellant production.

The hybrid hard module-soft expandable approach presents important advantages afforded by each structure type. The hard shell enables equipment and utility systems to be pre-integrated and checked out prior to launch, readily accommodates external berthing interfaces and other penetration seals, and applies proven technology (Fig. 20.18 and Table 20.17).

The soft deployable section conserves launch volume for companion payloads while expanding deployed volume functionality and crew comfort.

Fig. 20.18 Evolutionary growth lunar base configuration, the Hopping Crane is attached to the Lander, CLAM deployable module and HabLog modules attached with inflatable connecting tunnels (credit: SICSA)

20.4.15 Lunar Outpost Scenario 12.1

The Constellation Outpost consists of elements assembled from multiple launches, a so-called Class II construction. Once the lander has reached the lunar surface, an 'All-Terrain Hex-Limbed Extra-Terrestrial Explorer' mobility system, called 'Athlete', is used to offload, assemble and construct the outpost.

The habitat concept is based on mobile and modular pressurized elements that can be adapted to the designated location on the lunar surface. Three types of pressurized elements are used for the configuration: a small 12 m³ pressurized Lunar Electric Rovers (LER), a large 56 m³ cylindrical habitat module, and an inflatable 9 m³ external airlock (Fig. 20.19 and Table 20.18). The airlock accommodates two suits and has a deployable porch. It is stowed for launch and deployed on the lunar surface (Howe et al. 2010).

Fig. 20.19 Lunar Electric Rovers and a habitat module with inflatable airlock carried by an Athlete mobility system (credit: NASA)

Table 20.18 Project Summary 'Lunar Outpost Scenario 12.1'

Design Objective	Establishment of a lunar 'core' capability with redundant habitation volumes that can allow growth and flexibility
Structural Concept	Class II, development of a heavy-lift transportation system
	Cylindrical habitat with end domes and truss frame on top and bottom
	Dimensions: 56 m³ habitat, 12 m³ LER, 9 m³ inflatable airlock volume
Deployment Concept	Inflatable Airlock and foldable EVA porch
	Deployed: 1 m / deployed with EVA porch 4 m+
Habitability Aspects	- Extension of crew operations on the lunar surface
	- Modular growth capabilities
	- Adaptability to environmental and operational requirements

Constellation Outpost; NASA; 2010; Design Study

20.4.16 X-HAB

X-Hab is a structural concept prototype, developed by ILC Dover in frame of inflatable habitat development studies under the NASA ESRT project as part of the Constellation Program for NASA (Fig. 20.20 and Table 20.19).

The "mid-structure" expandable space habitat is comprised of two rigid end-caps and a flexible middle segment. The flexible body is stored between the rigid end-caps in the flight configuration, in a 1 m segment, which is to be deployed on the building site to reach a length of 5 m. Besides the logistics, storage is designed to be positioned in the rigid end sections. The flexible mid-segment shell is made from Vectran. The pressure bladder of the structure is designed to operate with 9 psi internal pressure and the bladder structure is manufactured from a specialized coated fabric in order to contain the inflation gas. The deployable floor unit is a light weight rigid structure included in the delivery package of the habitat system (ILC Dover 2010).

Fig. 20.20 Deployment Testing of the X-Hab Lunar Habitat at NASA LaRC (credit: NASA / ILC Dover)

Table 20.19 Project Summary 'X-HAB'

Design Objective	Development of lightweight inflatable structures that would enable a mid-structure expandable space habitat to function on the lunar surface
Structural Concept	Structural concept based on a hybrid (rigid/inflatable) system
	Rigid Metal Elements, Flexible Membranes
	Dimensions: L: 1.0 m (Packed); L: 5.0 m (Deployed), 9 psi
Deployment Concept	Deployment of the structure along the longitudinal axis by inflation
	Industrial-terrestrial manufacture, assembly and pre-packaging of elements, automated in-situ deployment
Habitability Aspects	- Improved habitability by greatly increasing the habitat volume to number of crew ratio;
	- Usable in packed and deployed modus;
	- Development of know-how on hybrid system design & testing

X-Hab Lunar Habitat ; ILC Dover ; 2010 ; Prototype built

20.4.17 Minimum Function Habitat

The Minimum Function Habitat is the product of a design study by a partnership of ILC Dover, Hamilton Sundstrand, and SICSA under contract to NASA in support of NASA's Constellation Program (ILC Dover 2010).

The habitat was designed to accommodate a crew of 4 astronauts for 16 days on the lunar surface. The assessed structure is a 'bare bones' habitat element that incorporates the benefits of flexible materials (Fig. 20.21 and Table 20.20). Besides multifunctional outfitting schemes for diverse operational concepts, analyses on systems-material-component design were made during the study. Designs include cylindrical modules for different uses (e.g. storm shelter, crew quarters), providing more than 5 m³ of habitable space for each crew member (NASA 2009, ILC Dover 2010).

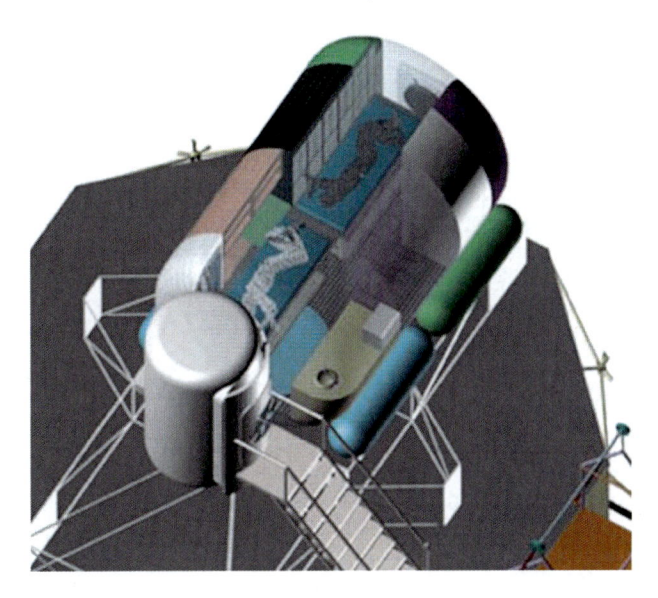

Fig. 20.21 Minimum Function Habitat with Inflatable Pod Option (Image: ILC Dover)

Table 20.20 Project Summary 'Minimum Function Habitat'

Design Objective	The assessment of a 'basic habitat element' that incorporates the benefits of flexible materials
Structural Concept	Deployable modules, based on flexible materials in combination with rigid structures
	Dimensions: D: 3.0 - 4.5 m, L: 4.5 m (Deployed)
Deployment Concept	Inflation driven deployment of flexible modules integrated within a hybrid structural system
Habitability Aspects	- Extensive analyses on habitation system components (e.g. expandable habitat, inflatable airlock, deployable radiator, etc.);
	- The study produced valuable know-how on deployable habitat architecture

Minimum Function Habitat; ILC Dover in partnership with Hamilton Sundstrand and SICSA; 2009; Design Study

20.5   Resume and Outlook for Innovative Applications for Lunar Habitation

As human spaceflight gained the ability of extra vehicular activities, the first generation of spacesuits marked a new era of space habitats that are deployable, inflatable and tailored to human scale. Utilizing the products of research and technology, suit designers created deployable structures composing different forms of materials like nylon, Teflon, vinyl and many others. Continuous development of these 'minimal habitats' provided valuable know-how on inflatable structures, taking the available materials, construction and operation techniques into focus. The latest spacesuit designs introduce hybrid systems, as their deployable habitat counterparts do. These optimize the lightness and flexibility of the inflatable components and the structural integrity of the rigid elements within a composite system.

Once on the Moon, astronauts made the best out of their limited surface time under extreme conditions of the lunar surface. Restriction to the walking distance was overcome by integrating a deployable surface vehicle into the mission. Apollo 15, 16 and 17 crews enjoyed an extended exploration radius of app. 10 km (Eckart 1999) provided by the Lunar Roving Vehicle. Foldable and light, the Lunar Rover emerged as a key asset of human exploration of the Moon. With its crews in spacesuits, the Lunar Rover system presented a combination of deployable systems, integrated into a surface exploration platform.

Currently, lunar exploration system architectures are highlighted by systems capable of supporting surface crews for various mission profiles. The efforts toward augmented capabilities indicate the need for innovative structural and operational systems. The deployable habitation systems, therefore, are among the critical assets that can be integrated into the missions. Gaining a momentum through the design and development studies, deployable systems for lunar habitation will be improved in the coming era of spaceflight with tighter budgets and more demanding goals than ever, marking their advantages ranging from transportation to operation in lunar exploration missions.

Deployable structures have the advantage of minimizing volume and weight during launch while at the same time maximizing living space at a remote destination. The integration of adaptable and flexible structures can further lead to an even more sustainable concept of a lunar base by integrating the idea of reusing materials and available resources. Further technical-functional as well as social-spatial adaptability can improve habitability and suitability for its inhabitants in a specific environment over a certain period of time (cf. Häuplik-Meusburger at al. 2009).

Acknowledgements. The authors wish to thank the following people for their assistance in preparation of this chapter (in alphabetical order): Manuela Aguzzi (EAC Cologne/ALTEC SPA), Olga Bannova (SICSA), Scott Howe (Plugin-Creation), Daniel Lee, Susmita Mohanty (Earth2Orbit), Giorgio Musso (Thales Alenia Space), Polina Petrova (HB2, VUT Vienna), Stephen Ransom (Liquifer Systems Group) and Andreas Vogler (Architecture and Vision).

References

Abramov, I.P., Skoog, A.I.: Russian Spacesuits. Springer-Praxis Books (2003)

Baker, D.: Lunar Roving Vehicle: Design Report. Spaceflight 13, 234–240 (1971)

Bigelow, Online 1999-2010, http://www.bigelowaerospace.com/

Cadogan, D., Stein, J., Grahne, M.: Inflatable Composite Habitat Structures for Lunar and Mars Exploration. Acta Astronautica 44(7-12), 399–406 (1999)

Cadogan, D.: Overview of Inflatable Airlock & Related Technologies: Technical Briefing for the Future In-Space Operations Colloquium. ILC Dover, April 25 (2007)

Cohen, M.M.: Mobile Lunar and Planetary Bases (AIAA 2003-6280). In: AIAA Space 2003 Conference & Exposition, American Institute of Aeronautics and Astronautics, Reston, September 23-25, pp. 2003–6280 (2003)

Eckart, P.: The Lunar Base Handbook. Mc Graw Hill, New York (1999)

Freeland, R.E., Bilyeu, G.D., Mikulas, M.M.: Inflatable Deployable Space Structures Technology Summary. IAF-98-1.5.01 (1998)

Gruber, P., Haeuplik, S., Imhof, B., Oezdemir, K., Waclavicek, R., Perino, M.: Deployable structures for a human lunar base. Acta Astronautica 61(1–6), 484–495 (2007)

Häuplik, S.: Space and Society. In: AIAA 2003-6242, AIAA Space 2003 Conference & Exposition, Reston, Virginia, USA, September 23-25, pp. 2003–6242 (2003)

Häuplik-Meusburger, S., Sommer, B., Aguzzi, M.: Inflatable Technologies: Adaptability from Dream to Reality. Acta Astronautica 65, 841–852 (2009)

Häuplik-Meusburger, S.: Architecture for Astronauts: An Activities based Approach. Springer Praxis Books (2010)

Howe, S., Howe, J.W.: Mobitat: Mobile Planetary Surface Bases, White paper, submitted in response to the NASA Exploration Systems Enterprise, Request for Information RFI0421(2004)

Howe, S.A., Spexarth, G., Toups, L., Howard, R., Rudisill, M., Dorsey, J.: Constellation Architecture Team: Lunar Outpost 'Scenario 12.1' Habitation Concept. In: Proceedings of the Twelfth Biennial ASCE Aerospace Division International Conference on Engineering, Science, Construction, and Operations in Challenging Environments (Earth & Space 2010), Honolulu, Hawaii, March 14-17 (2010)

Kennedy, K.J.: The Vernacular of Space Architecture. In: Howe, A.S., Sherwood, B. (eds.) Out of This World: The New Field of Space Architecture, ch. 2, pp. 7–21. American Institute of Aeronautics and Astronautics, Reston (2009)

Kuhn, M.: Archplus, Textile Architektur. Glossar Zum Leichtbau 107, 66–69 (1991)

ILC Dover, Website on Habitat Concepts (2011), http://www.ilcdover.com/Habitats-and-Shelters/

L'Garde (2010), http://www.lgarde.com/

Mohanty, S., Imhof, B., van Susante, P.J.: European Lunar Base Concepts (SAE 2003-01-2652). In: 33rd International Conference on Environmental Systems (ICES), Vancouver, British Columbia, Canada, Warrendale, Pennsylvania, USA, July 7-10, pp. 7–10 (2003)

Musso, G., Buffa, D.: Inflatable Habitat Study Program, Presentation at the 3rd European Workshop on Inflatable Space Structures, ESA-ESTEC, October 10 - 12 (2006)

Musso, G., Lamantea, M., Buffa, D.: Lunar Inflatable System, Presentation at the 4th European Workshop on Inflatable Space Structure, ESA-ESTEC, May 16-18 (2008)

NASA [GRIN], Great Image In NASA - Inflatable Station Concept (1961), GPN-2003-00106, online 2010, http://grin.hq.nasa.gov/ABSTRACTS/GPN-2003-00106.html

NASA [History] Marshall Space Flight Center History Office, Huntsville Alabama, http://history.msfc.nasa.gov/

NASA, Lunar Stay Time Extension Module (STEM), Final Report, NASA Langley Research Center, Hampton, Virginia, Goodyear Aerospace Corporation; N67-33706 (1965)

NASA, Early Lunar Shelter Design and Comparison Study, NASA, George C. Marshall Space Flight Center, Washington D.C., US. N72-75858 (1967)

NASA, Concepts for Manned Lunar Habitats. NASA, Langley Research Center, Hampton, Virginia, US, NASA-TM-104114 (1991)

NASA, Reference Mission Version 3.0, The Reference Mission of the NASA Mars Exploration Study Team (June 1998), http://ares.jsc.nasa.gov/

NASA, International Space Station History, TransHab Concept (2003), online 2011, http://spaceflight.nasa.gov/history/station/transhab/index.html

NASA, The Apollo Lunar Roving Vehicle. Goddard Space Flight Center, online 2005, http://nssdc.gsfc.nasa.gov/planetary/lunar/apollo_lrv.html

NASA, Final Presentation on Minimum Functionality Habitation Element, Lunar Surface Systems Workshop, US Chamber of Commerce, Washington DC, US (2009)

NASA, Space Suit Evolution - From Custom Tailored to Off-The-Rack, online 2010, http://history.nasa.gov/spacesuits.pdf

NASA, Human Space Flight Exploration Imagery (1989), online 2007, http://spaceflight.nasa.gov/gallery/images/exploration/lunarexploration/html/s89_26097.html

NASA, Mars Exploration Rovers Website, online 2011, http://marsrover.nasa.gov/mission/spacecraft_edl_airbags.html

Ozdemir, K.: A methodical approach to the transfer and the integration of design knowledge from terrestrial extreme environment structure designs to inhabited space structure design concepts. PhD Thesis, Vienna University of Technology. Institute for Architecture and Design, Vienna (2009)

Portree, EVA and the Soviet Manned Lunar Mission Plan (copyright 1997-1998), online 2010, http://www.myspacemuseum.com/evasov.htm

Project Apollo - NASSP Website, online 2011, http://nassp.sourceforge.net/wiki/Moonbases

Roberts, M.: Inflatable Habitation for the Lunar Base. In: Second Conference on Lunar Bases and Space Activites of the 21st Century, NASA, published by the Lunar and Planetary Institute, Houston (1988) LPI Contribution 652

SATC, Space Architecture Technical Committee for education, research and commercial aerospace architecture publications (2002), online 2011, http://www.spacearchitect.org

SICSA, Houston, TX, US. SICSA Outreach: Inflatable Space Structures, Sasakawa International Center for Space Architecture, Houston, TX, US, vol. 1(7) (May – June 1988)

Skoog, Å.I., Abramov, I.P.: Russian Spacesuits. Praxis Publishing Ltd, Chichester (2003)

Smith, A.: Mechanics of Materials in Lunar Base Design. In: Benaroya, H. (ed.) Applied Mechanics of a Lunar Base. Applied Mechanics Review, vol. 46(6), pp. 268–271 (1993)

Vogler, A.: Modular Inflatable Space Habitats. In: First European Workshop on Inflatable Space Structures, ESA-ESTEC, Noordwijk, The Netherlands, May 21-22 (2002)

Young, A.: Lunar and Planetary Rovers. The Wheels of Apollo and the Quest for Mars. Springer-Praxis Books, Heidelberg (2007)

World News, World News Network: Aleksei Leonov (2011), http://wn.com/Aleksei_Leonov

21 Curing of Composite Materials for an Inflatable Construction on the Moon

Alexey Kondyurin

School of Physics, University of Sydney, Australia

21.1 Where Will We Live While on the Moon?

The Moon is the nearest celestial body to our Earth. Humans first visited the Moon in 1969. However, only short term stays have been possible. In order for long-term missions to be possible, large pressurized constructions are needed. The 15-20 m^3 Altair habitat planned in the Constellation Program and the 6.65 m^3 pressurized crew compartment volume realized in the Apollo program are insufficient. Hundreds of cubic meters per crew member are required for living area, working area, greenhouse with sufficient plants and animals for food, air and water recovery and storage. Projects discussing such large metal constructions delivered from Earth were proposed since the first flight on Moon. However, this method is not realistic: such large construction cannot be launched from Earth (too big mass and size), large construction cannot be landed on the Moon (too big inertia), building of large constructions on the Moon requires a long presence of workers (workers need pressurized cabins, life support system, water, and food that needs large construction to deliver and to keep it), delivering of separate blocks to one place with landing rockets (accuracy of landing of a few meters is too complicate, or it needs Moon's tracks to collect all landed blocks). Therefore, a moon base should be constructed on the Moon itself.

There is only one way to get sufficient volume of the pressurized crew compartment: an inflatable construction. The soft shell of an inflatable construction can be prepared on Earth, folded and transported in a small container to the Moon. Then the shell is inflated to a sufficiently bigger volume than the container. The shell materials should be soft to unfold easily and light to provide a low mass and high durability of the construction.

American and Russian space agencies have an extensive experience in inflatable constructions. The history of inflatable space structures started from the

"Echo", "Explorer", "Big Shot" and "Dash" balloon satellites and "Vostok" airlock in the 1960s (Wilson 1981). Based on the success of the balloon satellite flights new projects utilizing inflatable structures for antennas, reflectors, Moon's and Mars's houses and bases, airlocks and modules based on light polymer films were proposed from these years, for example (Kato et al. 1989; Cassapakis and Thomas 1995; Guidanean and Williams 1998; Sandy 2000; Semenov et al. 2000; Veldman and Vermeeren 2002; Simburger et al. 2002; Pappa et al. 2003; Belvin 2004; Ruggiero and Inman 2006). From these times inflatable structures based on new materials were developed and used for space applications.

However, inflatable structures haven't had a wide application in space exploration because of high risk of damage of the soft inflatable shell. The use of inflatable construction in a high vacuum environment needs a durable frame for the wall. Since the first inflatable construction flights, several methods of rigidization have been discussed: rigidization due to chemical reaction of a soft polymer matrix by thermal initiation of the reaction, by UV-light initiation and by inflation for a gas reaction; mechanical rigidization due to a stressed aluminum layer in the deployed shell; foam inflation; passive cooling below T_g of the material; and evaporation of liquid from gel. In some cases a combination of hard and rigidizable structures was developed. All of these methods were tested in laboratory experiments on Earth. Only one real mechanism of rigidization was successfully tested in real space conditions - Aluminum stressed layers. However, this method is not suitable for large construction.

The best way of rigidization is a chemical reaction of a polymer matrix impregnated by fiber filler, which gives a durable composite material. These materials were tested under superior conditions including real space flight experiments. These materials are used for a wide number of space constructions now on Low Earth Orbits (LEO) and Geostationary Earth Orbits (GEO). These materials have long life-time (up to 20 years) under free space conditions and are certified by all space agencies. Therefore, it appears that a chemical reaction rigidization process is the best method to use on the Moon surface.

For the creation of the construction frame, the fabric impregnated with a long-life matrix (prepreg) is prepared in terrestrial conditions, which, after folding, can be shipped in a space ship to the Moon (Kondyurin 1997; Briskman et al. 2001). The curing process should be slow to keep the prepreg soft. On the Moon's surface the prepreg is carried outside and unfolded by, for example, inflating an internal pocket. Next the chemical reaction of matrix curing should be initiated. The reaction can be initiated with high temperature or UV light irradiation. The required temperature for the chemical reaction can be achieved with solar irradiation or with internal heaters inserted in the prepreg. If the curing reaction requires UV light for the initiation, sunlight or UV lamps can be used to illuminate the prepreg. After complete curing the construction can be pressurized and fitted out with apparatus and life support systems.

However, the curing technology for the composite material on the Moon's surface is not yet developed. The curing process in terrestrial environments is different than it will be on the Moon. The prepreg cannot be placed in a thermobox for precise temperature cycle, as done on Earth. The prepreg cannot be kept at Earth atmospheric pressure to prevent evaporation of active components. The composite material cannot be tested after curing to be sure of the strength and exploitation characteristics of the material, as is usually done for construction materials on Earth. Instead, the curing technology on the Moon's surface will use different principles of composition, curing and testing than on Earth. It should be well developed and proved.

21.2 What Are Problems Associated with Curing a Composite Material on the Moon?

Polymer matrices are very sensitive to lunar conditions, which include high vacuum, cosmic rays, sharp temperature changes, low gravity, and high meteorite flux. The virgin pressure on the Moon's surface is 2×10^{-12} Torr (2.7×10^{-9} Pa) at night (NASA fact sheets, http://nssdc.gsfc.nasa.gov/ planetary/factsheet/moonfact.html). The composition of the Moon's atmosphere is (particles per cubic cm): Helium 4 – 40,000; Neon 20 – 40,000; Hydrogen – 35,000; Argon 40 – 30,000; Neon 22 – 5,000; Argon 36 – 2,000; Methane - 1000; Ammonia - 1000; Carbon Dioxide – 1000. But the pressure in the area near the space ship or other constructions sent from the Earth is much higher due to degassing of the construction walls and exhaust gases of spaceship rocket engines. Even so, the real pressure near the spaceship is likely to be 10^{-3}-10^{-7} Pa, which is lower than the evaporation pressure of polymer matrix components.

The intensity of UV and X-ray irradiations on the Moon's surface is the same as on GEO and LEO. Vacuum ultraviolet (VUV) radiation is a part of solar spectra of irradiation. The intensity of VUV light is low, but the effect of VUV light on polymers is significantly higher than visual and UV light. The level of VUV light at 121.6 nm wavelength estimated at Earth's orbit is about 4×10^{11} photons per cm^2 per second. The Sun's radiation flux is 0.75 mkW/cm^2 in VUV for wavelengths of 100-150 nm and 11 mkW/cm^2 in UV for wavelengths of 200-300 nm. The level of X-rays in the Earth's orbit equals to 2.3×10^{-9} W/cm^2 for wavelengths of 0.1-0.8 nm and 1.43×10^{-10} W/cm^2 for wavelengths of 0.05-0.4 nm. The highest flux of X-rays is directed from Sun and less from stars. These intensities of UV light and X-rays are significant for destruction of polymer materials.

The Moon does not have a magnetic field and dense atmosphere Therefore all high energy particles directed towards the Moon from deep space and from the sun reach the Moon's surface. The energy spectrum of electron and ion fluxes is complex. An investigation of the cosmic particles on the Moon surface has been done during Apollo missions. But a detailed investigation of charged particles including continuous monitoring with time is proceeding mostly in LEO and partially in GEO. The energy spectrum and flux of charged particles depends on the kind of particle, altitude, longitude-latitude, seasons, and solar and geomagnetic activity. The full energy spectrum of charged particles ranges from few eV to 10^{12} eV. The number density of electrons with energy of 0.1 eV at LEO altitude of 400 km is totally 10^5 part./cm^3 (night side) and 10^6 part./cm^3 (day side). The electron flux at GEO mission is typically 10^9 part./cm^2/sec for electrons with energy of 0-12 keV. A typical number of electron density of 1.12 part./cm^3 at average energy of 1.2×10^4 eV and an ion density of 0.236 ion/cm^3 with average energy of 2.95×10^4 eV are measured on GEO missions. The high energy ions are mostly hydrogen ions (90%) and Helium ions (9%) with less than 1% of all other elements. These ion and electron fluxes are measured under the Van Allen radiation belts (at altitude more that 65,000 km) whereas GEO and LEO are at 40,000 km and 300 km altitude respectively. The actual ion and electron fluxes at the Moon surface are higher than at LEO and GEO. Such fluxes of high energy particles cause significant structural transformations in polymer materials.

The average total solar irradiation at Earth's distance from Sun equals to 1362-1367 W/m^2 in dependence on solar activity. The solar irradiation level depends on season (position of Earth on its solar orbit) and it can vary from 1316 W/m^2 at minimal solar energy flux (summer solstice) to 1428 W/m^2 at maximal solar energy flux (winter solstice). The temperature of the construction wall depends on day/night lunar time, altitude position on the Moon, orientation to the Sun, shadowing of lunar rocks, absorption and emission indexes of the wall surface and internal heat sources. The temperature of the wall surface on the Moon is expected to be in wide range from -150 to +150°C. This temperature range is much wider than any temperature ranges applicable to polymers on Earth.

The lunar gravity (g=1.6 m/s^2) is lower than Earth (g= 9.8 m/s^2). The influence of microgravity on polymerization is based on the exclusion of convection and sedimentation processes in curing polymers. In polymers with high concentration of fibers, such as epoxy composite materials, the flow of uncured matrix is low and the gravity doesn't influence on the polymerization process. However, if large construction with uncured matrix is under high gravity, the uncured matrix components could flow down so that the fiber/matrix ratio increases in the top part of the construction. This effect limits the size of the prepreg to about 3 m on Earth. On the Moon the limitation of the construction height due to flow of liquid resin will be less or absent.

Meteorite flux on the Moon's surface is higher than on Earth surface due to absence of Moon's atmosphere. Meteorite flux has a strong effect on brittle materials. It can cause cracking and mechanical destruction of the material and the whole construction. In the case of soft polymers the influence of meteorite flux is negligible because a collision of a meteorite particle with the surface of liquid polymer does not lead to crack formation. Erosion of polymer due to meteorite flux corresponds to 0.1 nm/year in deep space. This is not significant for 20-30 years of exploitation of the composite. At short times of polymerization the probability of meteorite attack is very low. So, the influence of meteorite fluency can be ignored during curing. However, after curing the construction is solid and a protection against meteorites should be considered.

Lunar regolith contains of 40% oxygen, 20% silicon, 12% iron, 8.5% calcium, 7.3% aluminum, 4.8% magnesium, 4.5% titanium, 0.33% sodium, 0.2% chromium, 0.16% manganese, 0.11% potassium and traces of other elements of S (540 ppm), C (200 ppm), N (100 ppm), H (40 ppm), He^4 (28 ppm), He^3 (0.01 ppm). These elements and their oxides do not have specific interactions with proposed polymer matrices and so cannot disturb the chemical reaction of curing. However, lunar dust can cover the uncured prepreg, if the unfolding construction touches the lunar soil. The stacked dust particles can play a role of additional filler in the surface layer of the prepreg so, that this can change the exploitation characteristics of the composite.

Thus, the Moon's environmental conditions such as vacuum, cosmic rays and temperature changes are the most critical factors, which have significant influence on polymers and their curing processes. To decrease the effect of space factors, the projects of inflatable constructions with curable prepreg between two hermetic shells were proposed (Cadogan and Scarborough 2001). In such projects, the uncured prepreg is covered on two sides: an external cover protects against space factors and an internal shell is used for the inflation. However, the inflation of the hermetic shell in high vacuum occurs under very low pressure in the inflating bag. This inflating pressure is close to the vapor pressure of solid polymers and is lower than the vapor pressure of the components of uncured prepreg. Therefore, when the uncured prepreg is isolated with two hermetic bags and placed into vacuum, the inflation can begin solely due to the vapor pressure of the prepreg. In this case, the inflation is uncontrolled and can break the inflating bag. Such damages of the inflating constructions were observed in real flight experiments on LEO.

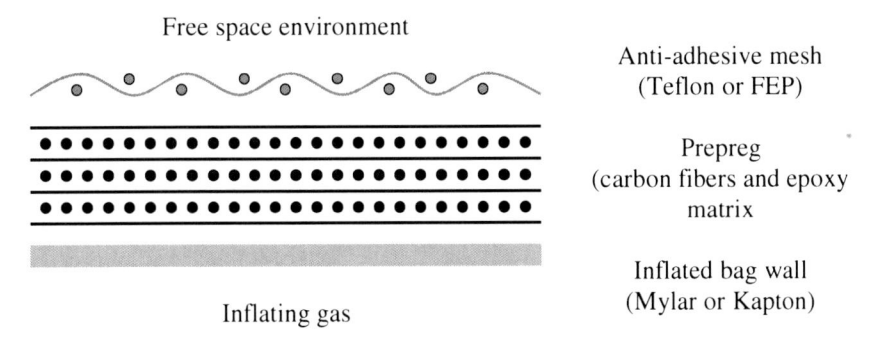

Fig. 21.1 The wall of the inflatable construction with inflatable bag, curable prepreg and anti-adhesive mesh. The prepreg is in direct contact with space environment

We propose a new geometry for the construction wall where the prepreg is exposed directly to free space (Fig. 21.1). The prepreg is covered on one side with the internal hermetic bag used for the inflation. The other side of uncured prepreg is open to the space or covered with rare mesh to prevent a sticking of prepreg layers. The evaporation of uncured prepreg components is free so that does not increase the inflating pressure. This inflation will be under control. But the prepreg should be cured directly in free space environment. Therefore, all factors of the lunar environment influence on curing of the prepreg must be investigated.

21.3 Can We Cure a Liquid Composition in Vacuum?

Most polymer composites certified for space applications are based on epoxy resin compositions with carbon, organics or glass fibers. The curing includes the chemical reaction of the epoxy group with amines, amides, anhydrides, metal-organic complexes or carboxyl acids as hardeners. A large number of different epoxy matrix compositions were developed since the discovery of epoxy resins for construction materials. In the curing reaction two or more components of low molecular mass take part and form crosslinks between molecules. For example, some general schemes of reactions are following:

primary amine hardener

$$R_1 - CH - CH_2 + H_2N - R_2 \rightarrow R_1 - CH - CH_2 - HN - R_2$$
$$\underset{O}{\diagdown \diagup} \qquad\qquad\qquad \underset{OH}{\mid} \qquad\qquad (21.1)$$

secondary amine, amide and urethane hardeners

$$\overset{R_3}{\underset{R_1 - CH - CH_2}{\mid}} + \overset{}{HN - R_2} \rightarrow R_1 - CH - CH_2 - \overset{R_3}{\underset{\mid}{N}} - R_2 \qquad (21.2)$$
$$\underset{O}{\diagdown \diagup} \qquad\qquad\qquad\qquad\qquad \underset{OH}{\mid}$$

tertiary amine hardener

$$R_1 - CH - CH_2 + HO - R_2 \rightarrow R_1 - CH - CH_2 - O - R_2$$
$$\underset{O}{\diagdown \diagup} \qquad\qquad\qquad\qquad \underset{OH}{\mid}$$

$$\vdots$$

$$\underset{R_3 \quad R_4 \quad R_5}{\overset{N}{\diagup \mid \diagdown}} \qquad\qquad\qquad (21.3)$$

These reactions require the presence of two or more components in the definite concentrations. When uncured prepreg is vacuumed, the components evaporate into the vacuum. The rate of evaporation depends on the molecular mass of the components, intermolecular interactions with neighbour molecules, and temperature. If the prepreg is thick enough, the evaporation rates depend on diffusion of the components to the surface of the prepreg. Because the components have different natures, the evaporation rates of the components are different. Therefore, after some time the concentrations of the components will be different than initial. The curing reaction then cannot be completed due to the deficit of the highly volatile components. Therefore, the prepreg cannot be cured and the required exploitation characteristics of the composite are not achieved.

When the evaporation rate of a component is too high, a cavitation effect is observed. The cured matrix contains bubbles and forms even foam. The mechanical strength of the cured composite goes down. This effect was observed in real flight experiment in LEO.

When the evaporation rates of all components are high, the amount of polymer matrix could decrease significantly. This could leave the fibers naked and the composite would then loose its mechanical strength.

American, European and Russian space agencies have certified tests for outgassing of the volatile components. Following these standards, the samples of polymer is vacuumed up to 10^{-3} Pa at temperature of 125°C for 24 hours. The total mass loss (TML), recovered mass loss (RML) and collected volatile condensable material (CVCM) are measured. For example, the ECSS-Q-70-02A ESA Standard requires, that acceptable materials must show RML < 1 % and CVCM < 0.1 %. However, this test cannot be applied to uncured prepreg because at high temperature the curing chemical reaction proceeds continuously during the test. On the other hand, the prepreg will be not kept uncured on the Moon for a long time, but it will be cured at such temperature. These outgassing standards do not reflect a realizable situation for curable materials. New standards and requirements for curing materials must be developed.

Experimental and theoretical investigations of the evaporation kinetics and curing kinetics in vacuum at various temperatures showed that the compositions curable under high vacuum conditions are found (Kondyurin et al. 2001, 2004, 2006, 2009, 2010; Kondyurin 2001, 2011; Kondyurin and Lauke 2003; Kondyurin and Nechitailo 2009). As a result of these investigations, a number of general rules for the selection of curable composition can be summarized:

1. The molecular mass of components must be as high as possible. However, the viscosity of the matrix should be low enough to keep the prepreg sufficiently elastic to be unfoldable;

2. The molecular mass of all components must be as close as possible in order to prevent the development of deficits of any active components.

3. A solvent must be excluded from any stage of the prepreg preparation. Even small traces of the solvent or other low molecular components can lead to a foam structure of the matrix when curing in vacuum.

4. Multifunctional components with 3 and more active groups in one macromolecule are more preferable than macromolecules with 2 active groups. One link of macromolecule to the hardener at the beginning of the curing process decreases the evaporation rate significantly, when other active groups can react late to provide required mechanical strength.

5. The geometry of the prepreg must permit low molecular components to evaporate freely. Any closed volumes and covers must be excluded. If some area of the prepreg is covered, the evaporated fractions can form cavities and foam structure. The thickness of the polymer matrix in prepreg is limited by the volatility of the components. In cases, where thick walls of the construction are required, the prepreg with a 3D complex structure of channels free to space must be used.

Using these rules, the following compositions were successfully cured under high vacuum conditions:

- Epoxy resin ED-20 with Triethanolaminetitanate
- EDT-10 epoxy composition
- ASPM1 prepreg, Alcatel Space
- Polypox epoxy resin with triethanolamine (TEA)
- Epilox epoxy resin with triethanolamine (TEA)
- SE 70 prepreg, Gurit

This list is not complete. Following these rules and experimental investigations, anyone can find compositions curable in high vacuum that will be suitable for specific curing regimes, fibers, exploitation conditions, and price.

However, this long list does not mean that all epoxy resins can be cured in high vacuum. An example is MAP epoxy resins (France). These resins are certified by ESA for space applications. The composition contains a number of low molecular weight active components. These components play a significant role in formation of dense polymer networks, providing strong mechanical properties and low outgassing of cured composite. The composite materials based on this composition excellently passed the outgassing and environmental tests for the ESA standards. However, the low molecular weight components in the uncured composition evaporate into vacuum much more quicker that the curing reaction proceeds. Finally, the polymer matrix disappears from the prepreg in vacuum. Curing is impossible.

In the case of adhesion joints, the polymer matrix must be separated with small dots or lines with open spaces between them. An example of adhesion joint is shown on Fig. 21.2. The adhesive lines provide mechanical strength and the spaces between these lines provide free evaporation of volatile components from the adhesive. The epoxy resin adhesive can be cured in high vacuum for this geometry. Such "space glue" technology can be used on the Moon and as well as in

LEO. For example, the Space Shuttle thermoprotection coating could be repaired with epoxy glue lines where the thermoprotecting ceramic plates are glued to the Shuttle wing even in Earth orbit.

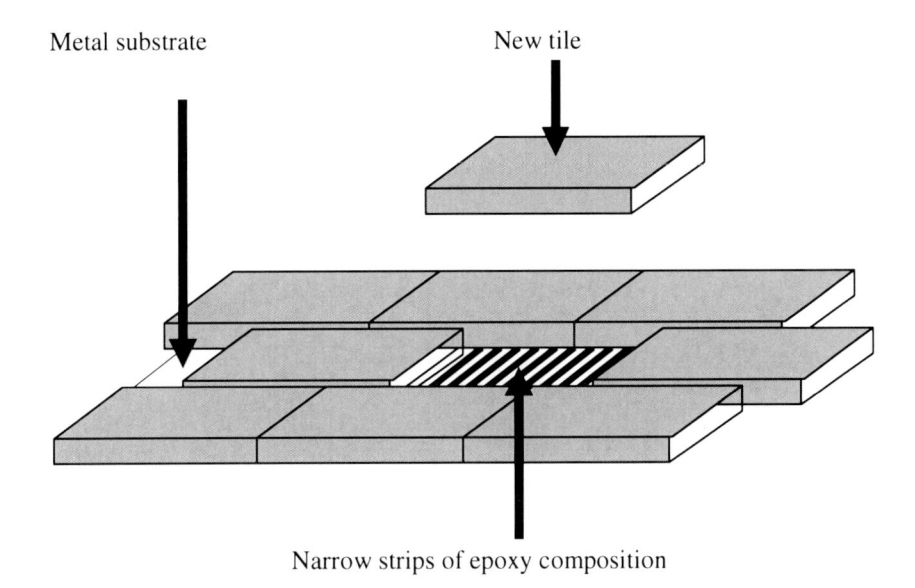

Narrow strips of epoxy composition

Fig. 21.2 "Space glue" is using for repair of thermoprotecting tiles under high vacuum. Narrow strips of the epoxy glue prevent bubbling of the liquid composition in vacuum.

Thus, the high vacuum of lunar environment is not a problem if the composition of the polymer matrix is selected and used correctly.

21.4 How Can Cosmic Rays Help with Curing?

Cosmic rays are a major destructive factor in space environment for all kinds of materials, including polymers. The destructive effect is based on energy transfer from the incoming space particle to an atom or electron of the polymer macromolecule. High energy transfer causes destruction of macromolecules so that atoms leave their positions and move far from mother macromolecule. At this movement, all bonds of the sputtered atom in the macromolecule break and a number of valence electrons, which formerly belonged to the sputtered atom and its neighbours, become unpaired. Therefore, one sputtered atom gives a number of free radicals proportional to the number of valence electrons of the atom. Cosmic ray particles have much more energy than is needed to break the chemical bonds in macromolecules. Thus, after passage of one cosmic ray particle of high energy, the polymer has a number of free radicals.

The free radicals are chemically very active and take part in a chain of chemical reactions:

the reaction of moving the free radical

$$R_1\text{--*CH--CH}_2\text{--R}_2 \rightarrow R_1\text{--CH}_2\text{--*CH--R}_2 \tag{21.4}$$

$$R_1\text{--*CH--CH}_2\text{--R}_2 + R_3\text{--CH}_2\text{--CH}_2\text{--R}_4 \rightarrow$$

$$R_1\text{--CH}_2\text{--CH}_2\text{--R}_2 + R_3\text{--*CH--CH}_2\text{--R}_4 \tag{21.5}$$

when two free radicals meet, they form a double bond

$$R_1\text{--*CH--CH}_2\text{--R}_2 + R_3\text{--*CH--CH}_2\text{--R}_4 \rightarrow \tag{21.6}$$

$$R_1\text{--CH=CH--R}_2 + R_3\text{--CH}_2\text{--CH}_2\text{--R}_4$$

and crosslink

$$R_1\text{--*CH--CH}_2\text{--R}_2 + R_3\text{--*CH--CH}_2\text{--R}_4 \rightarrow \tag{21.7}$$

$$R_1\text{--CH--CH}_2\text{--R}_2$$
$$|$$
$$R_3\text{--CH--CH}_2\text{--R}_4$$

a free radical could break a hydrocarbon backbone

$$R_1\text{--*CH--CH}_2\text{--CH}_2\text{--R}_2 \rightarrow R_1\text{--CH=CH}_2 + \text{*CH}_2\text{--R}_2 \tag{21.8}$$

and oxygen-containing backbone

$$R_1\text{--*CH--O--CH}_2\text{--R}_2 \rightarrow R_1\text{--CH=O} + \text{*CH}_2\text{--R}_2 \tag{21.9}$$

The products of the reactions (21.6)-(21.9) are stable groups. The reaction (21.7) gives crosslinking of the polymer. This reaction has the same strengthening effect as the curing reaction. The reactions (21.8) and (21.9) break the macromolecule, as on opposing reaction for the curing. The ratio of crosslinking to scission reactions depends on the chemical structure of the macromolecules.

The epoxy group can also take part in reactions with free radicals:

$$R_1-CH-CH_2 + {}^*R_2 \rightarrow R_1-CH-CH_2-R_2 \rightarrow \qquad (21.10)$$
$$\overset{\diagdown\diagup}{O} \qquad\qquad \overset{|}{O^*}$$

$$R_1-CH-CH_2-R_2 + {}^*R_3 \rightarrow R_1-CH-CH_2-R_2$$
$$\overset{|}{O^*} \qquad\qquad\qquad \overset{|}{O-R_3}$$

Therefore, the cosmic rays can cause crosslinking of the macromolecules or breaking of macromolecules. The Bisphenol A and F resins are mostly crosslinkable substances under irradiation. Therefore, the compositions based on such resins will be crosslinked under cosmic rays. The effect of crosslinking was observed in laboratory experiments on simulation of cosmic irradiation and in stratospheric flight experiment (Kondyurin et al. 2001, 2004, 2006, 2009, 2010; Kondyurin 2001, 2011; Kondyurin and Lauke 2003; Kondyurin and Nechitailo 2009).

Other components of the polymer matrix could belong to the classes of crosslinkable or scission molecules. Experimental testing of the components under simulated cosmic rays must be used to check it.

The crosslinking effect was observed in the following compositions, which were tested under plasma and stratospheric flight conditions:

- Epoxy resin ED-20 with Triethanolaminetitanate
- EDT-10 epoxy composition
- ASPM1 prepreg, Alcatel Space
- Polypox epoxy resin with triethanolamine
- Epilox epoxy resin with triethanolamine
- SE 70 prepreg, Gurit

These compositions can be used now.

Thus, the cosmic rays from Universe and from the Sun can play a role as an additional hardener, which accelerates the curing effect for a correct choice of the polymer matrix components.

21.5 Curing in Lunar Frost

The curing reaction for all epoxy resin composites is temperature dependent. The curing kinetics of epoxy resin composition proceeds with acceleration and deceleration (Begishev and Malkin 1999):

$$\frac{dC}{dt} = -Ck_1(1 + Ck_2)(1 - Ck_3) \qquad (21.11)$$

where C is the concentration of the epoxy groups, t is time of the reaction, k_1 is the reaction rate, k_2 is the acceleration constant, and k_3 is the deceleration constant. All constants depend on the temperature of curing. The curing rate constants are sensitive to additives, initiators, catalyst and inhibitors of the reaction. The curing kinetics for the proposed composition must be analyzed in detail to predict the curing reaction on the Moon.

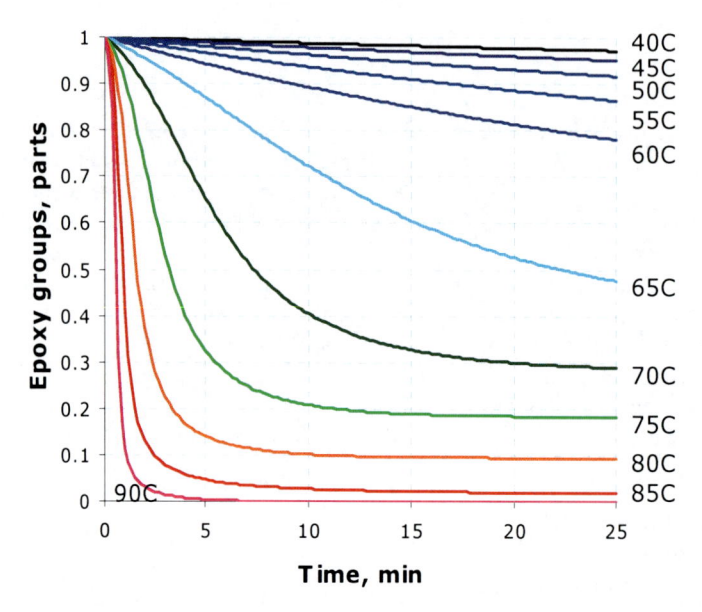

Fig. 21.3 Curing kinetics of SE70 Gurit prepreg based on epoxy composition at different curing temperature. The kinetics was measured by using FTIR spectra

For example, the curing kinetics of Gurit prepreg SE70 is described with the following equations:

$$k_1 = 9.68 \cdot 10^{11} \exp\left(-\frac{12021}{T + 273}\right) \tag{21.12}$$

$$k_2 = 1.55T - 92.6 \tag{21.13-14}$$

$$k_3 = 63 \exp\left(-\frac{T}{15}\right) + 0.8$$

where T is the temperature in °C. The calculated curing kinetics of the composition is shown in Fig. 21.3. Such prepreg can be stored for a long time at room temperature or cooler. After unfolding, the prepreg should be heated by sunlight and cured.

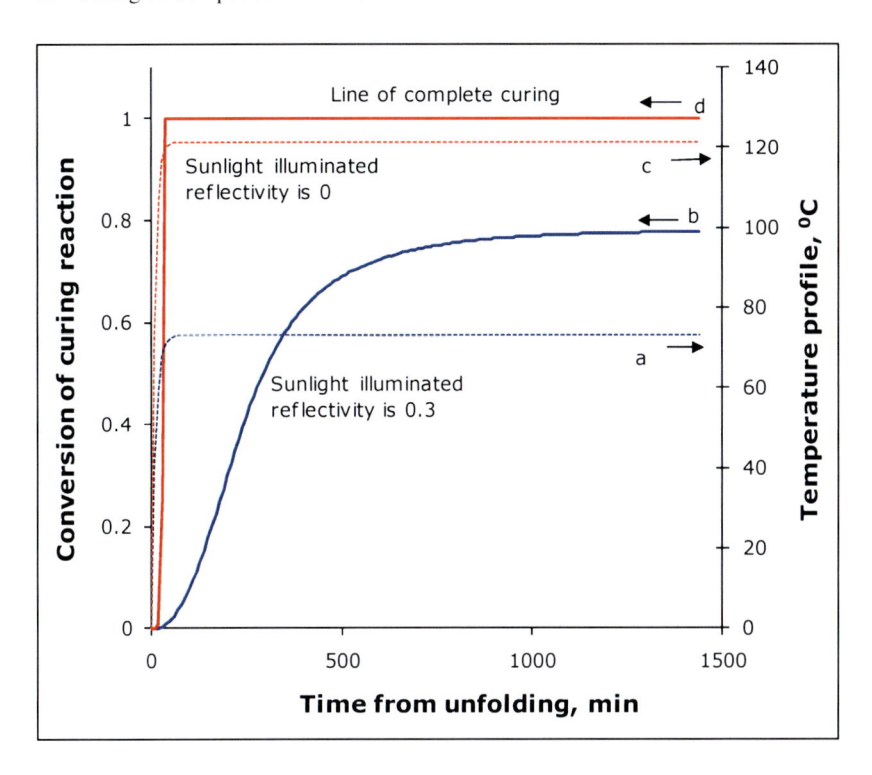

Fig. 21.4 Calculated temperature (a and c) and conversion (b and d) of the curing reaction in a wall illuminated by sunlight. Bottom (red) curves are for a partially reflective surface of the wall (a and b); top (red) curves are for a non-reflective surface of the wall (c and d). The curing reaction is complete after 40 min for the non-reflective wall, while the reaction for a partly reflective wall is still not complete after 24 hours

When the wall is illuminated by sunlight, the temperature elevates up to equilibrium, when the absorbed energy becomes equal to the radiated energy (Fig. 21.4). If all energy from the Sun is absorbed, then the temperature reaches 120°C and the curing reaction is completed after 40 min. When part of the Sun's energy is reflected, the equilibrium temperature depending on the reflectivity can be significantly lower, than is required for the curing. In this case, a long time is needed to complete the curing.

However, present plans for a Moon's base are to place the base in one of the craters near the North or South Poles of the Moon, where there is the highest probability to find water. Solar heating is less available there than on equator. Total shadow or sliding sunlight cannot heat the prepreg enough for the curing reaction. Internal heaters and power sources will be required. The temperature regime can be calculated via the dependence on heat flux (Fig. 21.5). A short intensive

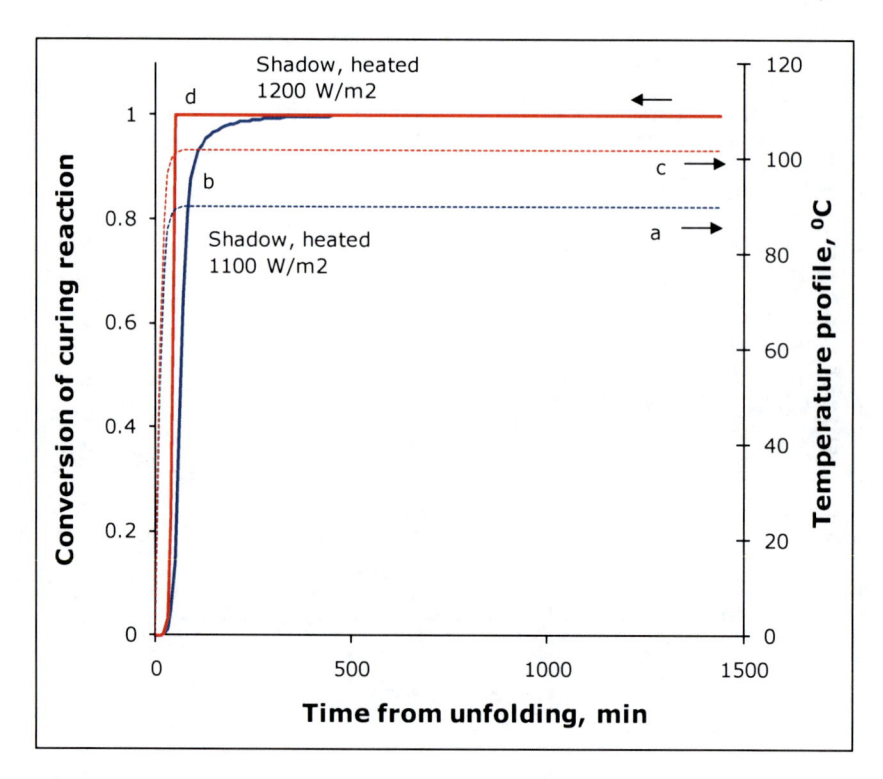

Fig. 21.5 Calculated temperature (a and c) and conversion (b and d) of the curing reaction in the wall in shadow. Bottom (red) curves are for the wall heated with 1.1 kW/m² (a and b); top (red) curves are for the wall heated with 1.2 kW/m² (c and d). The curing reaction is complete after 40 min for 1.2 kW/m² of heater power and after 6-8 hours for 1.1 kW/m² of heater power

heating event is more effective than a low heat flux for a long time. A short effective heating could be done with chemical sources of energy. Curing with internal heaters based on carbon fibers at low environmental temperature (-40°C) was successfully tested. The amount of power needed for a complete curing depends on the size of the construction, its geometry, and the required temperature profile.

Thus, after completing the curing, the durable frame of the Moon base can be pressurized and fitted with equipment and crew.

21.6　The Constructions We Take

The environment of the Moon does not permit usual building activities, where blocks of materials are transported, carried out, lifted, jointed, isolated and painted. No one will do these in a spacesuit. There is no robotic system to do this now. Therefore, the inflatable construction should be provided in a ready-to-use statement after deploying and curing. The maximal mass of the construction delivered from Earth depends on the parameters of the space carrier. Modern space systems can lift 18-20 tons to LEO, from where a second space carrier can deliver it to lunar orbit and land. Assuming 20 tons mass and 1 bar internal pressure, the thickness of the construction wall needs to be not less than 5 mm for composites based on Zylon fibers and epoxy resin matrix (Kondyurina et al. 2006). The construction has to be attached with hermetic ports, windows, heating and pumping systems, electronics and others. Estimates can be done for a cylindrical construction with diameter of 10 m and length of 80 m. The internal volume of the habitat is about 6300 m^3. This is about 1000 times bigger than 6.65 m^3 of pressurized crew compartment in Apollo Moon module, and about 300 times bigger than the 15-20 m^3 of planned Altair habitat in Constellation Program. Such a big volume is suitable for a small crew and life support system for a continuous base on the Moon's surface.

The cured construction can be protected against radiation and meteorites with a regolith covering, as proposed in modern Moon base projects. Similar inflated and polymerized constructions can be placed near first and jointed with hermetic pressurized ports to form a net of habitats. Then, the industrial, storage, and unpressurized constructions can be built and connected.

References

Begishev, V.P., Malkin, A.Y.: Reactive Processing of Polymers. Chemical Technology Publishing, Toronto (1999)

Belvin, W.K.: Advances in Structures for Large Space Systems. AIAA paper 2004-5898 (2004)

Briskman, V.A., Yudina, T.M., Kostarev, K.G., Kondyurin, A.V., Leontyev, V.B., Levkovich, M.G., Mashinsky, A.L., Nechitailo, G.S.: Polymerization in microgravity as a new process in space technology. Acta Astronautica 48(2-3), 169–180 (2001)

Cadogan, D.P., Scarborough, S.E.: Rigidizable materials for use in Gossamer Space Inflatable structures. AIAA paper 2001-1417 (2001)

Cassapakis, C., Thomas, M.: Inflatable structures technology development overview. AIAA paper 95-3738 (1995)

Guidanean, K., Williams, G.T.: An inflatable rigidizable truss structure with complex joints. AIAA paper 98-2105 (1998)

Kato, S., Takeshita, Y., Sakai, Y., Muragishi, O., Shibayama, Y., Natori, M.: Concept of inflatable elements supported by truss structure for reflector application. Acta Astronautica 19(6/7), 539–553 (1989)

Kondyurin, A.V.: Building the shells of large space stations by the polymerisation of epoxy composites in open space. Plasticheskie Massy 8, 25 (1997) (in Russian); translated in Int. Polymer Sci. and Technol. 25(4), T/78

Kondyurin, A., Kostarev, K., Bagara, M.: Polymerization processes of epoxy plastic in simulated free space conditions. Acta Astronautica 48(2-3), 109–113 (2001)

Kondyurin, A.: High-size space laboratory for biological orbit experiments. Advanced Space Research 28(4), 665–671 (2001)

Kondyurin, A., Lauke, B.: Space Environmental Effects on the Polymerisation of Composite Structures. Report ESA Contract No. 17083/03/NL/SFe (2003)

Kondyurin, A., Lauke, B., Richter, E.: Polymerization Process of Epoxy Matrix Composites under Simulated Free Space Conditions. High Performance Polymers 16, 163–175 (2004)

Kondyurin, A., Lauke, B., Vogel, R.: Photopolymerisation of composite material in simulated free space environment at low Earth orbital flight. European Polymer Journal 42, 2703–2714 (2006)

Kondyurin, A.V., Komar, L.A., Svistkov, A.L.: Modelling of curing of composite materials for the inflatable structure of a lunar space base. Journal on Composite Mechanics and Design 15(4), 512–526 (2009)

Kondyurin, A.V., Nechitailo, G.S.: Composite material for Inflatable Structures Photocured under Space Flight Conditions. Cosmonautics and Rockets 3(56), 182–190 (2009) (in Russian)

Kondyurin, A.V., Komar, L.A., Svistkov, A.L.: Modelling of curing reaction kinetics in composite material based on epoxy matrix. Journal on Composite Mechanics and Design 16(4), 597–611 (2010)

Kondyurin, A.: Direct Curing of Polymer Construction Material in Simulated Earth's Moon Surface Environment. Journal of Space Craft and Rockets 48(2), 378–384 (2011)

Kondyurina, I., Kondyurin, A., Lauke, B., Figiel, L., Vogel, R., Reuter, U.: Polymerisation of Composite Materials in Space Environment for Development of a Moon Base. Advances in Space Research 37, 109–115 (2006)

NASA fact sheets, http://nssdc.gsfc.nasa.gov/planetary/factsheet/moonfact.html

Pappa, R.S., Lassiter, J.O., Ross, B.P.: Structural Dynamics Experimental Activities in Ultralightweight and Inflatable Space Structures. J. of Spacecraft and Rockets 40(1), 15–23 (2003)

Ruggiero, E.J., Inman, D.J.: Gossamer Spacecraft: Recent Trends in Design, Analysis, Experimentation, and Control. J. of Spacecraft and Rockets 43(1), 10–24 (2006)

Sandy, C.R.: Next generation space telescope inflatable sunshield development. ILC Dover, Inc., ILC paper (2000)

Semenov, Y., Efremov, I., Blagov, V., Cherniavskiy, A., Kravchenko, Y., Tziganko, O., Medzmariahvili, E., Kinteraya, G., Bedukadze, G., Datashvili, L., Djanikashvili, M., Khatiashvili, N.: Space Experiment REFLECTOR on Orbital Station MIR, Noordwijk, The Netherlands. European Conference on Spacecraft Structures, Materials and Mechanical Testing, ESTEC (2000)

Simburger, E.J., Matsumoto, J., Lin, J., Knoll, C., Rawal, S., Perry, A., Barnett, D., Peterson, T., Kerslake, T., Curtis, H.: Development of a multifunctional inflatable structure for the powersphere concept. AIAA paper 2002-1707 (2002)

Veldman, S.L., Vermeeren, C.A.J.R.: Inflatable structures in aerospace engineering - an overview. ESA paper (2002)

Wilson, A.: A history of balloon satellites. J. of the British Interplanetary Society 34, 10–22 (1981)

22 Natural Resources of the Moon and Legal Regulation

Lotta Viikari

University of Lapland, Rovaniemi, Finland

22.1 Introduction

Things may be considered as resources of some sort to the extent that they are perceived to have a beneficial potential. In this sense, a wide variety of phenomena may be regarded as resources. Hence, the category of natural resources of the Moon includes the unpolluted environment, for instance. A more limited concept of lunar natural resources would restrict them to mineral resources and other material substances that can be derived from the Moon. Humans have found a variety of such resources on the Moon that they are eager to exploit (technology permitting). There may still be also new resources to be found, such as minerals, energy sources, organic and non-organic substances. To avoid undue speculation on an already complicated issue, this chapter will concentrate on the legal regulation of human activities involving resources of the more traditional type, i.e., known resources relating directly to the soil of the Moon.

Legally speaking, the Moon is an area which does not fall under national sovereignty of any state. It is thus what is called a "global common" (the other "global commons" are Antarctica, the high seas and the deep seabed). Activities on the Moon are governed by public international law, within which the most important instruments in this connection are the 1967 Treaty on Principles Governing the Activities of States in the Exploration and Use of Outer Space Including the Moon and Celestial Bodies (hereinafter "the Outer Space Treaty" or "the OST") [adopted 27 Jan. 1967 by UNGA Res. 2222(XXI), opened for signature 27 Jan. 1967, in force 10 Oct. 1967, 610 UNTS 205] and the 1979 Agreement Governing the Activities of States on the Moon and other Celestial Bodies (hereinafter "the Moon Treaty") [adopted 5 Dec. 1979 by UNGA Res. 34/68, opened for signature 18 Dec. 1979, in force 11 July 1984, 1363 UNTS 3], both negotiated under the auspices of the United Nations. Pursuant to Article II of the Outer Space Treaty, space – "including the Moon and other celestial bodies"– remains totally beyond the scope of any national sovereignty. Accordingly, the natural resources of the Moon lie completely outside national jurisdictions.

In the following, I will first introduce some questions of general nature related to different approaches to lunar resource activities. Next, the historical development of the legal regulation pertaining to such activities is presented. Constituent elements of the Moon Treaty are examined in detail, as well as reactions of the international community to the Treaty. Finally, the current status and future prospects of the Moon Treaty are considered after a short assessment of another UN instrument, the 1996 Declaration on International Cooperation in the Exploration and Use of Outer Space for the Benefit and in the Interests of All States, Taking into Particular Account the Needs of Developing Countries (hereinafter "Space Benefits Declaration") [UNGA Res. 51/122, 13 Dec. 1996].

22.2 General Questions

The resources of the Moon cannot be exploited without a corresponding diversity of actors being affected and, conversely, no single actor can determine the outcomes of exploration and utilization of the resources. In addition to the type of actors, the legal regulation of the use of the resources depends on the ideological and political approaches adopted, as well as on the nature of the resources themselves, their intended use, the environmental impacts of this use, etc. Some of these highly interlinked factors are briefly examined below in order to facilitate better understanding of the development of the international law of outer space for lunar resource activities.

22.2.1 Freedom versus Communalism

As regards the ideological framework, there are two obvious lines that one can follow in lunar resource utilization. One option would allow for complete freedom of exploration, which would naturally entail the "first come, first served" principle. This means that those who have the means – technology, in particular – would benefit the most (or even exclusively) from common resources. The other extreme option would entail exploration on a completely communal basis. While satisfying the spirit inherent in the concept of Common Heritage of Mankind the latter option may, however, not spur technological development or generate the financial and other investments necessary for the utilization of such resources. Clearly, it is not in line with the current strong trends towards privatization and commercialization, as illustrated by the major difficulties in reconciling the notion of profitability demanded by the private sector with that of equitable sharing advanced by many states. Feasible solutions are likely to lie somewhere in between the two extremes.

22.2.2 World Politics

The ideology of the international governance of any natural resources is closely connected to and further generates international political tensions. Due to differences in technological development, geographic location, political traditions, etc., states easily form like-minded groups to further their common interests against other states. This is also the case with space activities. States have widely ranging opinions regarding both different types of activities and their extent – not least where activities involving the use of natural resources of the Moon are concerned. Accordingly, in the process leading to the adoption of the Moon Treaty, world politics played an essential role and various state groups emerged. For instance, developing states aimed at preventing the leading industrialized countries from monopolizing the natural resources. The question of new mineral resource exploitation was especially acute for the developing countries which derived a significant amount of income from exporting the same minerals that might be mined in outer space. The disputes provoked by the New International Economic Order (NIEO) that was proposed by Third World countries further exemplify the underlying nature of these distributional issues (Rosenau 1983). The two other prominent groups in the Moon Treaty negotiations were the western industrialized countries and the socialist bloc of Eastern European states.

22.2.3 Types of Natural Resources

The type of resources in question is also obviously of fundamental importance when discussing the issue of their exploitation. Lunar natural resources are relatively recently discovered resources that are to be distributed for the first time. It is generally agreed that they are common property resources – whether by definition of the Common Heritage of Mankind or CHM, common property of mankind, or being used for the benefit of mankind (Jasentuliyana 1986). Furthermore, the status of natural resources depends to a large extent on their properties, in particular on whether the resources are renewable, non-renewable or inexhaustible. The lunar resources of the type that are the focus of the current examination are nonrenewable: they may be used up completely over time. Environmental resources, on the other hand, may not be physically exhaustible, but their quality could nevertheless decline. Another consideration is that even those natural resources of outer space that are in fact exhaustible have either been so abundant or so difficult to exploit in the past that we may not have viewed them as exhaustible (Brown Weiss 1983). Today, however, it is becoming obvious that more rigorous rules must be applied to the utilization of non-renewable resources irrespective of where they are in order to prevent their depletion and to preserve them for future generations. This is evident also in the development of principles of environmental law, the principle of sustainable development and the precautionary principle, above all (Viikari 2008).

22.2.4 Use of the Resources

There are different ways to use outer space natural resources, necessitating different kinds of regulation. There might be need to distinguish between industrial exploitation and scientific research, for example (although all lunar resource activities in the foreseeable future are likely to have at least some scientific functions, too; Sterns and Tennen 2006). In another sense, resources may be used for military or peaceful purposes. "Use" may also focus on preservation and/or controlled development. Where the former is concerned, one needs to make another threshold distinction: whether the preservation is active or passive. Passive preservation equals non-activity (such as neglect and abandonment) and requires little or no capacity. Active preservation, on the other hand, usually requires high scientific and legal professionalism for planning, a legislative authority to implement plans, and a police authority to enforce them. Above all, it requires decisions about what should be subjected to preservation and what should be used. Such decisions need to be based on a wider plan for development of the resources; without sound planning there can be no sustainable criteria for preservation (O'Donnell 1999). It is only in the case of total preservation that a detailed plan and criteria are not needed, but such a case excludes development of the resources. There is no substantial movement to prohibit all uses of resources with respect to outer space or the Moon; quite the contrary. Therefore, there is a need to develop a comprehensive plan for balancing the numerous interests among all the possible types of use of lunar resources.

22.2.5 Actors in Resource Utilization

When discussing the regulation of natural resources, in addition to the more general ideological framework referred to above, one has to take into consideration the particular type of possible actors in the field. These may be private, governmental, international or intergovernmental entities, to name the most obvious alternatives, and it is clear that their objectives and activities vary significantly. The type of actors has major relevance when designing any organizational model for the use of the resources. Moreover, one has to take into account that there may be intense competition among as well as within the different groups of potential resource users, which also calls for solutions. For any actor, an essential prerequisite for the development of a resource is that it has some assurance that its efforts will not be subject to interference by other actors. As will be shown below, this has been a focal concern in the regulation of lunar resource activities.

22.2.6 Environmental Concerns

The environmental impacts of any resource utilization are of great importance, although they are often disregarded until they become extremely severe. This is unfortunately especially the case where remote areas of substantial natural wealth are concerned – the Moon being a prime example. An additional problem may be a relatively low level of scientific knowledge about the possible impacts of activities taking place in the fragile environment of outer space. Highly conflicting opinions about the environmental consequences of space resource activities exist. Nevertheless, it is obvious that the use of lunar resources may affect the environment in various ways. It may cause chemical contamination or pollution in the form of space debris, for instance. In addition to possible areas of decay on the Moon, the activities may harm the surrounding outer space environment. Resource activities in space can be an environmental hazard during every phase of operation.

22.3 Development of the Legal Regulation

22.3.1 Short History of Space Law

Space law has from its very outset had to govern activities carried out in a new and distinctive area, the nature, dimensions and risks of which exceed all previous experience. It has also needed to accommodate a climate of fierce intergovernmental rivalry: as late as 1969, when the first lunar landing enabled the extraction of samples from the Moon, only the United States and the Soviet Union could truly be referred to as space nations. At that time, space could be regarded as a mere Cold War playground for power politics. By the end of the next decade, Japan, China, India and the European Space Agency had also acquired the ability to build and launch satellites into orbit. After that, the number and kind of participants in the exploration and use of outer space have increased rapidly. Today, every country possesses and utilizes at least some form of space applications, e.g., weather forecasting, remote sensing and satellite telecommunications. Science and technology have reached unprecedented degrees of development and the shift of space programs from the public to the private sector is quickly transforming the focus of activities "from scientific investigation and national prestige to considerations based on a return on investment" (Sterns and Tennen 1999).

Accordingly, the international law of outer space is facing more challenges than ever. The new space applications and the increasing participation by both private entities and new space-faring nations call for better elaborated rules. Much of the UN space law negotiated in the 1960s and 1970s must be reconsidered, as it often remains silent on, or sometimes is even counter-productive to, the orderly development of the present reality. Large numbers of treaty provisions, perhaps even entire treaties, are rendered obsolete by the emerging needs of the new spacefaring society. There is also a danger that the fast evolving domestic and bilateral regulations on space issues will prove difficult to fit into a multilateral framework anymore.

One important area prompting the need for extensive new legislation is the use of the natural resources of space. The place where such activities are likely to be undertaken first is the Moon; hence the particular need for regulating lunar resource utilization. The UN space treaties never offered much guidance in this respect. At the dawn of Space Age, exploitation of space resources still loomed far ahead and the drafters of the first legal instruments in the field saw little immediate need for focusing attention on regulating such activities. In general, the early space law authors were divided into two categories with regard to the Moon and other celestial bodies, i.e. those in favor of the permissibility of occupation and those against it (Fasan 1981). This dispute was solved in practice when the unanimous UN Resolution 1721(XVI) of 20 December 1961 declared that "celestial bodies are free for the exploration and use by all states [---] and are not subject to national appropriation" (para. 1.b).

The next phase in the development of the international law of outer space was the adoption of instruments of a legally more binding character: the United Nations space treaties. As long as space activities were beyond the interests and capacities of most countries (and of limited utility even to the spacefarers), space treaties could be drafted and ratified in a series of easy compromises. In such a situation, a consensus method permitted the negotiators to use ambiguous language (later defined by national interpretations) that rarely could be challenged by application and conflict. Another fact contributing to the relatively easy beginning of space law was that space law started from the "Charter" 1967 Outer Space Treaty (referred to also as the "Principle Treaty", the "Constitution of Space Law", the "Space Treaty", etc.) and consequent treaties were formulated as new problems emerged. Such was the process for the conclusion of the first four space treaties: the OST was followed by the elaboration of more detailed space law to cover assistance to astronauts, their return and the return of space objects (hereinafter "the Rescue Agreement") [1968 Agreement on the Rescue of Astronauts, the Return of Astronauts and the Return of Objects Launched into Outer Space, 672 UNTS 199], liability for damage from space objects (hereinafter "the Liability Convention") [1972 Convention on International Liability for Damage Caused by Space Objects, 961 UNTS 187], and registration of space objects (hereinafter "the Registration Convention") [1976 Convention on Registration of Objects Launched into Outer Space, 1023 UNTS 15].

The first four space treaties make no specific mention even of the words "exploitation" or "natural resources". They do refer to "use", however, which may well be interpreted to include exploitation. During the debates on the 1967 OST, the representative of France actually suggested, in an effort to clarify the meaning of the word "use", that it was the equivalent of "exploitation". The other delegates accepted this view, including illustrations of such use, such as meteorological research, telecommunications and extraction of Moon minerals (Christol 1991). Similarly, it may be argued that the frequent phrase in space treaty jargon "outer space, including the Moon and other celestial bodies" also refers to the natural resources found there. However, this interpretation has not received support in practice (Gorove 1995). The Rescue Agreement, the Liability Convention and the Registration Convention obviously do not have any immediate bearing on space resource utilization, as they only regulate aspects of space activities for the limited areas for which they have been drafted. Nor does the OST contain any principle that would explicitly regulate the activities of exploring and exploiting natural resources of outer space, let alone specific institutional arrangements for managing such activities. The fifth (and so far last) of the space treaties, the 1979 Moon Treaty, went farther in the attempt to define and develop the provisions of the OST with relation to the Moon and other celestial bodies – too far, it seems, as the Moon Treaty has been ratified by only a handful of countries.

The international community has not been able to produce any other legally binding instruments in the field of space law (by consensus or otherwise) since the Moon Treaty. Later negotiations on the further codification of space law have proceeded slowly, resulting thus far in nothing but a few sets of principles adopted through General Assembly Resolutions. One of these, the so-called Space Benefits Declaration, will be briefly examined below.

22.3.2 UN Space Treaties and Natural Resource Activities

22.3.2.1 The Outer Space Treaty

Even though it was not the aim of the Outer Space Treaty to establish specific norms for the utilization of natural resources of the Moon or other celestial bodies, a few of its provisions nevertheless have relevance to these issues. Besides, the Moon Treaty is essentially based on and supplements provisions of the OST, hence the two cannot be examined in isolation from each other. Moreover, the fact that the Moon Treaty is not generally adopted by states makes the Outer Space Treaty in practice the most important regulator of lunar resource activities at the moment.

The OST was an important step not only because it was the first instrument to create legal rules in space where none previously existed, but also because it rejected application of the res nullius principle to the exploitation of celestial bodies. The acceptance of the *res communis* principle was anything but clear before the first UN General Assembly Resolutions and the OST, as both principles had strong proponents among international lawyers. There was also a third school taking part in this heated scientific debate, namely, the one that preferred a completely unique legal status for the Moon. The main factors working in favor of the exclusion of territorial sovereignty and of the adoption of the res communis concept were, firstly, the extreme competition between the USA and the USSR in the conquest of space, in which neither could be sure of its own victory. Secondly, during that same period a great number of countries became independent and the Third World, also influenced by the East-West confrontation, started playing an important role on the international political scene. For these reasons, it was quite untenable to defend a legal regime allowing a single nation to establish sovereign rights in outer space (Oosterlinck 1997).

Accordingly, the OST embraced the notion that outer space is the "province of all mankind", that activities in outer space are to be conducted in the interest and for the benefit of all humankind, and that outer space should be developed for the benefit of humankind (Art. I). The OST has since become the framework for building cooperation and peace in space relations through the following guiding principles: 1) international law, specifically the Charter of the United Nations [done 26 June 1945, in force 24 Oct. 1945, 1 UNTS xvi], extends to space and celestial bodies; 2) outer space is free for exploration and use by all states; 3) celestial bodies cannot be subject to national appropriation or claims of sovereignty; and 4) celestial bodies shall be used only for peaceful purposes (Arts I-II). These are noble goals, but one has to admit that the OST is merely a declaration of fundamental principles with only generally worded provisions that do not provide much help in actually implementing the ideals in practice. The vagueness of the OST reflects various political compromises during the drafting process. As mentioned, at the time of the preparation of the instrument only a very limited number of states were actually capable of using space and only a small part of the present technology for such use had been invented. Consequently, the OST contains no specific reference to private activities in outer space, nor to "cosmic mining", for instance.

The fundamental basis for any resource exploitation activities in space is the freedom of exploration and use that the OST declares as one of its basic principles as well as the prohibition of "national appropriation" of outer space (the Moon and other celestial bodies included) by any means. Although the OST declares outer space to be the "province of mankind", this principle in itself does not impose any specific treaty requirement regarding how benefits are to be shared, for instance. Under Article VIII, states retain jurisdiction and control over registered objects launched into outer space and bear the corresponding responsibility (Art. VI) and liability (Art. VII) for space activities. In accordance with Article VI, the activities of non-governmental (private) entities in outer space are authorized, regulated and supervised only by the appropriate responsible state. Moreover, the OST embodies some general principles related to the commitments of states to avoid harmful contamination of outer space, including the Moon and other celestial bodies (Art. IX). The "appropriate international consultations" provided for removing potential conflicts in space activities of States Parties (Art. IX) also apply to resource activities. Admittedly, these concepts remain at a rather general level but they do provide an important starting point for the further elaboration of obligations and for a more developed control mechanism for resource utilization on the Moon and elsewhere in outer space.

22.3.2.2 The Moon Treaty

When the work to conclude a treaty on the legal status of the Moon and its natural resources began in the United Nations Committee on the Peaceful Uses of Outer Space (UNCOPUOS) in 1970, there existed a wide range of differing opinions about the subject. The controversies centered around the use of the natural resources of the Moon and the general scope of the Treaty. Some thought that space resources could be lawfully and freely exploited, while others viewed such activity as unlawful appropriation. Some made a distinction between the activities of states and private entities, others between scientific and commercial purposes. Some commentators supported the CHM principle, while others showed great interest in (less communal) analogies between the exploitation of outer space resources and deep seabed activities or the high seas. Some were in favor of applying the CHM principle only to the Moon but not to its natural resources (Christol 1984). There were also differing opinions about whether the Moon Treaty would apply to any areas other than the Moon (Oosterlinck 1997).

At that time, the US had just landed a man on the Moon (on July 21, 1969) and the Soviet Union had obtained lunar soil samples, which led to the (false) prediction that exploitation of the Moon's natural resources was beginning. Although some states (such as the USSR and Japan) were of the opinion that an international agreement on the use of space natural resources would be premature (Matte 1977), generally speaking, the importance of a Moon Treaty seemed evident. Despite the otherwise widely varying aspects of the issue, there was a strong will for continuing to ensure that the Moon would be used only for peaceful purposes (Galloway 1998). However, lunar exploration by the two superpowers soon ceased and the issue was no longer primarily one of providing space law in tandem with space science and technology (which had been the case with the previous space treaties spun from articles of the 1967 OST). Instead, the main motive was to establish a concept of CHM which would eventually require an international regime with jurisdiction over resource exploitation.

It was when UNCOPUOS finally adopted a draft text of the Moon Treaty in 1973 that ideological differences first flared up. Some of the deliberations witnessed the USA and the USSR on one side and the developing countries on the other. Others, though far fewer, reflected the superpower confrontation (Wadegaongar 1984). For a long time, the deliberations in UNCOPUOS concentrated on the issue of whether the concept of CHM should govern the legal status of the Moon and its resources and whether certain institutional arrangements should be included in the Moon Treaty (Kopal 1999). Not surprisingly, the developing nations requested that the principle of the CHM be applied to the Moon and its natural resources. Other groups were reluctant to accept this (Kopal 1986). Several legal commentators predicted that the conflicting opinions and interests of the industrial nations and the developing states were too divergent to be reconciled (Finch and Moore 1981). This seemed at first to be a false prediction, for the General Assembly finally secured sufficient agreement to adopt by consensus the provisions of the Moon Treaty on 9 December 1979.

The Moon Treaty was opened for signature on 18 December 1979 and for ratification in the following year. By that time, however, a different climate of opinion had developed and the provisions for an international regime governing the exploitation of lunar resources again provoked opposition (particularly in the USA). The general ambiguities and uncertainties concerning the CHM, which gave rise to a variety of interpretations and definitions, were the main factor contributing to the US Government´s decision to put off ratification (Galloway 1998). The other space powers also declined to ratify. Eventually, it took as long as five years for the Moon Treaty to obtain even the required five ratifications in order to become binding on the ratifying nations. The first states to ratify the Treaty were Chile, the Philippines, Uruguay, the Netherlands, and Austria; the last lodged the fifth instrument of ratification with the UN Secretary-General on 11 July 1984, thereby causing the Moon Treaty to enter into force. This record is in sharp contrast to that of the four other UN-formulated space treaties, which encountered no such delay – or outright rejection.

After the more than three decades that have passed since its conclusion, the Moon Treaty still has been ratified by thirteen states only (Australia, Belgium, Kazakhstan, Lebanon, Mexico, Morocco, Pakistan, and Peru, in addition to the first five countries); that this number excludes the major space powers is obviously evidence of a serious problem. The positions of the major states / state groups that shaped the Moon Treaty (the USA, the USSR and the developing countries) will be examined in the following in order to try to understand the reasons for this general non-acceptance. The Moon Treaty has not been amended later. The 1996 Space Benefits Declaration, which will also be examined later, did point the way in which the CHM concept seems to be evolving, however.

USA

The resource distribution philosophy of the USA with respect to outer space was grounded in a "freedom of outer space" ideology. According to this doctrine, no state may claim or acquire exclusive sovereign rights to areas of pristine space. Barring some agreement to the contrary, states may, however, use space resources as long as there is reasonable regard for the rights and activities of others (Marko 1993). It was also concluded that the right of a state to exploit the natural resources of the Moon "was recognized in Article I of the 1967 Outer Space Treaty, and may be considered to have existed well in advance of that time, too" (American Bar Association Report to the House of Delegates, 18 April 1980, quoted in Christol 1984). The USA was of the opinion that if a further agreement dealing with space resources were to be concluded, it should cover all celestial bodies, as it expected that it would soon be feasible to travel beyond the Moon (Oosterlinck 1997). Originally, the USA, along with several Third World nations, supported the Common Heritage concept: as late as in 1972 the USA introduced a draft article for the Moon Treaty embracing the CHM (Galloway 1998). Later, during the negotiations, this support eroded, however. The US reconsidered its position and started emphasizing instead its concern about the threats to national economic and security interests that the Moon Treaty could pose.

Perhaps the foremost of the various sources of that concern was the view that the meaning(s) attached to the concept of CHM by many countries were hostile to the system of free enterprise and thereby contrary to the interests of "advanced" states with free-market economies. The USA feared that the CHM regime would serve as a considerable disincentive to development (Christol 1984). Furthermore, the notions in the Moon Treaty of "orderly development" and "equitable sharing" were seen as imposing a substantial "tax" on lunar development for the benefit of nations that put nothing at risk. The USA also feared that in imposing a commitment to negotiate an "international regime" and requiring authorization for the removal of lunar natural resources the Moon Treaty could result in a "de facto moratorium" on resource-related activities in space. In its opinion, this would not have deterred the Soviet Union from moving forward with resource development disguised as scientific investigation, however (Nash 1989). Still, even as late as in 1980, the US Department of State indicated that the Moon Treaty was considered by the USA as providing "the best possible structure for regulating activities which governments may now or in the future engage in on the Moon or elsewhere in space" (Christol 1984). Ultimately, however, as a result of a storm of protest over the Treaty by a legion of space interest groups, the Senate failed to vote on ratification (Supancana 1998).

Soviet Union

The Soviet Union was the first nation to propose the conclusion of an International Agreement Regarding the Moon and Other Celestial Bodies, but its draft of 1969 did not contain any detailed provisions relating to the exploitation of space natural resources. Soviet experts did call attention to the desirability of the use of the natural resources of the Moon and other celestial bodies (Christol 1984), but for a long time the USSR maintained fierce resistance to the CHM concept. They argued, among other things, that such a principle would threaten state sovereignty and that it would be erroneous to apply civil law concepts such as "common heritage" to relations between states (Wadegaongar 1984). According to Soviet scholar Y.M. Kolossov, "far from corresponding to the fundamental principles of space law [---] [the CHM] tends to revise international law in its entirety, including its basic provisions of respect for national sovereignty, which is fundamental to the safeguarding of each state´s vitally important and legitimate interests" (quoted in Christol 1984).

In the Soviet view, national appropriation of the Moon or parts thereof was forbidden but not the appropriation or use of its natural resources (Matte 1977). Likewise, the USSR opposed the concept of a moratorium on the exploitation of space resources. It was concluded that, taking into account the "exploration", "use", and "activities" as used in the 1967 OST, it was permissible to use lunar natural resources and that this utilization should allow for the realization of "profit" and for the "practical use" of such materials (Christol 1984). In fact, the Soviet Union long took the position that space resources should be totally and freely exploitable by the nations that find and develop them (Reynolds and Merges 1989). In addition to opposing any form of international control of natural resources in space, the USSR opposed the application of any parts of a possible Moon Treaty to celestial bodies other than the Moon (Spitz 1989).

The Soviet opposition was one of the principal reasons why the conclusion of the Moon Treaty was delayed until 1979 (Cheng 1996). When consensus on the final text of the Treaty was finally achieved in UNCOPUOS, this was largely due to the fact that the Soviet Union had accepted Brazil´s proposal that the CHM concept should apply only to the Moon Treaty and the developing countries had given up their demand for a moratorium on exploitation of lunar resources prior to the establishment of an international regime (Galloway 1998). The Soviets even joined in the consensus on the need to negotiate about an international regime (including a future international organization to implement it). However, they contributed considerably to the rejection of the moratorium on space resource activity pending the implementation of the CHM principle, thereby again demonstrating their strong commitment to the practical exploitation of the Moon and its natural resources for national ends (Christol 1984). The USSR liked to style itself as a friend of the developing world (favoring in this instance, however, rather unfriendly solutions vis-à-vis the developing countries´ aspirations to benefit from the natural resources of space), so the refusal of the USA to ratify the Moon Treaty probably made it politically easier for the Soviets to refrain, too. Besides, it hardly would have been even economically feasible for only one of the two superpowers (as the only space nations of the time) to join the space resource regime of the Moon Treaty, as they were also the only ones considered capable of taking part in the international governing of space resources and of financing this governing (Reynolds and Merges 1989).

Developing States

A third major actor in the Moon Treaty negotiation process was the group of developing nations. Together, they formed an influential counterpart (already due to the vast number of these countries) to the actual spacefaring nations. Some developing states showed more interest and activity than others. Argentina, for instance, was very active in the early development of space law and even delivered its own Draft Moon Treaty. It was originally the Argentine draft of 1970 that introduced the Common Heritage concept with respect to the natural resources of the Moon and other celestial bodies. The draft also suggested innovatively that the regime for governing natural resources in space could be different for materials used in their place of origin and for those brought back to Earth. According to the Argentine draft, the benefits from the latter should be shared between all peoples (Matte 1977).

Obviously, the developing countries were motivated by a concern that the natural resources of space would be exploited to their disadvantage. The eventual utilization of space resources appeared to be open only to technologically advanced countries and, moreover, the developing nations felt that such exploitation would further widen the gap between the "have" and "have-not" countries. On the other hand, they saw in the vast new resources an opportunity to become indirect beneficiaries of resource exploitation on the basis of the principle of equitable justice and the idea that the resources may be regarded as part of the Common Heritage of Mankind (Gorove 1991). It was obvious that the developing countries could only realistically hope to fulfil their expectations of sharing in the benefits of space resource activities through an international regime in which they had a substantial voice. During the drafting of the Moon Treaty, some developing nations also persistently argued for an immediate moratorium on resource exploitation (Gorove 1995). Those efforts, however, were rejected.

22.4 Constituent Elements of the Moon Treaty

The Moon Treaty is a twenty-one article document which applies to the Moon and other celestial bodies in the solar system, excluding the Earth. The compromise between the Soviet view of limiting the Treaty to the Moon only and the US idea of a comprehensive celestial body regime was that the Treaty covers all celestial bodies but only until such time as specific treaties or agreements would be set up to deal with other celestial bodies (Art. 1). The Moon Treaty proclaims that "the moon and its natural resources are the common heritage of mankind" (Art. 11.1). Any exploration or exploitation of the Moon is to be carried on for the benefit of all countries, without regard to their level of development (Art. 4.1). In carrying out their activities, states are to be guided by "principles of cooperation and mutual assistance" (Art. 4.2). Examples of such cooperation are the sharing of scientific samples and exchanges of personnel on lunar missions (Arts 6.2, 6.3), the right to visit others´ lunar installations (Art. 15.1), prompt disclosure of mission logistics and mission data to the UN Secretary-General (Art. 5.1) and disclosure of the results of each mission, especially results concerning natural resources or phenomena that might endanger human life (Art. 5.3). It is further required that the

Secretary-General "as well as the public and the international scientific community" be informed about any natural resources discovered on the Moon (Art. 11.6). Although cooperation and good-will are required in the exploitation of outer space, the Treaty allows states to retain control over their lunar missions (Art. 12.1). All states are free to conduct scientific investigations, including space-vehicle use, personnel and equipment placement and movement, and the establishment of lunar space stations (Arts 8.2, 9.1). Freedom of scientific investigation must be taken without discrimination and on the basis of equality and in accordance with international law (Art. 11.4).

The Treaty further proclaims that the Moon is not subject to national appropriation as a result of any claims of sovereignty (Art. 11.2). The purpose is to ensure that the Moon and its resources cannot become the property of any country or other entity (Art. 11.3). The Treaty also states that the Moon is to be used exclusively for peaceful purposes (Art. 3.1). This prohibition includes a ban on military bases, weapons tests, nuclear weapons and the threat of force (Arts 3.2, 3.4). However, military personnel are allowed as long as they engage solely in scientific work or any other peaceful purpose (Art. 3.4). The Moon Treaty further requires that lunar exploration occur in accordance with the UN Charter and the entire system of international law (Art. 2). Finally, the Treaty provides that a conference is to be declared ten years after the Treaty has come into force or after it has been in force for at least five years and one-third of the members of the Treaty call for a meeting (Art. 18). However, the UN General Assembly decided not to take any action regarding the revision of the Moon Treaty on 9 December 1994, when the 10-year period had elapsed – obviously because of the "purely hypothetical character of the activities concerned" (Hofmann 2005).

The Moon Treaty also requires that disputes be settled in a peaceful manner. A state that believes that another party is violating the Treaty may request consultations with that party (Art. 15.2). Any party to the Treaty may join the consultations, and all must try to achieve a mutually acceptable resolution to the controversy. If these negotiations fail, the disputing parties may attempt any other appropriate method of peaceful resolution. The UN Secretary-General may be consulted in the event that no resolution can be reached (Art. 15.3). However, the Moon Treaty does not provide for mandatory jurisdiction before an adjudicatory body or any other compulsory process in the event that negotiations fail.

On balance, the Moon Treaty provides for regulation which goes further than any previous space treaty in expressing the aim of more advanced utilization of outer space. Most importantly, it was the first international treaty to incorporate provisions on the exploitation of the natural resources of the Moon and other celestial bodies. It should be further emphasized, that while Article 11 refers to the Moon only, Article 1 provides that the Moon Treaty also applies to other celestial bodies within the solar system, as well as to orbits around the Moon or other celestial bodies or the trajectories to or around them. The wording of Article 11 emerged as a compromise solution to the conflicting views of developing states and industrialized countries. On the one hand, the Moon and its natural resources are classified as the CHM (para. 1) and States Parties to the Treaty undertake to establish an international regime to govern exploitation of the natural resources of the Moon when such exploitation is about to become feasible (para. 5). On the other hand, the right of States Parties to explore and use the Moon without discrimination of any kind "on the basis of equality and in accordance with international law and the terms of this Agreement" is explicitly confirmed in the same article (para. 4). Thus, the main principle in the instrument is that there exists a right of free research in and use of outer space that has to be exercised in the interests of all countries.

In the following, I shall sketch out in more detail the most important provisions regarding the use of space resources as regulated by the Moon Treaty: those concerned with the freedom of exploration and use, the CHM, the non-appropriation principle, the international regime envisioned in the Treaty, and environmental questions pertaining to resource activities. Each section also examines some of the problematic aspects of the articles in question. Most of these topics are highly interconnected, thus a slight overlapping in the examination is unavoidable.

22.4.1 Freedom of Exploration and Use

Free exploration and use of the Moon (and other celestial bodies) as the fundamental starting point for the regulation of all space activities is confirmed in various instruments of space law. Also, Article 4 of the Moon Treaty provides that the Moon is the "province of all mankind" and its exploration and use are to be for the benefit of all countries. This freedom includes the erection and maintenance of installations (such as space stations) on the Moon and the right to use the area required for the installation (Arts 8.1, 9.1). States retain ownership of equipment and facilities placed on the Moon (Art. 12.1). There is a right to collect and keep scientific samples (Art. 6.2); the Moon Treaty puts only a "moral" obligation on those who are carrying out such activities to make samples available to others by providing that "States Parties shall have regard to the desirability of making a portion of such samples available to other interested States Parties and the international scientific community for scientific investigation". Furthermore, under the same Article, states may use Moon substances in the course of scientific investigations "in quantities appropriate for the support of the mission".

There are certain problems connected with the freedom of exploration and use of the Moon, however. For instance, in the case of permanent space structures on a particular site, the use of the underlying surface would in practice amount to appropriation (Matte 1977); yet, according to Article II of the OST one cannot appropriate space or celestial bodies. Furthermore, it is obvious that for the establishment (and possibly for the maintenance as well) of such an installation, building material will be needed and it is likely to be material from the celestial body in question that is used as far as possible. If such material included metals or oxygen, for instance, the installation might become partly self-supporting. Despite permission to use the substances "in quantities appropriate for the support of [a scientific] mission", it can be difficult to see how this kind of large-scale activity would any longer be such exploration and use as is provided for in the Treaty; rather, it would more likely resemble industrial exploitation of space mineral resources (Fasan 1981). However, with respect to minerals or other resources that are absolutely essential for maintaining an established structure or space station, the approach could be more generous. As we may well expect huge installations on the Moon in the future, this problem, too, should be discussed in detail before actual legal problems emerge. In any case, it seems unequivocal that space minerals can be extracted for research purposes as long as the extraction does not interfere with the researching rights of other states and the natural balance of the celestial bodies is not altered (Miklódy 1980). The above outlines may, however, not be applicable in the scope of economic research, as it can be seen as an essential prerequisite for industrial exploitation.

22.4.2 *Common Heritage of Mankind*

An analysis of current space law as embodied in the OST and the Moon Treaty indicates that while all states have an equal right to explore and use space resources, that right must be exercised in a non-prejudicial way. Even before the particular term "common heritage of mankind" was introduced by the Moon Treaty, the doctrine had become part of the law of outer space in the 1967 OST, which provides that outer space shall be for the benefit and in the interests of all countries, irrespective of the degree of economic or scientific development and shall be the province of all mankind (Art. I). The CHM doctrine is also articulated in the preambular reference of the OST to the common interest of all mankind in the progress of the exploration and use of outer space for peaceful purposes and to the propositions that the exploration and use of outer space shall be carried out for the benefit of all peoples irrespective of the degree of their economic or scientific development. Similarly, the Liability Convention and the Registration Convention both speak in their preambles of the "common interest of all mankind in furthering the exploration and use of outer space for peaceful purposes".

The Common Heritage of Mankind principle of the Moon Treaty Article 11.1 and other concepts of the type seem to establish and guarantee equality of opportunities for both the first-comers and the less developed states. However, even if the freedom principle offers all states freedom of access, exploration and use, it is mostly only certain developed countries which are capable of exercising these rights in practice. For the vast majority of countries, it is impossible to use and explore space freely or obtain benefits from such activities, at least without cooperation with other, usually industrialized, countries (Nakamura 1997). The phrase Common Heritage of Mankind can be regarded as mere confirmation of the very general concept of the OST declaring space the "province of all mankind" (Finch and Moore 1981). It has been suggested that the CHM of the Moon Treaty was only an "attempt to move into language more commonly used in international law", in which case the distinction between the two expressions would be only declaratory (Finch 1980). The legal content of the CHM would thereby equal that of the "province of all mankind", which has been described rather generally as international cooperation on exploration and use of outer space without discrimination against any state and the duty to take the interests of all other states into account (Nakamura 1997).

However, the CHM does seem to be given a bit more substance by the Moon Treaty. This becomes apparent from the stipulation that the concept finds its expression in other provisions of the Treaty, particularly in the obligation of the parties to establish an international regime to govern the exploitation of natural resources when such exploitation is about to become feasible (Arts 11.1, 11.5). As a complementary obligation, the States Parties need to inform the Secretary-General of the UN as well as the public and the international scientific community, to the greatest extent feasible and practicable, of any natural resources they may discover on the Moon (Art. 11.6). To make these binding provisions concerning the CHM even more realistic, the Treaty also spells out the main purposes of the international regime: (a) the orderly and safe development of natural resources, (b) their rational management, (c) the expansion of opportunities in relation to them and (d) an "equitable sharing" by the Parties in the benefits derived from the resources, whereby the interests and needs of the developing countries as well as the efforts of those countries which have contributed to the exploration must be given special consideration (Art. 11.7). Finally, the Treaty makes clear that any activity with respect to the natural resources of the Moon must be carried out in a manner compatible with the purposes of the international regime, even prior to its actual establishment (Art. 11.8). Thus, even though its meaning is not precisely determined, the Common Heritage concept of the Moon Treaty clearly envisages at least the establishment of an international regime for the exploitation of space natural resources.

The common use of the natural resources of the Moon and other celestial bodies also implies their rational utilization on the basis of equality of states without any discrimination as well as prohibition of exceptional rights in favor of any state (Art. 11.4). In accordance with Articles 12 and 15, states retain exclusive jurisdiction and control over their facilities and other states are obliged to avoid interference with the normal operations of such facilities. Rational utilization must be carried on by taking account of the interests of both the international community as a whole and states individually and, moreover, in the interests of not only the present but future generations as well. The stipulations of Article 4.1 that in the exploration and use of the Moon "due regard shall be paid to the interests of present and future generations as well as to the need to promote higher standards of living and conditions of economical and social progress and development" provide for the concepts of intra-generational equity (equity between states) as well as inter-generational equity (consideration of the needs of future generations), which also further emphasizes the principles of cooperation and mutual assistance. However, the same terms do not seem to require equal sharing of every advantage, profit or benefit that can be received from space activities. Nor do they prescribe, for instance, that exploration of resources may be undertaken only by joint ventures of all countries with equal participation.

Moreover, even the general goal of equitable sharing of the benefits derived from space natural resources among "all States Parties", which is defined to be one of the main purposes of the international regime (Art. 11.7), seems highly restrictive in practice. The language of the Article would appear to limit "mankind" (at least in reference to the exploitation of the natural resources of celestial bodies) merely to that portion of humankind which is party to the Moon Treaty – i.e., at the time the mere thirteen ratifying countries. Even within this limited community, special consideration is to be given to certain groups, namely, developing countries and countries contributing to the exploration of the Moon. However, justifying a distinction between developing countries that are party to the Moon Treaty and those that are not (especially within the framework of the concept of CHM) seems somewhat difficult. Special consideration for countries contributing to the exploration of the Moon appears to be more easily justifiable in terms of the expenses required for such activities, but even here the language of the Treaty is ambiguous: it only speaks about countries contributing to the exploration of the Moon and makes no explicit distinction between lunar exploration in general and exploration connected to exploitation of the Moon´s natural resources (Rosenfield 1980). At least where the consideration of "all mankind" is concerned, limiting it only to the States Parties to the Moon Treaty would appear to be in violation of Article I of the Outer Space Treaty, which makes the Moon available to "all countries" and "all mankind". Obviously, these problems stem from the poor wording of treaty provisions, whose drafters took it for granted that the result of their work would be universally adhered to.

22.4.3 Non-appropriation

According to the Outer Space Treaty, it is every bit as clear a principle as the freedom of use and exploration that one cannot appropriate space or celestial bodies (Art. II). Article 11.3 of the Moon Treaty repeats the rule: "Neither the surface nor the subsurface of the moon, nor any part thereof or natural resources in place, shall become property of any State, international intergovernmental or non-governmental organization, national organization or non-governmental entity or of any natural person". This would logically also forbid the prerogatives attached to the right of ownership (exchange, sale, purchase, transfer, lease, etc.). Exploitation of the natural resources of the Moon (although they are deemed CHM) is permitted through an international regime. However, exploiting most natural resources is practically impossible without appropriating them: things which are consumed by being used will cease to exist after such use, common heritage or not. Thus, if the CHM is understood as common property, parts of the common heritage of outer space will disappear through the exploitation of celestial natural resources (Fasan 1995).

Another ambiguity lies in the OST´s reference to "national" appropriation, which would seem to exclude international organizations, for instance, from the scope of the ban on appropriation. Some proponents of space commercialization have attempted to argue away the application of the non-appropriation principle to private entities in general or at least to favored projects in particular (Sterns and Tennen 1999). However, it is widely accepted today that non-appropriation in the Moon Treaty is meant to prohibit ownership of the Moon by any entities, the only exception being the international regime to be established when exploitation becomes feasible (Gorove 1995).

Even though the Moon Treaty prohibits ownership by prohibiting national appropriation, and the surface and subsurface of the Moon cannot become property, there are still numerous activities which are usually associated with appropriation and property rights that are not prohibited but in fact explicitly allowed. These include the placement of personnel, space vehicles, equipment facilities, stations and installations on or below the surface of the Moon, as well as structures connected with its surface or subsurface (Art. 11.3). This allows for the establishment and (under Art. 12) ownership of the infrastructure necessary for commercial exploitation programs as well. The right to collect and remove Moon samples for scientific investigation is also allowed, and the States Parties may even use substances on the Moon for the support of their scientific missions (in appropriate quantities) (Art. 6.2). It needs to be emphasized that the use of the Moon, its minerals and other substances in scientific investigations is not even restricted to scientific inquiry but is also permitted for any other activity supporting the mission. Combined with the freedom of scientific investigation (Art. 6.1) and the fact that the nature and extent of such investigations may, over time, be quite far-reaching, this provision can actually allow for very substantial use of natural resources (Christol 1984). In practice, the "scientific research" provision may even be used to engage in illegal mining and use of resources, because the Treaty does not define "scientific investigations" or "appropriate quantities" in any way. Therefore, even though the space-resourced states were not granted a status allowing them to assert national sovereignty over the Moon and other celestial bodies, they were allowed rather exclusive use of the substances identified above in the course of "scientific investigations" – even to the extent that the prohibition against appropriation of resources may seem somewhat illusory and practically negated.

Moreover, all this concerns only celestial body materials in situ; i.e., the principles of non-appropriation (and common use) apply to the resources of the Moon only as long as they have not been extracted from the lunar surface or subsurface. The expression "natural resources in place" in Article 11.3 and the general meaning of the Article testify to the fact that the ban on appropriation only concerns the exclusion of sovereignty, not of possession. It has been suggested (particularly by developing nations) that the phrase "outer space, including the moon and other celestial bodies" already in the OST was meant to include their resources so that the ban of appropriation would apply to space resources in situ as well. Such interpretations of the non-appropriation principle have been rejected by UNCOPUOS, however (Gorove 1995). Also, the drafters of the 1967 OST envisaged only immovable concessions; as soon as a portion of the soil is removed, they regarded it as movable (Matte 1977). If minerals are separated they pass into the ownership of those who extract them; this interpretation has been supported by both the USA and USSR. It has also been pointed out that when these superpowers first returned rocks and other samples from manned and robotic missions to the Moon in the late 1960s, prior to the Moon Treaty, no objection to the ownership of the materials by the state which had collected them was presented (Sterns and Tennen 1999).

Evidently, after their extraction lunar resources may be considered as property – or at least the non-appropriation principle would no longer prohibit their extraction, sale, exchange, etc. This means that the principle of Common Heritage as applied to celestial bodies and as applied to their resources is quite different. As regards celestial bodies, "common heritage" is used in connection with the principle of non-appropriation to ensure that celestial bodies cannot be subjected to the sovereignty of any state (or even public property in the form of ownership); in the case of the resources of celestial bodies, the principle envisages appropriation in that non-appropriation only applies to natural resources in place, for, once removed, these may be regarded as property (Dekanozov 1981). However, it remains an open question how states will interpret these concepts once large-scale extraction of lunar resources becomes feasible. Besides, in this respect the Moon Treaty also seems to permit somewhat questionable outcomes: it does not sound tenable that resources to which an entity never had any rights while in outer space could become the possession of that entity merely as a result of the fact that they have been extracted – regardless of the legality of this extraction.

22.4.4 International Regime

According to Article 11.5 of the Moon Treaty, "States Parties to this agreement hereby undertake to establish an international regime, including appropriate procedures to govern the exploitation of the natural resources of the moon as such exploitation is about to become feasible". The characteristics of the regime of Article 11 have for the most part already been discussed, in particular while examining the CHM concept. However, it should be further emphasized that by Article 11.5 of the Moon Treaty states only "undertake to establish" the regime in the future prior to the exploitation of celestial natural resources and that no such regime is established by the Treaty. All sentences refer to future time. Furthermore, the establishment of the international regime has been made contingent on the feasibility of natural resources exploitation but the Article does not elaborate the criteria underpinning the phrase "when such exploitation is about to become feasible" or indicate how a decision on such feasibility is to be made and who would be competent to make it. Article 11, though referring to "appropriate procedures" to be included in the international regime, does not specify such procedures in any way and does not even mention the establishment of a special institutional machinery for ensuring the application of the future international regime. Moreover, to "undertake to establish" may mean that such a task is performed successfully by states at a certain time in the future – but it may also mean that the undertaking fails. And as the US, for instance, has emphasized, the "acceptance to join in good faith negotiations of such a regime in no way institutes acceptance of any particular provision which may be included in such an agreement" or an "obligation to become a Party to such a regime regardless of its contents" (Finch and Moore 1981). Nevertheless, it should be noted that Article 26 of the Vienna Convention on the Law of the Treaties [done 23 May 1969, in force 27 Jan. 1980, 1155 UNTS 311] prescribes: "every treaty in force is binding upon to the parties to it and must be performed by them in good faith". Also, an agreement to agree is to be performed in good faith.

As regards the contents of the regime, the Moon Treaty defines the purposes of the regime (Art. 11.7) but gives no further details. The vague goals of the international regime include the development and rational management of the Moon's resources as well as "expansion of opportunities" in their use. The primary focus of the regime will be "equitable sharing" in the benefits derived from those resources by all states, but the Treaty fails to define "equitable". No system of international taxation of any profits is required, nor is mandatory transfer of technology. The Moon Treaty only insists that, in setting up the regime, the interests of developing countries and the efforts of those states that have contributed to the exploration of the Moon are to be given special consideration.

On balance, it is obvious that the "international regime" of the Moon Treaty is a very flexible concept. It may consist of a set of rules to govern resource exploitation and management or it may go beyond that and create an international organization for the purpose (Haanappel 1986). The organization may take the form of an enterprise or be a mere licensing authority. It could also function as one of the actors (Brooks 1998). The regime could, in principle, be developed through international consent multilaterally, bilaterally or regionally (although the provisions of the Moon Treaty clearly refer to multilateral concepts). In Article 18, it is only stated that 10 years after the entry into force of the Moon Treaty the question of the review of the Treaty shall be considered by the UN General Assembly, in addition to which, at any time after the Treaty has been in force for 5 years, the UN Secretary-General shall at the request of one-third of the States Parties to the instrument, and with the concurrence of the majority of them, convene a Conference of States Parties to review the Treaty. This review conference "shall also consider the question of the implementation of the provisions of Article 11.5". The international regime can thus be planned by States Parties any time after the Treaty has been in force for five years, or as a result of action in ten years by the General Assembly. The Treaty does not indicate that a Conference of the States Parties must necessarily create a regime. More likely, at least the first few meetings called would only discuss the need for a regime at that time and a suitable type of regime. In any case, the States Parties are committed to enter into some kind of negotiations about the "international regime" when resource exploitation is about to become feasible and to consider the resources of the Moon and other celestial bodies as being part of the Common Heritage of Mankind. All subsequent agreements on the issue must be compatible with this notion (Supancana 1998).

22.4.5 Environmental Issues

The juridical nature of the resources of the Moon is defined partly by their international protection and partly by the common use involving their appropriation in accordance with specific criteria in the interests of both the international community and individual states for the benefit of the present and future generations. Thus, international protection of lunar resources, as part of the protection of the environment of all of outer space, comprises measures for both their preservation and utilization. Already Article IX of the OST obligates States Parties to conduct exploration of the Moon (as well as other celestial bodies) "so as to avoid their harmful contamination". The Article also speaks about "consultations" to be conducted in case of potentially harmful interference with activities of other States Parties. The Moon Treaty elaborates further the provisions of the Outer Space

Treaty in this respect as well; in fact, it is the most advanced of the UN space treaties in an environmental sense. Article 7 not only provides for taking measures to prevent the disruption of the existing balance of the Moon´s environment, "whether by introducing adverse changes in that environment, by its harmful contamination through the introduction of extra-environmental matter or otherwise" (para. 1) but also envisions the designation of areas of the Moon having special scientific interest as "international scientific preserves for which special protective arrangements are to be agreed" (para. 3).

Although the Moon Treaty here represents an advance in developing environmental space law, it neither formulates definitions of such notions as "adverse changes" and "harmful contamination" nor establishes standards for how space activities are to be conducted. Consequently, the obligations regarding environmental protection in the Moon Treaty, like those in the other space treaties, are ambiguous and hardly legally enforceable in practice (Viikari 2008). Besides, the Moon Treaty expressly permits the removal and collection of samples of the lunar surface and subsurface (Art. 6.2), the landing of objects on and launching of them from the Moon (Art. 8.2.a), and placing and freely moving personnel, vehicles, equipment, facilities, stations, and installations on or below the lunar surface or subsurface (Art. 8.2.b). All such activities involve obvious potential for environmental degradation of the Moon.

Furthermore, also the Moon Treaty fails to establish any system of sanctions. However, States Parties have the obligation to inform the UN Secretary-General of the measures being adopted for the prevention of environmental harm in the Moon (Art. 7.2); this could partly increase the effectiveness of the duty to prevent disruption of the environment. Even more importantly, States Parties are obligated to make prior notification ("to the maximum extent feasible") of placements of radioactive materials on the Moon (as well as of the purpose of such placements). They must also inform the UN Secretary-General of any phenomena in space that could endanger human life or health (Art. 5.3) and report to him or her "concerning areas of the Moon having special scientific interest" which might be worth designating as international scientific preserves (Art 7.3). Such reporting requirements constitute a refinement vis-à-vis the OST, pursuant to which states parties only have to provide the UN Secretary-General (as well as the public and the international scientific community) with information about the nature, conduct, locations and results of their space activities "to the greatest extent feasible and practicable" (Art XI). However, there is, for instance, no indication of the manner in which the information must be conveyed to the scientific community and the public, nor any standards for determining when human life or health may be endangered such as to trigger the reporting requirement.

22.5 Failure of the Moon Treaty

Even during the last negotiation session in the process of drafting the Moon Treaty a number of states continued to voice their opposition to the adoption of the Common Heritage concept and/or stressed the need for a moratorium. Nevertheless, a draft text was unanimously accepted containing both the notion of Common

Heritage and provisions that can be interpreted to imply permission for states to start their operations on the Moon before an international regime has been established. This withdrawal of objections probably resulted at least from the fact that a great number of states had vigorously criticized the lack of progress in the implementation of the rules of the OST and, consequently, there appeared to be a strong will to achieve a breakthrough on at least one issue expanding the OST that had been discussed for many years. Another explanation may be the influence of the two most powerful decision-makers in space law at the time, the US and the Soviet Union, who apparently were in agreement on both the need to complete a draft without further delay and (more or less) the provisions it should contain (Reynolds and Merges 1989). However, the Moon Treaty later met with tremendous international resistance to ratification by the governments of the potential parties to it. To date, there are still only 13 states that have ratified the Treaty (and an additional 4 signatories, all from the early 1980s: France, Guatemala, India, and Romania). The countries that have refused to ratify include the USA, the Soviet Union / Russian Federation, most of the European Union nations and, of course, a great number of developing countries, too.

This failure has been explained by the very different nature of the Moon Treaty and the entire negotiation process when compared with earlier developments in the international law of outer space. The earlier space treaties were based on current problems and were able to draw on factual information from scientists and engineers that could be closely interlinked with legal solutions. The Moon Treaty was quite correctly based on the view that the Moon contains valuable natural resources, but the problem proved to be that the drafters of the instrument were trying to formulate binding rules decades before actual capabilities of exploiting such resources existed. The interest in regulating the new resources was based on historical fears of colonialism rather than on factual reports identifying the Moon´s natural resources, the technology required for exploitation and the funding of such activities. Furthermore, there was a lack of communication between the space industry and the diplomatic agencies that negotiated the Treaty (Reynolds and Merges 1989). Consequently, the Moon Treaty, on the one hand, is more specific in its aims than, for instance, the OST; on the other, its solutions still rely on general principles, such as the concept of CHM and the international regime, which can be interpreted in various ways according to different political and economical assumptions (Galloway 1998). They leave considerable uncertainty as to the type of regime that would be established, for instance.

The main reason for states´ reluctance to sign and ratify the Moon Treaty proved to be the CHM provision and the vaguely drafted international regime of Article 11, although the potential disadvantages that the states feared could actually only arise from a subsequent accord on the implementation of that regime (requiring separate approval or disapproval), and not directly from the Moon Treaty itself. It was suggested that both free-enterprise countries and socialist states should be able to "live very comfortably within the CHM principle" because of the provision of Article 11.7.d prescribing that as much special consideration is to be accorded to those countries "which have contributed either directly or indirectly to the exploration of the Moon" as is to be given to the developing countries. Moreover, even though the CHM principle can sustain the hope for a sharing of resource benefits, it does not seem to set preferences as to the forms and means of political or economic organization and production. (Christol 1984). Apparently, such considerations were, however, not enough to calm the doubts of the space powers and convince them that they, too, would be beneficially served by the CHM principle. The reservations of the Soviet Union and the USA about the definition of Common Heritage effectively terminated any chance that the current Moon Treaty would be ratified. The two most important factors that led to the unfortunate lack of support – and the main controversies connected with the international regime of the Moon Treaty – are the possible moratorium on exploitation of natural resources and the poor incentives for commercial exploitation. Both of these questions will be taken up briefly in the following.

22.5.1 Commercial Exploitation

Utilization, signified by the term "exploitation", may include an element of commerce, but the conditions stipulated to be the main purposes of the international regime of the Moon Treaty will determine whether commercial aspirations will be compatible with that regime. The objectives expressed in Article 11.7 do not necessarily prohibit the exploitation of the natural resources of outer space in a commercial way. However, it is quite another matter whether a commercial organization would be able to exercise its commercial function within the limits of these objectives, in particular, the special attention given to developing countries; the concept of Common Heritage may be also unable to accommodate a regime based on commercial principles. While western economists emphasize the need for advanced property rights to ensure orderly and efficient development of space resources, many developing country economists still stress the need for equitable sharing of space resources and opportunities in order to avoid "economic colonization of space along neo-imperialistic lines" (Reynolds and Merges 1989).

The fact that the political-economical philosophies supported by different countries vary so widely introduces considerable uncertainty with regard to the anticipated course of policy. This obviously has contributed to the significant hesitation to ratify the Moon Treaty, particularly on the part of the states most interested in the eventual exploitation of space resources (van Traa-Engelman 1993). It is quite obvious that one of the main shortcomings of the Moon Treaty is the fact that it seems to accept the exploitation of natural resources forthwith on an experimental basis, but the rules governing actual exploitation will only be established later, by someone else. This approach is more likely to lead to non-exploitation of the resources since no actor (governmental or private) would invest large amounts of money in such an uncertain venture. The main purposes of the international regime to be established, as stated in Article 11.7, make it even more doubtful that substantial investments will be made (Oosterlinck 1997). However, the Moon Treaty is already rendered so obsolete by many factors that its prohibitive effect is fast diminishing, in particular as technological developments are soon likely to make resource activities possible in practice.

22.5.2   The Moratorium

Another factor that has hampered further ratification of the Moon Treaty is the question whether and in what form exploitation of lunar resources could at all be initiated before an international regime along the lines of Article 11 has been established. Since the adoption of the Moon Treaty, most authors have agreed that the instrument cannot be interpreted as creating any kind of a moratorium on the exploration, experimental exploitation or possibly even commercial exploitation of the Moon´s natural resources (Cheng 1996). There is no explicit mention of any such moratorium in the Treaty and, moreover, language specifically calling for a moratorium was rejected (at least twice) during the negotiations (Finch and Moore 1981).

Nevertheless, it has also been suggested that the Moon Treaty implies a moratorium on the exploitation of lunar resources until the international regime is established (Wassenbergh 1997). The idea of a legal moratorium pending the establishment of the international regime can be based on 1) the Common Heritage principle, in which a moratorium is "implicit"; 2) Article 11.5, according to which a regime is to be negotiated once exploitation is about to become feasible, i.e., the regime thus seems to precede exploitation; and 3) Article 6.2, which limits the use of celestial resources to "scientific investigations" and in "appropriate quantities" only (Marko 1993). For obvious reasons, private enterprises in particular have expressed anxiety about a possible moratorium (van Traa-Engelman 1993), whether based on a legal restriction against mineral exploitation or even on a de facto restriction resulting from the ambiguity of the Moon Treaty. Potential investors are understandably reluctant to make the large investments required if there is a possibility that their efforts will later be nullified by an unacceptable regime. Moreover, it has been feared that a party to the Treaty may not be able to make use of the resources due to a failure of the States Parties to agree on the international regime (or even because of a desire of some states not to agree), while, simultaneously, states not party to the Treaty would be under no restraint with regard to exploration and exploitation of space resources (Spitz 1989). On balance, it is quite irrelevant why countries believe in a moratorium and whether their arguments have any legal foundation; what matters is that they do think this way, and as long as they fear a moratorium, they will not sign the Moon Treaty.

22.6   The UN Declaration on Space Benefits

The United Nations defined and adopted the concept of benefits sharing already in the 1960s General Assembly resolutions on outer space and, subsequently, as part of the fundamental Article I of the OST, according to which exploration and utilization of outer space is to be for the benefit and in the interests of all countries. However, there have always been various interpretations of the meaning of the concept. Already in the 1960s, the developed countries agreed that Article I.1 of the OST set out limitations and obligations regarding the use of outer space but did not diminish their inherent rights to determine how they share the benefits derived from their space activities. The developing countries, on the other hand, believed that the benefit-sharing principle was not only an appeal to but an obligation on all states to conduct their activities in space on a cooperative international basis (Supancana 1998).

Also, the implications of the Moon Treaty´s natural resource exploitation regime have been a central topic of discussion ever since the negotiating process in the 1970s. Even if there is no moratorium established, the aspirations of less developed nations have been forwarded. The Treaty clearly states that the resources of celestial bodies are to be regarded by the States Parties – even prior to the establishment of the international regime – as CHM and that all activities undertaken by the Parties must be compatible with the purposes of the proposed regime, including the requirement of the expansion of opportunities to use space resources and of "equitable sharing" in the benefits derived from the resources. Of course, benefit sharing may be called a principle of space policy rather than a legal obligation, but the majority of nations surely also had a fair expectation of some real benefit sharing from the exploration of space by industrialized countries. At the same time, however, it is very clear that there is as yet no detailed machinery agreed upon regarding the international space resource regime nor any indication of how the principle of equitable sharing is going to be translated into concrete benefits, inasmuch as no detailed rules for such sharing are stipulated by the Moon Treaty. In light of recent developments in the UN, the establishment of any arrangements to improve the situation does not seem likely, at least not in the near future.

The issue of the utilization of space resources was addressed in more detail by the UN General Assembly in what is called the Space Benefits Declaration of the 1996. This instrument provides an authoritative, current interpretation of the cooperation principle (including benefit sharing) in Article I of the OST. The item was placed on the UNCOPUOS agenda originally by the developing countries in 1986, when they again became serious about integrating the issues of outer space into their concept of the New International Economic Order (Benkö and Schrogl 1997). They wanted to make clear once and for all the legal nature of the principles of OST Article I and elaborate the judicial content and legal consequences of the principles, preferably by the adoption of a new international space law convention (Kopal 1995); otherwise, to the vast majority of countries, the noble goals of the Article would simply remain "just another text with very little practical consequence"(Thaker 1997). In 1991, a group of developing countries presented a draft "Principles Regarding International Cooperation in the Exploration and Utilization of Outer Space for Peaceful Purposes". This draft was rich in NIEO language aiming at forced cooperation and contained as its key element automatic transfer of financial and technological resources from the North to the South. The industrialized states refused to even discuss such a proposal and, consequently, the developing countries had to present a revised draft for international space cooperation, the tone of which was far less aggressive (Benkö and Schrogl 1996). The actual negotiations finally started in 1995.

The industrialized countries emphasized that they were already making considerable efforts voluntarily (e.g., through broad multi- and bilateral programs with developing countries consisting of participation in space projects as well as considerable amounts of financial help to build satellite utilization capabilities in those countries) and that cooperation had to be beneficial to both sides (Benkö and Schrogl 1997). They were of the opinion that the existing international legal framework had proved sufficient to further international cooperation and that there was no need for additional rules and regulations. These countries stressed the sovereign right of each state to choose with whom it cooperated and that it would not be reasonable to obligate them to cooperate with all states indiscriminately, especially since there already existed close cooperation between states on a voluntary basis; it was suggested that the elaboration of a rigid new framework for cooperation could even paralyze the existent flexible cooperation. Furthermore, the industrialized states pointed out that numerous developing countries had already established their own space activities (which they had often obtained by using the financial and personnel resources of the industrialized countries) (Thaker 1997). Consequently, the French-German counterproposal to the developing countries in 1995 was based on two main considerations: states are free to determine all aspects of their international cooperation and they should choose the most effective and appropriate mode of cooperation in order to allocate resources efficiently. Already by the following year, the developing countries had given up their revolutionary redistribution-minded goals and all member states of UNCOPUOS reached an agreement on the basis for a liberal regime grounded in the French-German proposal (Benkö and Schrogl 1997).

The Space Benefits Declaration starts with a preamble stressing equally the demand for more international, mutually beneficial cooperation (para. 8) and the fact that international cooperation is already well established (para.7). The basic policies to be served by this cooperation are described as "(a) promoting the development of space science and technology and of its applications; (b) fostering the development of relevant and appropriate space capabilities in interested states; and (c) facilitating the exchange of expertise and technology among states on a mutually acceptable basis" (para. 5). Contractual terms in cooperative ventures should be fair and reasonable and they should be in full compliance with the legitimate rights and interests of the parties concerned (para. 2). The Declaration makes it clear that spacefaring nations must not forget to integrate developing countries into space exploration and that particular attention should be given to their benefit and interests (para. 3). This provision is not, however, intended to force cooperation but instead focuses on existent space cooperation. Furthermore, cooperation should be conducted in the modes that are considered most effective and appropriate by the countries concerned (para. 4). Finally, although the focus of the declaration is on the promotion and fostering of international cooperation on an equitable and mutually acceptable basis, states are free to determine all aspects of their participation in such cooperation (para. 2).

The Space Benefits Declaration attempts an important compromise regarding the Common Heritage provision by offering a means to share benefits while recognizing market principles. Importantly, for the first time in UN space law this Declaration appears to place commercial space activities on the same level as state activities. It has been hoped that it could mark the end of the North-South debate on the issue of forced cooperation and transfer of resources. Some commentators have optimistically suggested that the Declaration would encourage the space powers to conduct their activities for the benefit of all countries on an equitable and mutually acceptable basis, and stimulate Third World countries to develop their own efforts in the field of space activities. All this could also make many countries more equal and alluring partners in cooperation, on terms which the space powers are ready to accept, and even lead to effective regional cooperation between the developing countries themselves (Benkö and Schrogl 1997). Other commentators have pointed out that it should be self-evident (as a principle of civilized nations) that the space powers educate the other states about what they are doing, share the scientific and technical knowledge, and assist the non-space powers in developing indigenous capability, so that they may participate in space activities and receive adequate benefits themselves (Wassenbergh 1997).

At first sight, the principles of the Declaration indeed appear to provide a means for the equitable sharing of benefits, especially as the description of policies is similar and complementary to the formulation expressed in the Moon Treaty for the international regime policies. However, the principles also show an obvious "tendency of softening the formerly harsh and economically restricting approach of the economic element of the common heritage conception" (Hobe 1999). The fact that this is done so clearly in conformity with free-market principles seems to lead in practice to a situation where the developing countries may not be in a position to expect much after all. Quite contrary to the initial expectations of the developing countries, the Declaration actually affirms that no requirements of a legal nature can be derived from OST Article I and that the Article is to remain a principle of moral and philosophical value only. Thus, principles of the Space Benefits Declaration add neither new obligations for spacefaring nations nor new rights for developing countries. In fact, it seems that even if the original concept of benefit sharing in space law did reflect some kind of genuine "common heritage" thinking, any vestiges of such may have been eliminated by the Space Benefits Declaration. It has even been suggested that benefit sharing should be litigated to find out whether or not it exists anymore (O'Donnell 1999).

On balance, the Third World seems to have failed to revive its NIEO aspirations through the Space Benefits Declaration. The Declaration focuses on interstate cooperation but fails to obligate states to enter into such cooperation. The text of the principles of the Declaration is vague and general and, moreover, essentially reiterates what already exists in other space law instruments. The way the principles are formulated keeps them on a completely non-committal level and as such they merely reflect good intentions on the part of states, with no need to translate these into action. (Wassenbergh 1997). It does not seem very likely that the developing countries, after their defeat in UNCOPUOS during the negotiations of these principles, would push for a codification of the Declaration in a new space treaty any time soon. It can be argued that the unanimous adoption by the General Assembly gives the Declaration certain binding effect (Thaker 1997). However, such merely academic considerations will obviously not help the developing states much in practice. It is still too early to say what – if any – effect the Space Benefits Declaration will actually have on enhancing international cooperation in the exploration and use of outer space.

22.7 The Moon Treaty Today

The Moon Treaty suffers from a chronic – very likely fatal – lack of adherents: thirteen ratifying states, none of which is a major spacefaring country, are obviously not enough to make the Treaty relevant in practice. If a space mission is conducted by an entity which does not include a State Party to the Moon Treaty – as is most likely the case – the provisions of the instrument may be wholly inapplicable. Initially, only five ratifications were considered necessary for the entry into force of the Moon Treaty because of its apparently general approval by the international community: the text of the Treaty was adopted in the UN General Assembly without a vote. However, instead of the general international legal regime it was planned to establish, the Moon Treaty has gained exceptionally few adherents. During the past decades, the original definitions of problems for which the legal remedies of the Treaty were formulated have become even far more outdated. New scientific, technological, political and commercial trends have developed to the extent that we would need a current assessment of the questions connected with the Moon Treaty in general and with the use of lunar natural resources in particular. Possibilities for revision of the Moon Treaty have been discussed in the past years within the UNCOPUOS, for instance. It may well be that the Treaty is so severely handicapped that it cannot survive the present demands in any way. Indeed, it has often been proposed that a completely new international agreement be drafted to replace the Moon Treaty altogether (e.g., Hobe 2005).

On the other hand, it has been suggested that the expansion of commercial enterprises from industrialized states, in the absence of effective international regulations, could give new impetus for the less developed states to reconsider the desirability of ratifying the Moon Treaty (Sterns and Tennen 1999). The reasoning here is that the Moon Treaty would then offer at least some protection for the developing countries and without it there would be little hope for an equitable arrangement anyway. However, it has been pointed out that, even though the Moon Treaty was supposed to be a tool for achieving equity in space resource distribution and it contains several provisions which in theory would help correct economic imbalance and social injustice, certain aspects of the Treaty are in fact potentially disadvantageous to developing countries. For instance, the uncertainty of the regime that is to be created may work to their detriment as well. It is also true that the entire definition of equity is open to subjective interpretation and therefore the "equitable sharing" concept might be invoked by the developed countries to justify excluding those countries that do not invest and participate in a particular lunar mining project, for instance (Marko 1993).

At least thus far there has been no rush of less developed states to adopt the Moon Treaty. Nevertheless, it has recently received some new States Parties. As many as four states joined the Treaty within the past decade: Kazakhstan in 2001, Belgium in 2004, Peru in 2005, and Lebanon in 2006. The UN has repeatedly encouraged new states to adhere to the Moon Treaty. In 2008, some States Parties to the Treaty presented at a session of the Legal Subcommittee of the UNCOPUOS a Joint Statement on the benefits of adherence to the Moon Treaty (Joint Statement 2008). The statement stresses that the Moon Treaty represents "a mutual commitment to seeking a multilateral solution for the exploitation of the natural resources of celestial bodies in accordance with the general principles of outer space law" (para. 8). It further provides that "the States parties to the Moon Agreement encourage States that have signed but not yet ratified the Agreement, as well as other States, to become parties to it, in particular considering their possible involvement in forthcoming missions or projects aimed at exploring celestial bodies" (para. 9). However, as long as the controversy about Article 11 continues, it seems that at least the USA (and therefore many others) will not ratify the Moon Treaty, which leaves the prospects for a true "common heritage" of space resources rather poor. The obvious dilemma of the Moon Treaty is that without the Common Heritage provision it would be largely just a reiteration of previous treaties but with it, it looses the appeal to major space powers whose support it desperately needs (Spitz 1989). The ambiguity of Article 11 makes the Moon Treaty less alluring to other states as well.

Barring adoption by the major space powers, the Moon Treaty is unlikely to play a major role in the future. It is of course at least a theoretical possibility that the few states that actually have already ratified the Treaty may become major space powers. However, even without any further adherents the Moon Treaty does have importance as one of the first international documents to embody concepts identified with the NIEO. The Treaty was many years in the making and thus reflects a careful balancing of the international interests of that time. Besides, the economical thinking represented by the developing states has not faded away; it has perhaps only taken somewhat different forms. Moreover, even many western economists and lawyers believe that an international organization will be required to allocate development rights to lunar resources (e.g., Kerrest 2005). Despite its low number of ratifying states, the Moon Treaty is likely to serve as the basis for any proposals for future agreements on space resources. Consequently, it is important to learn from the experience of the Moon Treaty and carefully explore the alternative organizational forms a space resource allocation system might assume.

A noteworthy attempt to interpret the Common Heritage of Mankind concept of the Moon Treaty was contained in a resolution adopted by the International Law Association in 2002 (ILA Resolution 2002). Pursuant to the resolution, "the common heritage of mankind concept has developed today as also allowing the commercial uses of outer space for the benefit of mankind". Accordingly, the resolution refers to the need of "certain adjustments" to Article 11 of the Moon Treaty "concerning the international regime to be set up for the exploitation of Moon resources which will make it more realistic in today´s international scenario". The conclusion of the ILA was that the current reality necessitates such an interpretation of the CHM which also implies the possibility of commercial space activities for the benefit of mankind. Thus the idea of equitable sharing of benefits and resources seems to be evolving into less rigid forms of economic cooperation (Hobe 2005). As long as the requirements incorporated in the concept of the CHM are complied with, even private entities could be allowed to carry out commercial lunar resource operations (Tronchetti 2009). In a similar manner, some States Parties to the Moon Treaty have stressed their view that the Treaty "does not propose a closed and complete mechanism [rather], it adopts an intelligent approach". The Moon Treaty "does not preclude any modality of exploitation, by public or private entities, or prohibit the commercialization of such resources, provided that such exploitation is compatible with the principle of a common heritage of mankind" (para 7.e, Joint Statement 2008). If such a view is generally accepted, there should be no need to abandon the Moon Treaty: its shortcomings may be amended by more market-oriented interpretations instead. Even the less developed states appear to be increasingly receptive to such an approach to the CHM concept (Tronchetti 2009).

22.8 Conclusion

After several decades of significant scientific and technological advances, exploitation of lunar resources has evolved into a matter of increasing interest worldwide. The first and thus far most important manifestation of this development was the Moon Treaty of 1979. The Moon Treaty specifically refers to the exploitation of the natural resources of the Moon and other celestial bodies, declares them to be the Common Heritage of Mankind and envisages the establishment of an international regime as soon as exploitation is about to become feasible. However, as the number of actors interested in outer space and having the resources to explore it multiplied over the decades, the process of producing space law became increasingly complicated. Consequently, there has not been much progress in developing the legal regime for the exploitation of lunar resources since 1979.

In the absence of up-to-date regulation, many important questions of space resource exploitation remain without adequate answers. One difficult question is likely to be the use of space natural resources, especially after their extraction, e.g., the status of lunar resources in the implementation of space defense systems. According to OST Article IV, the Moon and other celestial bodies are totally demilitarized and are to be used exclusively for peaceful purposes, but in the area around the Earth only the placement of nuclear weapons and weapons of mass destruction is prohibited. Are lunar materials in Earth orbit under the peaceful-purposes regime for the Moon and other celestial bodies or subject to the near-Earth partial demilitarization only (Goldman 1986)? Under the Moon Treaty, the legitimacy of the use of lunar materials could engender a similar debate, as Article 3.2 prohibits "any threat or use of force or any other hostile act or threat of hostile act on the moon".

An at least equally important yet also unresolved issue is mineral exploitation for commercial purposes. In more general terms, it may be necessary to draw distinctions between use which is by its nature exclusive and use which can be shared simultaneously, use which involves the resource being consumed and use which leaves it available for further use, use which involves some kind of deterioration or transformation and use which has no such effect, etc. (Jenks 1965). In particular, there should be some distinction between the waste or contamination of natural resources and use that may even contribute to resource and nature conservation (Goldman 1986). Furthermore, lunar resource activities are extremely capital-intensive industries and not all States will possess the requisite technology, personnel and financial means to exploit the resources to an equal extent at the same time. The very issue affects some countries more than others, depending on economic and political interests. In economic terms, some countries would need the revenue of these new minerals more than others. Then again, countries mining the same minerals in their own territory may suffer from the new competition. Politically, the issues related to the utilization of natural resources of outer space are very much inclusive of all nations: the distribution of profits from such resources is a topic that has divided opinions considerably.

Demands made regarding the resources of the Moon have originated in economic and security interests or interests of states or even private entities (usually to utilize the resources in one way or another); on the other, they derive from interests of the so-called global commons and interests of future generations (aiming rather at the preservation of resources and management of the environment). On another level, demands are being made by geographically, technologically or otherwise disadvantaged countries, as well as by leading world powers and technologically advanced states. The actors include governments, international organizations and private entities, all of whom have to interact in order to exploit the common property resources (Jasentuliyana 1986). Countries with superior technological capacity are likely to claim a "right of access" to the resources (and have in fact done so), whereas less advanced countries are more inclined to defending a right to conserve and manage the resources with a view to future exploitation. The latter demands are often treated as being of a lower order of priority, i.e., mere "political" and "ideological" considerations, in contrast to the more demonstrable "needs" of the technologically advanced countries (Pinto 1983). The increase of public-private partnerships and multinational cooperation, as well as the general gradual shift from exploration to exploitation in space activities add yet another layer to the discussion (Trepczynski 2010).

Yet, at present, the resources of the Moon do not have much immediate bearing on the daily lives of the vast majority of people. Consequently, those whose welfare might actually be threatened by the proposed solutions to questions regarding these resources are unlikely to be aware of such threats. Even if they become conscious of them, they will most probably feel that there is not much they can do about the situation. Moreover, the issues associated with these remote resources are mostly very technical, involving levels of information and comprehension attained only by specialists. For these reasons, the politics of the new resources are unlikely to be affected by much intervention from the general public. They are issues that tend to be debated in scientific discourse rather than in the inflammatory rhetoric of politics. On the other hand, it is also possible that the limitations of scientific knowledge will result in a situation where scientists lack sufficient tools for dominating the debate on space resources and politics will take over the discourse. Also the advent of increasing scarcities and a growing awareness of lunar resources will in all likelihood foster increasing politicization of the issues. Even then, however, the resources will remain located in a global common and their utilization will still pose numerous questions which cannot be addressed without a commitment to establish scientific proof (Rosenau 1983).

On balance, if lunar resource activities are pursued, complex questions of law, politics, economics and technology cannot be avoided. Unfortunately, the present regulatory framework is far from being adequately equipped for answering questions related to such activities although they are soon likely to emerge on the international political scene. Pressure is rapidly building on the international community to establish a more acceptable and in every aspect more distinctive legal regime applicable to the exploitation of space resources and, in particular, one under which the resources can also be utilized for commercial purposes. In all likelihood, various models of regulation will be needed, depending on the particular circumstances of the resources, their scarcity, intended use of the resources, the location of activities, and the type of projects and actors, for instance (Sterns and Tennen 2006). It does not require much insight to discern that these new resource issues are not likely to evolve along traditional lines; instead, an innovative, flexible and open-minded approach is needed in looking for solutions. Whatever forms the new regulation will take, it will have to be responsive to the realities of the space sector of today and, preferably, even of tomorrow.

References

Benkö, M., Schrogl, K.-U.: A new approach for the debate on "space benefits" in the UNCOPUOS. In: Proceedings of the 38th Colloquium on the Law of Outer Space (Oslo, 1995), pp. 293–299. American Institute of Aeronautics and Astronautics, AIAA (1996)

Benkö, M., Schrogl, K.-U.: History and impact of the 1996 UN Declaration "Space Benefits". Space Policy 13, 139–143 (1997)

Brooks, E.: Dangers from asteroids and comets: relevance of international law and the space treaties. In: Proceedings of the 40th Colloquium on the Law of Outer Space (Turin 1997), pp. 234–263. AIAA (1998)

Brown Weiss, E.: Conflicts between present and future generations over new natural resources. In: Dupuy, R.-J. (ed.) The settlement of disputes on the new natural resources, Workshop (The Hague 1982), pp. 177–191. Martinus Nijhoff Publishers, The Hague (1983)

Cheng, B.: Studies in international space law. Oxford University Press, Oxford (1996)

Christol, C.Q.: Modern international law of outer space, 2nd printing. Pergamon Press, New York (1984)

Christol, C.Q.: Space law: past, present, and future. Kluwer Law and Taxation Publishers, Deventer (1991)

Dekanozov, R.V.: Juridical nature and status of the resources of Moon and other celestial bodies. In: Proceedings of the 23rd Colloquium on the Law of Outer Space (Tokyo 1980), pp. 5–8. AIAA (1981)

Fasan, E.: Some legal problems regarding the Moon. In: Proceedings of the 23rd Colloquium on the Law of Outer Space (Tokyo 1980), pp. 9–11. AIAA (1981)

Fasan, E.R.: Space law, planets and Gobriel´s Ius Spatiale. In: Palyga, E.J. (ed.) International Space Miscellanea, Warsaw, pp. 68–74 (1995)

Finch, E.R.: 1979 United Nations Moon Treaty encourages lunar mining & space development. In: Proceedings of the 22nd Colloquium on the Law of Outer Space (Munich 1979), pp. 123–124. AIAA (1980)

Finch, E.R., Moore, A.L.: The 1979 Moon Treaty encourages space development. In: Proceedings of the 23rd Colloquium on the Law of Outer Space (Tokyo 1980), pp. 13–18. AIAA (1981)

Galloway, E.: The United States and the 1967 Treaty on Outer Space. In: Proceedings of the 40th Colloquium on the Law of Outer Space (Turin 1997), pp. 18–33. AIAA (1998)

Goldman, N.C.: Space activities: transforming space law. In: Proceedings of the 28th Colloquium on the Law of Outer Space (Stockholm 1985), pp. 227–231. AIAA (1986)

Gorove, S.: Developments in space law. Martinus Nijhoff, Dortrecht (1991)

Gorove, S.: Space resources and developing nations – a legal assessment. In: Palyga, E.J. (ed.) International Space Miscellanea, Warsaw, pp. 97–103 (1995)

Haanappel, P.P.C.: Comparison between the law of the sea and outer space law: exploration and exploitation. In: Proceedings of the 28th Colloquium on the Law of Outer Space (Stockholm 1985), pp. 145–148. AIAA (1986)

Hobe, S.: Common Heritage of Mankind – an outdated concept in international space law? In: Proceedings of the 41st Colloquium on the Law of Outer Space (Melbourne 1998), pp. 271–285. AIAA (1999)

Hobe, S.: ILA Resolution 1/2002 with regards to the Common Heritage of Mankind principle in the Moon Agreement. In: New Developments and the Legal Framework Covering the Exploitation of the Resources of the Moon, IISL/ESL Space Law Symposium (Vienna 2004), Proceedings of the 47th Colloquium on the Law of Outer Space (Vancouver 2004), pp. 536–544. AIAA (2005)

Hofmann, M.: Recent plans to exploit the Moon resources under international law. In: Proceedings of the 47th Colloquium on the Law of Outer Space (Vancouver 2004), pp. 425–434. AIAA (2005)

ILA Resolution. In: 70th Conference of the International Law Association, New Delhi (2002) (January 2002)

Jasentuliyana, N.: Balancing the conflicting demands in legislating common property resources of the oceans and space. In: Proceedings of the 28th Colloquium on the Law of Outer Space (Stockholm 1985), pp. 149–150. AIAA (1986)

Jenks, W.C.: Space law. Stevens, London (1965)

Joint Statement on the Benefits of Adherence to the Agreement Governing the Activities of States on the Moon and Other Celestial Bodies by States parties to the Agreement. UNCOPUOS Legal Subcommittee, 47th session (Vienna 2008), Agenda item 6 "Status and application of the five United Nations treaties on outer space". A/AC.105/C.2/L.272

Kerrest, A.: Exploitation of the resources of the high sea and Antarctica: lessons for the Moon? In: New Developments and the Legal Framework Covering the Exploitation of the Resources of the Moon, IISL/ESL Space Law Symposium (Vienna 2004), Proceedings of the 47th Colloquium on the Law of Outer Space (Vancouver 2004), pp. 530–535. AIAA (2005)

Kopal, V.: Analogies and differences in the development of the law of the sea and the space law. In: Proceedings of the 28th Colloquium on the Law of Outer Space (Stockholm 1985), pp. 151–155. AIAA (1986)

Kopal, V.: Progressive development of space law and concept of Common Heritage of Mankind. In: Palyga, E.J. (ed.) International Space Miscellanea, warsaw, pp. 107–114 (1995)

Kopal, V.: What kind of institutional arrangements for managing space mineral resource activities should be done in a foreseeable future? In: Proceedings of the 41st Colloquium on the Law of Outer Space (Melbourne 1998), pp. 12–22. AIAA (1999)

Marko, D.E.: A kindler, gentler Moon Treaty: a critical review of the current Moon Treaty and proposed alternative. J. Nat. Resources & Envtl. L 8, 293–337 (1993)

Matte, N.M.: Aerospace law: from scientific exploration to commercial utilization, Toronto (1977)

Miklódy, M.: Some remarks on the question of the rights of possession of mineral resources of the celestial bodies. In: Proceedings of the 22nd Colloquium on the Law of Outer Space (Munich 1979), pp. 179–180. AIAA (1980)

Nakamura, M.: Review of Article 1 of the Outer Space Treaty. In: Proceedings of the 39th Colloquium on the Law of Outer Space (Beijing 1996), pp. 132–137. AIAA (1997)

Nash, M.: Contemporary practice of the United States relating to international law. In: Reynolds, G.H., Merges, R.P. (eds.) Outer space – problems of law and policy, pp. 110–114. Westview Press, Boulder (1989)

O'Donnell, D.J.: A new institution is proposed to manage space resources: the Metanation in space. In: Proceedings of the 41st Colloquium on the Law of Outer Space (Melbourne 1998), pp. 23–31. AIAA (1999)

Oosterlinck, R.: Tangible and intangible property in outer space. In: Proceedings of the 39th Colloquium on the Law of Outer Space (Beijing 1996), pp. 271–283. AIAA (1997)

Pinto, M.C.W.: Settlement of disputes concerning new natural resources: new resources – old problems. In: Dupuy, R.-J. (ed.) The Settlement of Disputes on the New Natural Resources, Workshop (The Hague 1982), pp. 19–23. Martinus Nijhoff Publishers, The Hague (1983)

Reynolds, G.H., Merges, R.P.: Outer space – problems of law and policy. Westview Press, Boulder (1989)

Rosenau, J.N.: New natural resources as global issues. In: Dupuy, R.-J. (ed.) The Settlement of Disputes on the New Natural Resources, Workshop (The Hague 1982), pp. 25–34. Martinus Nijhoff Publishers, The Hague (1983)

Rosenfield, S.B.: Article XI of the Draft Moon Agreement. In: Proceedings of the 22nd Colloquium on the Law of Outer Space (Munich 1979), pp. 209–212. AIAA (1980)

Spitz, S.: Space law – Agreement Governing the Activities of States on the Moon and Other Celestial Bodies. In: Reynolds, G.H., Merges, R.P. (eds.) Outer Space – Problems of Law and Policy, pp. 115–116. Westview Press, Boulder (1989)

Sterns, P.M., Tennen, L.I.: Institutional approaches to managing space resources. In: Proceedings of the 41st Colloquium on the Law of Outer Space (Melbourne 1998), pp. 33–45. AIAA (1999)

Sterns, P.M., Tennen, L.I.: Private enterprise and the resources of outer space. In: Proceedings of the 48th Colloquium on the Law of Outer Space (Fukuoka 2005), pp. 240–252. AIAA (2006)

Supancana, I.B.R.: The international regulatory regime governing the utilization of Earth-orbits. Rijkuniversiteit te Leiden, Leiden (1998)

Thaker, J.S.: The development of the Outer Space Benefits Declaration, vol. XXVI, pp. 537–558 (1997)

Trepczynski, S.: Is a new look necessary in the age of exploration and exploitation? In: 30th anniversary of the "Moon Agreement": retrospect and prospects, Space Law Symposium (Vienna 2009), Proceedings of the International Institute of Space Law 2009, pp. 513–522. AIAA (2010)

Tronchetti, F.: The exploitation of natural resources of the Moon and other celestial bodies: a proposal for a legal regime. Martinus Nijhoff Publishers, Leiden (2009)

van Traa-Engelman, H.L.: Commercial utilization of outer space: law and practice. Martinus Nijhoff, Dordrecht Boston (1993)

Viikari, L.: The environmental element in space law: assessing the present and charting the future. Martinus Nijhoff Publishers, Leiden (2008)

Wadegaonkar, D.: The orbit of space law. Stevens, London (1984)

Wassenbergh, H.A.: The international regulation of an equitable utilization of natural outer space resources. In: Proceedings of the 39th Colloquium on the Law of Outer Space (Beijing 1996), pp. 132–137. AIAA (1997)

23 The Property Status of Lunar Resources

Virgiliu Pop

Romanian Space Agency, Bucharest, Romania

23.1 Introduction – Water from the Moon

In 1993, Celine Dion released a song whose lyrics spoke of the seemingly imposs-ible tasks of getting "water from the moon" in order to earn someone's love. As late as two decades ago, Earth's natural satellite was being viewed as the ultimate desert – a "vast, lonely, forbidding type of existence or expanse of nothing" and as a "magnificent desolation", as described by its visitors Frank Borman and Buzz Aldrin (Chaikin 1994, pp. 121, 211). As years passed, the Moon turned, under the scrutinizing eyes of the scientists, from a presumed bone-dry desert into a watery treasure chest, capable of quenching the NewSpace entrepreneurs' thirst for rocket fuel. The idea of orbital depots supplied with water extracted from the Moon be-came feasible, and companies were born, such as the Shackleton Energy Compa-ny, with an eye on mining the ice-water-rich Lunar south pole (Wall 2011).

The technical issues of mining the Moon's riches have been explained in other chapters of this volume. In what follows, we will examine the legal implications of extracting lunar resources, and of owning land on the Moon.

23.2 Extraterrestrial Real Estate – Is the Moon for Sale?

Until quite recently, the press abounded of stories about a man in the United States who "owns" the Moon and much of the Solar System, and who is more than happy to share his extraterrestrial wealth with everybody – for a fee. Throughout the three decades of its existence, Dennis Hope and his "Lunar Embassy" catered for literally millions of people from all walks of life, such as attorneys, doctors, edu-cators, but also for hundreds of celebrities counting Hollywood stars, politicians from many countries, and space travelers. The advent of the global communica-tions era had ushered an otherwise obscure novelty business into an undeserved spotlight, making it the leader of the "extraterrestrial real estate business".

The claim that the Moon belongs to a man self-styled "The Head Cheese" may be trivial; one cannot however ignore the seriousness of the problem behind this humorous mask. In an era where a return to the Moon is being envisaged by NASA and private enterprise is poised to play a major role in opening the high frontier, the question of extraterrestrial ownership is not a Byzantine discussion. While in the following sections the importance and problematic nature of property rights in outer space will be outlined, one needs first to critically analyze and dismantle the issue of so-called "extraterrestrial real estate".

The first reason for invalidating the "Lunar Deeds" is the non-appropriation principle of space law. Article II of the 1967 Outer Space Treaty – the main document governing the conduct of States and their subjects in regard to the extraterrestrial realms - forbids the "national appropriation by claim of sovereignty, by means of use or occupation, or by any other means" of outer space, including the Moon and other celestial bodies.

The second rationale is the plurality of claims lodged for the Moon and other celestial bodies. In the United States, lunar real estate was being peddled as early as the 1890s, when, according to journalist R. D. Whytock (quoted in Reno Evening Gazette 1929), "unscrupulous money-gleaners" from New York were known for selling "lots on Luna... and scores of other commodities that would appeal to the credulous public". And in 1910, lecturer Frank Dixon (quoted in The Daily Northwestern 1910) was complaining that "there is no legal relief against the bunco steerer. Any man can organize a real estate company to sell lots on the planet Mars and get a charter in the state of New Jersey."

On June 15th, 1936, A. Dean Lindsay of Ocilla, Georgia, presented himself before a Pittsburgh Notary Public and made original claims for -

"[a]ll of the property known as planets, islands-of-space or other matter ... located in all the region visible (by any means) ... from the city of Ocilla, Ga, together with all ... matter (except this world ...) visible from any other planet, island-of-space or other matter."

In a letter sent to the Clerk of Superior Court in Ocilla, Lindsay said he realized these great holdings have no private owner, hence he decided to lay claim to them so that henceforth they will always have one. Accompanying the letter were the deeds sent for record, together with the due amount for recording them (Ocilla Star 1937). And recorded they were on June 28th, in Deed Book 11, pages 28 and 29, at Irwin County Courthouse in Ocilla.

Throughout the following years, many more claims to extraterrestrial properties were lodged all though the world, leading to the establishment of entities such as the "Nation of Celestial Space" or the "Interplanetary Development Corporation". We have examined elsewhere the long history of lunar salesmen and overlords; what matters is that, if wolves and dogs have howled at it since time immemorial, many actual claims precede Dennis Hope's one.

The third reason for invalidating the claims presented above is the lack of *corpus possidendi*. In plain language, claiming does not mean owning. In the acquisition of possession, two concurrent elements - "the mind" and "the body" are required. One is insufficient without another; there must be "both an intention to take the thing and some act of a physical nature giving effect to that intention" (Reid 1996, p. 103).

The first element required is the *"animus possidendi"*, the intention to possess. The desire to own the Moon, even a tiny slice, is hardwired into the human mind.

There is no child who has not raised, at least once, a hand towards the Moon and stars, trying to grab them. However, lunar salesmen can not own the Moon just because they want to. They lack the second element required in the acquisition of possession, namely the "*corpus possidendi*"; without an act of physical nature giving effect to the intention to take the thing, animus is insufficient. The Scottish jurist Stair (1693, II.i.18) has explained this in very illustrative terms: "if any act of the mind were enough, possession would be very large and but imaginary". An imagination as large as the Universe, in the case of the purported extraterrestrial salesmen, who present a very valid *animus*, but no *corpus* at all.

All the arguments above – the non appropriation principle, the fact that a mere claim is not tantamount with ownership, and the host of unsubstantiated claims by countless wannabe landlords -- invalidate the "Lunar Deeds". Neither Dennis Hope, nor his predecessors or copycats, own the Moon. [Note: for a more elaborate analysis of the subject of extraterrestrial real estate, see Pop (2006)]

23.3 The Commons Regime – Everybody's and Nobody's

But who, after all, does own the Moon? Article I of the Outer Space Treaty proclaims that the extraterrestrial realms – outer space, including the Moon and other celestial bodies – shall be "free for exploration and use by all States without discrimination of any kind, on a basis of equality", and that "there shall be free access to all areas of celestial bodies". The exploration and use of the extraterrestrial realms is declared as being the "province of all mankind". This norm effectively establishes among the States Parties an open access and free use regime on the Moon, making it a public good whose owner is everybody and nobody.

Far from being Orwellian doublethink - that is, simultaneously accepting two mutually contradictory beliefs -, this state of affairs hails from the "bundle of rights" theory of property. Property is an embodiment of three attributes - *jus utendi* (the right to use), *jus fruendi* (the right to enjoy the fruits) and *jus abutendi* (the right to "abuse" or "dispose" of one's own good). The commons regime is built around *jus utendi* (use is permitted, hence "everybody owns the Moon") yet forbids *jus abutendi* (title is denied, hence "nobody owns the Moon"). As to *jus fruendi* – i.e, to collect the "fruits" of the extraterrestrial realms - this will be addressed later in this study.

The current incarnation of *corpus juris spatialis* is not strictly an open access regime, containing several regulations. According to Article I of the Outer Space Treaty, the freedom of exploration and use of the extraterrestrial realms pertains to States. Article VI of the same Treaty requires non-governmental entities to obtain authorization from the appropriate State Party in order to carry out activities in the extraterrestrial realms, and to consent to being continually supervised by same. States bear international responsibility for national activities carried out in outer space and on the celestial bodies, whether these are performed by governmental entities or by private enterprise. The regime is therefore a hybrid of *res communis* (whereby ownership pertains to the community as a whole, and every member has the non-exclusive right to use the property) at international level – and *res publica* (whereby property is centrally enclosed and cannot be used without permission from the community) at municipal level, given the need for a nationally issued license.

As explained by French jurist Pothier – quoted by the U.S. Supreme Court in *Geer v. Connecticut* (1896), res *communis* originates from "all those things which God had given to the human race", this community being different from the "positive community of interest, [existing] between several persons who have the ownership of a thing in which each have their particular portion". This original community was -

> "'a negative community,' which resulted from the fact that those things which were common to all belonged no more to one than to the others, and hence no one could prevent another from taking of these common things that portion which he judged necessary in order to subserve his wants. Whilst he was using them, others could not disturb him; but when he had ceased to use them, if they were not things which were consumed by the fact of use, the things immediately re-entered into the negative community, and another could use them".

In a nutshell, *res communis* implies freedom of an actor to use a good; as long as one uses this good, one may not be obstructed by another, yet this latter is free to use the good as soon as the first has ceased to use it.

Under space law, *jus utendi* is not absolute; the right of using the Moon and other celestial bodies has to be carried out, under the provisions of Article IV of the Outer Space Treaty, "exclusively for peaceful purposes". Article IX of the same pact requires the State Parties to use the principle of co-operation and mutual assistance as a guide in the exploration and use of the outer space and celestial bodies. The State Parties need to conduct all their activities in the extraterrestrial realms "with due regard to the corresponding interests of all other States Parties", according to the same legal norm. The Outer Space Treaty elaborates, up to a certain degree, the mechanism for accommodating the interests of the other members of the public. Thus, Article IX requires States Parties to publicize, "to the greatest extent feasible and practicable" the nature, conduct and locations of their space activities. If an activity by a State Party or one of its nationals is likely to cause "potentially harmful interference" with the space activities of other States Parties, the first State Party is required by the same tenet to "undertake appropriate international consultations" prior to carrying out such activity. In the same time, should a State Party to the Outer Space Treaty rightfully fear that a space activity of another State Party could harmfully affect its own activities, the first State is entitled to request appropriate consultations on this subject. While the Treaty makes no mention of it, its clear "non-interference" principle implicitly creates a "first-come, first-served" regime.

Whereas several rules are established and ought to be respected by the State Parties, most of these remain at the stage of principle – as recognized by the document's title – "Treaty on Principles Governing the Activities of States in the Exploration and Use of Outer Space, Including the Moon and Other Celestial Bodies". Many open access regimes are self-regulatory, and where the text of the law is silent, custom is bound to develop.

23.4 Finders Keepers – Property Status of the Extracted Planetary Resources

We promised above to address the issue of *jus fruendi* – i.e, the right to collect the "fruits" of the extraterrestrial realms, under the current, commons regime enshrined in the Outer Space Treaty.

It is to be remarked *ab initio* that the legal treatment applied to real estate and substances removed thereon is different. As C. Sweet (1882, p. 259) says,

"[w]hile unsevered, minerals form part of the land, and as such are real estate. When severed, they become personal chattels".

Whereas the immovables examined above are subject to the *lex situs* of outer space, extracted resources are movables, subject mainly to the *lex domicilii* of the person who caused their removal.

Is the conversion of immovables into movables by way of extraction, allowed in the light of the non-appropriation principle of the Outer Space Treaty? The legal text is silent as to the permissibility of appropriating natural materials in the course of exploration and use of the celestial bodies. In the absence of a specific norm clarifying the ownership of lunar resources and samples, most scholars draw a clear distinction between the appropriation of outer space and celestial bodies, and the appropriation of materials thereon. Article I of the Outer Space Treaty consecrates the freedom of scientific investigation, exploration and use of the extraterrestrial realms, whereas the "residuary rule of presumptive freedom of action" (Lauterpacht 1975, p. 220) proclaims that what is not prohibited is permitted.

Indeed, even Hugo Grotius, a strong opponent of national appropriation of the high seas, did not discourage the appropriation of the resources found there. In support of this view, he found it appropriate to cite the play *Rudens* ("The Rope") by Titus Maccius Plautus: -

"[W]hen the slave says: - the sea is certainly common to all persons - the fisherman agrees; but when the slave adds: - then what is found in the common sea is common property -, he rightly objects, saying: - But what my net and hooks have taken, is absolutely my own" (Grotius 1608).

A similar logic is found in John Locke's (1690, Sect. 30) Second Treatise of Civil Government, where -

"by virtue thereof [of the original law of nature], what fish any one catches in the ocean ... is, by the labour that removes it out of that common state nature left it in, made his property who takes the pains about it".

By their public domain character, celestial bodies are also akin with the seashore under Roman law, common and accessible to all by the law of the nature (Justinian, II.I.1). While, as a *res communis*, the seashore was incapable of ownership (Thomas 1975, p. 75), not the same regime was valid for the "pebbles, precious stones and the like which are found on the seashore"; these "at once become the finder's property by the law of nations" (Justinian, II.I.18). By analogy, the moon rocks "found" on the moon would become the finder's property.

Whereas the Moon Agreement contains a more restrictive regime, as long as the space powers do not ratify the document, their private enterprises are "entitled to acquire and retain space resources" for their own disposition "without limitation on possible profit", as agreed by Stephen Gorove (1985 p. 227).

In December 1993, Russia set an important precedent by commercializing lunar material retrieved by a Soviet probe. Sotheby's auctioned three small particles of regolith weighing about one carat (200 mg), who sold for US$442,500 - i.e. US$2.2 million per gram (Arthur 1998). As no objections were been voiced by third States, it can be stated that, as an attribute of ownership, the right to commercialize extraterrestrial material has entered into customary international law. Indeed, Gyula Gal finds worthy of mentioning "some facts of international practice", namely that both USA and USSR collected and returned lunar samples – objects that "were appropriated by U.S. and Soviet authorities respectively and have been owned by them without objection from the international community." (Gal 1996)

It is safe, therefore, to conclude that although fee simple ownership is prohibited, under the Outer Space Treaty regime the conversion of immovables into movables by way of extraction is allowed in the light of the non-appropriation principle, a distinction being made between the appropriation of outer space and celestial bodies and the appropriation of materials thereon. While lacking fee simple ownership over the land thereof, private actors are entitled to explore and exploit the natural resources of the moon. Ownership of extraterrestrial products vests in those who sponsored their removal, through the labor invested in seizing them and, as an attribute of ownership, the right to extract and commercialize extraterrestrial material is part of customary international law.

The status quo of the commons regime is not very stable, being challenged on two fronts. On the left, *res communis* is bordered by the Common Heritage of Mankind regime, whereby users have to share with the community the benefits accrued from the use of the commons. On the right, the supporters of property rights want the enclosure of the lands whereby they lay. We will next examine these two paradigms.

23.5 The Common Heritage of Mankind – Reaping without Sowing

At the beginning of 2011, on the occasion of the World Water Day, the Socialist president of Venezuela Hugo Chavez (quoted by Chinea 2011) offered his two cents on the matter of extraterrestrial water: -

"I have always said, heard, that it would not be strange that there had been civilization on Mars, but maybe capitalism arrived there, imperialism arrived and finished off the planet ...Careful! Here on planet Earth where hundreds of years ago or less there were great forests, now there are deserts. Where there were rivers, there are deserts".

The post-Sputnik and post-colonial era could not escape the materialist conception of history, where the world and its extraterrestrial surroundings are the scene of the class struggle between the antagonistic spacefaring and non-spacefaring nations, developed and developing states, the haves and have-nots.

Karl Marx, Friedrich Engels and many other communists sought an end to capitalism and the establishment of an egalitarian society. The 20th century saw the establishment of several Communist States, and the adoption of several Socialist principles into the mainstream. In the 1970s, the Marxist ideas received escape velocity through the work of the United Nations, who oversaw the drafting of the "Agreement Governing the Activities of States on the Moon and Other Celestial Bodies" (known as the Moon Agreement), effectively planting the Marxist standard in the lunar soil. Article 11.1 of this legal document proclaims that "[t]he moon and its natural resources are the common heritage of mankind", whereas Article 11.3 contains a facial prohibition of landed property in outer space:-

"Neither the surface nor the subsurface of the moon, nor any part thereof or natural resources in place, shall become property of any State, international intergovernmental or non-governmental organization, national organization or non-governmental entity or of any natural person"

Besides prohibition of title, the Moon Agreement contains another key tenet, namely an egalitarian distribution of benefits. As an attribute of property, *jus fruendi* embodies the right to enjoy the income (fruits) derived from an asset. Under the Moon Agreement regime, this cannot be fully enjoyed, as Article 11.7.d of the said document provides for –

"[a]n equitable sharing by all States Parties in the benefits derived from [the natural resources of the Moon], whereby the interests and needs of the developing countries, as well as the efforts of those countries which have contributed either directly or indirectly to the exploration of the moon, shall be given special consideration".

These provisions are in resonance with the 1848 Manifesto of the Communist Party, whereby Karl Marx and Friedrich Engels called for the

"[a]bolition of property in land and application of all rents of land to public purposes" (Marx and Engels 1848).

In the 1979 report of the Independent Commission on International Development Issues, the organism chaired by Willy Brandt (quoted in Vicas 1980, p. 303) considered that "'Global commons' is a neat catchword, but hardly appropriate", because -

"It connotes villagers in medieval England who have the right to pasture their cattle in the village commons. The space analogy is nations 'pasturing' their satellites in the global commons. The term connotes something of free access to outer space, but none of the distributional aspects of the 'common benefit' or 'common heritage'".

Indeed, whereas *res communis* offers free access, it does not entail a share of the benefits. The villager pasturing his cow in the village commons needed not share the meat and milk with the other villagers, even if originated from the grass grazed from a common pool. In contrast, under the CHM regime, the lunar miner has to share with all other humans what his equipment extracted from the Moon.

Lenin (1917) defined Socialism as the "social ownership of the means of production and the distribution of products according to the work of the individual"; in his view, Socialism will "ripen into Communism, whose banner bears the motto: 'From each according to his ability, to each according to his needs'. The tenets of the Moon Agreement hold middle ground between Socialism and Communism, providing for a share of the lunar benefits from each according to his ability, to some according to their work and to some according to their needs.

Dave Anderson (2002) laments the whole society having moved toward a culture of entitlement over the years, an ethos whose motto is "[w]eaken the strong to strengthen the weak". Keith Urbahn (2005) criticizes as well the "culture that promotes widespread dependence on government handouts". In his view, a society of entitlement rates justice "not by how much the government encourages those who succeed, but by how much it rewards those who don't". In his view, this equates with a "license for mediocrity", denying individual responsibility and creating convenient excuses for failure. Instead, as explained by John Locke (1690, Chap.5, Sect. 34) more than three centuries ago, the world was given -

> "to the use of the industrious and rational, ... not to the fancy or covetousness of the quarrelsome and contentious. He that had as good left for his improvement, as was already taken up, needed not complain, ought not to meddle with what was already improved by another's labour: if he did, it is plain he desired the benefit of another's pains, which he had no right to, and not the ground ... whereof there was ... more than he knew what to do with, or his industry could reach to".

An egalitarian regime, which not only prohibits private landed property but also calls for an 'equitable' *jus fruendi* offers no incentive to exploit the extraterrestrial ores when the investors have to share the benefits with free riders who believe in a culture of entitlement. Together with Locke, we believe that law ought to provide for the "industrious and rational" and not for the "quarrelsome and contentious".

23.6 Mine Is Better Than Ours – Homesteading the Final Frontier

What distinguishes the *res communis* from full property rights is its lack of marketability. As explained by Robert P. Merges and Glenn H. Reynolds (1997, p. 121), rights can be separated in two broad classes, namely usufruct – "a right to continued use for a limited time" and fee – "a more permanent interest that can be traded, devised, or otherwise transferred". According to the two authors,

> "For some purposes, the usufruct may prove to be valuable in the space environment, but generally we have in mind a fee interest, more specifically a right akin to the fee simple of Anglo-American law. The fee interest has the advantages of predictability ... and flexibility"

In 2004, the US President's Commission on Implementation of United States Space Exploration Policy recommended that -

"Congress increase the potential for commercial opportunities related to the national space exploration vision ... by assuring appropriate property rights for those who seek to develop space resources and infrastructure". (Aldridge 2004, p.33)

This *lex ferenda* proposal would represent an important shift from the *res communis* approach consecrated in the Outer Space Treaty. [Note: *Lex ferenda* is a Latin phrase meaning "future law", in the sense of "what the law should be" - as contrasted to *lex lata* - "the current law"].

Private property rights have long been promoted by space advocates as the most appropriate – if not the only way for developing and settling outer space. In the tradition of the American Frontier, whose settlement was encouraged by governmental plans of privatizing the public domain, the proponents of private property in space seek a similar privatization of the extraterrestrial realms. The frontier is inexorably linked with the idea of individualism and private ownership, of transforming *res nullia* and *res publica* into private property.

In 1976, Gerard K. O'Neill published "The High Frontier", a book where he sketched out a course for expanding humankind into outer space on a permanent and sustainable basis. A decade prior, Gene Roddenberry, the creator of "Star Trek", a successful TV show, had dubbed outer space "the final frontier", while in 1965, the Gemini 5 astronauts chose a Conestoga wagon as their flight emblem, true to Ralph Waldo Emerson's (1862) guidance – "Hitch your wagon to a star".

The association between space colonization and the American Frontier is a recurrent theme in much of the pro-space ethos, complementing the earlier view of the American continent as a "New World". At the end of the 19th Century, historian Frederick Jackson Turner (1893) explained American development through "[t]he existence of an area of free land, its continuous recession, and the advance of American settlement westward". According to historian Barbara Tuchman (1976), the frontier and private property rights are deeply connected, in her view -

"[t]he open frontier, the hardships of homesteading from scratch, the wealth of natural resources, the whole vast challenge of a continent waiting to be exploited, combined to produce a prevailing materialism and an American drive bent as much, if not more, on money, property, and power than was true of the Old World from which we had fled."

In a 2003 testimony before the U.S. Congress, Rick Tumlinson, founder of the appropriately named Space Frontier Foundation, pleaded for "the right to own new land in space", asserting the crucial need to "begin putting in place the rights of those who explore and develop such new 'lands' in space to own them" in order for these to live up to their potential of great sources of wealth:-

"Throughout history, it has been the ability to gain and hold land which has driven [the explorers and developers] forth, and given them the will to carve new human domains out of wilderness. Space is no different. If people are going to invest their wealth and lives in opening the frontier, they should have the right to pass what they have done down to the next generations. When the time is right, the US should stand up and recognize that in space, the same rights to won property exist as on Earth" (Tumlinson 2003, p.16).

Most of the resources of outer space are, in practical terms, unlimited; given their abundance, it is illogical to forbid their private appropriation. In 1968, Arthur C. Clarke and Stanley Kubrick, the creators of "2001: A Space Odyssey", estimated that the number of people who had ever lived on Earth matches the number of stars in our galaxy, hence, possibly

"there is enough land in the sky to give every member of the human species, back to the first ape-man, his own private, world-sized heaven - or hell". (Clarke and Kubrick 1999, p. 7).

Perhaps the best arguments for the privatization of the extraterrestrial realms have been brought by John Locke, long before the start of the Space Age. In his view – to which we adhere -, privatization enhances the common heritage, provided there is still enough left for the others:-

"[M]en had a right to appropriate, by their labour ... as much of the things of nature, as he could use: yet this could not be much, nor to the prejudice of others, where the same plenty was still left to those who would use the same industry" (Locke 1690, sec.37).

Locke did not justify greed, bearing in mind the interests of the fellow humans. Yet, as shown in the previous section, this interest cannot be the entitlement to the work of the other, but the right to use one's "industry" for homesteading the common heritage. In his correct reasoning, he who proceeds at appropriating land by the means of his labour "does not lessen, but increase the common stock of mankind", because enclosing an acre and using it yields more essential products than if the land were left "waste in common". The privatization is seen, thus, as an active administration of the public trust stemming from the common heritage in its original sense:-

"And therefore he that incloses land, and has a greater plenty of the conveniences of life from ten acres, than he could have from an hundred left to nature, may truly be said to give ninety acres to mankind".

23.7 Conclusion – and a Caveat

We have seen that, although we cannot purchase land on the Moon from the Lunar Embassy, the Outer Space Treaty regime does give us the right to use the extraterrestrial realms on a non-exclusive basis and conduct without "harmful interference." A leftward move towards the Common Heritage of Mankind regime would be detrimental to the development of space; a refutation of this principle does not mean, however, that the developing world will, or should, be left behind in the space era. China, India and Brazil are living proofs that a developing country can, through its own efforts, join the spacefaring club. Instead of freeloading on the efforts of the older spacefarers, have-not nations should pool their financial resources into a common space agency or into regional ones, and proceed at exploiting the riches of outer space for themselves. Indeed, they need not build the infrastructure themselves if they can buy commercial space services on the global market. The rallying cry of Marxism – "Proletarians of all countries, unite; you have nothing to lose but your chains" should evolve into "Countries of the world unite – you have nothing to lose but the chains of gravity".

On the contrary, the frontier paradigm has proven its worth on our planet, and it most likely will do so in the extraterrestrial realms. Privatization of the public lands is likely to transform the lunar desert in the same manner as it transformed the 19th century United States. Space is indeed a new frontier calling for individualism rather than collectivism, and its challenges need to be addressed with a legal regime favorable to property rights. A rightward move securing such a regime is the only means of opening the extraterrestrial realms to settlement, given the reluctance of most entrepreneurs to invest money in an endeavor without having the security that they will enjoy the benefits. Given the abundance of extraterrestrial resources, it would be nonsensical to forbid their private appropriation. Securing effective space property rights would be a small price to pay, and more beneficial to humankind, compared to the alternative of keeping the extraterrestrial realms undeveloped. Whereas appropriation of the extraterrestrial realms is outlawed in the current incarnation of the international law of outer space, law is a dynamic phenomenon and it may evolve towards a regime supportive of property rights in outer space. A shift from the commons regime may be in the cards, given the official support of the Aldridge commission for property rights. Until this shift happens, the non-appropriation principle remains nonetheless the *lex lata*.

The support for property rights in outer space expressed here ought to come with a caveat: Private property, even in its terrestrial form, is an eroded concept. Throughout the last century, the "owner" has involved into a mere "privileged user" of a certain object. The State taxes that object and the benefits derived thereon, and can confiscate that object by means of eminent domain. Fee simple ownership, the epitome of property, has been drawn into extinction by the gradual advance of the society into private matters. It is therefore essential to bear in mind that, even if formally allowed, landed private property rights in outer space would still be subject to taxation, eminent domain and other State powers. Support for property rights in outer space needs to be complemented by support for property rights on earth. [Note: an earlier version of this chapter has been published as Pop (2010). For a more elaborate analysis of property rights in outer space, see Pop (2008)].

References

Aldridge, E.C.: A Journey to Inspire, Innovate, and Discover: Report of the President's Commission on Implementation of United States Space Exploration Policy. US Government Printing Office, Washington DC (2004)

Anderson, D.: Changing a culture of entitlement into a culture of merit. The CPA Journal (November 2002),
http://www.nysscpa.org/cpajournal/2002/1102/nv/nv8.htm

Arthur, C.: Jewellery's final frontier. The Independent (May 11, 1998)

Chaikin, A.: A Man on the Moon. Penguin, London (1994)

Chinea, E.: Chavez Says Capitalism May Have Ended Life on Mars. Reuters (March 22, 2011), http://www.reuters.com/article/2011/03/22/us-venezuela-chavez-mars-idUSTRE72L61D20110322

Clarke, A.C., Kubrick, S.: Foreword to 2001: A Space Odyssey. Orbit, London (1999)

Emerson, R.W.: American Civilization. The Atlantic Monthly IX(54), 502 (1862)

Gal, G.: Acquisition of property in the legal regime of celestial bodies. In: Proceedings of the Colloquium on the Law of Outer Space, vol. 39, p. 45 (1996)

Connecticut, G.: Supreme Court of the United States. 161 U.S. 519 (1896)

Gorove, S.: Private rights and legal interests in the development of international space law. Space Manufacturing 5, 226 (1985)

Grotius, H.: The freedom of the Seas, or the right which belongs to the Dutch to take part in the East Indian trade. Justinian's Institutes (1608)

Lauterpacht, H.: International law ('the collected papers'), vol. 2. Cambridge University Press, Cambridge (1975)

Lenin, V.I.: The Tasks of the Proletariat in Our Revolution. Priboi Publishers, St. Petersburg (1917)

Locke, J.: The Second Treatise of Civil Government, London (1690)

Marx. K., Engels, F.: Manifesto of the Communist Party, London (1848)

Merges, R.P., Reynolds, G.H.: Space resources, common property, and the collective action problem. New York University Environmental Law Journal 6, 107 (1997)

Ocilla Star, Records claims to heavens, Moon and stars. Ocilla Star (July 15, 1937)

Pop, V.: Unreal Estate: The Men who Sold the Moon. Exposure Publishing, Liskeard (2006)

Pop, V.: Who Owns the Moon? Extraterrestrial Aspects of Land and Mineral Resources Ownership. Springer Science+Business Media BV, Dordrecht (2008)

Pop, V.: Planetary Resources in the Era of Commercialisation. In: Bhat, S.B. (ed.) Space Law in the Era of Commercialisation, pp. 57–71. Eastern Book Company, Lucknow (2010)

Reid, K.G.C.: The law of property in Scotland. The Law Society of Scotland/Butterworths, Edinburgh (1996)

Reno Evening Gazette, City bunco men's games fail in country. Reno Evening Gazette, p. 9 (1929)

Stair, J.: The Institutions of the law of Scotland, Edingurgh (1693)

Sweet, C.: A dictionary of English law. Sweet, London (1882)

The Daily Northwestern, Wants square deal. The Daily Northwestern, December 6, p. 5 (1910)

Thomas, J.A.C.: The Institutes of Justinian - text, translation and commentary. North-Holland Publishing Company, Amsterdam (1975)

Tuchman, B.: On our birthday—America as idea, Newsweek (July 12, 1976)

Tumlinson, R.: Testimony before the Senate Committee on Commerce. Science and Transportation (October 29, 2003)

Turner, F.J.: The significance of the frontier in American history. In: Proceedings of the State Historical Society of Wisconsin (December 14, 1893)

Urbahn, K.: Putting the torch to a culture of entitlement. Yale Daily News (March 23, 2005)

Vicas, A.G.: The New International Economic Order and the emerging space regime. In: Proceedings of the Symposium Space Activities and Implications: Where From and Where to at the Threshold of the 80's, October 16-17, p. 293. McGill University, Montreal (1980)

Wall, M.: Mining the Moon's Water: Q & A with Shackleton Energy's Bill Stone. Space.com, (January 13, 2011), http://www.space.com/10619-mining-moon-water-bill-stone-110114.html

24 Telecommunication and Navigation Services in Support of Lunar Exploration and Exploitation

Marco Cenzon[1] and Dragoş Alexandru Păun[2]

[1] Aviospace S.r.l.
[2] Sofiter System Engineering S.p.A., Torino, Italy

24.1 Introduction

A structured approach to the design of Space Exploration Systems is fundamental to creating a coherent and sustainable global exploration effort. In fact an extensive robotic and human settlement on the Moon will be achieved only if suitable services will be provided in support of the lunar exploration and exploitation activities. This is even more valid when taking into account the fact that, settlements on the Moon will not be completely autonomous from Earth and therefore will rely on its support in terms of strategic resources and consumables.

Among the different services, like crew and cargo transportation to and from the Moon and industrial logistics for lunar resources on Earth, communication and navigation systems will represent the backbone of any architecture from the early stages of exploration up to the utilization phase. Implementation of a Lunar Navigation and Communication (LNC) System represents a floor capability of any lunar utilization infrastructure by providing basic service to lunar assets. Since the timescale of human lunar exploration extends for several decades, the demand of NavCom services is expected to grow alongside the complexity of the entire lunar exploration system of systems. This chapter presents an analysis into what the future needs of human lunar exploration will be and proposes a system designed to fulfill those needs.

24.2 Lunar Activities and NavCom Evolution

The Exploration Program is still in its formulation stages, however envisioning an architecture that would support such a complex effort as that of exploring and utilizing the Moon requires well founded assumptions to be made. Several studies have recently been performed with the aim of identifying the possible shape of future exploration of the Moon (Culbert et al. 2010; Stanley et al. 2005; Hufenbach and Leshner 2008). Their collective conclusion underlines that a sustainable and economical exploration of the Moon and Mars can only be accomplished by a flexible phased approach that takes into account the various associated complexities. Although these studies take into account the large periods of time inherent to human space exploration, the exploitation of lunar resources remains a subject reserved for the end of the exploration timeline. Due to the complexity of the lunar exploration effort and of the NavCom segment in particular, an extended view of the exploration timeline is illustrated in Fig. 24.1 (based on ESA Aurora; http://www.esa.int/esaMI/Aurora/). This revised timeline also takes into account lunar utilization.

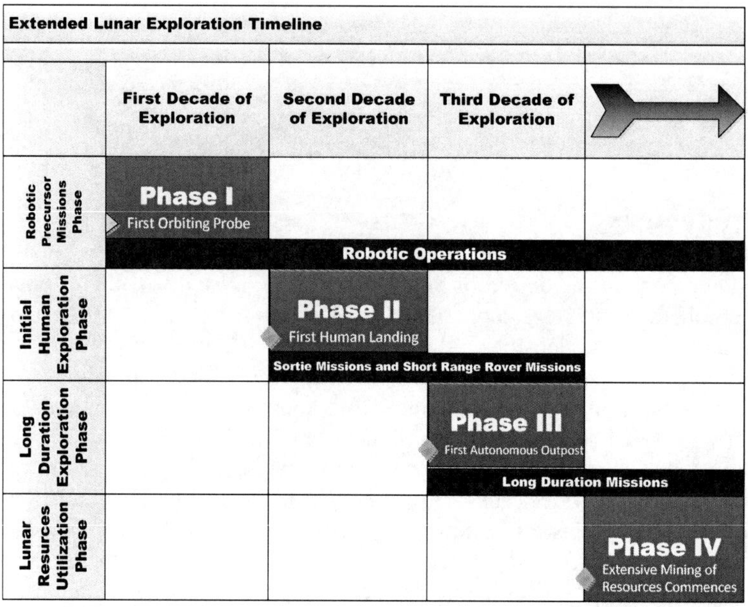

Fig. 24.1 Extended View of the Lunar Exploration Timeline

A phased approach to the exploration of the Moon also implies a flexible, phased and modular design of the NavCom architecture. Such a design would minimize for each phase the infrastructure to the strict minimum for each phase costs and therefore speed up construction time.

In the phased approach, the Robotic Precursor Missions would form the first stage in the lunar exploration effort. This phase is expected to last for up to one decade and its main objectives would include but not be limited to accurate mapping, resource prospecting, environment characterization and deployment of infrastructure (Schrunk et al. 2008; Culbert et al. 2010). Typical robotic end users of NavCom services would include for example scientific instruments in orbit, or in small rovers or landers.

Due to the typical requirement of minimizing payload mass, among others in scientific instruments that have a long operational life, the need of minimizing the resources allocated to the communication subsystem of each instrument arises. A possible solution to this problem, also underlined by Culbert et al. (2010), Bhasin et al. (2008) would present itself in the form of a LNCS capable of storing and relaying the scientific data from a multitude of sources placed on the lunar surface. This solution could provide total lunar coverage although not in a continuous mode and would minimize the needs of direct and consequently of high power communication with Earth for each instrument or vehicle on the Moon.

A major milestone in the human exploration effort is represented by a new Human Lunar Landing which would mark the beginning of the second exploration phase, namely the Initial Human Exploration Phase. This phase would be primarily characterized by a series of Sortie Missions, Small Pressurized Rover Missions and Pressurized Habitats and is expected to last for approximately 8 to 10 years. Robotic elements would still play a major role in the overall strategy as human presence will be limited in both time and range. They would perform science alongside humans as well as aid in the build-up of infrastructure. According to current plans and ideas, crews would land and explore the South Polar Region using rovers in excursions not exceeding one lunar day (14 earth days) (Culbert et al. 2010). Additional users of NavCom services for this phase would be:

- several small servicing robots (part of the human/robotic partnership),
- small pressurized rovers,
- power infrastructure,
- In-Situ-Resource-Utilization demonstration plants,
- heavy mobility systems,
- astronauts during EVAs.

A set of two LNCS's could provide continuous coverage of an area with a 500 km radius around the South Lunar Pole (Culbert et al. 2010; Bhasin et al. 2008). For missions taking place in locations without direct access to communication with Earth, mission specific capability needs to be ensured. The NASA SCAWG report (SCAWG 2006) proposes the launch of assets that could serve specific mission needs but that can nevertheless meet a long duration role in a larger NavCom architecture.

The initiation of long duration human missions would mark the beginning of the third phase. This phase would take place in the presence of one or more lunar outposts deployed in advance by the use of dedicated robotic infrastructure. Typical end users of LNC services for this phase would be dedicated habitat(s) and its/their associated science and technological module(s) and enhanced robotics. Already supporting South Polar Region activities and already complemented by spacecraft launched with the task of supporting several sortie missions, the LNCS's now forming an almost complete constellation around the Moon can provide adequate communication and navigation support for the ongoing operations taking place on the lunar surface.

The final and longest phase would commence once lunar mining/harvesting and associated processing and manufacturing operations begin. This phase would be executed by a large number of vehicles moving on the surface of the Moon as well as in lunar orbit and being utilized primarily in a teleoperated or completely autonomous operation mode. High bandwidth capability enables remote manipulation of vehicles and the return of video, audio and telemetry links covering the entire lunar surface. As such, the constellation charged with providing NavCom services needs to support high data transmission rates. SCAWG (2006) predicts that by the year 2030, a single spacecraft would be capable of transmitting in an 8 hour interval a quantity of data equivalent to the entire quantity present in the Planetary Data System today.

24.3 Requirements

The LNC systems will have to satisfy specific high level requirements. These are shown and explained in this section, dividing between general NavCom requirements (R-x) and specific navigation and communication requirements (NAV-x and COM-x). Typical driving requirements, against which the LNC implementation have to be designed, derive from this chapter rationales. The whole set of requirements refers to the fourth phase of the Extended Lunar Exploration Timeline (see Figure 24.1).

R-1 The LNC architecture shall provide communication and navigation services to support lunar activities

Implementing such an LNC infrastructure will unburden upcoming missions by eliminating the need of individual unique NavCom relay capabilities and by providing a standard shared LNC infrastructure. The goal is to reduce the total dependency or at least the total demand for Earth based NavCom services.

R-2 The architecture shall follow a step-wise phased evolutionary approach to enhance performance and capabilities

A fundamental aspect of this effort is the ability to evolve capabilities until reaching global, persistent high precision navigation and full communications connectivity among lunar elements and Earth.

The success of the final lunar resources exploitation initiative will rely on the initial lunar missions and their ability to succeed in providing needed services for both robotic and human elements. These missions can be aided and enabled by a navigation and communication infrastructure that can evolve in capability to support extended lunar operations. This evolution can be schematized in Phases of Lunar Exploration which are very important in determining the characteristics of the services. The LNC system generations will replace each other in order to follow the overall evolution in accordance with the exploration/utilization timeline. In fact the needs of NavCom generally increase with human presence and the scientific return of the mission and decreases with the level of autonomy of each element. Far ahead achievement of continuous moon-resources utilization would provide a basis of capability to support further exploration of the solar system, including Mars and beyond.

R-3 The LNC system shall provide service to elements in Low Lunar Orbit (LLO) as well as on its surface

Moon resources utilization will require a large number of elements both on the surface (e.g. outposts, processing plants, rovers, robotic machinery, etc.) and on orbit (hoppers, orbital carries, orbital bases, spaceports, etc.). The two classes present peculiar and different ways to interface with an LNC system both in terms of needs and technical implementations.

R-4 The LNC system shall be independent from the architecture of other systems to prevent continuous design changes inherent in tightly coupled systems.

R–5 The LNC system shall provide service to elements including when the Earth is not in direct view as on the lunar far-side.

This requirement clearly states that lunar exploration/exploitation must not be constrained by the availability of a direct view of the Earth. NavCom orbital relays have to be foreseen. This need is found in identified objectives that require operations out of direct contact with Earth such as the ability to communicate with Earth while on the far side of the lunar surface, at the bottom of craters or operations that benefit from in-situ navigation support.

R-6 The LNC system shall provide full and continuous Moon coverage even if with reduced performance.

Any location on the Moon is potentially an area where activities will take place (e.g. mines). Many identified areas of interest are currently in lunar depressions that will make navigation hazardous and communication impossible without a relay system. With respect to providing an extended presence at the Moon, the missions would require real-time navigation support and communication relays that are in the lunar locale. Up to now it is clear that at the beginning the lunar poles will see more activities due to the presence of the first lunar outposts, but the human/robotic settlements will evolve and spread toward specific sites of interest that are rich in material resources and scientific perspectives as of yet undefined. Continuous full coverage is required to ensure support to any activity. The requirement is relaxed by not asking for nominal performance every time and everywhere. These Reduced and Nominal performances shall be specified.

COM-1 The Communication System shall provide a Safety Service available at any time.

To what concerns human tended missions, it is mandatory to foresee a Safety Service to be used in case of emergency. Such a contingency mode shall be at least double failure tolerant (in addition to what has provoked the emergency) and be able to support human crew survivability and control of critical elements. The contingency mode is assumed to operate with reduced performance.

COM-2 The Communication System shall provide a Short-haul link among all lunar elements and a Long-haul link between these elements and Earth.

The Short-haul link refers to communications among lunar elements on the lunar surface and eventually elements in LLO, while the Long-haul refers to communications between lunar elements and the Earth.

COM-3 The Communication System shall provide Short-range communication on the lunar surface over distances up to 10 km.

Continuous communication links among surface elements, with EVA astronauts and robotic systems, and with external surface systems, shall be provided within a specified range.

COM-4 The Communication System shall maintain a radio-quiet zone in the Shielded Zone of the Moon (SZM) for radio astronomy below 2 GHz and above 3 GHz.

The far side of the Moon is a peculiar area which is not "polluted" by manmade electromagnetic signals, especially in radio frequency. This presents a chance to perform scientific radio astronomy and research. In order to preserve the shielded zone for radio-astronomy, the only frequencies allowed on the far side range between 2 and 3 GHz (Noreen et al. 2005). The LNC system shall take into account these limitations.

NAV-1 The Navigation system shall localize or allow navigation of any lunar surface element with a precision higher than approximately 100 m.

Several studies conducted in recent years give an estimation of surface mobility accuracy. NASA programs (Stadter et al. 2006; Gutiérrez-Luaces 1997) are planning a general surface operations accuracy of 30 m. In (Stadter et al. 2006) the foreseen accuracy using a fleet of three navigation satellites is about 10 m for a fixed user and 100 m for a mobile user. In other studies the static position of a mobile rover will be estimated with an error of less than 10 m computed in a few minutes of waiting time. In this case roving navigation requires periodic stops to obtain in-situ static position fixes at approximately every 30-60 min because the image data is not taken while roving (NASA 2007). The studies done by the Surface Exploration team of the MIT/Draper Lab group (Chabot 2005) suggests that the envisioned surface operations on both the Moon and Mars will require beyond line-of-sight navigation (and communication) coverage with 100 m absolute position accuracy. In conclusion basing on the bibliography it is reasonable to think about a surface navigation precision of few tens of meters to be collocated in three to four decades from now.

NAV-2 The Navigation system shall localize or allow navigation of any lunar orbital element.

NAV-3 The Navigation system shall guarantee time synchronization with a maximum error of 10 ns.

Precise synchronization is fundamental for navigation purposes as to perform range measurements. Well synchronized on board clocks enable pseudo-noise sequence transmission that is synchronous with an established local time such that individual clocks differ in knowledge by no more than a certain error from each other. Such a capability is necessary to enable a usable one-way radiometric service. Therefore though the LSE and the LNCS have both on orbit and on ground USOs, they need to be periodically synchronized depending on the stability of their clocks. To achieve an accuracy of about 100 m, using only one-way range measures, it is necessary to synchronize the clocks with an error around 10 ns, this guarantees range accuracy of 10 m without considering any other source of uncertainty. The same consideration can be found in the NASA architecture where no more than 10 ns of error are expected (NASA 2007).

NAV-4 The Lunar Navigation system once fully settled shall be nominally autonomous with respect to the Earth ground segment.

Traditional navigation is accomplished via radiometric measurements from a GEE. This is certainly feasible for lunar elements, but it has some severe limitations when compared to in-situ options without any interaction with Earth (Stadter et al. 2006). The first advantage of this is that it greatly simplifies the calculations, as the additional complexities of the Earth- Moon legs are no longer present. Because the total range is lower, the quality of the measurements is improved as well. Finally, since the user receives the measurements in real time, navigation data can be used to guide critical maneuvers such as ascent, descent, rendezvous, and orbit insertion. Long term lunar resource utilization would be too expensive and constraining if full reliability on Earth is considered (e.g. ground antennas). The Earth infrastructure should be dedicated to other purposes. Once fully operative the LNC system shall rely on Earth antennas and computers just in case of contingency. The natural conclusion is that the LNC system shall implement an in-situ navigation service on a dedicated navigation channel.

24.4 LNC High Level Architecture

The technical analysis presented hereafter in this chapter has been largely derived from the project work activity performed during the fourth edition of the SEEDS activities (http://www.seeds-master.eu/ for more details) (SEEDS 2008-2009; SEEDS 2009). The results have been revisited and presented in the frame of this book.

The reference LNC architecture consists of four segments.

- A constellation of Lunar Navigation and Communication Satellites (LNCS).
- Lunar Surface NavCom Terminals such as local communications terminals for the surface Wireless LAN or providing tracking and time services by working as a beacon of the Orbitography Beacons Network (OBN).

- Lunar Users (LU) adopting a product line of interoperable radios and navigation units. Four basic types with decreasing levels of capability are: fixed base radios for large elements (e.g. habitat or mines), radios for orbital elements, mobile user radios for surface rovers and robots, and EVA radios for EVA suits.
- The Ground based Earth Elements (GEE) including new and existing antennas, the Operations Center managing LNC satellites and the Mission Operations Center which deals with the end-to-end lunar navigation. This later could also be allocated to the Lunar Base.

Figure 24.2 schematically shows what is going to be illustrated in the following part of the chapter.

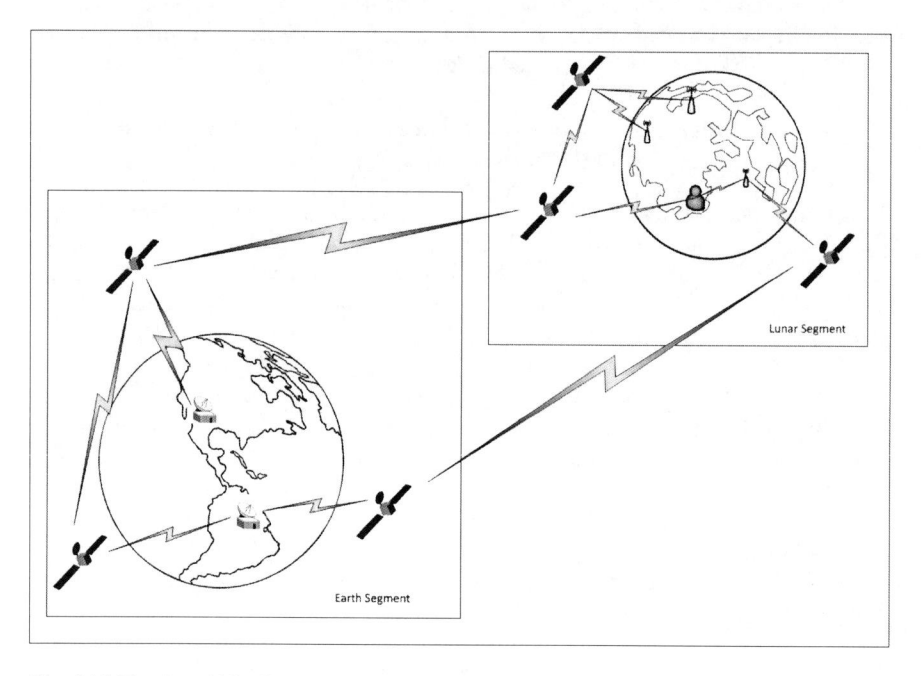

Fig. 24.2 Top Level NavCom Architecture

24.5 Orbital Mechanics for LNC

The navigation and communication functions will be coupled in the same orbital system. Each LNC satellite (LNCS) will guarantee both positioning and communication services under its region of visibility, and the constellation will continuously cover the entire lunar surface. However lunar orbits possess peculiarities that must be carefully considered.

High-altitude circular orbits, higher than 1200 km over the Moon surface are unstable because of the third body effects given by Earth's proximity which affects the orbital dynamics of the satellite. Within tens of days the orbit would collapse on the Moon or insert the spacecraft in a hyperbolic trajectory.

Low-orbiting satellites encounter problems as well. The Moon's gravity field is highly non-spherical (e.g. masscons), and the perturbation of the high order elements of its gravitational potential is significant, and causes satellites to crash on the surface. The instability of the orbits poses a real problem in terms of satellite control and therefore satellite lifetime. However the improved knowledge of this complex non-spherical field shows the existence of families of orbits that maintain long-term stability. Those discoveries enable the consideration of "frozen" orbits [Frozen Orbits are considered those which keep constants (at least cyclically) some or all their orbital parameters.]. The utilization of those orbits reduces the need of actuation, improving performances and increasing the life of the satellite. In reference (Todd and Lieb 2005) it is concluded that frozen orbits would avoid the need for any active deterministic orbit control for 10 years under the influence of gravitational and solar radiation pressure accelerations. This characteristic makes frozen orbits interesting candidates for lunar orbiters with low maintenance costs. An initial goal of defining frozen orbits for any LNC architecture becomes problematic when the semi major axis and eccentricity are limited by coverage constraints. Considering that the Moon orbit is inclined with approximately 5° with respect to the ecliptic plane, orbits close to the equator will face a shadowing period with every revolution. Possible solutions for avoiding this daily shadowing on the satellite given by the presence of the Moon are using an inclined lunar orbit or a high radius orbit.

As suggested by the NASA Space Communication Architecture Working Group (SCAWG 2006), the LNC architecture shall support missions that may land anywhere on the Moon. LNCS will be deployed as needed to provide coverage for individual missions and remain available for the future LNC constellation and science missions. An initial LNCS could cover some sortie areas, while an updated constellation with some LNCS's would provide continuous coverage to multiple sites, including the lunar far-side. Finally the complete constellation will continuously cover the whole surface.

Multiple satellites in the same orbit provide system-level reliability improvement against stochastic failures within a single satellite. Moreover multiple satellites improve navigation and also provide greater geometric diversity (if present in different planes). Accordingly to Schier et al. (2005), continuous lunar global coverage can be achieved with 5 or more satellites but the 5 satellite case requires 5 planes with higher deployment risk and launch constraints. In addition, the loss of one satellite equals with the loss of an orbital plane, in consequence it is not a robust, fault tolerant solution. Several good candidates exist for global lunar coverage with 6 satellites in polar or inclined orbits. Eight satellites are needed for higher data rates and volume.

Several possible constellations have been considered and analyzed combining different orbits (e.g. Halo, equatorial, polar, etc.) against several performance indexes. The performed assessment has identified as most promising the option proposed by D. Folta. An analytical study of the lunar problem, verified with a full potential model including third body effects, has been performed by Folta and Quinn (2006). Its result is a constellation placed in frozen orbits providing continuous and global coverage of the entire lunar surface. Such a constellation should also reside at an altitude allowing coverage of most spacecraft in LLO. This constellation can be considered as reference for a suitable option. The constellation configuration is composed of two orbital planes with two orbits each. On each orbit there are two spacecraft equally spaced (the entire set of orbital parameters is listed in Table 24.1. The orbits have an 18-hour period and are frozen.

Table 24.1 LNCS's orbital parameters

LNCS	Plane	Orbit	a [km]	e	i [°]	Ω [°]	ω [°]	v [°]
1	1	1	8049	0.4082	45	0	90	0
2	1	1	8049	0.4082	45	0	90	180
3	1	2	8049	0.4082	45	0	270	0
4	1	2	8049	0.4082	45	0	270	180
5	2	3	8049	0.4082	45	180	90	132
6	2	3	8049	0.4082	45	180	90	228
7	2	4	8049	0.4082	45	180	270	132
8	2	4	8049	0.4082	45	180	270	228

Figure 24.3 shows the constellation configuration; while Fig. 24.4 was captured using an STK® simulation and shows how the constellation satisfies the full coverage requirement.

A wider study on eclipses has been performed on the chosen constellation (Fig. 24.5). The maximum eclipse duration that each satellite experiences is 4195 s (slightly exceeds one hour).

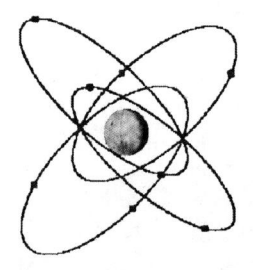

Fig. 24.3 LNC Constellation Configuration

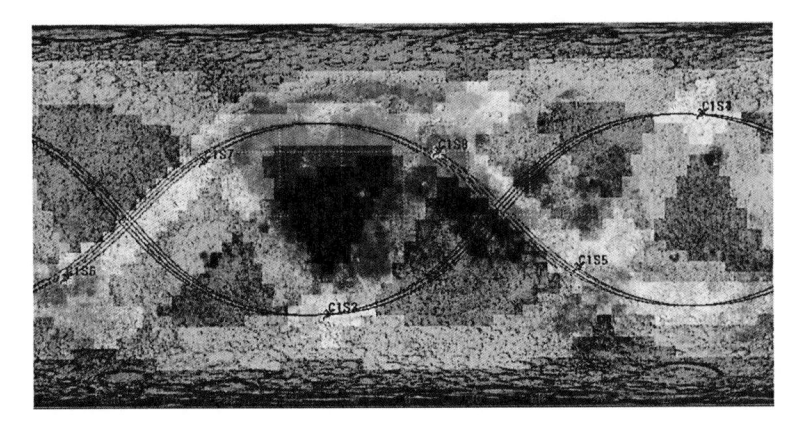

Fig. 24.4 Coverage analysis for LNC Constellation performed with STK®. Colours refer to the number of satellites in view, from one shown with the lighter colour to four satellites with the darker one. The lines are the LNCS's ground tracks.

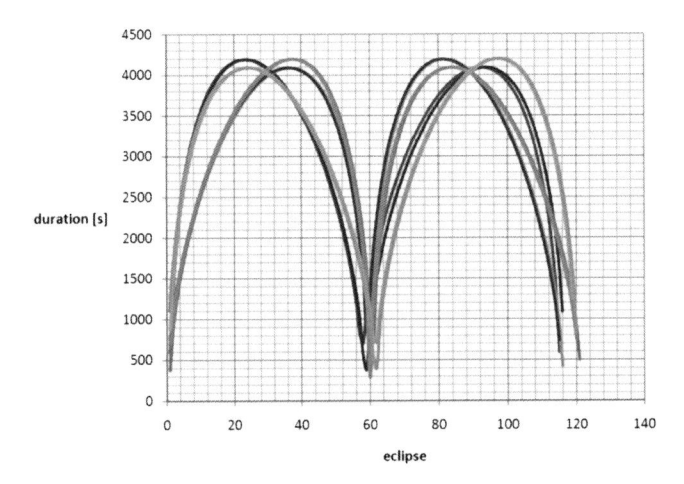

Fig. 24.5 LNC Constellation Eclipse Analysis. Each color refers to a different satellite.

These orbital features have been used to design and size the NavCom system as it is going to be shown.

24.6 Communication Architecture

The lunar elements cannot operate in a fully autonomous mode with the major reason involved being safety. In addition, TT&C and other data such as scientific data collected during the mission need to be transmitted to Earth. Thus every mission needs to communicate with the Ground Segment even if just for telemetry and commands. Starting from the System Requirements and Functional Design, a high-level communications scenario has been created.

Communication system design starts with an evaluation of the most common rates suitable for the planned mission. The considered technological horizon implies the usage of technologies available in the next years (up to ~2030). In any case in which a future compression algorithm is used (e.g. H.265), the assumption of a real time coding has been made. This evaluation is consistent with Moore's Law (*"The number of transistors per integrated circuit doubles every month"*), assuming the law valid for the next 20 years and applying it to the computational power.

The communication data rate could be assumed to be proportional to the complexity of the lunar systems (crew presence, number of elements, etc.) and to the overall scientific return. Table 24.2 is an attempt to estimate the needs of a standard manned element in terms of communication.

Table 24.2 Data communication needs for standard elements.

Data source		Data Rate
Digital Commands	Astronaut	≤ 500 kbps
	Robotic element	≤ 50 kbps
Technical Data	Astronaut	5 kbps
	Robotic element	500 kbps
Speech	Astronaut	8 kbps
Video VCD	Helmet-cam, CD quality, H.265 compression	320 kbps
Video DVD	Videoconference and medium definition, H.265 compression	1.25 Mbps
Video HD	High quality video, H.265 compression	4 Mbps
Data	Uplink	10 Mbps
	Downlink	5 Mbps

With current state of the art technology it seems reasonable to adopt a switched-packet network architecture using the IPv6 protocol. This choice implies major flexibility of the network and the usage of well-known and proven standards; nevertheless an overhead on the data transmission rates needs to be considered. Overhead refers to the amount of additional data needed to manage communications and this has been estimated to be 15%. The data rate is one of the major drivers in the design of communications systems. Based on Table 24.2 it can be assumed that each LNC node will be able to produce 10 GB/day of data; the overhead to be added increases the overall figure to 11.5 GB/day. A satisfactory value for channel rate could be 10 Mbps. At this rate all the data could be transferred in less than 3 hours, leaving the channel available for mission-tailored contents.

The global scenario foresees four different communication segments described below and in Fig. 24.6.

- Lunar Element Internal Communication (IC): communication inside the same element, or between elements constituting a physical unit (attached by a physical link). The internal communication system acts as a backbone for the gateways needed to connect the several sub-networks.
- Short-Range Communication (SRC): communication on the lunar surface over distances up to 10 km, providing continuous connection among lunar elements, EVA astronauts and robotic systems.
- Long-Range Communication (LRC): communication on distances exceeding 10 km, or when SRC mode is not available (i.e. in case of adverse terrain morphology). This kind of communication includes the communication to space with LNCS. Satellites are most advantageous way to establish LRC. Communication links are foreseen between all satellites in order to transmit data between two indifferent points of the lunar surface. Signals are sent to one of the satellites, where they are regenerated by repeaters (amplification, re-shaping and re-timing) before being sent back to the Moon or to the next satellite.
- Toward Earth Communication (TEC): communication involving Earth. It manages the communication between Moon elements and GEE. It foresees external lunar orbiting systems, such as a station orbiting in EML1, acting as an intermediate relay for communication from/to Earth. This station is assumed to be operative in the fourth phase of the Moon exploration/utilization scenario.

The IC segment can be for instance constituted by a double-ring optical fiber topology that ensures redundancy and reliability; two optic fibers per channel (ring) would allow reaching rates in the order of 20 Gbps. Such an architecture is enough to provide both internal communications and backbone capabilities to gateways towards external networks. Two physically connected elements must communicate to each other as if they were the same element (e.g. habitable pressurized module and laboratory) having the same bandwidth available. It is even possible to integrate a wireless solution, in order to eliminate cables and harness considerably simplifying the systems.

Short-Range Communications SRC system has to provide a continuous connection among not physically connected elements presumably in a mesh-network configuration and EVA sub-systems (human and robotic) in a star-network configuration. EVA astronauts will be provided with a digital command and audio channel to receive data and speech, and with a technical data and video channel to broadcast back physical or technical information and a helmet-quality video. EVA rovers and robots will be provided with digital commands and medium quality video for remote control and communications. A reliable and continuous communications channel therefore is of paramount importance ensuring mission success and crew safety. Considering the reference time frame, 3GPP Long Term Evolution (LTE) is a possible solution for SRC. It foresees the usage of the same technology for both inter-element and EVA communications. LTE is able to provide a data channel in the same order of data rate magnitude with WiMax, and is oriented to a degree of mobility similar to UMTS. Total required data rate could be of 172 Mbps for inter-element communication and 50 Mbps in upload from EVA network elements. LTE offers bigger rates than necessary and the remaining capacity

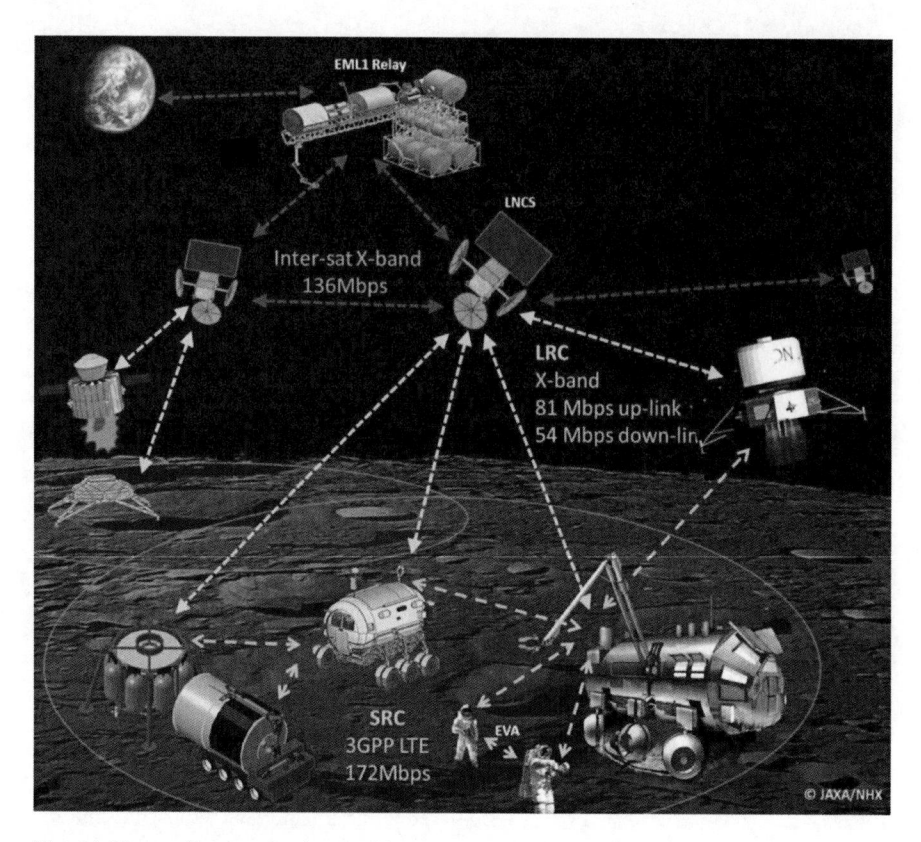

Fig. 24.6 Lunar Communication Architecture example.

may be used to broadcast additional data. For example in mesh networks the nodes are also forwarding the traffic of other peers in order to sustain the network. The area visible from an antenna is named radio horizon. Since on the Moon there is no atmosphere, the radio horizon overlaps with the optical horizon neglecting terrain irregularities. Choosing an antenna height of 7.5 m to be deployed on top of a trail or a natural headland will allow communications with equipment placed on the ground at distances of up to 5 km, and 10 km with another antenna placed at the same height (limit due to the Moon roundness not considering the terrain obstacles than could be seen even with 7.5 m masts).

In order to estimate the power needed by lunar elements for SRC, a link budget has been performed using data from commercial devices available on the market for LTE-like technologies (mainly UMTS and WiMax). Considering the three antennas transmitting at the same time (each one transmitting over a 120° range), the total amount of power needed for 10 km inter-element communications is 7.6 W. The link budget has been computed also for EVA communications at a distance of 7.5 km and in this case the transmission power is of 4.3 W. Link budgets for EVA communications led to results being similar to ones one could expect for a mobile telephony network on Earth; this is not surprising if one considers the topology of the EVA/robotic rovers communications network. Each module shall be able to emit a total of about 12 W of RF power; considering the DC/RF conversion acting with 50 % efficiency, each module shall allocate about 24 W for SRC.

Considering instead long-range communication (LRC) at distances farther away than 10 km, the choice of using satellite for communication is the expected result of an iterative trade-off process. The link budget is the overriding tool to size satellite communications channels but documenting the theory behind these computations is not a subject of this chapter. Before analyzing the single LRC links, data rates for LRC have been considered assuming that:

- Code Division Multiple Access (CDMA) is used for communications
- A single satellite link has to provide communications for many lunar elements, passing through only one element. The aggregated data rates are considered to be transmitted to Earth: this kind of communications requires the widest bandwidth, so it has been chosen to size the channel.
- Data are divided in three priority classes:
- A: safety-critical. This class includes all the communications needed for the safety of the crew or critical for mission survivability
- B: mission video. This class includes all the data and videos needed to control the mission (e.g. telemetry and commands)
- C: mission data. This class includes scientific data (e.g. high definition video)

Table 24.3 reports the considered 4-elements aggregated data rates for long-range and Earth communications. The division between onboard storage and downlink will depend on the relative data rates and the number of lunar users to be serviced.

At this point, it is important to distinguish between the two kinds of satellite-aided communications. Single Hop Communication means that both the transmitter and the receiver lie under the same satellite beam hence there is only one spacecraft acting as relay. In Multi-Hop Communication the transmitter and receiver do not lie under the same beam due to the distance between them. In this case several satellites are involved as relays in the communication and attention should be given to the switchover when satellites disappear over the horizon or mobile elements get out of the visibility zone. This second scenario highlights the need for inter-satellite links able to transfer information all around the Moon. In this sense, LNCS's will be the central nodes of an extended lunar IP network able to provide connectivity to several locations on the surface.

The S-band has been assigned to the following links:

- Emergency communications in the lowest part. This channel is used in contingency conditions, working with omnidirectional antennas on the ground at degraded data rates.
- Satellite communications with the surface. This channel transmits audio, video, telemetry, telecommand and data to/from the caravan.
- Emergency TT&C for the satellites. This channel is an Earth-LO channel, intended to be used as an emergency channel for satellite telemetry and telecontrol.

Table 24.3 LRC data rates required

Direction	Priority	Type	Rate [kbps]	Number of Chan-nels	Total Rate [kbps]	Channel Rate [Mbps]	Channel rate plus overhead [Mbps]	Overall bandwidth (2 trains) [Mbps]
Downlink	A	Speech	8	4	32	3,03	3,5	7
		Astronaut digital commands	0,5	4	2			
		EVA rover digital commands	50	1	50			
		Element digital commands	50	4	200			
		Robotic rover digital commands	50	5	250			
	B	Robotic rover technical data	500	5	2500	7,5	8,6	17,2
		Rover robotic video	1250	5	6250			
		Video	1250	1	1250			
	C	Video HD	4000	2	8000	13	15	30
		Data	5000	1	5000			
Uplink	A	Speech	8	4	32	5,3	6,1	12,2
		Astronaut technical data	5	4	20			
		EVA rover technical data	500	1	500			
		Element technical data	500	4	2000			
		Robotic rover technical data	500	5	2500			
		Robotic digital commands	50	5	250			
	B	Video (helmet)	320	2	640	11,9	13,7	27,4
		Element video	1250	4	5000			
		Video	1250	5	6250			
	C	Video HD	4000	2	8000	18	20,7	41,4
		Data	10000	1	10000			

- Surface communications. This bandwidth has been allocated for short-range communications (SRC) on the lunar surface.
- Navigation services.

The X-band has been instead allocated to the following links:

- Emergency communications in the lowest part. This channel will be allocated for the same usage as the S-band one.
- Inter-satellite communications. Satellites transmit on this band with CDMA in order to route the data from/to the lunar surface. Part of this band has been allocated to nominal satellite telemetry and telecommand.

Links between satellites do not have to be compliant with radio-astronomy regulations, since antennas are not pointed towards the lunar surface. For the same reason, and to save frequencies in S-band, an inter-satellite channel is allocated in X-band.

- EML1 communications. This part of X-band has been allocated to transmissions between the constellation of satellites and the orbiting relay system placed in EML1.

A trade-off has been performed to determine the best way to transfer data from the lunar surface to Earth. Different link budgets were calculated, considering propagation and attenuation in the atmosphere. The chosen solution foresees usage of an EML1 relay between the LNCS and Earth.

The frequency plan must be compliant with International Telecommunication Union (ITU) regulations, already present elements (e.g. GEE) and respect the far side as a shielded zone in order to preserve it for radio-astronomy (unless eventual safety-critical situations arise). The considered frequencies for Moon communication are shown in Fig. 24.7.

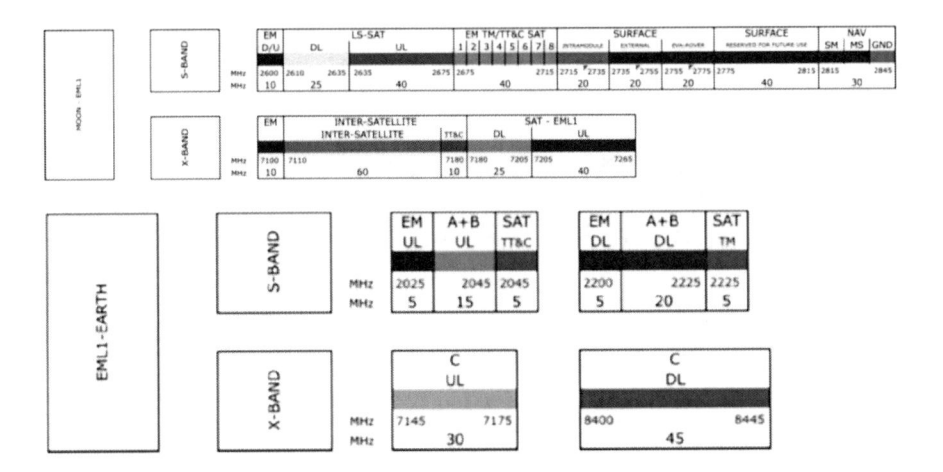

Fig. 24.7 Lunar Communication System Frequency Plan.

A suitable modulation scheme and codes could be DVB-S2 (ETSI 2005). DVB-S2 supports four different modulations: Quadrature Phase Shift Keying (QPSK) (2 bps/Hz) up to 32-PSK (5 bps/Hz). To compensate for the higher error probability due to higher compression factors the DVB-S2 standard supports error correction mechanisms foreseeing the use of concatenated Low Density Parity Check (LDPC) and Bose, Chaudhuri, Hocquenghem (BCH) codes.

Table 24.4 reassumes the major link budget parameters for the various LRC and TEC links that have been utilised for the LNC system design presented here.

Table 24.4 LRC and TEC Link Budgets.

		LNCS ↑ LS	LNCS ↓ LS	LNCS ↕ LNCS	EML1 ↑ LNCS	EML1 ↓ LNCS
Tx	Power [W]	10	7	11	30	30
	Frequency [MHz]	2655	2625	7140	7240	7200
	Antenna Diameter [m]	1.5	1	1	1	2
	Antenna Efficiency [%]	55	55	55	55	55
	Gain [dB]	29.81	26,19	34.88	34.99	40.97
	Pointing Accuracy [deg]	0.2	0.2	0.2	0.2	0.2
	EIRP [dBW]	39.29	34.13	44.74	49.21	55.02
Channel	Max Distance [km]	11,133	11,133	16,028	78,580	78,580
Rx	Antenna Diameter [m]	1	1.5	1	2	1
	Antenna Efficiency [%]	55	55	55	55	55
	Gain [dB]	26.28	29.71	34.88	41.02	34.95
	Pointing Accuracy [deg]	0.2	0.2	0.2	0.2	0.2
C/N_0	C/N_0 [dBHz]	80.51	78.74	82.28	80.04	78.75
Modulation	Margin [dB]	1	1	1	1	1
	Efficiency [bps/Hz]	2.242	2.242	2.242	2.242	2.242
Bitrate	Required [Mbps]	81	55	136	81	54
	Achievable [Mbps]	92	61	138	82	61

Some studies about desert sand impact on communications systems and several extrapolations were performed in order to roughly evaluate the possible impact of lunar regolith on the overall communication design since it has been observed that regolith moves due to electrostatic fields, or can be simply raised by moving vehicles or engine plumes. A good example for communication link requirements through a dust-laden channel was the operation Desert Storm performed by U.S. Army in 1990-91 in the Iraqi desert (Comparetto 1993), which has shown that signal attenuation due to sand or dust storms should be considered in ground-supported satellite communications systems. The maximum particle size observed in desert dust storms is on the order of 0.2 mm or approximately 50 times smaller than the minimum signal wavelength in the millimeter wave regime. Regolith is about ten times smaller than the desert sand being in the order of tens of micrometers. The behavior of particles with less than 1mm of radius is almost constant. Therefore it is assumed that for a 3 GHz frequency the attenuation will be about 1/3 of the 10 GHz. This leads to an attenuation of about 0.04 dB/km. Furthermore, Apollo mission's heritage gave some results (Dietz et al. 1972), obtained directly on the Moon. These results, reported in the following, highlighted a correlation between lunar soil temperature and signal losses and they have stressed that the radio absorption length is a power absorption skin depth (the inverse of the radio absorption coefficient). In conclusion surface and satellite communications in the millimeter wave regime through a dust laden path can be assured via an adequate link design and that the link budget must take into account just 0.1 dB/km losses in the worst case scenario.

24.7 Navigation Architecture

The lunar sky will optimistically see hundreds of vehicles going up and down, orbiting around and going back and forward towards Earth. These vehicles would be cargos for extracted resources or supplies, orbital stations, crew vehicles, hoppers and many others. This big variety should be based on the same navigation infrastructure able to support any kind of mission, including orbital maneuvers and fixed and especially mobile surface elements navigation as well. We are referring to a large variety of different systems such as rovers, outposts, drillers, trains, mobile machineries, "walking" astronauts and so on.

The Navigation problem is often associated to the Guidance (GN) referring to both the localization and trajectory planning and decision functions.

In general, localization is the problem of determining the position of a system from sensor data, in a known environment. Localization is a critical issue both in mobile robotics and human systems. In fact if the mobile element does not know where it is; it cannot effectively plan movements/maneuvers, locate objects, or reach goals. Therefore the navigation subsystem should be capable of providing the necessary accuracy during the path/trajectory, in order to ensure that the objective is reached. The navigation system would be fault tolerant with at least two independent measurement types per mission phase, integrating multiple sensors for all phases (e.g., cruise, landing, and roving). New technologies (sensors, inertial navigation systems, cameras, computer processors, and image processors) will make the future trips on the Moon easier for astronauts avoiding any kind of damage to themselves and to the systems.

Considering for just a while only the Lunar Surface Elements (LSEs) and not orbital spacecraft, different GN operations modes can be identified for the navigation of LSEs. The GN modes are defined in such a way to subdivide all the possible movements envisaged for the elements into a range-classifying definition which refers to possible different implementations and strategies. In fact, while moving towards a target, the relative distance decreases progressively. In these cases there is a switch in the operations modes when the target of the elements enters in the field of view of the GN subsystems (sensors), useful and available in a shorter range. An accurate knowledge of the position could be acquired in the proximity of targets by means of simple observations and matching different types of maps. On Earth the perception of the surrounding environment is sufficient to locally ensure positioning and navigation. Table 24.5 aims to roughly identify surface GN modes in terms of range and the accuracy required.

Table 24.5 Reference GN operative modes.

GN modes	Description	Range	Accuracy (95%)
Long Distance Navigation	The absolute positioning of the LSE on the lunar surface, including the capability of reaching foreseen far regions.	>= 100 km	1 km
Far-Field Navigation	Capability of reaching prefixed sites for operations.	100 km - 1 km	100 m flatlands / 50 m unevenness
Near-Field Navigation	Approach of a near targets	1 km - 50 m	1 m
Proximity Operations	Relative movement of LSE during planned or contingencies operations in the same area. They include also assisted landing or ascending of vehicles.	50 m - 1 m	0.1 m
Close Up Operations	High accuracy operations like docking, mating and inspection.	1 m - 0.01 m	1 cm

The GN system is strictly related to the availability of maps of a certain accuracy for the lunar surface; its implementation has to be designed taking into consideration the map details and resolution. Theoretically any mobile element is safe when it is sure that any obstacle can be avoided; but approaching the obstacle, at a certain point the distance will be below the global positioning system accuracy plus the resolution of the map, at this point local GN (using a different set of sensors and equipment) shall replace the global positioning system. The accuracy and precision needed for the generation of maps that will be used for long and medium distance lunar navigation should be primarily related to the type of terrain in which the navigation is done. For example over lunar seas a lower resolution could be accepted. At present the Lunar Reconnaissance Orbiter (LRO) mission has locally achieved a meter accuracy map of the Moon. Although this resolution would be enough to ensure the derivation of maps of enough accuracy to ensure safe navigation, a global coverage is needed and neither one of the past missions has gained a complete dataset using the same level of quality. Following missions should gather precise, various and complete data of the Moon. NASA created the Lunar Geodesy and Cartography Working Group (LGCWG) and by 2009 this working group had started working on laying the ground for a unified manner of collecting and processing topographic data and suggesting these standards to the different teams from around the world that are currently working on missions to the moon. It can be considered that the available map resolution in 2050 will be lower than 10 m (horizontally and vertically) and that it shall be available a wide variety of map scales.

24.7.1 Absolute Positioning

After this introduction of the GN function, the following paragraphs will be discussing only the absolute positioning capability (only Long Distance and Far-Field Navigation for surface elements) since this is one of the main aspects of the LNC architecture (see Fig. 24.6 for communication and Fig. 24.11 for navigation). Several solutions have been investigated by taking into consideration several parameters such as the achievable accuracy, coverage and complexity. Finally the proposed implementation for the positioning solution foresees a synergy of four different systems (see Fig. 24.8):

1. Satellite aided radio positioning which would also ensure the communication function based on a constellation of LNC Satellites (LNCS).
2. Celestial Navigation looking at the celestial bodies directly from on board the elements as it is often already done for orbital spacecraft.
3. Correlation with Digital Elevation Maps (DEM) using on board sensors (lasers, cameras, or radars) to acquire images of the surrounds (e.g. skylines or radio images) and processing them to match with the stored DEMs. [In the NASA's navigation architecture (Schier 2008) the celestial navigation using stars will be coupled with the correlation with DEMs in a unique camera, extracting information of star position and sky-line shape from the same images].The system will work by combining maps taken by satellites in orbit around the moon with images taken by the ground elements on the lunar surface.
4. Inertial navigation using IMUs onboard the elements.

Of course depending on the complexity and the resources available on board the LSEs not all four techniques can be implemented. Such an integrated architecture will be:

- *Robust* with respect to failures and therefore safer and more reliable. In fact, the navigation system will be fault tolerant with at least two independent measurement types. Safety and reliability are improved with redundancies of the positioning techniques. For example a system-wide risk such as Coronal Mass Ejection (CME) is being considered as a credible hazard that could cause interference preventing the radiometric-based navigation system from working.
- *Accurate*, because by combining different concurrent measures it is possible to reduce the absolute positioning error by means of sensor fusion algorithms which permit data integration thus improving the performance of the entire navigation system [Sensor fusion is the combining of sensory data or data derived from sensory data from disparate sources such that the resulting information is more accurate, more complete and more dependable than would be possible when these sources were used individually].
- *Flexibility* of the entire system.

For the scope of this chapter only the satellite aided radio positioning is shortly treated because the other three methods are less correlated in the whole LNC

Fig. 24.8 Absolute Positioning Integrated Architecture.

system and could be better described separately, it is sufficient to say that the same satellites will provide both communication and navigation services.

Satellites are the most common solution for modern navigation or localization; it is usually implemented on Earth by means of radio tracking. The authors are proposing a GN system adopting Doppler and Range measurements for the global localization over the lunar surface (see Argos and Cospas-Sarsat). The choice is also in line with both ESA and NASA navigation architectures (Hufenbach and Leshner 2008), which comprise of the implementation of both radio measurements from satellites around the Moon. In NASA's architecture (SCAWG 2006; NASA 2007), the Lunar Relay Satellites include a transponder that enables support for both one-way and two-way on-demand tracking and time transfer by the users. The two-way capability eliminates the burden on some users to provide an USO. The Doppler positioning with only one satellite requires several measures in order to reconstruct the Doppler curve (depending on the orbits). Therefore in this case there is the need to wait for some satellite displacement along its orbit (which may not be a problem as there is a continuous link between LSE and LNCS, i.e. a continuous localization/positioning). The LSEs will always be in view of at least one LNCS, their One-way transmitted signals in down-link guarantee Doppler, range and differential Doppler measures. It is also possible to imagine using differential positioning with respect to stations on the moon surface (such as OBN). By implementing both Doppler and range measurements at the same time (or using more satellites), it is possible to instantaneously determine the absolute position without waiting time, because there are no ambiguous solutions.

In the One-way link approach the radio sources originate the forward links, and the user receives and tracks the signals from any radio sources in view. From these in-view links, the user collects one-way radiometric measurements for its own processing (see Table 24.6 for a reassume of the Doppler positioning solutions).

Table 24.6 Link direction features

Link direction		Satellite equipment	Surface equipment	Processing allocation	Definition
One-way	Up-way	Rx, USO	Tx, USO	LNCS	Localization
	Down-way	Tx, USO	Rx, USO	LSE	Navigation
Two-ways	Repeater on ground	Tx, Rx, USO	Tx, Rx	LNCS	Localization
	Repeater on orbit	Tx, Rx	Tx, Rx, USO	LSE	Navigation
	Repeater on ground and on orbit	Tx, Rx, USO	Tx, Rx, USO	LNCS LSE	Navigation / Localization

To what regards radar positioning, the common glossary depends on the directions of the link. When the user is only receiving and calculating its position, commonly the process is called navigation (cf. GPS, TRANSIT). When instead an external system is determining the user position and then sending it to the user, the process is called localization (cf. Radar, Argos, Cospas-Sarsat). Both services are part of positioning and radio-determination satellite services (RDSS).

Considering the required equipment, impact of frequency stability, number of users and transmission continuity it seems that the best solution for the directionality of Doppler measures is the Down-way. The LSEs can evaluate their position by themselves by carrying a radio receiver. Therefore most of the mobile elements have to include a receiving pack: semispherical antenna, Ultra-Stable Oscillator (USO) and dedicated hardware and software to perform positioning. This method allows architecture expandability, since the signals of the LNC satellites can be used by an unlimited number of LSEs. The choice also implies the need to make the positioning computation onboard the surface elements. In itself, this is not a heavy constrain but the design of the surface elements must consider this.

Different solutions could be envisaged:

- One-way measures where the LSE transmits in an up-link instead of receiving the navigation signals; in this case the localization could be performed directly on the LNCS and then sent down by means of the telecommunication system,
- Two-way measurements with the repeater on the LSEs which allows more accurate positioning performance,

Unfortunately for these two options the number of LSEs obtaining the position data would be related to the number of channels on the satellite's receiver. In the case of several LSEs not being linked with any LNCS, they can nevertheless be positioned by using onboard means to estimate their relative position with respect to the other LSEs or natural landing marks.

Orbitography processing is necessary to compute the orbit of the LNCS. The LNCSs positioning are greatly improved when using their constant tracking. Therefore the LNCSs should be continuously or periodically tracked, depending on the stability of their orbits and the accuracies required, their ephemerides have to be established via tracking. Instead of using Direct Tracking from Earth (radio or laser links) or relay satellites (e.g. GPS or TDRSS) the use of an Orbitography

Beacons Network (OBN) composed by navigation beacons spread on the lunar surface has been conceived. [The ephemeris is a predicted orbit based on Doppler observations previously acquired by tracking stations]. Direct Tracking from Earth should be avoided because it constrains the constant use of Earth stations. The use of other relay satellites does not eliminate the tracking problem due to the fact that the relay satellites have to be tracked nevertheless. Therefore, a complete autonomous lunar navigation system requires the implementation of a surface network of orbitography beacons and this appears to be the best solution. This network of beacons is adopted to localize the satellites and characterize their orbits by means of one-way signals in up-link. OBN signals will be received by the satellite and subsequently this data will be assimilated in orbit determination models to keep permanent track of the satellite's precise position. Since the position of every beacon is unknown, the orbitography will follow a three step evolutionary process, see Fig. 24.9. In the first phase of the mission, the LNCS's position will be computed with the help of GEEs (e.g. ESTRACK or DSN). In the meanwhile the LNCS's will start to localize the beacons; this process will last until each beacon's position is known with the desired accuracy. In the following phase the satellites will use the reference beacon network for their own positioning without the need of any Earth antennas. From this moment on, the GEE station use is discontinued and the LNC satellite aided positioning system will be completely independent from Earth.

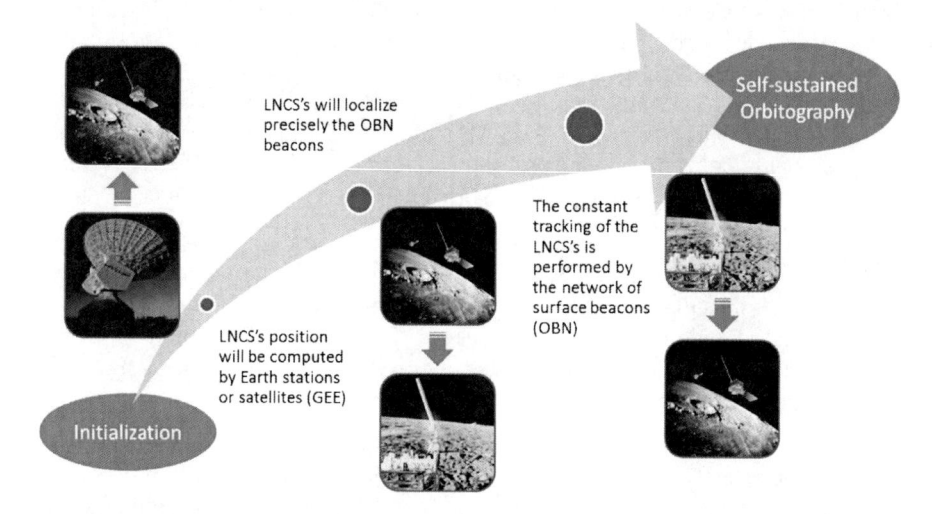

Fig. 24.9 Orbitography Evolution

The beacons synchronization can be in charge of the LNCSs themselves. The orbitography beacons will be incrementally placed opportunely spreaded by means of autonomous landers, or will be dropped by the LSEs. Adding new beacons to the network, the navigation accuracy for orbit determination, landing site targeting, and rendezvous improves and latency decreases.

A link budget shows that the beacon transmitter requires 25 W in radio frequency, and so at least 50 W for transmission considering the efficiency of the electronic system. Considering that it may be required that the beacons work also during the lunar night, the most promising power source are RTGs, because they are simple, reliable, long-life devices and allow complete autonomy of the beacons. From literature it is possible to foresee an End of Life (after 15÷20 years) energy density of 3.5 We/kg for a ostensible future RTG (Bennet 2006). Therefore considering the power demand, an RTG of 15 kg should be sufficient to sustain the power needs of a beacon. The OBN beacon will be very simple and light since it is a simple radio transmitter.

The OBN design requirements stem from the constraints on the knowledge of the satellite orbits. Taking into consideration the satellites coverage in the worst case scenario ($9 \cdot 10^6$ km^2 at periselene), the approximate minimum number of beacons can be estimated to be approximately 10, sufficient to give continuous service to all the LNCS's. It is very likely that more beacons are necessary to track each LNCS with the required precision. With a simple geometrical consideration it is possible to verify that especially with the satellite at the periselene and an eccentricity of 0.6, the LNCS can see both one polar region and a portion of the equatorial belt (refer to Fig. 24.10).

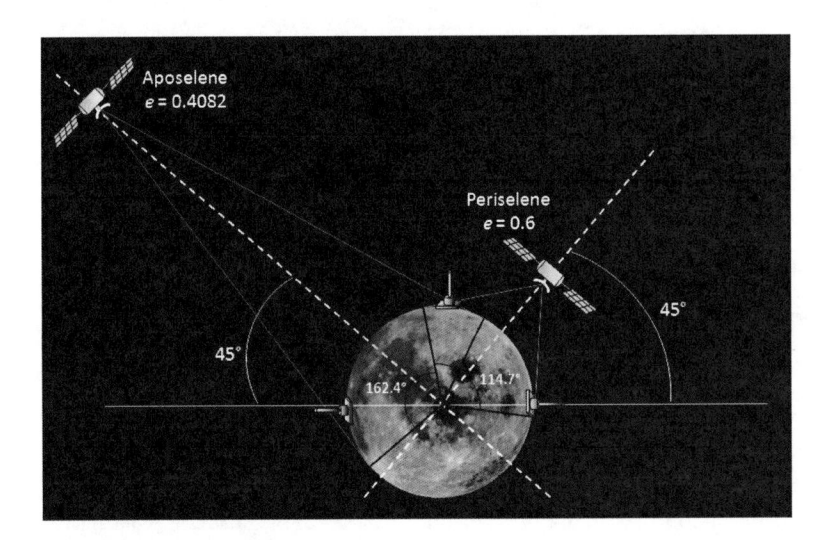

Fig. 24.10 Visibility between OBN and LNCS's.

The satellite navigation system (Fig. 24.11) can be schematized with the following steps:

1. The beacons transmit their signal to the satellites.
2. The satellites receive the signal from the beacons, and performing Doppler measures they are able to determine their position and orbit.

3. The satellites transmit a signal that is received by the LSEs. There the LSE compute from Doppler and range measurements, the position of the LSE with respect to the satellite.
4. From this data, it is then possible to compute the LSE position with respect to the available lunar maps.

In case of several satellites being visible from the LSE, or in the case of performing several possible measurements during one satellite pass, it will be possible to improve their accuracy in positioning, so the LSE can be able to simultaneously correlate, track, and collect more one-way radiometric data (range and Doppler) depending on the number of satellites concurrently in view. It is recommended to perform only Doppler navigation on the LNCS, assuming that they will have continuous access to at least one OBN beacons, therefore they can continuously track themselves.

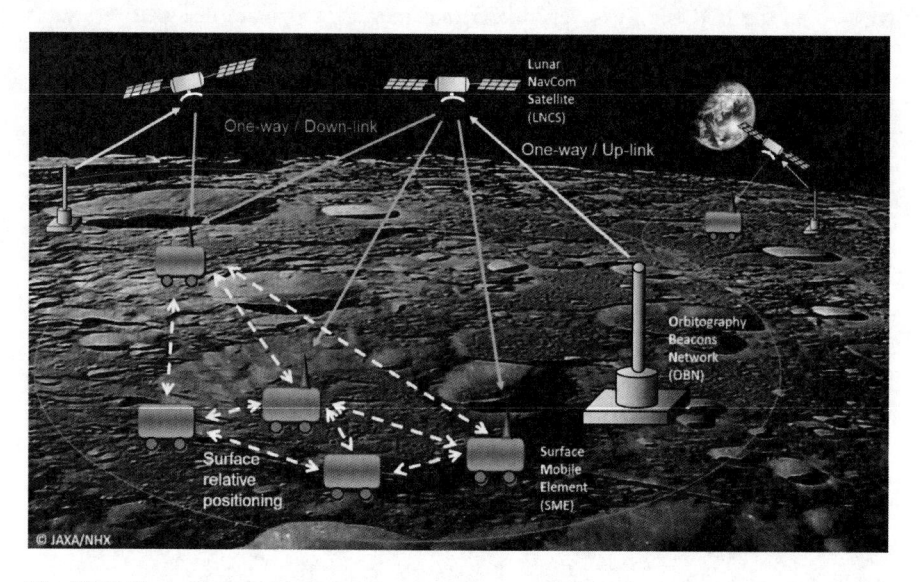

Fig. 24.11 Navigation Architecture

The LSEs can self-evaluate their position by carrying a radio receiver. Therefore most of the mobile elements have to include a receiving pack consisting of: semispherical antenna, Ultra-Stable Oscillator (USO) and dedicated hardware. In this way the architecture allows expandability, since the signal of the satellites can be used by unlimited number of LSEs. The choice also involves the need to make the positioning computation onboard the LSEs. This is not a heavy constrain but the design of the surface elements has to consider this.

Table 24.7 resumes the characteristics of the radio links involved in the satellite aided positioning system among Earth, the OBN, the navigation satellites and the mobile elements.

The satellite aided positioning system is similar to the old TRANSIT system. The Navy Navigation Satellite System, also known as TRANSIT, was the world's first operational satellite navigation system. Transit was originally conceived in the early 1960s to support the precise navigation requirements of the Navy's fleet ballistic missile submarines. With the same concept but with new more recent technologies and without the disturbance of the atmosphere it is possible to ensure for navigation a Doppler positioning accuracy of less than 100 m by means of the orbit segment.

Table 24.7 Navigation links features

	Earth → LNCS	OBN → LNCS	LNCS → LSE
Range / Range rate	Range rate and Angular from GEE	Range	Range Range rate
One-way / Two-way	Two-way/ Repeater on orbit	One-way/ up-way	One-way / down-way
Modes of Observation	Simultaneous point positioning	Simultaneous point positioning	Point positioning mode Multi-receiver techniques (if OBN in concurrent view)
LNCS equipment	Tx, Rx	USO, Rx	USO, Tx, Rx
Beacons equipment	-	USO,Tx	-
LSE equipment	-	-	USO, Tx, Rx
Band	X	S	S

Assuming the reference time frame belongs to the lunar headquarters base (i.e. South Pole), the synchronization of all LSEs would be made in three steps (Fig. 24.12):

1. the first step is the synchronization of the LNCS's with the base and among themselves,
2. the second step is the synchronization of the clocks on board the LSEs element in direct link with the orbital segment,
3. the third step is the synchronization of the remaining LSE elements by means of local wireless communication systems.

Each link in the chain involves a loss of precision, but the final error time shall be compliant with the precision requirements.

Frequency-Division Multiple Access (FDMA) has been considered to be most suitable for satellite aided localization purposes. The positioning signals will be structured as it is already done in present systems such as ARGOS, GPS, Cospas-Sarsat and other similar systems. They will contain LNCS ephemerides, locations, clock models, and other ancillary data required to process the radiometric tracking.

The signal has a pure carrier preamble of a certain period (160 ms similar to the ARGOS system) in order to improve precision in the frequency reading to the maximum extent, plus a set of phase modulated data on the carrier. The data part of the signal could contain:

- an identification code (ID) of the satellite (even if the channel belonging to different subjects is divided in frequency),
- the satellite ephemerides,
- the signal time of departure (TOD),
- information for the time correction of the clocks,
- a Pseudo-Random Noise sequence.

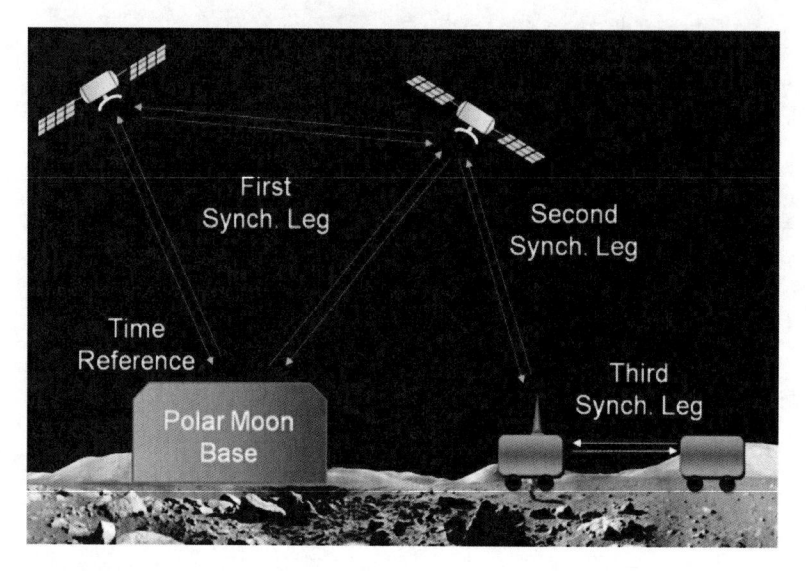

Fig. 24.12 Lunar Time Synchronization Legs

Transmitted signals from LNCS for LSE navigation can be structured as shown in Fig. 24.13.

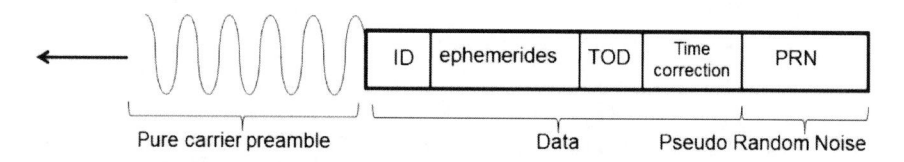

Fig. 24.13 Navigation Signal Structure

The Time of Arrival (TOA) will be extracted from the Pseudo-Random Noise sequence (PRN) by means of correlation with a copy of the same PRN sequence on the receiver, these codes are in general long (e.g. 1023 bit for GPS).

The same approach is followed for the LNCS tracking, when the beacons are the transmitter of the positioning signals. In this case the transmission is in up-link instead of down-link. The transmitted signal from OBN for LNCS tracking is simpler due to the lack of range measurements between the OBN and the LNS (see Fig. 24.14).

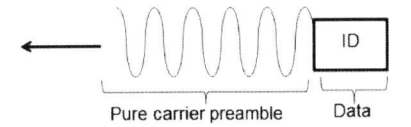

Pure carrier preamble   Data

Fig. 24.14 Orbitography Signal Structure

The choice of a FDMA implies that each channel needs a bandwidth that is wide enough for the transmission of the positioning message and the Doppler signal. The channel bandwidth has to consider the Doppler shift of the frequencies (Fig. 24.15), and this movement depends of course on the selected orbits. Doppler shift expected in the LCN system, in the worst case (periselene), is around $3.4 \cdot 10^6$ Hz/Hz with 1 hour of visibility at Time of Closest Approach (CTA) of 45°.

The frequency allocation performed in conjunction with the communication service study has assigned 20 MHz of bandwidth from 2815 MHz to 2835 MHz for Navigation purposes. The band is therefore been split in two distinguishing between down-link among LNSCs and LSEs and up-link between OBN and LNCS (see Table 24.8).

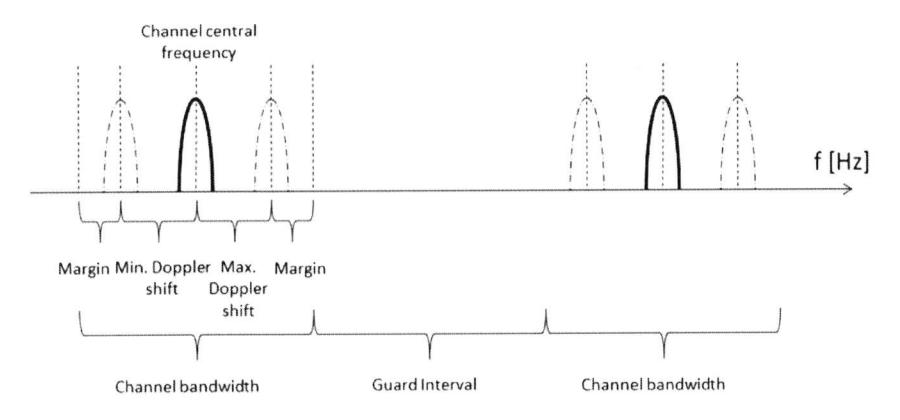

Fig. 24.15 Navigation signal frequency analysis.

Table 24.8 Positioning link frequency allocation.

	Central frequency [MHz]	Band with [MHz]
OBN → LNCS	2820	10
LNCS → LSE	2830	10

For the selected LNC constellation the maximum Doppler shift is of around 10 kHz. The total bandwidth required for a single channel is therefore the sum of the channel itself and the guard interval (Table 24.9).

Table 24.9 Positioning link signal frequencies.

Maximum Doppler Shift [kHz]	Bandwidth margin [%]	Band with [kHz]	Guard interval margin [kHz]	Guard interval [kHz]	Total channel [kHz]
9.588	50	28.917	10	38.917	67.834

Given that each satellite has its own channel for the navigation message, with 8 satellites a band of 541 kHz is needed. This means that more than 9 MHz are still available for other purposes, such as the synchronization signals. The available band is wide enough to allow a Frequency Division Multiple Access (FDMA) method for coordinating the signal of the different satellites and beacons.

Table 24.10 Positioning Error Budget

	Error Source	Error contribution	Design Constraints
USO	Drift (long term)	3 m	$2.5 \cdot 10^{-5}$ Hz/s
	Noise (short term)	7 m	Stability for Doppler measures; < ~10 ns @1s for range measures
LNCS's orbits (simultaneous point positioning) or OBN beacons position (Multi-receiver techniques) knowledge		25 m	beacons number, distribution and precision of localization
Hardware	Frequency reading (Doppler measure)	10 m	Not Available
	PRN correlation (time measure)		Synchronization error ~33 ns
LSE speed		Negligible	Good inertial and odometry navigation
Others	(e.g. relativistic effect, DEM accuracy, etc...)	5 m	-

Therefore the satellites could transmit their positioning signal continuously (maximum duty cycle of 100 %) without problems due to the synchronizations of the transmissions and increasing the number of radar measurements for the surface navigation, which means a better final accuracy of the absolute localization system.

The satellite aided navigation is expected to perform localization with an accuracy of 50 m. Table 24.10 aims to collect the major considerations about the final accuracy estimation, distinguishing the contributions of the various sources of position errors.

Table 24.10 shows how the errors can be minimized by using actual technologies or future ones. Only the knowledge of the LNCS orbital parameters is strictly related to the proposed architecture for the satellite aided navigation. The adoption of multiple receiver techniques (e.g. translocation) can reduce the impact of the problem but it is still present. At the end of these considerations the authors believe that an absolute positioning accuracy of the LSE of around 50 m is achievable. This implies severe constraints on the definition of the OBN which has to be analyzed to check if it would be able to allow navigation of the LNCS with an error of around 25 m.

24.8 LNC Satellites

Up to now the NavCom payloads of the LNCS have been roughly defined, but it is missing a concept study of the LNCS's able to support and integrate such payloads. All previous considerations give important inputs for development of the satellite design. The satellite shall host three antennas for telecommunications, one pointing towards the Moon, the others towards space. Another antenna towards the Moon is needed for the Guidance and Navigation system; in Stadter et al. (2006), it has been estimated that the on-board navigation system would require 1.5 kg of mass and 14 W of power based on current technologies. A final omnidirectional antenna is needed for telecommunication in case of contingency. Since the communications antennas are large (about 1 meter diameter), foldable antennas have been considered in order to save space inside the launcher fairing. Typically, satellites are composed of a payload module able to host the payloads and by a service module that hosts the other subsystems (e.g. propulsion, power generation and management). Figure 24.16 shows a possible LNCS configuration.

Since there are two orbital planes, separated by a $90°$ angle and it is very demanding to change the inclination of an orbital plane. Two different launches could be able to bring 4 LNC satellites each time into one of the two orbital planes. Thus each launcher carries 4 satellites and transports them from Earth to a temporary circular lunar orbit, with an inclination of $45°$. The last stage of the launcher performs a boost which brings the satellites into an Earth circular waiting orbit 700 km high. The 4 satellites would be mounted on a so-called service module, which brings the satellites into Lunar Transfer Orbit (ΔV of 3.2 km/s) and performs a final boost to inject all satellites into a circular Moon parking orbit (radius of 12000 km) contained in one of the two final orbital planes, with a ΔV of 180 m/s.

Once in the Moon's parking orbit, the satellites are separated from the service module. Afterwards, all the maneuvers are performed by the satellites themselves. This Initial Acquisition Mode considers all the activities needed to get a completely autonomous positioning system. In this period the satellite has to work in

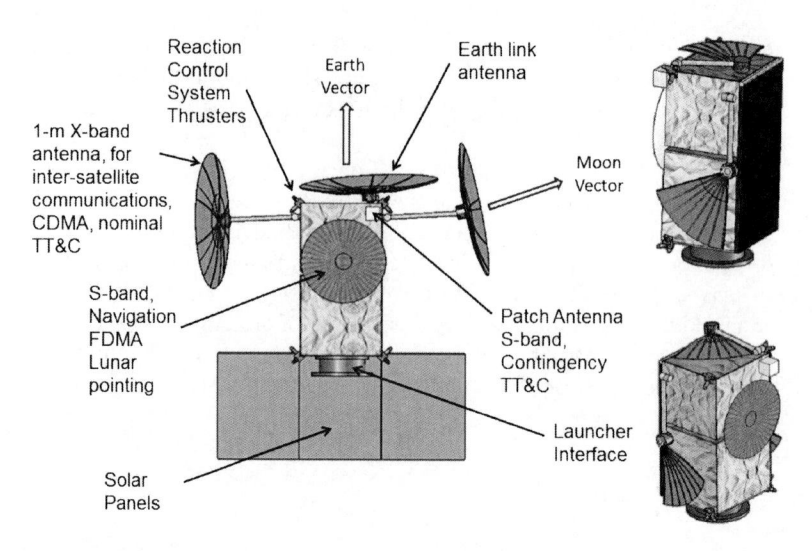

Fig. 24.16 LNCS configuration - On the right the LNCS in represented in a folded configuration

collaboration with both the GEE and OBN on the Moon surface, in order to calibrate the Doppler navigation system. Each satellite brings itself from the Moon's parking orbit to the final one. In particular, satellites have to correct the inclination error and perform a Hohmann transfer from the Moon's parking orbit to the circular lunar orbit for which the radius is equal to the final apogee (i.e. 11335 km). The satellites perform the final boost to enter the permanent elliptical orbit. The first and third satellites perform this maneuver at the same time. The last two satellites perform this maneuver together 9 hours later. As explained earlier the chosen orbits are stable, which means that their parameters change their value in a periodic way. As a consequence, the amount of ΔV dedicated to station keeping is assumed to be negligible. The overall amount of ΔV that is considered for the design of the satellite propulsion subsystem is 182 m/s. In the nominal scenario the remaining propellant will be used at the end of lifetime to put the satellites either into a graveyard orbit or into a controlled trajectory which leads to a safe crash on the Moon surface.

Two additional operational modes have been considered:

- The Stand-By Mode regards the inactive satellite period. It generates basic housekeeping telemetry and awaits incoming telecommands but takes no action to control the spacecraft. The satellite is typically in this mode before it separates from the launcher and in the first seconds after separation.
- The Safe Mode is activated after a very serious anomaly has been detected. The objective of Safe Mode to keep the satellite in a safe state; the satellite aligns the solar panel toward the sun and points the antenna toward Earth.

The Attitude and Orbital Control System (AOCS) uses sun sensors, star trackers and reaction wheels as its primary sensors and actuators. The satellites are not spin-stabilized but 3-axis controlled by means of four reaction wheels, placed in a tetrahedrical configuration in order to guarantee the best control in all directions and 12 5-N thrusters. The thruster's position in the satellite has been chosen in order to avoid thermal problems and antenna damage. Due to configuration and payload interference, just 12 thrusters could be mounted in such a way to guarantee rotation around the three body axes and motion in the direction of the main engine only. The set of small thrusters can also control the thrust vector and avoid any misalignment. Thrusters could be either chemical or electrical but for simplicity they have been considered to be hydrazine thrusters.

The satellite has to nominally provide 240 W for the correct functioning and solar arrays have to charge batteries during the period of illumination (assuming the efficiency of the battery is about 0.65 and the charging efficiency of 0.9). To this value a 10% for the harness losses plus a 15% for the margin is added. The total amount of power to be produced by the solar arrays is 365 W. Considering a conversion efficiency of 0.26, value achievable with common GaAs solar arrays, a Sun incidence angle of 15 degrees, and a degradation per year of 2.5% with a lifetime of 20 years the total solar array area required is 2.1 m^2. Considering a discharge efficiency of the regulator of 0.85 it has been estimated that 10 kg of batteries should be sufficient. The required minimum radiator area is of approximately 0.8 m^2.

When the mass of the solar array, battery, radiator and of all other equipment is available it is possible to have the complete mass budget. The value of the LNCS total mass is reasonable at about 200 kg with dimensions of around 1 m^2 meter top view section and less than 2 m high.

References

Bennet, G.L.: Space Nuclear Power: Opening the Final Frontier. In: 4th International Energy Conversion Engineering Conference and Exhibit (IECEC), Paper, AIAA-2006-4191, June 26-29, AIAA, San Diego (2006)

Bhasin, K.B., Warner, J.D., Anderson, L.M.: Lunar Communication Terminals for NASA Exploration Missions: Needs, Operations Concepts and Architectures. In: 26th International Communications Satellite Systems Conference (ICSSC), Paper AIAA-2008-5479, June 11, AIAA, San Diego (2008)

Chabot, T.: Integrated Navigation Architecture Analysis for Moon and Mars Exploration. Thesis, Advisor – How JP, Massachusetts Institute of Technology - Dept. of Aeronautics and Astronautics (June 2005)

Comparetto, G.: The Impact of Dust and Foliage on Signal Attenuation in the Millimeter Wave Regime. Journal of Space Communication 11(1), 13–20 (1993)

Culbert, C., Gonthier, Y., Mongrard, O., Satoh, N., Seaman, C., Troutman, P.: Human Lunar Exploration: International Campaign Development. In: 61st International Astronautical Congress 2010 Prague, International Astronautical Federation, Paper IAC-10.A5.2.10 (2010)

Dietz, R.H., Rhoades, D.E., Davidson, L.J.: APOLLO Experience Report - Lunar Module Communication System, Technical note TN D-6974, Manned Spacecraft Center, Houston – Texas (1972)

European Telecommunications Standards Institute (ETSI) ETSI Standard TR 102 376 V1.1.1 Digital Video Broadcasting (DVB) User guidelines for the second generation system for Broadcasting, Interactive Services, News Gathering and other broadband satellite applications (DVB-S2). Standard, European Telecommunications Standards Institute (ETSI), 650 Route des Lucioles F-06921 Sophia Antipolis Cedex – France (2005)

Folta, D., Quinn, D. (2006) Lunar Frozen Orbits. In: Astrodynamics Specialist Conference and Exhibit, Keystone – Colorado, AIAA 2006-6749, Paper, AIAA/AAS, August 21-24 (2006)

Gutiérrez-Luaces, B.O.: Radio Astronomy Use of Space Research (Deep-Space) Bands in the Shielded Zone of the Moon. In: Yuen, J.H. (ed.) The Telecommunications and Data Acquisition Progress Report 42-129, NASA Code 315-90-31-00-01, NASA JPL, Pasadena – California (1997)

Hufenbach, B., Leshner, R.B.: The NASA-ESA Comparative Architecture Assessment. In: Joint Annual Meeting of LEAG-ICEUM-SRR, Study Report, Cape Canaveral, Florida, October 28-31 (2008)

NASA, NASA's Lunar Communications & Navigation Architecture. In: Technology Exchange Conference, Presentation, NASA Technology Exchange Conference, Galveston, Texas, November 14-15 (2007)

Noreen, G.K., Cesarone, R.J., Deutsch, L.J., Edwards, C.D., Soloff, J.A., Ely, T., Cook, B.M., Morabito, D.D., Hemmati, H., Piazzolla, S., Hastrup, R., Abraham, D.S., Sue, M.K., Manshadi, F.: Integrated Network Architecture for Sustained Human and Robotic Exploration. In: IEEE Aerospace Conference, Paper, Big Sky, Montana, March 5-12 (2005)

Stadter, P.A., Sharer, P.J., Kantsiper, B.L., DeBoy, C., Finnegan, E.J., Napolillo, D., Duven, D.J., Kirby, K.W., Gramling, J.J.: Lunar Navigation and Communication System Implementation Concept. IEEEAC paper #1214, Version 3, Updated December 13 (2006)

Stanley, D., Cook, S., Connolly, J., Hamaker, J., Ivins, M., Peterson, W., Geffre, J., Cirillo, B., McClesky, C., Hanley, J., Davis, S., Falker, J., Pettit, D., Bailey, M., Bellinger, F., Craig, D., Derleth, J., Reuther, J.: NASA's Exploration Systems Architecture Study. Study final report, NASA-TM-2005-214062 (2005)

Schier, J.S., Rush, J.J., Williams, W.D., Vrotsos, P.: Space Communication Architecture Supporting Exploration and Science: Plans and Studies for 2010-2030. In: 1st Space Exploration Conference: Continuing the Voyage of Discovery, NASA, Paper AIAA-2005-2517, Disney's Contemporary Resort Orlando, Florida, January 30- February 1 (2005)

Schier, J.S.: NASA's Lunar Space Communication and Navigation Architecture. In: 26th International Communications Satellite Systems Conference (ICSSC), San Diego, California, Paper AIAA-2008-5476, June 10-12 (2008)

Schrunk, D.G., Sharpe, B.L., Cooper, B.L., Thangavelu, M.: The Moon - Resources, Future Development, and Settlement, 2nd edn. Springer/Praxis Publishing, Chichester (2008)

Space Communication Architecture Working Group (SCAWG) NASA Space Communication and Navigation Architecture Recommendations for 2005-2030. SCAWG Final Report (May 2006),
https://www.spacecomm.nasa.gov/spacecomm/
programs/system_planning/default.cfm

Space Exploration and Development Systems Master (SEEDS) - 4th Course edn. Design Reports, Torino – Toulouse – Bremen, Consorzio per la Ricerca e l'Educazione Permanente (COREP), Torino (2008-2009)

Space Exploration and Development Systems Master (SEEDS) - 4th Course Edition. ALICE: Advanced Lunar Itinerant Caravan for Exploration. In: Workshop Executive Summary of the International Post Graduate Master Course, Consorzio per la Ricerca e l'Educazione Permanente (COREP), Report, Torino (2009)

Todd, A.E., Lieb, E.: Constellations of Elliptical Inclined Lunar Orbits Providing Polar and Global Coverage. In: AAS/AIAA Astrodynamics Specialist Conference, Lake Tahoe, California, August 7-11 (2005)

25 A Laser Power Beaming Architecture for Supplying Power to the Lunar Surface

Henry W. Brandhorst, Jr.

Carbon-Free Energy, LLC, Auburn, USA

25.1 Infrastructure Support to Lunar Commercial Activities

In order to undertake space exploration and utilization of the moon for resources, a complete commercial infrastructure must be in place. Too often only a single point of departure is taken to be the total rationale for the commercial undertaking. However, insofar as the potential lunar resources and energy requirements are concerned, a fully-functional infrastructure for maintaining a continuous flow of materials to and from the moon is essential. Otherwise single-point activities will have infrastructure costs that don't permit profitability.

There are several options for a lunar commercial infrastructure but all have a common initial condition: everything needed to start the lunar activity is on the surface of the earth. Substantial effort must be made to emplace robots, other machines, chemicals and other essential materials on the lunar surface. The traditional approach to accomplishing that goal is to develop large chemical rockets in the 70-130 MT range to perform a direct ascent to low lunar orbit (LLO) with the cargo and lander. The lander and cargo (power, materials, chemicals, robots etc.) would then descend to the lunar surface and disgorge its cargo at the targeted site. All parts of this approach would be used only once. It is possible that the lander stage could return to LLO with products (propellant, solid materials, etc.), but would not be able to transport that new cargo to another location for profit.

This single-use type of architecture is expensive and will involve the development of specialized vehicles that support each endeavor on the lunar surface. Each will have their own launch vehicles, landers, robots etc. in order to conduct their work. To base this entire infrastructure on chemical propulsion does not appear to be cost-effective due to the multiplicity of vehicle types and the dependence on either hydrogen-oxygen or hydrocarbon-oxygen propulsion systems. Furthermore, the infrastructure must contain more elements to ensure success, therefore, an architecture based primarily on electric propulsion for the transit between low earth orbit (LEO) and LLO will be described (Brandhorst 2008, O'Neill et al. 2006).

25.1.1 Architecture Components

The architecture for this concept has three major aspects: A LEO propellant depot, a Solar Electric Propulsion (SEP)-based tug capable of carrying lunar cargo, structure elements, and a lunar landing stage capable of carrying 22 MT of cargo to the surface (O'Neill et al. 2006). With the exception of the earth-to-orbit launch vehicles, all these elements are reusable and are sent into a low earth orbit of about 500 km using the emerging range of COTS and commercial launch vehicles. The LEO propellant depot will be able to be refueled. Robotic assembly and other operations are extensively employed. Each will be described in turn.

25.1.1.1 Earth to LEO Launch Vehicles

Low cost launch vehicles are vital to lowering the cost of utilizing lunar resources and creating a lunar infrastructure to an economically viable level. In contrast to previous studies (NSSO 2007), this architecture uses existing launch vehicles as well as those that will be emerging in the next five to ten years under the NASA Commercial Orbital Transportation System (COTS) program. None of these vehicles are the heavy-lift launch vehicles that have been the baseline for launching large space systems into GEO. This architecture could benefit from a heavy-lift launch vehicle as will be shown later. However, the current launch vehicle families may be adequate (albeit with more launches) to fulfill the requirements of a robust lunar commercial infrastructure.

In the past, earth-to-LEO launch costs below $100/kg were required to meet the overall economic feasibility of a lunar commercial architecture. Approximate 2001 launch costs (Futron Corporation Report 2002) to LEO shown in Fig. 25.1 are around $4,000/kg with the Atlas, Delta, Proton and Zenit vehicles. The new generation vehicles being developed offer the promise of $1,000/kg and some payload capacity above 50 MT with the heavy-lift vehicle ranging from 70-130 MT. With emerging commercial markets and opportunities for space tourism occurring over the next two decades, significant cost reductions will naturally take place. At a minimum, the learning curve will come into play and costs of components will continue to decrease as production innovations occur. However, caution must be used in interpreting these data as they are somewhat out of date and the cost of a launch depends on many factors. Some of these factors are the amount of lift capacity actually needed, the volume of the payload and the urgency of launch.

As can be seen from Fig. 25.1, the current stable of launch vehicles around the world displays a wide range of both cost (price) and mass delivered to LEO. Furthermore, LEO is defined by U.S., European or Russian launch providers anywhere from 185 km to 400 km circular with inclination of 5.2°, 28.5° or 51.6° depending on launch site. However, it is noted that launch vehicle capacity to LEO is increasing and payload capacity of 20 MT and more are available today. Capacities up to 50 MT and costs of $1,000/kg may be forthcoming in the next decade.

However it is important to note that the cost/price to LEO is substantially cheaper than that for a vehicle going into Geosynchronous Transfer Orbit (GTO). Because of the increased energy required, the payload decreases as physics

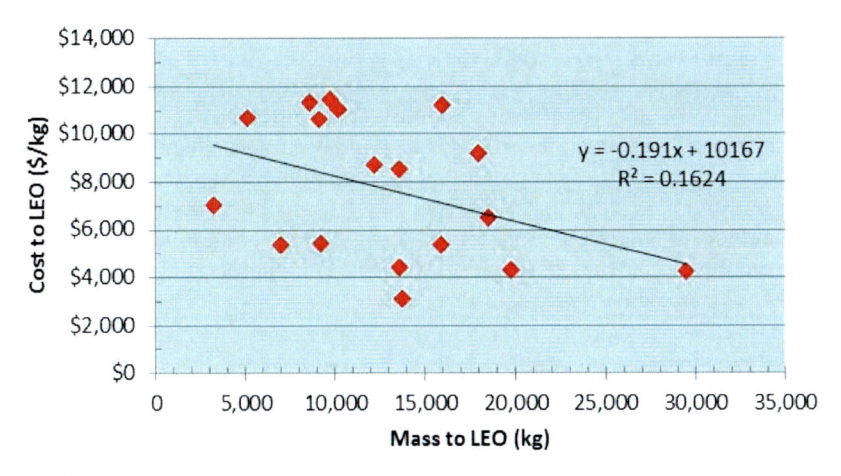

Fig. 25.1 Launch Costs to LEO

requires. In addition the spacecraft must also carry an additional propulsion system – usually a chemical "kick:" motor. However, as satellite power to mass level increases, it makes the use of electric propulsion attractive for reaching the final GEO orbit.

With that in mind, Fig. 25.2 shows the 2001 launch costs (Futron Corporation Report 2002) for payloads to GTO. Now the impact of physics is clear – cost/kg of payload has increased substantially (about a factor of 2.0 – 2.5) and the mass delivered has also dropped by the same ratio. Thus the cost of an equivalent mass to GTO is about 4 to 5+ times greater than to LEO. In addition, of the mass placed into GTO, only about 45% of it is payload, the rest is the propulsion system. Thus the cost now increases by 8 to 12-fold over a LEO launch.

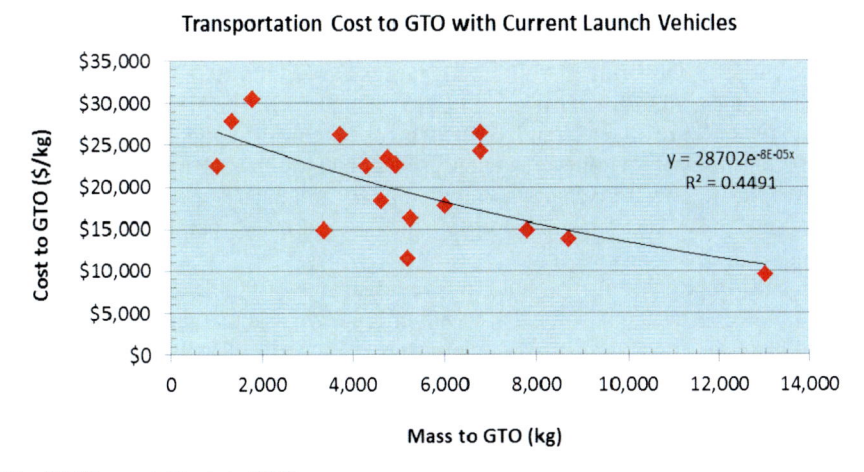

Fig. 25.2 Launch Costs to GTO

This has been the dilemma that has stymied concepts like a space-based solar power system from the very beginning of the concept. The most often prescribed solution to this barrier has often been a huge heavy-lift launch vehicle, but the cost to develop such a vehicle has been prohibitive without other commercial launch opportunities. The architecture proposed herein overcomes that limitation and permits the use of current or emerging near-term vehicles but does not exclude development of a heavy lift launch vehicle. In a later section we will make use of a 70 MT vehicle to emplace the main LEO to LLO transportation vehicle in LEO. However, multiple launches of a smaller vehicle with appropriate capacity could be used in place of the heavy lift launch vehicle.

25.1.1.2 Propellant Depot/Staging Area

The propellant depot/staging area will be launched into a low earth orbit of between 400 and 500 km with inclination dependent upon the launch site. From the Kennedy Space Center that would yield an inclination of 28.5°; for Kourou, French Guiana, the inclination would be approximately 5.2°. The launch vehicle to be used would depend upon the mass of the depot/staging area. For the purposes of this architecture, the depot would have basic satellite services of power, energy storage, attitude control, fuel storage tanks and capability for autonomous rendezvous and robotic fueling of other vehicles.

These latter capabilities were demonstrated in 2007 by the U.S. Orbital Express flight. On two separate occasions, one of the satellites, Astro, autonomously transferred hydrazine fuel to the NextSat satellite. The process was then repeated in reverse. During those linkages, battery and circuit board transfers using a robotic arm were also demonstrated. In addition, Europe's Automated Transfer Vehicle (ATV) Jules Verne successfully docked with the International Space Station on April 3, 2008. This fully automated docking of the 19,000 kg ATV represents an additional demonstration of the availability of these essential technologies for broad use. The design of this station will depend critically on the type of propellant to be used and options will follow.

Fuel resupply will occur through regular launches that will be bringing the lunar hardware to this orbit. Based on the successful Orbital Express flight, the depot's fuel tanks will refilled by either ullage recompression or pumped transfer.

Initially the propellant would be xenon (or some other noble gas) for use in Hall or ion thrusters. However, xenon used in electric propulsion systems is expensive, requires well-insulated storage and may ultimately be in limited supply when large scale electric propulsion systems are routinely operating in the earth, cis-lunar and deep space regimes.

Thus the depot will also be able to take advantage of the next step in low-cost fuels for electric propulsion systems. One of the key fuels may be ammonia. Ammonia has been used in the 1 MW-class Pulsed Inductive Thruster (PIT) demonstrated in the 1970s (Hrbud et al. 2002). An updated version of the PIT, called the Faraday Accelerator with Radio-frequency Assisted Discharge (FARAD), is currently under development. Both of these are pulsed systems that use current flowing through an inductive coil to accelerate plasma.

PIT and FARAD can use many different propellants beside Xe or other rare gases. These include NH_3, CO_2, and H_2O. Specifically, ammonia has been used with a nominal 50% efficiency in early tests. Ammonia can be stored at room temperature under slight pressure (like propane). It is widely used for agricultural fertilizer and safety and handling procedures are well known. For a comparison of cost, xenon costs around US$3,000/kg today while anhydrous ammonia costs only US$0.50/kg even at today's inflated prices. Hydrazine is also a candidate but is a more challenging health hazard.

25.1.1.3 LEO to LLO Solar Electric Propulsion Vehicle/Tug

The tug is designed for five round trips from LEO to lunar orbit, where a chemically fueled lunar lander will be released from the tug to deliver the cargo to the surface (O'Neill et al. 2006). Each round trip will take less than one year and will result in 22 MT of cargo delivery to the lunar surface. Thus, 110 MT of cargo will be delivered over the five year mission. Two launches per year will be needed for launch vehicles whose payload is less than 70 MT to deliver all the required materials to LEO for the five-year mission. A single launch is all that would be required with the 70 MT vehicle. The same solar array and tug will be reused five times over the five-year mission. A nominal 600 kW, 2,000 m^2 solar array will be needed to direct-drive the Hall-effect thrusters, which use xenon as their propellant. No solar array of this size has ever been built, so its design is speculative and dependent upon the array technology used. Figure 25.3 shows a schematic of one possible configuration. As noted above, other propellants may be used with other thruster systems in the future.

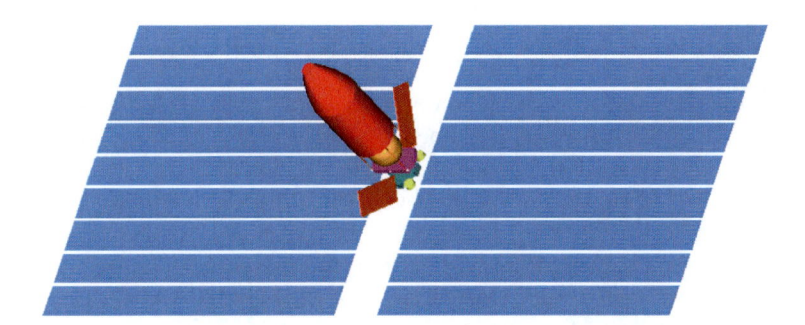

Fig. 25.3 Conceptual Drawing of a 600 kW SEP Tug for Lunar Transportation

In order for a vehicle like this to be successful and have the capability of delivering that size of payload to the lunar surface, low mass, and high efficiency solar arrays are an absolute necessity. Over the past two decades, a concentrating solar array technology using line-focus Fresnel lenses focusing onto multi-junction solar cells, with the latest embodiment called the Stretched Lens Array (SLA), has been developed and flown on a mission to a comet in 1998. This array uses a series of line-focus arched lenses which evolved into the SCARLET array that performed flawlessly for the full thirty-eight-month mission on NASA's Deep Space 1 probe, shown schematically in Fig. 25.4. SCARLET (acronym for Solar Concentrator Array using Refractive Linear Element Technology) employed silicone Fresnel lenses of a unique design. The refractive elements are configured such that the red and blue wavelengths are effectively merged onto the solar cell so there is no loss of the solar spectral distribution due to chromatic aberration. Monolithic triple-junction (GaInP/GaAs/Ge) cells were placed in the focal lines of the SCARLET lenses. The SCARLET array powered both the spacecraft and the ion engine on Deep Space 1 and performed as predicted on this highly successful mission (Murphy 2000).

720 ENTECH Lenses on
2.5 kW SCARLET Solar Array
on NASA/JPL Deep Space 1

Fig. 25.4 Artists' Concept of the Deep Space 1 Satellite

This solar array that concentrated sunlight about 8-fold, was able to keep cell temperatures low because small area cells were used. Thus with a substrate made from high thermal conductivity graphite fibers the cell temperatures were only a few degrees warmer than were full-area cells on a conventional planar solar array. However, that array was mounted on rigid panels which reduced the mass advantage of the SLA approach. Thus the next step was to develop a lightweight substrate that would meet the needs of high power electric propulsion missions. Figure 25.5 shows a demonstration model of the lightweight panel. The solar cells are illuminated by sunlight underneath the middle of each lightweight silicone lens. The black material is the carbon-fiber composite thermal rejection substrate. The lenses are erected with carbon fiber arches. They retain their shape over about one meter in length. In GEO, the solar cells in this array design reach a temperature of only 70°C in contrast to a conventional rigid array that operates at 60°C as confirmed by the earlier Deep Space 1 flight with the SCARLET array. The specific mass of this SLA blanket assembly is only 1.4 kg/kW using 31% efficient, 100μm thick triple junction GaAs-based solar cells. If the emerging 33% efficient Inverted MetaMorphic cells (IMM) reach production and are feasible, the mass would drop to only 1.1 kg/kW.

Fig. 25.5 Lightweight SLA Blanket Demonstration Panel

However, the blanket assembly is only one part of the solar array. The blanket must be stored, deployed and pointed at the sun to be effective. The size of the solar array proposed for this tug exceeds the state-of-the-art by 20-fold, thus substantial development of these large arrays must proceed in a stepwise fashion, complemented by space testing and validation at each step. With this in mind, an initial 30 kW design has been produced based on a lightweight, reliable, and extremely simple deployment system called the Rolled-Out Solar Array (ROSA). This array is made from carbon composites, deploys under its stored strain energy and is very rigid and accurate in its pointing capabilities.

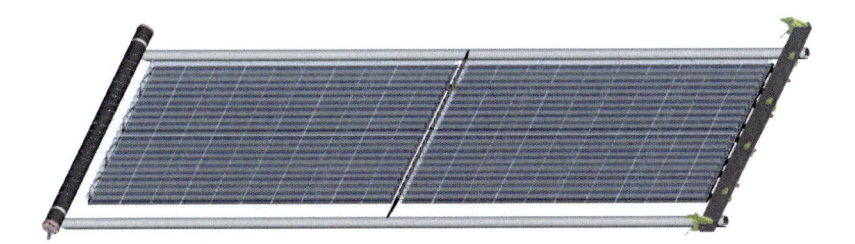

Fig. 25.6 Schematic of 15 kW SOLAROSA wing

Figure 25.6 shows a single 15 kW wing of this array, now called "SOLAROSA" (Stretched Optical Lens Architecture with Roll-Out Solar Array). Using this deployment approach, the mass of the SOLAROSA array using the 100μm thick triple junction solar cells increases to 2.6 kg/kW and with the IMM cells the mass becomes only 2.15 kg/kW. This value of alpha (kg/kW) is the lowest ever achieved for solar electric propulsion systems. In contrast, today's rigid panel planar solar arrays have an alpha of 12.5 kg/kW. A flexible blanket solar array that has flown to Mars has an alpha of 7.7 kg/kW, so the SOLAROSA array represents and enabling capability.

The SOLAROSA array for this mission will require more shielding on the solar cells than for a normal GEO mission, since the SEP tug will spiral slowly through the Earth's radiation belts during each trip from LEO to the Moon and back. To determine the optimal amount of shielding, the total mission radiation environment must be defined. Because the electron and proton radiation fluences vary widely with orbital altitude and inclination, the trajectory information must first be calculated. A spreadsheet model has been generated based on orbital energetics to estimate the spiral trajectories, in addition to other relevant mission parameters.

Table 25.1 shows the typical inputs and outputs of this model. Each round trip from LEO to the Moon includes an outbound spiral trajectory of the "heavy" tug (with lunar lander, lunar lander chemical fuel, cargo, and xenon propellant for the complete round trip) and a return trip of the "light" tug (without the lunar lander, its chemical fuel, cargo, and xenon used during the outbound trip). The minimum days in sunlight for both the outbound and inbound trips can be estimated with the rocket equation, but this ignores two important effects: the Earth's shadowing of the array during a portion of each spiraling orbit and residual atmospheric drag in LEO. These two important effects are treated in the spreadsheet model by a stepwise integration over time to calculate the actual spiral trajectories (the real days in Table 25.1).

Drag is modeled based on the worst-case atmospheric density (ρ) versus altitude curve from the European Space Agency's outstanding online tool known as SPENVIS (http://www.spenvis.oma.be). The orbital-average projected area (A) of the rotating solar array is combined with a worst-case drag coefficient (C_D) of 4, and with the local orbital velocity (V) to complete the drag calculation using the usual formula:

$$D = \frac{1}{2}\rho V^2 C_D A \qquad (25.1)$$

Because drag acts on the solar array for the full orbit while the Hall-effect thrusters only provide thrust for the illuminated portion of each orbit, the orbit-average thrust to drag ratio is a critical parameter.

Figure 25.7 shows this ratio as a function of altitude with solar array performance parameterized in terms of areal power density. Note that array performance is important, especially for relatively low starting altitudes, since the thrust to drag ratio must be obviously be well over 1.0 to enable the tug to move outward rather than falling back to Earth. The SOLAROSA array will produce between 300 and 400 W/m^2 when equipped with 33+% efficient solar cells that are currently in production.

Table 25.1 Solar Electric Propulsion (SEP) Lunar Tug Spreadsheet Model

Input Parameters										
Power	Array Specific Power	ISP SEP	SEP Power Efficiency	ISP Chem	SEP Delta-V LEO to LLO	Chem Delta-V LLO to Surface	Cargo	Mass Ratio LLO to Surface	Array Areal Power	Array Area
600,000 W	300 W/kg	2,500 Sec	60%	400 Sec	8,000 m/s	2,000 m/s	22,000 kg	1.67	300 W/sqm	2000 sqm

LEO to LLO												
Dry SEP Tug Mass	Dry Lunar Lander Mass	Chemical Fuel for Lunar Descent	Solar Array	Xenon Start	Cargo	Total Start Mass	Total End Mass	Xenon Used	Minimum Days in Sunlight	Real Days	Plane Change Days	Total Days
7,000 kg	3,000 kg	16,641 kg	2,000 kg	24,372 kg	22,000 kg	75,013 kg	54,116 kg	20,897 kg	202 days	246 days	11 days	257 days

LLO toLEO												
Dry SEP Tug Mass	Dry Lunar Lander Mass	Chemical Fuel for Lunar Descent	Solar Array	Xenon Start	Cargo	Total Start Mass	Total End Mass	Xenon Used	Minimum Days in Sunlight	Real Days	Plane Change Days	Total Days
7,000 kg	0 kg	0 kg	2,000 kg	3,475 kg	0 kg	12,475 kg	9,000 kg	3,475 kg	34 days	41 days	2 days	43 days
Thrust	Thrust	M-Dot	Min Sun Days Out	Min Sun Days Back			Total Xe Used	24,372 kg			RT Time	300 days
0.0294 kN	29.4 kN	104 kg/day	202 days	34 days			Difference	0 kg			RT to Outbound time ratio	1.17

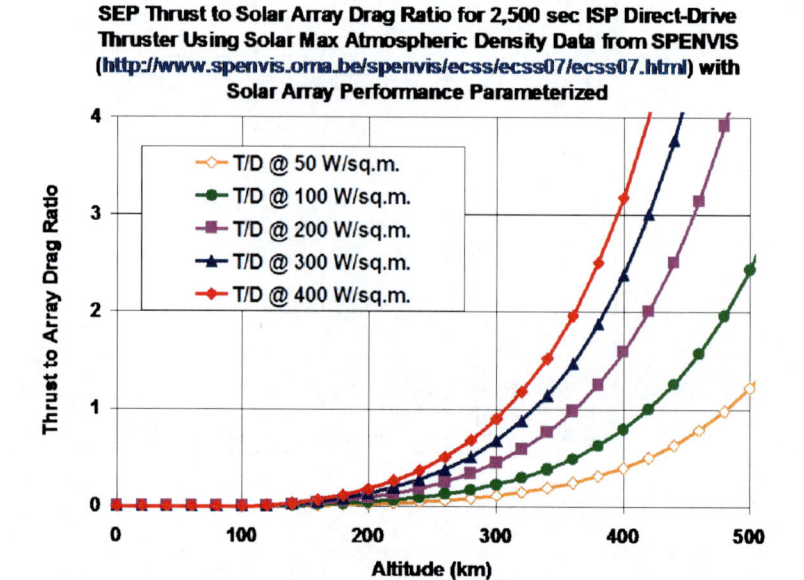

Fig. 25.7 SEP Tug Thrust to Drag Ratio in LEO

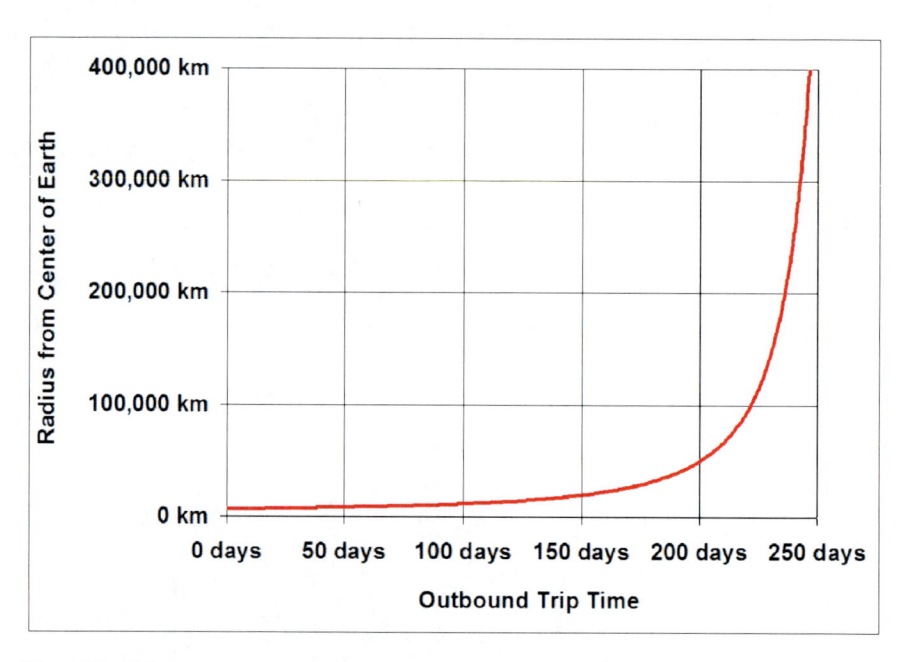

Fig. 25.8 SEP Tug Outbound Trajectory

It is important to also note that the cell efficiency increases about 10% with the 8X concentration level.

With drag and Earth shadowing properly treated, Fig. 25.8 shows the calculated outbound trajectory from LEO to low lunar orbit (LLO) for the SOLAROSA-powered SEP tug. The return spiral has the same curve shape as Fig. 25.8, but takes only 1/6th as long because of the lighter mass of the returning tug.

The trajectory information for all five round trips is used to calculate the equivalent 1 MeV electron fluences as a function of solar cell cover glass thickness, and the corresponding solar cell power degradation for triple-junction cells, using again the European Space Agency's powerful online tool, SPENVIS, (http://www.spenvis.oma.be). The SPENVIS results are used to determine the optimal solar cell shielding for the SLASR array, with typical results summarized in Fig. 25.9.

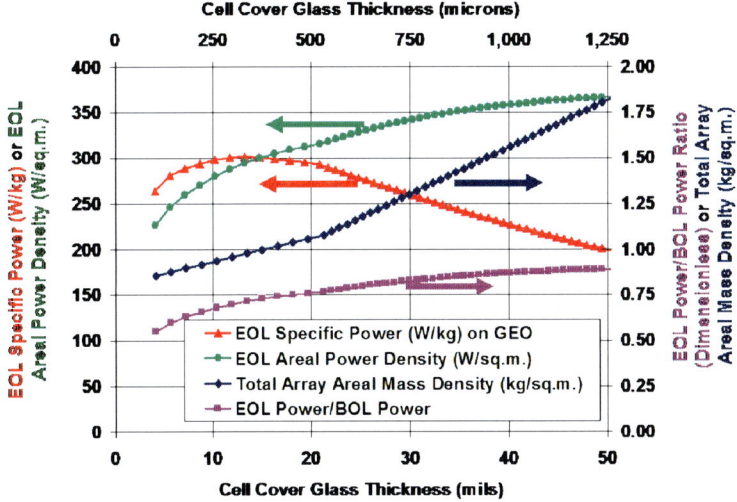

Fig. 25.9 Cell Shielding Optimization for the Mission

Note that the end-of-life (EOL) specific power has a peak value of 300 W/kg corresponding to an optimal 300-375 µm (12-15 mils) of cell cover glass. These results are based on a future multi-junction cell with room-temperature efficiency at 8 AM0 suns of 39% in 2015. That goal is consistent with current development programs as noted above. Today's triple junction solar cells have demonstrated 31+% room temperature efficiency, and multi-junction cells have averaged a 1% absolute efficiency gain per year for the past 20 years. As noted above, the emerging IMM cells have reached 33% efficiency with expectations of reaching 39% in the next five years. These values are remarkable and were not envisioned a decade ago.

The other major factor that concentrator solar arrays bring to space is their lower cost. The major factor in solar array cost is the solar cells themselves. In a planar solar array, the solar cells account for about 40% of the array cost, or about $400/W. For the concentrator array, because of the 8x concentration level, about $1/8^{th}$ the area of solar cell material is needed. This makes the concentrator solar cell less expensive because 12 concentrator cells are produced on a single wafer instead of only 2. This substantially improves the yield of cells/wafer. In addition, because the solar cells are narrower than the radiator/substrate, a cover glass that completely encapsulates the solar cells and interconnections can be used. This allows the solar array to operate at voltages above 1000V without arcing, even in the presence of micrometeoroid bombardment (Brandhorst and Best 2005). As observed in that reference, the lens material even acts as a meteoroid bumper and shatters the incoming particle before it can damage the solar cells. The bottom line is that the SOLAROSA array at 30 kW can be produced for 50% the cost of a conventional rigid panel, planar solar array. The numbers used to determine this value were based on the costs of solar cells, structures and mechanisms plus the sustaining engineering and recurring labor costs plus the costs of asssembly, integration, testing and shipping for both types of arrays.

25.1.1.4 The LEO to Lunar Surface Solar Electric Propulsion Tug Mission

With the previous sections as background, the total mission concept can be described (O'Neill et al. 2007). The propellant depot/staging area was briefly described in section 25.1.1.2. This depot would be established using the existing stable of launch vehicles or the COTS vehicles being developed by NASA. This depot would not need the 70-130 MT heavy lift launch vehicle. This facility would contain the propellant storage and transfer facility, storage locations for materials and robotic vehicles used to prepare and load the SEP tug. The mission scenario assumes that initial launches of the SEP tug from this depot/staging area would be used to establish a lunar base for the production of propellants or other products from lunar resources. This facility can be located either at a polar location or anywhere else on the surface of the moon.

If these missions to the surface are planned carefully, product from the commercial facility can be uploaded to LLO via spare landers (assuming adequate propellant supplies can be generated and the vehicle is reusable. Furthermore, this SEP tug can also deliver satellites to lunar orbit that will beam power to any surface location. That application will be described in later sections. The exact mixture of missions to send materials to and from the lunar surface or to emplace satellites in lunar orbit has not been established. One location where hydrogen and oxygen created from water on the lunar surface can be taken is to a propellant depot in the Earth-Moon L1 libration point. This location seems to be favored as a launching point for crewed space exploration missions.

The first launch of the SEP tug mission components to LEO would require a 70 MT launch vehicle or two vehicles on the order of 40 MT. Assuming a single vehicle launch, that launch would contain the following: the SEP tug itself, the Xe fuel for the first mission to LLO (and return to LEO), the chemically-fueled lander and 22 MT of cargo that will be transferred to the lunar surface by the lander.

Upon assembly of the vehicle, it would be sent of on its first mission to LLO which will take about 257 days as shown in Table 25.1. The site of this depot is assumed to be in a 28.5° inclination. Plane changes are made at apogees near the moon to save fuel. The lander would take the 22 MT of cargo to the appropriate lunar surface location and remain there. The SEP tug would return to the LEO propellant depot/staging area over a subsequent 43 days. Thus the total mission time for this initial step in establishing a commercial base on the moon is 300 days.

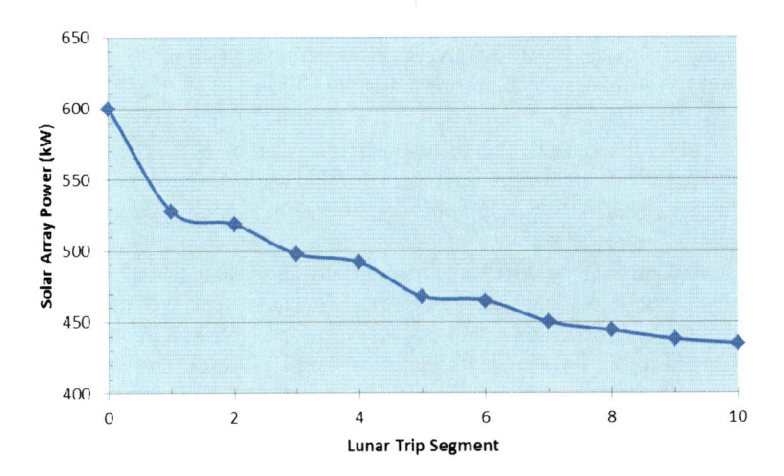

Fig. 25.10 Power Loss of SOLAROSA during Trip Segments

The solar arrays will be damaged by the transit through the Van Allen radiation belts around Earth. Figure 25.10 shows the expected degradation of the solar array over the five round-trip missions. The return segments (even numbers) show less damage because of the shorter trip time to LEO from LLO because the vehicle is lighter. If it is carrying cargo back to LLO these numbers will change. The initial outbound trip causes the greatest amount of damage to the solar array. Subsequent transits, either inbound or outbound cause less and less damage due to the logarithmic nature of particulate damage to solar cells. This damage curve is typical of today's 31% efficient multijunction solar cells in commercial production. They do not relate to future cells and their radiation damage characteristics. Those cells are expected to be substantially more damage-tolerant. The performance characteristics of the SOLAROSA arrays shown in Table 25.1 uses the maximum loss in power due to radiation damage after the fifth round trip mission for all the trips.

After the first mission, only the following materials would have to be brought to LEO from Earth: new xenon fuel (24 MT), a new fueled lander (20 MT), and the 22 MT of cargo. As noted above, these payloads could be taken to LEO with a mixture of Atlas- and Delta-class vehicles or the emerging COTS vehicles depending on the lowest cost options. They would be sent to the LEO propellant depot/staging area and integrated onto the returned SEP tug. It is clear that other activities may be taking place at this depot, hence these cargo elements may be carried on several different vehicles over a period of time.

The one year round trip mission time should not be an impediment to establishing a lunar commercial presence. Although we are accustomed to chemical transportation that can significantly reduce trip times, the costs are very high for this alternate approach. In a previous study (O'Neill et al. 2006) this mission scenario was directly compared to the same five round trips based solely on chemical propulsion systems. In that study the total launch mass and associated launch cost for delivering 110 MT of cargo to the lunar surface in support of exploration missions was calculated based on a base cost of $10M/MT. Future launch costs may be lower. For the conventional chemical system, the rocket equation was used to estimate propellant mass for the LEO to LLO trip, assuming a specific impulse of 400 sec (the same as for the chemical lander rocket in Table 25.1) and a ΔV of 4 km/sec (half the value for SEP in Table 25.1). The chemical-based mission scenario launch-only cost was $6.8B based on mass assumptions only while the SEP SOLAROSA tug mission scenario, launch-only costs were $3.5B for a saving of over $3B.

Thus the advantages of the SEP tug mission are very clear and it is easy to understand how such a vehicle can become a workhorse for the Earth-Moon vicinity. A wide range of commercial and government mission scenarios can easily be developed for an infrastructure that relys on SEP vehicles as the baseline carrier. To name a few: large debris removal (e.g. spent upper stages), removal of GEO satellites to disposal orbits, maintenance (e.g. refueling) and repair on satellites equipped for such servicing.

With this baseline delivery system established, we will next move on to discussing the means of delivering power to any part of the lunar surface from orbiting vehicles. The emplacement of these vehicles is enhanced by the use of the SEP tug and infrastructure described above.

25.2 Lunar Power Beaming Using Lasers

One of the key issues that will affect the lunar exploration and commercialization is the ability to provide electric power to any surface location. This power should be available through daylight times and especially during the extended lunar night times. Although nuclear reactor power systems and radioisotope power sources are options for continuous power, those options will not be discussed here. It is the purpose of this chapter to explore the advantages of a solar-powered, electrically propelled spacecraft to support the delivery of cargo to the lunar surface. This scenario was described above. A second capability of such a vehicle is that it can also emplace solar-powered spacecraft into orbits where they can use laser beaming to provide power to any location on the lunar surface. Thus such a vehicle or vehicles can meet several diverse mission needs and do it for less cost than can conventional chemically-fueled mission scenarios.

As described above, the major benefit of a SEP vehicle is that it can deliver more payload mass to any lunar location for less cost than one using chemical propulsion. In addition, SEP allows orbital adjustment to permit a range of orbital characteristics to fit the application. One disadvantage is that it takes longer to reach the moon, but this is not a limiting factor for this power beaming mission. Of course careful planning of launches, cargo transfer and uploading and staging of the vehicles is essential.

Although the baseline mission was to deliver 22 MT to the lunar surface once per year, other capabilities can be realized with a different vehicle but with smaller payload capability. The ability of this vehicle to deliver 42 MT to LLO far exceeds the needs of other mission scenarios – such as satellite emplacement around the moon.

25.2.1 Lunar Orbital Vehicle Characteristics

Because of the capability of the SEP tug (46 MT to LLO), it is quite feasible that several power beaming satellites could be emplaced in a single mission. However, no detailed design of the laser beaming spacecraft was attempted. However, for a nominal satellite of 5-6 MT such a large vehicle could carry seven or eight to LLO. This seems to be rather excessive. Therefore, instead of using the SEP tug described above (or even a vehicle half or one-fourth that size); a second option was chosen (Brandhorst et al. 2006).

In this option, the entire beaming spacecraft is launched into LEO at the depot where it is fueled with Xe. The satellite has 100 kW of SOLAROSA arrays, multiple laser beamers operating at wavelengths in the 0.800 – 0.900 μm wavelength range. These arrays have a specific mass in excess of 300 W/kg hence weigh less than 333 kg. Multiple lasers are used to keep power transmission distances short and thermal issues distributed and manageable.

The 5,000 kg conceptual spacecraft uses multiple Hall thrusters each able to operate at power levels between 10 and 50 kW. At maximum power, the thrusters will produce a total thrust of ~5 N. The mission begins at an altitude of ~500 km with thrusting occurring only during the sunlight period of the orbit in the tug case. This spacecraft is estimated to weigh about 5000 kg and could easily be sent to LEO by the Delta 2 (7920/5) launch vehicle. For this case, the total mission time to the moon was about 89 days beginning at a 28.5° inclination. Lower radiation damage will take place to the solar arrays due to the shorter duration in the Van Allen belts.

25.2.2 Lunar Orbital Results

Many options exist for orbits around the moon and the selection depends upon the application. In this case, an equatorial Molniya-type orbit with an apogee of 30,000 km and a perigee of 500 km meets the desire to provide power to locations near the equator. In this orbit, the satellite can beam power to locations within 45° north (N) and south (S) of the equator when the satellite is in sunlight and the site is in darkness. Obviously if the site is in sunlight, laser power beaming won't be necessary. A previous study examined both microwave and laser beaming from the L1 and L2 positions (Little and Brandhorst 2004). Those results showed power beaming over that distance was not feasible for microwaves and only marginally effective with lasers. This section will examine the benefits of much lower lunar orbits and two beaming satellites.

A wide range of potential orbital parameters exists for an equatorial satellite that can view locations between 45° N and S. An initial survey was performed using Satellite Tool Kit® v.7.1 that encompassed circular and elliptical equatorial orbits with apogees as high as 55,000 km. Based on examination of these orbits, an orbit of 500 km perigee and 30,000 km apogee was chosen because it was the best match of long coverage times to multiple lunar surface sites over a two year period and a somewhat short time with no coverage whatsoever. The orbital condition required that there would be laser power beaming when the site was in shadow and the satellite was in sunlight. Of course, beaming when the site and the satellite were illuminated is possible but not particularly feasible. This is so because the solar array on the surface site must track the sun during the day, and the beaming satellite is not in a sun-synchronous orbit, so the surface array won't be able to receive power from the satellite except intermittently. Thus we set the guideline that there is no power beaming done when the site is in sunlight.

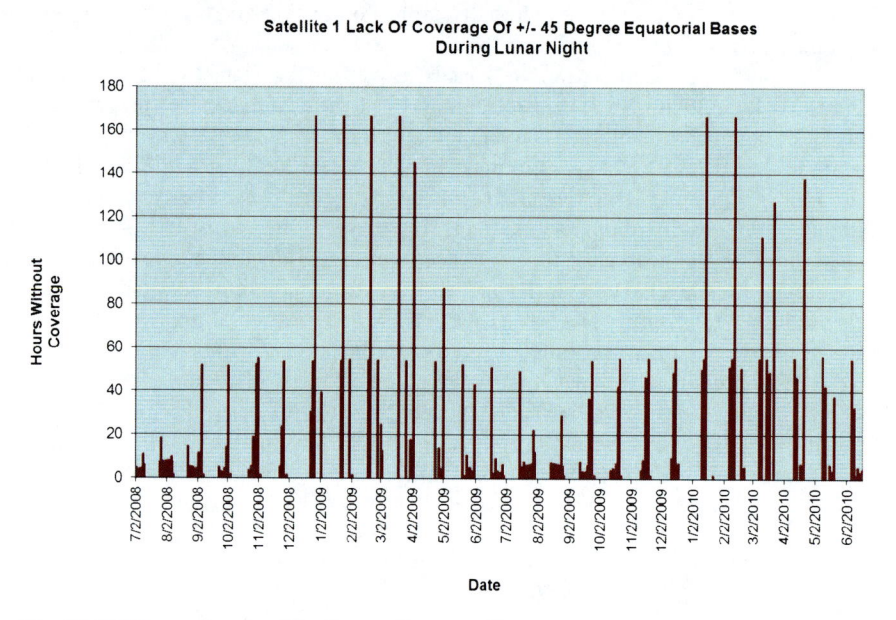

Fig. 25.11 Times when Satellite Cannot Beam to Site

An arbitrarily start time of July 1, 2008 was used and continued the mission through June 30, 2010. Figure 25.11 shows the hours without coverage for sites at ±45° N and S from the equator for a single satellite. There was little difference between North or South locations, so the site at 45° N will be used for the analysis. Of course, sites closer to the equator will receive more coverage, but the details won't be presented here.

Inspection of this chart shows the maximum duration with no coverage by this one satellite for these sites to be 164 hours or nearly 7 days. This would require a massive storage system, assuming the night-time power requirement is the same as in day-time. Over the two year period, there were only ten times when the dark time exceeded 84 hours.

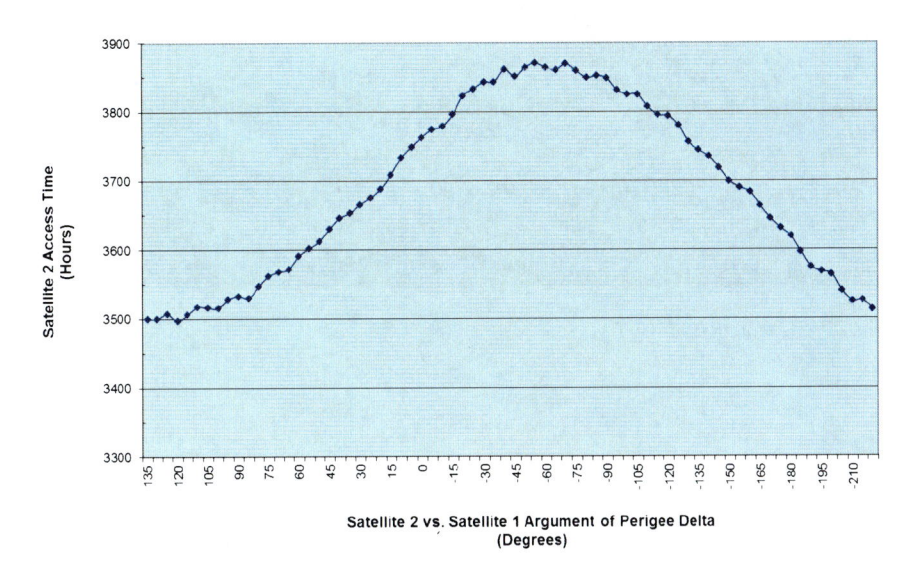

Satellite 2 Base Access Time vs. Satellite 1 Argument of Perigee Delta

Satellite 2 vs. Satellite 1 Argument of Perigee Delta (Degrees)

Fig. 25.12 Satellite 2 Access Times with Satellite 1 Fixed

In order to reduce the dark period length, a second satellite was added. However, one of the critical issues is the angular relationship between the first satellite and the second one. Accordingly, a plot was made of access time of the second satellite to the 45° N site as shown in Fig. 25.12. This figure indicates that the total access time ranges from 3500 hours to about 3870 hours when there is a 55° offset between the two. Configuration of the satellites is shown in Fig. 25.13.

This analysis showed that the times when no satellite is able to cover the lunar sites at 45° N had dropped dramatically. The maximum time without coverage is 84 hours (3.5 days) as shown in Fig. 25.14. Furthermore, there are only eight of these periods over the two-year span. In addition, in many cases the duration drops to 54 hours (2.25 days). Both these options represent significant mass savings for a lunar site. It should be remembered that sites closer to the equator will have shorter "dark" times.

A plot of both the two satellite access times as well as the maximum duration of any single dark time for elliptical, equatorial orbits with apogees up to 55,000 km showed that the 30,000 km apogee condition is appropriate for this mission to generally achieve the best of both options. The dark times have been substantially reduced – to generally less than three days.

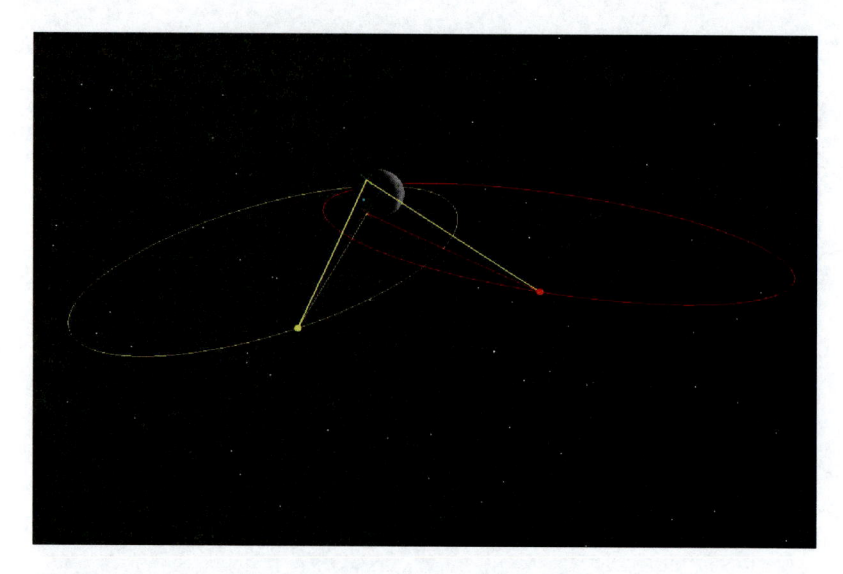

Fig. 25.13 Depiction of the Two Satellite Orbits

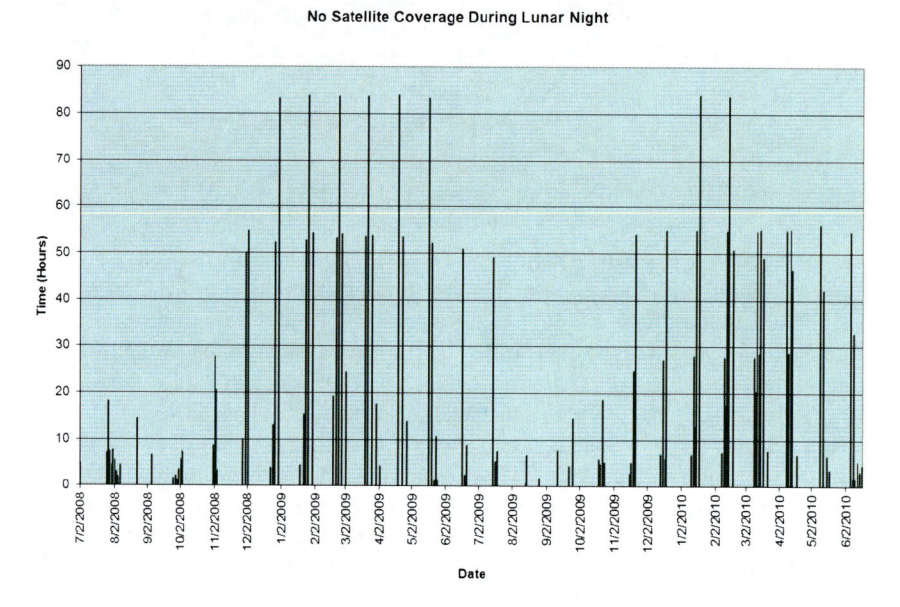

Fig. 25.14 Times when Neither Satellite Views Site

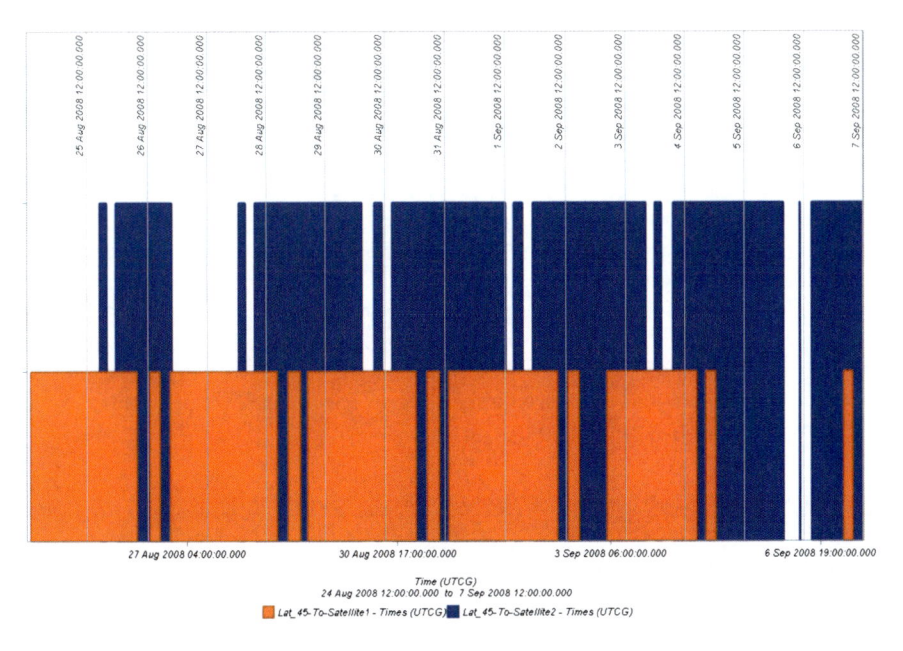

Fig. 25.15 Two Satellite Overlapping Coverage

In addition, there are many times when both satellites view the site over the same time period. This suggests the option of having increased power at these times for increasing the activities at the site. However that would require use of two separate arrays on the surface. Although it isn't certain what activities would be conducted at night, it would seem reasonable that propellant production from the regolith or water from the dark craters could easily take advantage of this "bonus" in power. Figure 25.15 shows a period where there is a major coverage of the site by both satellites. The exact amount of power delivered will depend on the orbital conditions of each satellite as well as the base location. This chart also shows that there are some very short periods of darkness as shown above in Fig. 25.14.

25.2.3 Solar Cell Selection for the Lunar Surface Receiving Array

Before discussing the results of laser power beaming, the type of solar cell to be used in the lunar surface array must be discussed. Although the high efficiency of the multijunction solar cells is attractive, physics prevents their use. For the multijunction solar cell to function correctly, all of the junctions must be illuminated. Some layers will receive more light, and hence create more potential current, but because the layers are connected in series, the junction with the least amount of current will control the output of the device. Because lasers emit a monochromatic beam, multijunction solar cells cannot be used as at least one of the junctions will not receive any photons and in effect, be "in the dark". Thus the output current of such a cell will approach zero.

There are two types of single junction solar cells that would be suitable as laser receivers: silicon and GaAs. Of course, with modern processes for solar cell fabrication, new devices with a mixture of Group III and V materials can be produced. The only requirement will be that the maximum power point and the wavelength of the laser at the cell operating temperature be in harmony. There are also two lasers that fit one or the other of these cells: a 1.06 μm laser and a solid state laser in the 0.800 to 0.900 μm wavelength range. Both these laser wavelengths have been suggested for beaming from GEO to Earth (Brandhorst and Forester 2003, Brandhorst and O'Neill 2002) as there are windows through the atmosphere in these bands. The 1.06 μm wavelength will only work with the Si cell, and it is close to the band edge where absorption in the cell is not ideal. Additionally, the band gap of Si is about 1.11 eV. This low band gap results in a high temperature coefficient of power and the cell voltage is low. These factors work against Si as the best choice for this application.

ENTECH SLA Using
EMCORE GaAs Cell
Tested Outdoors
September 12, 2005

Fig. 25.16 SLA Module with GaAs Cells

The laser version of the SLA uses single junction GaAs solar cells instead of the multijunction solar cells for the reasons noted above. For infrared laser light in the wavelength range of 0.800-0.850 μm, GaAs is an ideal photovoltaic cell. To demonstrate the achievable conversion efficiency of the laser version of SLA, a number of prototype SLA moduleswere developed. Figure 25.16 shows one of these SLA modules during outdoor testing under sunlight irradiance. Of course, under sunlight irradiance, the SLA performance is significantly lower with single-junction GaAs cells than with triple-junction GaInP/GaAs/Ge cells. The net SLA/GaAs module efficiency as tested in natural sunlight of 22% is much lower than the typical 30% net SLA module efficiency measured under terrestrial sunlight with a triple-junction cell, but that was expected because the GaAs cell has only a single junction.

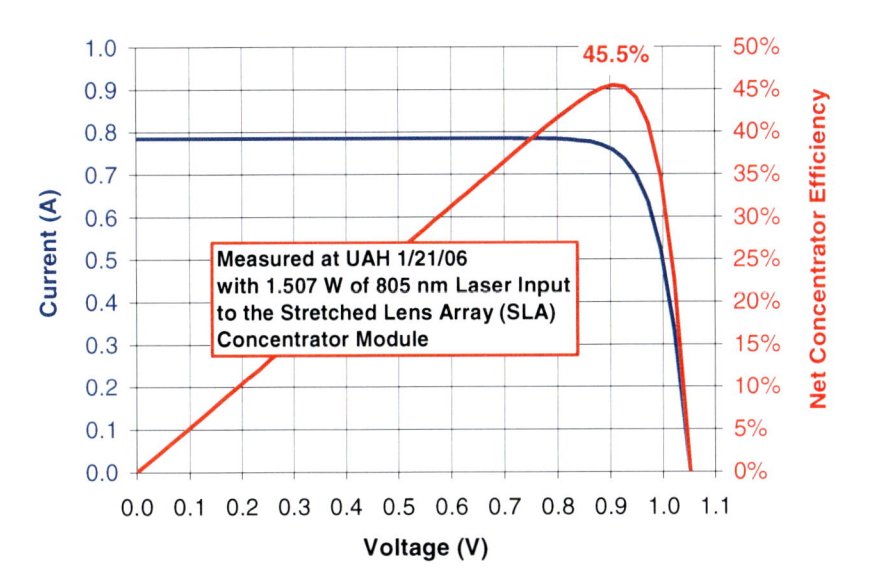

Fig. 25.17 SLA/GaAs Module under Infrared Laser Illumination (1.597 W @ 805 nm)

This SLA module equipped with GaAs cells was then exposed to a laser. The shape of the I-V curve was the same as sunlight testing, but that the overall net SLA module conversion efficiency had more than doubled for the monochromatic laser light compared to sunlight. The module reached an efficiency of 45.5% with a laser operating at 0.805 μm as shown in Fig. 25.17. The 45.5% net SLA module conversion efficiency corresponds to the product of about 50% cell efficiency times about 92% lens optical efficiency. The 50% cell efficiency corresponds to an operating cell temperature only slightly higher than room temperature, because the measurements were made indoors in an air-conditioned room. The much higher SLA module efficiency under laser irradiance is due to the module's much better current response for laser light (about 0.52 A/W) than for solar irradiance (about 0.25 A/W), since the laser wavelength is very near the peak spectral response of the GaAs cell. Previous results[7] also indicate that the peak efficiency of the GaAs solar cell occurs at approximately 840 nm for a cell at room temperature.

The lens was removed from another lens/cell combination and laser measurements were made on the prism-covered GaAs cell directly. Without the spiked irradiance profile produced by the mini-dome lens, the cell alone achieved much higher performance levels, with conversion efficiency above 56% with a laser operating at 0.835 μm. The fill factor improved to 84-85% for the I-V curve, even at a higher current level than for the solar test. This 56% cell conversion efficiency is very good, considering the age (16 years), vintage (1988), and intended use (solar) of the cell in this test (O'Neill 2004).

These tests suggest that the maximum efficiency for conversion of laser light to electricity with GaAs cells can be as high as 60% with several improvements. The antireflection coating on the GaAs cell can be adjusted to completely eliminate any reflections at the laser wavelength. This would add about 3% to the output. Additionally, increasing the laser wavelength to about 0.850 μm will increase performance another 5%. It must be kept in mind that this 8% increase just balances the 8% loss in the stretched lens, so a laser module overall efficiency of 55% seems reasonable.

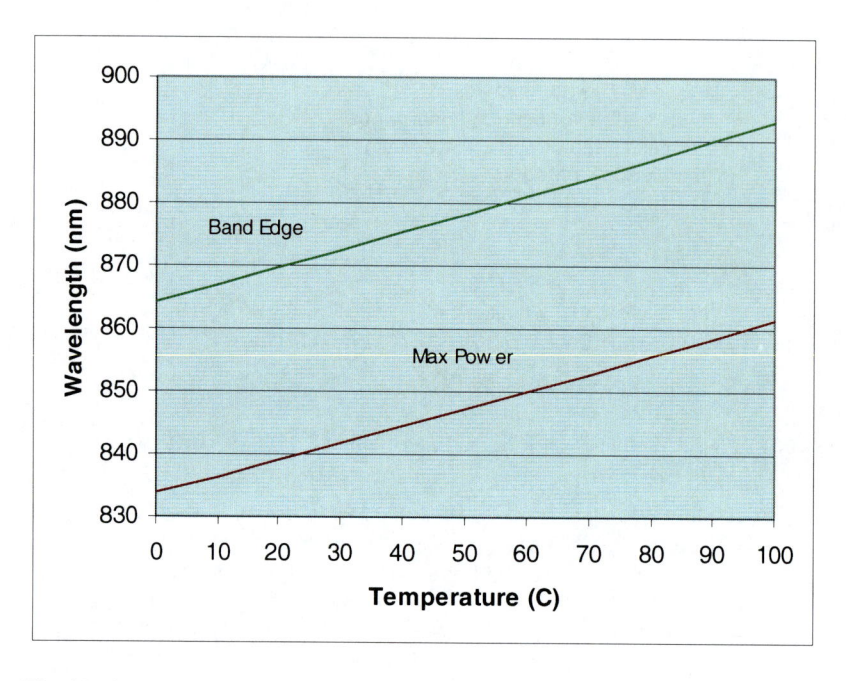

Fig. 25.18 Calculated GaAs Solar Cell Performance

The operating temperature of the GaAs cell on the lunar surface will be an important factor in determining the best laser wavelength. At an operating temperature of 60°C, the peak power would occur at 850 nm wavelength as shown in Fig. 25.18. If the operating temperature reaches 90°C, the peak power wavelength shifts to 860 nm. These calculations show that testing to date has not reached the full potential efficiency of a well-designed cell and system. It is also an analysis that must be performed if cells made from Group III-V materials are created.

Because the GaAs cell will operate at high efficiency under laser irradiation, the exact operating temperature is expected to be lower than for conventional cells responding to sunlight. However, the expected operating temperature on the lunar surface will be dependent on the mounting and the laser beam intensity. For example, Olsen et al. (1991) demonstrated a conversion efficiency of 52% at 25°C with a laser intensity of 100 mW/cm^2 and a wavelength of 0.806 μm.

25.2.4 *Laser Power Beaming Results for Equatorial Lunar Locations*

With the previous test results that show that efficiencies above 45% can be achieved with GaAs solar cells and laser wavelengths in the 0.800-0.900 μm wavelength range we can now turn to the issues of beaming from a satellite to the lunar surface. Laser beaming from the satellite in the elliptical orbits described above yields a complex beam pattern and intensity. For the 45° N site, the solar arrays will be tilted at the latitude. The satellite will rise over the horizon and move about 180° in azimuth. At the same time it will increase in elevation to 45° as shown in Fig. 25.19. This chart provides the combined beam incidence angles for this site. Because the moon is only slightly tilted on its axis, these angles will remain essentially constant. It is assumed that the surface array would track the source in one axis as it moves in azimuth. This will fulfill the need to have a ±2° angle of incidence on the lens in azimuth. With proper design of the array, no north-south adjustment will be necessary.

As the satellite rises over the horizon, it will be at a low, ~500 km altitude. The laser beam will be very intense and the coverage area will be small. As it moves toward its apogee of 30,000 km, the beam intensity will drop and the beam area will expand. Figure 25.20 shows the variation of laser beam intensity in terms of equivalent AM0 sunlight from a 4 m^2 transmitter aperture. This chart is for a three day period in the month of August, 2009. In this calculation, a laser power level of 40 kW, and a wavelength of 850 nm, which is nearly ideal for GaAs cells has ben assumed.

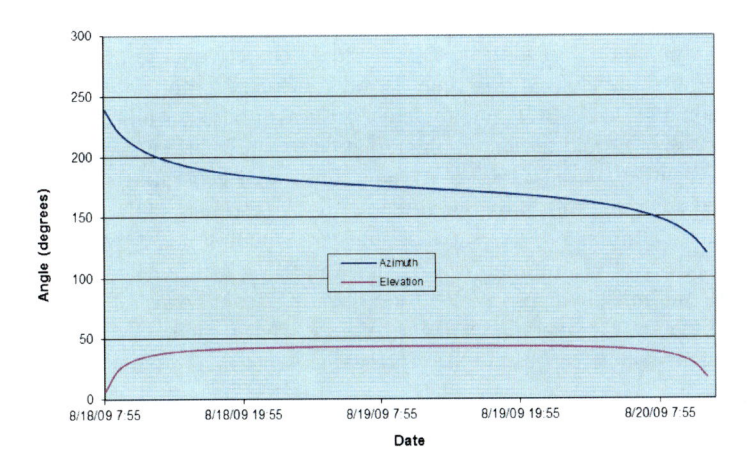

Fig. 25.19 Laser Beam Incidence Angles

It is important to note that the intensity of the laser beam is only about 0.2 equivalent suns over most of its traverse and peaks at only 2 "suns" for a very short time. Thus temperatures of the GaAs solar array on the lunar surface are

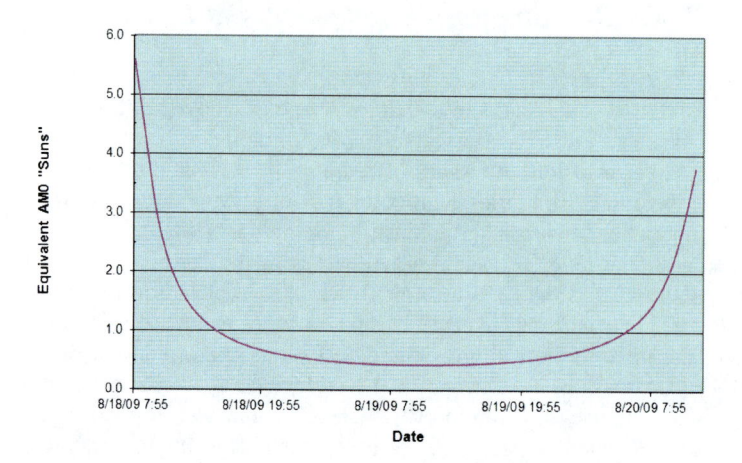

Fig. 25.20 Equivalent Solar Intensity of Laser Beam

expected to be within the bounds shown in Fig. 25.18. A second beam with a planar array could increase power to the site without causing array temperatures to rise too high.

As the satellite moves toward its apogee, the diameter of the laser beam will increase. If we assume a 60 kW surface Stretched Lens Array that uses GaAs cells, we can compute the maximum diameter of the laser beam relative to that area as shown in Fig. 25.21. Thus in this case also, the 30,000 km apogee does not cause the laser beam to exceed the area of the nominal 60 kW solar array.

Finally, what is the amount of laser power that could be delivered to the site at 45° N from a single satellite over this time period? As noted before, the power output of the nominal 100 kW BOL satellite solar array will be determined by the transit time to this orbit around the moon, the radiation damage suffered while traversing the Van Allen belts around the Earth, the efficiency of generating the laser beam, and the thermal conditions of the satellite in orbit. An estimate of the amount of laser power that this initial 100 kW satellite could provide will include a worst-case 10% loss in power traversing the Van Allen belts, the laser beam production efficiency of 50% and mirror losses of 12% for a resulting beam power of 40 kW. This assumes that the array is operating at a normal deep space temperature.

In addition, the characteristics of the Stretched Lens Array on the lunar surface will be important. Because the power will be beamed to the surface at night, the solar array will be cold initially. As the laser beam intensity reaches a maximum level of 0.2 equivalent suns, the array temperature will rise to no more than 28°C depending on surface features, tilt angle and other considerations. We are assuming a 45% conversion efficiency of the laser beam (Brandhorst and O'Neill 2002) as determined by tests on GaAs solar cells at various temperatures. This is a conservative value and can be increased by the cell design improvements described above.

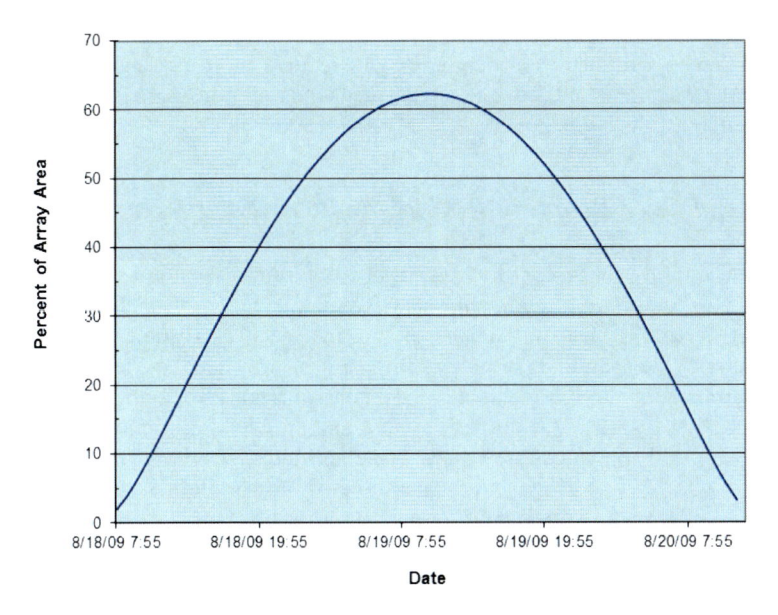

Fig. 25.21 Percent of 60 kW GaAs Solar Array Area

Given these assumptions, Fig. 25.22 shows the power that could be delivered to the lunar site for a 50% conversion efficiency of satellite power to laser beam. This also is a conservative value as the efficiency of laser beam production in solid state devices is continually improving. In this case, 18 kW could be supplied to this surface site. Beaming from a second satellite as shown before will further increase this power level as noted above if a planar array is used. Thus, one single satellite can supply 30% of the daytime power at night.

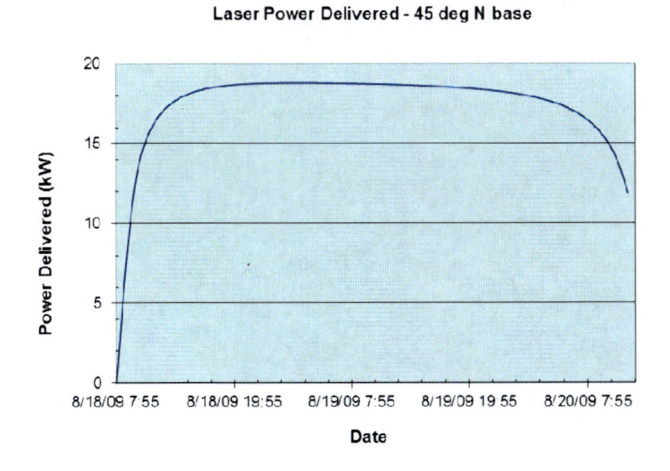

Fig. 25.22 Laser Power Delivered to a 45° N Site

These calculations represent a "worst case" to a site located at either 45° N or S. Any other site within those boundaries will receive more energy over the same period of time. Let's next examine delivery of laser power to lunar sites located from either of the poles down to 45° N or S.

25.2.5 Laser Power Beaming Results for Polar Locations

All of the assumptions made for the equatorial laser power beaming case still apply to this polar case. A similar approach was taken to produce the results (Brandhorst et al. 2007). The initial part of the analysis was to determine the orbital parameters. Two satellites in polar elliptical orbit was used as the baseline for the polar application. Using the same analysis as in Sect. 25.2.2 above, these satellites were offset by ~180°. A range of orbits ranging from 500 x 5,000 km to 500 x 30,000 km. All these results will be presented later. However, the 500 x 5,000 km orbit with ~7.5 hr orbital time gave the most power and the least times without power – albeit when two satellites were in orbit. For both satelliltes, their apogee was over the south pole. The same will be true for satellites with apogees above the north pole. However, for a single satellite, this orbit produces the least viewing time.

For the rest of the system, the laser operated at 0.850 μm laser with an aperture of 1.5 m^2 (1.38 m dia). This condition increases beam size on the lunar surface as compared to the equatorial case. Because of the 180°offset between the two satellites, there were only a limited number of times when the lunar array could see both satellites. Therefore the surface system is a single junction GaAs tracking array. An alternate option is to use a fixed array. However this option will reduce the power delivered to the surface. It should not be forgotten that these beaming satellites are not constrained to supply power to one single location, but multiple sites.

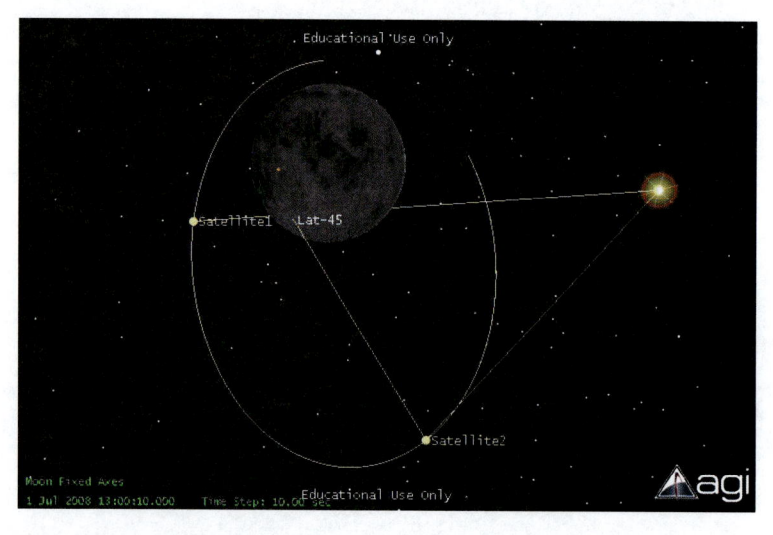

Fig. 25.23 Polar Orbital Depiction

An example of this orbit is shown in Fig. 25.23. In this depiction, a single satellite is beaming to a 45° S location. Thus management of power delivery to the surface will require substantial organization and management of orbital resources. Figure 25.24 shows the orbital conditions of each of the two satellites with both azimuth and elevation results for the elliptical orbits with perigee at 500 km and apogee at 5,000 km. With two satellites, this orbit produced the highest power to the surface for the widest range of locations. It is also important to note that these orbits are some of the so-called "frozen orbits" that are exceptionally stable with time.

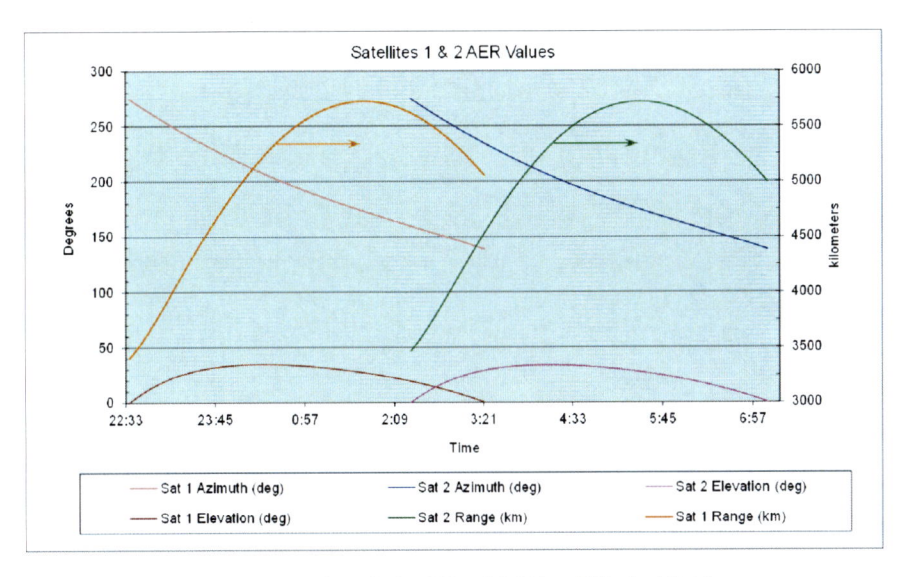

Fig. 25.24 Satellite Parameters for a Polar 500 x 5,000 km Elliptical Orbit

Aximuth angles for either satellite varies from 135° to 275° and the elevation changes from zero degrees (the horizon) to about 35°. Thus the single tracking array will have about 140° angular tracking with minimal elevation changes. In fact, a single axis tracking system with a fixed 15° tilt would lose only about 4% of the maximum power over the total elevation span.

The range to the satellites varies from about 3,400 km to 5,700 km. This occurs because the perigee of the satellite is essentially opposite the site. With these parameters, there is not a period of more than 15 – 20 minutes where the base (when in the dark) is not covered by at least one satellite. As the base transitions from shadow to sunlight, there is a period where, when in penumbra, there is no access for approximately 1.5 hours. Overlapping coverage peaks out at about 1.4 hours.

As can be seen in Fig. 25.25, polar orbits give excellent access times to lunar sites from the pole to ~30° N or S. As might be expected, the view time of the satellite with the 5,000 km apogee is the lowest time for a single satellite as would be anticipated. This requires the deployment of the second satellite. Both satellite access times are comparable and, as expected, the access time depends on satellite altitude.

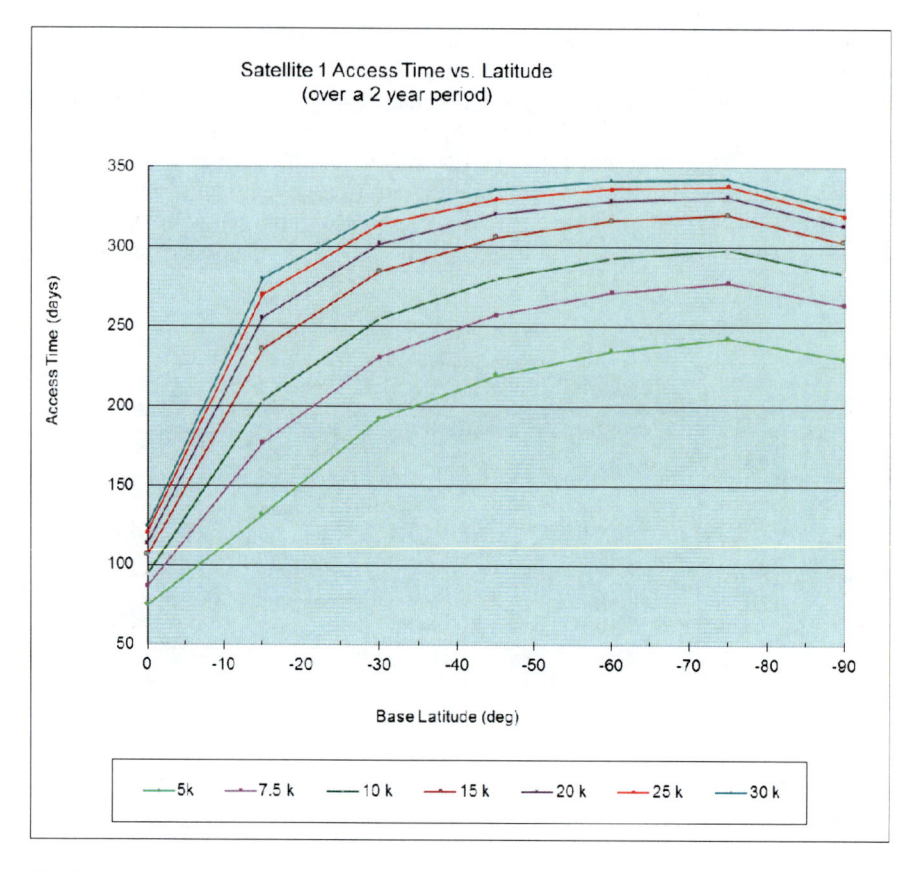

Fig. 25.25 Access Times for Various Orbits

The deployment of the second satellite ensures almost total coverage – the time with no laser illumination drops to only 1.5 hours. The higher apogee provides more access but the longer beam distance reduces power received. For sites from either pole to 45° N or S, access times for a single satellite reach form 320 to 340 days for apogees 20,000 km to 30,000 km.

As before, the first factor that is vital to determine is the laser beam intensity received at the surface. Figure 25.26 shows these results. For this condition, the peak intensity is just over 1.9 suns. Althugh higher than hoped for, it is still within the operating temperature range of the GaAs solar cells in the SLA tracking array. The total beaming time is about five hours out of the 7.5 hour orbital period.

The second factor is to determine if the laser beam size exceeds the area of the array. Figure 25.27 shows that this is not an issue and only 40% of the GaAS array area is covered.

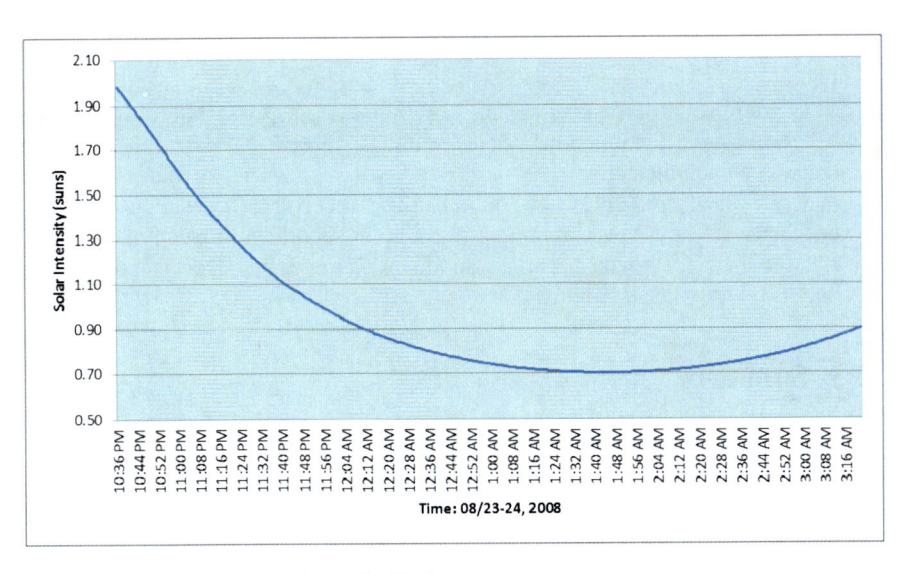

Fig. 25.26 Laser Beam Intensity at the Surface

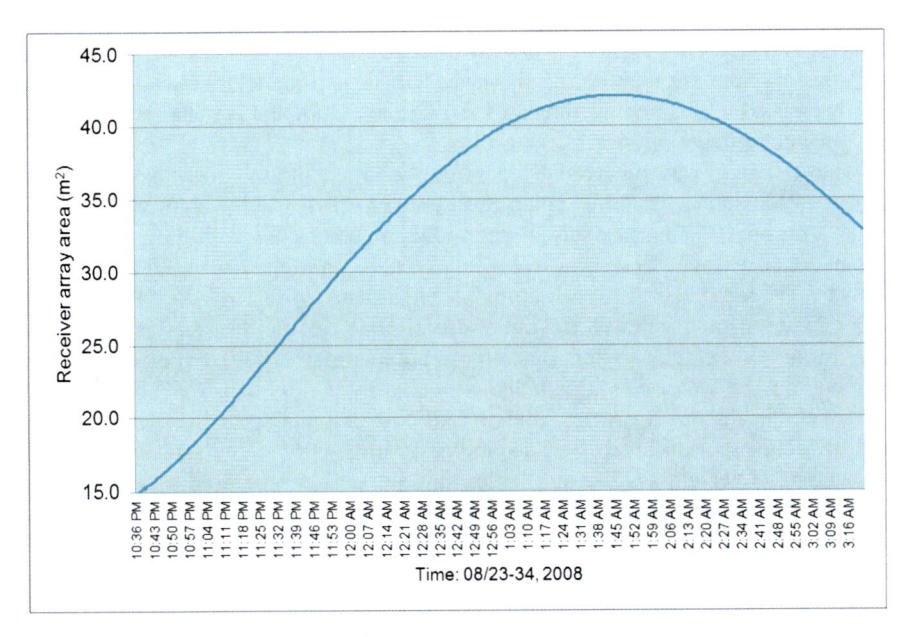

Fig. 32.27 Beam Coverage for 62 m² GaAs Array

With a tracking array, power to the surface is essentially constant at ~16.8 kW per satellite. This results from a 50% conversion of power on the satellite into the laser beam. Using some of the conservative numbers given above for the conversion of laser light into electricity of 45% and it includes other losses as well in the power processing chain. The GaAs SLA surface array is 62 m² in area.

Thus, except for the extreme poles this system will provide adequate power for the night times. The surface array used will produce about 15 kW in sunlight, so a virtual constant source of power for any site from the pole to 45°N or S is feasible. Neither receiving array area nor laser beam intensity is excessive and the beaming parameters can be adjusted further.

Finally, with two satellites, the longest time a receiver at 45° does not receive power is only 1.5 hours maximum, less for a polar site. This *substantially* reduces storage and in effect makes storage an insignificant issue. Hence laser power beaming is a very plausible option.

25.3 Summary

An architecture for transporting large quantities of mass (22MT) to the lunar surface at an affordable cost using a reusable space tug has been described. Although detailed cost studies have not been completed, it is apparent that unique features of this architecture have eliminated the major high cost barrier of previous studies. The first feature is a reusable space tug powered by solar electric propulsion that makes one round trip per year to the moon. It can operate for five years without solar array replacements. This architecture replaces the direct ascent, chemically-boosted non-reusable approach to deliver cargo to the moon. The second feature that this architecture may not need a heavy lift launch vehicle – even though one has been used in this architecture. A LEO fueling depot and staging area provides the final component of the architecture.

The SEP tug was powered by a refractive concentrator array based on the stretched lens concept and is called SOLAROSA. It is a lightweight, high power solar array when using present-day multijunction solar cells with 33% efficiency. It can achieve a specific power over 400 W/kg, or an alpha (α) of 2.5 and above. This can be compared to today's common satellite arrays at 80 W/kg ($\alpha = 12.5$). It can operate at high voltages to "direct drive" Hall electric thrusters and the concentrating lens serves as a micrometeoroid bumper also. The radiation loss in traversing the Van Allen radiation belts will certainly be less than 10% because the solar cells in this array are 1/8th the area of conventional space solar cells due to the 8X concentration level. Because the cells are small in area and the radiating area large, cover glasses that are larger than the cell can be used to seal the exposed surfaces and prevent arcing at high voltages. Careful design of the radiator area and the high efficiency of the cells help keep the array close to conventional GEO arrays in size.

When implemented, it is likely that this infrastructure will lead to other commercial undertakings such as retrieval, repair and refurbishing of existing satellites, cargo delivery to the L1 and L2 libration points and a staging base for interplanetary missions that use solar electric propulsion. The use of solar electric propulsion systems for this endeavor will likely usher in a new era in space transportation and enable existing missions to be conducted at much reduced costs.

Also included was the use of this solar electric propulsion vehicle to emplace satellites in lunar equatorial orbit that would then beam laser power to bases at 45° N or S of the equator. These two satellites in lunar equatorial orbit can provide substantial power to these bases depending on the total power available on the satellites. A nominal power level of 100 kW was assumed. It was necessary to carefully space the satellites to achieve maximum power beaming time. Over a two year period, there were only eight times when the lack of view time of the 45° N location reached 84 hours (even though the satellites were receiving solar energy). Over the rest of the two year mission, the times when the satellites did not view the sites dropped to less than 54 hours. Thus the need for energy storage on the lunar surface drops dramatically with this approach.

The surface array was assumed to be made of GaAs cells in a Stretched Lens Array called SOLAROSA with a surface power in the day time of 60 kW. The power delivered by one satellite was 18 kW and there are many times when both satellites are in view of the site. This offers the opportunity of further increasing the power to the site; however the surface array would have to be a non-concentrating array. With this planar array, the power delivered to the surface could then double to 36 kW.

The laser used in this application was tuned to the 0.820 to 0.870 μm range to maximize the efficiency of the GaAs cells. The wavelength range is important because the band gap of GaAs shifts with temperature and to achieve maximum performance the wavelength must be matched to the maximum power point of the cell at its operating temperature. Direct measurement of the efficiency of GaAs cells illuminated by laser wavelengths in the 0.805 and 0.835 μm ranges showed efficiency of 46.6% even though the cells were designed for sunlight operation. Changes in the cell design, cover glass and other aspects were projected to lead to efficiencies above 50% in the near term.

A laser efficiency of 50% conversion of electric energy into laser energy was used. However, advances in the efficiency of diode lasers in this wavelength range are expected to make this option increasingly attractive for providing power to the lunar surface.

In order to power the other areas of the moon, a second study was performed using two satellites in polar lunar orbit. These two satellites were displaced from each other by 180°. With a tracking surface array, the times when the site(s) are not receiving power from the laser beam drops to only 1.5 hours. This significantly reduces the need for energy storage and is a great mass savings. These satellites were in 500 x 5,000 km orbits, and were thus able to maintain virtually constant power to any site from either pole to ±45° N or S. Thus four satellites could provide power to any lunar location with minimal need for storage. The other benefit of these polar orbits is that they are so-called "frozen orbits" whose parameters don't change due to gravitational influences thus minimizing the need for substantial amounts of propellant. In this case about 17 kW could be sent to the surface continually. Depending on the need and location of the surface sites, these elliptical orbits can also cover the back side of the moon and help provide communications to those locations as well as providing power needed at those sites.

Thus it appears possible to provide power to virtually any lunar location at any time of the year with a reusable, cost-effective architecture. The future will determine how useful the moon will be in space exploration missions, but the architecture described herein will offer significant benefit to both exploration and commercial efforts that may result. With reasonable costs, a wide range of comercial efforts will be undertaken and will transform our use of the space environment.

References

Brandhorst, H.W.: A Cost-Effective Architecture for Delivering Space Solar Power Systems. In: 26th International Symposium of Space Technology and Science, Hamamatsu City, Japan, June 1-8 (2008)

Brandhorst, H.W., O'Neill, M.J.: In-Space and Terrestrial Solar Array Technologies for Beamed Laser Power. In: World Space Congress, paper #IAC-02-R.2.08, Houston, TX, USA (2002)

Brandhorst, H.W., Forester, D.R.: Effects of the Atmosphere on Laser Transmission to Gas Solar Cells. In: 54th International Astronautical Congress, paper #IAC-03-R-3.08, Bremen, Germany, September 29-October 3 (2003)

Brandhorst, H.W., Best, S.T.: Hypervelocity Impact Testing of Solar Cells in a Plasma Environment. In: 9th Spacecraft Charging Technology Conference, Tsukuba, Japan, April 4-8 (2005)

Brandhorst, H.W., Rodiek, J.A., Crumpler, M.S.: A Solar Electric Propulsion Mission for Lunar Power Beaming. In: 57th International Astronautical Congress, paper #IAC-06-C3.4.2, Valencia, Spain, October 2-6 (2006)

Brandhorst, H.W., Rodiek, J.A., Crumpler, M.S.: A Solar Electric Propulsion Mission with Lunar Power Beaming. In: Rutgers University Symposium on Lunar Settlements, June 4-8 (2007)

Futron Corporation Report. Space Transportation Costs: Trends in Price per Pound to Orbit 1990-2000 (September 6, 2002)

Hrbud, I., Lapointe, M., Vondra, R., Dailey, C.L., Lovberg, R.: Status of Pulsed Inductive Thruster Research. In: El-Genk, M.S. (ed.) AIP Conference Proceedings of Space Technology and Applications International Forum – STAIF 2002, Albuquerque, NM, February 3-6, vol. 608, pp. 627–632. American Institute of Physics, Melville (2002)

Little, F., Brandhorst, H.W.: An Approach for Lunar Power – 24/29. In: 4th Int. Conf. on Power from Space SPS 2004 Together with the 5th Int. Conf. on Wireless Power Transmission WPT5, Granada, Spain, June 30-July 2 (2004)

Murphy, D.M.: The SCARLET Solar Array Technology Validation and Flight Results. In: Deep Space 1 Technology Validation Symposium, Jet Propulsion Laboratory, Pasadena (2000)

NSSO, Report to the Director, National Security Space Office: Space-Based Solar Power as an Opportunity for Strategic Security (2007),
http://www.nss.org/settlement/ssp/library/
final-sbsp-interim-assessment-release-01.pdf
(released October 10, 2007)

Olsen, L.C., Huber, D.A., Dunham, G., Addis, F.W.: High Efficiency Monochromatic GaAs Solar Cells. In: 22nd IEEE Photovoltaic Specialists Conference Record, Las Vegas, NV, USA, pp. 419–424 (1991)

O'Neill, M.J.: Laser Power Transmission Employing a Dual-Use Photovoltaic Concentrator at the Receiving End, Phase I Final Report, NASA SBIR contract No. NNM04AA80C; Phase II Final Report, NASA SBIR Contract No. NNM05AA11 (2004)

O'Neill, M.J., Piszczor, M., Brandhorst, H., Carpenter, C., McDanal, A.J.: Stretched Lens Array (SLA) Solar Electric Propulsion (SEP) Space Tug: SLA-SEP Offers Multi-Billion Dollar Savings Delivering Lunar Exploration Cargo. In: 4th World Conference on Photovoltaic Energy Conversion, Waikoloa, HI, May 7-12 (2006)

O'Neill, M.J., Piszczor, M., Brandhorst, H., McDermott, P., Lewis, H.: The Stretched Lens Array (SLA) for Solar Electric Propulsion (SEP). In: 20th Space Photovoltaic Research and Technology (SPRAT XX) Conference, Cleveland, OH, September 25-27 (2007)

26 Building the First Lunar Base – Construction, Transport, Assembly

Werner Grandl

Consulting architect, Tulln, Austria

26.1 Introduction

Since the space age began less than 60 years ago, the exploration of our moon has been one of the prime goals of space technology and astronautics. If we decide to expand human civilization into space and to use the resources of the solar system, the establishment of a permanently occupied lunar base should be the first step. Prospecting and exploitation of lunar resources such as Helium-3 and scarce materials on a cost- effective basis could be a test bed for further space enterprises in the solar system up to moons of Jupiter. Among various possibilities of scientific research, such as geology, artificial ecosystems and human physiology and space medicine, the moon is an ideal location for optical, infra-red and radio astronomy. Last not least the effort to establish a lunar base would start a tremendous economic stimulus for terrestrial economies (ESA 2003).

During the last decades various proposals for lunar stations have been made. The early NASA concepts in the 1960ies preferred cylindrical modules, using the payload capacity of the Saturn V launcher (Gatland 1981). Recognising the problems of cosmic rays and micrometeorites some authors like Krafft Ehricke, Isaac Asimov et al., proposed to locate lunar settlements in artificial or natural cavities under the lunar surface, e.g. lava tubes, using inflatable balloons to prepare the construction site (Asimov and McCall 1974; Smolders 1986; Bogen 1993).

Studies of inflatable structures on the lunar surface were made e.g. by Vanderbilt et al. and Novak et al., while Chow and Lin proposed a lunar base built of double-skin membranes, filled with structural foam (for details see Benaroya 2002). In 2006 Petra Gruber and Barbara Imhof presented a study on bionic (biomimetic) structures (Gruber and Imhof 2007).

26.2 Design Criteria for an Initial Lunar Base

Any building on the lunar surface is penetrated constantly by high-velocity micrometeorites and hit by cosmic rays. The 11-year cycle of solar flare eruptions can cause lethal danger for humans. The severe temperature cycles, when the sun shines undimmed for 14 days and disappears for another 14 days, can cause material fatigue and brittle fractures on exposed structures and materials.

The average lunar surface temperature is assumed to be about -170 °C in the lunar night and +130 °C during the day (Rükl 1990).

The fundamental technologies to erect a lunar base are available since the 1970ies. To minimize risks and costs we assume to use only current and proven technology.

The most important *design criteria* we postulate as follows:

- Maximum safety for humans during construction and habitation; easy escape and rescue in case of damage;
- Structural redundancy and reliability of the entire design;
- Shelter against micrometeorites, cosmic rays, solar flares and temperature cycles;
- Ease of construction: the number of launches from earth and extravehicular work of astronauts should be minimized; robotic vehicles and remote control should be preferred;
- Use of lunar material (regolith), as an in situ resource, especially for shielding;

Do e.g. inflatable (pneumatic) structures fulfil the above criteria? During inflation a pneumatic construction may easily be damaged by small meteorites and cosmic rays. Because of extremely high or low temperatures brittle fractions may occur in the pneumatic skin before it is covered with regolith. To cover inflatables with a 1.0 m thick layer much lunar material and a big crane is necessary. In case of meteorite impact the bubble will deflate immediately. Last not least equipment and furniture can just be put into the habitat through the airlocks. Inflatables are no proven technology under the conditions of space yet.

To start any long-term human presence on the moon we need first of all a "site hut" like on terrestrial building sites, easy to assemble and modular.

In the long run architects and engineers can use various advanced methods for design and construction of lunar buildings.

26.3 A Modular Lunar Base Design

We propose to build the initial lunar station of six cylindrical modules, each one 17 m long and 6 m in diameter. Each module is made of aluminium sheets and trapezoidal aluminium sheeting and has a weight on earth of approx. 10.2 tons, including the interior equipment and furnishing. The outer wall of the cylinders is built as a double-shell hull, stiffened by radial bulkheads. Eight astronauts or scientists can live and work in the station, using the modules as follows:

- 1 Central Living Module;
- 2 Living quarter Modules, with private rooms for each person;
- 1 Laboratory Module for scientific research and engineering;

- 1 Airlock Module, containing outdoor equipment, space suits, etc.;
- 1 Energy Plant Module, carrying solar panels, a small nuclear power source and communication antennas;

The usable living and working area is about 270 m² (Energy Plant Module not included).

During assembling the first two or three modules can be used immediately as a shelter for astronauts. After assembling, every module is divided from each other by airlocks or fire doors, which is essential for safety and rescue operations.

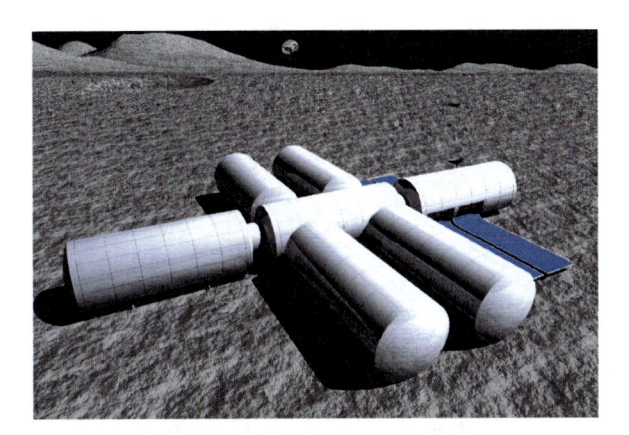

Fig. 26.1 Modular Lunar Base

Figure 26.1 shows the initial stage of six modules with one additional module (to the left).

To protect the astronauts from micrometeorites and radiation, the caves between the two shells of the outer wall are filled with a 0.63 m thick layer of regolith in situ by a small teleoperated digger vehicle (Fig. 26.2). Using lunar material for shielding the payload for launching can be minimized.

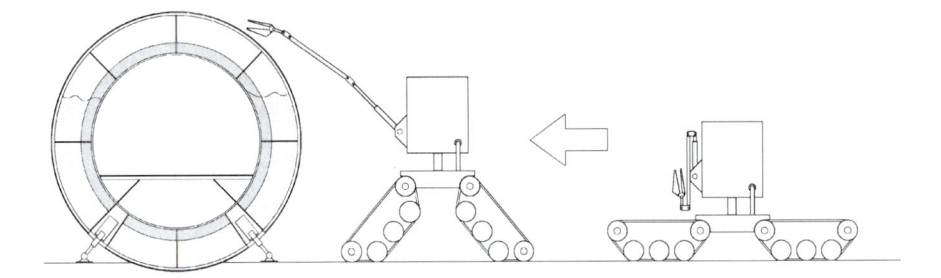

Fig. 26.2 Teleoperated Digger Vehicle, filling the caves with regolith

As a thermal insulation a layer of 0.25 m foamglass is used (Fig. 26.3). The entire outer wall of the cylinders provides a thermal conductivity of approx. 0.0942 W/m²K.

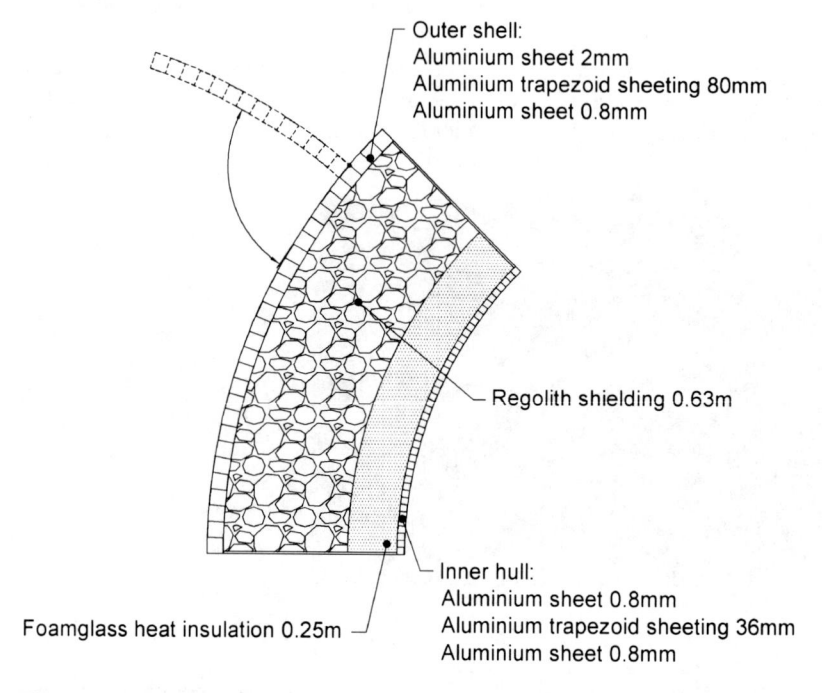

Outer shell:
Aluminium sheet 2mm
Aluminium trapezoid sheeting 80mm
Aluminium sheet 0.8mm

Regolith shielding 0.63m

Inner hull:
Aluminium sheet 0.8mm
Aluminium trapezoid sheeting 36mm
Aluminium sheet 0.8mm

Foamglass heat insulation 0.25m

Fig. 26.3 Section of the double-shell cylinder wall

The light-weight structure of thin aluminium sheets and trapezoidal sheeting reduces the payload mass for launching.

In the recent study "Design and Construction of a Modular Lunar Base"(Grandl 2010) the author has compared the proposed double-shell structure (Fig. 26.4) with single-shell designs and inflated structures focussing on thermal protection during the lunar night.

Assuming an external temperature of -170°C, and +21°C and 60% humidity of air inside the habitat the calculations in this study showed the advantages of the double-shell system (Table 26.1).

On the inner surface of the pneumatic skin condensed water will occur and freeze. The temperature difference between inner surface and internal air is approx. 27°C.

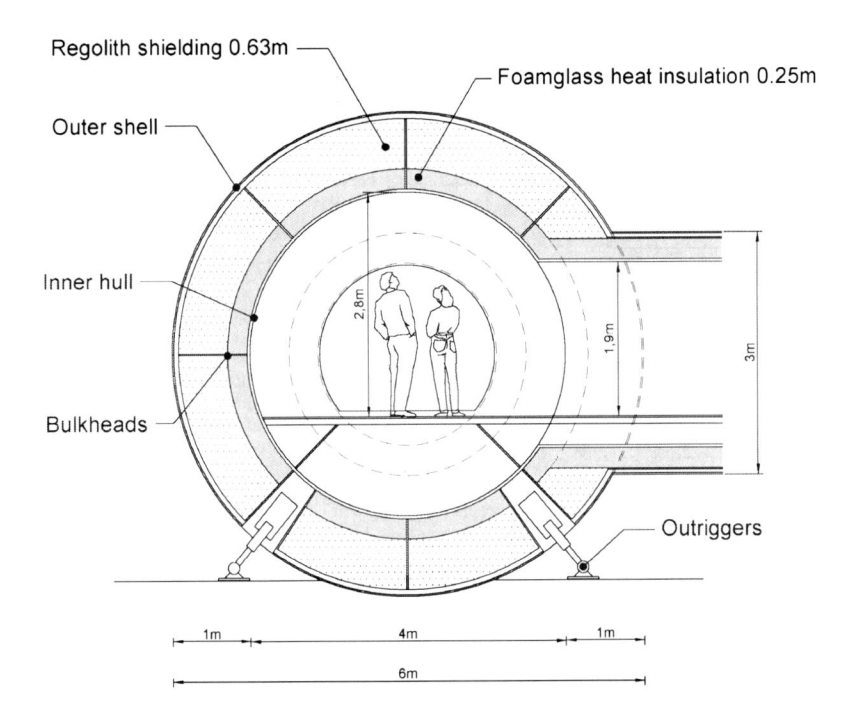

Fig. 26.4 Section of a cylindrical Module

Table 26.1 Thermal protection and temperature distribution (Grandl 2010)

Structure	Regolith shilding (m)	Thermal insulation (m)	Thermal conductivity (W/(mK))	Inner surface temperature (°C)	Internal temperature (°C)
double-shell	0.65	0.25	0.0942	19.2	21.3
single-shell	0.70	0.08	0.3375	13.0	20.9
inflatable	1.00	-	1.1800	- 6.7	21.0

A single-shell hull with 0.08 m foamglass insulation, covered with 0.7 m regolith will cause condensed water on the inner surface at 60% humidity of air. In this case the temperature difference between the inner surface and the internal air will be approx. 8°C -uncomfortable for humans. The calculations demonstrate the conclusive advantage of the proposed double-shell structure, which provides a comfortable climate inside the habitat. The temperature difference between inner surface and air is about 2°C -similar to a terrestrial building (Grandl 2010).

26.4 Launching, Transport and Assembly

The proposed light-weight aluminium design by using thin metal sheets and trapezoidal sheeting and stiffened by radial bulkheads enables us to reduce the payload
mass of one cylinder to 10.2 tons, including the furniture mainly made of carbon
fibre material. As an example a modified ARIANE 5 ESC-B launcher was assumed (Fig. 26.5). This launcher will be able to put a payload of 12 tons into geostationary transfer orbit (GTO).

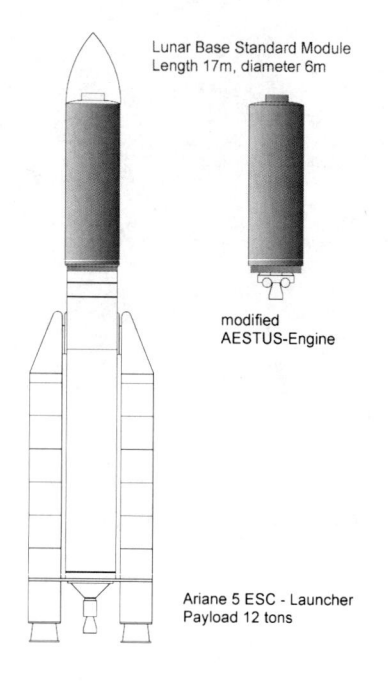

Lunar Base Standard Module
Length 17m, diameter 6m

modified
AESTUS-Engine

Ariane 5 ESC - Launcher
Payload 12 tons

Fig. 26.5 ARIANE 5 launcher with standard
module payload and a modified fairing

For the flight from low earth orbit to lunar orbit the modules can either be pro-
pulsed by a small rocket engine (AESTUS engine) or be moved by a "space
tug"(Gatland 1981).

The landing and assembling of the modules on the lunar surface is controlled
and teleoperated by the astronauts in the Lunar Orbiter. For finishing the astronauts can descend to the lunar surface using a small Manned Lander.

To establish the initial Modular Lunar Base on the moon eleven launches are
necessary:

- first flight: Lunar Orbiter, a small manned spaceship for two astronauts;
- one flight: Manned Lander and docking module for the orbiter;
- one flight: Teleoperated Rocket Crane (TRC) (Fig. 26.6);
- six flights: Lunar Base modules;
- one flight: machinery, teleoperated digger vehicle;
- one flight: scientific equipment, lunar rover, etc.;

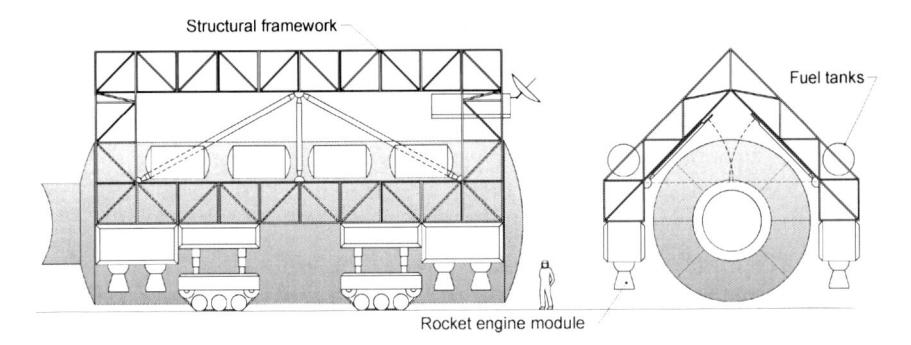

Structural framework

Fuel tanks

Rocket engine module

Fig. 26.6 Teleoperated Rocket Crane (TRC), landing and moving a module on the lunar surface

To land, move and assemble the modules on the lunar surface the Teleoperated Rocket Crane (TRC) is used. This vehicle will be assembled in the lunar orbit and is built as a structural framework carrying rocket engines, fuel tanks and crawlers to move on the lunar surface. After connecting a module to another one the TRC can start and return to lunar orbit. The building site just needs to be roughly prepared because the modules can be adjusted by outriggers (Fig.26.4.).

26.5 International Cooperation, Future

Due to the modular design, which provides standard weights, measures and sizes for all components of the Lunar Base Project, international cooperation is easily possible.

The Ariane 5 launcher -as an existing rocket- was taken as an example. The modules can also be launched by US, Russian, Indian or Chinese rockets with similar payload capability.

The modular design enables the station to be enlarged by stages, finally becoming an "urban structure" for astronauts, scientists and space tourists within the next decades. Both designs, the Modular Lunar Base and the TRC can also be used to land on other celestial bodies, like the moons of Jupiter.

In a second step of lunar settlement after some years of accommodation and the establishment of mining and industry on the moon, advanced structures, like inflatables or the use of lava tubes will be a challenge for architects and engineers.

References

Asimov, I., McCall, R.: Unsere Welt im All. Verlag C.J. Bucher AG, Frankfurt/M (1974)

Benaroya, H., Bernold, L., Chua, K.M.: Engineering, Design and Construction of Lunar Bases. Journal of Aerospace Engineering 15, 33–45 (2002)

Bogen, W.: Creating habitable volumes from lunar lava tubes. In: 11th SSI- Princeton Conference on AIAA-Space Manufacturing 9, The High Frontier, pp. 247–251 (1993)

ESA-report, Moon: the 8th continent. Human Spaceflight 2025, Final Report of the Human
    Spaceflight Vision Group (2003),
    http://esamultimedia.esa.int/docs/exploration/
    StakeholderConsultations/Moon_The_8th_Continent.pdf
Gatland, K.: Space Technology. Salamander Books Ltd., London (1981)
Grandl, W.: Lunar Base 2015 Stage 1 Preliminary Design Study. Acta Astronautica 60,
    554–560 (2007)
Grandl, W.: Design and Construction of a Modular Lunar Base. In: Lunar Settlements,
    pp. 509–517. CRC Press, Taylor & Francis Group (2010)
Gruber, P., Imhof, B.: Transformation: structure/ space studies in bionics and space design.
    Acta Astronautica 60, 561–570 (2007)
Rükl, A.: Mondatlas, Artia, Praha, Czech Republic (1990)
Smolders, P.: Living in Space. Airlife Publishing Ltd., England (1986)

27 Advanced Systems Concept for Autonomous Construction and Self-repair of Lunar Surface ISRU Structures

H. Benaroya, S. Indyk, and S. Mottaghi

Rutgers University, USA

27.1 ISRU for Manned Space Exploration and Settlement

In-Situ Resource Utilization is viewed by most as the basis for a successful manned exploration and settlement of the solar system. It is as its name implies, the "living off the land" that is a necessity for any untethered manned activity. The goal of ISRU is to allow settlements on the Moon and Mars and beyond to live with minimal if any dependence on the shipment of raw materials from Earth. The existence of a broad spectrum of elements on the lunar surface and on Mars gives hope that in principle it is possible to build a sustainable infrastructure on both these bodies that is based on local resources.

In the 1981 NASA report by Williams and Hubbard, a methodology for the selection of potential lunar resource sites is examined. They point to the major need for autonomous systems for data collection and analysis upon which intelligent decision can be made about the usefulness of a potential resource site. This study examined five major topics – lunar resources, geological studies, geophysical studies, geochemical instrumentation, and robotics. Three distinct resource types were identified:

(i)   bulk soil with no beneficiation beyond simple size sorting,
(ii)   identification of a concentration of a specific mineral, and
(iii) identification of a resource created by geologic processes unique to the Moon.

In the last category would be resources such as solar wind hydrogen recovery and the recovery of lunar polar ice. As summarized in this chapter, to use a lunar resource it must be mined, beneficiated and refined and then manufactured into desirable products. The need for automated systems is clearly identified. Automated systems must have mobility, in situ material characterization, manipulation and analysis. The authors of this chapter estimated that a 30 year lead time would be needed to proceed through the needed R&D to create the needed systems. Had we started this process when this report was written in 1981, we would have been ready today with the needed technologies to proceed.

In the proceedings edited by Mendell in 1985, there are many chapters on all aspects of the creation of lunar bases and what space activities in the 21st century might look like. Chapter 7 discusses lunar materials and processes. Contributors identified the available lunar raw materials needed for manufacture, explored possible lunar mining technologies, suggested a beneficiation technique for the lunar mineral ilmenite, explored technologies and material properties for lunar utilization, suggested drilling techniques, microwave-based beneficiation, among other relevant topics.

A workshop report edited by Fairchild and Mendell in 1988 investigated potential joint development of the key technologies and mechanisms required to enable the permanent habitation of space. In addition to NASA and the US Department of Energy, cosponsors were United Technologies, Kraft Foods, Large Scale Programs Institute, and Disney Imagineering. The objectives of the workshop were fourfold:

(1)   to develop collaborations between space planners and the non-aerospace industrial sector;

(2)   to obtain opinions from this industrial sector on the key issues facing the space program;

(3)   to formulate priority recommendations for space technology developments, and

(4)   to develop strategies for private investments in space development.

The working groups in the workshop developed recommendations in the following areas:

(i)   construction, assembly, automation, and robotics
(ii)   prospecting, mining, and surface transportation
(iii)   biosystems and life support
(iv)   materials processing, and
(v)   innovative ventures whereby a framework can be created to accelerate private space technology development.

It was concluded that a successful and vigorous civilian space program requires both NASA and the private sector collaboration.

The 1998 workshop report edited by Duke has as its principal purpose the examination of high-priority, near-term applications of in situ planetary resources that could lower the construction costs of human outposts on the Moon and Mars. Of particular interest was the use of indigenous material for the initial stages of planetary outpost installation. An example is the production of propellants. Also a high priority of this workshop was to link ideas to program applications, that is, create interfaces between technology innovators and mission designers.

With the 2004 declaration of the Bush vision for the manned return to the Moon by 2020, NASA embarked on an assessment of capabilities known as the capability roadmaps in key technical areas. One of these was the ISRU capability roadmap of 2005 chaired by Sanders and Duke. This report is an excellent and detailed overview of the ISRU capability. As stated in its introduction, potential space resources include water, solar wind implanted volatiles, metals and minerals, atmospheric constituents, solar energy, regions of permanent light and darkness, the vacuum and zero gravity of space, and trash and waste from human crew activities. The following five high criticality-to-mission success/cost areas strongly affected by ISRU were identified:

(1) transportation
(2) energy/power
(3) life support
(4) sustainability (repair, manufacture, construction), and
(5) commercialization.

The early incorporation of ISRU into missions and architectures would lead to reduction in mass, cost, risk, increase mission flexibility and mission enhancement and general capabilities. This report also provided timelines and decision points for the evolution of manned missions to the Moon and Mars. The following key capabilities were identified:

- Lunar/Mars regolith excavation and transportation
- Lunar oxygen production from regolith
- Lunar polar water/hydrogen extraction from regolith
- Mars water extraction from regolith
- Mars atmosphere collection and separation
- Mars oxygen/propellant production
- Metal/silicon extraction from regolith
- In situ surface manufacture and repair
- In situ surface power generation and storage
- Lunar/Mars surface cryogenic fluid liquefaction, storage, and transfer.

Also identified in this report were critical interdependencies of ISRU with other roadmap capabilities. Examples include power and propulsion, human health and support systems, and scientific instruments and sensors. Numerous gaps were identified in ISRU capabilities development. Dust mitigation, reduced-gravity effects, specialized sensors, for example. Also critical are the abilities to estimate the resource and technical risks associated with the incorporation of ISRU into the missions. Resource risk refers to the possibility of not locating sufficient resources and technical risk refers to uncertainties in the processes of extraction and beneficiation.

In the paper by Duke et al. (2006), an excellent and substantial discussion was provided on all aspects of the manned return to the Moon. An extensive and detailed section was written on the development and use of lunar resources.

27.2 Advanced Concept Description

The next step for manned exploration and settlement is a return to the Moon. Such a return requires the construction of structures for habitation as well as for manufacturing, farming, maintenance and science. The most challenging of these is the construction of structures that can be used for habitation, although the other mentioned applications each offer unique challenges to the design engineers and, until now, the astronaut construction teams that must erect the structures.

The use of freeform fabrication technologies by autonomous mini-robots can form the basis for habitable lunar structures using primarily in-situ resources and solar/nuclear power. Over a six to twelve month period prior to the arrival of astronauts such a system can erect a near-habitable structure.

Every aspect of this concept and goal has serious technical challenges. Currently we do not have the suite of capabilities needed to autonomously erect structures on the Moon using local resources. (We cannot even do this on Earth, where the challenges are less severe.) Some minimal capabilities exist in robotics (Novara et al. 1998) and the processing of regolith simulant. For example,

- prospecting strategies for lunar resources with some ideas as to where certain elements are located (Taylor and Martel 2003);
- carbothermal reduction of oxides for elemental extraction and zone refining for obtaining high-purity metals (Sen et al. 2005, Balasubramaniam et al. 2010);
- extraction systems and transportation (Siegfried and Santa 2000) as well as the discovery of water ice in significant quantities (Sridharan et al. 2010);
- oxygen, metals, silicon and glass refining on the Moon (Landis 2007), extraction of magnesium and copper using a surfactant and water in supercritical carbon dioxide (Wang et al. 2008);
- using a regolith-based substrate in the fabrication of solar cells directly on the lunar surface (Horton et al. 2005);
- production of structural elements via a geothermite reaction in a mixture of lunar regolith simulant and aluminum powder (Faierson et al. 2010);
- growing plants – lunar and Martian agriculture (Kozyrovska et al., Davies et al. 2003);
- ISRU technologies development for Mars (Moore 2010), for life support systems (Sridhar et al. 2000), propellant production (Hu et al. 2007, Holladay et al. 2007).

Structures for manned habitation must be designed to protect against the extreme lunar surface environment. Some argue that manned structures be buried – and they will be one day as the infrastructure is created on the Moon that is capable to engineer underground structures – but initial forays to the Moon will be modest given our limited capabilities to build on such a hostile environment. Many grand and exciting visions such as Fig. 27.1 have been created for how cities on the Moon will look but they invariably depict second and third generation facilities. As wonderful as it is to behold these creations, it is more important to be able to formulate a framework for the erection of the first lunar structure for habitation.

The question is the following: Given the costs associated with bringing material to the Moon (assuming no Space Elevator exists before man returns to the Moon), how do we design and construct habitable structures on the lunar surface in a way that is feasible from mass and energy constraints and minimizes astronaut construction time and risk? What is needed is a small "machine" that can be sent to the Moon and can build structures utilizing primarily in-situ resources and can operate via solar power in conjunction with nuclear power as backup.

Fig.27.1 This lunar settlement artist's concept is by NASA engineer Gary itmacher. Courtesy NASA.

Prospecting to identify the most fruitful resource locations will be essential. Depending on the specific 3D printing that is being used, prospecting may be as simple as identifying a small region set a distance away (to minimize the amount of dust that gets kicked into suspension and increase dust mitigation complications caused by excavation). Or it may be as complex as identifying the regions that have the highest and most widespread concentration of the mineral of interest that can then be collected and refined further.

Many tactics for scientific exploration may be easily converted for use in prospecting. Barfoot et al. (2010) provide an excellent description of the robotic requirements for a system to prospect ground ice on Mars. Sensors that could also be carried over to lunar prospecting include Ground Penetrating Radar (GPR), Light Detection and Ranging, (LIDAR), and Stereo Camera. Kisdi and Tatnall (2011) pose an interesting method of scouting and identifying regions of interest through a biologically inspired honeybee method. Not all methods of prospecting would depend on purely robotic means. Fong et al. investigate the operation requirements for a joint robotic and human scientific exploration that could be easily adapted to a joint robotic-human prospecting expedition.

Fundamental systems/capabilities that will be required for prospecting are mobility, navigation, autonomy, communications, and prospecting sensors.

27.3 The "Grand Vision"

The "grand vision" is the design of freeform fabrication machines that operate almost completely under solar power. One can envision such machines doing most of their work during the lunar day – perhaps during the night as well if battery or nuclear technologies are also included in the equation.

To analyze this prospect a study is proposed comprising of the following aspects:

1. select a benchmark lunar structure for analysis and which is to be built using the autonomous technologies proposed;

2. establish structural strength of "blocks" built from in-situ/regolith material (what can be expected from such a process ideally and realistically, knowing the kind of efficiencies one can obtain on Earth);
3. determine energy/power requirements and feasibility (can current and antic-ipated solar energy conversion technology provide the needed power to drive such a machine, and if not, what are our other options);
4. rate of construction as function of above (how fast can the most likely such robot build a facility that can then be organized for human habitation);
5. examine limitations on the possible complexity of such a structure (how com-plicated can it be made and can the fabrication process also include holes for pipes and power lines, for example);
6. examine the transferability of the technology to the Martian surface (clearly solar power will not be sufficient – nuclear battery technology is a viable power source for robots on Mars).

Others have suggested sintering regolith, utilizing microwaves, and some form of automated habitation construction, but technical issues still require resolution. The rate at which such an autonomous system can operate is low. It may take many months for a team of autonomous systems to erect a simple igloo-like structure of the kind proposed below.

In order to develop and evaluate this concept, we need a realistic lunar structure upon which to apply our ideas. Thus, what we propose has two distinct components:

1. conceptual and actual design of a lunar structure – a benchmark lunar surface structure, and
2. assessing the issues that need to be resolved for the design of autonomous construction mini-robots, as described herein.

27.4 A Benchmark Lunar Surface Structure for Autonomous Construction

Key environmental factors affecting lunar structural design and construction are: 1/6 g, the need for internal air pressurization of habitation-rated structures, the re-quirement for shielding against radiation and micrometeorites, the hard vacuum and its effects on some exotic materials, a significant dust mitigation problem for machines and airlocks, severe temperatures and temperature gradients, and nu-merous loading conditions – anticipated and accidental. The structure on the Moon must be maintainable, functional, compatible, easily constructed, and made of as much local materials as possible.

Cast regolith has been suggested as a building material for the Moon. The use of cast regolith (basalt) is very similar to terrestrial cast basalt. The terms have been used interchangeably in the literature to refer to the same material. It has been suggested that cast regolith can be readily manufactured on the Moon by melting regolith and cooling it slowly so that the material crystallizes instead of turning into glass. Virtually no material preparation is needed. The casting operation is simple requiring only a furnace, ladle and molds. Vacuum melting and casting should enhance the quality of the end product. More importantly, there is terrestrial experience producing the material; but it has not been used for construction purposes yet.

Cast basalt has extremely high compressive and moderate tensile strength. It can easily be cast into structural elements for ready use in prefabricated construction. Feasible shapes include most of the basic structural elements like beams, columns, slabs, shells, arch segments, blocks and cylinders. Note that the ultimate compressive and tensile strengths are each about ten times greater than those of concrete.

Table 27.1 Typical properties for cast regolith

Property	Units	Value
Tensional Strength	N/mm^2	34.5
Compressive Strength	N/mm^2	538
Young's Modulus	kN/mm^2	100
Density	g/cm^3	3
Temperature Coefficient	10^{-6}/K	7.5 – 8.5

Cast basalt also has the disadvantage that it is a brittle material. Tensile loads that are a significant fraction of the ultimate tensile strength need to be avoided. The fracture and fatigue properties need further research. One possibility is to reinforce it with high tensile strength materials such as locally manufactured magnesium or aluminum or with carbon nanotubes (Alford et al. 2006).

It should also be feasible to use cast regolith in many structural applications without any tensile reinforcement because of its moderately high tensile strength (Table 27.1). However, a minimum amount of tension reinforcement may be required to provide a safe structure. The reinforcement could be made with local materials as well. These local materials would also have to be derived from ISRU manufacturing and refining processes that have yet to be developed and tested.

Cast regolith is most suited for use in structures that are dominated by compression. However, using prestressed applications will offer a wide variety of shapes and structures. Prestressing tendons can be made from lunar materials.

Since it is extremely hard, cast regolith has high abrasion resistance. This is an advantage for use in the dusty lunar environment. It may be the ideal material for paving lunar rocket launch sites and constructing debris shields surrounding landing pads. The hardness of cast basalt combined with its brittle nature makes it a difficult material to cut, drill or machine. Such operations should be avoided on the Moon. Production of cast regolith is energy intensive because of its high melting point. The estimated energy consumption is 360 kWh/MT.

The structure in Fig. 27.2 and the related analysis and discussion is from a paper by Ruess et al. (2006). The preliminary analysis and design of an autonomous system that is capable of constructing such a structure on the surface of the Moon is a goal. This structure will be ready for the first astronaut team that arrives on the Moon to fit it with systems necessary for human habitation. In essence, this structure is envisioned to be a "shell" into which pipes, wiring, windows (if any), and equipment can be installed by the astronaut team. The appropriate volumes will be already in the structure. Eventually, in principle, we can anticipate that almost all of the structure and life support equipment can be locally fabricated and installed by autonomous robots.

A preliminary analysis of this structure is summarized here and this will form the basis for evaluating the autonomous construction method proposed.

The structure will be human-rated, meaning that it will be shielded and can be pressurized upon a human presence. The presence of a structural shell on the Moon, awaiting human arrival, has enormous implications on the logistical planning of man's return to the Moon. All the volume that would normally be allocated for bringing structural materials to the Moon can now be replaced with other items. This leads to an enormous saving in time and money.

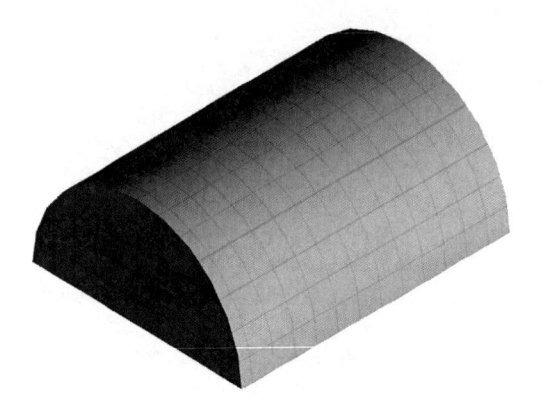

Fig. 27.2 Rendering of a lunar habitat "igloo" module

Determining the dimensions of a lunar base habitat is a challenging task. Numerous factors like crew size, mission duration and function of the base as an industrial or scientific outpost influence the necessary habitat size. Hence, a global approach considering the necessary habitable volume per person will be pursued. Habitable volume is interpreted as free volume, excluding volume occupied by equipment or stowage.

As demonstrated by the Gemini missions, relatively short duration missions of up to two weeks can be endured by a person restrained to a chair most of the time. The habitable volume per crewmember in Gemini was 0.57 m^3. Currently, the NASA ManSystems Integration Standards (NASA STD 3000) recommend a minimum habitable volume at which performance can be maintained for mission durations of four months or longer of about 20 m^3. Despite this recommendation, a design volume (living and working areas) of 120 m^3 per person for a lunar habitat has been recommended, based on research of long-term habitation and confined spaces. This value is about equivalent to the volume per crewmember onboard the International Space Station.

The regolith upon which the lunar structure is built needs to be understood from a foundation engineering view. The bulk density of regolith ranges from 0.9 to 1.1 g/cm^3 near the surface and reaches a maximum of 1.9 g/cm^3 below 20 cm. The average is at 1.7 g/cm^3.

- The porosity of the regolith surface is about 0.45.
- Cohesion of undisturbed regolith is $c = 0.1$ to 1.0 kN/m^2.
- The friction angle is about $30°$ to $50°$.
- The regolith's modulus of subgrade reaction is typically 1000 $kN/m^2/m$.
- The compressibility ranges from compressibility index $C_c = 0.3$ (loose regolith) to 0.05 (dense regolith).
- Interparticle adhesion in the regolith is high. It clumps together like damp beach sand.

Typical properties on Earth are: bulk density for sandy soil is 1.5-1.7 g/cm^3 with 2.65 g/cm^3 for most mineral soils; porosity typically ranges from less than 0.01 for solid granite to more than 0.5 for peat and clay; cohesion is in the range of almost zero to about 25 kN/m^2 with friction angles in the range of $8°$ to $45°$; the compressibility index values are from 0.05 to 0.06 for a loose state and 0.02 to 0.03 for a dense state.

Figure 27.3 shows how the space within the arch will be divided into the different functional areas. Figures 27.4-27.6 depict the various loads on the structure, both internal and external.

The loads for the regolith cover assume the regolith can be placed uniformly on the structure. If instead loose soil is simply heaped upon the top of the structure, the resulting load will be trapezoidal, not uniform.

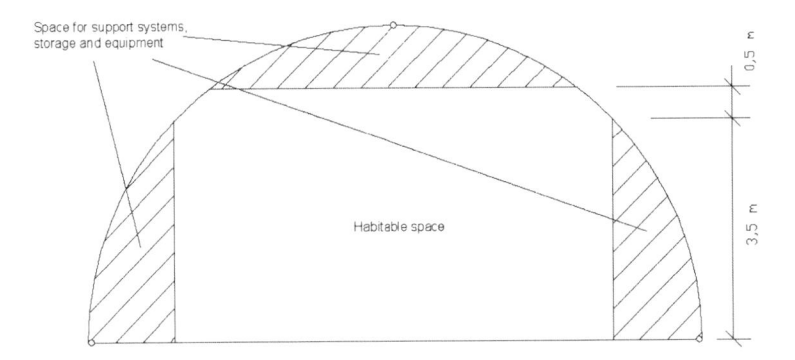

Space for support systems, storage and equipment

Habitable space

0,5 m

3,5 m

Fig. 27.3 Space use within the proposed structure.

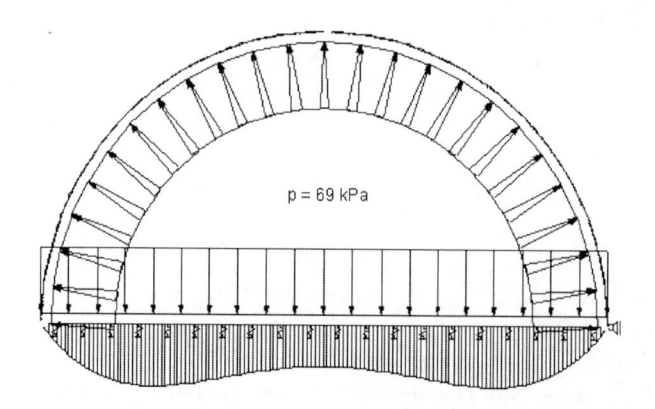

Fig. 27.4 Internal pressure loads and the bending moment for the circular arch.

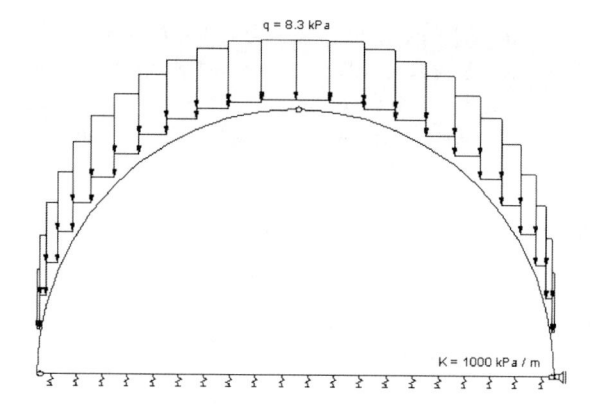

Fig. 27.5 Regolith cover load

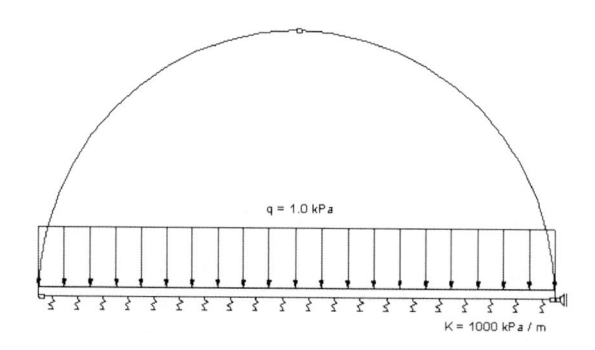

Fig. 27.6 Floor loads

Most of the loads described above may act at the same time. There are also a number of different scenarios that the designer needs to account for: starting with construction stages, the structure being initially pressurized with the regolith not yet on top of it, next the regular operational mode with all loads acting, and finally a planned or accidental decompression. The maximum effect on the structure has to be found using load combinations. For each scenario only the loads that increase the stresses in the structure are to be included. Self-weight is always present. Four combinations were used to find the maximum stresses in the members:

1. internal pressure plus floor loads;
2. regolith cover plus installation loads;
3. all loads;
4. half the regolith cover (during construction).

Two main conclusions result from the preliminary structural analysis. First, the arch segments can have a uniform cross-section. It is possible but not necessary to adjust the arch cross-section to the distribution of internal forces since these are almost uniform. Second, in order to get an efficient cross-section for the tie it has to be adjusted to the distribution of internal forces. The bending moment has the shape of a parabola, so it was decided to give the tie a similar shape. Figure 27.7 shows the principal shape of the tie/floor/foundation.

Fig. 27.7 Front view of the general tie shape

The regolith foundation will need to be sintered before the structure is fabricated and will result in a higher modulus of subgrade reaction and therefore lower deflections of the tie. It does not affect the arch deflections. Calculations show that the modulus of subgrade reaction would have to be increased about tenfold to get in the range of desired deflections. This is very likely not possible to be achieved by sintering the regolith. Some additional reinforcing is needed. More research data is needed for this topic.

27.5 A Conceptual Development Work Plan

An engineering evaluation of the use of freeform fabrication technologies as the basis for building autonomous mini-robots that can, as a team, build a first lunar structure for habitation, using in-situ resources and primarily solar power, over a six to twelve month period prior to the arrival of astronauts can be comprised of the following aspects:

1. an examination of the structural strength of "blocks" built from in-situ/regolith material (what can be expected from such a process ideally and realistically, knowing the kind of efficiencies one can obtain on Earth);
2. an examination of the energy/power requirements and feasibility (can current and anticipated solar energy conversion technology provide the needed power to drive such machines, and if not, what are other options for power);
3. establish the rate of construction as function of above (how fast can the most likely ensemble of such robots build a facility that can then be organized for human habitation);
4. what is the appropriate mix of mini-robots, that is, how many different functions are needed and what is an optimal ensemble;
5. how complicated a lunar structure can be erected (how difficult is it for the fabrication process to include holes for pipes and power lines, for example);
6. address briefly the transferability of this technology to the Martian surface.

Our vision is that an optimally grouped set of mini-robots will work as a team to autonomously erect a habitable volume for finishing by the first astronaut team on the Moon. Issues that will be resolved in a preliminary study is the appropriate mix of robots by function, energy/power demands as a function of rate of construction, and redundancy needs for reliability. A point of study will be to examine whether solar and other radiation that hits the lunar surface can be factored into the "curing" process. Also, the hard vacuum on the Moon can be used to some advantage.

The appropriate mix of robots refers to the fact that different free-form manufacturing processes will be needed. We know that different parts of the structure require different construction approaches. For example, ballistic particle manufacturing may be appropriate for creating the foundation layer. For such ejection-type systems, a filtering process is needed so that relatively uniform sequences of regolith particles are fed. After a preliminary layer is prepared, some additional sintering may be required, perhaps via microwave as has been suggested in the literature.

One can envision the following sequence of events. One or several mini-robots (size of a lawn mower) are landed on the Moon, perhaps six to twelve months before the arrival of humans. Their purpose is to prepare the site and erect a simple structure that the arriving astronauts can then fit for habitation. This means installing (in a modular and easy way) life support and maintenance equipment.

Once the mini-robots land, they begin to prepare the regolith, first smoothing out the site, some sintering with attention to reinforcement (perhaps using carbon nanotubes). Once the site is prepared, the mini-robots begin to build the lunar structure in layers, a structure that has been designed *a priori*, leaving open volumes for mechanical and electronic equipment that is to be placed subsequently by astronauts. Of course, there would be a large volume, as shown in the previous figures, for habitation. We anticipate that it will be possible within a reasonable timeframe to have much of the equipment manufactured locally and robotically installed.

Studies will determine what kind of reinforcing the ISRU-based structure requires, possibilities being a glass fiber-reinforced or a nano-composite matrix.

Some studies have shown that such a layered manufacturing process can be achieved in a number of ways. One promising approach is based on microwave sintering. Much of the reinforcing material can be found in the regolith. Some suggestions have been made for autonomous construction on the Moon, but no study exists for autonomous construction using freeform manufacturing technologies utilizing in-situ resources, or for the erection of the first habitable structure on the Moon via such technologies.

Rapid prototyping processes are a relatively recent development. The first machine was released onto the market in late 1987. While rapid prototyping is the term commonly applied to these technologies the terminology is now a little dated, reflecting the purpose to which the early machines were applied. A more accurate description would be layered manufacturing processes. An alternative term is freeform fabrication processes. These processes work by building up a component layer by layer, with one thin layer of material bonded to the previous thin layer. There are a number of different processes. The main ones are:

- Stereolithography
- laser or microwave sintering
- fused deposition modeling
- solid ground curing
- laminated object manufacturing
- ballistic particle manufacturing and
- three-dimensional printing,

all of which have appeared on the market.

Some ballistic particle manufacturing technology use piezoelectric pumps that operate when an electric charge is applied, generating a shock wave that propels particles. All these processes essentially start with nothing and end with a completed part. Rapid prototyping processes are driven by instructions that are derived from three-dimensional computer-aided design (CAD) models. CAD technologies are therefore an essential enabling system for rapid prototyping. Other enabling technologies are for mini-rovers, energy beams, solar power systems, and structural reliability/durability.

The processes use different physical principles, but essentially they work either by using lasers or microwaves to cut, cure or sinter material into a layer, or involve ejecting material from a nozzle to create a layer. Many different materials are used, depending upon the particular process. Materials include thermopolymers, photopolymers, other plastics, paper, wax, or metallic powder, for example. The processes can be used to create models, tooling, prototypes, and even in some cases to directly produce metal components. As such, one may view these capabilities and laying the groundwork for self-repairing systems.

In the end, we envision a team of such freeform manufacturing mini-robots that will work as a team, with the group comprising the skill set of capabilities needed to construct all the parts of the structure. As an example, the ballistic particle manufacturing robot can be placed on a ridge to project particles to the higher locations of the lunar structure.

We believe that freeform 3D printing technologies today offer the advantage that they can create full size structure for habitation.

27.6 3D Printing

3D printing, also called Additive Fabrication, Rapid Prototyping, Layered Manufacturing, and Fabbing is a relatively new manufacturing process. As one of its names implies, it is an additive process, as opposed to a subtractive process. In a subtractive process you start with a block of material and remove parts of it to arrive at the final shape. In an additive process you instead start with nothing and add material. The way this is done, as another name of the process implies, is layer by layer. A three dimensional CAD model of a part is split into thin layers, and the piece is built from the bottom layer to the top.

There are several advantages to this process, all of which are important in solving the problems in construction on the Moon and Mars mentioned in the first section of this chapter. 3D printing can utilize a wide variety of materials, ranging from powdered metal to sand. The process also limits waste material, since only material actually utilized to make a part is used – there are no scrap pieces as in subtractive methods. With the severely limited resources found on the Moon and Mars, these advantages have a very positive impact. Also, 3D printing allows for the creation of very complex geometry, as shown in Fig. 27.8. Since the part is built layer by layer, there are virtually no limits on how complex the part can be (the same cannot be said about conventional methods of manufacturing). This complex geometry capability can be utilized to create crevices for wiring, plumbing, etc. in the structures.

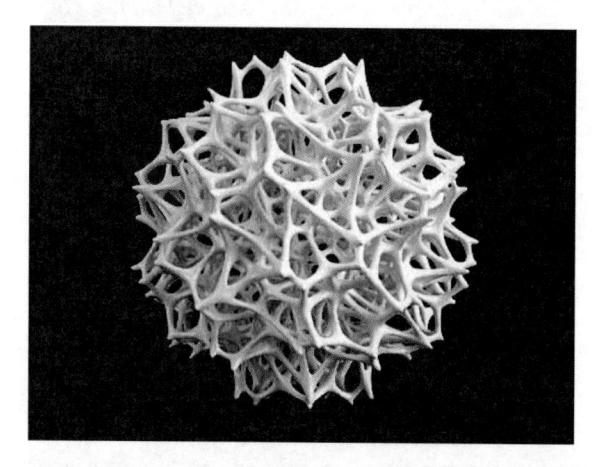

Fig. 27.8 An example of the complex geometry possible with 3D printing
http://www.psfk.com/2009/01/shapeways-3d-printing-marketplace.html

27.6.1 D-Shape

Another type of 3D printing technology that can have an application in the construction of primary bases on the Moon and Mars is D-Shape created by Dini Engineering. This technology, instead of melting powdered metal, uses a sandy material and a special inorganic, non-toxic binder. A layer of sandy material is deposited and then a movable nozzle deposits the binder, "gluing" the sand particles together to form a solid layer. A new layer of sand is deposited, and the process is repeated.

The appeal of this technology, when it comes to construction on the Moon and Mars, is that it is capable of producing parts up to 6m x 6m x 12 m in size. Potentially, the building platform can be removed altogether, and robots can deposit the binder. This way the size of the parts could be as large as it needs to be. Another positive aspect of this technology is that regolith found on the Moon and Mars could be used as the "sandy material." This possibility is actually already under consideration by Dini Engineering and several research and development companies in Europe. Figure 27.9 demonstrates the versatility of this technology. The machine requires 40 kW peak power.

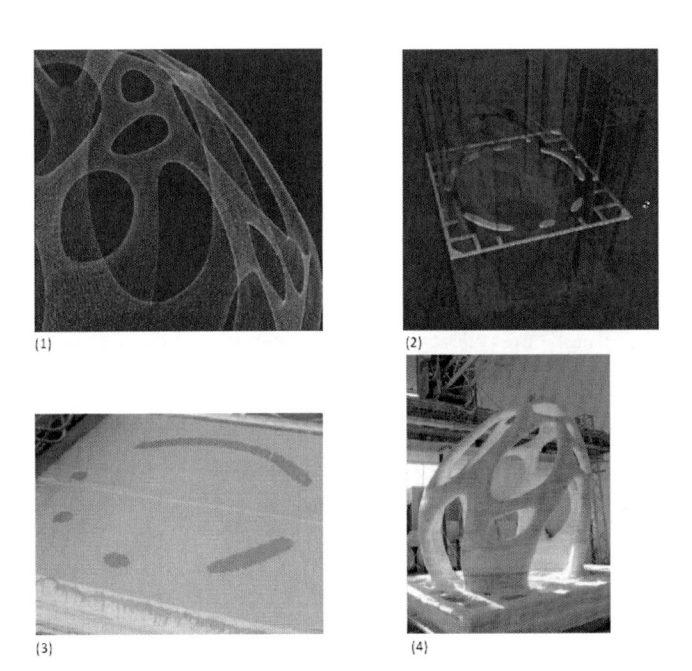

Fig. 27.9 The build process for D-shape: Start with a 3D CAD model (1), take a thin layer of the model (2), deposit binder to create the layer (3), keep adding layers until final product is built (4), (d-shape.com)

The parts that this technology makes have the strength of structural concrete, so D-shape can build bases with structural integrity as well. The problem with this concept, however, is that while structural concrete is strong in compression, it is relatively weak in tension. Since bases on the Moon and Mars would need to be internally pressurized, they need high tensile strength. Another setback of D-shape is that since the structures it builds are made out of bound sand, they will be susceptible to abrasion. Both of these issues can be fixed by utilizing Metal Spray technology.

27.6.2 High Velocity Oxy-Fuel Metal Spray Technology (HVOF)

HVOF Metal Spray is a technology developed by Sulzer Metco. It works by mixing a powdered metal with a fuel like propylene, propane, hydrogen, natural gas, or kerosene inside of a special "gun". The metal/fuel mixture is then ejected from the gun at high velocity and ignited via an oxidizer. The semi-molten metal impacts the target surface, deforms plastically, and forms an even coating of metal. This process is demonstrated in Fig. 27.10. The coatings made with HVOF can be thicker than with other metal spray methods, and they are denser as well. The technology can use material such as H13 Tool Steel (described above), Grey Iron (Iron, Silicon, and Carbon), Ductile Iron (Iron, Silicon, Carbon, Manganese, Magnesium, Phosphorus, and Sulfur) and many other types of alloys. As with H13 Tool Steel, Grey Iron and Ductile iron are mostly composed of materials that can be found on the Moon and Mars (mostly Iron and Silicon) and they are wear resistant.

The way that HVOF would solve the drawbacks of D-shape mentioned before (low tensile strength and susceptibility to abrasion) is by adding a thick layer of materials such as H13 Tool Steel on top of the structures created by D-shape. This would not only provide resistance to elevated temperatures and abrasion, but it would also give the base extra structural integrity. Steel is much stronger in tension than structural concrete, so the extra layer of steel would provide the tensile strength necessary to internally pressurize the structures built on the Moon and Mars.

To date, current 3D printing effectively deposits a single material. Depositing of multiple materials through a single machine would increase the effectiveness of the structural erection. Or, perhaps, if the medium is raw regolith, an additional sintering "head" could be incorporated to sinter the regolith as it is deposited. The process is further enhanced by including a method of laying wire for electrical systems or ducts for future ventilation systems. Similarly, an increase in the resolution of the layer deposition could also enable miniature mechanisms that could be useful to be included during the construction of the structure.

If we are to assume that transport to the Moon is readily and economically accessible, then the question is what needs to be done in order to prepare a machine such as the D-shape so that it can operate on the Moon. Debilitating environmental conditions include: extreme temperatures and temperature gradients, abrasive dust, vacuum, meteorite impacts, and radiation. All these factors would negatively impact the operation of any machine on the surface of the Moon.

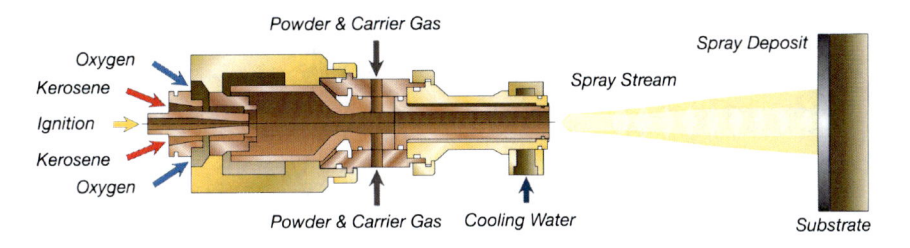

Fig. 27.10 Diagram of HVOF Metal Spray (Courtesy Sulzermetco).

We know that the following machine element component designs need radical changes for operation on the Moon: seals (high and low temperature effects, regolith dust), bearings (high and low temperature effects on lubricants, microscopic dust particles), radiation effects on electrical and electronic components, micrometeorite impacts and general abrasion to machine parts. Options include knitted mesh seals and graphite-metal bearings – these are produced for machines in extreme environments on Earth. They could be redesigned for a space/lunar rating. Electrical and electronic components will need to be shielded and hardened against radiation. These technologies are available but will need to be customized for the lunar surface environment.

27.7 Concluding Thoughts

We have presented a conceptual overview of how an automated ISRU effort can create near-habitable structures in advance of the arrival of astronauts on the Moon and eventually Mars and beyond. Clearly we have extrapolated from currently available technologies in three dimensional printing. These technologies offer the promise that they can be adapted to the hostile lunar environment and can be powered by solar and nuclear generation systems envisioned for the Moon. We have not much discussed the robotics needed for such an automated ISRU system. It appears to us that the printing technologies are further advanced than are the robotics. Serious challenges exist along the way to the types of systems proposed here. But there can be no settlement of the Moon and beyond without such systems.

Acknowledgments. Prof. Haim Baruh and the NASA Space Grant Consortium for their interest and generous support. Pem Lasota for our past work together, some of which is referred to above. Florian Ruess for our past work together, some of which is referred to above.

References

Alexander, P., Dutta, D.: Layered manufacturing of surfaces with open contours using localized wall thicknesses. Computer-Aided Design 32, 175–189 (2000)

Alford, J.M., Mason, G.R., Feikema, D.A.: Free Fall Plasma-Arc Reactor for Synthesis of Carbon Nanotubes in Microgravity. Review of Scientific Instruments 77 (2006)

Balasubramaniam, R., Gokoglu, S., Hegde, U.: The reduction of lunar regolith by carbothermal processing using methane. Int. J. Mineral Processing 96, 54–61 (2010)

Barfoot, T., Furgale, P., Stenning, B., Carle, P., Thomson, L., Osinski, G., Daly, M., Ghafoor, N.: Field Testing of a Rover Guidance, Navigation & Control Architecture to support a Ground-Ice Prospecting Mission to Mars Robotics and Autonomous Systems (2010) (in press)

Benaroya, H., Ettouney, M.: Framework for Evaluation of Lunar Base concepts. Journal of Aerospace Engineering 5(2), 187–198 (1992a)

Benaroya, H., Ettouney, M.: Design and Construction Considerations for a Lunar Outpost. Journal of Aerospace Engineering 5(3), 261–273 (1992b)

Benaroya, H.: Reliability of structures for the Moon. Structural Safety 15, 67–84 (1994)

Benaroya, H.: Economic and technical Issues for lunar development. Journal of Aerospace Engineering 11(4), 111–118 (1998)

Benaroya, H., Bernold, L., Chua, K.M.: Engineering, design and construction of Lunar Bases. Journal of Aerospace Engineering 15(2), 33–45 (2002)

Benaroya, H.: An overview of Lunar Base structures: past and future. In: AIAA Space Architecture Symposium, pp. 1–12. AIAA, Houston (2002)

Braun, B.: Faserverbundkunststoffe (FVK) als tragende Struktur. Tech. rep., University of Stuttgart, Institute of Structural Design, KE (2003)

Cohen, M.M.: Selected precepts in lunar architecture. Tech. rep., 53rd International Astronautical Congress, The World Space Congress (2002)

Criswell, M.E., Sadeh, W.Z., Abarbanel, J.E.: Design and performance criteria for inflatable structures in space. In: SPACE 1996, pp. 1045–1051. ASCE (1996)

Davies, F.T., He, C., Lacey, R.E., Ngo, Q.: Growing Plants for NASA – Challenges in Lunar and Martian Agriculture. In: Combined Proceedings International Plant Propagators' Society, vol. 53, pp. 59–64 (2003)

Duke, M.B.: Workshop on Usin. In: Situ Resources for Construction of Planetary Outposts. Lunar and Planetary Institute, 98-01 (1998)

Duke, M.B., Benaroya, H.: Applied mechanics of lunar exploration and development. Applied Mechanics Review 46(6), 272–277 (1993)

Duke, M.B., Gaddis, L.R., Taylor, G.J., Schmitt, H.H.: Development of the Moon. Reviews in Mineralogy & Geochemistry 60, 597–656 (2006)

Eckhart, P.: The Lunar Base Handbook. McGraw-Hill, New York (1999)

Eichold, A.: Conceptual Design of a crater Lunar Base. In: Proceedings of Return to the Moon II, pp. 126–136. AIAA (2000)

Ettouney, M., Benaroya, H.: Regolith mechanics, dynamics and foundations. Journal of Aerospace Engineering 5(2), 214–229 (1992a)

Ettouney, M., Benaroya, H., Agassi, N.: Cable structures and lunar environment. Journal of Aerospace Engineering 5(3), 297–310 (1992b)

Faierson, E.J., Logan, K.V., Stewart, B.K., Hunt, M.P.: Demonstration of concept for fabrication of lunar physical assets utilizing lunar regolith simulant and a geothermite reaction. Acta Astronautica 67, 38–45 (2010)

Fairchild, K., Mendell, W.W. (eds.): Report of the In Situ Resources Utilization Workshop, NASA Conference Publication 3017 (1988)

Graf, J.C.: Construction operations for an early Lunar Base. In: SPACE 1988, ASCE, pp. 190–201 (1988)

Happel, J.A.: The design of lunar structures using indigenous construction materials, Master of Science in Civil Engineering, University of Colorado (1992a)

Happel, J.A.: Prototype Lunar Base construction using indigenous materials. In: SPACE 1992, pp. 112–122. ASCE (1992b)

Happel, J.A.: Indigenous materials for lunar construction. Applied Mechanics Reviews 46(6), 313–325 (1993)

Holladay, J.D., Brooks, K.P., Wegeng, R., Hu, J., Sanders, J., Baird, S.: Microreactor development for Martian in situ propellant production. Catalysis Today 120, 35–44 (2007)

Horton, C., Gramajo, C., Alemu, A., Williams, L., Ignatiev, A., Freundlich, A.: First demonstration of photovoltaic diodes on lunar regolith-based substrate. Acta Astronautica 56, 537–545 (2005)

Hu, J., Brooks, K.P., Holladay, J.D., Howe, D.T., Simon, T.M.: Catalyst development for microchannel reactors for Martian in situ propellant production. Catalysis Today 125, 103–110 (2007)

Johnson, S.W., Chua, K.M., Carrier III, W.D.: Lunar soil mechanics. Journal of the British Interplanetary Society 48(1), 43–48 (1995)

Karalekas, D., Antoniou, K.: Composite rapid prototyping: overcoming the drawback of poor mechanical properties. J. Mat. Proc. Tech., 526–530 (2004)

Kisdi, A., Tatnall, A.R.L.: Future Robotic Exploration Using Honeybee Search Strategy: Example Search for Caves on Mars. Acta Astronautica (2011), doi:10.1016/j.actaastro.2011.01.013

Kozyrovska, N.O., Lutvynenko, T.L., Korniichuk, O.S., Kovalchuk, M.V., Voznyuk, T.M., Kononuchenko, O., Zaetz, I., Rogutskyy, I.S., Mytrokhyn, O.V., Mashkovska, S.P., Foing, B.H., Kordyum, V.A.: Growing pioneer plants for a lunar base. Advances in Space Research 37, 93–99 (2006)

Landis, G.A.: Materials refining on the Moon. Acta Astronautica 60, 906–915 (2007)

Mendell, W. (ed.): Lunar Bases and Space Activities of the 21st Century, Lunar and Planetary Institute (1985)

Moore, C.L.: Technology development for human exploration of Mars. Acta Astronautica 67, 1170–1175 (2010)

Novara, M., Putz, P., Marechal, L., Losito, S.: Robotics for lunar surface exploration. Robotics and Autonomous Systems 23, 53–63 (1998)

Reynolds, K.H.: Preliminary design study of lunar housing configurations. In: NASA Conference Publication 3166, NASA, pp. 255–259 (1988)

Ruess, F.: Structural Analysis of a Lunar Base. Master's thesis, Universität Stuttgart / Rutgers University (May 2004)

Ruess, F., Schänzlin, J., Benaroya, H.: Structural Design of a Lunar Habitat. J. Aerospace Engineering 19(3), 133–157 (2006)

Sanders, G.B., Duke, M.: NAS. In: Situ Resource Utilization (ISRU) Capability Roadmap Final Report (2005)

Sen, S., Ray, C.S., Reddy, R.G.: Processing of lunar soil simulant for space exploration applications. Materials Science and Engineering A 413-414, 592–597 (2005)

Siegfried, W.H., Santa, J.E.: Use of Propellant from the Moon in human exploration & development of space. Acta Astronautica 47, 365–375 (2000)

Spudis, P.D.: Harvest the Moon. Astronomy 6, 42–47 (2003)

Sridhar, K.R., Finn, J.E., Kliss, M.H.: In-situ Resource Utilization Technologies for Mars Life Support Systems. Adv. Space Res. 25(2), 249–255 (2000)

Sridharan, R., Ahmed, S.M., Das, T.P., Sreelatha, P., Pradeepkumar, P., Naik, N., Supriya, G.: 'Direct' evidence for water (H_2O) in the sunlit lunar ambience from CHACE on MIP of Chandrayaan I. Planetary and Space Science 58, 947–950 (2010)

Taylor, G.J., Martel, M.V.: Lunar prospecting. Adv. Space Res. 31(11), 2403–2412 (2003)

Taylor, L.A., Meek, T.T.: Microwave Sintering of Lunar Soil: Properties, Theory and Practice. Journal of Aerospace Engineering (ASCE) 18(3), 188–196 (2005)

Wang, T., Debelak, K.A., Roth, J.A.: Extraction of magnesium and copper using a surfactant and water in supercritical carbon dioxide. J. of Supercritical Fluids 47, 25–30 (2008)

Williams, R.J., Hubbard, N.: Report of Workshop on Methodology for Evaluating Potential Lunar Resource Sites, NASA Technical Memorandum 58235 (1981)

28 Moon Dune – Bacillithic Cratertecture

Magnus Larsson and Alex Kaiser

Species of Spaces, London, UK

28.1 Introduction

The potential for microbial life to adapt and evolve in environments beyond its planet of origin should be assessed. Little is currently known regarding the consequences when earthly microbial life is transported into space or to other planets, where the environment is very different from that of Earth. The findings from such studies will determine whether life on Earth is strictly a local planetary phenomenon or can expand its evolutionary trajectory beyond its place of origin (NASA 2003).

While the exogenetic panspermia theory – the hypothesis that life has been distributed by meteroids, asteroids, and planetoids to extend throughout the universe – remains heavily debated, the proposition that all known elements can be considered a form of cosmic dust is widely held to be true. "Dust," writes Joseph Amato,

"is found within all things, solid, liquid, or vaporous. With the atmosphere, it forms the envelope that mediates the earth's interaction with the universe" (Amato 2000).

Indeed all terrestrial elements with an atomic mass above that of hydrogen were probably formed in the core of stars via stellar nucleosynthesis and supernova nucleosynthesis events. But of course dust doesn't stop there. It is found beyond the earthly sphere, throughout infinite expanses of interstellar space, a "ghostly residue of unspooled stars, astronomical structures reduced to mist" (Manaugh 2008). Elements are indivisible particles of dust; gas and dust clouds are primary precursors for planetary systems – as well as for human beings. A priest quoting *Genesis 3:19* from her King James Bible – *Dust thou art, and unto dust shalt thou return* – is essentially correct. As Carl Sagan put it, we are all "made of stellar ash" (Sagan 1980).

In 1967, a television camera landed on the surface of the Moon as part of the unmanned lunar lander Surveyor 3. Two and a half years later, on 20 November 1969, Apollo 12 astronauts Pete Conrad and Alan L. Bean carefully recovered the camera. When it was examined by NASA scientists back on Earth, specimens of *Streptococcus mitis* were found to still be alive inside the camera. It was clear that the bacteria had entered the camera before the launch of the lander, and so had survived for 31 months in the vacuum of the Moon's atmosphere, despite immense monthly temperature swings and a complete absence of water (Jones 1995).

Perhaps we shouldn't be all too surprised. Bacteria are also star stuff, but in many cases of a much sturdier kind than fragile human beings. 40 million bacterial cells will typically be found in a gram of soil, while a millilitre of fresh water will contain about a million bacterial cells. It has been estimated that there are approximately five nonillion (5×10^{30}) bacteria on Earth (Whitman et al. 1998) forming a biomass that exceeds that of all plants and animals combined (Hogan 2010). Some of those bacteria are classified as extremophiles, organisms (within all three domains of *Bacteria*, *Archaea*, and *Eucarya*) able to survive in extreme environments – in near-boiling water, sub-zero temperatures, and at nuclear reactor radiation levels (Carey 2005). Extremophiles constitute a core research area within astrobiology, as they expand the number and types of extraterrestrial locations that may be targeted for exploration (Cavicchioli 2002).

Despite their impressive resilience, bacteria grow in colonies that depend on the prevalence of resources such as food. If the population becomes too dense, they suffer. The brilliant science writer Philip Ball once remarked that "for many of us, this sounds a little bit like home" (Ball 2009).

It is a home that is certainly becoming increasingly crowded. Since the end of the Bubonic Plague around the year 1350, the world population has experienced continuous growth. While annual births have reduced since their peak in the late 1990s, and the death rate is expected to increase in years to come, current projections show a continued population increase that could see some 36.44 billion people inhabit the planet in the year 2300 (UN 2004).

Add to such predictions the perceived increase in resource-related threats to our ecosystem, the explosive nature of current international relations, humanity's relative fragility in the wake of crises such as the recent tsunami in Japan and its subsequent nuclear explosions, the perils associated with potential future outbreaks of significant pandemics, and the different dangers of astronomical magnitudes associated with near-Earth object collisions, and it becomes significantly harder to remain optimistic about the survival of our species on this planet alone.

In this context, advocates of space exploration have viewed settlement of the Moon as an initial and logical step in the expansion of humanity beyond the Earth. Already in 1950, Arthur C. Clarke argued that humanity has a choice between expanding into space or death in the wake of cultural and biological stagnation (Clarke 1950). Interviewed in 2001, theoretical physicist and cosmologist Stephen Hawking offered his view:

"I don't think the human race will survive the next thousand years, unless we spread into space. There are too many accidents that can befall life on a single planet. But I'm an optimist. We will reach out to the stars" (Highfield 2001).

Seven years later, in a speech delivered in honour of NASA's 50th anniversary, Hawking focused in on lunar settlement:

"The most obvious [site of a human colony] is the Moon. It is close by and relatively easy to reach. (…) On the other hand, the Moon is small and without atmosphere or a magnetic field to deflect the solar radiation particles, like on Earth. There is no liquid water, but there may be ice in the craters at the north and south poles. A colony on the Moon could use this as a source of oxygen with power provided by nuclear energy or solar panels. The Moon could be a base for travel to the rest of the solar system" (Hawking 2008).

One particular idea for how to design and build such a base is the topic of this chapter. *Moon Dune* is a proposal that brings us back to the cosmic dust and the hardy *Streptococcus mitis*, linking the incredibly small with the unfathomably vast. While scientists and design teams have offered many settlement proposals that entail the use of existing lunar resources, the present text argues that one possible way of using the Moon's regolith as an *in-situ* building material might have been overlooked, and speculates that these particles can be solidified using microbial precipitation.

In brief, *Moon Dune* seeks to explore the opportunities for a novel lunar architecture that utilises a new (and currently theoretical) material based on a process that adds star stuff to star stuff in order to create habitable crater structures on the Moon. While many pieces of this conceptual jigsaw puzzle remain to be both found and put in place at the time of writing, we still believe the project to be a valid addition to the debate on lunar resource utilisation, adding a discursive reflection to the debate without necessarily providing answers to all questions that might be raised. It is an unresolved hypothesis in need of much work, further experimentation, and empirical testing. As Louis Pasteur would have it, fortune favours the prepared mind: our ambition here is primarily to initiate an international discussion and raise awareness of a progressive design response to the challenge of *in-situ* resource utilisation on the Moon, an attempt to add a new *architectural* dimension to that preparation in the context of future lunar living.

28.2 ISRU and Lunar Materials

Moon Dune is based on a terrestrial architecture that pushes its analogous space architecture towards the realms of synthetic biology and beyond, consiliently bridging the biological, digital, and architectural worlds in order to create a foundation for future strategies that could be based on biological computing and programmable built environments. The scheme harnesses the emergent behaviours of cellular agents, thereby creating a *radically modular* design strategy that opens up a much wider array of possible constructed forms and spatial arrangements, while potentially offering a resource-efficient way of sustainably transporting and utilising biologically manufactured, renewable building materials at the molecular scale.

According to NASA, *in-situ* resource utilisation (ISRU) – the idea of living "off the ground" in an outer space context, the strategy of putting local raw materials to practical use as humanity expands outward into the solar system – will "enable the affordable establishment of extraterrestrial exploration and operations by minimizing the materials carried from Earth and by developing advanced, autonomous devices to optimize the benefits of available in-situ resources" (NASA 2008).

This is important because lunar structures deployed for life support would have to be massive, calling for huge amounts of thrust, fuel, and rocket building, in order for us to survive the near-vacuum of the Moon and be able to create a lunar colony of any size. Using ISRU to lower the mass needed to be brought to the Moon is one way of reducing the environmental and financial expense needed to create a permanent human presence there. Harvesting materials from and processing the surface of a planetary body is a more efficient method compared to sending entire readymade habitats out into space. The Moon's regolith can already handle the lunar environment, and ISRU requires the launch of less rockets than would be needed for prefabricated earth structures (Hsu 1995). Alternative, less heavy construction methods are also crucial if we are to lower the cost of a Moon habitat: a normal cement truck on Earth weighs between 9,000 and 14,000 kg and can carry roughly 18,000 kg of concrete; sending it to the Moon involves a truly astronomical expense. Furthermore, sending less building materials into space opens up the opportunity to carry more science equipment instead, as well as to explore and potentially colonise more sites.

The taxonomy of lunar-based structures is divided into three structure classes. This is largely a theoretical construct – in real life, we would almost always end up with a hybrid of multiple classes. In class I can be found pre-integrated hard shell modules ready to use immediately upon delivery. Class II holds prefabricated kits-of-parts that can be deployed through surface assembly after delivery. Class III is the ISRU class that we are interested in here: structures derived from local materials, with integrated Earth components (Cohen and Benaroya 2008). In this latter class, schemes have been suggested that derive all or almost all of their raw materials from lunar resources, albeit at a high cost, the idea being that construction materials for landing/launching pads, radiation shields, thermal wadis, and so on could be produced *in situ* from lunar regolith.

Building blocks ranging from bricks to tiles to ceramic layers on the Moon surface itself have all been proposed in the past, using methods such as self-propagating high-temperature synthesis (SHS), also known as combustion synthesis (Shafirovich 2010). At the NASA/Marshall Space Flight Center, the Habitat Structures project has been developing materials and construction technologies to support development of *in-situ* structures, including proposals to develop extruded concrete and inflatable concrete dome technologies based on waterless and water-based concretes, development of regolith-based blocks with potential radiation shielding binders including polyurethane and polyethylene, pressure regulation systems for inflatable structures, production of glass fibers and rebar derived from molten lunar regolith simulant, and development of regolith bag structures (Bodiford et al. 2006).

In-situ-manufactured lunar materials generally refer to products made from regolith such as lunar concrete, sintered or vitreous masonry, and bulk-regolith applications (radiation shields, roads, berms, and so on) (Cohen and Benaroya 2008). A number of research efforts have been aimed at producing construction materials from the elements available on the Moon, including lunar base construction using concrete (Lin 1987), modules made from lunar concrete (Namba et al. 1988), lunar cement/concrete for orbital structures (Agosto 1988), sulfur-based construction materials (Leonard and Johnson 1988), different cement-based materials (Yong and Berger 1988), and lunar brick designs (Strenski et al. 1990).

In 1998, Gracia and Casanova extended Leonard and Johnson's efforts by suggesting a concept for sulfur-based concrete, in order to overcome the lack of water on the Moon (sulfur is readily available on the Moon) (Gracia and Casanova 1998). Other proposed materials and methods include waterless concrete (Toutanji et al. 2006), an extruded concrete system called Contour Crafting (Khoshnevis et al. 2005), inflatable concrete domes known as Binishells (Bini 1967), and lunar cement made from anorthite using steaming (Horiguchi et al. 1998). Using existing landforms such as lava tubes, underground channels through which lava flows during volcanic eruptions, has also been proposed as potential sites (Daga et al. 1990). Two architectural suggestions that we will return to in the section on construction methods are worth mentioning here: Nader Khalili's domes coated with glaze and vitrified *in situ* by intensely hot incendiary sources inside the structures, and Alice Eichold's cable-suspended habitat shelters inside existing lunar craters (Cohen and Benaroya 2008).

One interesting proposed material is lunar cast basalt, an alternative that would allow for very strong lunar buildings by the fusing of raw lunar soil – rich in silicon and aluminium – at about 1,550 K, which would then be allowed to cool and solidify into a crystalline material stronger than Earth cast basalt, which is similar in strength to cast iron (Hsu et al. 2005). However, each of these technologies and materials has limitations that would have to be worked around. Melting regolith, for instance, takes a lot of solar power (combustion is impossible on the Moon, nuclear power may be too bulky, and the great conductivity of molten regolith makes electric heating dangerous). A very large dish would be needed to cast the regolith quickly. So are there other, less energy-intensive options?

One historical precedent proposal is perhaps of particular importance to this discussion. In 1984, 18 university teachers and researchers got together with a larger group of attendees from universities, Government, and industry on the Scripps campus of the University of California at San Diego (UCSD) for a summer study program that looked at what space resources can be used to support life on the Moon and exploration of Mars. The participants "explored the use of space resources in the development of future space activities and defined the necessary research and development that must precede the practical utilization of these resources" (McKay et al. 1992).

The findings resulted in a report in which Karl R. Johansson is reported to have "studied the literature" and

"found several bioprocesses for the beneficiation of lunar and asteroidal materials by the action of microorganisms. Notably, the extraction of metals by (1) oxidation-reduction reactions, (2) acid leaching, (3) pH alteration, (4) organic complexing, and (5) cellular accumulation of metals due to the action of bacteria on minerals. (…) The process of microbe-enhanced vat leaching, which is used terrestrially to concentrate copper ores, might be applicable to extracting common lunar metals like iron and manganese from lunar rocks and soils."

An accompanying caption, headlined
  "Bacterial Processing of Metal Ores," goes on to explain that although

"most concepts of processing lunar and asteroidal resources involve chemical reactors and
techniques based on industrial chemical processing, it is also possible that innovative techniques might be used to process such resources. (...) Bacteria are already used on Earth to
help process copper ores. Advances in genetic engineering may make it possible to design
bacteria specifically tailored to aid in the recovery of iron, titanium, magnesium, and aluminum from lunar soil or asteroidal regolith" (McKay et al. 1992).

  While Johansson is careful to point out that all of these bioprocesses would require "stringent radiation and temperature controls in closed aqueous environments having elements in which the Moon is deficient, like carbon, nitrogen, and
hydrogen," his proposition points to an alternative and potentially more sustainable manufacturing process through microbial processes. Arguably, one of the main
space exploration questions is whether a viable lunar architecture can be designed
that avoids huge development and maintenance costs while remaining flexible and
efficient. Drawing on an existing speculative terrestrial analogue project that also
uses microbial action as an alternative material strategy, we will now attempt to
draw up an initial framework for such an architecture.

28.3 *Aggrerosion* Tactics: Arenaceous Architecture

In a mineral world like ours, all design is fundamentally about aggregation and
erosion. Giving shape is a matter of deciding where elements will aggregate and
matters densify, and where they will erode or remain as voids. Architecture at its
most fundamental is an act that follows simple laws of arithmetic: we add something to contract a space, we subtract something to expand a space. At its core, architecture is about the manipulation – at different scales – of landscape.

  Solitary and minuscule, a grain of sand may seem insignificant. But this splin-
ter of rock is nevertheless an elementary particle within the material history of
architecture. Take a poetic look at almost any city: what you will see is a conglo-
meration of sand piled into buildings, frozen grains stacked high into the sky, an
urbanly orthogonal version of our deserts' rolling dunescapes, streets and avenues
lined with *castles made of sand.*

  Without sand there would be no brick, no concrete, no glass – even wooden
structures are sanded down to smoothen their edges. Sand is an incredibly renewable material: one *billion* grains of sand come into existence around the world
every second through a cyclic process that sees entire mountain ranges weather
and release tiny splinters. Some of those fragments lithify (from *lithos*, Greek for
'stone' or 'rock'), into a clastic sedimentary rock, a sandstone. As that sandstone
weathers, new grains break free. The majority of quartz sand grains are derived
from the disintegration of older sandstones; perhaps half of all grains of sand have
been through six cycles. A typical mountain will be lowered by a few millimetres
every year (Welland 2009).

  That amounts to a lot of sand. Dry areas cover more than one-third of the
earth's land surface, and desertification – "the diminution or destruction of the

biological potential of the land" – is a major threat on all continents, affecting more than 100 countries in the world. Some estimates suggest that the livelihoods of 850 million people are at risk, spread out across 35 percent of the Earth's land surface (Grainger 1990).

Dune – Arenaceous Anti-Desertification Architecture, the terrestrial architecture project presented by one of the authors to an international audience in 2009, is based on the idea that this awe-inspiring desert landscape, these immense masses of sand, can be manipulated into habitable structures through a strategically controlled local solidification process whereby sand is turned into sandstone via microbially induced carbonate precipitation (MICP) (Le Metayer-Levrel et al. 1999, Nemati and Voordouw 2003, DeJong et al. 2006, Whiffin et al. 2007).

This is a rapidly developing area. The first of the works cited above investigated how the natural process of biomineralisation, more specifically the production by carbonatogenic bacteria of biocalcin (surficial protecting coatings) on limestone surfaces, is feasible within the frame of industrial constraints, leading to "the manufacturing of biological mortars and cements". It also studied different ways through which the nitrogen cycle's metabolic pathways can be stimulated.

The second paper compared the enzymatic formation of calcium carbonate ($CaCO_3$) in unconsolidated porous media, showing that an increase in enzyme (urease) concentration enhances the rate of $CaCO_3$ production, while increases in reactants (urea and calcium chloride) increases the quantity of produced $CaCO_3$. An increase in temperature enhanced both the production rate of $CaCO_3$ and the extent of conversion (furthermore, plugging studies with Berea sandstone cores in a core-flooding system confirmed the effectiveness of $CaCO_3$ in reducing permeability).

The third study (which will later be described in further detail) achieved MICP using *Bacillus pasteurii,* an aerobic bacterium that we shall return to, which was introduced to sand specimens in a liquid growth medium amended with urea and a dissolved calcium source; subsequent cementation treatments were shown to increase the cementation level of the sand particle matrix.

The final study evaluated Microbial Carbonate Precipitation (MCP) as a soil-strengthening process by injecting a a 5 m x 55 mm sand column with bacteria and reagents under conditions that were realistic for field applications.

The combined outcome of these and other Earth experiments in microbial geotechnology point toward engineering potentials for carbonate precipitation and other applications of microbiological methods in order to improve the mechanical properties of soil – the processes underpinning microbial construction methods. While these are naturally occurring phenomena, an architectural exploitation of such processes as design and construction methods (the controlled "growing" of stone structures using *aggrerosion* tactics) is a novel *architectural* speculation, a newfangled technology with staggering potentials (Larsson 2011).

The procedure establishes a radical shift in structural thinking, away from both pre-fabricated and common *in-situ* construction, towards the localised biocementation of granular materials. Once we are able to control the solidification of sand into sandstone – the addition and subtraction of densifications mentioned above – we can investigate the possibilities of a completely monolithic architecture based not on components that are attached to each other in order to create a structural system, but on the binary densification of aggregate matter: the sand is either turned into stone, or it is not.

The geological term 'arenaceous' means "consisting of sand or sandlike particles". Used from the mid-17th century, its roots lie in the Latin *arenaceus,* from *arena (harena):* sand. In biology, it is also used to describe animals or plants that live or grow in sand. As the *Dune* scheme is literally built from sand using organisms that live and grow in sand, calling it arenaceous seems highly appropriate.

The mountain-to-sand cycle described above gives us a glimpse of one of nature's rudimentary rules: that erosion follows aggregation follows erosion. Harnessing this perpetual pattern could potentially be a way of making exceptionally sensitive physical interventions in the landscape: exploring the poetic qualities of infinitesimal building blocks – and how they can be utilised in innovative ways to establish an understanding of (near-)microscopic stacking, packing, heaping, massing, mounding, piling, sheafing, and so on – could lead to an entirely new taxonomy of buildings, series of new architectural typologies that utilise Aeolian forces to define their exterior while harnessing the power of biocementation processes to sculpt their interiors.

Putting aggregation and erosion together was the starting point for turning *Dune – Arenaceous Anti-Desertification Architecture* into a study in *aggrerosion,* a design strategy based on the accumulation and reduction of granular materials. As a building material, grains of sand can be employed across a gradient of conditions: granular mass, dynamic medium carried by Aeolian forces, compressive membrane, solid stone, and so on. The granularity allows for both structural and constructional inventions: the sand can for instance be used both as primary ingredient of the finished structure and as its formwork during construction. If we solidify parts of a dune to hold it in place, the non-solid parts will continue moving, making it possible to plan for Aeolian forces to excavate (part of) the structure for us. A sandy desert is thus a construction site waiting to be deployed: Aeolian forces become our extended, invisible hand in the design process – the wind carries the material to the site, and then carries the excess material away from it again.

As soon as we begin to view them in this way, the sand dunes are conceptually transformed into readymade buildings. All we need to do is solidify the sand wherever we need solid surfaces, and then excavate the sand we don't need – or have the wind excavate it for us. Using the existing sand masses as our base material, we can sculpt arches, gaps, caves, pockets, and other patterns straight into the dune. As long as we take care to design *with* the aggregation rather than *against* it, we can simply allow saltation (a type of particle transport) to propel the grains in place, and then, once the sand has aggregated into a beneficial form, use an intelligent strategy for how to solidify it, petrify it, freeze it into a solid state that speaks of that one moment in time. This is the novel idea at the heart of *Dune – Arenaceous Anti-Desertification Architecture.*

28.4   Project Analogue: Dune – Arenaceous Anti-Desertification Architecture

The *Dune – Arenaceous Anti-Desertification Architecture* proposal (henceforth abbreviated to *Dune*) is the scheme out of which *Moon Dune* was born, and therefore the analogue project on which the present proposition is based. This terrestrial idea has been described in some detail elsewhere (Larsson 2011); what follows is a condensed outline of the scheme covering enough detail to establish its obvious connection to the project at hand.

The starting point for *Dune* is an environmental threat that we have already touched upon: desertification. While sand dunes cover only about one fifth of our deserts, the edges of those extreme areas are good places to introduce barriers of greenery in order to keep the sands in place, counter soil degradation, and stop the dunes from migrating. The idea of a "Green Wall for the Sahara" was first proposed by former Nigerian president Olusegun Obasanjo in 2005. The initiative originally called for 23 African countries to come together in order to plant trees across a 7,000-by-15 kilometres wide stretch just south of the Sahara desert – a total area of 105,000 km^2, or 10.5 million hectares (AU/CEN-SAD, 2009).

Dune is an architectural speculation that grew out of the notion that an architectural support structure could be constructed for this Great Green Wall for the Sahara and Sahel Initiative (GGWSSI) through the creation of a network of solidified sand dunes in the desert. The final outcome is a habitable anti-desertification structure made from the desert itself, a sand-stopping device made out of sand. The spatial pockets created within the resulting solid, sedimentary rock structure would help retain scarce water and mineral resources, while also serving as programmable spaces – a habitable wall straddling an entire continent, binding villages, people, and countries together.

Organising grains of sand into novel spatial structures through microbially induced carbonate precipitation (MICP) using the microorganism *Bacillus pasteurii*, also known as *Sporosarcina pasteurii*, an aeorobic bacterium pervasive in natural soil deposits (Le Metayer-Levrel et al. 1999, Nemati and Voordouw 2003, DeJong et al. 2006, Whiffin et al. 2007) could create a very narrow and roughly 6,000 kilometre long pan-African city as a mitigatory measure against the Sahara's shifting sands.

In 2007, professor Jason DeJong at the University of California at Davis proposed a sustainable alternative to the often toxic chemicals that are injected into soils to make them withstand the phenomenon of *liquefaction*, which can cause more damage during an earthquake than its tremors: use *Bacillus pasteurii*, natural bacteria that live between sand grains and in soils, and that cause calcite (the most stable polymorph of $CaCO_3$) to precipitate, a process that glues the grains together and turns loose sand into solid rock. "Starting from a sand pile, you turn it back into sandstone," DeJong explained, before making clear that there are no toxicity problems, that the treatment could be applied after the construction of a building, and that the structure of the soil doesn't change as the void spaces between the grains are filled in (The Engineer, 2007).

Chemically, in the medium containing urea and calcium chloride, the *Bacillus pasteurii* produce enzyme urease that hydrolyses urea by the following reaction:

$$(NH2)2 + 3H2O \rightarrow 2NH+4 + HCO-3 + OH- \qquad (28.1)$$

The bacteria hydrolyse urea to ammonium, an enzymatic reaction that increases pH, whereby hydrocarbonate is produced, which in turn precipitates calcium as calcium carbonate under high pH, which binds the grains of sand together (Kucharski et al. 2005, Ivanov and Chu 2008).

Inject sand with cultures of *Bacillus pasteurii*, feed them well and provide them with oxygen, and they will solidify loose sand into sandstone. DeJong and his colleagues experimented with sterilized sand and bacteria, and were able to control and monitor nutrients, oxygen levels, and other variables to determine exactly how the bacteria hardened their sand specimens. The microbes were added in a liquid growth medium amended with urea and a dissolved calcium source (nutrients are required by microorganisms for cellular material – carbon and minerals – as well as for a source of energy). Subsequent cementation treatments were passed through the specimen to test the increase in cementation level of the sand particle matrix. The results for the MICP-cemented specimens were compared with those of gypsum-cemented equivalents, and were assessed by measuring the shear wave velocity with bender elements.

The experiments indicated that the MICP-treated specimens exhibit a non-collapse strain-softening shear behaviour, with a higher initial shear stiffness and ultimate shear capacity than untreated loose specimens – a behaviour similar to that of the gypsum-cemented specimens, which represent typical cemented sand behavior. SEM microscopy showed formation of a cemented sand matrix with a concentration of precipitated calcite forming bonds at particle-to-particle contacts, while x-ray compositional mapping confirmed that the observed cement bonds were comprised of calcite.

For MICP to be effective, the microorganism must be capable of CO_2 production paralleled by a pH rise in the surrounding environment to an alkaline level that induces precipitation of calcium carbonate. This forces calcium and carbonate dissolved in the water to combine and form crystals of calcium carbonate – calcite – the same natural cement that binds together sandstone, as well as manmade concrete. Aerobic microorganisms capable of consuming urea as an energy source (such as *Bacillus pasteurii*) are particularly good candidates because they provide two sources of CO_2 – respiration by the cell and decomposition of urea. Furthermore, cells of *Bacillus pasteurii* do not aggregate, which ensures a high cell-surface-to-volume ratio, an essential condition for efficient cementation initiation (DeJong et al. 2006).

How long does the microbial part of the construction process take? According to DeJong's experiments, factors critical to the success of the microbial treatment include pH, oxygen supply, metabolic status, concentrations of microbes, ionic calcium in the biological and nutrient treatment flushes, and the timed sequence of injections. While the treatment time is dependent on numerous factors such as microbial concentration, reaction kinetics, and soil characteristics, DeJong found the maximum shear wave velocity versus time value to be 1,700 minutes, and the relative amount of cementation resulting from each treatment to reach its peak towards the fifth or sixth injection. This would indicate that an initial sandstone surface could be created within the dune in approximately 28 hours, and that the structure could reach its optimal structural strength in a single week (1,700 x 6 = 10,200 minutes) (DeJong et al. 2006). In private conversation with one of the authors, however, DeJong has indicated that variations are likely to occur, and that "the more patient you can be, the better" (DeJong 2008).

Different methods for flushing the bacteria through the sand could be investigated, as could different sand qualities, in order to find ways of articulating the material properties and characteristics further. The possibilities for research within this area are vast. Once this new way of manufacturing sandstone in order to create new architectures had been established, the next step in the development of *Dune* was to define potential strategies and methods for how to actually construct the microbial sandstone structures *in situ*. Three indicative construction options were outlined: pneumatic balloon precipitation, injection pile precipitation, and surface precipitation.

The first method involves filling a pneumatic structure with a mixture of *Bacillus pasteurii* and the necessary nutrient medium, and then allowing a sand dune to wash over the vessel. Once the wind has sculpted the sand dune into an optimal shape for the solidification process to begin, the microbial solution is distributed through specifically designed apertures in the skin of the structure, solidifying the sand surrounding it into structurally compressive surfaces. Following the solidification, once the interior of the resulting sandstone structure has reached optimal strength, the balloon structure could be recycled for the construction of the next stretch of the scheme. This method could be viewed as similar to that of the application of spraycrete, the high-strength polymer coating designed predominately for the re-surfacing of old concrete and tarmac (Fig. 28.1).

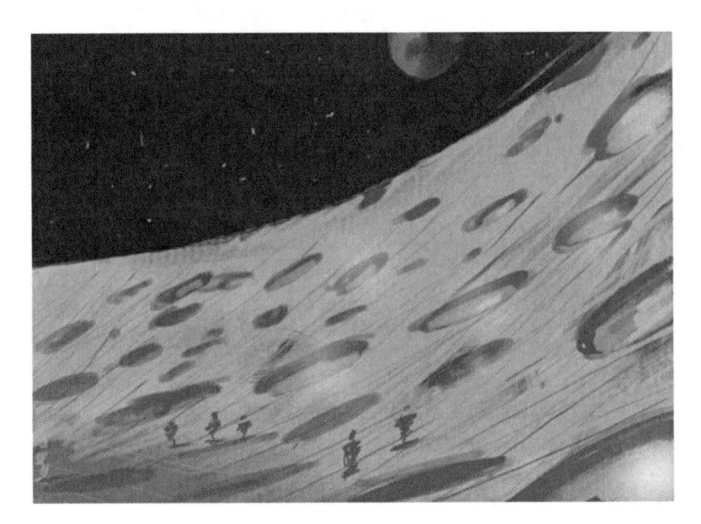

Fig. 28.1 A pneumatic structure is filled with a mixture of *Bacillus pasteurii* and a nutrient medium, sand is allowed to wash over the vessel, the microbial solution is distributed through specifically designed apertures in the skin of the structure, and the sand is solidified into structural compressive surfaces

The second alternative is to use injection piles, normally used for grouting processes. These piles would be pushed through the dune, after which a first layer of bacteria would be distributed through the piles to solidify an initial surface within the dune. The piles would then be pulled up in a controlled fashion, creating almost any conceivable (structurally sound) shape along the way, with the loose sand acting as a jig or mould before being excavated either by hand or by the wind to create our spaces (Fig. 28.2).

This method would be in line with one particular system that has been developed specifically for the purpose of using calcium carbonate as a laboratory cementation agent: the Calcite *In-Situ* Precipitation System, or CIPS. This involves injecting a proprietary chemical solution that causes the precipitation of calcite crystals within the pore fluid and on the surfaces of constituent sand grains (Ismail et al. 1999a). Studies using CIPS have shown that the solidification level and cementation rate can be altered by using different chemical formulations or multiple solution flushes (Ismail et al. 1999b).

Fig. 28.2 Injection piles are pushed through the dune, and a first layer of bacteria distributed to solidify an initial surface within the dune. The piles are then pulled up, creating almost any conceivable (structurally sound) shape along the way, with the loose sand acting as a jig or mould before being excavated

The final construction method would involve solidifying just the surface of the dune rather than creating solidified subsurface spaces, using much simpler distribution methods that might perhaps include the use of machines similar to today's fertilizer spreaders – or even agricultural robots capable of driving themselves using GPS maps and electronic sensors – and to then connect these surfaces to the loadbearing sandstone structure.

If this use of controlled microbial precipitation is not original enough, the idea of working directly within the material volume itself certainly is. This strategy has been imaginatively compared to

"a kind of *infection of the earth*... a vast 3D printer made of bacteria (that) crawls undetectably through the deserts of the world, printing new landscapes into existence over the course of 10,000 years" (Manaugh 2009).

Using the existing sand dunes as granular readymade structures is arguably a radical approach to the sourcing of local materials: a delicately devised construction method that adjusts the composition of the dunescape interior just enough to create a structurally valid composition within it – the kind of *light-touch* technique that makes even more sense in an outer space context.

Needless to say, transferring terrestrial microbial precipitation to a lunar setting is no straightforward matter. Nevertheless, the process can be broken down into its constituent parts and studied as a series of scientific and engineering tasks. First, an appropriate strain of bacteria needs to be singled out. Within the emergent field of microbial geotechnology, the two most promising methods of improving soil characteristics are bioclogging (the reduction of the hydraulic conductivity of soil and porous rocks through microbial grouting *in situ*), often using the microbial production of water-insoluble polysaccharides, and biocementation (the generation of particle-binding materials that increase the shear strengths of soil through structural microbial grouting *in situ*), using processes similar to DeJong's: bacteria-mediated transformations of sand to sandstone (or similar materials such as for instance limestone) using soil bacteria, urea, and calcium ions. Due to the extreme conditions on the Moon, however, we are unlikely to find a microorganism that fits the bill perfectly on this planet – we will probably have to re-engineer the bacteria towards extremophile characteristics. The starting point, however, should be clear: The most suitable microorganisms for large-scale construction and environmental problems are facultative anaerobic and microaerophilic bacteria, and the urease-producing microorganisms are from the genera *Bacillus*, *Sporosarcina*, *Sporolactobacillus*, *Clostridium*, and *Desulfotomaculu* (Ivanov and Chu 2008).

The starter culture of this microorganism then needs to be kept alive and safely transported, together with reactants, to the Moon. An inflatable breeding module either in orbit or on the lunar surface could provide an ideal environment for bacterial reproduction. Nitrogen could possibly be harvested on the Moon (lunar soil brought back by Apollo astronauts contained nitrogen, one of the six most abundant elements in the Universe) or would have to be brought from Earth. Energy could come either from the Sun's rays or through helium-3 (the non-radioactive isotope abundant on the Moon that might fuel future nuclear fusion processes and possibly power the Earth). And as NASA's biomass production chamber at Kennedy Space Center has shown, bio-regenerative life support environments could be used to control not only aerobiosis and a high production of biomass, but also the controlled provision of oxygen and condensed water (Wheeler et al. 1996).

A pressurized tank could be used to release a solution of bacterial culture into the regolith, initiating the solidification process and beginning to outline our first habitable biological machine (Fig. 28.3).

Fig. 28.3 A pressurized tank releases a solution of bacterial culture into the regolith, initiating the solidification process and beginning to outline our first habitable biological machine

28.5 Habitable Biological Machines

In September 2007, a study was published that showed microbes that cause salmonella had returned from space even more virulent and dangerous than before they travelled millions of kilometres in orbit. While the experiments, carried out by microbiologists at Arizona State University, only studied microgravity growth conditions (the gravity on the Moon is about 16.7% of that on Earth), they established that the morphology of the bacteria changed during flight, indicating the formation of biofilms, which are associated with increased bacterial virulence (Wilson et al. 2007). In their abstract, the scientists state that a

"comprehensive analysis of both the molecular, genetic, and phenotypic responses of any organism to the space flight environment has never been accomplished because of significant technological and logistical hurdles. Moreover, the effects of space flight on microbial pathogenicity and associated infectious disease risks have not been studied" (Wilson ibid).

This highlights the nascent nature of the scientific research on microorganisms in space.

We have already seen that in the late 1960s, terrestrial specimens of *Streptococcus mitis* survived for 31 months in the vacuum of the Moon's atmosphere, and that certain extremophile organisms thrive in physically or geochemically extreme conditions that are detrimental to most life on Earth. We don't know whether some such (or other) organisms are in themselves detrimental to life on this planet, and so safeguarding the Earth from potential back contamination will remain the highest planetary protection priority in future exploration missions of any kind. While the issue of contamination is inextricably linked to the topic at hand – we are after all suggesting the use of terrestrial microorganisms in an extraterrestrial context – it lies outside the scope of the present text. Space contamination is of principal and undeniable importance; *Moon Dune* is based on the assumption that the necessary processes can be developed to turn the operations involved into a safe technology in this respect.

Having said that, the probability that terrestrial life can grow in the harsh environment on the lunar surface is rather low, and the current planetary protection policy for the Moon related to forward contamination falls within the not-so-strict Category 1. With lack of liquid water, large temperature extremes, a nearly non-existent atmosphere, and intense ultraviolet (UV), galactic, and solar cosmic radiation, making our bacteria survive for long enough on the lunar surface is likely to be a larger problem than the risk of contamination. On the one hand, experiments carried out on NASA's Long Duration Exposure Facility (LDEF) have indicated that even after six years in space, a large fraction of spore-forming bacteria (*Bacillus subtilis*) will survive if they are not directly exposed to solar UV radiation (Horneck et al. 1994), suggesting that bacteria can be delivered to the surface of the Moon for instance by robotic spacecraft. Indeed, it has been shown that changes in the bacteria's extracellular fluid composition increases their growth in space flight (Brown et al. 2005). On the other, without further engineering bacterial growth on the Moon remains unlikely, though in non-UV-exposed regions such as the permanently-shadowed south polar region of the Moon or below the lunar surface it might be possible even with today's technologies (Glavin et al. 2010).

Although the lunar surface environment may represent a worst-case scenario for the survival of microorganisms and even terrestrial organic matter, several earlier proposals have suggested that the use of bacteria could form part of a lunar *ecopoiesis*, the "fabrication of a sustainable ecosystem on a currently lifeless, sterile planet" (Haynes 1990). This neologism, created by the late Canadian geneticist and biophysicist Robert Haynes to support his vision of terraforming entire extraterrestrial planetary environments in order to support a biological evolution that can proceed independent of further human intervention, has been the topic of several papers, though rarely if ever with the present scheme's aim of implementing the initial seeding of microbial life on the Moon surface in order to open up new architectural possibilities and create a new building material through humanly controlled bacterial growth of habitable structures inside of the lunar regolith. Rather, they have focused on areas ranging from the support and enhancement of plant growth in lunar greenhouses (Lytvynenko et al. 2005; Kozyrovska et al. 2005; Kozyrovska et al. 2010), via regolith biomining (Dalton and Roberto 2007) to space-based pharmaceutical research in low-gravity environments (Horneck et al. 2010). Other processes could include sewage and solid waste processing, pharmaceutical bioreactors (the production on demand of drugs from a "bacterial library"), and biological manufacture of products and implements at a smaller scale than buildings, such as tools, utensils, and instruments.

Any microbial ISRU would need to be based on strategies for 1) transporting the bacteria to the target celestial body, 2) growing them *in situ*, 3) supplying them with nutrients to keep them alive, and 4) deploying them during the construction phase. While this process presents many challenges (most microbes would need elements that the Moon doesn't have in large supplies, such as hydrogen, carbon, and nitrogen), the advantages are potentially vast: with a safe technology in place, rather than coping with the enormous costs of delivering building materials or entire compounds to the Moon, vials of microbes could be transported instead: a renewable resource that can be grown *in situ*. Rather than carrying materials, we could carry microorganisms: machines with which to sculpt the materials. Rather than basing construction on human work in an environment where any EVA (Extra Vehicular Activity)-based missions are difficult, bacterial micronauts could do the work for us. Rather than focusing on *in-situ* control of the construction phase, the work could be monitored and supervised remotely through robotic agents. We shall return to these concerns in the forthcoming discussion of construction methods.

In November 2010, NASA officials announced that a satellite had begun conducting astrobiology experiments in low-Earth orbit. Covered in solar panels, the Organism/Organic Exposure to Orbital Stresses (O/OREOS) satellite is no bigger than a breadbox. Inside it, two types of microbes commonly found in salt ponds and soil – *Halorubrum chaoviatoris* and *Bacillus subtilis* – as well as samples of organic molecules, are used in two experiments. In the first one (Space Environment Survivability of Live Organisms, or SESLO), the dried and dormant astrobugs are rehydrated, fed dyed liquid nutrients, and grown in three sets of colonies – a change in their colour would indicate healthy appetites, helping us understand how microbes react to weightlessness and cosmic radiation. In the second (Space Environment Viability of Organics, or SEVO), four types of organic molecules that exist throughout our galaxy are contained in microenvironments that mimic space and planetary conditions, exposing the compounds to solar ultraviolet (UV) light, visible light, and cosmic particles to measure changes in their molecular stability (Jaggard 2010).

If successful, these experiments will confirm two hypotheses that are particularly important to us here: 1) that strains of microbes can be transported in a dormant, dried state and revived *in situ* to perform their duties, and 2) on a more pragmatic level, that similar low-cost satellites can be used to conduct all kinds of space-based research. Indeed, the entire process of using bacterial precipitation in space could theoretically be carried out and tested using an analogous laboratory in orbit. An alternative to orbital space flight would be a vacuum chamber on Earth.

Bacteria are very resilient, and known to have survived tremendously inhospitable conditions on this planet. When the first bacteria colonised the Earth almost four billion years ago, it was by our current standards a hostile place. The air was hot and full of noxious chemicals such as sulphurous gases released by volcanoes. No ozone blocked out the sun's ultraviolet radiation, and there was no oxygen to breathe. Furthermore, nuclear radiation from decaying U235 was about fifty times as abundant as now (Lovelock 1988). While sending today's bacteria outside of the Earth's protective atmosphere and into much harsher environments might be a great challenge, recent advances primarily within the emergent field of synthetic biology are highly promising.

Not only is it now comparatively easy to envisage a near future in which it is possible to engineer microbes to control features that would make it easier for them to survive and thrive on the Moon, including the ability to grow without some terrestrial elements (in December 2010, it was shown that the strain GFAJ-1 of the *Halomonadaceae* can vary the elemental composition of its basic biomolecules by substituting arsenic for phosphorus to sustain its growth (Wolfe-Simon et al. 2010)). Extrapolating from a 2011 article in Nature (Tamsir et al. 2011), we can plausibly deduce that multicellular computing – the writing of software to control the creation of *genetic circuits* in microbes – "pathways of genes, proteins, and other biomolecules that the cells use to perform a particular task, such as breaking down sugar and turning it into fuel," to quote an interview with co-author Christopher Voigt in MIT's Technology Review, is around the corner (Bourzac 2011).

Voigt and his colleagues have already managed to create basic circuit components in the bacterium *Escherichia coli*, and are now "trying to make the cell understand where it is and what it should be doing based on its understanding of the world" (Bourzac *ibid*). This understanding is aided through giving bacteria the senses of touch, sight, and smell. Equipped with this, the cells can be programmed on an abstract level like robots and made to perform complex, coordinated tasks.

The implication for architecture is breathtaking. The construction of new or augmented organisms lays the groundwork for the construction of new or augmented materials, and the construction of new or augmented materials suggests new ISRU possibilities. Once the materials we use become biologically programmable, the very nature of the discipline is bound to change irrevocably: while interdisciplinary research has already begun to yield *metamaterials* (artificial materials engineered to have properties that go beyond those found in nature), the use of computational processes to control cellular compositions essentially gives us the power to precisely configure the built environment, expanding vastly our capacity to use programmable bacteria as building elements. Soon this new generation of synthetic biologists will be able to control the biological machines that control cells, and when that happens, a whole new realm of architecture opens up, based on "the design rules by which simple logic can be harnessed to produce diverse and complex calculations by rewiring communication between cells" (Tamsir et al. *ibid*).

Legendary French architect Le Corbusier enjoyed talking about the house as a machine for living in, Ray Kurzweil named one of his books The Age of Intelligent Machines, while leading theorist, author, and founding editor of Wired magazine, Kevin Kelly, pronounced the future of machines to be biology (Le Corbusier 1923; Kurzweil 1990; Kelly 1995). Once we learn to create novel materials from computer-controlled synthetic ecosystems, architects can start designing *intelligent biological machines for living in.* Habitable biological machines built from a new sub class of reconfigurable, computational biomaterials that are able to fundamentally change their properties and characteristics beyond traditional notions of self-assembly, in accordance with chemical signal inputs received from a human designer; a spectacular design strategy falling somewhere between *in silico* and *in vivo;* sculpting with code in real time on a microbial scale.

An architect's technical drawing might soon contain new sets of instructions: instead of material specifications, perhaps timings for the release of quorum molecules within the bacterially controlled material volume; instead of dimensions, spatial configurations of the bacterial colonies; instead of surface treatments, programmed instructions detailing the precise mapping of where microbe "robots" will release, emit, repair, cover, coat, glaze, deposit, decompose, etch, polish, grind...

Architecture takes its cues from prevailing scientific and cultural movements, and at the moment, we're working within a veritable explosion of new biology. This naturally shapes today's (or at least tomorrow's) architecture much as the explosion of data and computers shaped architecture throughout the 20th century. While, as Kurzweil has pointed out, utilising full-scale nanoengineering to construct macroscale objects at the molecular scale is "still considered a middle to late 2020s technology" (Kurzweil 2010), there is no reason why we shouldn't start planning for the building of biological machines on the Moon.

28.6 Introducing *Bacillith*

Author and design theoretician Bruce Sterling has called Branko Lukić "the best design-fiction designer in the world" (Lukić 2010). Some of that design fiction is based on a perhaps slightly whimsical, yet not impossible, imaginary materials, such as an anticipation of the recently Nobel-lauded carbon allotrope graphene, an extension of which Lukić turns into the "incredibly thin and incredibly strong" theoretical ideal material *thinium* (Lukić *ibid*).

In the narrative world, there is a long and proud history of such fictional materials, from the super-hard *adamant* (the material that the fictitious flying island is made of in Jonathan Swift's Gulliver's Travels, which takes on magnetic properties allowing its hovering ability), via Isaac Asimov's invention *thiotimoline*, a fabricated compound with chemical bonds that project back to the past and into the future, through to the *balthorium* used in the Russians' doomsday device in Stanley Kubrick's Dr Strangelove (Swift 1726; Asimov 1948; Kubrick 1964).

Science is also no stranger to imagined materials: the success of Russian chemistry professor Dmitri Ivanovich Mendeleev's periodic table in 1869 (as opposed to that published independently by Julius Lothar Meyer the following year) came partly from his leaving gaps in the table for corresponding elements that had not yet been discovered. In the years following Mendeleev's publication, the gaps were filled as chemists discovered more chemical elements. While the last naturally occurring element to be discovered was francium in 1939, the periodic table has since grown with the addition of synthetic and transuranic elements (Atkins 1995; Ball 2002).

In some cases, scientists appear to have been inspired by human designs in their search for natural elements: the carbon allotrope fullerene, and in particular the C_{60} molecule of that allotrope, was named after American engineer, author, designer, inventor, and futurist Richard Buckminster Fuller: it is now known as buckminsterfullerene, or buckyballs. Consisting of 60 carbon atoms, this molecule closely resembles a spherical version of Fuller's geodesic dome, and the trio Kroto, Curl, and Smalley was awarded with the 1996 Nobel prize in chemistry for their discovery of the fullerene (Encyclopædia Britannica 2011). Fullerenes have since been found in nature (Buseck et al. 1992) as well as in outer space (Cami et al. 2010). In other cases, as with Lukić's *thinium*, designers and architects have conceived of objects and structures from imagined materials, or rather materials not yet discovered or created, in order to think and design beyond the confines of the known world.

This has historically been the case with several proposed materials for lunar construction, including the imaginary aggregate building material *lunarcrete* (or *mooncrete*), first proposed by Larry A. Beyer of the University of Pittsburgh in 1985, consisting of regolith mixed with water and cement (Beyer 1985). An alternative to that proposal tried to bypass the water by instead using molten sulphur (Omar and Issa 1994). Other theoretical lunar materials include sintered regolith (loose particles that have been collected, placed into forms, compressed, and heated via microwaves or solar energy to sinter or fuse into bricks, blocks, and other shapes, similar to terrestrial masonry, and, if pre-stressed, even into beams and slabs); lunar glasses (in some areas up to 40% of the regolith is glass, and even the regolith itself – if melted and cooled rapidly – forms glass); and cast regolith (a material very similar to terrestrial cast basalt) (Ruess et al. 2006).

Indeed, synthetic basalt was manufactured in ancient Mesopotamia, possibly as early as in the early second millennium B.C. (Stone et al. 1998) – the same Mesopotamia where the first artificial brick was fired around 3000 B.C. (Weston 2008). In terrestrial architecture, those are early examples of building with anthropic rocks, that is, rocks that have been made, modified and/or moved by humans. This fourth class of rocks (supplementing geology's long-identified igneous, sedimentary, and metamorphic groups) was proposed by US geologist Dr James Ross Underwood, Jr, who further proposed that anthropic rocks be subdivided into three sub-classes: anthropogenic rocks (those made by humans), anthropotechnic rocks (those modified by humans), and anthropokinetic rocks (those moved by humans) (Cathcart 2011).

On 20 July 1969, one peculiar anthropic material that is anthropogenic, anthropotechnic, and anthropokinetic was created on the Moon, as Neil Armstrong and crew fired up the ascent engine of their Apollo 11 Lunar Module *Eagle* in order to blast off the lunar surface and re-dock with the Command Module. The rockets must have eroded and melted some lunar regolith, leaving behind the first *in-situ*-created, anthropic lunar rock in history (Cathcart *ibid*).

That method of manufacturing Moon materials may seem somewhat violent, but is essentially analogous to *in-situ* sintering, lunar glass melting, and regolith casting, all of which require very high temperatures. The creation of the theoretical material proposed here, *bacillith,* would call for a much less forceful process: the addition of terrestrial bacteria to the lunar regolith.

When Neil Armstrong stepped off the *Eagle* in 1969, one of the first things he noticed was the nature of the surface of the Moon: how fine the sediment was. Though there are outcroppings of rock and crater ridges, this sediment – the regolith – is the most prominent surface type on the Moon, filled with ferrous metals. This lunar dust (and sand) is very dangerous to breath in, sticks to space suits, and easily clogs machinery. If the surface is disturbed, as for instance during a space craft landing, the regolith particles can be accelerated to incredible speeds, the blast of a lift-off potentially even driving some of the dust to circle the Moon before settling down again on the surface, due to the lack of air resistance and relatively modest pull of gravity (Miller and Coit 2008).

George P Merrill first defined the term regolith (which combines the two Greek words rhegos (ρηγος), "blanket," and lithos (λίθος), "rock") in 1897, writing that this "entire mantle of unconsolidated material, whatever its nature or origin, it is proposed to call the regolith" (Merrill 1897).

Lunar regolith is a 4.6 billion-year-old layer of loose, heterogeneous material covering solid rock. It includes dust, soil, broken rock, and other related materials, and covers nearly the entire lunar surface – bedrock is exposed really only on very steep crater walls and occasional lava channels. The regolith was formed by the impact of large and small meteoroids and the steady bombardment of micrometeoroids, as well as solar and galactic charged particles breaking down surface rocks. In mare areas, the regolith is generally four to five meters thick, while the older highland regions are covered in ten to 15 meters of regolith (Heiken et al. 1991). The thicknesses vary, however, extending from the lunar surface to depths ranging from centimetres to several hundred meters (Fritz et al. 2009).

Furthermore, the impact of micrometeoroids, at times travelling faster than 96,000 km/h, generates enough heat to melt or partially vaporise some dust particles, welding them together into glassy, jagged-edged agglutinates reminiscent of terrestrial tektites (Mangels 2007).

The top 150 mm or so of the Moon's regolith is composed of loosely compacted fine soil particles. Below this surface the density increases rapidly, with depths below one meter achieving relative densities greater than the value 95, "Very dense," on Peter Eckart's scale (Eckart 1999, Ruess et al. 2006). Grain size in the regolith ranges from sub-micron particles near the surface, to larger, millimeter-sized grains below the surface (Fritz et al. *ibid*). Since the heavy bombardment ceased about 3.8 billion years ago, the upper several meters of the Moon have been modified by micrometeorite impacts. That regolith may be much finer grained than typical regolith as it developed on hundreds of meters of fine-grained material. Should this be the case, the physical properties of the regolith – such as porosity, thermal conductivity, shear and bearing strength, angle of repose, and tribology – is hard to predict. A finer grain size provides a much larger surface area for a given mass of regolith, which could enhance adsorption of H2O and other volatiles and their reaction with regolith grains (Taylor et al. 2004).

The continual ultraviolet radiation and micrometeoroid bombardment on the Moon's lit side has pulverised the lunar surface into a powdery texture and left it with a significant electrostatic charge, a rough, jagged microstructure, and a high surface energy – the regolith providing a unique granular material whose properties are little understood (Fritz et al. *ibid*). While the jaggedness of the regolith grains has been shown to cause lunar concrete to require more mixing water than well-rounded terrestrial sands, it also increases the bond between cement paste and aggregate, providing increased strength (Lin et al. 1986). The fine-grained nature of this layer of unconsolidated debris actually makes it a good ingredient for any concrete-like *in-situ*-manufactured material, as this property provides a substance with a minimal amount of air gaps within the cementitious material. The finer the grains, the more homogeneous the final *bacillith* will be, and homogeneity usually provides more strength. Furthermore, vaccum welding of small grains to larger ones could potentially turn the regolith into a stronger material than terrestrial concrete. This regolith, the bulk of which is a fine gray soil with a bulk density of about 1.5 g/cm3 (Meyer 2003), is our aggregate, the first part of the three that make up the recipe for *bacillith*.

The other two basic ingredients follow those of the *Dune* project: a bacterium producing calcite or a similar mineral (analogous to the *Bacillus pasteurii* used on Earth), and a solution of water and nutrients. The water would either be supplied from the moon, or by combining oxygen with hydrogen produced from lunar soil (Ruess et al. *ibid*). Experiments with lunarcrete have shown that steam might be used to cure a dry aggregate/cement mixture, a method that might be re-engineered to work with bacteria as well (Herbert 1992). Some of the necessary nutrients (such as urea) could be produced by recycling human waste in space.

The main difference from lunarcrete is that rather than manufacturing cement by beneficiating lunar rock with a high calcium content, we grow a biological binder in the form of bacteria *in situ*, leaving us with less human/robotic work in terms of harvesting cement to be carried out on the Moon itself (though admittedly more work caring for the microorganisms).

Some practical challenges have already been addressed in discussions about lunarcrete. The casting would require a pressurised environment, as casting in a near-vacuum would simply result in the water (or nutritional solution in the case of *bacillith*) to evaporate. Proposed solutions have included the use of steam injection of a premixed aggregate/cement blend (Ruess et al. *ibid*), and the use of an entire pressurised concrete fabrication plant that could produce pre-cast lunarcrete blocks (Bennett 2002). The lack of tensile strength (analogous to that of terrestrial concrete) could be overcome with lunar glass (Ruess et al. *ibid*) or Earth-imported kevlar (Bennett *ibid*).

It has been argued that lunarcrete will absorb gamma rays, and that it would require less energy than production on the Moon of steel, aluminium, or brick (Bennett *ibid*). This is likely to be the case with *bacillith* as well. As with lunarcrete, a *bacillith* building would not be an airtight material, and so would have to either be given a coating of epoxy or some similar polymer, or be combined with a substructure such as a pneumatic interior vessel.

As is the case with lunarcrete, whether or not a *bacillith* structure would be able to withstand the micrometeor bombardment remains to be seen. The performance of any material used for lunar construction needs to be examined in light of the fact that structures that would be unsuitable on Earth may be adequate for the reduced-gravity lunar environment (Chow and Lin 1989). Even if we were only able to produce thin panels of *bacillith*, there would still be some merit to the concept: while unusable on Earth, on the Moon they could still be strong enough to make cheap and sturdy unpressurised structures, such as garages for lunar vehicles.

What about the third ingredient in the *bacillith* recipe, the producer of our binding agent, the bacteria? The authors are not micro- or geobiologists; our primary interest lies not in finding the right microbial workforce, but in considering how the theoretical material resulting from their efforts can be applied in the context of space architecture. Facilities have been developed for microbiologists to study the adverse effect of extraterrestrial conditions on microorganisms over the last 60 years, furthering our understanding of the physiological requirements for survival

– a good review of this history is in (Olsson-Francis and Cockell 2010a); we remain hopeful that the scientific community can rise to the challenge of identifying, or re-engineering, the ideal extremophile nominee.

Specialists would need to produce a bacterial candidate that can withstand and thrive in an environment featuring strong radiation, a temperature difference of - 155 to 100 °C, a gravity of 0.17g, an irregular and weak magnetic field, an atmospheric pressure that approaches vacuum, igneous surface materials, no or close to no annual precipitation, and no organic life. In particular, we need bacteria that produce as much binding agent material as possible (while being starved of the oxygen the *Bacillus pasteurii* of the *Dune* project use to metabolise the calcium lactate into calcite) using the least amount of water.

Henk Jonkers of Delft University of Technology has carried out interesting experiments during which he has created a self-healing biological material by packing concrete with calcite-producing bacteria that can take on a dormant spore state for long periods (up to 50 years at least), without nutrients or water, before being activated, a technique Jonkers compares to seeds waiting for water to germinate (McAlpine 2010).

In recent news, the cyanobacterium *Anabaena cylindrica*, commercially used as a natural fertilizer in rice paddies, has been proposed as an organism that could potentially be suitable for space *in-situ* resource use due to its biomining capabilities in environments that call for extreme survival measures (Olsson-Francis and Cockell 2010b).

In another recent study, scientists have discovered three new multicellular marine species that appear to have never lived in aerobic conditions, and never metabolized oxygen (Danovaro et al. 2010).

Put these three advances together, and we may be getting closer to fulfilling the final part of the *bacillith* recipe. If the calcite-producing property of *Bacillus pasteurii* was combined with a spore-forming bacterium (maybe an oligotrophic organism capable of growth in nutritionally limited environments, or maybe a lithoautotrophic microbe deriving its energy from reduced compounds of mineral origin) that can survive in aerobic conditions and withstand the radiation levels on the Moon, we might have found our preferred exophile.

28.7 From Terrestrial and Lunar Cave Dwelling to *Cratertecture*

As Robert Heinlein famously observed, the Moon is a harsh mistress (Heinlein 1966). It is also comparable to a 4-billion-year-old desert (Phillips 2006). While the Moon does have an atmosphere (composed of elements that include helium, argon, sodium, and potassium), it is extremely thin and too weak to provide sufficient protection for human beings. The thin lunar atmosphere means that sunlight strikes the lunar surface with an intensity unknown on Earth. It also creates radiation and micrometeorite impact concerns for humans, structures, and equipment deployed on the surface. Three radiation sources affect the Moon: galactic cosmic rays, solar wind particles, and, in particular, solar flare particles (which are highly detrimental to equipment). Micrometerorites are abundant to say the least: about 30 000 zap pits per year will be formed on a given square meter of the lunar surface, and any Moon structure must be able to absorb and withstand this type of constant bombardment. The 28.5 days it takes for the Moon to spin about its axis not only impacts on circadian rhythms: during the fourteen-day lunar days and nights, equatorial temperatures can range from 400 K to 100 K (127 °C to -173 °C), with rapid temperature changes at sunset and sunrise. In the polar regions, many crater floors are in permanent shadow. These areas, with their constantly cold (about 80 K, or -193 °C) temperatures, are our best hope for locating permafrost (water) on the Moon (Seybold 1995).

This extreme change in temperature throughout the day-night cycle is one strong argument for subsurface dwelling, as at a depth of one meter beneath the surface, the temperature is a nearly constant 238 K, or -35 °C (Heiken et al. 1991). Another justification for working beneath the surface of the regolith is of course the use of natural rock mass as radiation protection for habitats and workspaces, and as protection from micrometeorites, somewhat analogous to how the *Dune* project offers protection from desert sandstorms on Earth. A third reason could be planetary protection considerations: containing the primary human habitation and work activities within the confines of subsurface environments is very much in line with recent approaches to forward contamination, which call for localisation and zoning of human impact in the same vein that for instance the Antarctic is currently being protected as a *Special Region* (Boston 2010). A fourth aspect that speaks in favour of using the existing regolith is that a habitat constructed "organically" from regolith facilitates a "blending in" to the landscape, similar to terrestrial rock dwellings – a more integrated, but also less temporary, approach. A fifth and final point is the typological/programmatic benefits of working directly with the Hakka-like crater structures, as outlined in the construction section above.

Whether natural crater cavities or artificial cave structures are used, the challenges of living in a subsurface environment may be easier to overcome than developing methods to ameliorate the effects of radiation as experienced on the lunar surface. The problem of protecting humans on the surface in space suits even for limited durations is intractable enough (cf. Carr and Newman 2008; Zeitlin et al. 2006; Hodgson 2001).

The use of *bacillith* construction techniques would allow for existing lunar (crater) caves to be structurally secured, while new cave-like structures could be created where needed (Fig. 28.4). This would provide a degree of randomness and unpredictability in morphological terms, but a level of control when choosing the site. We can use the natural structures on the Moon to integrate our structure into the existing landscape, but are not necessarily dependent on existing craters or other landforms. The fluid nature of *bacillith* means *Moon Dune* habitats could retain the formal language of the lunar environment, or turn into any form that could be made with more common terrestrial materials such as concrete or adobe: the architecture could be orthogonal rather than sinuous, asymmetric rather than symmetric, more like a cave or more like a bunker, and so on.

Fig. 28.4 The use of *bacillith* construction techniques allows for existing craters to be structurally secured, while new cave-like structures can be created where needed

A series of pockets of sponge-like spaces could be created around the rim of the crater, *bacillithic* voids lined with pneumatic structures in which to gather for a conversation, a meal, a workout session, a prayer. As the microbes close some of the gaps in between the grains of regolith to turn it into *bacillith*, yet another space will be added to this architectural lunar refuge. While the space architect gives up a certain level of formal control to nature as the bacteria solidify the lunar dust, she gains a boundless beauty, the traces of bacteria being harnessed to sculpt selected parts of the Moon into a new, habitable environment.

A good review of some differences between lunar and terrestrial construction is found in Cohen and Benaroya (2008): The hard vacuum on the lunar surface means some materials (including many plastics) are not chemically stable on the Moon. Construction in vacuum also poses the challenge of outgassing (loss to vacuum) of oil, vapours, and lubricants – including, potentially, water from *bacillith* or any other concrete-like material. Vacuum tribology means a drill bit might fuse with the regolith it drills, while blasting creates gas overpressures making it hard to predict explosion results (particles could even go into orbit around the Moon). Habitats need to contain hardshell or inflatable pressure vessels to counter this vacuum. While at 1/6-g, the stability of human and robotic movements is impaired, the same partial gravity also means a given structure typically offers six times the weight-bearing capacity on the Moon as it would on Earth; however, under reduced gravity artificially compacted regolith will be less consolidated and provide less containment than undisturbed regolith. The different properties of regolith and terrestrial soil pose unique problems for excavation, trenching, backfilling, and compacting activities. On top of this, shielding from radiation exposure and episodic solar flares constitutes a tough design challenge for any lunar architecture. Shielding is also needed to protect against the steady flux of micrometeoroid particles. Structures aimed to support human habitation must also withstand internal pressures. Furthermore, the lunar thermal environment poses several challenges, including the structural threat of material fatigue, as well as issues with human thermal comfort, which has led to proposals for ISRU-based heat sinks for energy storage (Sullivan 1990). It is also understood that lunar regolith is very difficult to excavate.

And those are considerations to account for before the actual construction even begins. The concept of health and safety takes on a new meaning in an environment hinged on spaceflight and EVA. The slow lunar diurnal cycle has direct implications for human (and systems) performance as well as the quality of life. The on-site construction team would by necessity have to be kept small, and would operate in pressure suits and via tele-operated machines. Fine, abrasive dust would work its way into every interface. Pair this with the remoteness of lunar construction sites and the high cost of transporting machines, materials, and (wo)men into space, and we have a situation that suggests lunar structures should be designed for ease of construction (Cohen and Benaroya *ibid*).

While detailed strategies for applied construction methods clearly lie outside the scope of this chapter, the considerations listed above remain at the heart of our thinking: *Moon Dune* seeks to offer an architecture that minimises both construction time and costs. While using (and thus only having to transport) renewable bacteria for the *bacillith* is one way of coming to terms with the material part of the latter, using existing landforms is another, aimed at the former.

Pragmatic alternatives for *in-situ* lunar construction using *bacillith* might be the topic of papers to come. Here it might be sufficient to mention some of those methods and precedents we have considered so far, including the proposed permanent domed Moon oasis of Alice Eichold, and Nader Khalili's extension of terrestrial techniques developed in antiquity into dome-shaped tied-arch structures using lunar regolith (Eichold 1996; Khalili 1989). *Bacillith* could be used to construct blocks, bricks, or slabs – possibly reinforced with cast lunar basalt or similar. It could be deployed as a surface treatment, injected into the regolith, robotically distributed above or below ground. It could be used to support and make safe existing lunar landforms, or to solidify accumulations of regolith that have been piled into architectural volumes and spaces.

Lunar base development has previously been categorised into stages, usually by the degree of human presence on the moon (Eckart 1999, Toklu 2000, Ruess et al. 2004) – here, we follow both that categorisation and, in particular, an alternative one outlined in the paper A Review of Technical Requirements for Lunar Structures (Jablonski and Ogde 2005), which uses a structure-based definition to focus on technical construction requirements.

In the first (2010-2020) of their three general phases of lunar construction, Jablonski and Ogde introduce non-habitable support and shelter structures for scientific equipment during initial unmanned robotic or manned missions. In the second phase (2020-2030), structures for medium-length stays (up to several months) are developed. In the third and final phase (2030-), long-term structures serving many different purposes – including habitats, observatories, laboratories, production plants, and permanent lunar bases – are constructed.

Though Jablonski and Ogde view Phase One structures as being built on Earth and transported to the Moon (where they are to be either automatically deployed or set up by robots or humans), constructing an initial set of experimental *bacillith* structures seems a viable alternative, in particular if construction is robotic and remotely controlled. Since this phase does not call for the same amount of radiation shielding and air leakage control as those to come, a surface structure could be conceived, made from fine regolith and bacteria – a method closely following the terrestrial analogue of the *Dune* project. Lunar lander laboratories could hold the bacteria until deployed by robots. The final construction could be carried out using a modified large-scale rapid prototyping technology such as a smaller (1-2 m printing width) version of the existing D_Shape printer, coupled with a rover that collects and lays down the regolith. A temporary mould could be inflated, the regolith propped up against it, and the solidification carried out using the bacterial process.

At least since Robert Hooke experimented with them in the context of the rebuilding of St Paul's Cathedral in the 17th century (Jardine 2001), catenary arches have been favoured as compressive systems with optimal loadbearing capability in architecture. In formal terms, we imagine *Moon Dune's* Phase One structures as being catenary half-igloo shapes, perhaps similar in shape to hollowed-out barchan (crescent-shaped) dunes on Earth.

The D_Shape system was developed by Italian inventor Enrico Dini as a terrestrial building system, and suggested as a potential construction technology for a lunar outpost in a paper presented at the 61st International Astronautical Congress in Prague in 2010, following a general study programme contract awarded by the European Space Agency in 2009 to allow for a consortium to assess the concept of using 3D printing technology to build lunar habitats from regolith. The authors (with knowledge from space technology development, 3D printing at building scale, complex architectural design, and robotics) verify that the printing process works properly in vacuum and that the reticulation process takes place using a regolith simulant. Furthermore, the vacuum tests carried out show that evaporation or freezing of the "ink" can be prevented by adopting a proper injection method – a similar concern as that of bacterial survival and performance in the case of *bacillith* (Ceccanti et al. 2010).

The month-long human stays of Phase Two could perhaps utilise a combination of construction on the lunar surface and a subsurface engineering within an existing landform, such as a bowl-shaped crater with a smooth rim and a small diameter, falling within the ALC class of Wood and Andersson's morphologic classification system (Wood and Andersson 1978). If the crater is located close to one of the lunar poles, it will benefit in terms of sun exposure from the "peaks of eternal light," remaining in sunlight almost at all times, with the sun revolving very close to the horizon at an elevation of about ±1.5 degrees, thus allowing for optimal use of solar arrays for electrical power (De Weerd et al. 1998). A crater with an existing lava tube or similar would be ideal, as this would offer the opportunity to test different ways of creating habitable *bacillith cratertectures* both above and below ground. The lava tube or crater cavity could be vitrified *in situ* using *bacillith* and pressure sealed to create a subsurface habitat, in side of which supporting surface structures could be created either the same way as described above (employing the D-Shape technology), or using gabions (regolith-filled fabric cylinders) and (perhaps again inflatable) mould sections to keep the regolith in place during the bacterial solidification process (Fig. 28.5).

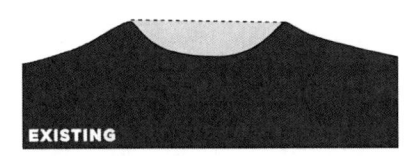

Fig. 28.5 The lava tube or crater cavity is vitrified *in situ* using *bacillith*, and pressure sealed to create a subsurface habitat, inside of which supporting surface structures are created using large-scale 3D printers

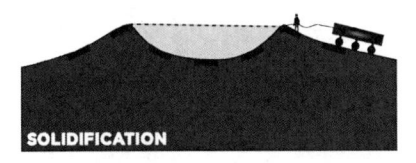

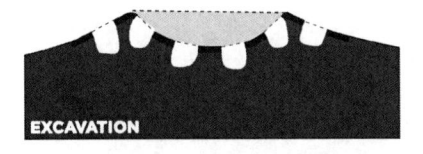

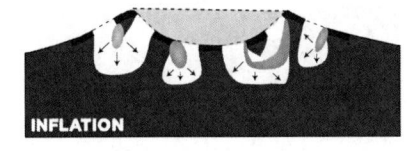

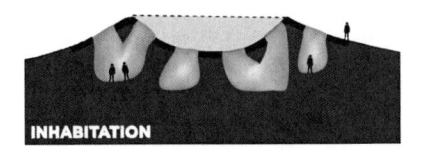

Any *bacillith*-based architecture (as indeed most lunar ISRU-based projects) is likely to employ a hybrid strategy, with an unpressurised shield structure – which could be an existing landform secured with *bacillith* – protecting a pressurised interior. This allows for a separation of the sealing capability (the pressurisation technology) of that interior from the thermal, mechanical, and radiation-shielding functions of the exterior.

An inflatable module could provide the pressurised shell containing the environment's breathable environment, with the external shell made from non-pressurised *bacillith*. While the porosity and permeability of *bacillith* might be engineered to provide sufficient sealing capability, for a Phase Two habitat constructed for up to ten people to conduct science and investigate possible locations for a permanent lunar base, such a development would probably not be economically viable. Furthermore, the internal pressures of a monolithic *bacillith* system would create strong tensile stresses in the intrinsically fragile shell – removing this pressure reduces the loads, which reduces the amount of material that needs to be consolidated, which reduces the amount of bacteria that needs to be grown and distributed (Ceccanti et al. *ibid*).

Fig. 28.6 Inflatable insertions allow for deployment once the cave is completed to achieve a pressurised and habitable volume

Inflatable insertions allow for deployment once the cave is completed to achieve a pressurised and habitable volume (Fig. 28.6). Alternatively, the entire inflatable insertion could be used as initial mould for the regolith structure to wrap around, similar to the inflatable moulds of our Phase One structures (Fig. 28.7). In Wolf Hilbertz's *Ice City* project in Fargo, North Dakota (1973), for instance, inflatables were sprayed over with ice to create temporary structures (Manacorda and Yedgar, 2009). Similar practices are employed in the production of temporary concrete shelters that inflate before being cured hard (Hooper 2005).

To a large extent, architecture is about the shielding of interior space from the exterior environment, the forces of nature, and about the selective screening of sensory data (Hall 1966). As we know, this is particularly relevant in a lunar context with regards to radiation, extreme temperatures, pressure, micrometeorites, and so on. In order to answer to the main purpose of a Phase Two shelter – to protect the crew (and then their equipment) from radiation and micrometeoroids – a typical wall thickness of one to two meters would be called for (depending on orientation and safety margin).

Fig. 28.7 The entire inflatable insertion can be used as an initial mould for the regolith structure to wrap around

Such a wall section would also provide for an adequate thermal insulation. Three main structural loads would have to be countered: gravity, moonquakes, and thermo-elastic loads between parts of the structure that are sunlit and those that are shaded, including the interior and exterior surfaces of the walls (Ceccanti et al. *ibid*).

The long-term structures of Phase Three call for permanent bases able to house many people comfortably for extended periods of time, and for an expandable approach that allows the capacity of the architecture to be increased. In this phase, processing and production plants will be built to further develop the *bacillith* production (and other lunar-based material and construction technologies). The *in situ* resources will be used for more than simple regolith-based shielding, creating habitable spaces with a more advanced radiation shielding that can supply comfortable living conditions (as opposed to mere survival), as the inhabitants will use the architecture for a longer time. We envision this stage as being formally and programmatically based on a strategic exploitation of the entire range of craters up to the TYC class of the morphologic classification system, taking full advantage of terraced craters (or parts of craters) with crenulated rim crests, large flat floors, and a diameter of 30 to 175 kilometers (Wood and Andersson *ibid*). This would be ISRU on a massive scale, providing the basic infrastructure for a further colonisation of heavenly bodies further out in space.

A *cratertecture* on that scale actually has a terrestrial analogue, albeit one that was never constructed. In 1971, a team of architecture students led by Buckminster Fuller started designing *Old Man River's City*, a single-community "dwelling machine" for 125,000 inhabitants. The outer surface of the rim of this 1,500-meter-diametre mega structure would be terraced to provide 25,000 earth-sheltered garden homes with exceptional views of the Mississippi River scenery outside East St. Louis, Illinois, while the inner surface would be terraced for communal use. A "moon-crater-shaped" design, its optimum-efficiency ethos called for "the whole terraced crater structure, inside and out" to "be of thin-wall reinforced concrete" (Fuller 1981).

That's not the first time craters have been used as readymade buildings, nor is it the first case of centripetal/centrifugal, circular building typologies. French architect Jean-Louis Chanéac proposed an entire city made from craters in his *Villes Crateres* (1963-1968) project (Busbea 2007). *Eco-city 2020* is a proposal designed by architecture studio AB Elis Ltd for the rehabilitation of the Mirniy industrial zone in Eastern Siberia, Russia. In Yemen, the seaport city of Aden lies in the crater of an extinct volcano, while in Arizona, artist James Turrell has been working since 1972 with turning the Roden Crater into a monumental work of art (Cook 1999). Perhaps the best terrestrial analogue in programmatic terms are the vernacular dwellings that make up the walled villages of the Hakka people of southern China: large, circular, multi-family, communal living structures sometimes called *tulou*, designed to be easily defensible. The most representative and best preserved examples are probably the Fujian Tulou (UNESCO 2008).

The best example on the Moon is probably the Clavius Base, by an ingenious designer we have already met: visionary science fiction author Arthur C Clarke.

A lunar settlement in Clarke's fictional *Space Odyssey* universe, the base, located at the Clavius crater, is featured in both the novel and film versions of 2001: A Space Odyssey. As depicted in Stanley Kubrick's screen version, Clavius Base features some surface features (a landing pad and control tower, together with other ancillary support structures), though the vast majority of the base is located beneath the Lunar surface (Kubrick 1968). This story also features an example of inter-crater communication, as following his meeting with officials at Clavius, the protagonist Dr. Heywood R. Floyd departs for the crater Tycho on the (fictional low-altitude rocket craft) Moonbus.

At the scale of a Phase Three habitat, a section through the crater would reveal such inter-crater infrastructure nodes together with circulatory communication and services channels running around the entire rim, with communal areas toward the centre of the crater. Interior terraces and cavities accommodate pressurised inflatable interiors, with enough horizontal surfaces to hold interplanetary school and playground areas, spacecraft launching pads, permanent habitation pods, space tourism hotel rooms, interlevel ramps, lunar laboratories, bacterial growth tanks, greenhouse structures, solar panel arrays, and so on. As the floors of large craters are often cluttered with smaller craters and there are many examples of smaller craters breaking into larger ones (North 2000), secondary crater structures could be conceived to support functions of the primary mother crater.

But the programming of the *cratertecture* is perhaps a later concern. First, we need to show that *bacillith* is a valid ISRU building material. In order to do so, we are currently devising two experiments: one that will seek to establish the basic properties of the new material when grown in a controlled laboratory (vacuum) environment here on Earth, one that will use a technology similar to that of the O/OREOS satellite in order to test the actual making of the first-ever piece of *bacillith* in an extraterrestrial environment. Both of these experiments will be outlined in further detail in forthcoming papers.

Falling somewhere in between ISRU classes II and III, *bacillithic cratertecture* could be a way of simplifying and accelerating the construction of lunar habitats while limiting cost. By working within the regolith itself we can minimise the amount of material we need to bring to the Moon and the amount of material we need to bring back. Using existing craters as the foundation of our architecture could help cause as little (actual and visual) damage as possible while optimising our exploitation of the existing lunar assets, in particular if situated within the polar regions.

From an initial crater, over time a nodal network of more and more craters is added to the infrastructure as more and more people – astronauts, scientists, tourists, construction workers, adventurers, or just normal families – go to live in the newly constructed Moon bases. Slowly, over time, new craters are added to the old, their rims and crater floors lighting up at night, the pattern of a massive extraterrestrial city glimmering in the dark cold of space: our first space colony, our first urban *cratertecture*, our first society on a different planetary body, built from minuscule Earth organisms and billion-years-old lunar landforms.

28.8 Moon Dune: The Final Frontier

It is relatively easy to visit the Moon. Even in the late sixties of the Apollo mission, lunar living was just three days' spaceflight from the Earth. What's considerably more complicated is to set up a lunar base where humans can live, work, and breathe.

And still that's precisely what several space agencies are planning to do. While the US plan of the Constellation program (to return manned missions to the Moon by 2020) has now been cancelled, NASA has envisioned a lunar outpost at one of the poles by 2024; the European Space Agency is working towards a permanently manned Lunar base by 2025, and Russia is seeking to establish a permanent base on the Moon a few years later than that. The space race used to be about sending a man to the Moon. Today it is about sending construction workers there to start building the first lunar capital.

The first buildings of that capital will probably not be made from *bacillith*, or lunarcrete, or lunar basalt, or any other ISRU material. Those initial planetary outposts will be composite, pre-integrated surface structures designed to cater for exploration rather than habitation.

And yet with a little planning it seems feasible to incorporate ISRU strategies right from the beginning as we begin to write the next chapter in the history of our space colonisation. We need to find a way of living off the land in space. Air, water, and electricity are the key requirements of any human habitation on any planetary body. Since it would be prohibitively expensive to provide supplies from the Earth, these would probably have to be produced on the Moon itself. The raw materials are there, though they may not be easily extractable.

Water poses a particular challenge: while a "significant amount" (about 110 litres) of water was found when NASA's Lunar Crater Observation and Sensing Satellite (LCROSS) was deliberately crashed into the Cabeus crater near the Moon's south pole in late 2009 (Amos 2010), it has historically been viewed as entirely dry. To launch one kilo of water into low Earth orbit costs about 20,000 euro (Vogler and Vittori 2008). It has been pointed out that if lunar concrete existed naturally on the Moon, it would make more sense to mine it for water than it would to mine the polar regolith (Cohen and Benaroya 2008).

In its current state, the *bacillith*-making process requires water either obtained from the Moon or transported from Earth, water that would also evaporate more quickly in the thin atmosphere, potentially resulting in a weaker material than its Earth analogue. To prevent the water from evaporating from the mortar, epoxy binders could be added to the *bacillith*, or it could perhaps be preset in a pressurized environment to prevent evaporation (Eckart 1999); a pressurised construction dome has already been suggested for lunarcrete construction (Langlais and Saulnier 2000).

Perhaps paradoxically, the fact that *Moon Dune* is based on the deliberate displacement of a terrestrial microorganism in an extraterrestrial context might be the key to finding a way around the water issue. While experiments have shown that cyanobacteria can grow in regolith (Brown et al. 2008), and while some astrobiologists have gone so far as to say that we won't be able to colonise the Moon (or Mars) without the use of microbes (Choi 2010), the task of using bacteria as microbial construction workers and material element will have to be performed in phases. As the environmental conditions evolve on the Moon, the capabilities of the organism needs to be adjusted, probably through the introduction of a newer species to continue (or inactivate) a task associated with a previous organism, or alternatively by initiating a new, secondary objective. Some of the harsh conditions that the initial species would have to tolerate include micrometeorite bombardment, radiation exposure, severe shifts in temperature – and the dehydration discussed above.

While we have seen that there are approximately five nonillion bacteria on Earth (Whitman et al. 1998), and only a fraction of them have been studied, it is unlikely that we will find a perfect match that fits the bill perfectly and ticks all of our boxes. Even though some species can survive a vacuum, there is still a substantial difference between an organism surviving and an organism proliferating. Capability of spore formation, for instance, is a defensive mechanism for the organism: just because the bacterium has the capacity to protect itself doesn't necessarily mean it is capable of reactivation or indeed any kind of environmental interaction.

This means we will have to engineer our microbial workers. We will have to prepare them for the task at hand. Fortunately, with the budding field of synthetic biology, we are beginning to understand how to run such a bacterial boot camp.

The most exciting topic within architecture today is biology. This is not surprising: arguably, the most interesting evolution of society today is happening in biology, and in particular within the synthetic field that allows never-before-seen habitable biological machines to be built from a new sub class of reconfigurable, computational biomaterials that are able to fundamentally change their properties and characteristics beyond traditional notions of self-assembly, in accordance with chemical signal inputs received from a human designer. As pointed out above, this could eventually lead to a spectacular design strategy falling somewhere between *in silico* and *in vivo*: real-time programmed sculpting of microbe materials.

The resulting biological machines, moreover, would double as living, reprogrammable memory storage units. Their performative properties could be programmed to change with incoming data; they could save and play back previous configuration patterns for control purposes; they could even host "internal debates" within their own material structures to inform responses to incoming parameters: Chris Voigt's NOR gates can be combined to perform any logical operation, and some of his bacterial circuits hack into existing bacterial quorum-sensing communication systems, based on the local density of their population, allowing the cells to increase the quality of the computations performed by "voting" on an output (Bourzac 2011).

The construction of completely new organisms lays the groundwork for the construction of completely new materials. The construction of completely new materials suggests the possibility of completely new architectures, which can in turn provide a basis for the creation of new kinds of urban realities; alternative and innovative cities; perfect acts of landscape architecture on the surface of a planetary body other than our own. In the context of space colonisation, this makes a lot of sense. And when fused with the large-scale *in-situ* utilisation of resources that have been turned into readymade buildings over the course of eons, it seems to make even more sense: *Moon Dune* is situated on a 4.6-billion-year-old construction site on which impacts of larger and smaller meteoroids together with a steady bombardment of micrometeoroids, as well as charged particles unremittingly breaking down surface rocks, have prepared the landscape for our *cratertecture*.

Many challenges are of course left to overcome before *Moon Dune* and its methods become a viable alternative to existing, better-documented technologies. Some would argue it is too optimistic a proposal, but then some argued it was too optimistic to go to the Moon in the first place as well. All acts of design are inevitably and inherently about change, and therefore radically optimistic. Architecture is all about opportunities: old buildings give way to new, structures are razed and new ones take their place, one city turns into the foundations of the next. The moment we stop believing in the possibility of making the world – on this planet and elsewhere – a better place is the moment we give up on architecture. "You never change things by fighting the existing reality," said R. Buckminster Fuller. "To change something, build a new model that makes the existing model obsolete."

Moon Dune is one such new model. this stage, it is little more than an architectural idea with an inspiring Earth analogue, but with a little help from nonconformist supporters and some highly skilled synthetic biologists, our *bacillithic cratertecture* just might go beyond the final frontier, and end up making the existing model obsolete.

Acknowledgements. The authors wish to thank Prof. Viorel Badescu both for his invitation to submit this chapter, and for his graceful patience (verging on the angelic) while waiting for it to get finished. Chris Hatherill and Super/Collider for sparking the initial transition from terrestrial to extraterrestrial. Geoff Manaugh at BLDGBLOG for continual inspiration. Tor Edsjö for a timely Rudofsky reminder that was cut at the last minute. Enrico Dini for welcoming us in Pisa. Jason DeJong at UC Davis and Paul Todd at TechShot for agreeing to be our referees. Madelaine Levy and Thidaa Roberts for remaining our partners despite insane sleeping patterns. Finally, we are particularly thankful to Regina Peldszus for her inspiring and committed support.

References

Agosto, W., Wickman, J., James, E.: Lunar Cement/Concrete for Orbital Structures. In: Proceedings of SPACE 1988; Engineering, Construction, and Operations in Space, pp. 157–168. American Society of Civil Engineers, New York (1988)

Amato, J.: Dust: A History of the Small and the Invisible, p. 3. University of California Press, Berkeley (2000)

Amos, J.: Moon's water is useful resource, says Nasa, bbc.co.uk, October 22 (2010), http://www.bbc.co.uk/news/science-environment-11598813 (Viewed April 13, 2011)

Asimov, I.: The Endochronic Properties of Resublimated Thiotimoline (Astounding Science Fiction, Street & Smith) (1948)

Atkins, P.W.: The Periodic Kingdom. HarperCollins Publishers, Inc. (1995) ISBN 0-465-07265-8

AU/CEN-SAD (2009) Plan of Action for the Implementation of the Great Green Wall of the Sahara and Sahel Initiative. Draft for submission to the AU Executive Council. Addis Ababa, Ethiopia, February 1–3 (2009)

Ball, P.: The Ingredients: A Guided Tour of the Elements. Oxford University Press (2002)

Ball, P.: Branches, p. 61. Oxford University Press, Oxford (2009)

Bennett, D.F.H.: Concrete: the material – Lunar concrete. Innovations in concrete, pp. 86–88. Thomas Telford Books (2002)

Beyer, L.A.: Lunarcrete – A Novel Approach to Extraterrestrial Construction. In: Faughnan, B., Maryniak, G. (eds.) Proceedings of the Seventh Princeton/AIAA/SSI Conference on Space Manufacturing 5: Engineering with Lunar and Asterodial Materials, May 8-11, p. 172. American Institute of Aeronautics and Astronautics (1985)

Bini, D.: A New Pneumatic Technique for the Construction of Thin Shells. In: Proceedings of the First International Association for Shell Structures (LASS), International Colloquium on Pneumatic Structures, University of Stuttgart, Germany (May 1967)

Bodiford, M., Fiske, M., McGregor, W., Pope, R.: In Situ Resource-Based Lunar and Martian Habitat Structures Development at NASAMSFC. In: Proceedings: AIAA 1st Exploration Conference on AIAA Paper AIAA-2005-2704, Orlando, FL (January 2005)

Bodiford, M., Burks, K., Perry, M., Cooper, R., Fiske, M.: Lunar In Situ Materials-Based Habitat Technology Development Efforts at NASA/MSFC. In: NASA Marshall Space Flight Center, Earth & Space 2006: Engineering, Construction, and Operations in Challenging Environment, pp. 1–8 (2006), doi:10.1061/40830(188)70

Boston, P.J.: Location, Location, Location! Lava Caves on Mars for Habitat, Resources, and the Search for Life. Journal of Cosmology 12, 3957–3979 (2010)

Bourzac, K.: Software for Programming Microbes. Technology Review (January 5, 2011), http://www.technologyreview.com/computing/27025/page1/?a=f (viewed April 13, 2011)

Brown, I.I., Sarkisova, S.A., Garrison, D.H., Thomas-Keprta, K., Allen, C.C., Jones, J.A., Galindo Jr. C., McKay, D.S.: Bio-Weathering of Lunar and Martian Rocks by Cyanobacteria: A Resource for Moon and Mars Exploration. Lunar and Planetary Science XXXIX (2008), http://www.lpi.usra.edu/meetings/lpsc2008/pdf/1673.pdf (viewed April 13, 2011)

Brown, R.B., Klaus, D.M., Todd, P.: Buoyancy-driven fluid flow generated by bacterial metabolism and its proposed relationship to increased bacterial growth in space (2005), http://www.dtic.mil/ cgi-bin/GetTRDoc?AD=ADA431056&Location=U2&doc=GetTRDoc.pdf (viewed April 13, 2011)

Busbea, L.: Topologies: The Urban Utopia in France, pp. 1960–1970. The MIT Press, Cambridge (2007)

Buseck, P.R., Tsipursky, S.J., Hettich, R.: Fullerenes from the Geological Environment. Science 257(5067), 215–217 (1992)

Cami, J., Bernard-Salas, J., Peeters, E., Malek, S.E.: Detection of C60 and C70 in a Young Planetary Nebula. Science 329, 1180 (2010)

Carey, B.: Wild Things: The Most Extreme Creatures (article in Live Science), February 7 (2005), http://www.livescience.com/133-wild-extreme-creatures.html (viewed April 13, 2011)

Carr, C.E., Newman, D.J.: Characterization of a lower-body exoskeleton for simulation of space-suited locomotion. Acta Astronautica 62, 308–323 (2008)

Cathcart, R.B.: Anthropic Rock: a brief history. History of Geo and Space Sciences 2, 57–74 (2011)

Cavicchioli, R.: Extremophiles and the search for extraterrestrial life. Astrobiology 2(3), 281–292 (2002)

Ceccanti, F., Dini, E., De Kestelier, X., Colla, V., Pambaguian, L.: 3D Printing Technology for a Moon Outpost Exploiting Lunar Soil. In: 61st International Astronautical Congress, Prague, CZ, IAC-10-D3.3.5 (2010)

Choi, C.: Space Colonists Could Use Bacteria to Mine Minerals on Mars and the Moon. Scientific American (September 10, 2010),
http://www.scientificamerican.com/
article.cfm?id=space-colonists-could-use-bacteria
(viewed April 13, 2011)

Chow, P.Y., Lin, T.Y.: Structural engineering's concept of lunar structures. ASCE J. Aerosp. Eng. 2(1), 1–9 (1989)

Clarke, A.: Interplanetary Flight – An Introduction to Astronautics, ch. 10. Harper & Brothers, New York (1950)

Cohen, M.M., Benaroya, H.: Lunar-Base Structures. In: Howe, A., Sherwood, B. (eds.): Out of This World – The New Field of Space Architecture, pp. 179–204. American Institute of Aeronautics and Astronautics, Reston (2009)

Connor, S.: Cromwell's moonshot: how one Jacobean scientist tried to kick off the space race, article in The Independent, Sunday, October 10 (2004),
http://www.independent.co.uk/news/uk/
this-britain/cromwells-moonshot-how-one-jacobean-scientist-
tried-to-kick-off-the-space-race-535171.html
(viewed April 13, 2011)

Cook, E.: Roden Crater: The Art & Vision of James Turrell. 1978 – A Year in the Project (1999-2001. Enhanced and refined by Gail Cook, April 2001) (1999),
http://www.lasersol.com/art/turrell/roden_crater.html
(viewed April 13, 2011)

Daga, A., Daga, M., Wendell, W.: A preliminary assessment of the potential of lava tube-situated lunar base architecture. In: Johnson, S., Wetzel, J. (eds.) Engineering, Construction, and Operations in Space II, pp. 568–577. ASCE, New York (1990)

Dalton, B.P., Roberto, F.F. (eds.): Lunar Regolith Biomining Workshop Report (Report of a Workshop Sponsored by and held at NASA Ames Research Center, Moffett Field, CA, May 5-6, 2007, NASA/CP–2008–214564) (2007)

Danovaro, R., Dell'Anno, A., Pusceddu, A., Gambi, C., Heiner, I., Christensen, R.M.: The first metazoa living in permanently anoxic conditions. BMC Biology 8, 30 (2010)

DeJong, J.T., Fritzges, M.B., Nüsslein, K.: Microbially Induced Cementation to Control Sand Response to Undrained Shear. Journal of Geotechnical and Geoenvironmental Engineering 132(11), 1381 (2006)

DeJong, J.T.: Private correspondence with the author (March 12, 2008)

De Weerd, J.F., Kruijff, M., Ockels, W.J.: Search for Eternally Sunlit Areas at the Lunar South Pole from Recent Data: New Indications Found. In: 49th International Astronautical Congress, IAF 98-Q.4.07, Melbourne, September 28-October 2 (1998)

Eckart, P.: The Lunar Base Handbook. In: Larson, W.J. (ed.), McGraw Hill, Montreal (1999)

Eichold, A.: Conceptual Design of a Crater Lunar Base. Society of Automotive Engineers, Paper 961464 (July 1996)

Encyclopædia Britannica, "fullerene" (Encyclopædia Britannica Online) (2011), http://www.britannica.com/EBchecked/topic/221916/fullerene (viewed April 13, 2011)

Fritz, I., Kreppel, S., Crosby, K.M., Martin, E., Pennington, C., Frye, B., Monegato, J., Agui, J.: Repose Angles of Lunar Mare Simulants in Microgravity (2009), http://www.carthage.edu/dept/physics/flight/wsgc-proc2009.pdf (viewed April 13, 2011)

Fuller, R.B.: Critical Path, pp. xxxiv–xxxv. St. Martin's Press, New York (1981)

Glavin, D.P., Dworkin, J.P., Lupisella, M., Kminek, G., Rummel, J.D.: In Situ Biological Contamination studies of the Moon: Implications for Future Planetary Protection and Life Detection Missions (2010), http://ntrs.nasa.gov/archive/nasa/casi.ntrs.nasa.gov/20100036597_2010034233.pdf (viewed April 13, 2011)

Gracia, V., Casanova, I.: Sulfur Concrete: A Viable Alternative for Lunar Construction. In: Proceedings of the ASCE, SPACE 1998 Engineering, Construction, and Operations in Space, New York, pp. 585–591 (1998)

Grainger, A.: The Threatening Desert – Controlling Desertification. Earthscan, London (1990)

Hall, E.T.: The Hidden Dimension. Doubleday, Garden City (1966)

Hawking, S.: NASA's 50th Anniversary Lecture (NASA, Monday, April 28, 2008), http://www.spaceref.com/news/viewsr.rss.html?pid=27805 (viewed April 13, 2011)

Haynes, R.H.: Ethics and planetary engineering. 1. Ecce ecopoiesis: Playing God on Mars. In: Macniven, D. (ed.) Moral Expertise, pp. 161–183. Routledge, London (1990)

Heiken, G.H., Vaniman, D.T., French, B.M. (eds.): The Lunar Sourcebook: A User's Guide to the Moon. Cambridge University Press, New York (1991)

Heinlein, R.A.: The Moon is a Harsh Mistress. G.P. Putnam's Sons, New York (1966), http://rosuto.paheal.net/Books/Heinlein,%20Robert%20A%20-%20The%20Moon%20is%20a%20Harsh%20Mistress.pdf (viewed April 13, 2011)

Herbert, G.W.: Lunar concrete. In: Yarvin, N. (ed.) Archives: Space: Science, Exploration (1992), http://yarchive.net/space/science/lunar_concrete.html (viewed April 13, 2011)

Highfield, R.: Colonies in space may be only hope, says Hawking: The Telegraph, London (newspaper interview, October 15, 2001), http://www.telegraph.co.uk/news/uknews/1359562/Colonies-in-space-may-be-only-hope-says-Hawking.html (viewed April 13, 2011)

Hodgson, E.: CP 00-02 Phase I – A Chameleon Suit to Liberate Human Exploration of Space Environments. NASA Institute for Advanced Concepts (2001)

Hogan, C.: Bacteria. In: Sand, D., Cleveland, C.J. (eds.) Encyclopedia of Earth. National Council for Science and the Environment, Washington DC (2010)

Hooper, R.: Need a Building? Just Add Water (2005), http://www.wired.com/science/discoveries/news/2005/03/66872 (viewed April 13, 2011)

Horiguchi, T., Saeki, N., Yoneda, T., Hoshi, T., Lin, T.D.: Behavior of Simulated Lunar Cement Mortar in Vacuum Environment. In: Space 1998, pp. 571–576. American Society of Civil Engineers, Reston (1998)

Horneck, G., Bucker, H., Reitz, G.: Long-term survival of bacterial spores in space. Advanced Space Research 14, 41–45 (1994)

Horneck, G., Klaus, D.M., Mancinelli, R.L.: Space Microbiology. Microbiology and Molecular Biology Reviews 74(1), 121–156 (2010)

Hsu, H., Baruh, H., Benaroya, H.: In-Situ Resource Utilization in a Lunar Environment (Rutgers University NASA-SHARP, at GISS) (2005)

Ismail, M.A., Joer, H.A., Randolph, M.F., Kucharski, E.: Cementation of porous materials using calcite precipitation. University of Western Australia Geomechanics Group, Geotech. Rep. G1422 (1999a)

Ismail, M.A., Joer, H.A., Randolph, M.F., Kucharski, E.: CIPS, a novel cementing technique for soils. University of Western Australia Geomechanics Group, Geotech. Rep. G1406 (1999b)

Ivanov, V., Chu, J.: Applications of microorganisms to geotechnical engineering for bioclogging and biocementation of soil in situ. Rev. Environ. Sci. Biotechnol. 7, 139–153 (2008)

Jablonski, A.M., Ogde, K.A.: A Review of Technical Requirements for Lunar Structures. In: International Lunar Conference 2005 (2005), http://sci2.esa.int/Conferences/ILC2005/Manuscripts/JablonskA-01-DOC.pdf (viewed April 13, 2011)

Jaggard, V.: NASA's O/OREOS Dunking Bacteria in Space Rays (National Geographic News Watch article, November 23, 2010), http://newswatch.nationalgeographic.com/2010/11/23/nasa_ooreos_dunking_bacteria_space/ (viewed April 13, 2011)

Jardine, L.: Monuments and microscopes: scientific thinking on a grand scale in the early. Royal Society, Notes and Records Roy. Soc. London 55, 289–308 (2001)

Jones, E.: Apollo 12 Lunar Surface Journal. Surveyor Crater and Surveyor III, Corrected Transcript and Commentary: NASA (1995), (search for "little bacteria"), http://www.hq.nasa.gov/office/pao/History/alsj/a12/a12.surveyor.html (viewed April 13, 2011)

Kelly, K.: Out of Control: The New Biology of Machines, Social Systems and the Economic World. Addison Wesley/Perseus Books, Reading/New York (1994/1995)

Khalili, E.N.: Lunar Strcutures Generated and Shielded with On-Site Materials. Journal of Aerospace Engineering 2(3), 119–129 (1989)

Khoshnevis, B., Bodiford, M., Burks, K., Ethridge, E., Tucker, D., Kim, W., Toutanji, H., Fiske, M.: Lunar Contour Crafting – A Novel Technique for ISRU-Based Habitat Development. In: Proceedings of the 43rd AIAA Aerospace Sciences Meeting and Exhibit, Paper AIAA-2005-0538 (January 2005)

Kozyrovska, N., Zaetz, I., Lytvynenko, T., Voznyuk, T., Maria, M., Rogutskyy, I., Mytrokhyn, O., Lukashov, D., Mashkovska, S., Foing, B.: Microbial Community for Growing Pioneer Plants in a Lunar Greenhouse (2005), http://sci2.esa.int/Conferences/ILC2005/Manuscripts/KozyrovskaN-01-DOC.pdf (viewed April 13, 2011)

Kozyrovska, N.O., Zaetz, I.E., Burlak, O.P., Rogutskyy, I.S., Mytrokhyn, O.V., Mashkovska, S.P., Foing, B.H.: The Conception of Growing First Generation-Plants in Lunar Greenhouses. Космічна наука і технологія 16(2), 70–74 (2010)

Kubrick, S.: Dr. Strangelove (Columbia Pictures, 94 minutes) (1964)

Kubrick, S.: 2001: A Space Odyssey (MGM, 161 minutes) (1968)

Kucharski, E.S., Winchester, W., Leeming, W.A., Cord-Ruwisch, R., Muir, C., Banjup, W.A., Whiffin, V.S., Al-Thawadi, S., Mutlaq, J.: Microbial biocementation. Patent Application WO/2006/066326; International Application No.PCT/AU2005/001927 (2005)

Kurzweil, R.: The Age of Intelligent Machines. MIT Press, Massachusetts (1990)

Kurzweil, R.: How My Predictions Are Fairing (October 2010), http://www.kurzweilai.net/predictions/download.php (viewed April 13, 2011)

Langlais, D.M., Saulnier, D.P.: Reusable, Pressurized Dome for Lunar Construction in Space. In: Chua, K.M., et al. (eds.) American Society of Civil Engineers, Albuquerque, New Mexico, pp. 791–797 (2000)

Larsson, M.: Dune – Arenaceous Anti-Desertification Architecture. In: Badescu, V., Cathcart, R. (eds.) Macro-engineering Seawater in Unique Environments, Environmental Science and Engineering. Springer, Berlin (2011)

Le Corbusier: Vers Une Architecture (English version, Toward an Architecture. Translated by John Goodman, Getty Research Institute, Los Angeles (2007); first English translation (Towards a New Architecture 1927) (1923)

Le Metayer-Levrel, G., Castanier, S., Orial, G., Loubiere, J.F., Perthuisot, J.P.: Applications of bacterial carbonatogenesis to the protection and regeneration of limestones in buildings and historic patrimony. Sedimentary Geology 126, 25–34 (1999)

Leonard, R., Johnson, S.: Sulfur-Based Construction Materials for Lunar Construction. In: Proceedings of the ASCE, SPACE 1988 Engineering, Construction, and Operations in Space, New York, pp. 1295–1307 (1988)

Lin, T.D.: Concrete for Lunar Base Construction. Concrete International (ACI) 9(7) (1987)

Lin, T.D., Love, H., Stark, D.: Physical Properties of Concrete Made with Apollo 16 Lunar Soil Sample. In: Faughnan, B., Maryniak, G. (eds.) Space Manufacturing 6: Proceedings of the Eighth Princeton/AIAA/SSI Conference, May 6-9, pp. 361–366. American Institute of Aeronautics and Astronautics (1987)

Lovelock, J.E.: The Ages of Gaia, p. 123. W.W. Norton and Company (1988)

Lukić, B.: Nonobject. MIT Press (2010)

Lytvynenko, T., Zaetz, I., Voznyuk, T., Kovalchuk, M., Rogutskyy, I., Mytrokhyn, O., Lukashov, D., Estrella-Liopis, V., Borodinova, T., Mashkovska, S., Foing, B., Kordyum, V., Kozyrovska, N.: A rationally assembled microbial community for growing Tagetes Patula L. in a lunar greenhouse. Research in Microbiology 157, 87–92 (2005)

Manacorda, F., Yedgar, A.: Radical Nature: Art and Architecture for a Changing Planet, 1969-2009 (London, Koenig Books, co-edition with Barbican Art Gallery) (2009)

Manaugh, G.: Mineral Kinships, introduction to Maisel D, Library of Dust, p. 25. Chronicle Books, San Fransisco (2008)

Manaugh, G.: Sand/Stone BLDGBLOG (April 19, 2009), http://bldgblog.blogspot.com/2009/04/sandstone.html (viewed April 13, 2011)

Mangels, J.: Coping with a lunar dust-up (The Seattle Times) (2007), http://seattletimes.nwsource.com/html/nationworld/2003572876_moondust15.html (viewed April 13, 2011)

McAlpine, K.: For self-healing concrete, just add bacteria and food, New Scientist (September 1, 2010), http://www.newscientist.com/article/dn19386-for-selfhealing-concrete-just-add-bacteria-and-food.html (viewed April 13, 2011)

McKay, M.F., McKay, D.S., Duke, M.B.: Space Resources. Overview (NASA SP-509, Lyndon B. Johnson Space Center, Houston, Texas) (1992),
http://www.scribd.com/doc/43549497/
Space-Resources-Overview-1992-From-www-jgokey-com
(viewed April 13, 2011)

Merrill, G.P.: Rocks, rock-weathering and soils. MacMillan Company, New York (1897)

Meyer, C.: Lunar Regolith. NASA Lunar Petrographic Educational Thin Section Set, 46–48 (2003),
http://curator.jsc.nasa.gov/lunar/letss/Regolith.pdf
(viewed April 13, 2011)

Miller, J., Coit, D.: Lunar Property and Mining Rights (2008),
http://www.wpi.edu/Pubs/
E-project/Available/E-project-082708-
113454/unrestricted/LunarMiningRightsFinal.pdf
(viewed April 13, 2011)

Namba, H., Ishikawa, N., Kanamori, H., Okada, T.: Concrete Production Method for Construction of Lunar Bases. In: Proceedings of the ASCE Space 1988 Engineering, Construction, and Operations in Space, pp. 169–177. American Society of Civil Engineers, New York (1988)

NASA, Nasa Roadmap, Goal 6: Understand the principles that will shape the future of life, both on Earth and beyond (Responsible NASA official Lynn Rothschield) (2003),
http://astrobiology.arc.nasa.gov/roadmap/g6.html
(viewed April 13, 2011)

NASA, In-Situ Resource Utilization (NASA, Ames Research Center, page editor: Dino J, NASA official: Dunbar B) (March 29, 2008) (2008),
http://www.nasa.gov/centers/ames/research/
technology-onepagers/in-situ_resource_Utiliza14.html
(viewed April 13, 2011)

Nemati, M., Voordouw, G.: Modification of porous media permeability, using calcium carbonate produced enzymatically in situ. Enzyme and Microbial Technology 33, 635 (2003)

North, G.: Observing the Moon: The Modern Astronomer's Guide, p. 37. Cambridge University Press, Cambridge (2000)

Olsson-Francis, K., Cockell, C.S.: Use of cyanobacteria for in-situ resource use in space applications. Planetary and Space Science 58(10), 1279–1285 (2010a)

Olsson-Francis, K., Cockell, C.S.: Experimental methods for studying microbial survival in extraterrestrial environments. Journal of Microbiological Methods 80, 1–13 (2010b)

Omar, H.A., Issa, M.: Production of Lunar Concrete Using Molten Sulfur. In: Galloway, R.G., Lokaj, S. (eds.) Proceedings of the 4th International Conference on Engineering, Construction, and Operations in Space IV: Space 1994, Albuquerque, New Mexico, February 26-March 3, pp. 952–959. American Society of Civil Engineers, New York (1994)

Phillips, T.: The Mysterious Smell of Moondust (NASA Science, Science News) (2006),
http://science.nasa.gov/science-news/
science-at-nasa/2006/30jan_smellofmoondust/
(viewed April 13, 2011)

Ruess, F., Kuhlmann, U., Benaroya, H.: Structural Design of a Lunar Base. In: Maji, A., Malla, R.B. (eds.) Engineering, Construction, and Operations in Challenging Environments, pp. 17–23. ASCE, Houston (2004)

Ruess, F., Schaenzlin, J., Benaroya, H.: Structural Design of a Lunar Habitat. Journal of Aerospace Engineering 19(3), 138 (2006)

Sagan, C.: Cosmos: Abacus, London (2011 reprint of 1995 edition), p. 347 (1980)

Seybold, C.C.: Characteristics of the Lunar Environment (1995),
http://www.tsgc.utexas.edu/tadp/1995/spects/environment.html
(viewed April 13, 2011)

Shafirovich, E., White, C., Alvarez, F.: In-Situ Production of Construction Materials by Combustion of Regolith/Aluminum and Regolith/Magnesium Mixtures. Space Manufacturing, Center for Space Exploration Technologies Research, Mechanical Engineering Department, The University of Texas at El Paso, October 29-31, vol. 14 (2010)

Stevens, A.: Man's furthest aloft. National Geographic Magazine 69:59, 693–712 (1936)

Stone, E.C., Lindsley, D.H., Pigott, V., Harbottle, G., Ford, M.T.: From Shifting Silt to Solid Stone: The Manufacture of Synthetic Basalt in Ancient Mesopotamia. Science 280, 2091–2093 (1998)

Strenski, D., Yankee, S., Holasek, R., Pletka, B., Hellawell, A.: Brick Design for the Lunar Surface. In: Proceedings of the ASCE SPACE 1990 Engineering, Construction, and Operations in Space, pp. 458–467. American Society of Civil Engineers, New York (1990)

Sullivan, T.A.: Process Engineering Concerns in the Lunar Environment (AAIA Paper 90-3753) (September 1990)

Swift, J.: Travels into Several Remote Nations of the World. In: Gulliver, L. (ed.) Four Parts, First a Surgeon, and then a Captain of several Ships (1726) (reprinted edition: Gulliver's Travels. Penguin Classics, 336 pages (2011), The adamant island can be found in part III)

Tamsir, A., Tabor, J.J., Voigt, C.A.: Robust multicellular computing using genetically encoded NOR gates and chemical 'wires'. Nature 469, 212–215

Taylor, G.J., Neubert, J., Lucey, P., McCulloug, E.: The Uncertain Nature of Polar Lunar Regolith (Space Resources Roundtable VI) (2004),
http://www.lpi.usra.edu/meetings/roundtable2004/pdf/6040.pdf
(viewed April 13, 2011)

Taylor, L.A., McKay, D.S.: Beneficiation of lunar rocks and regolith - Concepts and difficulties. In: Article in Proceedings of the 3rd International Conference on Engineering, Construction, and Operations in Space - III: Space 1992, Denver, CO, May 31-June 4, vol. 1, pp. 1058–1069 (1992)

The Engineer. Editorial, Bacteria Help Protect from Quakes: The Engineer (February 23, 2007),
http://www.theengineer.co.uk/news/
bacteria-help-protect-from-quakes/298382.article
(viewed April 8, 2010)

Toklu, Y.C.: Civil Engineering in the Design and Construction of a Lunar Base. In: Chua, K.M., et al. (eds.) Space 2000, pp. 822–834. American Society of Civil Engineers, Albuquerque (2000)

Toutanji, H., Fiske, M., Bodiford, M.: Development of Lunar "Concrete" for Habitat Structures. In: Proceedings of 10th ASCE Aerospace Division International Conference on Engineering, Construction and Operation in Challenging Envirorments: Earth & Space 2006, League City, TX (March 2006)

UN, World Population to 2300 (United Nations Department of Economic and Social Affairs/Population Division) (2004), http://www.un.org/esa/population/publications/longrange2/WorldPop2300final.pdf (viewed April 13, 2011); The quoted figure is the "high scenario," based on extrapolations of current trends. The report itself states that "Any demographic projections, if they go 100, 200, or 300 years into the future, are little more than guesses," a fact that is further shown by an experiment in which the researchers held the total fertility indefinitely at its level between 1995 and 2000, producing "an unrealistic, and almost unimaginable world population of 134 trillion by 2300"

UNESCO, Examination of nomination of Natural, mixed and cultural properties to the World Heritage List - Fujian Tulou (CHINA) (200 Decision 32COM 8B.20) (2008), http://whc.unesco.org/en/list/1113 (viewed April 13, 2011)

Vogler, A., Vittori, A.: Space Architecture for the Mother Ship: Bringing It Home. In: Scott Howe, A., Sherwood, B. (eds.) Out of This World: The New Field of Space Architecture. AIAA, Reston (2009)

Welland, M.: Sand – The Never-Ending Story. University of California Press, Berkely (2009)

Weston, R.: Materials, Form and Architecture. Laurence King Publishing, London (2008), (1st edn. 2003)

Wheeler, R.M., Mackowiak, C.L., Stutteb, G.W., Sagerb, J.C., Yoriob, N.C., Ruffeb, L.M., Fortson, R.E., Dreschelb, T.W., Knottb, W.M., Coreyc, K.A.: NASA's biomass production chamber: A testbed for bioregenerative life support studies. Advances in Space Research 18(4-5), 215–224 (1996)

Whiffin, V.S., van Paassen, L.A., Harkes, M.P.: Microbial carbonate precipitation as a soil improvement technique. Geomicrobiology Journal 24(5), 417–423 (2007)

Whitman, W., Coleman, D., Wiebe, W.: Prokaryotes: the unseen majority. Proceedings of the National Academy of Sciences of the United States of America 95(12), 6578–6583 (1998)

Wilson, J.W., Ott, C.M., Hönerzu Bentrup, K., Ramamurthy, R., Quick, L., Porwollik, S., Cheng, P., McClelland, M., Tsaprailis, G., Radabaugh, T., Hunt, A., Fernandez, D., Richter, E., Shah, M., Kilcoyne, M., Joshi, L., Nelman-Gonzalez, M., Hing, S., Parra, M., Dumars, P., Norwood, K., Bober, R., Devich, J., Ruggles, A., Goulart, C., Rupert, M., Stodieck, L., Stafford, P., Catella, L., Schurr, M.J., Buchanan, K., Morici, L., McCracken, J., Allen, P., Baker-Coleman, C., Hammond, T., Vogel, J., Nelson, R., Pierson, D.L., Stefanyshyn-Piper, H.M., Nickerson, C.A.: Space flight alters bacterial gene expression and virulence and reveals a role for global regulator Hfq. PNAS 104(41), 16299–16304 (2007)

Wolfe-Simon, F., Switzer Blum, J., Kulp, T.R., Gordon, G.W., Hoeft, S.E., Pett-Ridge, J., Stolz, J.F., Webb, S.M., Weber, P.K., Davies, P.C.W., Anbar, A.D., Oremland, R.S.: A Bacterium That Can Grow by Using Arsenic Instead of Phosphorus. Science (2010), doi:10.1126/science.1197258

Wood, C.A., Andersson, L.: New morphometric data for fresh lunar craters. In: Proceedings of 9th Lunar and Planetary Science Conference, Houston, Tex., March 13-17, pp. 3669–3689. Pergamon Press, Inc., New York (1978) (A79-39253 16-91)

Yong, J., Berger, R.: Cement-Based Materials for Planetary Materials. In: Johnson, S. (ed.) SPACE 1988 Engineering Construction, and Operations in Space. American Society of Civil Engineers, New York (1988)

Zeitlin, C., Guetersloh, S.B., Heilbronn, L.H., Miller, J., Shavers, M.: Radiation tests of the extravehicular mobility unit space suit for the international space station using energetic protons. Radiat. Msmts. 41(9-10), 1158–1172 (2006)

29 Fundamentals of Modern Lunar Management: Private Sector Considerations

Mike H. Ryan and Ida Kutschera

Bellarmine University, Louisville, KY, USA

29.1 Introduction

Implicit in any assessment of management practice in space, the moon or even Mars is that a good reason exists for people to be present. The case for a human presence in space is fairly limited from a commercial standpoint but there some interesting viewpoints as to its present and future potential (Morris and Cox 2010). Limited development of commercial tourism combined with equally limited commercial activity in orbit to date does not make a convincing case for business. The justification for lunar commerce is even less convincing. Dr. David Livingston, noted space expert and host of the Space Show at TheSpaceShow.com, is a good example of the significant business concerns related to lunar enterprises. As he states:

> There is no commercial market for anything returning to the Moon that is real and not decades if longer off into the future. There is no cost effective, affordable transportation… All of what we hear is costly, has no commercial value and is of questionable sustainability…. Until a business case can be made that is real regarding the cis-Lunar environment, I remain skeptical and would consider this to be a major hurdle facing this type of plan (Livingston 2010, p. 334).

The practical limitations of current business models for the space environment do not reduce the need to address the possibilities for management practice on the moon. History has shown that often what seems a remote possibility is, in fact, much closer in terms of both need and actual practice. A series of breakthroughs in space transportation for example might make the potential of doing business in space both practical and an immediate concern. Therefore, an overview of the substantive issues related to managing in space or, in this case, a lunar environment makes sense. Whether individuals and firms operate on the moon in five years or twenty-five years is not germane. What is important is that issues critical for good business practices be considered before such activities take place rather than after problems begin to develop (Wellin and Elliott 2010).

29.2 Applied Management Principles

The application of management principles in a lunar context can be very challenging and straightforward at the same time. The ambiguity is a function of the inherent difficulties of the environment and the characteristics of the individuals needed to work there. Management practice is very clear for some situations. Other circumstances, especially those bringing together larger numbers of participants from different cultural and work backgrounds, may present more significant management problems. Regardless, many "interesting" problems faced by managers of people in space-based environments are inherently those brought to that situation by the participants, workers, or employees simply being human. Mary Roach, a science writer and author of *Packing for Mars* (Roach 2010), humorously underscores the essential elements of dealing with people in a space environment by suggesting that people will be the focus for many of the problems related to living and working in space related environments or places.

To the rocket scientist, you are the problem. You are the most irritating piece of machinery he or she will ever have to deal with. You and your fluctuating metabolism, your puny memory, your frame that comes in a million different configurations. You are unpredictable. You're inconsistent. You take weeks to fix. The engineers must worry about the water and oxygen and food you'll need in space, about how much extra fuel it will take to launch your shrimp cocktail and irradiated beef tacos. A solar cell or thruster nozzle is stable and undemanding. It does not excrete or panic or fall in love with the mission commander. It has no ego. Its structural elements don't start to break down without gravity, and it works just fine without sleep (Roach 2010, p. 15).

The opportunity for space-related management issues would be less without people involved in space (Jorgensen 2010). The absence of people working and living in non-terrestrial environments, however, would not eliminate managerial problems, concerns or consequences. It would merely move their proximate location to the office, laboratory or cubicle back on earth. Wherever humans live and work together, the issues and consequences of management practices will have to be addressed. Any discussion regarding possible management approaches to be used on the Moon should be constrained somewhat as the majority of the tasks to be conducted have terrestrial counterparts or analogs.

Managing on the Moon might be as much about adopting a different perspective on what to manage as how. Managers in remote, isolated, technically complex locations such as the Moon may find that their managerial model is much more like the parent of a bunch of very bright children. This is not to suggest that people would be behaving childishly, but that the variety and interaction of potential issues they will face is more complex than just "running the operation" (Ryan and Kutschera 2007, p. 45).

The practice of management is based on both experience and empirical research. Managing individuals requires ongoing adjustment on the part of managers as some things work well for managing some employees but not others. Sufficient reasons exist that some management approaches will be preferred in areas where information and/or experience is more limited, as in potential lunar operations. Differences in both how and what to manage will appear as more individuals begin to live and work on the moon as compared to their terrestrial counterparts. Many initial managerial assumptions for managers overseeing lunar employees will be subject to revision over time; most likely they will start with the best practices available and then adjust those practices as needed.

A significant number of existing organizations operate with inherently bad management practices, such as: badly conceived missions, fundamental planning mistakes, poor or missing organizational communications, inappropriate or misdirected reward and/or evaluation systems, as well as a host of things that could be and probably should be better than they are (Tenner 1996). Tenner's *revenge effect* is also a good way to remember the problems that arise from poor or incomplete planning efforts as in, *we intended... but what happened was....* Therefore, projecting both the use and misuse of current managerial practice that might be applied to a lunar environment includes the standard management issues compounded by space related issues, exacerbated by distance, extreme conditions, and very low fault tolerance. Engineers frequently chide less knowledgeable individuals about the rigors of designing things for space. By comparison, these difficulties may fade into the realm of "I remember when we only had to worry about keeping things airtight..." once large numbers of bright, headstrong, egocentric, and competitive individuals actually begin to live and work off planet. Managing the people, processes and business enterprises that may develop will require the best managerial techniques and people available.

Given the wide range of issues covered by management practice, only a comparative few can be explored within the current chapter. However, these issues, are critical for future successful lunar operations. The following concerns are either important in themselves or have the potential to impact a wide range of operational concerns for any lunar base, lunar business or lunar exploratory outpost.

29.3 Management Issue: Management by Default

The general factors that should be examined from a management perspective might be the same for a lunar facility as compared to one operating on earth but the actual application can vary tremendously. For example, planning, motivation, leadership, decision-making, teamwork, conflict resolution, employee selection, training, communication and culture are all components of the management environment regardless of location. Location imparts some essential differences in how management practice is implemented. What works well in one culture may be problematic in another. Culture alone makes direct insertion of business practices from one place to another more difficult than many managers are willing to admit. This seems to occur regardless of the direction of the cultural shift. Regardless of the organizations' point of origin, virtually all have some level of cultural insensitivity or myopia. It is often not deliberate but usually the result of simple oversight or overly optimistic assumptions about what will or will not work.

Adjustments must be made and practices adjusted to accommodate variations in what was expected versus how things will actually work. Presuming similar problems and issues will not be a part of space operations runs contrary to the body of management experience around the world. Unfortunately, awareness does not equal preparedness or avoidance of these sorts of management problems. Management problems, particularly problems related to culture, occur with disturbing regularity, even in organizations with significant experience and preparation. It is probably a forgone conclusion that such issues will find their way into space and on to the moon.

The surprising thing would be if no cultural clashes and no instances of operational problems occurred due to differences in areas such as culture, communication styles, and approaches to leadership and decision-making. The likelihood that such things will occur is what makes the issues of management so important. It is critical that management practices employed in these environments not be selected simply by default by which ever organization happens to be the primary operator. It is equally important that managerial activity not simply be placed on "autopilot" with an assumption that all is well if things are working technically. As more stakeholders from private sector move into space, technical concerns notwithstanding, the managerial focus must expand to consider the behaviors of the people finding their way into the workforce. Management practice should reflect what works or will work best in a given situation.

29.4 Management Issue: Management of Operational Transitions

Space-related activity in its present form is primarily driven by government-initiated or government-supported operations. The principal source of space activity in the United States has been the National Aeronautics and Space Administration (NASA) for half a century. In Europe, the European Space Administration (ESA) has served as the hub for the various national space organizations of its European members. The Japanese Aerospace Exploration Agency (JAXA) has operated in cooperation with NASA, ESA, and other national programs. Russia has operated its own national space administration with some variations in administrative structure since the beginning of the space-age during the 1950s. All of these organizations have engaged in both cooperative as well as competitive space ventures for decades. The preponderance of space experience as a consequence resides in government-operated or government-supported programs with quasi-civilian administrative control.

From a management perspective, this suggests two things. First, that most space programs tied to the various governments supporting those activities all enjoy the benefits incumbent with supporting significant bureaucracies. Benefits, of course, may be a sarcastic reference to the inherent problems that frequently develop in large-scale government programs over time. Private sector operations also encounter difficulties related to a bureaucratized environment. However, that environment cannot be generally sustained in the private sector in the absence of convincing profit or at least minimal cash flow. Governments seemingly sustain bureaucracies almost indefinitely without meeting the same sort of goals required in private sector ventures. Bureaucratic environments can create their own internal rules (rules that can be quite arbitrary) as to what constitutes success in contrast to what may be generally viewed as a successful activity. For example, budget expansion is seen frequently as success even when particular programs fail or specific goals are never met. In some cases, the quest for resources, influence and control can and does undermine good projects and diminishes the organization's ability to meet long-term goals and objectives. This is a criticism often aimed at NASA with some degree of accuracy. And, unhindered expansion may be a general concern any time an organization grows to a certain size or operates over a wide range of activities. Management cartoons such as Dilbert resonate across a wide audience of practicing managers for a reason.

Second, the pressure to maintain an organization's budget in the face of public scrutiny, financial vacillations, and the squabbles related to interagency and legislative conflict make long-term commitments to specific courses of action difficult and often impossible. Operational priorities within government agencies are often put aside to maintain the overall organizational budget in the short term. This Faustian bargain sacrifices long-term organization goals for short-term solutions to budget pressures. The inability to plan and execute an agreed-upon plan in the face of uncertain legislative priorities has proven problematic for more than one space agency.

Long-term development, critical to commercial and private sector space activities has often been put aside for other short-term considerations. The inability of government space organizations to consistently plan for long-term has, in the minds of many potential space entrepreneurs, rendered their government counterparts as unsuitable business or project partners. This situation has been exacerbated by governmental responses to private sector initiatives that frequently have been viewed as tactical roadblocks to limit or just outright prevent private sector operations in space. The apparent zero sum game played by various governments' space agencies has produced an almost impossible environment for business activity. One clear result of the zero sum game is the significant divide on what is the appropriate role and scope for private sector space operations. The private sector has limited interest in space operations and sees limited potential without markets, products or services from which revenue might be obtained. The absence of a robust private space sector has limited the speed with which technological innovation and new market opportunities allow the government to withdraw or minimize its role in supporting, subsidizing or driving space activity. This critical divide will create serious management problems for both government space organizations and developing private sector space companies. Neither government space agencies nor private sector companies can afford to operate without the other in the long-term. Both require the other's cooperation if space activity and lunar operations are to be successful and profitable.

29.5 Management Issue: Planning for the Extreme

Planning in general is the critical management component for long duration and/or remote activities. Planning is particularly important when transportation and communication are irregular, prone to disruption or subject to various interruptions. The inability to provide rescue or relief imparts a necessary condition of anticipating not just some possible problems to be overcome.

In some ways, outer space is fundamentally different from other endeavors in the business world. Space is exceedingly hard to get to, and once you put your space segment in place the physical environment is unrelentingly hostile. It is extraordinarily difficult to maintain space assets because you can't retrieve them or send a repair crew (at least not currently). The legal and regulatory environment for space differs in many ways from comparable activities on the ground. For example, there is no private property system for extraterrestrial real estate, and the international law governing salvage in space is more restrictive than that which applies to salvage at sea (Vedda 2009, p. 99).

Planning for any imaginable circumstance that might threaten operations is a necessary element for business operations in space and by extension the Moon. What backups for various activities need to be present? What skill sets must be redundant and available on a 24/7 basis at the actual remote location? The question of how much redundancy, safety, just-in-case capability ought to be present within a lunar operation is a planning exercise of some importance. And failure to consider what might otherwise be exceptionally remote possibilities could be fatal. These same questions apply to both ongoing lunar missions and to the support areas for such missions. Japan's 2011 earthquake and tsunami underscore quite clearly that events can occur well beyond those anticipated or projected. The best-prepared nation in the world in terms of earthquakes found itself facing unprecedented damage after combined effects of moving earth and water. More difficult were the problems involving damaged nuclear power plants all of which met Japan's stringent building and disaster codes. Even the best of planning has its limits in some circumstances. Yet, aid of all sorts reached this country in comparatively short order. It is difficult to imagine an equivalent response in the event of disaster on the moon. Even the notion of being able to respond might become unreasonable if a terrestrial disaster somehow interrupted the supply chain for earth-based resources.

Planning is a multi-layered endeavor. The obvious planning components are often over-emphasized over the more subtle issues related to actual implementation and execution. Fundamental measures of performance are often absent and whatever metrics are employed do not necessarily measure whether or not the actual plan was successful. Further complicating the situation is that activity can and often does occur at multiple levels simultaneously and can do so with little if any interaction among the levels. The probability of success in complex organizations falls dramatically as a direct function of the number of levels of planning employed. Unless the implicit values inherent in the plan as emphasized by the vision underlying the plan are embraced at each level in an organization, the likelihood that a consistent, recognizable planning process will be executed is virtually nonexistent. And without such a unified process the probability of serious problems developing within or related to any proposed lunar operation begin to become alarmingly high. One of the central tenets of planning is that if the first plan fails it is time for a new plan. It may be impossible for managers to "tweak" or recast the plan along an entire range of critical dimensions in uniformly unforgiving environments such as on the moon and in space.

The tools available for planning purposes are quite extensive. However, their application in specific organizations is often lacking due to the limited toolset employed. Alternatively, the expectations arising from any given toolset may also be perceived as more significant than they really are. For example, the difference between an assumed strategy (what we think we are doing) and a realized strategy (what we actually end up doing) often is simply whether or not people understand or even know what the objectives of any given plan are. People make their own assumptions in the absence of a clear well-articulated description of the plan and the implicit values that the plan is attempting to impart to people. Managers often will impose their own interpretation as to the implications of a specific company decision, organizational choice, individual action, or environmental situation and then assume that their interpretation is correct. NASA has seen this occur with two shuttle disasters. Frequently, when managers are confronted with information that conflicts with their perceptions of what is supposed to be going on, information is either changed or deleted so as to avoid cognitive dissonance. The implications for managers who are unable to process all the information or who lack a system for aiding in that process are almost always negative (Weick and Sutcliffe 2007).

29.6 Management Issue: Command and Control

The evolution of management theory generally points to a progression from more restrictive forms, or as some have termed, "industrial feudalism" (for example, feudal societies of Europe, the Hershey experiment, certain robber barons, Carnegie), with a capitalist or other individual having nearly total social control of a community, to less restrictive forms. It has been speculated that lunar organizations would start out more in the management style of early industrial-age organizations, with strong central control, and then evolve towards the more freewheeling organizations that are common on Earth today. This is linked to an initial situation where there would be central control of transport and vital resources such as power, water, and air (Ryan and Luthy 2003, p. 21).

No single issue has the potential to do more good or create more harm than how lunar operations would be managed in terms of command and control. As operations transition from predominately government-operated and controlled mission-driven activities to more enterprise-oriented business-driven activities, management considerations will become even more important to success. The problems likely to become the most troublesome from a managerial perspective are those that may occur in three general areas. The first would be in the transition from a more direct command structure to a civilian-based model. It is likely that most lunar operations would initially have significant government participation. Therefore, the most relevant currently available space-based operation, the International Space Station (ISS), may provide some guidance into the forms of government-sponsored space activities. Management of the facility is primarily operated from terrestrial-based locations with orbiting astronauts and cosmonauts taking their directions from earth-side mission controllers. Earth-based support will be crucial to

lunar activity success and earth-based support will be insufficient once large numbers of people are living and working on the moon. On-site management would
increasingly take the place of earth-based controllers, in part because the essential
lunar expertise would then be on the moon rather than the earth. Earth-based managers would become more advisory and provide support in much the same manner
that occurs with multinational organizations currently. Goal setting and long-term
planning would occur with lunar facility input but would probably happen in the
headquarters of the organizations paying for the lunar activity. Again, this approach is pretty much the standard operating model for organizations working in-
ternationally. The moon would not be that much different in that regard.

A second area of potential friction will occur during the transition from an
earth-directed, highly centralized command structure, to one more common to
commercial or scientific organizations. The issue of where ultimate control resides
and who makes the final decisions and in what areas could be quite troublesome.
A summary comparison of the typical military versus civilian command structure
suggests some clear areas of possible friction (see Table 29.1).

Table 29.1 Military versus Civilian Command Structures (Ryan and Luthy 2003, p. 23)

Military or Military-like Organizational Structures	Civilian or Civilian-like Organizational Structures
Often rigid rules as to	Often flexible rules derived by
• What will be done	• Consensus
• When it will be done	• Priority shifts allowed
• How it will be done	• Individual decides how task is to be done
Force of authority frequently derived from the position itself	Participative

Particular care to ensure an orderly transition from offsite management and
control to more onsite participation and direction will be problematic if not handled carefully. Potentially vexing from the point-of-view of both managers and
prospective employees may be a tendency to move operational control back and
forth between managerial/support positions depending on the circumstances. The
need for a final locus of control at critical times implies that a flight controller or
similar individual would be needed. Would things run best with a captain in
control model or with a city manager/city council style model of control? Again,
depending on the circumstances the answer could be either one. A crisis might
compel choosing a centralized authority approach while at all other times decisions would be made and implemented with a participative management approach.
The key would be in knowing what circumstances, situations, or events would
trigger the centralized authority model.

Finally, the third area for potentially serious management problems will arise
once a significant commercial presence is established. Business people are often
willing to sacrifice long-term considerations for short-term results. External pressure to do more with less, to make one's numbers, or to show consistent quarter to

quarter profits has undermined more than one business over time. The inherent problem of a manager or a firm willing to do "anything" to ensure profitability could undermine appropriate management practices. Bad managerial practices and poor managers are endemic across the globe. It would be naive to assume that such lapses would not find their way to the moon. The solution is to find ways to inculcate good practices into the very fabric of the lunar management. Individuals unable or unwilling to manage appropriately should be removed as soon as possible. Lunar employees should not be placed in positions of responsibility if during their selection and training it becomes evident that they do not have the salient characteristics to manage people, activities or organizations at an optimal level.

29.7 Management Issue: Basic Human Resource Management (See: Kutschera and Ryan 2010)

29.7.1 Selection

Recruitment and selection will be challenging tasks that are essential for the success of business driven activities on the Moon. The actual recruiting, i.e. attracting qualified candidates to apply for open positions, might not be as problematic since one anticipates enough interest from highly qualified individuals to work "abroad" (even, or especially, if that means working in space). However, the actual selection process will prove more demanding and difficult as it is absolutely crucial that the "right" people are hired.

What profile do the "right" people to work successfully on the Moon have? Of course certain medical requirements need to be met by a successful applicant. The recruitment criteria for The Canadian Space Agency for example include points such as standing height, visual acuity, and blood pressure (Seedhouse 2010, p. 18). The medical selection requirements for Mars missions include genetic screening for disease, having undergone appendectomy, having undergone gall bladder removal, having an above average bone density, genetic screening for radiation resistance, and screening for kidney stones (Seedhouse 2010, p. 174). The key issue is to have mission candidates that are unlikely to suffer avoidable disabilities or have increased resistance to known space hazards. Obviously, the physical minimum requirements for those individuals who want to work on the Moon will have to be re-defined and standardized to meet the unique working conditions of that particular location.

A potential hiring manager will look at two components beyond the physical requirements when making a selection decision, namely skill and attitude. First, professional expertise will be a definite selection condition as the applicant needs to have certain skills and abilities relevant to the open position. Unlike the employment models from ISS for example where every crew member is a fully-trained astronaut, not every single individual working on the Moon in the envisioned business model will need to be able to "fly the spaceship". However, individuals will still need to be versatile and able to fulfill more than one function as the total number of employees will still be limited. The second component of the selection process is attitude, and it is no less important for a successful applicant. Applicants should, with as much certainty as possible, be able to function in the extreme environment of the Moon. There won't be the luxury of trial periods after which any job applicant can easily be let go, as quitting or leaving one's job on the Moon is literally not going to be an option. Therefore, any job applicant needs to be carefully evaluated as to his or her ability to work in an extreme environment under extreme conditions. To some extent, even civilian applicants will be signing on for the duration of the assignment. The recruiting team most likely will rely on already developed requirement lists to develop their criteria for working on the moon. Again, Mars mission selection requirements are a starting point as they list both social skills and behavioral traits (e.g., social compatibility, emotional control, patience, introverted but socially adept, sensitive to the need of others, high tolerance for lack of achievement, self-confident without being egotistical, tolerance, agreeable and flexible, practical and hard-working, does not become bored easily, desire for optimistic friends, high tolerance for little mental stimulation, subordination of own interests to team goals) and crew compatibility traits (tactfulness in interpersonal relations, sense of humor, effective resolution skills, ability to be easily entertained) (Seedhouse 2010, p. 174).

Safety needs to be the top priority for every employee, no matter what individual goals ultimately are followed. Also, given that any moon-based infrastructure will be limited in space, employees need to be able to tolerate close quarters and unvarying close contact with co-workers. Managing individual differences is critical when people are operating in close proximity and have limited opportunity for reclusion. Individual differences include for example demographic factors, personality types, general work attitudes and national culture. There are examples of environments where individual differences play a critical role. These include Antarctic winter-overs, long duration polar expeditions, sustained submarine operations and even endurance yachting events. Essentially, wherever people are forced to endure each other's company for prolonged periods without relief, selecting for individual behavioral characteristics will become important.

There is much to be learned from the contemporary Antarctic selection programs for personnel, their on-site needs and support, and their reentry problems and experiences. Furthermore, a body of research literature on the subject has been building up for three decades on the emotional and mental health of Antarctic personnel. Both past and ongoing behavioral research of such human experience, combined with analysis of previous investigations of both foreign and space relocations, should provide immeasurable insight for the improvement of large-scale space deployment (Harris 2009, p. 253).

29.7.2 Training

What role should training play both for managers and other personnel if the individuals selected are assumed to have the requisite skills and knowledge for their specific job assignments? Training has two essential elements for personnel heading out for a lunar assignment. The first element is to acquaint every individual with the assorted operational and safety issues for the facility and its environs.

Learning what to do and what not to do during a variety of possible emergencies should constitute a significant part of the pre-departure training activity. Early Russian space missions found that tasks conducted in the weightless environment almost always took longer than training on earth accounted for. The mismatch in the actual time it took to do things versus the agreed upon expected time to accomplish experiments and other tasks contributed to increased stress and conflict among crews and ground controllers (Ivanovich 2008).

Practice for emergencies and assorted serious situations also needs to be repeated until a person's responses become second nature. For example, knowing where a critical switch, value, or tool was located even in the dark might make all the difference between having a *problem* or a *life or death situation*. Typical training could range from blind navigation exercises to practice with a wide variety of simulations from power outages, air loss, to simple water leaks. Simulations will need to convey a sense of seriousness and immediacy not unlike that created in ship and submarine disaster control exercises used by various navies. Managers cannot assume that the appropriately trained person will be in the necessary location at the time of an emergency. Therefore, they must make sure that everyone has a base level of training and sufficient experience to fill in any gaps that might occur, at least on a temporary basis. Making sure that everyone stays up to date with needed information and general facility training will be an ongoing management consideration. Other more specific training needs would include things such as spacesuit use, maintenance, and emergency repair.

Spacesuit training alone could take several weeks. Current designs for spacesuits are not intended for casual use. It is very likely that enhanced suit designs would be developed for various facets of lunar activity. Some suits would be intended for deployment on the lunar surface for exploration or construction. These suits would have to be built to deal with the rigors of the terrain while also being very reliable and easy to repair. Other suit designs might only be used to move from one part of the facility to another or from a transport to the facility. Such suits might not require the extra ruggedness of an exploration suit. Still, the need to practice tasks while wearing spacesuit gloves could be an important part of pre-departure training even for individuals unlikely to venture outside for prolonged periods. Familiarity with how it feels to do various tasks while encumbered with a spacesuit and the inherent loss of touch sensitivity when using gloves would be critical in preparing people for lunar assignments. Every person selected for a lunar-based assignment would be required to show a minimum level of proficiency in spacesuit operation and use before being cleared for transit to the moon.

The ability to operate within a spacesuit should not be assumed. For example, it is not uncommon for even experienced swimmers to behave somewhat erratically when learning to scuba dive. Losing one's mask, mouthpiece, or air supply are common events for which divers need to be prepared. Diving instructors frequently force students to experience these types of events in the comparative safety of the training environment. It is far better to see how an individual responds in a controlled situation with plenty of help around than to find out they are prone to panic in an actual problem situation. Some individuals are never comfortable underwater even with extensive training. Others are actually unable to adjust their responses to correspond to what is required while diving. Recreation divers often forget their training when they lose their air supply and break for the surface. Such actions potentially endanger the diver and others with the diver. Training alone does not necessarily ensure that when critical situations arise that an individual will respond appropriately. Experience with Russian spacesuit designs has many analogs for future lunar missions (Abramov 2003). People expected to function in spacesuits must be exposed to situations that test their ability to work within a suit. Those experiences must also create sufficient stress so to mimic real lunar emergencies. Training combined with appropriate reality-based activity is much more likely to reveal deficiencies in either the individual or their preparation. Preparing prospective employees for the unexpected would appear to be a critical link in all proposed lunar activities. Lunar managers would need to make preparation training and ongoing refresher training specific priorities for everyone involved as the unexpected may not be that uncommon on the Moon. Previous space training practice required extensive repetition until every possible action was virtually automatic. Business requirements in the future may preclude costs associated with extensive practice for low probability events. Eventually a balance will be found between preparing for possible events and preparing for probable events much as it has in commercial aviation.

In the past, there has been significant cost involved in the launch of a few people into space for just days. Consequently, every space activity, especially an EVA (extra-vehicular activity), was extensively rehearsed for weeks or months on Earth to maximize the efficiency for the actual time-constrained performance in space. We now live in a time when humans of several nations are currently living or planning to live in space. With ever-increasing frequency, they are being asked to respond to challenges without proceeding terrestrial practice. In the future, there will be even less opportunity to train for events in advance as humans move to explore and work in environments that we have yet to experience (Thomas and McMann 2006, p. 342).

The second element in essential training may be understanding group behavior and getting along with different people in a fairly rigid and isolated environment. Examples of difficulties that may be encountered range from cultural expectations to what constitutes standard work practices. Training for potential cultural conflicts and their resolution will be important for anyone working and living on the moon. It will also be important to recognize that operating and work practices vary across the range of prospective employees. At the absolute minimum, common nomenclature, measures, tools, warning colors and similar types of things must be standardized prior to putting large numbers of people in a lunar facility. Training with these common elements must be conducted so that when an individual uses a tool, checks a measurement or responds to a critical situation the potential for confusion is removed. Any lunar operation may well have participants from dozens of nations attempting to work collaboratively in a very complex and dangerous environment. Experience suggests that managers and workers will face serious and continuing problems in the absence of comprehensive cultural preparation, training and implementation. Problems such as: "facing unexpected language barriers requiring special expertise by their translators (e.g. technical terms having different meanings...); having long cross-cultural discussions with apparent agreement, only later to be surprised by their counterparts' different understanding.... (Harris 2009, p. 160)" Managers can expect to spend some time dealing with issues arising from cross-cultural misunderstandings. However, proper training should reduce the number of such incidents and their impact.

29.8 Management Issue: When Work and Living Must Co-exist

The practical problems of managing a workforce that will operate in comparative isolation and without many of the expected comforts of any modern community will begin after the selection and training process has occurred. Operating on the Moon will be an interesting managerial experience, if for no other reason than nothing can be taken for granted, not even the air. The entire experience will be inherently artificial with many choices constrained by what is needed to ensure survival. Even the manner of experiencing things as simple as dawn and dusk will have to be built into the operation of the lunar community. These sorts of things might be overlooked and are exactly the things with the potential to bedevil even the most experienced manager. For example, people generally do not pay much attention to the background noises of the forest, their community, or even their workplace. However, people take immediate notice if the noise is removed. Some individuals find it impossible to sleep without the familiar droning of a fan, and it is the normal sounds of the night or even passing traffic for others. A background environment made up of a living tapestry of sounds and sights, smells and colors, the coming and going of people and things is normal for most individuals. That tapestry will unravel on the Moon and everyone will notice the unraveling to some extent, at least initially. It will be more than just sleeping in an unfamiliar bed, room and building. All the things that individuals use subconsciously to evaluate and assess the state of their environment will change. Not everything will change to the same degree but some serious adjustments may be required. And the management of the facility will be expected to find solutions so that people will be able to maintain their productivity and focus. Managers will need to take care of these issues while also experiencing the same discontinuities.

29.8.1 Isolation

What does it take for people to want to live and ultimately enjoy living and working on the Moon over an extended period of time? What are the prerequisites for high job satisfaction, resulting in consistently high job performance? Lunar operations will no doubt be facing some of the same issues that other small communities face. Of course, the special physical circumstances that basically seal this community off will only intensify certain problems like, for example, those resulting from feelings of isolation and confinement (Ryan and Kutschera 2007, p. 46).

The management of people in isolation poses a number of special problems for managers. Simulations of long duration space voyages by the Russians have provides mixed results. Moscow's Institute of Biomedical Problems (IBMP) has conducted a number of isolation experiments since the late 1960s with reported instances of various psychological or interpersonal issues (see Table 29.2).

A summary of the most commonly reported psychological problems in these environments exemplifies how pervasive psychological disturbances are .... (Bishop 2010, p. 265)

Table 29.2 Reported Problems in Confined Environments (Bishop 2010, p. 266)

Reported Problems	MIR	US Space	SUBS	POLAR	SIMS
Interpersonal conflict	X	X	X	X	X
Somatic complaints	X	X	X	X	X
Sleep disturbances	X	X	X	X	X
Homesickness	X	X	X	X	X
Boredom, restlessness	X	X	X	X	X
Decrements in performance	X	X	X	X	X
Decline in group compatibility	X	X	X	X	X
Substance abuse	-	-	?	X	-
Communication breakdowns	X	X	X	X	X
Conflicts with mission control	X	X	?	X	X

Among capabilities of the individuals involved would be the following requirements for employees working and living within a lunar environment. Individuals would be expected to be adaptable and have almost an extreme tolerance for both confinement and austere living arrangements. Table 29.2 indicates, the range of potential psychological issues that might be experienced by any given individual is extensive. It is evident that individuals entering into prolonged periods of isolation would need to be emotionally stable (Spell 2010). In addition, a certain level of family stability would also be mandatory for those anticipating long stays. The fact that things have not always gone perfectly within simulations based on Earth suggests that even more care will need to be used in the selection, training and support of individuals chosen to spend prolonged periods in space or lunar environments. The pressures of isolation possibly compounded by difficulties of adjustment leading to increased friction among the various crewmembers may be expected as business interests arrive on the Moon. All sorts of problems will be encountered, unless specific provisions are made to allow both time and places for crewmembers to be alone, to vent their frustrations and to find intelligent ways to cope with the stress of their environment. Terrestrial examples are common where confined environments have the potential to create individual stress, such as ships at sea.

We can use the analogy of a sailing boat versus a cruise ship. On a sailing boat the crew is confined to cramped quarters; the extreme limitations of space are met by the limited circle of people one can interact with. On a cruise ship passengers are still confined in space, although the relatively larger vessel size makes it possible to 'escape' to a certain degree. Also, the amount of people on a cruise ship allows for interaction with people beyond a small group and there is still a perceived degree of anonymity and privacy (Ryan and Kutschera 2007, p. 46).

Communication will be critical to lunar employees. They will need to be able to interact with their family and friends frequently. The realities of long-distance communication should be considered or it will simply add to an individual's stress rather than alleviate stress. Picking the right time zones for respective work shifts may always be somewhat arbitrary. However, it may make sense as a good management practice to align on-duty and off-duty shifts to promote communication with family back on earth. Doing so would make it much easier for lunar workers to interact with family and friends. The longer the work rotation on the moon the more important it would become to ensure that employees remain mentally grounded by being connected to those that they value. The moon may never be a "normal" environment but at some point people must operate as if living and working there is normal. Maintaining a semblance of normal will require good communication capabilities with those back on earth as well as additional features such as health care, intelligent living arrangements, activities and features to build "community," and a range of decisions generally not the province of your average manager. Of course, being a manager on the Moon may itself require time for individuals to become comfortable, if it can ever be truly comfortable. It is certain that management on the Moon will not start out like managing in another country. Some areas of daily concern for lunar managers would only occur in emergencies for expatriate managers around the world. Fortunately, a body of research building upon the problems associated with isolation and working within stressful environments is developing.

29.8.2 Health Care

Few managers are acutely concerned with health care except as an issue related to benefits. In that context, a manager is often more concerned with the cost of health insurance to his/her employees and by extension to his/her organization than with the underlying health care process. Lunar-based managers will by necessity view health care and its associated technologies as a critical component in the overall operation of their facility. Managing a facility on the Moon will require self-sufficiency, particularly in health care, that exceeds the most rigorous requirements for long duration activities at remote locations on earth. Appropriate medical facilities and trained personnel on site would be absolutely essential. The question of how extensive the preparations should be can be debated to a point. For example, what medical specialties are virtually mandatory as compared to those a facility might like just in case? How many medical personnel are needed and what percentage of any medical specialty should reside in any single individual? Placing all the expertise in a single individual is not much help if that person is the injured party or is unavailable for whatever reason.

Telemedicine is an obvious adjunct to having trained personnel in place as it provides extensive capabilities for additional expertise and information. The problem is that communications can fail. The inability to communicate for any reason makes a medical emergency that much more dangerous, regardless of the amount of remote support available to individuals living and working on the Moon. Diagnostic equipment must also be present for earth-based specialists to determine what the medical problem involves and how to proceed. Industrial facilities incur a wide variety of accidents that require extensive medical procedures and support for injured personnel by their very nature. The ability to move critically ill or injured personnel is unlikely even with significant improvements in transportation technology. The time and stresses required to move an individual from the Moon back to an earth facility is itself a barrier to survival and recovery. Injuries that are survivable on earth will be terminal on the moon if the capability for treatment and support is only available on earth (Eckart 1996).

Therefore, lunar-based managers will be faced with the necessity of managing a self-contained health care operation staffed with sufficient personnel as to be able to handle emergencies without earth-based support in some circumstances. These types of requirements suggest not only medical staff but support staff capable of keeping diagnostic equipment properly calibrated and in good repair. Many good hospitals do not have the capability to completely maintain and/or repair their own x-ray, MRI, blood chemistry or similar equipment without outside support. Repair of equipment implies spare parts being available or that backup exists for critical care capability. All of these things will require managerial oversight, tracking, and regular checks to insure that the medical subsystems present on the moon work well and provide sufficient capacity for typical medical situations as well as the unexpected. Managers must remember that in this environment short-term cost savings decisions could well result in extraordinary problems in an emergency and significant loss of life.

29.8.3 Living Arrangements

Only in rare circumstances, such as in the off-shore oil industry, are general managers concerned about employee living arrangements. It is often necessary for managers in remote locations or where housing and/or work facilities are limited or extremely expensive to manage living arrangements. Managerial intrusions are generally limited, even in such circumstances. The Moon will require more significant and regular managerial oversight of habitation issues. Operational efficiency will be greatly affected by employee ability to rest and relax in off hours. The Moon will be a 24/7 operating environment and managers will be unable to ignore any issue that affects or might affect worker safety, morale or effectiveness. "All the necessary conditions to perpetrate a murder are met by locking two men in a cabin of 18 by 20 feet...for two months (as quoted in Seedhouse 2008, p. 112)." Even simple things such as getting a good night's sleep could easily find their way to a manager's desk. Seemingly inconsequential things such as background noises, smells, and employees private behavior could become part of a lunar manager's in basket. These types of issues would appear to be somewhat childish in nature and often beyond the concern of all but perhaps the lowest managerial level. However, that is not the case for Moon facilities. Each issue represents a substantial decision with significant consequences. The lunar facility, whatever its construction, remains a closed system, and individuals living and working within it are subject to a variety of constraints. Recirculated air and water, limits on available space, comparatively fewer choices in terms of what to do and where to go all combine to reduce the habitable quality of the facility.

Consequently, great care must go into creating ways for employees to have some measure of control without creating new problems. Not everyone enjoys the same food odors and lingering smells could be a serious source of friction among lunar residents. Food, particular food of one's own choosing will be an ongoing management problem. Not all requests can or should be accommodated, and to ignore this issue will be to ensure an ongoing morale problem. Managerial decision-making with regards to food choices, preparation and availability will be closely scrutinized. Privacy will be an even bigger issue. People do, on occasion, wish to be alone. Being alone can mean being "alone" or alone with another. Social politics notwithstanding it will be problematic to turn a blind managerial eye to relationship issues. People are social creatures and will interact with one another even in the most restrictive of environments. Working on the moon will not be different in that regard. Attempts to control, regulate or limit some level of interaction will fail. The failure to take privacy needs into account from a management perspective either in terms of individual needs or facility design will create an ongoing stream of management problems. It is better to address these issues upfront and provide sensible mechanisms for managers to deal with these issues among his or her colleagues and their employees. Questions regarding rotation policy for lunar employees should be addressed from the beginning. Explicit guidelines should be developed for extending lunar deployments, how they should be handled, and when extending an employee's tour is appropriate because of safety or other factors.

29.8.4 Community

Community as a concept is fairly simple. A group of people who work and live together come to know and appreciate all of the strengths and weaknesses of their respective group mates. Large numbers of individuals living, working and interacting for long periods of time in large communities is rare apart from some large expatriate communities. People seldom live and work in the same place. People who do live and work in the same place usually do so for limited periods and in some fairly restrictive environments as within military or science missions. The Moon would represent a different situation. The lunar facility would expand as time passed and evolve from a simple outpost to a much more diverse operation that may mirror a small city. This growing new city would encounter all of the issues that similar communities have experienced as they grew. Cities tend to be untidy places in many respects. They take on certain home-like characteristics that people enjoy. They reflect the character of their builders and the aesthetics of their inhabitants. So will the lunar community.

Much more care must go into design and construction of Moon communities due to the severity of the environment unlike its terrestrial counterparts. Failure to take into account concurrent basic human and hence community needs would cause many problems for managers interested in the overall effectiveness of operations within its parameters. How much space to devote to public areas, possible greenery, recreation and the creation of variability could be endlessly debated. Managers may decide in the long run that the needs for community outweigh some efficiency factors. Dorms will give way to apartments and apartments to custom-configured living spaces. Physical space on the Moon always may be at a premium and it makes sense to provide the best living and working environment possible. The goal would be to improve the livability of the habitation areas without significant loss of functional efficiency. The field of space architecture is rapidly developing in response to the unique needs of space-based operations (Howe 2009). Management requirements must be blended with human expectations to provide the necessary components for a reasonable lifestyle. You cannot pay people enough to be miserable for prolonged periods. Prospective employees that have options will be very resistant, regardless of the unique environment, to putting themselves in a place that is totally unsuitable by their standards. The addition to that unsuitability of extended absences from friends and family, one of the most significant social stressors, means that managers will face increasingly serious problems as the novelty wears thin for their employees. Managers in charge of this environment may find their own morale at risk as they attempt to manage their management tasks, their employees and themselves. Community will be a consideration that will be important to everyone.

29.8.5 Uncharted Territory

What lies beyond the boundaries of traditional management are the things that often do not get discussed. Relationships were described in a previous section in the context of privacy. However, managers could still pretend that they do not know what may be going on behind the closed door. Unfortunately, once large numbers of people work and live within a lunar facility, traditional limits on what a manager needs to know in terms of private information or even private behavior may change. The distinction between public and private may not be able to maintain the same boundaries on the Moon as it does on earth. The limited number of people available might suggest or even require extreme care in selection of individuals with no communicable diseases. Managers may need information that typically is protected under many countries' privacy laws, at least during the selection process and subsequent training and preparation phases. The limited size of the community might also require some management in terms of employee blood types. Much as the military needs specific health information from soldiers as a consequence of possible injury or wounds, so too would the health providers within the lunar facility. It would be unreasonable to not prepare for possible emergencies by having a balance of blood types even to the extent of limiting or monitoring rare blood groups. Working on the moon will preclude putting out a general call for help in the event of a blood emergency. Blood substitutes now becoming available might solve that problem or just make it more manageable. And, blood access is an example of a nontraditional issue that will not go away and will need to be addressed.

Certainly more controversial will be managerial issues as they relate to interpersonal relations between the people living and working in the lunar environment. Gender balance and the implicit issues regarding the sexual makeup of a crew, workforce, or community may become an ongoing managerial concern.

The issue of whether a crew should be all-male, all-female or mixed remains a contentious matter. Some have argued a female crew would exhibit preferable interpersonal dynamics and be more likely to choose non-confrontational approaches to solve interpersonal problems. Others have made a case for a mixed crew, claiming crews with women are characterized by less competition and seem to get along better. Evidence from Antarctic winter-over crews supports each of these arguments and suggests women, in addition to their mission function, would serve a socializing purpose. However, the introduction of a single female into a male group may have destabilizing effects because of sex issues - a topic that space agencies are notoriously reluctant to discuss (Seedhouse 2010, pp. 175-176)!

The subtext often absent in all of these discussions is the lack of attention given to how people interact in real environments rather than the simulated environments common to mission planning discussions. Planners are not overly naive in this regard and they expectedly try to avoid controversial topics. Many sensitive issues are either ignored, minimized or subsumed away as something to be dealt with at a later time as a consequence. The likelihood these issues will find their way to the desk of a manager who may not be as well prepared to deal with controversial and sensitive issues as she/he should be is high. Many of these seldom discussed topics will be exacerbated by the inherent complexity of multi-nation and multi-organization enterprises. A cornucopia of potentially incompatible international rules, regulations and laws will be added to this mix. Overlaying the legal framework will be a patchwork composed of religious and cultural expectations which in themselves are not necessarily consistent or compatible with the existing laws of many nations. It is possible to imagine all sorts of scenarios leading to bizarre outcomes if one were trying to accommodate all levels of complexity. For example, it might become necessary for managers to violate various discrimination laws or to seek exceptions to existing laws to manage what would otherwise be the managerial equivalent of an asylum. These are not issues or situations that most managers are prepared to deal with. The simple problem of managing may become a secondary consideration in the face of these types of confounding circumstances.

29.9 Conclusion: Lunar Management Will Be Hard – Very Hard

Each of the preceding areas poses special management problems. None of the managerial concerns identified, with the possible exceptions of those that hit religious or cultural hot buttons, are inherently so problematic as to encourage a competent manager to run screaming from the Moon if taken individually. That can be described as the good news. However, taken together and probably simultaneously, the management workload for even the most simple of lunar operations will be significant. Current managerial training does not train people to be the sort of managers that will be needed to successfully run a lunar base, community, business or businesses, and keep staff, employees, families, both working and living at peak efficiency and effectiveness. The military operates to an extent with some of these factors in mind but it is often criticized for its inability to implement policies consistently or successfully. The private sector has some places where these issues play out, however, none of these situations rival the magnitude required for any lunar operation of extended length. For example, the offshore oil industry uses Flotels, floating accommodation modules, for their employees which operate on a 24/7 basis. These modules frequently provide individual living compartments, often with extensive support capability, and represent a good analog for possible lunar facility equivalents.

The major space organizations have cadres of individuals well suited and exceptionally well trained for specific space missions. While their capabilities would be required for long-term space ventures, including those on the Moon, it is doubtful that their skills alone would be sufficient to manage all of the interrelated business, community, and individual issues. It is very clear that competent lunar managers will require a level of training and preparation well above what is available currently regardless of where a business person begins in terms of talent, skills, and training. It is also pretty clear that individuals beginning this training process will eventually need to morph into managers of exceptional flexibility and consummate skill. Whether the individuals that will be needed to start this process are unique or as ubiquitous as MBAs remains to be seen.

References

Abramov, I.P., Skoog, A.I.: Russian Spacesuits. Praxis, Chichester (2003)

Bishop, S.L.: Here to Stay: Designing for Psychological Well-Being for Long Duration Stays on the Moon and Mars. In: Benaroya, H. (ed.) Lunar Settlements. CRC Press, Boca Raton (2010)

Eckart, P.: Spaceflight Life Support and Biospherics. Kluwert/Microcosm, USA (1996)

Harris, P.R.: Space Enterprise: Living and Working Off World in the 21st Century. Praxis, Chichester (2009)

Howe, A.S., Sherwood, B. (eds.): Out of this World: the New Field of Space Architecture. American Institute of Aeronautics and Astronautics Inc., Reston (2009)

Ivanovich, G.S.: Salyut: The First Space Station. Praxis, Chichester (2008)

Jorgensen, J.: Human: the Strongest and the Weakest Joint in the Chain. In: Benaroya, H. (ed.) Lunar Settlements, pp. 247–259. CRC Press, Boca Raton (2010)

Kutschera, I., Ryan, M.H.: The Future Role of Human Resource Management in Non-Terrestrial Settlements: Some Preliminary Thoughts. In: Benaroya, H. (ed.) Lunar Settlements, pp. 87–99. CRC Press, Boca Raton (2010)

Livingston, D.: Turning Dust to Gold: Building a Future on the Moon and Mars, p. 334. Praxis, Chichester (2010) as cited in Benaroya H

Morris, L., Cox, K.J. (eds.): Space Commerce: The Inside Story By The People Who Are Making It Happen. An Aerospace Technology Working Group Book, United States (2010)

Roach, M.: Packing For Mars. WW Norton and Company, New York (2010)

Ryan, M.H., Kutschera, I.: Lunar-based enterprise infrastructure—hidden keys for long-term business success. Space Policy 23, 44–52 (2007)

Ryan, M.H., Luthy, M.R.: Management Architecture: Problems Facing Lunar-Based Entrepreneurial Ventures. J. Space Mission Architecture 3, 20–38 (2003)

Seedhouse, E.: Tourists in Space: A Practical Guide. Praxis, Chichester (2008)

Seedhouse, E.: Prepare for Launch: The Astronaut Training Process. Praxis, Chichester (2010)

Spell, C.S.: Mental Health Implications of Working in a Lunar Settlement. In: Benaroya, H. (ed.) Lunar Settlements. CRC Press, Boca Raton (2010)

Tenner, E.: Why Things Bite Back, Technology and the Revenge of Unintended Consequences. Vintage Books, New York (1996)

Thomas, K.S., McMann, H.J.: US Spacesuits. Praxis, Chichester (2006)

Vedda, J.A.: Choice Not Fate: Shaping a Sustainable Future in the Space Age. Xlibris Corporation, Bloomington (2009)

Weick, K.E., Sutcliffe, K.M.: Managing the Unexpected: Resilient Performance in an Age of Uncertainty. Wiley, San Francisco (2007)

Wellin, K., Elliott, E.: The Next Frontier: Commercialization of the Lunar Surface and CISLunar Space in the 21st Century. J. British Interplanetary Soc. 63(2), 53–60 (2010)

30 Highlights of Solar System Development on the 200th Anniversary of Men on the Moon

Yerah Timoshenko[*]

Rutgers University, Piscataway, New Jersey, USA

30.1 Introduction

Since humans returned to the Moon, we went from being an outpost on a barren rock orbiting Earth to being a nascent civilization on the Moon and Mars, with outposts on dozens of asteroids, outer planets and their moons. We went from a population of under a dozen to one approaching 300,000 extraterrestrials.

But our impact is larger than what one may expect of that many people. A city on Earth of that population is considered small, perhaps peripheral to the main avenues of power and influence. The 300,000 people who are today distributed throughout large sectors of the Solar System are all prime movers. Each of us has significant responsibilities. We oversee an infrastructure that is vast and very wealthy. We supply 80% of the energy needs of Earth and an increasing percentage of its raw material needs.

Space settlements, those in orbit around Earth, the Moon and Mars, as well as the many outposts and small cities on the Moon and Mars, are important to all humans because of the access these give us to resources otherwise unavailable to our industrial societies and for the multiple markets created. By settling the Moon and Mars, Earth sent into space its best and brightest in the arts and sciences and strengthened humanity's core positive achievements: democracy, individual rights, equal opportunities for individual achievements, and all that is inherent and is based on these principles. Humans explore because it is in our genes.

30.2 An Outline of Our Space History

Human civilization now encompasses the inner Solar System and this outward development from Earth has resulted in advances for humanity that could not have been predicted at the onset. We generally think *linearly* – we are best at extrapolating our current experiences and usually miss paradigm shifts due to unanticipated discoveries. In the year 1807, very few people anticipated the Wright Brothers'

---

[*] Yerah Timoshenko is a pseudonym for Haym Benaroya.

human flight a hundred years later. In 1869, only science fiction writers envisioned landing people on the Moon. Similarly, other great discoveries in physics, mechanics and electronics were not predicted and therefore the technologies to which those discoveries gave birth could not have been foreseen except perhaps by a tiny group of exceptional visionaries.

It took us longer to break our Earthly bonds than those who lived during the Apollo era had wished. While many who witnessed man's first steps on the Moon did not see the many achievements of the 21st and 22nd centuries, they understood the inevitability of what exists today in 2169 and had such keen imaginations that their minds' eyes fully saw how a spacefaring mankind would evolve. An example of the optimism coupled with the vision of the early 1960s was the Army Corps of Engineers study of the kinds of facilities that it would need to be able to build on the Moon for the coming human settlements. During the decade between the late 1980s to the mid 1990s, such studies had intensified both within NASA and outside the Government in industry and academe. Numerous of these studies discussed science on the Moon, the economics of lunar development, and the challenges for human physiology and psychology in space and on planetary bodies. An equally large literature on policy – economic and legal – developed the framework for our evolvement into a spacefaring species. All these disciplines and others related to the survival of living in space and low gravity were needed to plant humans on the Moon – and beyond – in a sustainable and viable way.

By the mid 1990s the political climate turned against a permanent return to the Moon and began to look at Mars as the "appropriate" destination, essentially skipping the Moon. The debate between "Moon First" and "Mars Direct" faded as it became clear that we did not have the technology and experience to send people to Mars for an extended stay until after we perfected our abilities to live outside the Earth cocoon. Physiological and reliability issues were unresolved for a trip to Mars. We are grappling with many of the same issues today.

Post-Apollo, many reasons were given why humanity needed a lunar base: lunar science and astronomy would benefit from the isolated and vacuum lunar environment. Space technology would be stimulated and much of that would have dual-use applications. The Moon would be an excellent test bed for the technologies required to place humans on Mars and beyond. Recovery of plentiful lunar resources would become economically viable once a basic infrastructure existed. Space could not be underestimated as a motivator for young people to study science and engineering. And there was growing concern that humanity needed an insurance policy against the possibility that life on Earth could be wiped out by a single collision with a relatively small asteroid. Settling the Moon and later Mars was starting to be viewed as the beginning of a long-range program to ensure the survival of the species. Of course, all these reasons were valid then and are in play now as humanity has found homes on multiple bodies in the Solar System.

While two hundred years is a long time against the scale of human life expectancy, it is a short time in the evolution of humanity. That it has taken us two hundred years to evolve from an Earth-based species to one that has hundreds of thousands living permanently in the inner Solar System is really remarkable since this is not a long time. The extreme environment on the Moon is very challenging to our engineers. Missteps can be fatal – we are on an unforgiving planet. There were many difficulties that had to be overcome so that we could achieve these ends. But the primary one was the economics of spaceflight. It was just too expensive to lift mass from the surface of the Earth into low-Earth orbit. Given the costs associated with spaceflight, the returns on investment demanded by private corporations as well as the taxpayers were difficult to overcome. Even with the benefits of dual-use technology, it was difficult to close the loop.

Figure 30.1 is of the historic document created by the US Army Corps of Engineering during the time of Apollo. At that time it seemed obvious that Apollo would lead to permanent settlements on the Moon. Then who else but the Army Corps of Engineers to build those facilities on the Moon? Even though Apollo was a geopolitical decision by the Kennedy Administration in response to the Soviet Union's surprise foray into space, many believed that success would lead to decisions to take advantage of the new capabilities and to then settle permanently on the Moon. Had that been done, we would likely be one hundred years ahead of our present capabilities.

A number of important issues had to be addressed and at least partially resolved as we evolved into a spacefaring civilization. Our colonies on the Moon had two general purposes: to learn how to survive on a non-terrestrial body that is naturally hostile to human life, and to begin to explore and use the resources of the Moon for human purposes. The second purpose led to mining and materials-processing operations, the synthesis of fuels from mined lunar hydrogen and oxygen and the

Fig. 30.1 Cover of NASA's Office of Manned Space Flight *Lunar Construction*, prepared by the Office of the Chief of Engineers, Department of the Army, April 1963. (Courtesy Al Smith and the U.S. Army Corps of Engineers)

development of industries that could benefit from the one-sixth Earth gravitational field and the hard vacuum. In-situ resource utilization was our lifeblood. We could have never survived without the ability of to mine and recover resources on the Moon and Mars and then be able to process those local resources into most of what we need to survive – from oxygen to building materials.

Figure 30.2 is an illustration of one of our earliest outposts on the Moon. Constructed from igloo-like modules that could house six people, it was possible to expand these facilities so that hundreds could be accommodated (Ruess et al. 2004).

Eventually structures and most facilities became subterranean because the surface is so toxic to living entities. With surface temperatures ranging between -153 °C to 107 °C, galactic and solar radiation, and regular micrometeorite impacts, early settlements seriously challenged our engineering skills. As soon as the infrastructure developed serious construction capabilities, we began digging our settlements to the point where they house thousands of people well beneath the surface. An example of such a facility is shown in Figure 30.3.

Fig. 30.2 The look of the igloo base on the Moon. From the scale of the figure, we can see that the shielding is not purely 3 m of regolith. This is a more advanced shielding material that is blended with the regolith. (Illustration by Andre Malok, by permission c 2007 The Newark Star-Ledger)

Our self-sufficient large lunar cities survive economically by exporting lunar minerals and finished products to Earth, and by servicing transportation, both commercial and military, between Earth and the emerging settlements on Mars, its moons, as well as early mining activity on the asteroids and the moons of the gas giants of the outer solar system. A variety of private and public-private partnerships created settlements with specifically economic purposes. Lunar development organizations of various pedigrees were created. Our present-day lunar ecosystem is a very complicated one. It includes plants that are able to grow in regolith with the aid of certain bacteria and minute amounts of water.

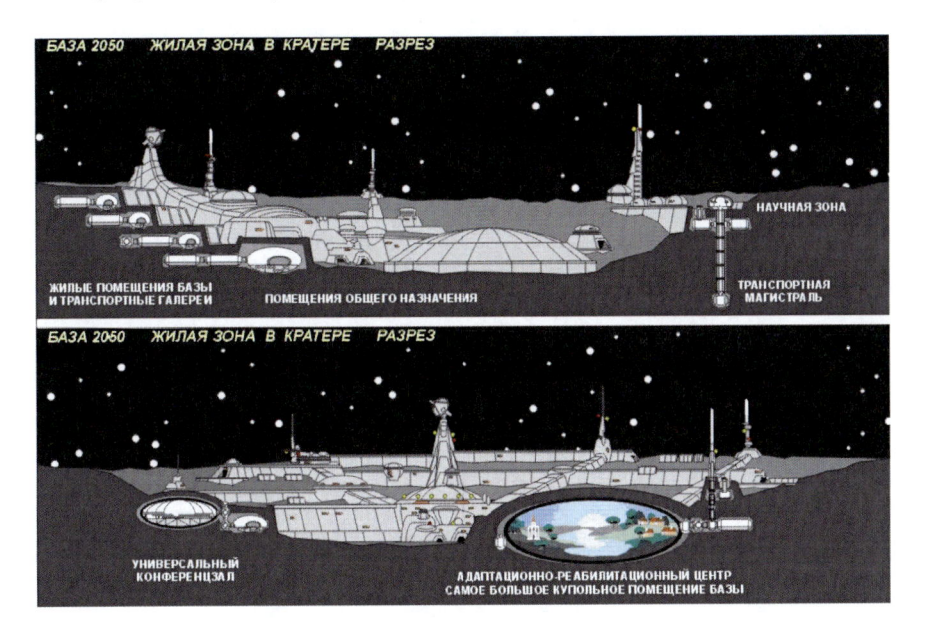

Fig. 30.3 Manned Base in Lava Tube. Energia-Sternberg Project. Side Views. The base is mostly within the lava tubes for shielding. In the top figure: on the left side are general purpose premises; the lower left are residential facilities; on the right side is the scientific zone and a transport line. In the bottom figure: in the lower left is a general conference hall, and in the lower right side are a rehabilitation center and the largest cupola-shaped premises on the base. (Courtesy Vladislav Shevchenko)

Along the way we had to address additional issues. These included the desire to avoid pollution and the rights of ownership. At some point a consensus evolved to ensure that the Moon, in particular, and the planets, in general, survived with integrity after colonies and industrial facilities began to be planted there. The Moon was our test – did we learn from the mistakes on Earth? One advantage of the long times it took us to return to the Moon was that it gave us a lot of time to think about how we would treat the Moon. For example, we realized that we did not want to strip mine the Moon for resources.

Therefore, the rare mining operation that erased the topology of the surface as it exhumed valuable ore was quickly viewed as an unacceptable form of mining to the population that now calls the Moon home. Many are eager to protect the stark beauty of this and all the new worlds we call home. We also believe that in the long-term such careful oversight of the Moon is economically advantageous. Working for short-term profits with long-term damage is viewed as an ineffective way to develop the Moon. What deserves saving on the Moon? Most agree that the Apollo sites and surrounding areas were worth preserving. A radius of no contact was established where no person and no vehicle could approach, not even overhead since rockets can destroy the footprints and locations of items on the surface (Hargrove 2008).

"The place where the two astronauts landed, resided and worked – a roughly 60 m^2 area named 'Tranquility Base' - is a unique Solar System physical location. Tranquility Base, and what was left behind there when the astronauts departed, should be preserved and protected for all, for all time (Chaikin 1998)."

It was also suggested that the UN declare the site a World Heritage Site (Rogers 2004). But there was little public support for the UN in the U.S. - the United States viewed Tranquility Base as historic for the world but still felt it to be a distinctive American accomplishment before the era of multinational space efforts.

"The Apollo programme truly widened the horizons of humanity at large. The image of an Earthrise over the lunar surface, taken by Apollo 8, brought home in a very visual way the relative position of Earth and demonstrated the fragility of our planet, 'space ship Earth'. It soon became a symbol for the environmental movement. The 'blue marble' photograph of a complete and only slightly cloud-covered Earth taken during Apollo 17 has become the icon of Earth in space. People were fascinated by these missions ... about a quarter of the human population at the time watched the live television broadcast as Neil Armstrong stepped onto the lunar surface (Spennemann 2004)."

To this day we support the value of the Apollo sites and have placed hemispherical glass domes over all of them. As "traffic" on the lunar surface grew, disturbed electrostatically suspended regolith began to migrate long distances, thus not only affecting all surface operations but also threatening historical sites. We have also protected the landing sites of the early probes on the Moon and on Mars.

There are sites on the Moon, Mars and some of the other bodies that have been viewed as worth preserving for their own sake .such objects, like works of art, have intrinsic value. The intrinsic value may be for scientific reasons or for aesthetic reasons. On the Moon, the harsh environment makes it difficult to erase mistakes, and there are no minor mistakes.

We have had several generations born on the Moon. While our physiology has changed infinitesimally our psychology is wholly lunar. We are seeing the evolution of mixed loyalties amongst Lunarians. While we retain a kind attitude to the planet of our great grandparents, our hearts and minds naturally have different perspectives on how our mother planet, the Moon, should be developed and evolve. My two children were born on the Moon.

Table 30.1 is taken from the book *Turning Dust to Gold* by Haym Benaroya. It is a chronology of milestones in humanity's evolution into a spacefaring species. The book is written as though a history book in the year 2169, the 200th anniversary of when Americans first stepped on the Moon with Apollo 11. Yerah Timoshenko (YT) is a fictional character in the book that occasionally leads the reader in reviewing aspects of this speculative but possible history.

Similarly, settlers and natives on Mars and the outer Solar System have their own views on development. What may have once been viewed as property of Earth-based interests, the lunar settlements have become entities in their own right, much like children who grow up to become individuals, unique and independent. In the same way that the Americas became independent of the mother countries, and properties developed by trading companies evolved into autonomous entities, cities on the Moon and Mars will demand independence from Earth in the not-too-distant future. At best, a confederation of equals may evolve between Earth and its former colonies.

Table 30.1 Chronology of milestones in humanity's evolution

1969	First men on the Moon
2009	Chandarayaan-1/Moon Mineralogy Mapper reveals H_2O molecules
2014	Space tourism reaches $1B threshold
2024	Humans return to Moon
2029	Permanent colony
2034	Humans land on Mars
2041	Permanent Mars colony
2046	Space elevator prototype construction begins over Earth
2049	Lunar space elevator construction begins
2059	Yerah Timoshenko's (YT) great-grandparents go to Moon
2060-61	YT's grandparents born on Earth (conceived on Moon)
2070	Fusion reactors go online on Moon, a year later on Mars
2084	First families on Moon
2089-90	YT's parents born on Earth
2094	First lunar Olympics
2099	First human birth on Moon
2115	YT's parents move to Moon
2119	YT born on Moon
2142-43	YT's boy and girl born on Moon
2169	The present – 200th anniversary of the first men on the Moon
Now	Terraforming of Mars
2179	150th anniversary of off-Earth permanent habitation

The historical summary provided by the table lists the key successes of a space-faring humanity during the past two hundred years. Space elevators were developed and became key components of our space civilization. As construction of the space elevators took place in orbit around the Earth (and Moon and Mars) a major infrastructure in orbit around these planetary bodies was erected. Once the space elevators became operational the pace of space development increased exponentially due to the drop in costs by several orders of magnitude.

While solar power is a key resource on the Moon, it is less so on Mars and the outer moons. Fusion-based power is well developed and supplements solar and nuclear power throughout the colonies.

All manner of human activity takes place on the Moon – witness the first Lunar Olympics of 2094. In that first such lunar competition, many hundreds of lunar inhabitants witnessed three-dimensional football, low-gravity swim competitions, and thirty meter pole vaulting. Hundreds of people from Earth visited the Moon for the games. Tourism significantly adds to the lunar economy, less so to that of Mars due to the added transit time – one way travel to the Moon takes about twelve hours, but between two and three weeks are needed to reach Mars depending on the time of year. Also noted is the ongoing terraforming activity on Mars the extent of which it is to be allowed is still being seriously debated.

The survival of living things in space and on extraterrestrial bodies was and is the key challenge to a spacefaring humanity. The first priority is of course living; the next priority is living with sufficient food and comfort, and the next priority is to live and enjoy life. The Apollo program, and Mercury and Gemini that preceded it, focused on the first priority. When Skylab, Mir and the ISS were built, there was more room. Astronauts could move around – they even had a bit of privacy.

These and more recent space stations became our nodes for material transport, for docking of transport ships that ferried people and goods between Mars, the Moon and Earth. The United States and The Russians were the backbone of manned space exploration and settlement in the 21st century, with the Chinese, Indians, Japanese and Europeans quickly evolving as equal partners in the space adventure. Many nations participated during the 21st and 22nd centuries in the development of the myriad of skills and technologies needed to settle space. By the early 22nd century, the distinctions were no longer by nationality, but rather by planetary body!

Was it comfortable? I would have been claustrophobic. The design of the chairs and the human/machine interfaces began to look beyond survival and into ease of use. Concepts for space stations and lunar settlements eventually took into account the appearance of the interior – what colors are stress reducers and make people at ease. With each new adventure, and each new generation of that adventure, designs added comfort and took into account the full human being. Layout design facilitated ease of human motility. There were significant efforts given to improving the social and organizational aspects of life in the settlements. Social and psychological issues – effects of stress, recreation and exercise, interpersonal dynamics in space, personal space, privacy, crowding, and territoriality – all began to be addressed.

Human physiology took the front row since being alive is a prerequisite for comfort and happiness. But human psychology was close behind. Just being alive was not enough. Once space travelers included people from throughout all of society's strata, not just engineers and scientists with years of training, the whole picture of human survival began to be considered. After all, space tourism had to be an enjoyable adventure, not just an exciting one.

Human physiological and psychological issues have been partially resolved but there exist lingering effects on humans who spend their whole lives on planetary bodies other than Earth. We have had to thrive in low gravity and within still very close quarters. This proved difficult for all of us and for some it proved insurmountable. Eventually, larger facilities were built on the Moon so that personal space became possible – but the space is still a far cry from that available on Earth. And then people began to settle Mars where distance from Earth and the Sun introduced new and profound psychological challenges. The Sun viewed form the surface of Mars is essentially just another star in the sky, unlike the view from the lunar surface where solar power is a major part of the energy and psychological equation. Mars feels much more isolated than the Moon – because it is!

A number of technologies have evolved significantly since the return to the Moon in 2024:

ISRU – in-situ resource utilization, the way by which we live off the land, operates continuously and is almost completely automated using a suite of robotic technologies. Every planetary body of interest to humanity is being developed for habitation autonomously by robotic ISRU-construction teams. Settlements are erected to welcome arriving astronaut pioneers who make the final installations. Robotic miners bring us our raw material needs. More of our systems have self-repairing capabilities.

The majority of our cities are underground where natural protection exists for the living and the inanimate from space's vacuum, radiation, micrometeorites, temperature extremes and severe gradients.

Space elevators eventually dropped by two orders of magnitude the cost of launching mass into orbit and then transporting it between points. These exist wherever there are a sizable number of settlers.

We have finally closed the loop on nuclear fusion power and have access to enormous quantities of Helium-3 fuel.

While our habitats are increasingly below ground, thus protecting us from radiation and micrometeorite impacts, we do spend time on the surface. Advances in gene therapy have helped shield us from radiation damage. The effects of low gravity have also been significant and gene therapy has helped in that regard as well. There is promising research on artificial gravity.

Currently the Moon has eleven cities with a total population of about 250,000 people, and Mars has four settlements with a total population of about 25,000 people. There are another 25,000 thousand people spread over our outposts on dozens of planetary bodies and asteroids throughout the Solar System.

30.3 Conclusion

I would like to end this brief essay on our spacefaring history by quoting the first man on the Moon – Neil Armstrong – from his interview in *Turning Dust to Gold* who responded to the following question as quoted here:

> How do we answer critics who say that space is too expensive and that there are numerous problems on Earth to take care of first?

Space exploration is expensive. Not nearly as expensive as many of our government enterprises: Defense is 30 times more expensive, Intelligence is 3 times more expensive, Health and Human Services is 38 times more expensive. NASA requires less than 1% of the U.S. national budget. NASA's responsibility is to develop options for future generations. If they do it properly, our grandchildren and great-grandchildren will benefit in many ways and it will have proven to be a very excellent investment.

References

Benaroya, H.: Turning Dust to Gold: Building a Future on the Moon and Mars. Springer-Praxis Books, Chichester (2010)

Chaikin, A.: A Man on the Moon, p. 200. Penguin Books (1998)

Hargrove, E.C.: The Preservation of Non-Biological Environments in the Solar System.2162.pdf. In: NLSI Lunar Science Conference (2008)

Rogers, T.F.: Safeguarding Tranquility Base: Why the Earth's Moon Base Should Become a World Heritage Site. Space Policy 20, 5–6 (2004)

Spennemann, D.H.R.: The Ethics of Treading on Neil Armstrong's Footprints. Space Policy 20, 279–290 (2004)

Ruess, F., Schänzlin, J., Benaroya, H.: Structural design of a lunar habitat. Journal of Aerospace Engineering 19(3), 133–157 (2004)

Author Index

Subject Index

Printed by Publishers' Graphics LLC
MO20120604